Molecular Beam Epitaxy and Heterostructures

NATO ASI Series

Advanced Science Institutes Series

A Series presenting the results of activities sponsored by the NATO Science Committee, which aims at the dissemination of advanced scientific and technological knowledge, with a view to strengthening links between scientific communities.

The Series is published by an international board of publishers in conjunction with the NATO Scientific Affairs Division

A	Life Sciences	Plenum Publishing Corporation
B	Physics	London and New York
C	Mathematical and Physical Sciences	D. Reidel Publishing Company Dordrecht and Boston
D	Behavioural and Social Sciences	Martinus Nijhoff Publishers Dordrecht/Boston/Lancaster
E	Applied Sciences	
F	Computer and Systems Sciences	Springer-Verlag Berlin/Heidelberg/New York
G	Ecological Sciences	

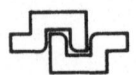

Series E: Applied Sciences – No. 87

Molecular Beam Epitaxy and Heterostructures

edited by

Leroy L. Chang

IBM T.J. Watson Research Center
Yorktown Heights, NY 10598
USA

Klaus Ploog

Max-Planck-Institute
D-7000 Stuttgart 80
FRG

1985 **Martinus Nijhoff Publishers**
Dordrecht / Boston / Lancaster
Published in cooperation with NATO Scientific Affairs Division

Proceedings of the NATO Advanced Study Institute on Molecular Beam Epitaxy (MBE) and Heterostructures, Erice, Italy, March 7-19, 1983

Library of Congress Catalog Card Number: 84-27312

ISBN-13: 978-94-010-8744-5   e-ISBN-13: 978-94-009-5073-3
DOI: 10.1007/978-94-009-5073-3

Distributors for the United States and Canada: Kluwer Boston, Inc., 190 Old Derby Street, Hingham, MA 02043, USA

Distributors for the UK and Ireland: Kluwer Academic Publishers, MTP Press Ltd, Falcon House, Queen Square, Lancaster LA1 1RN, UK

Distributors for all other countries: Kluwer Academic Publishers Group, Distribution Center, P.O. Box 322, 3300 AH Dordrecht, The Netherlands

FOREWORD

The NATO Advanced Study Institute on "Molecular Beam Epitaxy (MBE) and Heterostructures" was held at the Ettore Majorana Center for Scientific Culture, Erice, Italy, on March 7-19, 1983, the second course of the International School of Solid-State Device Research. This volume contains the lectures presented at the Institute.

Throughout the history of semiconductor development, the coupling between processing techniques and device structures for both scientific investigations and technological applications has time and again been demonstrated. Newly conceived ideas usually demand the ultimate in existing techniques, which often leads to process innovations. The emergence of a process, on the other hand, invariably creates opportunities for device improvement and invention. This intimate relationship between the two has most recently been witnessed in MBE and heterostructures, the subject of this Institute.

This volume is divided into several sections. Chapter 1 serves as an introduction by providing a perspective of the subject. This is followed by two sections, each containing four chapters, Chapters 2-5 addressing the principles of the MBE process and Chapters 6-9 describing its use in the growth of a variety of semiconductors and heterostructures. The next two sections, Chapters 10-11 and Chapters 12-15, treat the theory and the electronic properties of the heterostructures, respectively. The focus is on energy quantization of the two-dimensional electron system. Chapters 16-17 are devoted to device structures, including both field-effect transistors and lasers and detectors. This volume ends with Chapter 18 in the last section, which describes the metalorganic chemical vapor deposition as an alternative epitaxial process to MBE.

We would like to express our thanks to A. Gabriele of the Ettore Majorana Center for the excellent local arrangements, to H. Hoogervorst of the Martinus Nijhoff Publishers for her editorial support, and to M. Shaw and M. Powers for the preparation of the manuscripts.

L. L. Chang

K. Ploog

CONTENTS

THEORY OF HETEROSTRUCTURES

ELECTRONIC PROPERTIES OF HETEROSTRUCTURES

HETEROSTRUCTURE DEVICES

ALTERNATIVE EPITAXY

1

SEMICONDUCTOR SUPERLATTICES AND QUANTUM WELLS THROUGH DEVELOPMENT OF MOLECULAR BEAM EPITAXY

Leo Esaki

IBM Thomas J. Watson Research Center
Yorktown Heights, NY 10598

ABSTRACT: Advances in superlattices and quantum wells are reviewed in conjunction with the development of molecular beam epitaxy.

1. INTRODUCTION

In 1969, research on synthesized semiconductor superlattices was initiated with a proposal for a one-dimensional periodic structure consisting of alternating ultra-thin layers by Esaki and Tsu.[1,2] Two types of superlattices were envisioned: doping and compositional, as shown in the top and bottom of Figure 1, respectively.

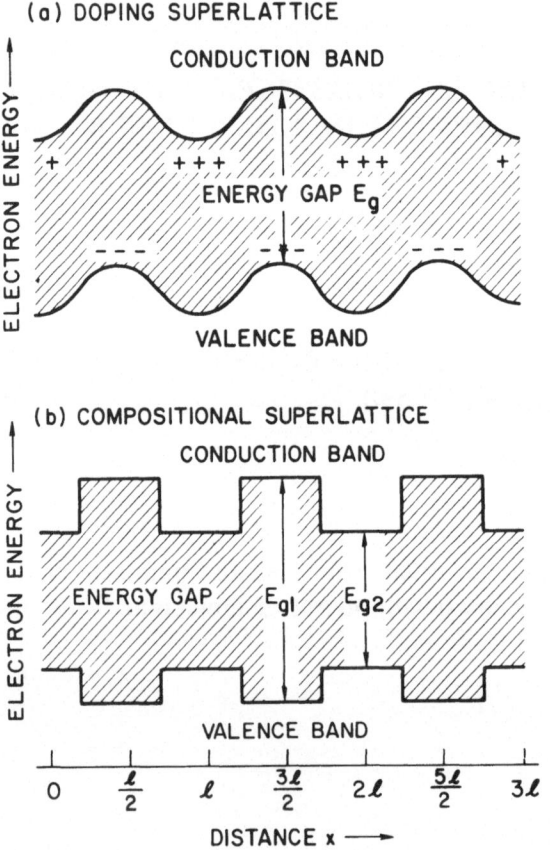

Figure 1 - Spatial variation of the conduction and valence bandedges in two types of superlattices: top, doping superlattice of alternating n-type and p-type layers; bottom, compositional superlattice with alternation of crystal composition.

The idea of the superlattice occurred to us while investigating the possible observation of resonant tunneling through double and multiple potential-barriers.[3] In general, if characteristic dimensions such as superlattice periods and widths of potential wells in semiconductor nanostructures are reduced to less than electron mean free path, the entire electron system will enter a quantum regime of the reduced dimensionality with the presence of nearly-ideal hetero-interfaces. Our effort for the semiconductor superlattice[4] is viewed as a search for novel phenomena in such a regime with precisely-engineered structures.

It was recognized at the beginning that, while the structure was undoubtedly of considerable interest, the engineering of such a crystal consisting of ultra-thin layers would be a formidable task. Nevertheless, the proposal of a semiconductor superlattice inspired a number of material scientists.[5-8] Indeed, steady improvements in thin-film growth techniques such as MBE or MOCVD during the last decade have made possible high-quality heterostructures having designed potential profiles and impurity distributions with a dimensional control close to interatomic spacing and with virtually defect-free interfaces in a lattice-matched case such as $GaAs-Ga_{1-x}Al_xAs$. Such great precision, indeed, has allowed access to a new quantum regime.

In this review, following remarks on MBE, I will survey significant milestones chosen from the research of superlattice and quantum wells occurring over the past fifteen years, including recent results in our efforts.

2. MBE

Since the era of selenium rectifiers, vacuum evaporation has been widely used for the preparation of semiconductor thin-film devices. For instance, in 1948, Shockley and Pearson[9] utilized evaporated films of germanium, silicon and copper oxide in an attempt to make field-effect thin-film amplifiers.

Since the 1940's, evaporated films of lead and tin chalcogenides were rather extensively investigated although superior epitaxy was not achieved until 1964 when Schoolar and Zemel[10] clearly demonstrated the growth of epitaxial PbS films on NaCl, using molecular beams generated from effusion cells. This work probably constitutes a precursor of the modern MBE technique.

The first methodical study on the growth of III-V compound semiconductors was reported by Günther[11] in 1958. With his "three-temperature" technique, stoichiometric compound semiconductor films were produced, although not epitaxial. With improved vacuum conditions, in 1968, Davey and Pankey[12] succeeded in growing epitaxial films on clean monocrystalline GaAs substrates using Günther's method. Around the same time, Arthur[13] investigated the kinetic behavior of the Ga and As species on GaAs surfaces to understand the growth mechanism, which laid the groundwork for the subsequent achievement of superior MBE films of GaAs and related III-V compounds by Arthur, LePore[14] and Cho.[15]

The wide use of MBE had to await the arrival of commercial vacuum equipment in the early 1970's. MBE[16] is basically a sophisticated version of vacuum deposition. The degree of sophistication solely depends on requirements of individual research in achieving its own objective. Because of vacuum deposition, MBE growth is governed mainly by kinetics of those beams reacted with crystalline surfaces, in contrast with other techniques such as liquid-phase epitaxy (LPE) and chemical vapor-phase deposition (CVD), which proceed near thermodynamic equilibrium conditions. Furthermore, because of the MBE process in the ultrahigh vacuum environment, a number of diagnostic methods such as reflection high-energy electron diffraction (RHEED), Auger electron spectroscopy (AES), secondary ion mass spectroscopy (SIMS), X-ray induced photoelectron spectroscopy (XPS=ESCA), etc., can be performed for in situ evaluation by incorporating necessary apparatuses in the system, together with a quadrupole mass analyzer for measuring beam intensities and an ion gun for surface cleaning. These powerful facilities for control and analysis, eliminating much of the guesswork, certainly give substantial advantages to MBE over other material preparation techniques. Salient features which MBE offers can be summarized as follows:

a.   High-purity monocrystals - the growth in ultrahigh vacuum and high purities of the beam fluxes.

b.   Formation of superfine structures with abrupt structural changes at the interfaces - relatively low growth temperatures preventing interdiffusion.

c.   Smooth, flawless surfaces for heteroepitaxy - the step-growth mechanism excluding any nucleation process.

d.   Ultrathin layers of controlled thicknesses - precise control of the beam fluxes and relatively slow growth rates.

e.   Formation of structures of complex profiles in terms of both alloy compositions and/or impurity concentrations.

f.   Fabrication of structures with predetermined uniaxial compres-
        sion or dilatation built-in in desirable regions, where the Γ_8
        degeneracy is lifted, thus the bandstructure is locally modified ---
        "bandstructure engineering."

Historically, the development of techniques for epitaxy, as well as
formation of desirable structures, has played an indispensable role in
the progress of modern semiconductor devices. The introduction of
MBE undoubtedly stimulated the imagination of scientists and engi-
neers in providing challenging opportunities for not only the fabrication
of substantially improved devices but also the preparation of unprece-
dented device structures.

3. MILESTONES

3.1 Superlattice Band Model and Experiment (1969-1972)

Esaki and Tsu[1,2] proposed the introduction of a superlattice potential
by a periodic variation of composition or impurities during epitaxial
growth. The techniques of thin films, then, were rapidly advancing
with the arrival of commercial ultrahigh vacuum equipment.

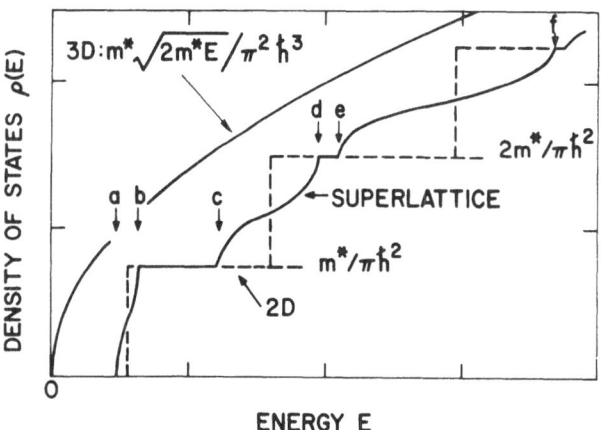

Figure 2 - Comparison of the density of states in a superlattice with
those in the three-dimensional (3D) and two-dimensional (2D) electron
systems.

        It was theoretically shown that such a synthesized structure
possesses unusual electronic properties of quasi-two-dimensional char-

acter. The introduction of the superlattice potential clearly perturbs the band structure of the host materials. Since the superlattice period is much longer than the original lattice constant, the Brillouin zone is divided into a series of minizones, giving rise to narrow subbands, separated by forbidden regions, analogous to the Kronig-Penney band model,[17] in the conduction band or the valence band of the host crystal. Figure 2 shows the density of states $\rho(E)$ for electrons in a superlattice in the energy range including the first three subbands: E_1 between a and b, E_2 between c and d and E_3 between e and f (indicated by arrows in the figure), in comparison with the parabolic curve for the three-dimensional electron system and the staircase-like density of states for the two-dimensional system.

The electron dynamics in the superlattice direction, with a simplified path integration method,[18] was analyzed for conduction electrons in narrow subbands of a highly perturbed energy-wave vector relationship. This calculation predicted an unusual current-voltage characteristic including a differential negative resistance.

In 1972, we[19] found that a MBE-grown GaAs-GaAlAs superlattice[20] exhibited a negative resistance in its transport properties, as shown in Fig. 3, which was, for the first time, interpreted in terms of the above-mentioned superlattice effect.

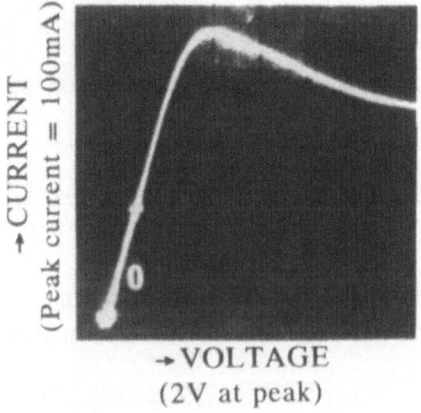

Figure 3 - Current-voltage characteristic for a 70Å period superlattice.

It is worthwhile mentioning here that, in 1974, Gnutzmann and Clauseker[21] pointed out an interesting possibility; namely, the occurrence of a direct-gap superlattice made of indirect-gap host materials, because of Brillouin-zone folding as a result of the introduction of the new superlattice periodicity, which was later reexamined by Madhukar.[22] The idea may suggest the synthesis of new optical materials.

3.2 Multibarrier Tunneling and Quantum Wells: Theory and Experiment (1973-1974)

Our superlattice concept arrived while seeking resonant tunneling. In 1973, Tsu and Esaki[23] computed the resonant transmission

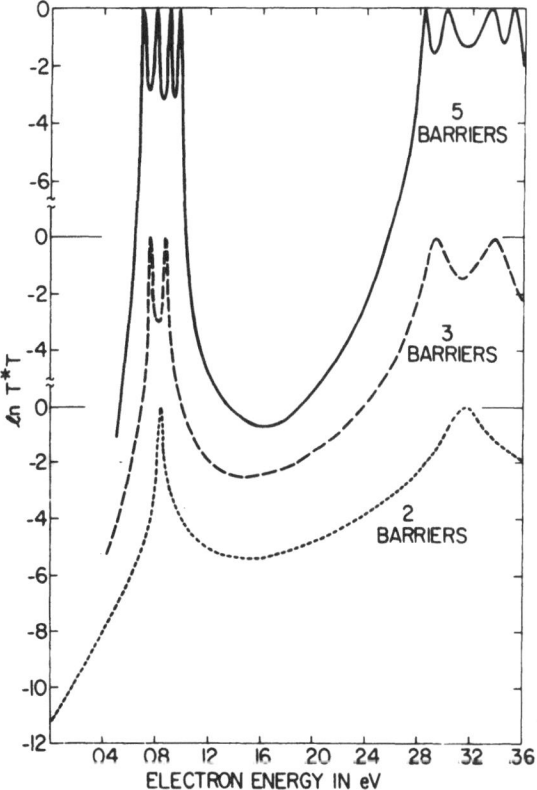

Figure 4 - Plot of ln T*T (transmission coefficient) vs. electron energy showing peaks at the energies of the bound states in the potential wells. The curves labelled "2 barriers," "3 barriers" and "5 barriers" correspond to one, two and four wells, respectively.

coefficient T*T as a function of electron energy for double, triple, and quintuple barrier structures from the tunneling point of view, as shown in Fig. 4, leading to the derivation of the current-voltage characteristics. Note that the resonant energies for the triple-barrier case consist of a doublet, and those for the quintuple barrier are a quadruplet. In the double-well case, each single-well bound state is split into a symmetric combination and an asymmetric one. The superlattice band model previously presented, assumed an infinite periodic structure, whereas, in reality, not only a finite number of periods is prepared with alternating epitaxy, but also the electron mean free path is limited. Thus, this multibarrier tunneling model provided useful insight into the transport mechanism.

In early 1974, Chang, Esaki and Tsu[24] observed resonant tunneling in double-barriers, and subsequently, Esaki and Chang[25] measured quantum transport properties for a superlattice having a tight-binding potential. The I-V and dI-dV versus V characteristics are shown in Fig.

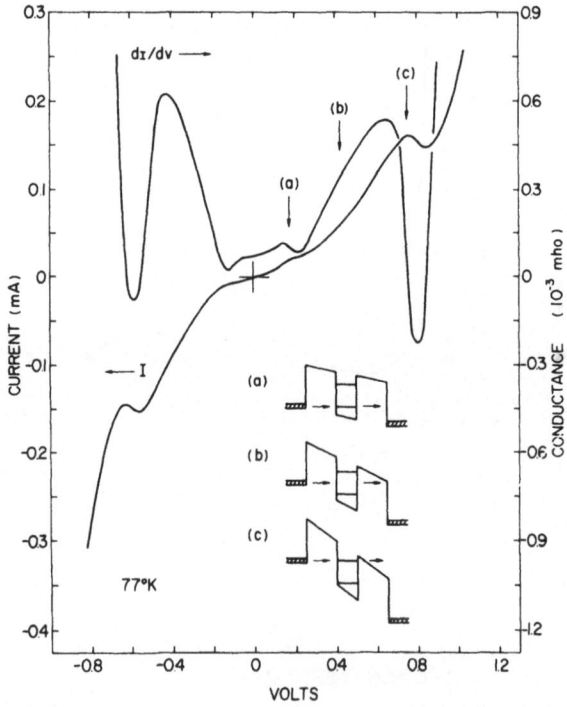

Figure 5 - Current-voltage and conductance-voltage characteristics of a double-barrier structure. Conditions at resonance (a), (c), and off-resonance (b), are indicated.

5 for a double barrier with a well of 50Å and two barriers of 80Å made of $Ga_{0.3}Al_{0.7}As$. The schematic energy diagram is shown in the insert where the two bound states, E_1 and E_2, are indicated. The resonance is achieved at such applied voltages as to align the Fermi level of the electrode with the bound states, as shown in cases (a) and (c). The energies of the bound states, referred to the original conduction band edge in the quantum well, can be obtained from such resonant curves: half of the voltages at the current peaks correspond to the bound energies. A large number of double barriers were MBE-grown with four different well widths, d_w: 35, 40, 50 and 65Å, for which the bound energies were measured. As shown in Fig. 6, although there is some understandable spread, measured values for the bound states have been found to agree with the calculated curves for E_1 and E_2.

This experiment, together with the quantum transport measurement for a tight-binding superlattice,[25] probably constitutes the first clear observation of man-made bound states in both single and multiple potential wells. Such achievement in superlattices and quantum wells can be viewed as a laboratory practice of elementary one-dimensional

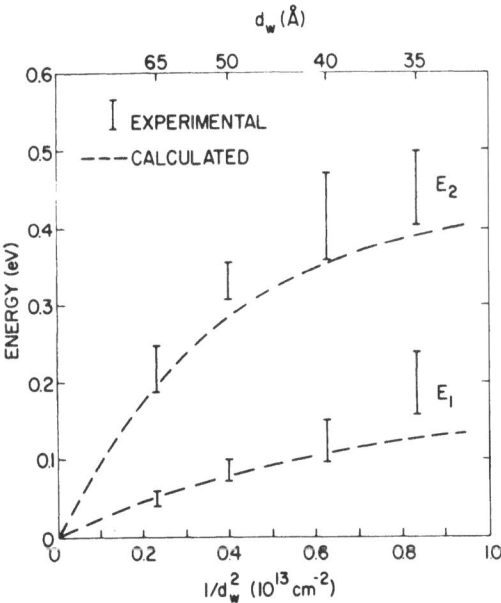

Figure 6 - Measured bound-state energies vs. $1/d_w^2$, where d_w: the well-width. The calculated curves are shown with dotted lines.

quantum physics described in textbooks (Do-It-Yourself Quantum Mechanics!). Sollner et al.[26] recently demonstrated dramatically-improved I-V characteristics in resonant double-barrier tunneling. Such an improvement clearly endorses the evolution of MBE for the last decade.

3.3 Optical Absorption for Quantum Wells and Superlattices (1974-1975)

Dingle et al.[27,28] observed pronounced structure in the optical absorption spectrum, representing bound states in isolated[27] and double quantum wells.[28] For the former, GaAs well widths in the range between 70Å and 500Å were prepared. The GaAs wells were separated by $Ga_{1-x}Al_xAs$ barriers which were normally thicker than 250Å. In low-tempereature measurements for such structures, several exciton peaks, associated with different bound-electron and bound-hole states, were resolved. For the latter study, a series of structures, with GaAs well widths in the range between 50Å and 200Å and $Ga_{1-x}Al_xAs(0.19 < x < 0.27)$ barrier widths between 12Å and 18Å, were grown by MBE on the GaAs substrates. The spectra at low temperatures clearly indicated the evolution of resonantly split discrete states into the lowest subband of a superlattice. In all experiments, in order to enhance the total GaAs absorption, as many as fifty (or eighty) GaAs layers were grown in a single structure, which, indeed, demonstrated the precision of MBE in fabricating thin and uniform layers. From analysis of the spectra, the electron and hole well depths were determined to be 85% and 15% of the total energy-gap difference, respectively.

3.4 Raman Scattering (1976-1980)

Manuel et al.[29] reported the observation of enhancement in the Raman cross section for photon energies near electronic resonance in $GaAs-Ga_{1-x}Al_xAs$ superlattices of a variety of configurations. Both the energy positions and the general shape of the resonant curves agree with those derived theoretically based on the two-dimensionality of the quantum states in such superlattices. Later, however, the significance of resonant inelastic light scattering as a spectroscopic tool was pointed out by Burstein et al.,[30] claiming that the method yields separate spectra of single particle and collective excitations which will lead to the determination of electronic energy levels in quantum wells as well as Coulomb interactions. Subsequently, Abstreiter et al.[31] and Pinczuk et al.[32] observed light scattering by intersubband single particle excitations, between discrete energy levels, of two-dimensional electrons in GaAs-Ga$_{1-x}$Al$_x$As heterojunctions and quantum wells.

Meanwhile, Colvard et al.[33] reported the observation of Raman scattering from folded acoustic longitudinal phonons in a GaAs(13.6Å)-AlAs(11.4Å) superlattice. The superlattice periodicity is expected to result in Brillouin-zone folding (as previously mentioned) and the appearance of gaps in the phonon spectrum for wave vectors satisfying the Bragg condition. Before this observation, Narayanamurti et al.[34] showed selective transmission of high-frequency phonons due to narrow band reflection determined by the superlattice period.

3.5 Modulation Doping (1978) and Subsequent Developments

It is usually the case that free carriers, electrons or holes created in a semiconductor by impurity inevitably suffer from impurity scattering. There are a few exceptions such as Si MOSFETs, where electrons or holes are induced by applied gate voltages. InAs-GaSb heterostructures are another example where electrons and holes are produced solely by electron transfer, as described later.

Now, in superlattices, it is possible to spatially separate carriers and their parent impurity atoms by doping impurities in the regions of the potential hills, as shown in Fig. 7. In the original article,[1] this concept was expressed in general terms as follows: "...if the superlattice structure is formed in such a manner that most scattering centers such as foreign atoms, imperfections, etc., are concentrated in the neighborhood of the potential hills, one can show that electrons would suffer less from such scattering center..."

In 1978, Dingle et al.[35] successfully implemented such a concept in modulation-doped GaAs-GaAlAs superlattices, as illustrated in the top of Fig. 7, achieving electron mobilities which far exceed the Brooks-Herring predictions. Modulation doping was performed by synchronizing the silicon (n-dopant) and aluminum fluxes in the MBE, so that the dopant was distributed only in the GaAlAs layers and was absent from the GaAs layers. Soon after, Störmer et al.[36] reported a two-dimensional electron gas at modulation-doped GaAs-GaAlAs heterostructures. Such heterostructures were used to fabricate a new high-speed field-effect-transistor[37,38] called MODFET, of which the

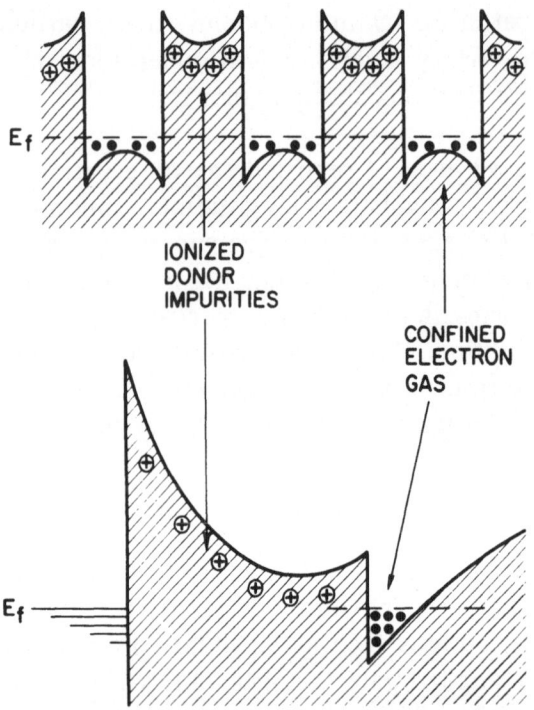

IONIZED
DONOR
IMPURITIES

CONFINED
ELECTRON
GAS

E_f

E_f

Figure 7 - Modulation doping for a superlattice (top) and a heterostructure with an attached Schottky junction (bottom).

energy diagram is shown in the bottom of Fig. 7. The device, if operated at 77K, apparently exhibits a performance three times faster than that of the conventional GaAs MESFET. Hall mobilities in the dark at 4.2K for such confined electrons recently exceeded 1000000 $cm^2/V \cdot sec$.[39,40]

Subsequently, a similar technique provided a two-dimensional hole gas at the hetero-interface.[41] Such a hole gas not only revealed characteristic quantum effects[42] but also was found useful for p-channel MODFETs.[43] More recently, Wang et al.[44] reported a hole mobility as high as $97000 cm^2/V \cdot sec$ at 4.2K and deduced a valence-band offset of $210 \pm 30 meV$ for $Ga_{0.5}Al_{0.5}As$-GaAs heterojunctions, corresponding to the conduction-band offset of 0.62 ± 0.05 energy gap difference.

3.6 Quantized Hall Effect (1980-1981) and Discovery of Fractional Filling (1982)

In 1980, v. Klitzing et al.[45] demonstrated the interesting proposition that quantized Hall resistance can be used for precision determination of the fine structure constant α, using two-dimensional electrons in the inversion layer of a Si MOSFET. Subsequently, Tsui and Gossard[46] found the modulation-doped GaAs-GaAlAs heterostructures desirable for this purpose, primarily because of their high electron mobilities, which led to the determination of α with a great accuracy such as $\alpha^{-1} = 137.035965(12)(0.089ppm)$.[47]

The quantized Hall effect in the two-dimensional electron or hole[42,48] system is observable at sufficiently high magnetic fields and low temperatures when there is little overlap in the density of state of neighboring Landau levels; in such a range of magnetic fields as to locate the Fermi level in the localized states between the extended states, the parallel component of resistance ρ_{xx} vanishes and the Hall resistance ρ_{xy} goes through plateaus, as shown in Fig. 8(a).
This surprising result can be understood by the argument that the localized states do not take part in quantum transport.[49] At the plateaus, the Hall resistance is given by $\rho_{xy} = h/e^2 i = \mu_0 c/2\alpha i \approx 25813$ Ω/i, where i is the number of filled Landau levels; h, Planck's con-

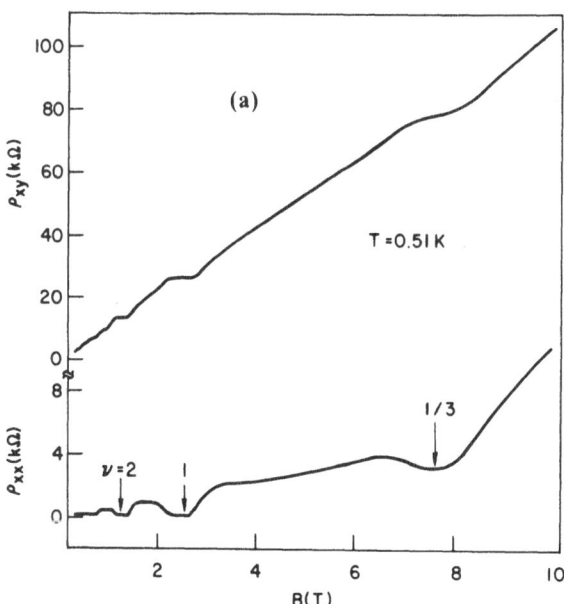

Figure 8(a) - Magnetoresistance ρ_{xx} and Hall resistance ρ_{xy} as a function of magnetic field at 0.51K for a heterojunction with an electron concentration of $0.6 \times 10^{11} cm^{-2}$.

stant; e, the electronic charge; μ_0, the vacuum permeability; and, c, the speed of light in vacuum.

Recently, Tsui, Störmer and Gossard[50] discovered a striking phonomenon: the existence of an anomalous quantized Hall effect, a Hall plateau in ρ_{xy} and a dip in ρ_{xx}, at a fractional filling factor of 1/3, as seen in Fig. 8(a), in the extreme quantum limit at temperatures lower than 4.2K. This discovery has spurred a large number of experimental and theoretical studies. Laughlin,[51] as an explanation of such a fractional filling, presented variational ground-state and excited-state wave functions which describe the condensation of a two-dimensional electron gas into a new state of matter, an incompressible quantum fluid. The elementary excitations of this quantum fluid are fractionally charged, and this elegant theory predicts a series of ground states characterized by the variational parameter m (m=3, 5...), decreasing in density and terminating in a Wigner crystal.

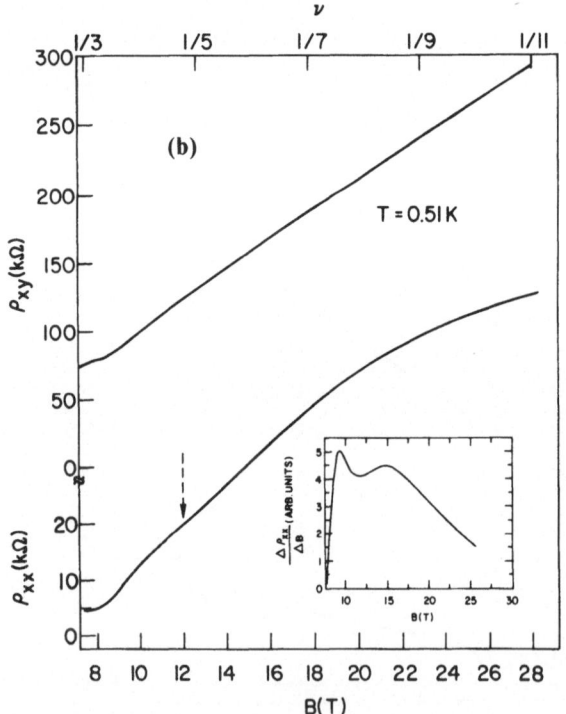

Figure 8(b) - High field magneto and Hall resistance for the same sample. The insert corresponds to the derivative with respect to the field of the magnetoresistance. The dip of the curve in the insert is indicated in the main figure by a broken arrow.

Mendez et al.[52] performed, as shown in Fig. 8(a) and (b), magnetotransport measurements at 0.51K (later 68mK)[53] and up to 28 T
for a dilute two-dimensional electron gas with a concentration of
$6 \times 10^{10} cm^{-2}$ (a mobility of 4.1×10^5 $cm^2/V \cdot sec$) in a GaAs-GaAlAs
heterojunction. The magnetoresistance indicated a substantial deviation
from linearity above 18 T and exhibited no additional features for
filling factors beyond 1/5, which suggested a transition to a crystalline
state. Meantime, Lam and Girvin[54] calculated the critical Landau-level
filling factor for transition from Laughlin's liquid state to a Wigner
crystal in comparing the energies of these states. The result appears to
be consistent with the experimental observation.

3.7 Variety of Heterojunctions and Superlattices and Relevant Topics

A major portion of the studies reported up to now have been
carried out with the $GaAs-Ga_{1-x}Al_xAs$ system. However, a variety of
other systems, notably InAs-GaSb(-AlSb), InAlAs-InGaAs,[55] InP-
lattice matched alloys,[56,57,58,59] Ge-GaAs,[60,61] CdTe-HgTe,[62,63,64]
PbTe-PbSnTe[65,66] and lattice-mismatched pairs of III-V
compounds[67,68] (strained-layer superlattices) have also been seriously
explored from scientific as well as technical aspects.

Semiconductor hetero-interfaces exhibit the abrupt discontinuity
in the local band structure, usually associated with a gradual band
bending in its neighborhood which reflects space-charge effects. According to the character of such discontinuity, known hetero-interfaces
may be classified into four kinds: type I, type II-staggered, type II-
misaligned, and type III, as illustrated in Fig. 9(a)(b)(c)(d): band
offsets (left), band bendings and carrier confinement (middle), and
superlattices (right). The conduction band discontinuity ΔE_c is equal to
the difference in the electron affinities of the two semiconductor. Case
(a), called type I, is applied to the GaAs-AlAs, GaSb-AlSb, GaAs-GaP
systems, etc., where their energy-gap difference $\Delta E_g = \Delta E_c + \Delta E_v$. On
the other hand, cases (b) and (c), type II, are applied to pairs of InAs-
GaSb, $(InAs)_{1-x}(GaAs)_x$-$(GaSb)_{1-y}(GaAs)_y$,[69] $InP-Al_{0.48}In_{0.52}As$,[59]
etc., where their energy-gap difference $\Delta E_g = |\Delta E_c - \Delta E_v|$ and electrons
and holes are confined in the different semiconductors at their heterojunctions and superlattices. Particularly, in case (c), type II-
"misaligned," the top of the valence band in GaSb is located above the
bottom of the conduction band in InAs by the amount of E_s, differing
from case (b), type-II "staggered," as shown in Fig. 9. In this classifi-

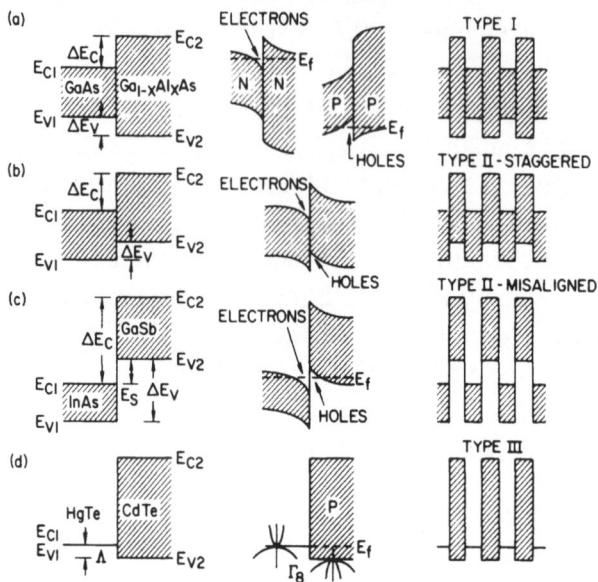

Figure 9 - Discontinuities of bandedge energies at four kinds of hetero-interfaces: band offsets (left), band bending and carrier confinement (middle), and superlattices (right).

cation, we add a unique member of the family, HgTe-CdTe,[64] as shown in Fig. 9(d), where HgTe is a zero-gap semiconductor due to the inversion of the relative positions of Γ_6 and Γ_8 edges. The Γ_8 light-hole band in CdTe becomes the conduction band in HgTe, where their energy difference Λ was determined to be 40 meV.

The bandedge discontinuities at the hetero-interfaces obviously command all properties of quantum wells and superlattices, and thus constitute the most relevant parameters for device design.[70] Recently, considerable efforts have been made to understand the electronic structure at interfaces[71] or heterojunctions.[72,73,74] Even in an ideal situation, the discontinuity provides formidable tasks in theoretical handling: Propagation and evanescent Bloch waves should be matched across the interface, satisfying continuity conditions on the envelope wave functions.[75,76] The fundamental understanding as well as the experimental determination for such important parameters as ΔE_c and ΔE_v, however, are still not satisfactory, even in the most-studied GaAs-Ga$_{1-x}$Al$_x$As system. Values of $\Delta E_c/\Delta E_g$, 85%, and $\Delta E_v/\Delta E_g$,

15%,[28] determined in 1975 apparently are to be revised to about 60% and 40%, respectively.[44,77,78,79,80]

Another breed called "n-i-p-i," the outgrowth of a doping super-lattice in the original proposal,[1,2] was pursued by Döhler[81] and Ploog et al.[82] As shown in Fig. 1, the periodic rise and fall of the bandedges is caused by a periodic variation of impurity doping. If this superlattice is illuminated, extra electrons and holes are attracted to minima in the conduction band and to maxima in the valence band, respectively. Thus, those extra carriers are spatially separated, resulting in anomalously long lifetimes. An interesting consequence of this fact is that the amplitude of the periodic potential is reduced by the extra carriers, leading to a crystal which has a variable energy gap. Döhler et al.,[83] indeed, observed in GaAs doping superlattices that the photon energies in luminescence, which represent the band gap, were varied from 1.3eV to 1.53eV by the laser excitation intensity.

Superlattice structures were also made of amorphous semiconductors. Abeles and Tiedje[84] pioneered the development of such structures consisting of alternating layers of hydrogenated amorphous silicon, germanium, silicon nitride, and silicon carbide.

The superlattice synthesis so far has been limited to the dual-constituent system. Esaki et al.[85] proposed the introduction of a third constituent such as AlSb in the InAs-GaSb system. Such a triple-constituent system leads to a new concept of man-made polytype superlattices, which offers an additional degree of freedom.

Some degree of the lattice-mismatch, however small, at hetero-interfaces is inevitable because of the joining of two different semiconductors. It is certainly desirable to select a pair of materials closely lattice-matched in order to obtain defect- and stress-free interfaces. However, heterostructures lattice-mismatched to some extent, 1 or 2%, can be grown with essentially no misfit dislocations, if the layers are sufficiently thin, because the mismatch is accommodated by uniform lattice strain.[86] On the basis of such premise, Osbourn[67] and his co-workers[68] prepared a number of strained-layer superlattices from lattice-mismatch pairs, claiming their relatively high-quality superlattices to be suitable for some applications. It is certainly true that, without the requirement of lattice matching, the number of available pairs for superlattice formation is greatly inflated.

Voisin et al.[87] made optical absorption measurements on GaSb-AlSb superlattices, of which the spectrum exhibits the two-dimensional density of states and pronounced free exciton peaks, as shown in the upper curve of Fig. 10. In comparison with the reported optical absorption spectrum[28] for the GaAs-GaAlAs case in the lower curve, it is noticed that exciton peaks 1 and 2 appear to be exchanged. Using an effective mass theory, such observation can be interpreted as the occurrence of the reversal of the heavy- and light-hole bands due to the strain effect induced by a lattice mismatch equal to 0.65%. Though this lattice mismatch implies that the present superlattice is rather modestly strained by Osbourn's standard, the pressure exerted on GaSb layers at the extreme (thinnest) limit is expected to be approximately 7.8×10^3 bars.

The structure is shown in the upper side of Fig. 11, where GaSb layers are made thinner than AlSb layers so as to observe the large modification of the GaSb bandstructure. The GaSb layers stretch in the direction shown with arrows as a result of a biaxial tensile stress,

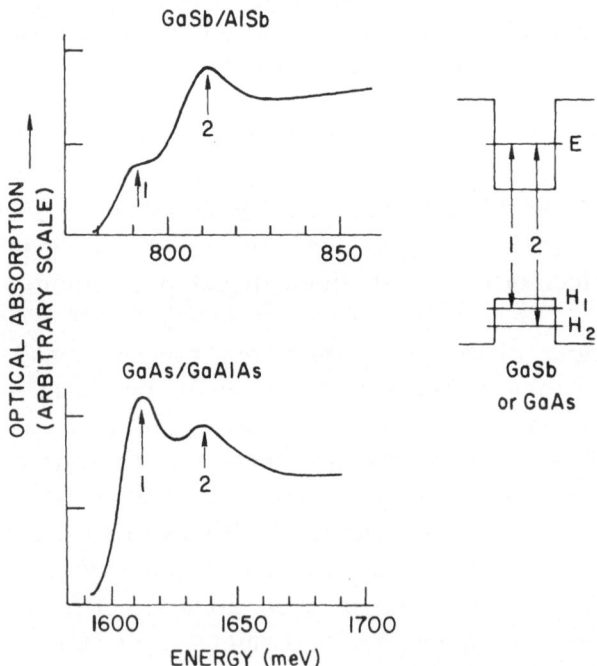

Figure 10 - Optical absorption vs. photon energy, comparing GaSb-AlSb and GaAs-GaAlAs. The relevant transitions in the quantum well are shown in the right.

while the AlSb layers are under a biaxial compressive stress. This can be analyzed in terms of the sum of a hydrostatic dilation plus a uniaxial compression for GaSb and a hydrostic compression plus a uniaxial elongation for AlSb. It is well known that the hydrostatic stress keeps the symmetry but changes the energy gap, whereas the uniaxial stress lifts the Γ_8 degeneracy. The result is schematically illustrated in Fig. 11 which indicates the ground valence band (GaSb) of this superlattice to be a light-hole state, despite its larger confinement energy.

3.8 InAs-GaSb Superlattices (1977-1982) and Quantum Wells (1982-)

In 1977, while searching for such a superlattice that the introduction of the periodic potential provides a greater modification to the

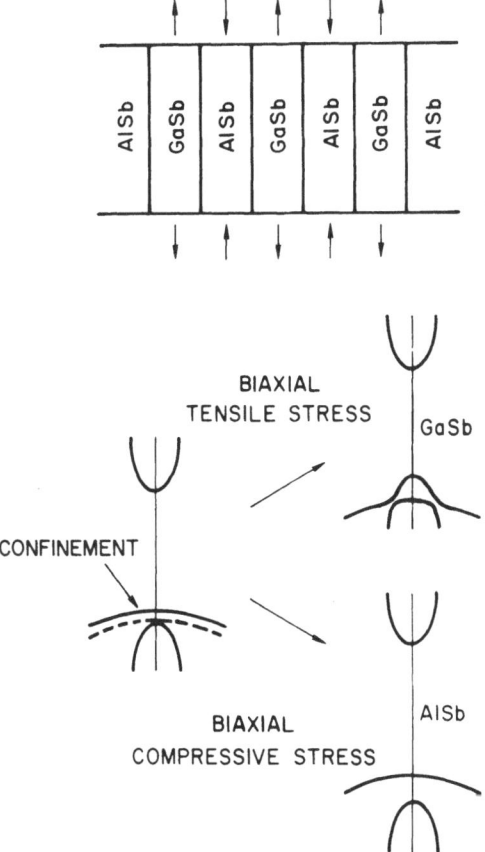

Figure 11 - Modification of the GaSb and AlSb band structure by built-in stress. The top shows the superlattice structure with the stress direction.

host bandstructure than that in the GaAs-AlAs system, the InAs-GaSb system was selected as a candidate, because of its extraordinary bandedge relationship at the interface, called type II "misaligned" in Fig. 9(c). It was observed that, in the study of $(InAs)_{1-x}(GaAs)_x$-$(GaSb)_{1-y}(GaAs)_y$ p-n heterojunctions,[69] the rectifying characteristic changes to nonrectification as both x and y approach zero, implying the change-over from the "staggered" heterojunction to the "misaligned" one. Such unusual nonrectifying p-n junctions are the direct consequence of "interpenetration" between the GaSb valence band and the InAs conduction band. At the hetero-interface, electrons which "flood" from the GaSb valence band to the InAs conduction band, leaving holes behind, produce a dipole layer consisting of two-dimensional electron and hole gases, as shown in the center of Fig. 9(c).

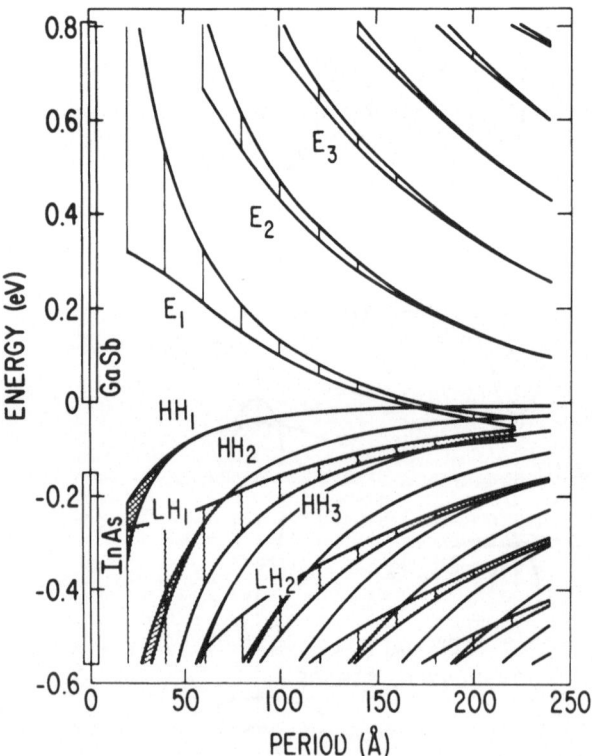

Figure 12 - Calculated subband energies and bandwidths for electrons (E_i) and light (LH_i) and heavy (HH_i) holes as a function of period, assuming $d_1 = d_2$.

First, Sai-Halasz et al.[88] made a one-dimensional calculation for InAs-GaSb superlattices, treating each host material in Kane's two-band framework. Subsequently, the LCAO band calculation[89] was performed, handling a large size of the primitive cell and ignoring charge redistribution at the interface. The calculated subband structure is strongly dependent upon the period. Figure 12 shows calculated subband energies and bandwidths for electrons and light and heavy holes as a function of period, together with the energy gaps of GaSb and InAs on the left, assuming $d_1 = d_2$, where the semiconducting energy gap is determined by the difference E_1-HH_1. This gap decreases with increase in the period, becoming zero at 170Å, as seen in the figure, corresponding to a semiconductor-to-semimetal transition. In those calculations, the misaligned magnitude, E_s (seen in Fig. 9(c)), was set at 0.15eV: a value which had been derived from analysis of optical absorption.[90] Recently, Altarelli[91] performed self-consistent electronic structure calculations in the envelope-function approximation with a three-band k•p formalism for this superlattice.

The electron concentration in superlattices was measured as a function of InAs layer thickness,[92] as shown in Fig. 13. The thickness of the GaSb layer is of secondary importance, since the energy gap is mainly determined by the ground subband in the conduction band of InAs. It is evident from Fig. 13 that the electron concentration exhibits a sudden increase of an order-of-magnitude in the neighborhood of 100Å This increase indicates the onset of electron transfer from GaSb to InAs: The transition from the semiconducting state to the semimetallic state, as the ground conduction subband, E_1, moves below the ground valence subband, HH_1. The observation is in good agreement with the theoretical prediction.

Far-infrared magneto-absorption experiments,[85,86] were performed at 1.6K for semimetallic superlattices with radiation near normal incidence to the layers. The transmission signal for each wavelength exhibits oscillations with increase in magnetic field. Figure 14 gives, as a function of the magnetic field B, the infrared energy positions of the transmission minima from such oscillations for the 120-80Å superlattice. The data indicate that the energies at which absorption maxima occur are directly proportional to B and all lines converge to -38±2 meV at zero magnetic field. Such absorption is interpreted as being due to interband transitions from H_1 to E_1 Landau levels illustrated in the inset of Fig. 14. If these transitions are assumed to occur

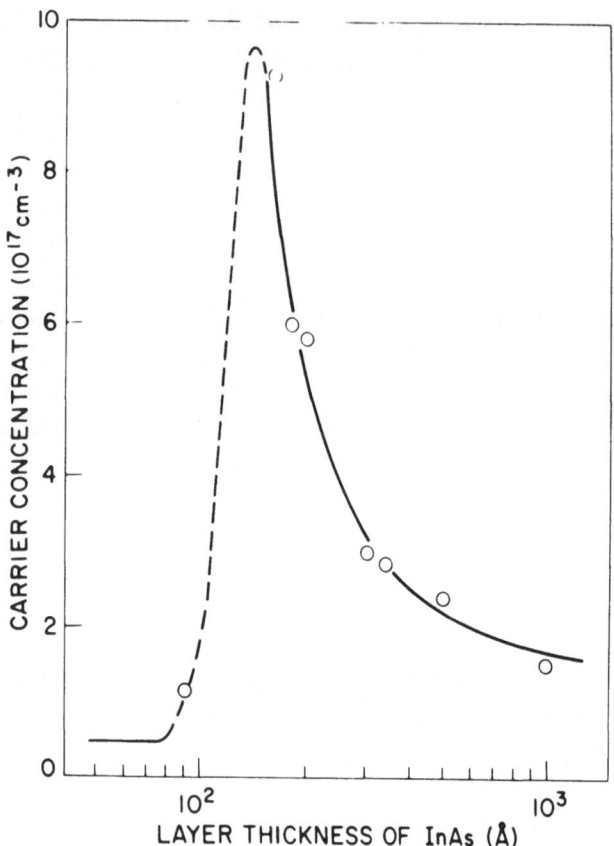

Figure 13 - Carrier concentration vs. InAs layer thickness to demonstrate the semiconductor-semimetal transition.

at a selection rule, $\Delta N=0$, the converged value should correspond to the negative energy gap of the semimetallic superlattice, E_1-H_1. The value is in good agreement with the calculation.

MBE-grown GaSb-InAs-GaSb quantum wells, as shown in Fig. 15(a), have been investigated, where the unique bandedge relationship allows the coexistence of electrons and holes across the two interfaces. Before an experimental approach, Bastard et al.[95] had performed self-consistent calculations for the electronic properties for such quantum wells, including the effect of magnetic fields, predicting the existence of a semiconductor-to-semimetal transition as a result of electron transfer from GaSb when the InAs quantum well thickness reaches a threshold, somewhat similar to the mechanism in the InAs-GaSb superlattices. Such a transition was confirmed by the experiment.[96] Figure

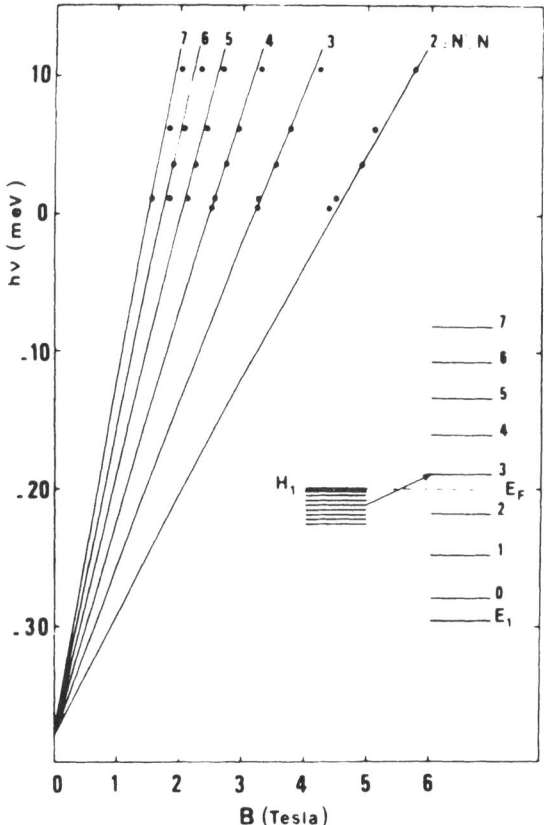

Figure 14 - Magneto-absorption vs. field B.   The insert shows the
Landau levels of E_1 and H_1 to illustrate interband transitions.

16 shows the electron concentration and mobility as a function of the
InAs layer thickness obtained from Hall measurements.   Two theoretical curves were drawn in the lower part of Fig. 16, assuming the misaligned magnitude, E_s, is 200meV and 175meV.   The theory apparently
predicts a faster decrease of carriers with decrease of well widths than
the observation.   Nevertheless, the data appear to indicate that E_s is
around 180meV, slightly larger than the value previously determined.
The calculations for electron mobilities[97] were made, which take into
account electron scattering by holes in GaSb and residual donors, N_i, in
InAs.   The results for $N_i = 0$ and 5×10^{16}cm^{-3} were shown with two
curves in Fig. 16.   The observed mobilities, particularly in thin wells,
are smaller than those calculated.   Scattering by the surface roughness
as well as possible charge centers in the neighborhood of the interfaces
may account for the discrepancy.

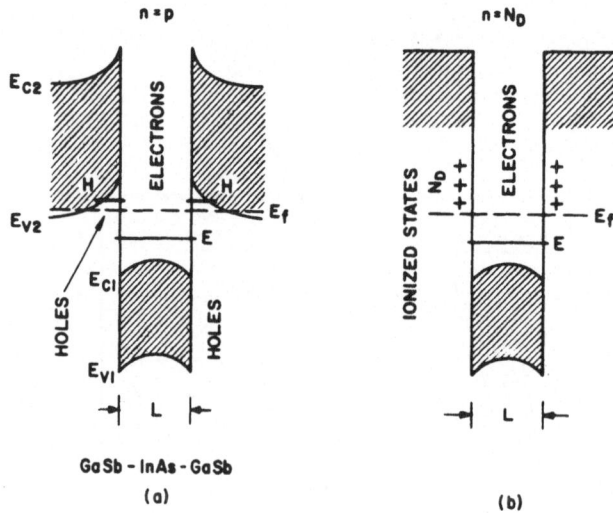

Figure 15 - Energy-band diagrams of GaSb-InAs-GaSb quantum wells. The left is for an ideal structure, whereas the right is the extreme case, where all electrons are supplied by ionized states, N_D.

The magnetoresistance ρ_{xx} and Hall resistance ρ_{xy} at 1.2K for a typical high-mobility sample with an InAs well width of 200Å (a mobility of $1.5 \times 10^5 \mathrm{cm}^2/\mathrm{V \cdot sec}$) are shown in Fig. 17, where the Landau level index, n, and the filling factor, i, in the quantized Hall effect are indicated. Figure 18 shows the standard plot of the Landau level index versus the inverse of magnetic fields of the maxima in ρ_{xx}, of which the slope was used to derive carrier concentrations of $9.0 \times 10^{11} \mathrm{cm}^{-2}$ at low fields and $7.5 \times 10^{11} \mathrm{cm}^{-2}$ at high fields.

There are a few surprises in this observation: No holes are evident in ρ_{xy} whereas the increase of ρ_{xx} is proportional to B^2 at low fields, suggesting a two-carrier conduction mechanism; there is a nearly 20% decrease of electrons at high fields; the most of all is the advent of anomalous peaks designated as A in ρ_{xx}, adjacent to ordinary peaks in Shubnikov-de Haas oscillations. Such anomalous peaks are plotted in Fig. 18 as black dots. Washburn et al.[98] made low-temperature measurements, as shown in Fig. 19, indicating a decrease of peak heights and then the total disappearance of the anomalous peaks at 19mK, as well as a shift of peak positions to lower magnetic fields with decrease in temperature.

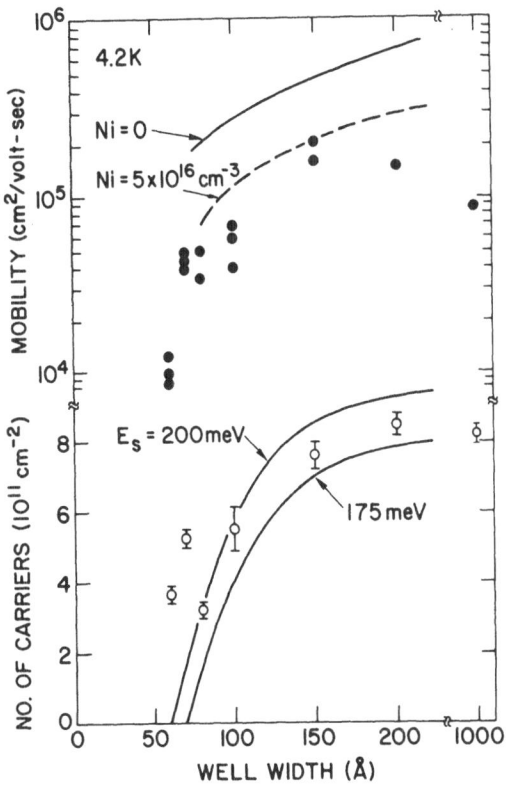

Figure 16 - Number of electrons and mobility as a function of the InAs well width. See text for theoretical curves.

Such phenomena appear to be too unusual to be interpreted in terms of the known effects at this time. Nevertheless, we can present the following considerations: First of all, the electron number, n, in the InAs layer, may be balanced not by the hole number, p, alone, but rather by $p + N_D$, where N_D is the concentration of ionized donor states in the neighborhoold of the interfaces, as shown in Fig. 15(b). The InAs well is also susceptible to unwanted modulation-doping from GaSb. It seems possible that p is a minor fraction of n, which means a rather large deviation from the theoretical calculations.[87]   Secondly, although these holes are contributing to conduction at low magnetic fields, they will form exotic electron-hole bound states such as excitons, possibly induced by a certain magnetic field strength, 5T or so, as shown in Fig. 20. The reduction of n may be indicative of the forma-

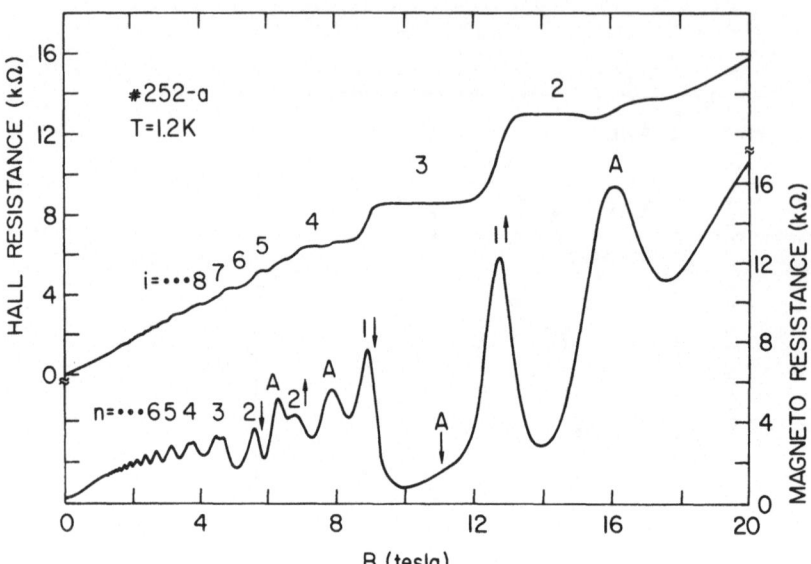

Figure 17 - Magnetoresistance and Hall resistance at 1.2K as a function of B for a high-mobility sample with an InAs well width of 200Å, where the Landau level index, n, and the filling factor, i, as well as anomalous peaks, A, are indicated.

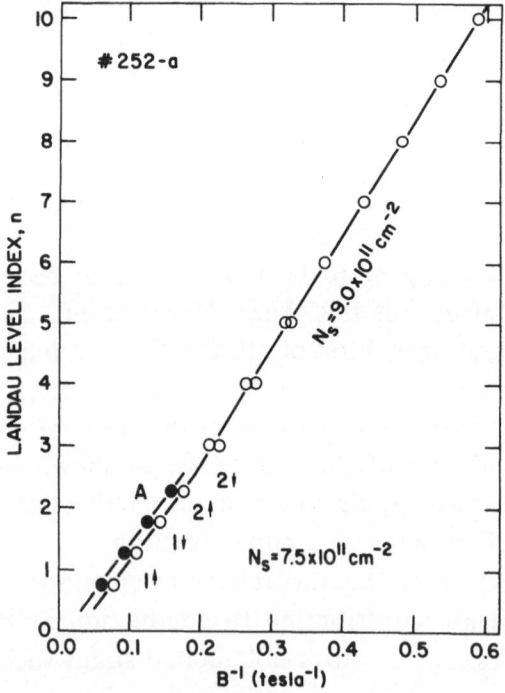

Figure 18 - Plot of the Landau level index vs. the inverse of magnetic fields of the maxima in ρ_{xx}, derived from Figure 17. The slope gives the electron concentration.

tion of such two-particle states which manifest themselves as anomalous peaks, but cease to contribute to conduction at very low temperatures. Along with this line, the binding energy for such an excitonic state is derived to be 6.4meV from the difference of the positions between the normal and anomalous peaks in Fig. 19. The obtained value is about three times the effective Rydberg of InAs (1.9meV), comparable to the hydrogenic binding energy[99] in the two-dimensional system, but larger than the previously calculated exciton energy[100] at no magnetic field condition.

MBE-grown AlSb-InAs-AlSb quantum wells[101] have also been studied. The electron concentration in the InAs well, however, appears to be susceptible to the exposure to moisture and its mobility is smaller than that in the GaSb-InAs-GaSb case, by, at least, an order of magnitude. A serious doubt about the integrity of the grown structure prevents the derivation of any conclusion from experimental results at this time.

4. CONCLUSION

We have witnessed remarkable progress of an interdisciplinary nature on this subject. Indeed, a variety of "engineered" structures exhibited extraordinary transport and optical properties; some of them, such as ultrahigh carrier mobilities, semimetallic coexistence of electrons and holes, etc., may not even exist in any "natural" crystal. Thus, this new degree of freedom offered in semiconductor research *through advanced material engineering* has inspired many ingeneous experiments, resulting in observations of not only predicted effects but also totally unknown phenomena such as fractional quantization which require novel interpretations.

The field is now so heavily proliferated that any future prediction is next to impossible. Nevertheless, one can safely make a few comments on the prospective research.
a) I would think that the route to one dimension is still wide open. It has recently become technically feasible to engrave ultrathin lines with dimensions less than 1000Å With such techniques and other means, a number of attempts have already been made for GaAs quantum well

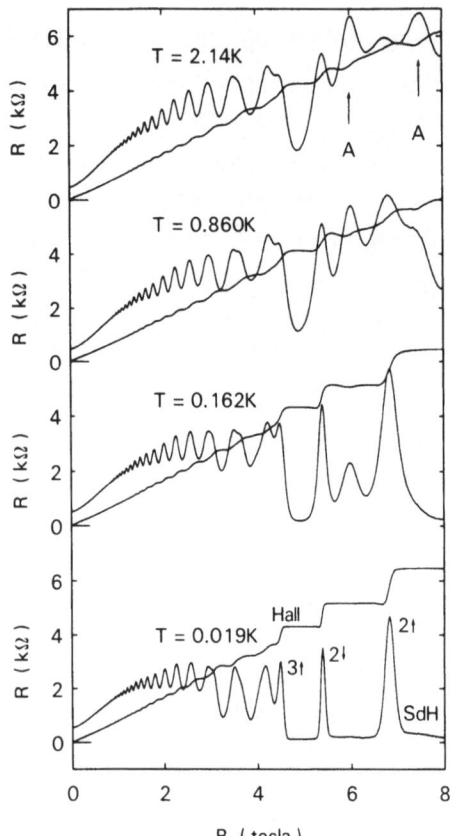

Figure 19 - Traces of the magnetoresistance and Hall resistance at four temperatures for the sample similar to that in Figure 17.

wires[102] or pinched accumulation layers in Si MOSFET.[103] A true one-dimensional electron gas, however, can be created with a combination of a superlattice potential and the surface inversion,[104,105] as schematically illustrated in Fig. 21. It is expected here that, using a superlattice either being p-type or having no carriers, surface electrons confined in two-dimensional quantum wells are generated either by an external field or by modulation-doping from a clad GaAlAs layer (not shown in the figure). The dimensionality between one and two will be exhibited, depending on the superlattice potential profile.

b) Our early efforts focussed on transport properties in the direction of the one-dimensional periodic potential. Later, however, transport of a two-dimensional gas in the layer plane prevailed. Former efforts apparently demand the stringent requirement for lateral uniformity as well as the low density of defects in grown wafers. With improved

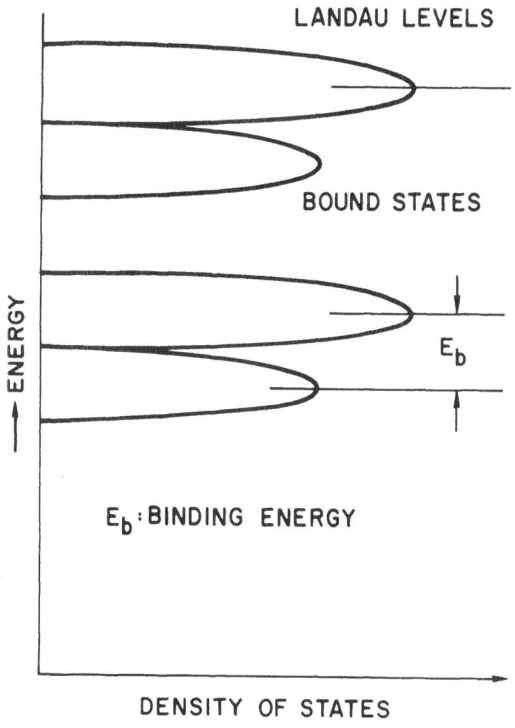

Figure 20 - Bound states associated with Landau levels.

techniques, more studies in this direction[26] will be made, expecting some interesting result.

c) The discovery of the fractionally quantized Hall effect is probably one of the most significant events in this field. More efforts will be needed for in-depth understanding of this unusual effect, including a transition to a Wigner crystal.

d) In two dimensions, there is no true metallic conduction:[106] all of the two-dimensional electron gas systems exhibit a logarithmic increase of the resistance as the temperature decrease,[107] which is characteristic of the weak localization. Thus, studies on the localization are inevitable in physics on the reduced dimensionality.

e) Finally, last but not least, still required are better theoretical treatment, high-quality hetero-epitaxy, and accurate measurement with regard to the bandedge discontinuities at the hetero-interfaces.

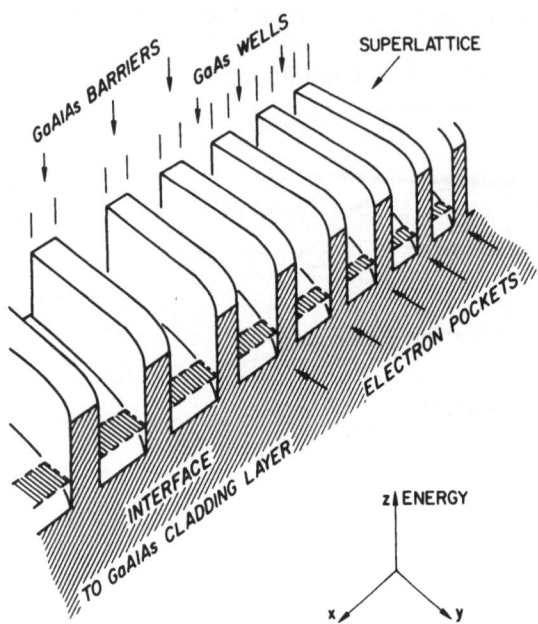

Figure 21 - The quasi-one-dimensional electron system on the inversion layer formed on the side surface of a superlattice crystal.

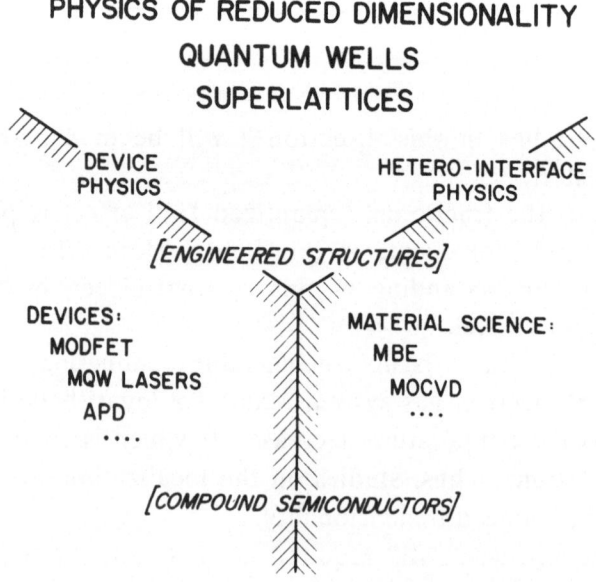

Figure 22 - Research in the interdisciplinary environment.

The interdisciplinary aspect of the research in this field is represented by a picture, shown in Fig. 22. The new frontiers of physics are explored in close cooperation with material science; such activities, in turn, give immeasurable stimulus to device physics, leading to novel semiconductor devices such as MODFET,[37,38] MQW-lasers,[108] advanced APDs[109,110] and real-space electron-transfer devices,[111] or provoking new ideas[112,113] for applications.

I hope this article, which cannot possibly cover every landmark, provides a flavor of the excitement in this field. Finally, I would like to acknowledge my many colleagues in our superlattice research for their contributions, and also the ARO's partial sponsorship throughout our investigation from the very beginning.

REFERENCES

1. L. Esaki and R. Tsu, IBM Research Note RC-2418 (1969).
2. L. Esaki and R. Tsu, IBM J. Res. Develop. 14, 61 (1970).
3. D. Bohm, *Quantum Theory* (Prentice Hall, Englewood Cliffs, N.J. 1951), p. 283.
4. L. Esaki, *Les Prix Nobel en 1973,* Imprimerie Royale P. A. Norstedt & Söner, Stockholm 1974, p. 66; Science 183, 1149 (1974).
5. A. E. Blakeslee and C. F. Aliotta, IBM J. Res. Develop. 14, 686 (1970).
6. L. Esaki, L. L. Chang, and R. Tsu, in Proceedings 12th International Conference on Low Temperature Physics, Kyoto, Japan, September 1970. (Keigaku Publishing Co., Tokyo, Japan), p. 551.
7. A. Y. Cho, Appl. Phys. Lett. 19, 467 (1971).
8. J. M. Woodall, J. Cryst. Growth 12, 32 (1972).
9. W. Shockley and G. L. Pearson, Phys. Rev. 74, 232 (1948).
10. R. B. Schoolar and J. N. Zemel, J. Appl. Phys. 35, 1848 (1964); J. N. Zemel, J. D. Jensen and R. B. Schoolar, Phys. Rev. 140A, 330 (1965).
11. K. G. Günther, Z. Naturforsch. 13a, 1081 (1968).
12. J. E. Davey and T. Pankey, J. Appl. Phys. 39, 1941 (1968).

13. J. R. Arthur, J. Appl. Phys. 39, 4032 (1968).
14. J. R. Arthur and J. J. LePore, J. Vac. Sci. Technol. 6, 545 (1969).
15. A. Y. Cho, Surf. Sci. 17, 494 (1964).
16. A. Y. Cho and J. R. Arthur, Proc. Solid State Chem. 10, 157 (1975); Molecular Beam Epitaxy, L. L. Chang and R. Ludeke, in *Epitaxial Growth*, ed. by J. W. Matthews (Academic, New York, 1975) Part A, Chapter 2.2; Molecular Beam Epitaxy, L. L. Chang, in *Handbook on Semiconductors*; (Series ed. T. S. Moss), Vol. 3, Materials, Properties and Preparation, (ed. S. P. Keller), (North-Holland, Amsterdam 1980) Chapter 9; Molecular Beam Epitaxy of III-V Compounds, K. Ploog, in *Crystals, Growth, Properties, and Applications*, (ed. by H. C. Freyhardt), Springer-Verlag Berlin-Heidelberg-N.Y., 1980, p. 73-162.
17. R. de L. Kronig, and W. J. Penney, Proc. Roy. Soc. A130, 499 (1930).
18. R. G. Chambers, Proc. Phys. Soc. (London) A65, 458 (1952).
19. L.L. Chang, L. Esaki, W. E. Howard, V. L. Rideout and J.F. Ziegler, presented at the IEEE Solid-State Device Research Conf., Edmonton, June 1972; L. Esaki, L.L. Chang, W.E. Howard and V.L. Rideout, presented at the 11th International Conf. Phys. Semicond., Warsaw, July 1972 (Proc. by PWN-Polish Scientific Publishers, Warsaw 1972, p. 431); L.L. Chang, L. Esaki, W.E. Howard and R. Ludeke, presented at the 19th National Vac. Sym., Chicago, Oct. 1972 (J. Vac. Sci. Technol. 10, 11, 1973).
20. L. L. Chang, L. Esaki, W. E. Howard, and R. Ludeke and G. Schul, presented at the Conf. Thin Film Phenomena, San Jose, March 1973 (J. Vac. Sci. Technol. 10, 655, 1973).
21. U. Gnutzmann and K. Clauseker, Appl. Phys. 3, 9 (1974).
22. A. Madhukar, J. Vac. Sci. Technol. 20, 149 (1982).
23. R. Tsu and L. Esaki, Appl. Phys. Lett. 22, 562 (1973).
24. L. L. Chang, L. Esaki, and R. Tsu, Appl. Phys. Lett. 24, 593 (1974).
25. L. Esaki and L. L. Chang, Phys. Rev. Lett. 33, 495 (1974).
26. T. C. L. G. Sollner, W. E. Goodhue, P. E. Tannenwald, C. D. Parker, and D. D. Peck, Appl. Phys. Lett. 43, 588 (1983).
27. R. Dingle, W. Wiegmann, and C. H. Henry, Phys. Rev. Lett. 33, 827 (1974).
28. R. Dingle, A. C. Gossard, and W. Wiegmann, Phys. Rev. Lett. 34, 1327 (1975).
29. P. Manuel, G. A. Sai-Halasz, L. L. Chang, C.-A. Chang, and L. Esaki, Phys. Rev. Lett. 25, 1701 (1976).

30. E. Burstein, A. Pinczuk, and S. Buchner, Physics of Semiconductors 1978, Institute of Physica Conference Series 43, London 1979, p. 1231.
31. G. Abstreiter and K. Ploog, Phys. Rev. Lett. 42, 1308 (1979).
32. A. Pinczuk, H. L. Störmer, R. Dingle, J. M. Worlock, W. Wiegmann, and A. C. Gossard, Solid State Commun. 32, 1001 (1979).
33. C. Colvard, R. Merlin, and M. V. Klein, and A. C. Gossard, Phys. Rev. Lett. 45, 298 (1980).
34. V. Narayanamurti, H. L. Störmer, M. A. Chin, A. C. Gossard and W. Wiegmann, Phys. Rev. Lett. 43, 2012 (1979).
35. R. Dingle, H. L. Störmer, A. C. Gossard, and W. Wiegmann, Appl. Phys. Lett. 33, 665 (1978).
36. H. L. Störmer, R. Dingle, A. C. Gossard, W. Wiegmann, and M. D. Sturge, Solid State Commun. 29, 705 (1979).
37. T. Mimura, S. Hiyamizu, T. Fujii and K. Nanbu, Jpn. J. Appl. Phys. 19, L225 (1980).
38. D. Delagebeaudeuf, P. Delescluse, P. Etienne, M. Laviron, J. Chaplart, and N. T. Linh, Electron. Lett. 16, 667 (1980).
39. M. Heiblum, E. E. Mendez, and F. Stern, Appl. Phys. Lett. 44, 1064 (1984).
40. E. E. Mendez, P. J. Price, and M. Heiblum, to appear in Appl. Phys. Lett.
41. H. L. Störmer and W. T. Tsand, Appl. Phys. Lett. 36, 685 (1980).
42. H. L. Störmer, Z. Schlesinger, A. Chang, D. C. Tsui, A. C. Gossard, and W. Wiegman, Phys. Rev. Lett. 51, 126 (1983).
43. H. L. Störmer, K. Baldwin, A. C. Gossard, and and W. Wiegmann, Phys. Rev. Lett. 44, 1062 (1984).
44. W. I. Wang, E. E. Mendez, and F. Stern, to appear in Appl. Phys. Lett.
45. K. v. Klitzing, G. Dorda, and M. Pepper, Phys. Rev. Lett. 45, 494 (1980).
46. D. C. Tsui and A. C. Gossard, Appl. Phys. Lett. 38, 550 (1981).
47. D. C. Tsui, A. C. Gossard, B. F. Field, M. E. Cage, and R. F. Dziuba, Phys. Rev. Lett. 48, 3 (1982).
48. E. E. Mendez, W. I. Wang, L. L. Chang, and L. Esaki, Phys. Rev. B, 28, 4886 (1983).
49. T. Ando and Y. Uemura, J. Phys. Soc. Jpn. 36, 959 (1974).
50. D. C. Tsui, H. L. Störmer, and A. C. Gossard, Phys. Rev. Lett. 48, 1559 (1982).
51. R. B. Laughlin, Phys. Rev. Lett. 50, 1395 (1983).

52. E. E. Mendez, M. Heiblum, L. L. Chang, and L. Esaki, Phys. Rev. B, 28, 4886 (1983).
53. E. E. Mendez, L. L. Chang, M. Heiblum, L. Esaki, M. Naughton, K. Martin, and J. Brooks, to appear in Phys. Rev. B.
54. Pui K. Lam and S. M. Girvin, Phys. Rev.B, 30, 473 (1984).
55. R. People, K. W. Wecht, K. Alavi, and A. Y. Cho, Appl. Phys. Lett. 43, 118 (1983).
56. K. Y. Cheng, A. Y. Cho, and W. R. Wagner, Appl. Phys. Lett. 39, 607 (1981).
57. M. Razeghi and J. P. Duchemin, J. Vac. Sci. Technol. B, 1, 262 (1983).
58. M. Voos, J. Vac. Sci. Technol. B, 1, 404 (1983).
59. E. J. Caine, S. Subbanna, H. Kröemer, J. L. Merz, and A. Y. Cho, to appear in Appl. Phys. Lett.
60. P. M. Petroff, A. C. Gossard, A. Savage, and W. Wiegmann, J. Cryst. Growth 46, 172 (1979).
61. C-A. Chang, Armin Segmüller, L. L. Chang, and L. Esaki, Appl. Phys. Lett. 38, 912 (1981).
62. J. N. Schulman and T. C. McGill, Appl. Phys. Lett. 34, 883 (1979).
63. G. Bastard, Phys. Rev. B, 25, 7584 (1982).
64. Y. Guldner, G. Bastard, J. P. Vieren, M. Voos, J. P. Faurie and A. Million, Phys. Rev. Lett. 51, 907 (1983).
65. H. Kinoshita and H. Fujiyasu, J. Appl. Phys. 51, 5845 (1980).
66. E. F. Fantner and G. Bauer, *Two-Dimensional Systems, Heterostructures, and Superlattices*, ed. G. Bauer, F. Kuchar and H. Heinrich, Springer Series in Solid-State Science 53, Springer-Verlag, 1984, p. 207.
67. G. C. Osbourn, J. Appl. Phys. 53, 1586 (1982).
68. G. C. Osbourn, R. M. Biefeld and P. L. Gourley, Appl. Phys. Lett. 41, 172 (1982).
69. H. Sakaki, L. L. Chang, R. Ludeke, C.-A. Chang, G. A. Sai-Halasz, and L. Esaki, Appl. Phys. Lett. 31, 211 (1977).
70. H. Kröemer, Surf. Sci. 132, 543 (1983).
71. M. L. Cohen, Advances in Electronics and Electron Physics, Vol. 51, 1, Academic Press, 1980.
72. W. A. Harrison, J. Vac. Sci. Technol. 14, 1016 (1977).
73. W. R. Frensley and H. Kröemer, Phys. Rev. B16, 2642 (1977).
74. J. Tersoff, to appear in Phys. Rev. B.
75. G. Bastard, Phys. Rev. B24, 5693 (1981).
76. S. R. White and L. J. Sham, Phys. Rev. Lett. 47, 879 (1981).

77. H. Kröemer, Wu-Yi Chen, J. S. Harris, Jr., and D. D. Edwall, Appl. Phys. Lett. 36, 295 (1980).
78. R. C. Miller, A. C. Gossard, D. A. Kleinman and O. Munteanu, Phys. Rev. B 29, 3740 (1984).
79. R. C. Miller, D. A. Kleinman, and A. C. Gossard, Phys. Rev. B29, 7085 (1984).
80. T. W. Hickmott, P. M. Solomon, R. Fischer, and H. Morkoc, to be published in J. Appl. Phys.
81. G. H. Döhler, Phys. Status Solidi (b)52, 79 and 533 (1972).
82. K. Ploog, A. Fischer, G. H. Döhler, and H. Künzel, in *Gallium Arsenide and Related Compounds 1980*, Institute of Physics Conference Series No. 56, edited by H. W. Thim (Institute of Physics, London, 1981), p. 721.
83. G. H. Döhler, H. Künzel, D. Olego, K. Ploog, P. Ruden, H. J. Stolz, and G. Abstreiter, Phys. Rev. Lett. 47, 864 (1981).
84. A. Abeles and T. Tiedje, Phys. Rev. Lett. 51, 2003 (1983).
85. L. Esaki, L. L. Chang, and E. E. Mendez, Jpn. J. Appl. Phys. 20, L529 (1981).
86. J. H. van der Merwe, J. Appl. Phys. 34, 117 (1963).
87. P. Voisin, C. Delalande, M. Voos, L. L. Chang, A. Segmüller, C. A. Chang, and L. Esaki, to appear in Phys. Rev. B.
88. G. A. Sai-Halasz, R. Tsu, and L. Esaki, Appl. Phys. Lett. 30, 651 (1977).
89. G. A. Sai-Halasz, L. Esaki, and W. A. Harrison, Phys. Rev.B18, 2812 (1978).
90. G. A. Sai-Halasz, L. L. Chang, J-M Welter, C.-A. Chang, and L. Esaki, Solid State Commun. 25, 935 (1978).
91. M. Altarelli, Phys. Rev. B28, 842 (1983).
92. L. L. Chang, N. J. Kawai, G. A. Sai-Halasz, R. Ludeke, and L. Esaki, Appl. Phys. Lett. 35, 939 (1979).
93. Y. Guldner, J. P. Vieren, P. Voisin, M. Voos, L. L. Chang, and L. Esaki, Phys. Rev. Lett. 45, 1719 (1980).
94. J. C. Maan, Y. Guldner, J. P. Vieren, P. Voisin, M. Voos, L. L. Chang, and L. Esaki, Solid State Commun. 39, 683 (1981).
95. G. Bastard, E. E. Mendez, L. L. Chang, L. Esaki, J. Vac. Sci. Technol. 21, 531 (1982).
96. E. E. Mendez, L. L. Chang, C-A. Chang, L. F. Alexander, and L. Esaki, to appear in Surf. Sci.
97. E. E. Mendez, G. Bastard, L. L. Chang, C-A Chang, and L. Esaki, Bull. Am. Phys. Soc. 29, 471 (1984).

98. W. Washburn, R. A. Webb, E. E. Mendez, L. L. Chang, and L. Esaki, to appear in Phys. Rev. B.
99. G. Bastard, Phys. Rev. B 24, 4714 (1981).
100. G. Bastard, E. E. Mendez, L. L. Chang, and L. Esaki, Phys. Rev. B 26, 1974 (1982).
101. C-A. Chang, E. E. Mendez, L. L. Chang, and L. Esaki, to appear in Surf. Sci.
102. P. M. Petroff, A. C. Gossard, R. A. Logan and W. Wiegmann, Appl. Phys. Lett. 41, 635 (1982).
103. A. B. Fowler, A. Harstein, and R. A. Webb, Phys. Rev. Lett. 48, 196 (1982).
104. H. Sakaki, Jpn. J. Appl. Phys. Lett. 19, L735 (1980).
105. The structure shown is the outgrowth of discussions with L. L. Chang and P. J. Stiles.
106. E. Abrahams, P. W. Anderson, D. C. Licciardello, and T. V. Ramakrishnan, Phys. Rev. Lett. 42, 673 (1979).
107. S. Washburn, R. A. Webb, E. E. Mendez, L. L. Chang, and L. Esaki, Phys. Rev. B 29, 3752 (1984).
108. W. T. Tsang, Appl. Phys. Lett. 39, 786 (1981).
109. F. Capasso, J. Vac. Sci. Technol. B1, 457 (1983).
110. T. Tanoue and H. Sakaki, Appl. Phys. Lett. 41, 67 (1982).
111. K. Hess, M. Morkoc, H. Shichijo, and B. G. Streetman, Appl. Phys. Lett. 35, 469 (1979).
112. J. J. Quinn, U. Strom, and L. L. Chang, Solid State Commun. 45, 111 (1983).
113. T. Nakagawa, N. J. Kawai, K. Ohta, and M. Kawashima, Electronics Lett. 19, 822 (1983).

KINETIC AND SURFACE ASPECTS OF MBE

B. A. Joyce

Philips Research Laboratories, Redhill, Surrey, England

ABSTRACT: Surface kinetic data relating to the growth of thin films of III-V compounds and alloys by MBE can be obtained using modulated beam techniques. Data acquisition and signal processing methods are described, and application to the study of Ga-As$_4$ and Ga-As$_2$ surface reactions presented in detail. The evaluation of composition control during growth of both mixed Group III and mixed Group V element alloys is also discussed.

Some consequences on film properties of the differing surface chemistry involved in growth from Group V element tetramers compared with dimers are considered in relation to possible reaction mechaniums. The evaluation of the surface crystallographic and electronic structure of differently reconstructed GaAs(001) surfaces using RHEED, ARPES and core level spectroscopy is briefly reviewed.

1. INTRODUCTION

The growth mechanisms involved in the preparation of thin films of III-V compounds and alloys by molecular beam epitaxy (MBE) have been studied extensively by the modulated molecular beam technique. This can provide detailed information on kinetic factors controlling the growth processes, which operate far from equilibrium. It is also important to have an understanding of the crystallographic and electronic structure of the surfaces involved especially during growth. This has been obtained principally by using a combination of reflection high energy electron diffraction (RHEED) with synchrotron- excited angle resolved photoemission (ARPES) and core level spectroscopy.

In this article we first describe the basic experimental details of data acquisition and signal processing used in the modulated beam technique and explain the methods of analysis. We then provide detailed experimental results for Ga-As$_4$ and Ga-As$_2$ reactions on GaAs(001) surfaces. These are used to deduce growth models for the two reaction systems and it is shown how the different surface chemistry involved in the two processes produces significant differences in film properties.

We next present a brief review of the results of a combined RHEED, ARPES and core level spectroscopy study of differently reconstructed surfaces grown in-situ by MBE. Surface structural models have been developed and the measured surface energy band dispersions compared with a calculated surface energy band structure.

Finally the kinetic processes occurring in the growth of mixed Group III element and mixed Group V element alloys are discussed, with particular emphasis on composition control and uniformity.

2. KINETICS OF III V COMPOUND GROWTH

We will treat GaAs as a typical example and consider growth from Ga + As$_4$ and Ga + As$_2$, but before describing results for these two systems, and the reaction models which have been developed, we will review briefly the experimental techniques which are available.

2.1 Experimental Methods

The basic experiment is, in principle, a very simple one: a beam of neutral atoms or molecules, having thermal velocities, and with an intensity in the range 10^{11} - 10^{16} atoms (molecules) cm^{-2} s^{-1} is directed at a substrate surface and the desorbing flux detected mass spectrometrically. The experiment is performed under UHV conditions ($\leq 10^{-10}$ torr) to minimise surface impurity effects and some provision may also be made for structural and compositional analysis of the surface, most frequently by RHEED and Auger electron spectroscopy (AES) respectively. A typical arrangement is shown in Fig. 1.

The major factor in surface scattering or desorption studies is to distinguish in the mass spectrometer between background signals and those produced by desorbing species. The technique used involves modulating either the incident beam or the desorbing flux and examining the mass spectrometer signal for a correlated response. The modulation and signal processing can be carried out in a number of ways, the complexity of which is largely determined by the nature of the desorbing species. The simplest method is to determine the response of the desorption signal to a step function change in the intensity of the

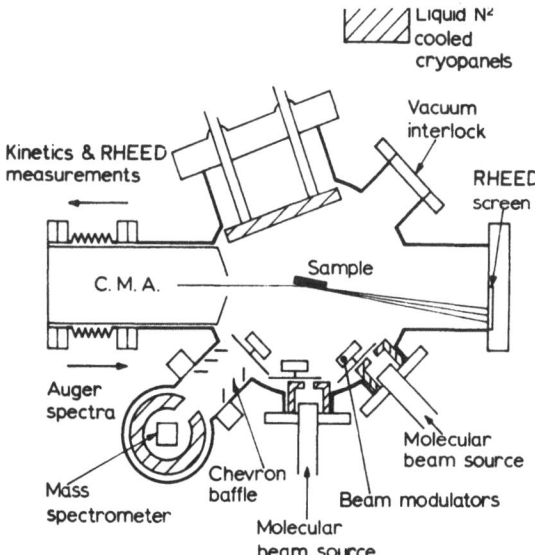

Figure 1 - Apparatus for modulated molecular beam measurements.

incident beam, produced by opening or closing a shutter. It is, however, only suitable for species which are readily condensible since otherwise it is not possible to obtain the required separation from background signals. Arthur[1] has successfully studied the desorption of Ga from GaAs in this way.

An improvement can be achieved by combining periodic modulation of the impinging beam with synchronous detection of the signals from the desorbing species. The amplitude and phase of the detected signal are compared with the amplitude and phase of the fundamental component of the modulated incident beam, but this only provides information at the fundamental (i.e., modulation) frequency. To extract details of surface processes measurements must therefore be made over a wide range of modulation frequencies, and time delays due to molecular flight times, velocity distributions, etc., must be accurately known. This method has been reviewed by Jones et al[2] and Schwarz and Madix.[3]

We have developed an alternative approach[4] to data acquisition and signal processing which still uses periodic modulation, but which has certain advantages over simple phase sensitive detection. Signal averaging is used to obtain statistically significant data, and by Fourier transforming the detected signal and the known time dependence of the incident flux, the attenuation and phase shift of the Fourier components of this flux due to all events between beam modulation and detection are obtained. Information on the surface process of interest is extracted by deconvolution. Since this is the technique which has been used to obtain most of the available kinetic results on surface reactions occurring in the growth and doping of GaAs films by MBE, we will consider it in rather more detail.

When analysing data from modulated beam experiments it is usually assumed that the behaviour corresponds to a time invariant linear system. If this is so, the response, $y(t)$, to an arbitrary stimulus, $x(t)$, is given by

$$y(t) = \int_0^\infty [h(\lambda)](t-\lambda)d\lambda \qquad (1)$$

where $h(\lambda)$ is the response of the same system to a unit impulse at time $-\lambda$. .An important consequence is that only those frequency components present in the stimulus $x(t)$ will be observed in the response, $y(t)$. This can be seen directly by Fourier transforming eq. (1) to give

$$Y(f) = [H(f)]X(f) \tag{2}$$

where $Y(f)$, $H(f)$ and $X(f)$ are the Fourier transforms of $y(t)$, $h(\lambda)$ and $x(t)$ respectively. The complex convolution integral is reduced to a simple product and it is obvious that an excitation at a frequency f results only in a response at the same frequency, but in general with modified phase and amplitude. If $H(f)$ is not a single process, but comprises a sequence of independent events, $H_1(f)$, $H_2(f)$, ..., $H_{(n-1)}(f)$, $H_n(f)$, they are commutative, and:

$$H(f) = \prod_{i=1}^{n} H_i(f) \tag{3}$$

In practice the overall transfer function $H(f)$ is a convolution of transfer functions representing several different processes, the most important of which are (i) the flight time of the velocity distributed molecular beam between modulation and the surface, (ii) the surface process of interest and, (iii) the flight time and velocity broadening between the surface and the detector. The transfer function representing the first process, $H_1(f)$ is given by the Fourier transform of:

$$h_1(\lambda) = 2\ell_1^4 \exp\left(-\ell_1/\lambda \, \alpha_1\right)^2 / \alpha_1^4 \, \lambda^5 \tag{4}$$

where ℓ_1 is the distance from modulator to substrate surface and $\alpha_1 = (2kT_F/M)^{1/2}$, where T_F is the source temperature and M the molecular weight of the effusing molecules. The required surface response function is $H_s(f)$ and that representing the third process, $H_2(f)$ is given by the Fourier transform of:

$$h_2(\lambda) = 2\ell_2^3 \exp\left(-\ell_2/\lambda\alpha_2\right)^2 / \alpha_2^3 \, \lambda^4 \, \pi^{1/2} \tag{5}$$

where ℓ_2 is the distance from substrate to detector and $\alpha_2 = (2kT_D/M)^{1/2}$, where T_D is the effective temperature of the desorbed species.

By combining eqs. (2) and (3)

$$H_s(f) = Y(f)\left\{[\prod_{i=1}^{n} H_i(f)]X(f)\right\}^{-1} = \frac{Y(f)}{X'(f)} \tag{6}$$

where $X'(f)$ may be regarded as an effective impulse function which produces response $Y(f)$ from the surface process $H_s(f)$. $X'(f)$ is calculated by transforming the known functions $h_i(\lambda)$ and the impulse function using a fast Fourier transform technique (5). Stages in this calculation are illustrated in Fig. 2. From a single experiment which gives values of $Y(f)$, values of $H_s(f)$ are therefore obtained at a number of frequencies, from which the following kinetic parameters can be obtained:

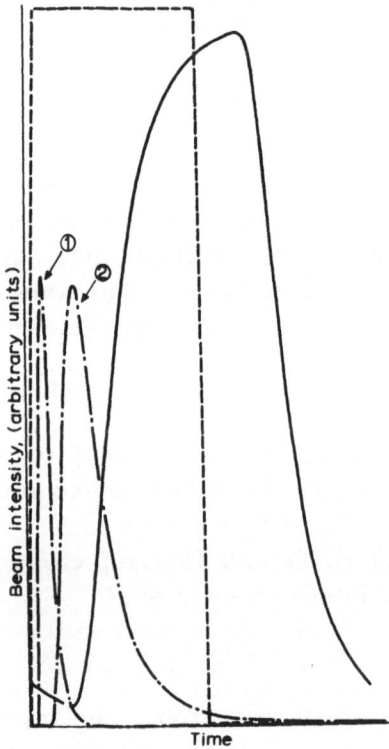

Figure 2 - Calculation of effective driving function at a fundamental frequency of 100Hz. Time domain representation: illustrating flight time delays and broadening introduced between modulator and surface and between surface and detector, (1) and (2) respectively.

(i)   thermal accommodation coefficients of molecules interacting with the surface.
(ii)   surface lifetimes and the energies of binding states.
(iii) sticking coefficients.
(iv)   orders of reaction.

By fitting the time or frequency domain curves given by eq.(5), with the single adjustable parameter T_D, to the known Maxwell-Boltzmann distribution function, the thermal accommodation coefficient, γ, can be obtained simply from

$$\gamma = (T_F - T_D)/(T_F - T_S) \tag{7}$$

where T_F has been previously defined as the temperature of the incident beam and T_S is the substrate temperature.

A molecule desorbing from a binding state of energy E and an associated surface lifetime τ at a substrate temperature T_S will give a transfer function H(f) in the frequency domain of:

$$H(f) = 1/(1 + i2\pi f\tau) \tag{8}$$

where

$$\tau = \tau_0 \exp (E/kT) \tag{9}$$

and τ_0 is a temperature independent vibrational term. τ can be calculated either from the phase shift produced at each frequency using the relationship:

$$\begin{aligned} \tan (\delta\theta)_f &= 2\pi f\tau \\ &= \mathrm{Im}H(f)/\mathrm{Re}H(f) \end{aligned} \tag{10}$$

where $(\delta\theta)_f$ is the measured phase shift at frequency f, or from the attentuation β_f, which is given by

$$\beta_f = [1 + (2\pi\tau f)^2]^{-1/2} = |H(f)| \tag{11}$$

If the residence time on the surface is long compared to the modulation period, a sticking coefficient can be defined as that proportion of the incident flux which reaches this state, and is measured as a frequency independent attenuation of the signal produced by desorbing molecules.

Finally, the order of the surface process is important for two reasons: firstly, for describing the kinetics and reaction mechanisms, but also because it can introduce non-linearity in the response if it is other than unity. This is illustrated in Fig. 3, which shows the relationship between surface concentration and desorption rate for a non-linear process. The response to a single frequency excitation will contain higher harmonics not present in the exciting signal, and their amplitude increases with increasing excitation amplitude. This invalidates the approach described here unless the amplitude of the excitation is small enough for the spurious harmonics to be of negligible amplitude.

In a practical modulated beam experiment this can always be checked by using a symmetrical excitation signal which therefore contains no even harmonics. The presence of even harmonics in the re-

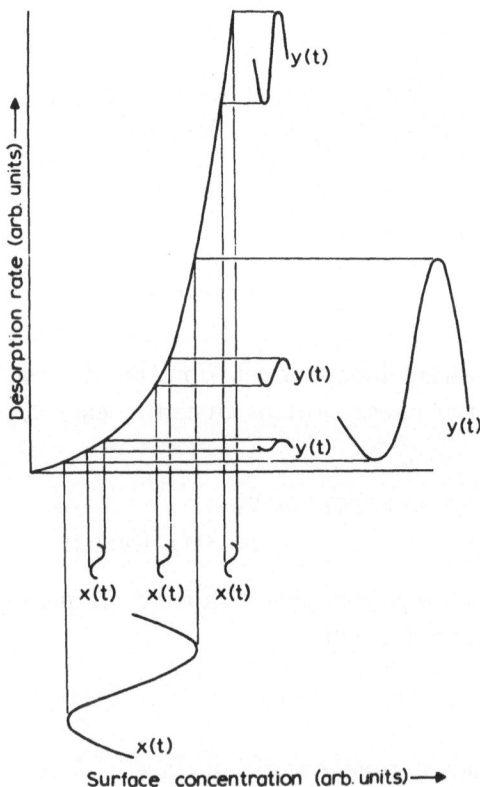

Figure 3 - Determination of reaction orders using small perturbations.

sponse then indicates non-linearity, so the intensity of the modulated beam must be reduced until they become negligible. The reaction order can then be determined by supplying additional material to the surface from a second, unmodulated source and measuring the response as a function of concentration.

2.2   Kinetic Models of GaAs Film Growth from the Application of Modulated Beam Techniques

Arthur[1] was the first to apply transient response techniques to study surface kinetics on GaAs. He measured the lifetime, τ, of Ga on (111) and ($\bar{1}\bar{1}\bar{1}$) GaAs over the temperature range 860K - 960K, and found it obeyed the usual Frenkel relationship ($\tau = \tau_0 \exp E_D/kT$), with $\tau_0 \approx 10^{-14}$s and $E_D \approx 2.5$eV. For T $\leq$ 750K, however, there was no desorption, the sticking coefficient of Ga was unity. The sticking coefficient of arsenic (as As_2) was effectively zero unless there was an excess surface population of Ga produced either by a separate flux or by heating the substrate above 775K, when thermal dissociation with arsenic desorption occurred. The important consequence of this work lies in its demonstration that the formation of films of III-V compounds from beams of the elements is kinetically controlled by adsorption of the Group V element, while the growth rate is determined only by the Group III element flux.

From this basis it has been possible to develop much more detailed growth models using the modulated beam methods described in the previous section (2.1), but it is important first to consider thermal effects occurring at a GaAs substrate, which can be investigated by similar techniques. We will restrict our discussion to the (001) oriented surface, which is polar and so may be terminated in either Ga or As atoms, or a combination of both. Above 600K a surface which is arsenic-rich will lose up to ≈ 0.5 monolayer of arsenic as As_2, leaving a Ga-rich surface. At temperatures $>$ 850K, the dissociative Langmuir evaporation of GaAs becomes significant;[6] it is congruent below $\approx$930K (i.e., the fluxes (J_i) leaving the surface are related by $J_{Ga} = 2J_{As_2}$). In this temperature range the evaporation rate of the compound is determined by the desorption rate of Ga, any surface arsenic evaporating as As_2. An upper limit to the evaporation rate of the compound is therefore set by the equilibrium vapour pressure of Ga over GaAs, which is similar to that over Ga,[6] and between 850K and 930K corresponds to evaporation rates from 0.01 to 1.0 monolayer s^{-1}.

Above 930K, As_2 is lost preferentially and the free Ga left on the surface aggregates to form liquid droplets.

We can now turn to the interaction of arsenic and arsenic + gallium on a GaAs surface, but it is important to define the species involved. Gallium is always monatomic, but an arsenic flux comprises either As_2 or As_4 molecules. Tetramers are produced from evaporation of elemental arsenic and dimers by evaporation of appropriate III-V compounds. This leads to a significant proportion of the Group III elements in the dimer flux ($\approx 15\%$ in the case of As_2 from GaAs) which makes accurate kinetic measurements difficult. It is therefore preferable to form the dimer beam from a two zone Knudsen cell,[7] the first stage of which is conventional and produces an As_4 beam, while the second, high temperature stage is optically baffled to maximise As_4 molecule collision with the hot surfaces. This arrangement can easily produce a conversion of $\geq 99.99\%$ As_4 to As_2.

Considering now As_2-Ga-GaAs interactions, Arthur[8] and Foxon and Joyce[9] have shown that the sticking coefficient of As_2 (S_{As_2}) is unity for a Ga adatom population of one monolayer. The latter authors also found that S_{As_2} increases linearly with the Ga adsorption rate, in fact reaching unity when $J_{Ga} = 2J_{As_2}$, i.e., stoichiometric GaAs will be formed provided $J_{Ga} < 2J_{As_2}$, any excess As_2 being lost by desorption. Above ≈ 600K however, additional surface processes become significant, as shown in Fig. 4, which shows the relative information available from modulating either the incident or the desorbing flux while maintaining a constant incident flux. With modulation of the desorbing flux, the detected signal is proportional to the total amount of As_2 leaving the surface, which is independent of temperature from 600K to 900K. Modulation of the incident flux however, produces a detected signal proportional to the amount of As_2 not chemisorbed, and this decreases with increasing temperature, i.e., the sticking coefficient of incident As_2 increases with increasing temperature. The total flux of As_2 molecules desorbing from the surface is thus made up of two parts: one from dissociation of the GaAs, which in turn creates a Ga surface population, and the other from incident molecules which are not chemisorbed. The sum is constant, but the ratio is a function of temperature, and it is clear the surface populations of gallium and arsenic present during growth will therefore depend on both substrate temperature and the relative flux intensities.

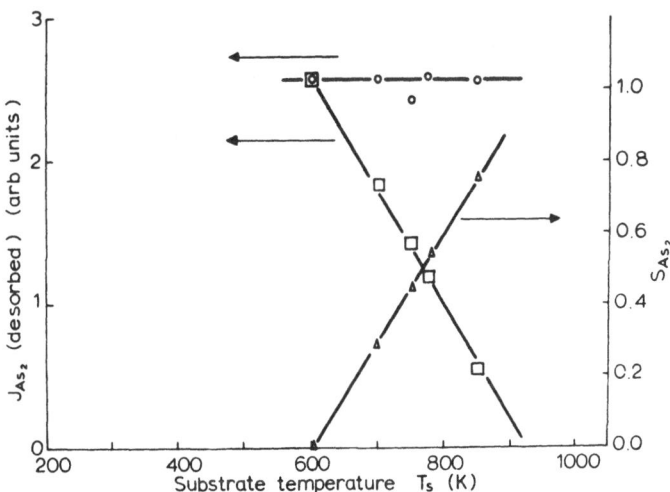

Figure 4 - The sticking coefficient of As_2 on GaAs (001) as a function of temperature. The various plots illustrate the relative information available from modulating either the incident or desorbing flux.

    Below $\approx 600K$ there is no measurable dissociation of GaAs but some incident As_2 molecules can associate on the surface to form As_4 before desorbing,[9] as shown in Fig. 5. The desorbing As_2 flux decreases monotonically with decreasing temperature, but the desorption rate of As_4 reaches a maximum at $\approx 450K$. The decrease at lower temperatures is an artefact arising from the use of a GaAs source to produce the As_2 This leads to the non-dissociative adsorption of some As_4 molecules on Ga atoms which arrive with the As_2 flux.

    The GaAs$_2$ interactions on GaAs are summarized in the growth model shown in Fig. 6. The basic process is a simple first order dissociative chemisorption of an As_2 molecule on a surface Ga atom, with the possibility of an association reaction to form As_4 at lower temperatures ($<600K$) and of some GaAs dissociation at temperatures above $\approx 600K$.

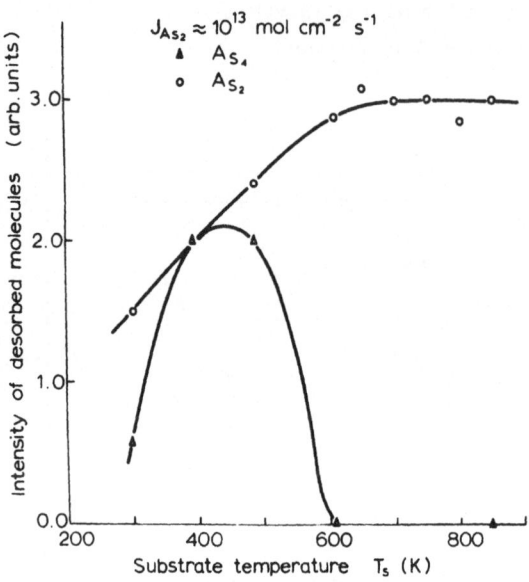

Figure 5 - Relative desorption rates of As_2 and As_4 for an incident As_2 flux as a function of temperature. Note the surface association of As_2 to form As_4 below 600K.

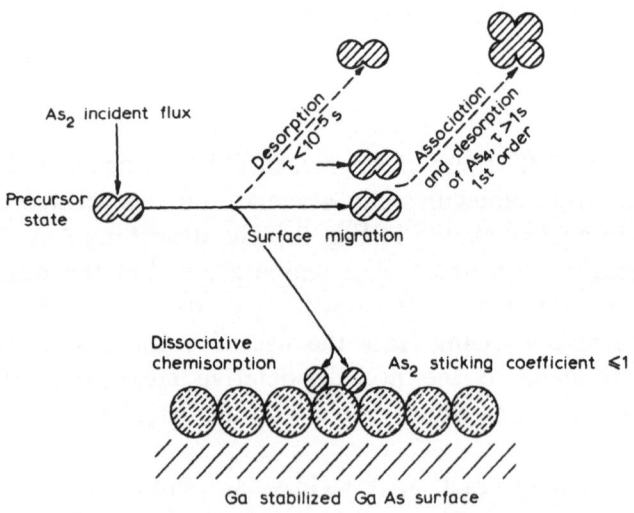

Figure 6 - Model of the growth of GaAs from molecular beams of Ga and As_2.

In contrast, processes involving an incident As_4 flux are significantly more complex,[10] as indicated by the results shown in Figs. 7 and 8. From the former we see that when Ga and As_4 beams interact on a GaAs surface the relative flux ratios strongly influence the As_4 sticking coefficient (S_{As_4}). For $J_{Ga} << J_{As_4}$, S_{As_4} is proportional to J_{Ga} $(S_{As_4} = J_{As_4}/4J_{Ga})$ and stoichiometric GaAs is produced. When $J_{Ga} \leq J_{As_4}$ however, S_{As_4} becomes independent of J_{Ga}, but never exceeds 0.5 (cf., S_{As_2}, which reaches unity under equivalent conditions). With As_4 moreover, in this region excess Ga is incorporated in the growing GaAs film, despite the fact that not more than half the As_4 supplied to the substrate surface is consumed.

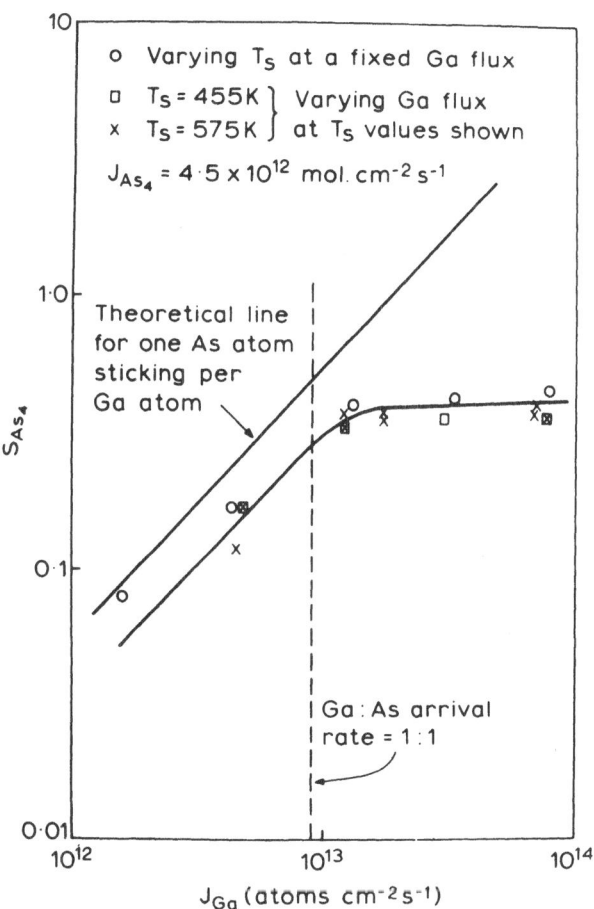

Figure 7 - The sticking coefficient of As_4 on GaAs (001) as a function of Ga flux.

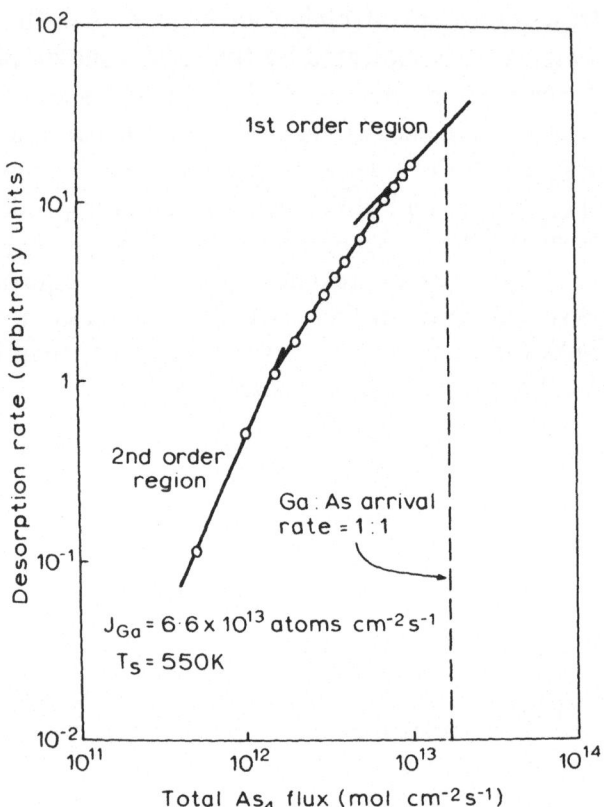

Figure 8 - Desorption rate of As$_4$ as a function of As$_4$ adsorption rate under Ga-rich surface conditions. The desorption rate is second order with respect to the incident flux at low fluxes, gradually becoming first order as the flux increases.

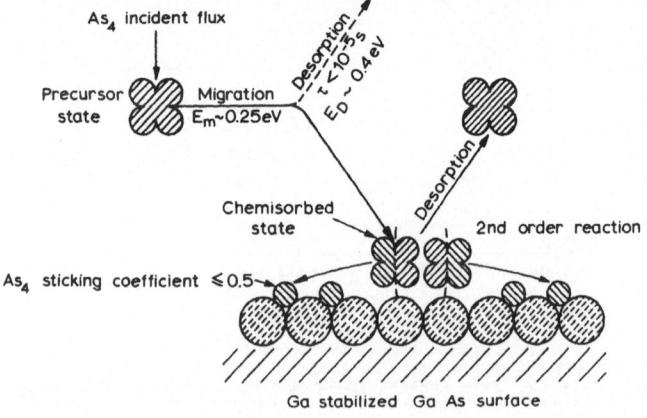

Figure 9 - Model of the growth of GaAs from molecular beams of Ga and As$_4$.

By measuring the desorption rate of As_4 as a function of the incident As_4 flux at a fixed Ga flux, we obtain the result shown in Fig. 8. At low fluxes (i.e., low surface concentrations) the desorption rate is second order with respect to the incidence rate, but becomes first order as the incident flux is increased. This does not imply a change of mechanism, it is the rate controlling step of the reaction which changes as the As_4 surface population increases.

Finally, at substrate temperatures $\leq 450K$, S_{As_4} is zero in the absence of a Ga surface population, but As_4 has a measurable surface lifetime, τ, given by $\tau = \tau_0 \exp (E_D/kT)$, where $\tau \approx 10^{-10}s$ and E_D, the activation energy of desorption, $\approx 0.4eV$.

The model for the growth mechanism of GaAs from Ga and As_4 beams which we have constructed from these results is shown in Fig. 9. The critical feature is the pairwise dissociation of As_4 molecules adsorbed on adjacent Ga atoms. From any two As_4 molecules four As atoms are incorporated in the GaAs lattice and the other four desorb as an As_4 molecule. This is consistent with the observed relationship $S_{As_4} = J_{As_4}/4J_{Ga}$ when $J_{Ga} \ll J_{As_4}$, i.e., when the surface population of Ga is small, and with $S_{As_4} = 0.5$ when the surface population of Ga is large. The model also explains the second order As_4 desorption rate at low relative incident rates, i.e., low surface concentration of As_4 molecules, since under these conditions the desorption rate will be determined by the probability of pairs of As_4 molecules being adsorbed on adjacent sites, which is simply proportional to the square of the adsorption rate. The change to first order at higher coverages reflects the change to a supply rate limited desorption rate.

2.3   Effects of Arsenic Species on Film Properties

The different growth mechanisms of GaAs films prepared from As_2 or As_4 might be expected to influence film properties, and in this section we will examine the available evidence. The crucial difference between the two reaction systems is that As_2 chemisorption involves a single Ga surface atom, while with As_4 there is an interaction involving adjacent pairs of Ga surface atoms. The steady state arsenic surface population should therefore be higher with As_2; with As_4 the maximum coverage will be $< 100\%$, since two adjacent sites must be occupied for interaction to occur, so that some proportion of single sites, estimated at $\approx 10\%$, will always remain unoccupied. It then seems reasonable to propose that non-occupied surface sites lead to (As) vacancy introduction, which will influence the deep level concentrations.

Neave et al[7] have conclusively demonstrated that the concentrations of three characteristic deep states (M1, M3 and M4) in MBE grown GaAs are substantially lower in films grown from As_2 than from As_4, all other conditions being identical, and this has been confirmed by Künzel et al.[11] The states involved are all electron traps in n-type material, and although they cannot unequivocally be related to native defects, the fact that they occur only in MBE-grown GaAs, with no dependence on the system used, strongly indicates this to be the case.

It has also been claimed[11,12] that the presence of sharp luminescence lines arising from defect induced bound excitons in the range 1.504eV to 1.511eV observed in low temperature photoluminescence spectra of MBE-grown GaAs could be attributed to the As_4 surface chemistry. When As_2 was used, the features were not present in the spectrum. However, Dobson et al[13] were not able to confirm this result, and observed the sharp emission structure with both As_2 and As_4 grown material but equally they found it could be absent in films produced from either species. Whether or not this spectral feature is observed probably depends on the free electron concentration and it is effectively screened out at values $\gtrsim 2 \times 10^{16}$ electrons cm $^{-3}$. It is therefore not certain that there is any substantiated effect of the growth chemistry on the low temperature photoluminescence spectra of MBE-produced GaAs films.

There is no doubt however, that there is a considerable influence on the ratio of near-band-edge to deep emission in the room temperature photo- luminescence spectra of $Ga_{0.8}Al_{0.2}As$ films, in that the band-edge-emission is increased at the expense of deep emission in As_2-grown films compared to equivalent As_4-grown layers.[14] This effect on the properties of $Ga_{0.8}Al_{0.2}As$ is even more apparent with $Ga_{0.8}Al_{0.2}As$-GaAs isotype double heterostructures, where if the alloy confinement layers are grown with As_2, the interface recombination velocity is decreased and the minority carrier lifetime increased with respect to equivalent As_4-grown structures. The species used to produce the GaAs layer is apparently not significant, but this probably relates both to the region where recombination predominantly occurs, i.e., close to the interface, and also to the diffusion length being considerably greater than the GaAs layer thickness, whichever species is used for growth.   Clearly, however, the concentration of recombination centres in the interface region is much lower when As_2 is used to grow the alloy film.

Finally, we can consider the influence of the arsenic species, and the associated surface chemistry, on the incorporation of an amphoteric dopant, germanium, in GaAs films.   On the basis of the proposed models, the relatively higher arsenic surface populaton obtained with As_2 should favour the incorporation of Ge as a donor (i.e., occupying a Ga site), and the degree of autocompensation should consequently be less with As_2 than As_4-grown films.   This was tentatively suggested by Neave et al,[7] but definitely confirmed by Künzel et al,[11] who observed almost a decade difference in the ratio at a growth temperature of 820K.

There is no doubt that the choice of arsenic species, As_2 or As_4, significantly influences film properties, and although the results are consistent with an explanation based on the surface chemistry models we have proposed, there is not yet definite proof.

3. CRYSTALLOGRAPHIC AND ELECTRONIC SURFACE STRUCTURE OF GaAs

In view of the apparent significance of surface chemistry for film properties, it is important to understand the surface structures, electronic and crystallographic, which are involved in the various processes. This information can be obtained from a combination of RHEED with synchrotron excited ARPES and core level spectroscopy, the surfaces being prepared in-situ by MBE.   The experimental arrangement we have used is illustrated in Fig. 10.   Since surface studies are the subject of a separate article, we will only present here a brief summary of those results relevant to surface kinetics, restricting our attention to (001)GaAs.

The characteristic feature is that this surface reconstructs, i.e., displays a lower symmetry than that produced by a simple termination of the bulk lattice.   In fact no clean, ordered (001)-1×1 structure related to either a cation or anion terminated surface exists.   The nature of the reconstruction is determined by the surface stoichiometry, since it is a polar surface and so can in principle be terminated by either a Ga

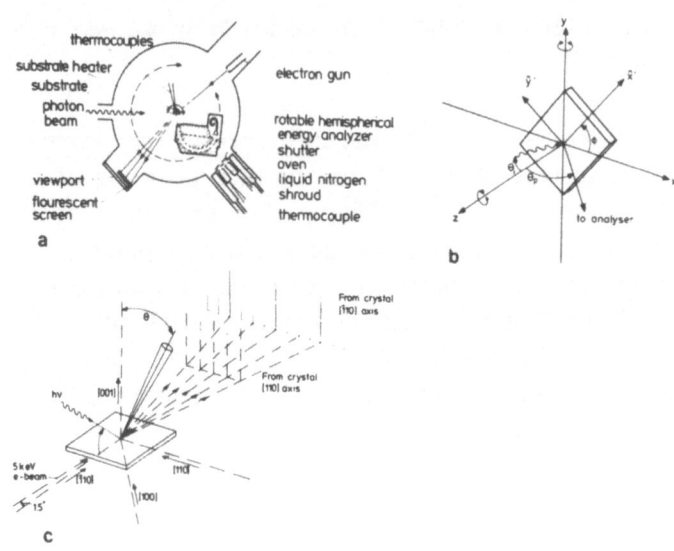

Figure 10 - (a) Horizontal cross-section of the experimental chamber; (b) angular relationships of incident radiation and photoelectron detector; (c) diffraction geometry related to the crystal surface (2×4 reconstruction) and photoelectron emission.

or As atom layer. In practice Ga-terminated surfaces have a very restricted existence region because any excess Ga quickly forms free metallic clusters. Furthermore, they tend only to be stable during growth, because any Ga atoms present in the surface when growth ceases provide ideal adsorption sites for arsenic species, arriving either from the vapour phase or by outdiffusion from the bulk. As-terminated surfaces are more stable and we will concentrate on the (001)-2×4, since it is the symmetry observed under most growth conditions, but we will also briefly discuss the (001)-C(4×4), which is produced when arsenic is present in excess.

3.1 The GaAs(001)-2×4 Surface

The symmetry is readily determined by observation of the RHEED pattern in two orthogonal <110> azimuths.[15] From a reciprocal lattice-Ewald sphere analysis of the length and width of the streaks in the RHEED patterns,[16] the main ordering direction is found to be [$\bar{1}$10], which displays two-fold periodicity, whereas in the orthogonal [110] direction, which is four-fold periodic, the ordering is much weaker. The primary reconstruction is therefore reflected in the two-fold periodicity. The appearance of curved streaks in RHEED patterns from azimuths between [110] and [$\bar{1}$10.] in fact confirms the presence of one-dimensional disorder boundaries, which are along the [$\bar{1}$10] direction.[17] The corresponding ARPES results show a two-fold periodicity with surface energy band dispersion along [$\bar{1}$10.], but no four-fold periodicity is observed directly.[16,18]

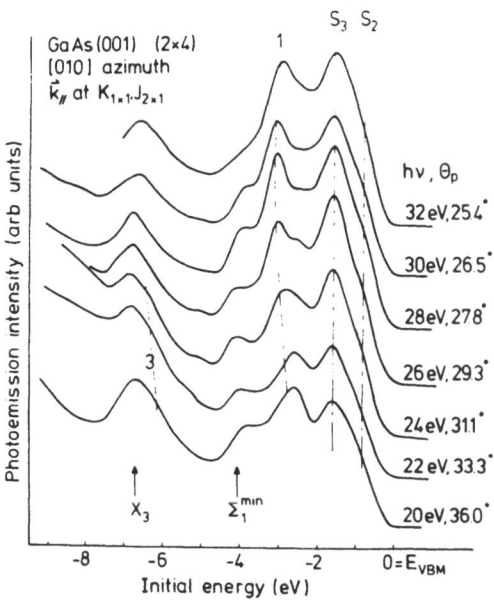

Figure 11 - Photoemission spectra for GaAs (001)-2×4, [010] azimuth. The photon energies and polar angles are related to keep $k_{\parallel}$ at the K_{1x1} symmetry point for $E_i = -1$eV.

ARPES experiments enable a direct determination to be made of the two-dimensional surface energy band dispersion $E = E(\bar{k}_{\parallel})$, where $\bar{k}_{\parallel}$ is the surface parallel wave-vector. For the GaAs (100)-2×4 surface results show a two-fold periodicity with energy band dispersion along [$\bar{1}10$], but four-fold periodicity is not observed. Because surface states do not disperse with $\bar{k}_\perp$ (the momentum perpendicular to the surface) for a given $\bar{k}_\parallel$, they can be distinguished from bulk states by choosing an appropriate combination of photon energies and polar angles. This allows us to vary $\bar{k}_\perp$ while keeping $\bar{k}_\parallel$ constant. A bulk transition which depends on $k_\perp$ will then disperse but a surface state transition will remain fixed in energy. This is shown in Fig. 11, where $\bar{k}_\parallel$ is kept fixed at the $K_{(1\times1)}$ symmetry point (see Fig. 12c, showing surface Brillouin zones (SBZ's) for (001) GaAs surfaces). Transitions 1 and 3 are from bulk states while S_2 and S_3 are from surface states. Alternative distinguishing procedures are to quench surface state emission by adsorption of foreign atoms, or to

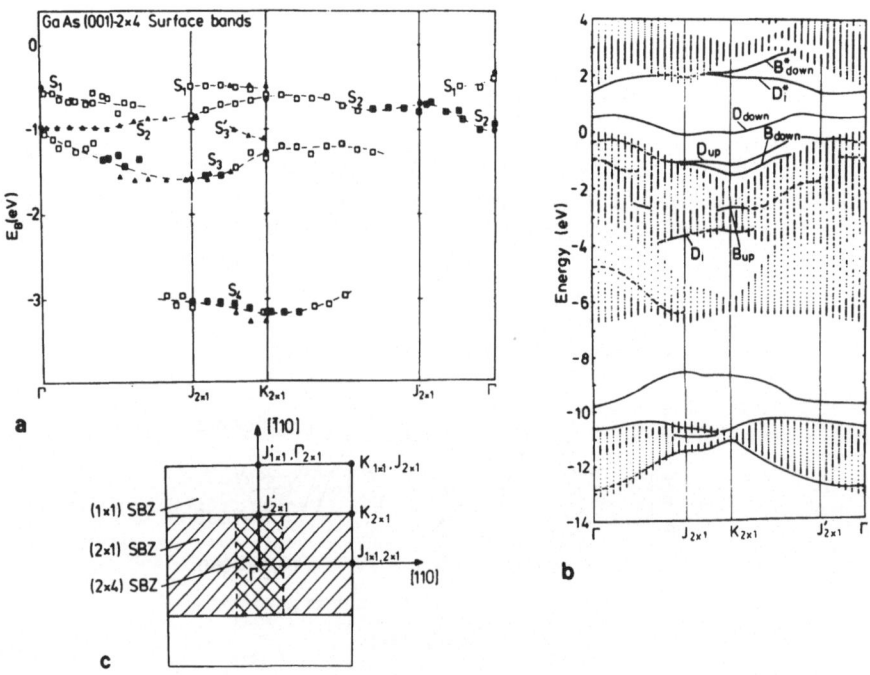

Figure 12 – (a) Experimental surface energy bands for GaAs (001)-2×4 along the symmetry lines of the 2×1 SBZ ; (b) Calculated surface band structure and projected bulk bands for the 2×1 asymmetric dimer model surface ; (c) SBZ's for the 1×1, 2×1 and 2×4 reconstructions.

compare spectra from different reconstructions of the same surface orientation, both of which can easily be achieved using the MBE approach. In general, the surface state emission is highly directional, with rather weakly dispersing surface energy bands.

The experimentally determined surface band structure[18] for GaAs (001)-2×4 is shown in Fig. 12a, within the framework of the 2×1 SBZ (Fig. 12c). This allows it to be compared directly with calculation, (Fig. 12b) which was carried out by the scattering theoretical method.[19] Using an empirical tight binding bulk Hamiltonian, surface states and resonances were derived from a layer, orbital and wave-vector resolved density of states. The essential feature of the surface model for this calculation is that the two-fold reconstruction is produced by the formation of a tilted (aplanar) As-As dimer bond. This gives rise to the band D_i near -3.5eV, and its antibonding counterpart D^*_i. The hybridisation and dimerisation leading to this is shown schematically in Fig. 13. Energy minimization calculations[20] have also shown the tilted As-As dimer to be a favourable configuration.

If now we compare the experimental and theoretical surface band structures we find very good agreement between the dimer bond related state D_i and the observed state S_4. The dangling bond states near the top of the valence bands are found both experimentally and theoretically, and angle of incidence-dependent measurements show them (S_1 and S_2) to be sp_z-like, but with some ($p_y - p_x$) character, indicative of tilted dangling bonds. There is not of course a one-to-one correspondence

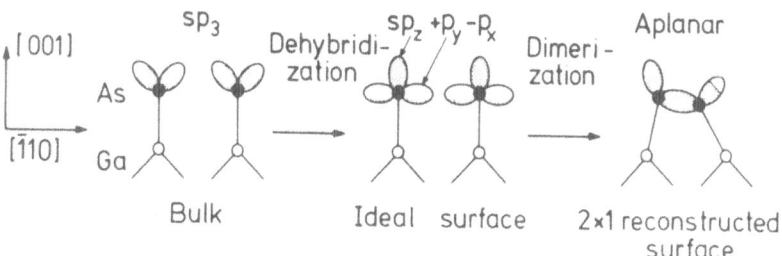

Figure 13 - Schematic diagram of the dehybridisation and dimerisation leading to the aplanar dimer structure.

between calculated and observed surface state energy locations, but this could not be expected with the limitations of the model used, which was based on 2×1, and not the actual 2×4 unit cell, in order to make the calculations tractable. From the combination of RHEED and ARPES data, however we are able to suggest a basis of the 2×4 unit cell, which is shown in Fig. 14. The two-fold periodicity in the [1̄10] direction is introduced by the asymmetric dimers, and the four-fold periodicity then merely reflects the phase sequence of dimer chains, giving rise to one-dimensional disorder boundaries along the [1̄10] direction.

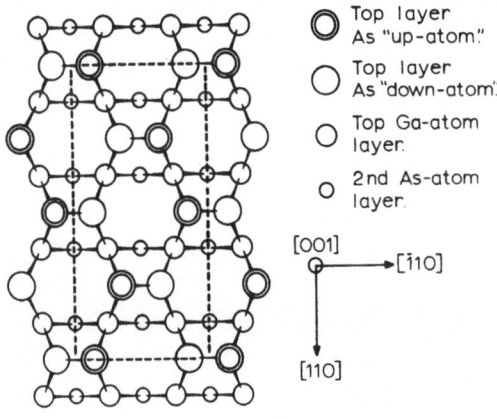

Figure 14 - Unit cell of the (001)-2×4 asymmetric dimer model.

Since we have seen significant differences between As_2 and As_4 growth in both surface chemistry and film properties, it is pertinent to ask if any effects are apparent in direct surface structure measurements. RHEED, in the absence of detailed intensity measurements is not sufficiently sensitive, but there are differences in the ARPES data, as shown in Fig. 15. The peak S_2 is due to emission from an As-derived surface state[18] and is sharper and more intense for Ga/As_2 than for Ga/As_4 growth. This indicates that the former surface is better ordered and has a slightly higher As concentration, consistent with the kinetic models, although of course not providing definite proof of them.

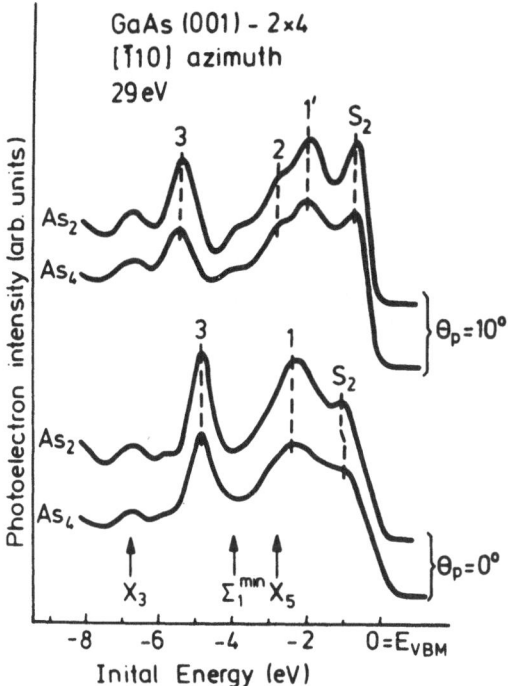

Figure 15 - AREDC's for the GaAs (001)-2×4 reconstruction prepared by Ga/As_2 and Ga/As_4 MBE growth.

3.2   The GaAs (001)-C(4×4) Surface

This surface symmetry is typically observed when a 2×4 reconstructed surface is cooled from the growth temperature in the presence of arsenic vapour (preferably As_2). Its significance lies in its potential for producing lowered Schottky barrier heights. It has been suggested[21] that the structure represents a perfect (i.e., complete monolayer) As-termination of the (001) surface, but from a combination of RHEED, angle resolved valence band spectra and surface sensitive As (3d) core level spectra we have been able to show that it is a chemisorbed structure, derived from excess As atoms trigonally bonded to the As-terminated (001) surface.[22]

The most important aspect of the photoemission results from the valence band region is the complete absence of the dimer-bond related band at $\approx$-3eV, which is the essential characteristic of the 2×4 surface.

In addition dangling bond states near the valence band maximum are different for the two reconstructions. It is, however, the core level spectra which in this case provide the critical evidence for the surface bonding. In Fig. 16a and b we compare the photoemission spectra of the As (3d) level from 2×4 and C(4×4) reconstructed surfaces taken at normal emission and hν=72.7eV. The final energy of the photoelectrons then corresponds to a small escape depth ($\approx$5.5Å) so the spectra are very surface sensitive. The spin-orbit split As (3d) level for the 2×4 spectrum is only slightly broadened by surface contributions on both sides of the bulk peak, but the C(4×4) spectrum shows an additional, well resolved doublet at 0.67 eV higher binding energy. The extra doublet coincides exactly with that measured for a film of elemen-

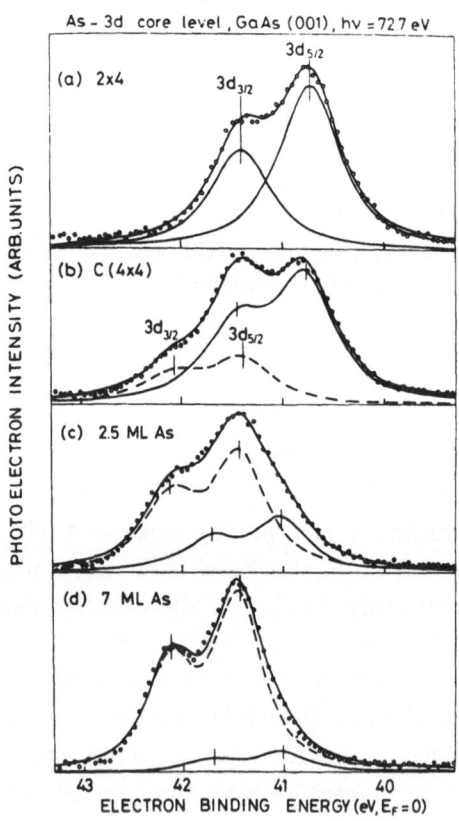

Figure 16 - Photoelectron As (3d) core level spectra for (a) GaAs (001)-2×4; (b) GaAs (001)-C(4×4); (c) and (d) amorphous arsenic film on a GaAs substrate.

tal, amorphous arsenic (Fig. 16c and d), which suggests that the bonding of As atoms is similar in the two cases. The shift to higher binding energy occurs because the bonding in GaAs is partly ionic, with charge displaced towards the As atom. This lowers the binding energy for As atoms in GaAs with respect to those in covalently bonded amorphous arsenic. There is no change in lineshape for the Ga (3d) level between the 2×4 and C(4×4) surfaces, indicating that the local environment of the Ga atoms does not change. It seems reasonable to suggest, therefore, that the C(4×4) structure is produced by chemisorption of additional As onto As atoms already present in the 2×4 reconstructed surface, and since the bonding in amorphous arsenic is trigonal, the core level spectra indicates it to be the same in the C(4×4) structure. RHEED results show that the same symmetry is maintained for a wide range of arsenic exposures (and hence coverages) and we have proposed[22] the type of unit cell shown in Fig. 17a and b for the C(4×4) reconstruction, corresponding to 25% and 50% additional As coverage respectively, while maintaining the same symmetry.

GaAs(001)-C(4×4)

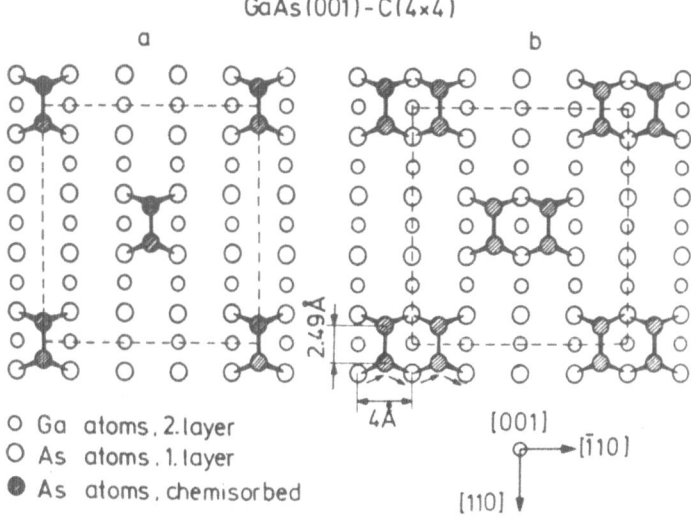

O Ga atoms, 2.layer
O As atoms, 1.layer
● As atoms, chemisorbed

Figure 17 - Models for the GaAs (001)-C(4×4) surface, based on a trigonally bonded As ad-layer (a) 25% coverage, (b) 50% coverage.

In this case we have observed that it is very much easier to obtain the C(4×4) reconstruction using As_2 than with As_4, which is again consistent with the model proposal that a higher surface arsenice population can be obtained with As_2.

4. KINETICS OF ALLOY FILM GROWTH

By regulating the composition of certain ternary and quaternary alloys of the types $III_x^{(a)} III_{(1-x)}^{(b)} V$, $III V_y^{(c)} V_{(1-y)}^{(d)}$ and $III_x^{(a)} III_{1-x}^{(b)} V_y^{(c)} V_{1-y}^{(d)}$, it is possible to adjust simultaneously their lattice constants and band gaps so that they are closely lattice matched to the underlying (binary) substrate ($\Delta a/a \lesssim 10^{-4}$) while having the appropriate optical and/or electrical characteristics. In this section we will discuss the kinetic parameters which influence composition for alloys with both mixed Group III and mixed Group V elements.

4.1 Alloys with Mixed Group III Elements

For simplicity we will confine our comments very largely to $Ga_xIn_{1-x}As$ and $Ga_xIn_{1-x}P$, but qualitatively all mixed Group III alloy systems are expected to behave similarly. Alloy growth reactions, insofar as the Group V element is concerned, are identical to those observed in binary compound growth, and the only problem is to establish those kinetic factors which can influence the ratio of the Group III elements. At comparatively low temperatures ($\lesssim 650K$) their sticking coefficients are unity so the alloy composition is determined simply by the flux ratio. However, to obtain adequate electrical and/or optical properties it is frequently necessary to use higher substrate temperatures and then several conflicting effects can occur.

The first is not in fact related to alloy composition as such, but rather to the thermal stability of the alloy. This is in effect controlled by the lesser stable of the two binary compound end members which comprise the alloy, the Group V element being lost preferentially from that compound.[23] The Group V flux necessary to ensure Group V stable growth at high temperature is therefore also determined by the lesser stable binary compound; e.g., in $Ga_xIn_{1-x}As$, it is the thermal stability of InAs which is the controlling factor.

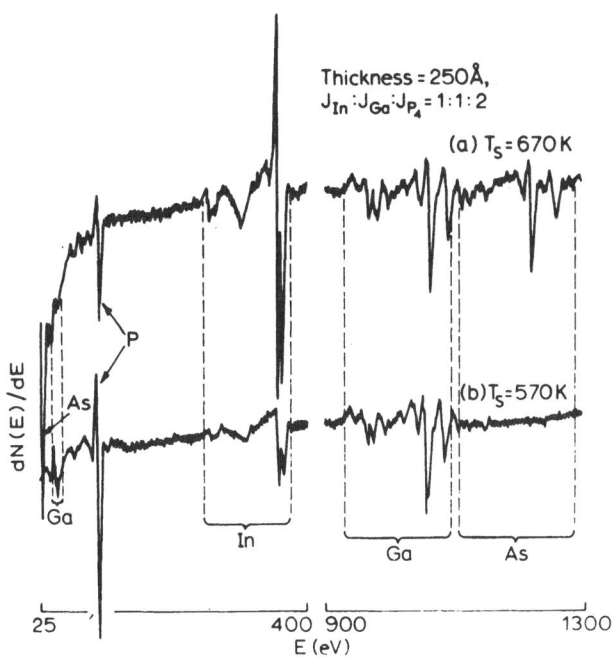

Figure 18 - Auger spectra from $Ga_{0.5}In_{0.5}P$ films on (001) GaAs substrates: (a) grown at 670K; (b) grown at 570K. Note the large In enrichment at the higher growth temperature.

A rather more serious problem is illustrated in Fig. 18, which shows Auger spectra for $Ga_{0.5}In_{0.5}P$ films grown on GaAs substrates.[23] For this composition the film is approximately lattice matched to the substrate, but the important result is that at the higher substrate temperature (670K) the surface is enriched with In. Similar effects have been reported recently[24] for $In_xGa_{1-x}As$ films lattice matched to InP substrates and grown at $\approx$ 770K where again the surface is In-rich but the substrate-layer interface region shows In depletion, and for $Al_xGa_{(1-x)}As$ films on GaAs substrates, which have a Ga-rich surface region.[25]

These results are consistent with a model of surface segregation within the Group III atom sub-lattice. Surface segregation is simply the enrichment of one species at the surface with respect to the bulk concentration for what is effectively a binary substitutonal solid. The two principal driving forces relate to differences in bond strength and atomic size. In thermodynamic terms, in the dilute limit, the free energy of segregation, ΔF_a, is given by

$$\Delta F_a = (F_s - F_s^0) - (F_b - F_b^0) \tag{12}$$

where F_b and F_s represent the free energy for a system with solute atoms in the bulk and surface respectively and F_b^0 and F_s^0 are the corresponding free energes for the pure solvent.

In our case, segregation is occurring during growth, so we should treat it on a two-dimensional basis. As each atomic layer grows, with a composition determined by the relative fluxes, the layer immediately beneath will adjust its composition by exchanging atoms with the new surface layer to minimise the total free energy. Any tendency to back diffusion by virtue of a concentration gradient will be reduced by the deeper potential well at the surface for the segregated atom. By using the hypothesis of surface segregation, the surface enrichment of In in $Ga_xIn_{(1-x)}As$ and $Ga_xIn_{(1-x)}P$ and of Ga in $Al_xGa_{(1-x)}As$ can be correctly predicted from the values quoted by Miedema[26] for the molar volume (V_m) and the surface tension of the solid at $OK(\gamma_s)$, which is a quantity closely related to the electron density at the interface between neighbouring atoms in the crystal.

The third high temperature effect is due directly to the different vapour pressures of the Group III elements over the alloy, which results in different loss rates and hence a composition different from the incident flux ratio. We can make a reasonable estimation of the loss rates from the vapour pressure data of the elements[27] since it is known[28] that the vapour pressure of the Group III element over the III-V compound is similar to that over the element itself, and for high temperature growth the alloy surface will be enriched in the more volatile Group III element. The results of the calculation are shown in Table 1. Typical MBE growth rates are approximately one monolayer per second ($\approx 1\mu m$ hr^{-1}), so that we would expect to observe a significant loss of In from In-containing alloys above $\approx 820K$ and of Ga from Ga-containing alloys above $\approx 920K$. Such effects are observed and the actual loss rates are close to the predicted values.[29]

Table 1. Approximate evaporation rates (ML s^{-1}) of Group III elements over III-V compounds and alloys containing them.

Temperature		Evaporation Rate, monolayers s^{-1}		
K	~°C	In	Ga	Al
820	550	0.03	negligible	negligible
870	600	0.3	negligible	negligible
920	650	1.4	0.06	negligible
970	700	8.0	0.4	negligible
1020	750	30	2	0.05

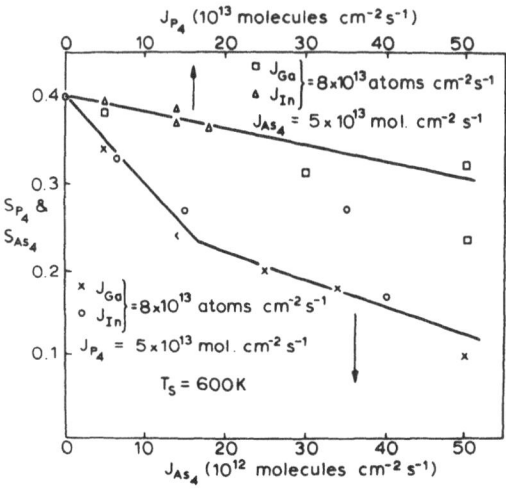

Figure 19 - The sticking coefficients of As$_4$ and P$_4$ as a function of the adsorption rates of P$_4$ and As$_4$ respectively, during growth of GaAs$_y$P$_{1-y}$ and InAs$_y$P$_{1-y}$. The sticking coefficient of As$_4$ is not strongly influenced by the P$_4$ flux, but S$_{P_4}$ is greatly reduced in the presence of an As$_4$ flux.

4.2 Alloys with Mixed Group V Elements

Three alloy systems with mixed Group V elements have been investigated, $GaAs_yP_{1-y}$,[30,31] $InAs_yP_{1-y}$,[31] and $GaAs_ySb_{1-y}$.[32] The important point is that in no case do the incorporation rates of the Group V species correspond directly to the relative fluxes of the two elements, or to the vapour pressures of Group V dimers in equilibrium with the alloy at the growth temperature. Kinetic factors are therefore of significance in controlling the compositions of such films.

Arthur and Lepore[30] first observed that for $GaAs_yP_{1-y}$ prepared from molecular beams of Ga, As_2 and P_2, the film contained approximately four times more As than P relative to the $As_2:P_2$ flux ratio arriving at the surface. Furthermore the ratio of the equilibrium vapour pressures of As_2 and P_2 over $GaAs_yP_{1-y}$ differs considerably from the solid phase composition ratio and the film composition obtained for a given $As_2:P_2$ ratio is quite different from the solid which would be in equilibrium with a vapour of equivalent flux ratio.

Foxon et al[31] have studied the interaction kinetics of As_4 and P_4 for both $GaAs_yP_{1-y}$ and $InAs_yP_{1-y}$ films. In this investigation particular attention was given to estimating accurately the As_4 and P_4 fluxes reaching the substrate surface. A beam-monitoring ion guage was calibrated for Ga, In, As_4 and P_4, and this enabled the sticking coefficient of As_4 and P_4 to be measured in-situ during growth, using modulated molecular beam measurements. The sticking coefficient of one tetramer, As_4 or P_4, was measured as a function of the arrival rate of the other at a fixed growth rate. The experiments were carried out with the substrate at 600K to avoid complications which would be caused at higher temperatures by the loss of As_2, P_2 and AsP molecules from the film, thereby creating Ga and In sites in the surface. The results in Fig. 19 show that the probability of incorporating As_4 molecules is much higher than that of P_4 molecules for both $GaAs_yP_{1-y}$ and $InAs_yP_{1-y}$ and that the nature of the Group III element is not important. Equivalent behaviour has been observed for $GaAs_ySb_{1-y}$ films[32] where the incorporation probability of Sb_4 is much higher than As_4. It has also been established[33] that for $GaAs_yP_{1-y}$ films the relative sticking coefficient of P_2 and As_2 are closely similar to those for As_4 and P_4.

It is clear therefore, that preferential adsorption occurs whether the Group V elements are supplied as dimers or tetramers, but the

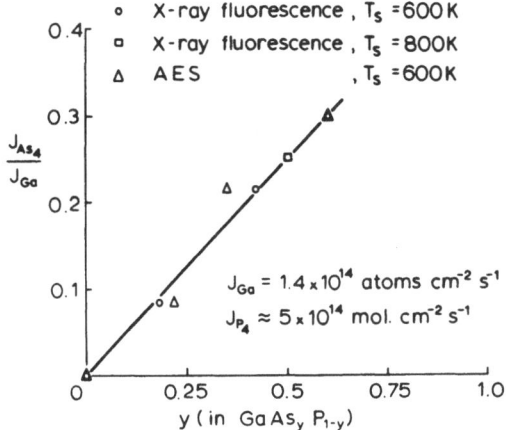

Figure 20 - The composition of $GaAs_yP_{1-y}$ films as a function of the ratio of As_4 to Ga fluxes for a constant P_4 flux.

mechanism is not known. It has been suggested[31,32] that it might be related to different surface lifetimes of the Group V element species, and although Foxon et al[31] found that the lifetime of As_4 on GaAs was greater than that of P_4 on GaP, the values appeared to be surface population dependent in a complex manner. We must conclude that there is no evidence of any simple relationships between individual surface lifetimes and relative probabilities of incorporation in a growing film.

Nevertheless, the relative sticking coefficient results do suggest a simple method for growing mixed Group V element alloy films of controlled composition. If the flux of the more reactive species is limited with respect to the Group III element flux and an excess amount of the other Group V element is provided, then alloy films of predetermined composition should be obtained. Such behaviour has been observed for $GaAs_yP_{1-y}$ films grown at both 600K and 800K from fluxes of Ga, As_4 and P_4 as shown in Fig. 20. The alloy composition is linearly related to the As_4:Ga flux ratio as the kinetic measurements predict and it is evident that the composition can be controlled in this manner. Similar behaviour is observed for $GaSb_yAs_{1-y}$ alloy films where the incorporation probability of Sb_4 is much higher than As_4[32].

Over a reasonable range of temperatures, relative sticking coefficients are not strongly temperature dependent and consequently there is no effect on film composition in that range. Taking $GaAs_yP_{1-y}$ as an example, however, Woodbridge et al[33] showed that while y was constant up to $\approx$ 800K, it decreased from 0.63 to 0.43 between 800K and 920K. This can be simply explained from the known kinetic behaviour. Below 800K, the incorporation rate of arsenic is approximately fifty times greater than that of phosphorus in the presence of a Ga flux, so vacant sites are filled preferentially by arsenic, and the composition is controlled by the gallium to arsenic flux ratio, with phosphorus supplied in excess. This situation begins to break down when the desorption of Group V element molecules becomes significant, which occurs in this case as the substrate temperature is raised beyond 800K. Since the vapour pressure of As_2 over GaAs is greater than that of P_2 over GaP[28,34] the desorption flux is mainly As_2, but because the incident arsenic flux (of As_2 or As_4) is restricted, the desorbing molecules will be increasingly replaced by phosphorus, if a sufficiently high phosphorus flux is maintained. The film therefore becomes increasingly phosphorus-rich as the substrate temperature is increased.

ACKNOWLEDGMENTS

The author is indebted to his many colleagues, but especially J. H. Neave, C. T. Foxon, P. J. Dobson and P. K. Larsen, for providing most of the results and much stimulating discussion.

REFERENCES

1. J. R. Arthur, J. Appl. Phys. 39 4032 (1968).
2. R. H. Jones, D. K. Olander, W. J. Siekhaus and J. A. Schwarz, J. Vac. Sci. Technol. 9 1429 (1972).
3. J. A. Schwarz and R. J. Madix, Surf. Sci. 46 317 (1974).
4. C. T. Foxon, M. R. Boudry and B. A. Joyce, Surf. Sci. 44 69 (1974).

5.   J. W. Cooley and J. W. Tukey, Math. Comp. 19 297 (1965).
6.   C. T. Foxon, J. A. Harvey and B. A. Joyce, J. Phys. Chem. Sol. 34 1693 (1973).
7.   J. H. Neave, P. Blood and B. A. Joyce, Appl. Phys. Lett. 36 311 (1980).
8.   J. R. Arthur, Structure and Chemistry of Solid Surfaces, ed. G. A. Somorjai (Wiley, New York, 1969) 46-1.
9.   C. T. Foxon and B. A. Joyce, Surf. Sci. 64 293 (1977).
10. C. T. Foxon and B. A. Joyce Surf. Sci. 50 434 (1975).
11. H. Künzel, J. Knecht, H. Jung, K. Wünstel and K. Ploog, Appl. Phys. A28 167 (1982).
12. H. Künzel and K. Ploog, Appl. Phys. Lett. 37 416 (1980).
13. P. J. Dobson, G. B. Scott, J. H. Neave and B. A. Joyce, Solid State Commun. 43 917 (1982).
14. C. T. Foxon, P. Dawson, G. Duggan and G. W.'t Hooft, Proc. 2nd Intern. Meeting on MBE, Tokyo, (Japan. Soc. Appl. Phys.) 81 (1982).
15. J. H. Neave and B. A. Joyce, J. Crystal Growth 44 387 (1978).
16. P. K. Larsen, J. H. Neave and B. A. Joyce, J. Phys. C. 14 167 (1981).
17. P. J. Dobson, J. H. Neave and B. A. Joyce, Surf. Sci. 119 L339 (1982).
18. P. K. Larsen, J. F. van der Veen, A. Mazur, J. Pollman, J. H. Neave and B. A. Joyce, Phys. Rev. B26 3222 (1982).
19. J. Pollmann and S. T. Pantelides, Phys. Rev. B18 5524 (1978).
20. D. J. Chadi, C. Tanner and J. Ihm, Surf. Sci. 120 L425 (1982).
21. R. Z. Bachrach, R. S. Bauer, P. Chioradia and G. V. Hansson, J. Vac. Sci. Technol. 19 335 (1981).
22. P. K. Larsen, J. H. Neave, J. F. van der Veen, P. J. Dobson and B. A. Joyce, Phys. Rev. B, 27, 4966 (1983).
23. C. T. Foxon and B. A. Joyce, J. Crystal Growth 44 75 (1978).
24. D. V. Morgan, H. Ohno, C. E. C. Wood, L. F. Eastman and J. D. Berry, J. Electrochem. Soc. 128 2419 (1981).
25. T. C. Chiang, R. Ludeke and D. E. Eastman, Phys. Rev. B25 6518 (1982).
26. A. R. Miedema, Philips Tech. Rev. 38 257 (1978/79).
27. R. E. Honig and D. A. Kramer, RCA Rev. 30 285 (1969).
28. C. T. Foxon, J. A. Harvey and B. A. Joyce, J. Phys. Chem. Sol. 34 1693 (1973).
29. C. T. Foxon, private communication.

30. J. R. Arthur and J. J. Lepore, J. Vac. Sci. Technol. <u>6</u> 545 (1969).

31. C. T. Foxon, B. A. Joyce and M. T. Norris, J. Cryst. Growth <u>49</u> 132 (1980).

32. C. A. Chang, R. Ludeke, L. L. Chang and L. Esaki, Appl. Phys. Lett <u>31</u> 759 (1977).

33. K. Woodbridge, J. P. Gowers and B. A. Joyce, J. Cryst. Growth <u>60</u> 21 (1982), quoting unpublished results of C. T. Foxon.

34. M. Ilegems, M. B. Panish and J. R. Arthur, J. Chem. Thermodynamics <u>6</u> 157 (1974).

THE APPLICATION OF THERMODYNAMICS TO MOLECULAR BEAM EPITAXY

R. Heckingbottom

British Telecom Research Laboratories
Martlesham Heath Ipswich IP5 7RE, England

ABSTRACT: Thermodynamics is most rigorously applied to systems at equilibrium. In its extension to MBE, the interaction with kinetics, the effect of chemical over-potential and the concept of local equilibrium are briefly described. The growth of GaAs is considered in detail, as experimental data is available for comparison. While in general in MBE, kinetic barriers to reactions are expected to confuse thermodynamic guidelines, it is found that under conditions where the best quality layers are obtained their impact is slight. Kinetically induced defects can be reduced to the part per billion range. Against this background the main competing (contaminating) reactions and doping processes are examined. Potential (elemental sulphur) and impractical (elemental zinc) dopants can be identified, without recourse to MBE, from existing thermochemical data. Kinetic barriers, when they occur, are more clearly revealed by the analysis. These barriers seldom involve the commonly assumed dopant incorporation step.

1. INTRODUCTION

It is fairly widely assumed that molecular beam epitaxy (MBE) usually takes place so far from thermodynamic equilibrium that thermodynamics has little relevance to the process. This is an assumption to be tested against experimental observations throughout the body of the paper. First though, it should be stated what can be expected from the application of thermodynamics as follows:

i. One can determine whether the desired reaction is possible or not under MBE conditions. This is trivial for a well-known reaction such as growth of GaAs, but for some doping reactions and III-V alloy growth the waste of much time and effort on reactions doomed to failure can be avoided.

ii. One can determine what competing reactions are also possible. This knowledge indicates which precautions are most important and may also aid in the selection of dopants which are more easily used.

iii. In the more detailed study of a particular reaction, the comparison of the thermodynamic framework with experimental data allows the identification of any kinetic barriers and hence leads to a clearer understanding of the reaction mechanism.

There is of course, as with any theoretical analysis, often a shortage of relevant data with which to quantify the predictions. Even here however, calculation of the thermodynamic framework helps identify the minium number of experiments needed to determine the most useful data.

Before proceeding to specific cases, it is worth considering briefly three general issues.

1.1 Thermodynamics and Kinetics

It is important to recognise that thermodynamics and kinetics are not opposed and exclusive but compatible and interrelated. Thus in reactions which proceeded facilely to equilibrium, they do so by reaction paths at clearly describable kinetic rates. It is the fact that kinetics often does not allow one to distinguish clearly whether equilibirum is being reached or not, particularly where say only the forward reaction can be followed readily experimentally, that leads to added understanding of the reaction when the thermodynamic framework is calculated.

Consider the fairly simple case of the solution of sugar in water. If we introduce far more sugar than can dissolve and wait long enough, the system will reach equilibrium with a large excess of solid and a saturated solution. The situation is dynamic however with sugar continuously dissolving and precipitating at equal and opposite rates. The position may be described

$$k_f[\text{Sugar}_{\text{solid}}] \Longleftrightarrow k_r[\text{Sugar}_{\text{solv}}] \tag{1}$$

where k_f and k_r are the forward and reverse rate constants and [ ] indicate concentrations, which can be used as approximations to activities. The actual forward and reverse rates are no doubt determined by the effective area of the interface between solid and solution and the temperature. If now a little more water is added, but not enough to affect the fact that the undissolved sugar is in large excess, more sugar will of course dissolve until the solution is again saturated. Note however that the forward rate will remain constant throughout and so will reveal nothing of the changes. It is the reverse rate which is reduced, through [$\text{Sugar}_{\text{solv}}$] , and so leads to a net dissolution current for a while. The extent of the reactions in the two situations is well described by the thermodynamic parameters for solid sugar and the saturated solution.

Though the example is a long way from MBE it is hoped that two principles are clear: a) determination of the kinetic details of a reaction (usually the most amenable to definitive study, like evaporation of solid into the vacuum) gives little or no indication of the point at which that reaction ceases to dominate in a more complex situation, e.g. growth, and (b) establishing a detailed kinetic fit with experiment, e.g. for the reverse reaction in (1), does not establish whether equilibrium is reached or not.

1.2 Thermodynamics and Overpotential

One of the reasons why thermodynamics has often been virtually ignored in MBE is that conditions at first glance look unsuitable. Thus reactants and products are each at their own temperatures, whereas much of the thermodynamics applies to a system at a single temperature. Experimentally however it has been found[1] that all arriving species quickly acquire the substrate temperature, and it is this that is used in thermodynamic calculations. Additionally much MBE occurs with a significant chemical over-potential for growth, e.g. supercooling of 80-100°C in the case of early GaAs studies. However while infinitesimal supercooling is the way to guarantee that equilibrium is reached, large supercooling does not preclude it. A simple example is provided by water flowing to mean sea level. The obvious "close to equilibrium" route is via say the mouth of the Amazon, but water falling spectacularly over a cliff into a Norwegian fjord reaches mean sea level just as surely. The only way to find out if equilibrium is reached is to examine the product. While it is quite clear that it is easy in general not to reach equilibrium in MBE, almost all the effort is spent in doing so with considerable success. Thus the best MBE layers are comparable with the best from the near equilibrium techniques of liquid and vapour phase epitaxies.

The explanation for this close approach to the equilibrium position lies in the slow growth rate in MBE. Thus an atom condensing on the substrate has typically one second to diffuse before incorporation in the bulk. For GaAs at 600°C and a sensible estimate of 1eV for the activation energy for surface diffusion, an atom will make $= 10^6$ site changes before incorporation. This number of moves presents plenty of opportunity for selection of the lowest free energy position. We may check the reasonableness of this estimate using the data of Nagata and Tanaka.[2]

They found the distance x diffused by Ga on GaAs at 550°C was 200Å under arsenic rich conditions and 1900Å under gallium rich positions. Using the equation

$$x = \sqrt{Dt} \tag{2}$$

where t is the time and D is the diffusion coefficient for Ga, and substituting the lower value for x with $t = 1$ gives $D = 4.10^{-12} cm^2 sec^{-1}$. D can then be expressed as

$$D = D_o \exp(-E/kT) \tag{3}$$

where E is the activation energy for suface diffusion, T is the substrate temperature and k is Boltzman's constant. Assuming $D_0 = a^2 \nu$ where a is the diffusion jump distance and ν is the surface atom vibration frequency, we have $D = 10^{-15} .10^{12}$ or 10^{-3} cm^2 sec^{-1}. For T = 823K, then E = 1.36eV in reasonable agreement with the initial guidline estimate. Using the larger value for x gives E = 1.05eV in even better agreement.

It has recently been suggested that this slow growth rate could lead to MBE allowing the system to approach equilibrium even more closely than in liquid phase epitaxy (LPE) in order to explain some results in GaAs/Ga$_{1-x}$Al$_x$As superlattice growth.[3] In general this is unlikely to be the case. Thus using the above experimental activation energy and raising the growth temperature to 900°C (more typical of LPE) increases D by a factor of 100 while LPE growth rates are greater than those in MBE by a factor of 50 or 60. Even more important is the fact that the activation energy for surface diffusion is likely to be significantly lower in LPE by about half the bonding energy of a Ga atom in Ga liquid. Hence at a given temperature D is likely to be much greater in LPE than that calculated from MBE measurements.

1.3 Partial or Local Equilibrium

Partial or local equilibrium is a more familiar concept than might appear at first sight. Thus all molecules of carbon, hydrogen and oxygen should react with further oxygen to form CO_2 and H_2O at equilibrium at normal temperatures and pressures. The examples of these molecules, that we class as food, do proceed to this equilibrium in the body. Other rather similar molecules in the stomach wall fortunately do not proceed to equilibrium to anything like the same degree, or we would not be able to use the digestive process.

Turning to growth of GaAs under MBE conditions we find an equally sharp divide. While it was shown in the previous section that a significant amount of surface diffusion (10^4-10^6 jumps per atom) is likely in MBE, it is crucial to the production of the sharp interfaces characteristic of MBE that bulk diffusion is negligible. Substituting the values from Goldstein[4] for the self diffusion of Ga in GaAs in equation (3) of $D = 1.10^7$ and E = 5.6eV gives a value for x of only $1.9 .10^{-5}$Å in 1 second at 600°C. Even after a growth run of about 2.5 hours (10^4 secs) x is only $1.9.10^{-3}$ Å, hence to a very good approximation atoms stay in the layer in which they are incorporated at the growing surface.

As a final aspect of this partial equilibrium it should be noted that not all reactions in the surface layer will necessarily proceed to equilibrium. In many reactions, such as the dissociation of molecular species, surface diffusion is unlikely to play a key role in the reaction path. It is in just such situations that consideration of both thermodynamic and kinetic aspects of reactions are required to unravel the reaction mechanisms.

2. GROWTH OF GALLIUM ARSENIDE

2.1 The Thermodynamic Framework

The basic thermodynamic framework for MBE growth of GaAs can be usefully considered in terms of an extended version of the vapour pressure curves presented by Arthur,[5] see Fig. 1. This treatment of GaAs serves to typify MBE growth of the binary III-V compounds in general and it will be considered almost exclusively, as the range of data available allows the approach to be judged in a fairly quantitative manner.

Three chemical equations need to be considered:

$$GaAs_{(s)} \Longleftrightarrow Ga_{(g)} + 1/2As_{2(g)} \tag{4}$$

with

$$K_4 = P_{Ga}P_{As_2}^{1/2} \tag{5}$$

$$GaAs_{(s)} \Longleftrightarrow Ga_{(g)} + 1/4As_{4(g)} \tag{6}$$

with

$$K_6 = P_{Ga}P_{As_4}^{1/4} \tag{7}$$

and

$$2As_{2(g)} \Longleftrightarrow As_{4(g)} \tag{8}$$

with

$$K_8 = P_{As_4}/P_{As_2}^2 \tag{9}$$

where subscripts (s) and (g) indicate solid and gas phase respectively, K_4, K_6 and K_8 are equilibrium constants for the respective reactions and P_{Ga}, P_{As_2} and P_{As_4} are the pressures in atmospheres of the respective species. Comparable reactions involving As are omitted for simplicity as it is never a dominant species for the temperature and pressure range considered.

The position at the typical growth temperature of 597°C ($10^3/T$ = 1.15) will now be considered in more detail by reference to Fig. 1. The first obvious point is that only the values of K are fixed at a given temperature - there are no unique values of equilibrium pressures. Thus the value of K_4 is $=9.10^{-17}$ atmos$^{3/2}$ which has boundary solutions close to P_{Ga} = 4.10^{-11} atmos and P_{As_2} = $7.5.10^{-12}$ atmos at one

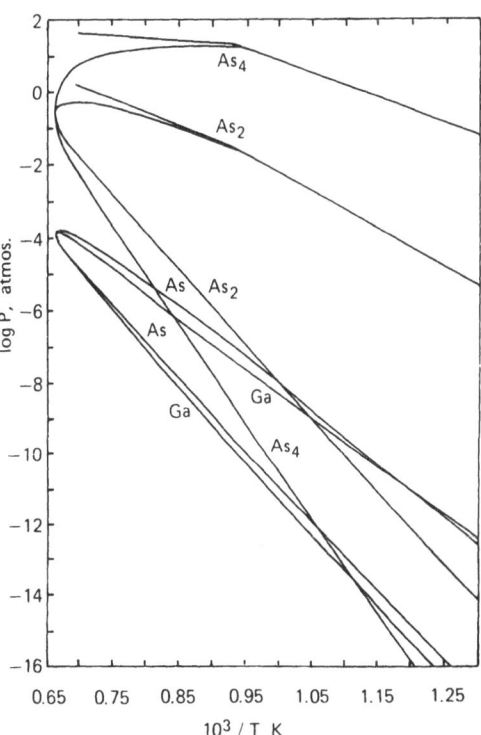

Figure 1 - Equilibrium vapour pressures of As, As$_2$, As$_4$ and Ga along the binary liquidus as a function of T^{-1}. Pressures of As$_2$ and As$_4$ over pure solid and liquid As are also shown.

extreme and to $P_{Ga} = 6.10^{-15}$ atmos and $P_{As_2} = 1.8.10^{-4}$ atmos at the other. Note that all values in this exercise have been taken from the figure and are only self-consistent to factor of 2 either way. From the point of view of the phase diagram of GaAs the former values correspond to Ga rich $GaAs_{(s)}$ and the latter to As rich material. These boundary values are set by the alternative reactions

$$Ga_{(g)} \Longleftrightarrow Ga_{(1)} \tag{10}$$

at the Ga rich side (where subscript (1) indicates the liquid phase) and

$$1/2As_{2(g)} \Longleftrightarrow As_{(s)} \tag{11}$$

at the As rich side. More exactly the pure elements in the condensed phases should be replaced by a saturated solution of GaAs in them, but in practice this is a small correction. It is clear that there is a continuous range of values of P_{Ga} and P_{As_2} consistent with $K_4 = 9.10^{-17}$ atmos.

In a similar manner it is found that $K_6 = 6.10^{-15}$ atmos$^{5/4}$. The alternative boundary values are $P_{Ga} = 4.10^{-11}$ atmos with $P_{As_4} 3.10^{-15}$ atmos on the one hand and $P_{Ga} = 6.10^{-15}$ atmos with $P_{As_4} = 1$ atmos on the other. Thus As_2 is the dominant arsenic species at MBE pressures, but at higher pressures, e.g. at the arsenic rich boundary of $GaAs_{(s)}$, it is As_4 that dominates at equilibrium. Inspection of these values shows clearly that all MBE conditions $0.5 < P_{As_4}/P_{Ga} < 20$ will give $GaAs_{(s)}$ close to the Ga rich boundary if equilibrium is approached, i.e. material similar to but not identical to that obtained from liquid phase epitaxy (LPE).

The boundary values of the arsenic species may be used via equation (8) to calculate a value of $K_8 = 4.10^7$ atmos^{-1}. The large values of P_{As_4} reached before the reaction

$$1/4As_{4(g)} \Longleftrightarrow As_{(s)} \tag{12}$$

occurs illustrates that the failure of excess As_4 molecules, i.e. all those above the concentration required to form $GaAs_{(s)}$, to condense is consistent with the thermodynamic framework and is not in itself proof of kinetic limitations. The latter comes from the failure of this excess As_4 to convert to As_2 under MBE conditions. The evidence for the effect of this failure on the resultant $GaAs_{(s)}$ will be examined in the next section. Here it is noted that P_{As_4} values typical of MBE conditions (10^{-9} atmos) are not sufficient to cause condensation of arsenic as in reaction (12) until the substrate temperature, $T_s \lesssim 180°C$. This region has been studied by Matsumoto[6] for MBE of GaAs on Si, SiO_2 and Al_2O_3.

Although the reaction was kinetically hindered at these low temperatures, excess arsenic was found in all layers grown at $T_s <$ 180°C. Significant kinetic limitations to the condensation of As_4, even on solid arsenic, have long been known and the successful growth of GaAs on GaAs (001) at temperatures $\lesssim 100°C$ by Joyce et al.[7] is seen to rely heavily on similar kinetic barriers.

Finally in this section, it will be recognised that typical MBE beam equivalent pressures are $\simeq 10^{-9}$ atmos - significantly greater than the equilibrium values at 597°C. For $P_{Ga} \simeq P_{As_2} \simeq 10^{-9}$ on the equilibrium vapour pressure plots of Fig. 1 a temperature of $\simeq 680°C$ is required. Hence the supercooling of a typical MBE system is $\simeq 80°C$. Under optimum growth conditions, the practical effects of this overpotential can be almost eliminated as discussed in section 1.2. A more detailed examination of this situation will now be given in the next section.

2.2 <u>Defect Concentrations in GaAs</u>

The defect concentration in a semiconductor is the most sensitive measure of the crystal quality. In the case of MBE gallium arsenide, it is clear that at growth temperatures of $\simeq 500°C$ the concentration of defects is sufficiently high that it dominates over all attempts to control the electrical properties by conventional impurity doping. As the growth temperature is raised through the range 500-650°C, progressively fewer unintentional defects remain. Although some lower temperature effects may also be caused by competing reactions (see section 3), the general trend is consistent with a large population of mistakes in crystal growth due to kinetic barriers to reaction gradually becoming less important as the temperature is raised.

At the higher end of the temperature range the number of defects can be extremely low. Deep traps are measured in parts per billion[8] and luminescence properties are excellent leading to some of the lowest threshold lasers known.[9] The only reasonable conclusion is that under these conditions thermodynamic equilibrium is approached as closely in MBE as it is in the near equilibrium techniques of LPE and VPE.

At slightly lower temperatures there is clear evidence that the use of the equilibrium species As_2 in the beam yields crystal with fewer defects than when As_4 is used.[10,11] This result gives important evidence of a specific kinetic barrier. As remarked earlier, the decomposition of As_4 to As_2 - reaction (8) - is hindered as determined by Foxon and Joyce[12] in modulated beam experiments. Here however it is seen that the crystal growth reaction (6) reverse can be kinetically hinered under conditions where the corresponding reaction (4) reverse is not. It is important to note however the small extent of the intrusion of the kinetic barriers. In the experiments reported[10,11] the additional 'mistakes' due to using As_4 were at concentration < 1ppm and as mentioned above, even this effect can be reduced by increasing the growth temperature.

The above results lead directly to consideration of the other main aspect of defect chemistry in GaAs - nonstoichiometry. In principle as the ratio P_{As_2}/P_{Ga} in the vapour is increased, as considered in the treatment of Fig. 1, the solid GaAs should become slightly richer in As. The change in stoichiometry to be expected is probably a part per thousand or less[13] at MBE growth temperatures, even over the full pressure range considered in Fig. 1. Over the limited pressure variations possible in MBE the situation can be illustrated by considering a change of 10^{-4} in As concentration in GaAs as P_{As_2}/P_{Ga} is varied from 1 to 20 and examining how closely this is matched by varying P_{As_4}/P_{Ga} from 0.5 to 10. By this comparison it is possible to assess whether the reaction important to crystal growth

$$1/4As_{4(g)} \Longleftrightarrow GaAs_{(s)} + V_{Ga} \qquad (13)$$

is kinetically hindered to any important extent, as compared with the analogous reaction

$$1/2As_{2(g)} \Longleftrightarrow GaAs_{(s)} + V_{Ga} \qquad (14)$$

where V_{Ga} represents a vacant gallium site in GaAs.

Foxon and Joyce et al.[12] showed that on a Ga(001) surface with the Ga rich (4x2) surface structure the sticking coefficient for As_4 reached a maximum of 0.5 instead of unity. This result shows a clear kinetic barrier but not one that is likely to interfere seriously with obtaining a crystal of the required stoichiometry. GaAs of good morphology is grown on GaAs(001) with the As rich (2x4) surface structure however and here the situation is more difficult to disentangle. The sticking coefficient of As_4 is expected to be very low on thermodynamic grounds (section 2.1) and so it is difficult to identify any additional kinetic limitations. To be specific, if thermodynamics is followed and a change of P_{As_4}/Ga from 1 to 10 gives an increase of As in GaAs of 10^{-4}, the sticking coefficient of As_4 falls from 0.25 to 0.0250025. If on the other hand, reaction (13) is effectively halted by kinetic barriers, so that reaction (6) always leads to GaAs of the same stoichiometry then the sticking coefficient of As_4 falls from 0.25 to 0.025. Clearly the difference will not be distinguishable by modulated beam techniques.

The only direct evidence on stoichiometry comes from recent work by Kunzel et al.[11] using Ge as a dopant. Doping will be considered in more detail in later sections. Here it is sufficient to note germanium on gallium sites gives n-type doping and this substitution is favoured by an increase in the concentration of V_{Ga}. Hence an interpretation of Ge doping allows assessment of reactions (13) and (14). It is clear from this work as shown in Fig. 2 that in the temperature range 500-600°C reaction (13) does proceed to an important extent, but there are kinetic barriers which prevent it being as effective as reaction (14). The effect is much more marked near 500°C and has almost disappeared by 600°C. It is not possible to distinguish whether the enhanced effect at 500°C is just the expected temperature effect on the rate constant or whether the difference is also enhanced by the increased surface arsenic concentration at the lower temperature. Both effects are likely. All these results refer to defect at concentrations at the 1ppm level. It is important to recognise that definitive comment on this point (reaction 14) has only become possible in the last year.

In summary therefore the effects of kinetic limitations on MBE growth of GaAs on the (001) surface can now be clearly defined. They are certainly detectable under typical MBE conditions, but are shown to be very small, and under optimum conditions, the resultant defect concentrations can be reduced to the parts per billion range. Against this background, several important doping reactions will be considered in section 4. Before this however one or two other aspects need to be considered.

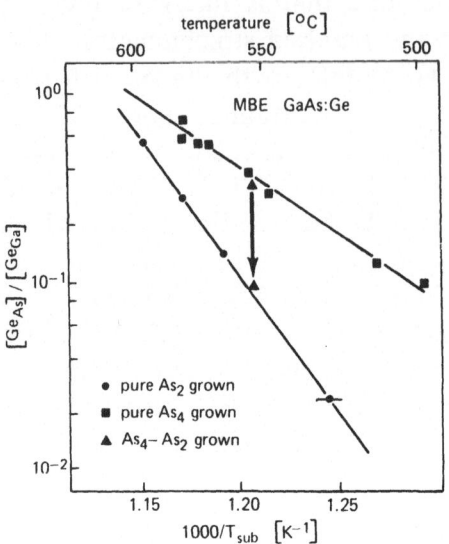

Figure 2 - Substrate temperature dependence of Ge autocompensation ratio in MBE n-GaAs:Ge layers; ■ samples grown from pure As_4 molecular beam species by evaporation of elemental arsenic. • samples grown from pure As_2 molecular beam species by incongruent evaporation of GaAs. ▲ samples grown from an adjustable As_2/As_4 ratio in a molecular beam. The arrow indicates the reduction of autocompensation by increasing the As_2/As_4 flux ratio.

2.3 Surface Effects

The surface chemistry of GaAs (001) presents a problem for quantitative analysis either thermodynamic or kinetic. Variations in gas phase conditions (pressure or flux ratio) cause corresponding changes in surface chemistry which are much larger and more complex than those occurring in the bulk (see section 2.2). In one sense the complexity, as evidenced by the series of surface structures which occur as either temperature or the arsenic/gallium flux ratio is varied, can be used to ensure some comparability between different experiments. The main problem however, is the general lack of certainty about the atomic positions and concentrations throughout the range of structures, and even moreso concerning the bond strengths involved. Thus it is difficult to be quantitative either about the free energy of a particular surface configuration (needed for thermodynamics) or likely energy barriers to specific reaction steps (needed for kinetics). Two of the simpler possible variations are illustrated for the specific case of arsenic vacancies in Figure 3.

In Figure 3a the arsenic vacancy concentration in the surface, $[V_{As}]_{surface}$, and the bulk $[V_{As}]_{bulk}$ are shown as a function of sample temperature - the concentrations shown are illustrative only. The situation applies only for a constant coverage, Θ, of arsenic on the surface. Note that this will involve a large increase in P_{As_2}/P_{Ga} over the temperature range shown - one not usually applied in MBE experiments as the values of P_{As_2} required could easily extend outside the MBE region. Figure 3b on the other hand shows the companion effect of variation in $[V_{As}]_{surface}$ at constant temperature as Θ is varied via P_{As_2}. Note particularly that around a surface phase transition one moves from the arsenic deficient end of an arsenic rich phase (many vacancies) to the arsenic rich end of an arsenic poor phase (few vacan-

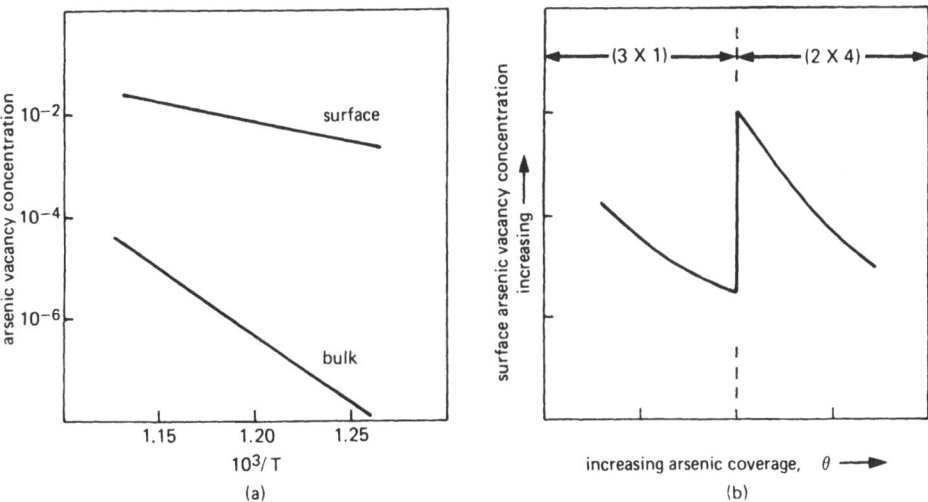

Figure 3 - Schematic representation of arsenic vacancy concentrations in GaAs:- (a) in the surface and bulk as a function of temperature for a fixed surface arsenic concentration Θ, and (b) in the surface as a function of Θ at a fixed temperature.

cies) as Θ decreases. In a typical range of experiments both effects are intermixed.

In the simplest situation it is an advantage of thermodynamics that the relationship between the gas phase and the bulk should be independent of the surface detail, as the latter represents part of the reaction path. Note however that the surface detail is expected to change with conditions and identification of this detail is seldom definitive, unless it can be compared with that to be expected at equilibrium. None-the-less marked changes in overall results (bulk defect population) as a function of surface changes, should be one of the more useful criteria for identifying kinetic barriers. In order to pursue this line one first needs to eliminate the less fundamental effects caused by starting growth runs or changing experimental conditions. At these points in the experiments one may observe induction periods which are simply connected with the limited supply of reactants in MBE. These limits obviously provide limitations which distort the thermodynamic framework. They must be allowed for first if one to use thermodynamics to help elucidate the more fundamental aspects of reaction mechanisms.

If the section on doping may be anticipated slightly, the case of a dopant atom interacting with V_{As} to occupy an arsenic site may be considered. Reference to Fig. 3a shows that an increase in substrate temperature will reduce the distribution coefficient $[V_{As}]_{surface}$ / $[V_{As}]_{bulk}$ but at the same time the increase in $[V_{As}]_{surface}$ is numerically larger and a dip in bulk doping level while these surface sites are filled is to be expected. It is not until the effects of these complications are identified that one can proceed to describe clear kinetic limitations, and the qualitative current understanding of the surface chemistry greatly hinders this process. Some related problems will be considered in section 4 on doping.

2.4 GaAs Summary

Sections 2.1 to 2.3 have considered the growth of GaAs on the (001) surface under MBE conditions. As the reaction is possible and we have ignored competing reactions, the application of thermodynamics has been restricted to improving the understanding of the reaction mechanisms. It is shown that on this surface the effects of kinetic barriers to growth can often be identified more clearly and also reduced to very small values. This provides a useful framework for considering the more complex doping reactions later. It should also be recognised however that the surface diffusion rates largely responsible for the growth of highly perfect crystals were a marked function of surface conditions. This means that the results cannot be transferred quantitatively to other surfaces of GaAs. It is well known even in the near-equilibrium techniques of LPE and VPE that crystal orientation effects on growth rate and doping can be quite marked. Experiment and analysis will be required to determine the situation in MBE.

3. COMPETING REACTIONS

One of the main possibilities for reactions competing with growth and doping of GaAs and other compounds in MBE is the formation of oxides from residual gas pressures of oxygen containing species. This reaction class has been treated most explicitly by Kirchner et al.[14] who considered the prospect of Ga_2O impurity in the molecular beam leading to Ga_2O_3 formation in the growing layer. They examine the reaction

$$4GaAs_{(s)} + Ga_2O_{3(s)} \Longleftrightarrow 3Ga_2O_{(g)} + 2As_{2(g)} \qquad (15)$$

which gives

$$K_{15} = P_{Ga_2O}^3 P_{As_2}^2 \qquad (16)$$

Using literature values for thermodynamic data they obtain

$$Log P_{Ga_2O\ max} = 17.52 - 2.8.10^{-4}/T - (2/3) \log P_{As_2} \qquad (17)$$

where $P_{Ga_2O\ max}$ is the maximum value, above which solid Ga_2O_3 will form. They choose the high value of 10^{-7} atmos for P_{As_2} and then calculate $P_{Ga_2O\ max} = 1.5.10^{-10}$ atmos at 600°C. They suggest that all GaAs grown below this temperature is liable to have Ga_2O_3 in it.

In most MBE systems a beam equivalence pressure of 10^{-6} torr for P_{As_2} is more typical ($1.3.10^{-9}$ atmos). It can be seen from Eq. (16) that this lower value for P_{As_2} means a higher value of P_{Ga_2O} and via Eq. (17) $P_{Ga_2O\ max}$ at 600°C then becomes $2.34.10^{-9}$ atmos ($1.8.10^{-6}$ torr). Again most MBE practitioners would claim that their beams were much purer than this. If however we consider 500°C, where the electrical properties of MBE grown GaAs are poor the analysis indicates a value of $1.25.10^{-10}$ torr for $P_{Ga_2O\ max}$ when $P_{As_2} = 1.10^{-6}$

torr. Reaction (15) could therefore be a serious contributor to the situation in GaAs at 500°C. One further general point is also worth noting. The point at which a separate phase of Ga_2O_3 forms corresponds to a saturated solution of O in GaAs. Oxygen impurities in GaAs corresponding to lower values of P_{Ga_2O} than $P_{Ga_2O\ max}$ could still be enough to affect the electrical properties of GaAs adversely.

Kirchner et al.[14] also consider the formation of Al_2O_3 in $Ga_{1-x}Al_xAs$ layers and recognise as have others before them,[15] that the oxygen incorporation problem is likely to extend to higher temperatures in this case. It is very likely that oxygen in $Ga_{1-x}Al_xAs$ contributed to the poor performance of $GaAs/Ga_{1-x}Al_xAs$ lasers when first fabricated in MBE layers. As with GaAs itself, the problem was solved by growing at higher temperatures. This change shifts the chemical equilibria favourably, but also reduced the effect of straightforward kinetic limitations, and it is fairly certain that both effects are necessary in practice.

The distinction between competing reactions and kinetic barrier effects can be easier to make where dopant impurities are involved. Thus one can choose a growth temperature where high quality material is obtained but problems occur with just one or two dopants. The most likely dopants to be susceptible to reaction with oxygen are the most electropositive ones. Magnesium is perhaps the best example. Kirchner et al.[14] calculate for the reaction

$$2GaAs_{(s)} + MgO_{(s)} \Longleftrightarrow Ga_2O_{(g)} + Mg_{(g)} + As_{2(g)} \qquad (18)$$

that $\log P_{Ga_2O\ max} = 26.53 - 5.42.10^4/T - \log P_{Mg} - \log P_{As_2}$ If the lower operating pressures of $P_{Ga} = P_{As_2} = 10^{-6}$ torr and $P_{Mg} = 10^{-12}$ torr (1 ppm doping) are used then $P_{Ga_2O\ max} = 1.3.10^{-9}$ torr at 600°C. This value ought to be avoidable and in recent work[16] this appears to have been achieved.

Reaction (18) is not the only one to consider however, More general sources of oxygen such as CO have been considered earlier[17,18] in explanation of the low electrical activity reported for Mg doping. For the straightforward reaction

$$Mg_{(g)} + CO_{(g)} \Longleftrightarrow MgO_{(s)} + C_{(s)} \qquad (19)$$

one may calculate from standard thermochemical data[19] that at 560°C

$$K_{19}^{-1} = P_{CO}P_{Mg} = 16.10^{-24} \text{atmos}^2$$

Thus for an attempt to dope at 5.10^{18}, $P_{Mg} = 1.3.10^{-13}$ atmos and $P_{CO} = 1.3.10^{-11}$ atmos ($9.7.10^{-9}$ torr). All conditions indicated are feasible for the early work on Mg doping.[15,17,18]

If the inclusion of CO_2 in the system is considered, the possibility of MgO formation is much greater for the reaction

$$Mg_{(g)} + CO_{2(g)} \Leftrightarrow MgO_{(s)} + CO_{(g)} \qquad (20)$$

at 560°C $K_{20}^{-1} = P_{CO_2}P_{Mg}/P_{CO} = 5.23.10^{-22}$ atmos.

Now even for $P_{Mg} = 1.3.10^{-15}$ (ppm doping) oxidation will be thermodynamically possible for all values of $P_{CO_2}/P_{CO} > 4.10^{-7}$. For sensible values such as $P_{CO_2} = 10^{-11}$ torr and $P_{CO} = 10^{-9}$ torr, significant oxidation of Mg might well be expected.

Two points need to be raised in interpreting these thermodynamic results. Firstly, they only serve to indicate possibilities - these competing reactions are just as likely to be subject to kinetic limitations as the intended reaction. A clear example of such a limitation will be identified in the case of S doping of GaAs. Secondly, some of the more striking results can be safely ignored simply because the flux of reactant is quickly exhausted. This point becomes clearer if we reexamine reaction (20). At 560°C if doping at 5.10^{18} is attempted ($P_{Mg} = 1.3.10^{-15}$ atmos) then $P_{CO_2}/P_{CO} = 4.10^{-9}$ will provide an overpotential for oxidation of Mg. With a value of $P_{CO} \simeq 10^{-12}$ atmos any value of $P_{CO_2} > 10^{-20}$ is sufficient. While it will probably be impossible to avoid higher pressures than this, one should note that the CO_2 will be exhausted by the time $10^{-5} P_{Mg}$ is oxidized. The effect can therefore be ignored at this level and probably tolerated at more realistic values of P_{CO_2} like 10^{-15} atmos even for a reaction coefficient of unity. It is clear from experimental results[16] that in GaAs growth the above problems can now be avoided in practice. The main guideline from analyses of this type is the probable advantage to be gained in growing materials containing electropositive elements like these, e.g. $Ga_xIn_{1-x}As$ and $Al_xIn_{1-x}As$ at temperatures as high as their dissociation pressures will allow, to minimize the risk of competing oxidation reactions. The reactions considered are by no means the only possible ones but they are among those most likely and illustrate the approach sufficiently.

4. DOPING OF GALLIUM ARSENIDE

The main doping reactions in gallium arsenide may be grouped as follows:

(a) substitution of gallium atoms by elements in Group 2 of the periodic table to give p-type doping (Zn, Cd, Mg, Be)

$$Zn_{(g)} + V_{Ga} \Longleftrightarrow Zn_{Ga}^- + h \tag{21}$$

where V_{Ga} is a vacant gallium site in GaAs, Zn_{Ga}^- is an ionized Zn atom on a gallium site in GaAs and h is a free hole in the valence band of GaAs. Mn also behaves this way.

(b) substitution of arsenic atoms by elements of Group 6 of the periodic table to give n-type doping (S, Se, Te)

$$1/2S_{2(g)} + V_{As} \Longleftrightarrow S_{As}^+ + e \tag{22}$$

where e is a free electron in the conduction band of GaAs.

(c) substitution of gallium atoms by group 4 elements to give n-type doping (Si, Ge, Sn, Pb)

$$Si_{(g)} + V_{Ga} \Longleftrightarrow Si_{Ga}^+ + e \tag{23}$$

(d) substitution of arsenic atoms by group 4 elements to give p-type doping (Ge)

$$Ge_{(g)} + V_{As} \Longleftrightarrow Ge_{As}^- + h \tag{24}$$

Si also behaves analogously in LPE.

In all cases a direct quantitative analysis of the thermodynamics of the reactions involves the enthalpies and entropies of defects such as V_{Ga} or $Zn^-{}_{Ga}$. These quantities are not known with any precision as was the case for surface defects discussed in section 2.3. However, there is one big difference from the surface case. The doped bulk GaAs can be obtained by other means such as LPE or VPE whereas the GaAs/vacuum interfaces cannot. As the thermodynamic state is not dependent on the route by which it is approached, one can use data from other growth techniques and experiments to describe it. The approach has been applied successfully to all the common dopants in GaAs by Heckingbottom et al.[18] Some of the most interesting aspects are considered in the remaining sections, together with some more recent results.

4.1 Sulphur Doping

Most of the thermodynamic analysis of doping of GaAs in MBE[18] served mainly to establish the validity of the approach, as the experimental results for most dopants already existed for comparison. The effective doping of GaAs using a molecular beam of elemental sulphur was however a sucessful prediction of the approach - up to that date only sulphur doping using PbS as a "captive" source had been tried.[21] The prediction has since been confirmed for beams of S_2 using an electrochemical cell incorporating Ag_2S as the source of sulphur.[22]

Successful doping with elemental sulphur serves to emphasize that the elemental vapor pressure is not an important consideration - except that a source which is compatible with bake-out procedures, and other aspects of UHV, is required. The doping reaction is described by reaction (22), with just one important additional aspect to consider. From equation (22) it follows that

$$K_{22} = [S_{As}^+] \cdot n/P_{S_2}^{1/2}[V_{As}] \tag{25}$$

where [ ] indicate concentrations of defects and n is the concentration of electrons e. If n is determined by reaction (22) then

$$n = [S_{As}^+] = K_{22}^{1/2}P_{S_2}^{1/4}[V_{As}]^{1/2} \tag{26}$$

It is found in practice however in vapor phase epitaxy[23] that although on cooling to room temperature $n = [S_{As}^+]$, the amount of $[S_{As}^+]$ is proportional to $P_{S_2}^{1/2}$ (actually P_{H_2S}, see equation (32)) present at the growth temperature. Hence at the growth temperature n must be constant and greater than $[S_{As}^+]$, so that

$$[S_{As}^+] = K_{22}P_{S_2}^{1/2}[V_{As}]/n \tag{27}$$

This observation that $n \neq [S_{As}^+]$ is typical of that for several dopants in both VPE and LPE of GaAs and at least two important models for explaining this behavior have been described.[24,25] The alternative explanation, that the discrepancy between experimental results and equation (26) is due to kinetic barriers has also been examined. While kinetic effects have been clearly identified, for example by comparing growth and doping on different faces of GaAs,[26,27] they typically lead to variations in growth rate or doping efficiency of factors of $\lesssim 5.$[27] It is also considered unlikely that kinetic limitations involving chlorides or hydrides (VPE) or Ga solutions (LPE) would all lead to the same basic modification, and an explanation in terms of an underlying aspect like a change in the relevant defect model is preferred. Equation (27) is therefore adopted in the following treatment. It remains to replace $[V_{As}]$ by an experimental parameter through the reaction

$$\frac{1}{4}As_4 + V_{As} \Longleftrightarrow 0 \qquad (28)$$

so that

$$K_{28} = [V_{As}]^{-1}P_{As_4}^{-1/4} \qquad (29)$$

and rearranging and substituting in (27) gives

$$[S_{As}^+] = K_{30}P_{S_2}^{1/2}/P_{As_4}^{1/4} \qquad (30)$$

where $K_{30} = K_{22}/n\,K_{28}$

Using equation (30) and the known doping achieved under VPE conditions one may now calculate the required conditions in MBE. In typical VPE growth[28] sulphur is introduced as H_2S in a large excess of H_2, allowing the equilibrium value of $P_{S_2}^{1/2}$ to be calculated from

$$\frac{1}{2}S_2 + H_2 \Longleftrightarrow H_2S \qquad (31)$$

giving

$$P_{S_2}^{1/2} = P_{H_2S}/K_{31}P_{H_2} \qquad (32)$$

where K_{31} can be calculated from standard thermodynamic data.[19]

For $P_{H_2S} = 10^{-7}$ atmos and $P_{H_2} = 1$ atmos a corresponding doping level of $\simeq 10^{17}$ cm^{-3} is obtained. Similarly the equivalent value of P_{As_4} in VPE is typically 10^{-3} atmos.[29] The resulting extrapolation to MBE conditions indicates that $[S_{As}^+] = 2.2.10^{17}$ is in equilibrium with

P_{S_2} = 4.2.10^{-24}atmos ($\equiv 3.2.10^{-21}$torr) at a typical growth temperature of 560°C. It should be noted that the answer is approximate as
(a) the VPE results apply at 750°C while the MBE is at 560°C, (b) the
value of P_{As_4} used in MBE is not at equilibrium, (c) it is doubtful
whether the dependence of P_{As_4} is as great as assumed in equation
(28)[25] and (d) only two terms equations were used in calculating
equilibrium constants. However if (c) applies it will serve to minimize
the effects of (b). It is most unlikely that any of these approximations
will affect the conclusion that sulphur doping should occur at useful
levels in MBE, as the result is used in the following way.

In MBE, straightforward control on doping is sought by control
of the growth rate and the arrival rate of the doping species. Thus in
growth of GaAs (001), where all the Ga condenses and so governs the
growth rate, the doping is given simply by (F_S/F_{Ga}) (Ga atoms cm^{-3} in
GaAs) where F_S, F_{Ga} represent the fluxes of sulphur and gallium atoms
respectively. If the flux ratio is approximated by the beam pressure
ratio, then noting that GaAs has 2.2.10^{22} atoms cm^{-3} of Ga, P_{Ga} =
10^{-6} torr and P_{S_2} = 5.10^{-12} torr will lead to $[S^+_{As}]$ = 2.2.10^{17}. The
thermodynamic calculations are used to establish that under such condi-
tions $[S^+_{As}]$ is stable and hence should form. It will be clear that the
operating pressure is $\simeq 10^9$ above the stability boundary and the approximate guideline is therefore quite sufficient. The experimental
results[22] confirm that this simple description is essentially correct at
560°C, with no evidence for any kinetic barriers to dopant incorporation to complicate the situation.

In a later study,[30] it was found that at higher temperatures
($\simeq 600°C$) sulphur incorporation is much reduced as the temperature is
increased, see Figure 4. The most likely reaction causing this observed
loss of sulphur is

$$2Ga_{(1)} + 1/2S_{2(g)} \Longleftrightarrow Ga_2S_{(g)} \tag{33}$$

where (as considered in section 2.1) $Ga_{(1)}$ is a good approximation to
the chemical potential of Ga in GaAs grown under LPE or MBE conditions.

Thus

$$K_{33} = P_{Ga_2S}/P_{S_2}^{1/2} \qquad (34)$$

and from available thermodynamic data,[19,31] the calculated values of K_{33} are $9.2.10^{10}$ atmos$^{1/2}$ at 560°C and $2.5.10^{10}$ atmos$^{1/2}$ at 650°C. Substituting the value of $P_{S_2}^{1/2}$ in equilibrium with $[S_{As}^+] = 2.2.10^{18}$ cm^{-3} ($4.2.10^{-22}$ atmos) in equation (34), at the lower temperature gives $P_{Ga_2S} = 1.89$ atmos. It is clear that the value will not change substantially through the range of growth temperatures up to 650°C. Hence essentially all the sulphur should be lost as Ga_2S, according to the thermodynamic framework, under all typical MBE conditions. Any observed doping must therefore be due to a kinetic barrier to the formation of Ga_2S on the GaAs (001) surface. To emphasize the completeness of this situation one may recast the figures in line with the assumption that virtually all the sulphur is converted to Ga_2S. Then for a beam pressure of $P_{S_2} = 5.10^{-11}$ torr, converted to $P_{Ga_2S} = 1.10^{-10}$ torr, equations (34) and (30) lead to a value of $[S_{As}^+] = 1.5.10^5$ cm^{-3} i.e., the approximation that all the sulphur is converted to Ga_2S is accurate to $\simeq 1.10^{-12}$.

In summary, therefore, there is only local equilibrium at 560°C. The reaction leading to the lowest free energy (Ga_2S formation) is essentially completely hindered kinetically while the route to the metastable position $[S_{As}^+]$ is kinetically facile. As the temperature is raised it is also found that the formation of Ga_2S is hindered by an increase in the P_{As_4}/P_{Ga} ratio. In fact, as shown in Figure 4, the temperature dependence of sulphur loss is similar to that for arsenic loss as measured by Panish[32], using the border between the arsenic stabilized GaAs (2x4) and GaAs (3x1) surface structures as a criterion. These results suggest that the situation is dominated by the surface arsenic population. As the arsenic coverage is reduced, an increasing surface gallium population becomes free to bond to sulphur to form Ga_2S via a much lower energy transition state. Here, as in the case of several other dopants, further doping experiments at different temperatures - with the Ga/As surface ratio held constant - are required to provide additional insight.

4.2 Tellurium and Selenium Doping

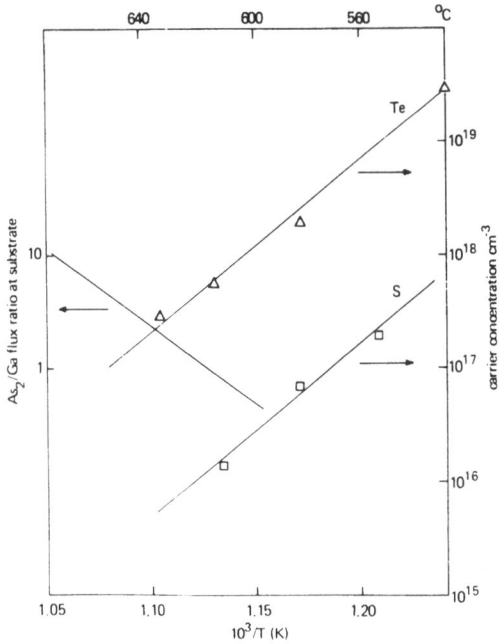

Figure 4 - The temperature dependence of S doping using an S_2 beam and Te doping using a beam of SnTe. Also shown is the temperature dependence of the GaAs (001) (2x4) As and GaAs (001) (3x1) RHEED pattern boundary as a function of As_2/Ga flux ratio.

Tellurium and selenium should also lead to doping via equation (22). Collins and Miller have recently studied doping of GaAs in MBE using SnTe[33] in the temperature range 540-640°C. Above 580°C the amount of Sn and Te in the growing layer is different, suggesting that here the dopants are acting independently. The temperature dependence of Te throughout the temperature range is identical with the sulphur results, as shown in Figure 4. This similarity supports the proposal that the rate controlling step is dependent on the Ga-As system. Collins and Miller argues that the increase in doping with arsenic pressure ruled out an explanation in thermodynamic terms. The effect is in the wrong direction in terms of equation (22), but they omitted to consider that the dominant effect of the arsenic could be on the Te loss mechanism by a reaction analogous to equation (33).

Recent work in our own laboratories using an electrochemical source for selenium has shown that dopant incorporation continues to higher temperature ($\geq 625°C$). This behaviour makes it a more useful dopant in practical terms but also indicates that the full explanation must be more complicated than that given above in terms of Ga and As only.

4.3 Zinc Doping

Zinc is an effective p-type dopant for GaAs in VPE and as a diffusion souce, and is incorporated as indicated in equation (21). Extrapolation from VPE conditions, using thermodynamic formalism,[18] showed that for a $P_{Zn} = 10^{-9}$ torr, a dopant concentration $[Zn^-_{Ga}] \leq 10^{12}$ cm^{-3} was predicted at $560°C$. The indicated sticking coefficient is thus $\simeq 10^{-7}$. The only reported experimental value[20] of the sticking coefficient under similar conditions is 5.10^{-7} at $500°C$. Zinc therefore well illustrates the principle that when the thermodynamic guideline indicates that the doping reaction itself will be ineffective, that is a sufficient condition. Ionized zinc or some similar energetic beam is needed to achieve useful doping at MBE pressures. Cd is similar to Zn but even less effective as a dopant.

4.4 Magnesium Doping

The original thermodynamic treatment of Mg doping of GaAs in MBE[18] was interesting due to the lack of VPE or LPE data to be used for extrapolation. Mg should be incorporated like Zn according to equation (21). It can therefore be compared with Zn, by considering the likely bonding in the Zn^-_{As} and Mg^-_{As} defects and using more basic concepts like electronegativity to estimate the difference in bond energies in the two cases. The results suggested that Mg was a borderline but usuable dopant with a sticking coefficient at $560°C$ of $\simeq 6.10^{-3}$. The uncertainty in the estimating procedure indicated that the value might be up to one order of magnitude larger or smaller. The result was however in conflict with the experimental data which suggested sticking coefficients of 10^{-4} - 10^{-5} (electrical results)[15,20] or unity (chemical analysis).[17] As discussed in section 3, these early experimental results were probably dominated by reactions involving oxygen containing contaminants.

In a recent paper[16] on Mg doping, these contaminant problems appear to have been avoided - for instance the doping increases as the growth temperature decreases, whereas the opposite would be expected if oxidation was important. At 560°C the sticking coefficient is measured as 3.10^{-3}, in excellent agreement with the calculated value. As in the case of sulphur doping however, the temperature variation of doping reveals a more complex situation[24] - see Figure 5.

The thermodynamic model predicts a reduction in doping as the temperature is increased, and the low sticking coefficient indicates that re-evaporation from the surface will be an important mechanism. This significant desorption from the surface was confirmed by experiment.[16] The evidence for kinetic complications stems from the magnitude of the temperature dependence of doping shown in Figure 5 and corresponding to 3.8 eV. The basis of the defect model used in calculation[18] however is that Mg is bonded to As in GaAs, so that a first approximation to the temperature dependence of surface loss (the negative of that for doping) should be given by the vapour pressure of Mg above Mg_3As_2. This vapor pressure was measured,[16] as Mg_3As_2 was used as a source of Mg, and it is also plotted in Figure 5. It is clear that the temperature dependence of Mg over Mg_3As_2 (and therefore to be expected in the thermodynamic model) is much less than that measured over GaAs.

As in the case of sulphur doping, it is suggested here that the reduction in surface arsenic concentration with increasing temperature, again plays a key role. Thus as the temperature in increased towards 600°C, the Mg environment changes from one resembling the Mg_3As_2 surface towards one resembling elemental Mg or an Mg/Ga alloy.

Correspondingly the evaporation rate should increase towards that given by the upper line in Figure 5. The rate of approach is given by the dotted line, which represents the sum of the Mg_3As_2 line and that for As surface loss shown in Figure 4. This combined temperature dependence is close to the observed value for doping. It should be stressed that this dominance of surface effects is kinetic. In the thermodynamic model the increase in free energy of surface atoms should increase the rate of incorporation into the bulk just as it speeds the evaporation so that the gas/bulk relationship remains approximated by the Mg_3As_2 plot. The simplest explanation in the practical situation is that there is only one unhindered incorporation mechanism, which is via

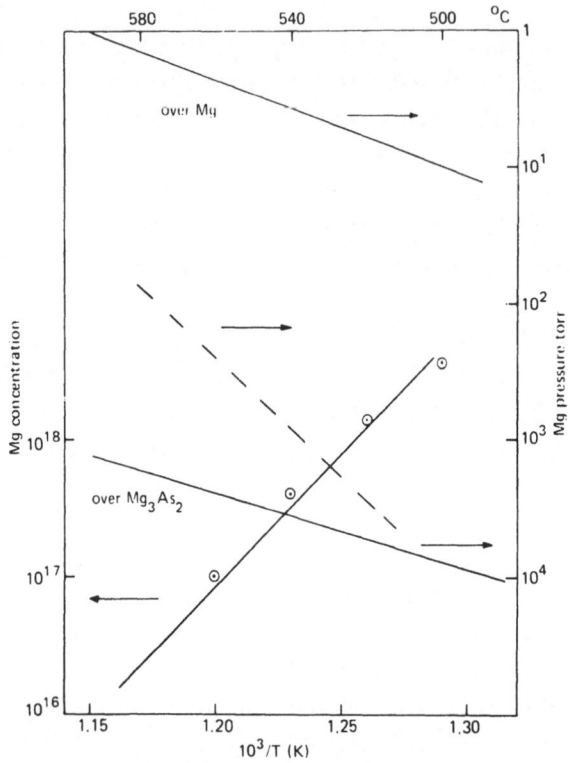

Figure 5 - The temperature dependence of Mg doping. Also shown (full lines) are the temperature dependence of the vapour pressure of Mg over elemental Mg and over Mg_3As_2 (taken as 10^3 x the beam pressure.[16] The dashed line results from adding the temperature dependence of Mg over Mg_3As_2 to that for the surface arsenic coverage in Figure 4.

the growing layer (as assumed in section 1.3), and that other options like direct diffusion into the bulk are insignificant. It is then clear that the dominant incorporation rate cannot increase as the free energy of surface Mg increases, since it is fixed by the Ga supply rate.

4.5 Germanium Doping

Germanium is incorporated in GaAs on both Ga and As sublattices according to equations like (23) and (24), respectively. Thus models have to explain not only the incorporation step but also the amphoteric doping. In practice the incorporation appears straightforward, with all the Ge being incorporated in MBE, with no evidence for any kinetic hindrance. The amphoteric doping is more complex.

Ge doping leads to p-type material in LPE and n-type material in VPE. In MBE either doping may dominate, depending on the growth conditions. Hurle[35] addressed the problem that the acceptor/donor (N_A/N_D) ratio only varies from 5.4 in LPE to 0.3 in VPE. On the other hand, based on reactions (23) and (24) and the relationship of the vacancy concentrations to P_{As_4} through equations (13) and (29), a variation in N_A/N_D of 10^5 is to be expected. To explain the experimental results Hurle proposed the existence of an additional acceptor defect complex $V_{Ga}Ge^-_{Ga}$ with

$$\frac{N_A}{N_D} = \frac{[Ge^-_{As}] + [V_{Ga}Ge^-_{Ga}]}{[Ge^+_{Ga}]} \tag{35}$$

and $[V_{Ga} Ge^-_{Ga}]/[Ge^+_{Ga}] = 0.3$ and independent of P_{As_4} throughout the n-type range. Heckingbottom and Davies[36] applied the Hurle model to MBE conditions and obtained very good agreement with the main features of the experimental results, as shown in Figure 6. Cho[37] found p-type GaAs with poor morphology $(P_{As_2}/P_{Ga} \leq 1)$ was obtained at 815K while more recently Kunzel et al.[11] showed that with more standard conditions $(P_{As_2}/P_{Ga} \geq 1)$ the p/n conversion point is approximately 880K (see Figure 2). These values compare well with the calculated p/n boundary at 833K for $P_{As_2}/P_{Ga} = 1$ and 860 for $P_{As_2}/P_{Ga} = 2$. The calculations emphasized that the equilibrium gas phase species is As_2 not As_4 in the MBE regime. It was also pointed out that strictly speaking Figure 6 cannot be used to obtain a quantitative guide to the nonstoichiometry of the GaAs, as for a constant growth rate increasing the P_{As_2}/P_{Ga} ratio also increases the departure from equilibrium. In practice however[11] it appears that the increase in over-potential with increase in P_{As_2} (or constant P_{As_2} with decreasing temperature) has no observable effect on the results and even P_{As_2} excess is partially effective - see section 2.2

In a recent paper Munoz-Yague and Baceiredo[38] also considered Ge doping of MBE grown GaAs from a thermodynamic standpoint. They restrict their treatment to uncomplexed defects and obtain an interesting correlation with their experimental results, though the calculated p/n conversion point is a little high at 650°C. They also make the interesting point that although the diffusion of atoms is severely restricted in the solid (in agreement with section 1.3), the diffusion of mobile defects such as V_{As} is probably not, so that a site

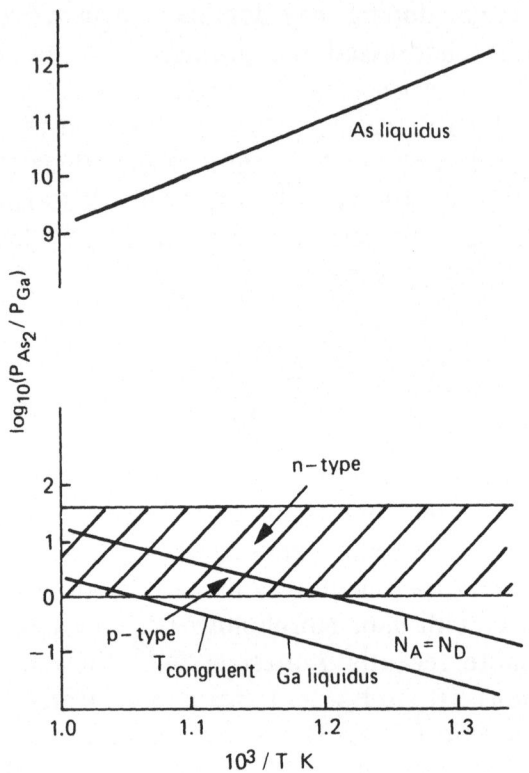

Figure 6 - P_{As_2}/P_{Ga} versus I/T for GaAs at the Ga liquidus, the As liquidus and the p/n boundary ($N_A = N_D$) for Ge doping, together with the MBE working range (hatched area).

exchange

$$Ge_{Ga} + V_{As} \Longleftrightarrow Ge_{As} + V_{Ga} \qquad (36)$$

is likely to be facile. It is difficult to see however how their model could explain VPE results - the dependence on P_{As_2} is too severe as Hurle first noted for all models of this type.

4.6 Silicon Doping

Silicon is in principle very similar to Ge as a dopant. It is how-ever much more useful in practice as incorporation is straightforward as shown in equation (23) and $N_D \gg N_A$ under most MBE conditions.[39,40] Measureable autocompensation has been detected at growth temperatues $\geq 625°C$ however[39] and at these higher tempera-tures the N_D/N_A was influenced by variation of P_{As_4} in significant excess ($P_{As_4}/P_{Ga} = 4-6$) in line with the discussion of section 2.2. The above results refer to the (001) surface, and under otherwise identical conditions p-type material results on the (001) surface.[41] Thermodynamically a change of orientation should have no effect but it is clear that the surface conditions vary greatly and are hindering the approach to equilibrium in some cases. No detailed thermodynamic analysis of Si doping has been reported, but it is known[38,42] that the p/n boundary lies close to the Ga liquidus in the 500-800°C range and so is similarly not far removed from the MBE regime.

4.7 Tin Doping

Tin is also incorporated as in equation (23) and forms n-type material. Growth at temperatures approaching 600°C results in high quality material[43-45] but also marked surface accumulation. Thus starting with a clean GaAs (100) surface, growth with tin doping leads to the characteristic doping profile shown in Figure 7. It has been shown[44] that, for the situation where reevaporation of tin is negligible, the profile is given by

$$\frac{dC_s}{dt} = J_{Sn} - KC_s \tag{37}$$

where C_s is the surface composition, t is the time from the start of tin doping and J_{Sn} is the flux of tin atoms. K has been interpreted as the rate constant of a surface rate limited incorporation step. At steady state C_s is constant and $dC_s/dt = 0$ giving $J_{Sn} = KC_s$. The situation is not reached however until C_s builds up sufficiently. In practice K is required to be $\simeq 10^{-3}$ and is a function of temperature and arsenic pressure.

The model is generally accepted and may be essentially correct but it is noted here that it is not yet proved, as the above mentioned fit is not a definitive criterion. Thus it could be assumed instead that the surface accumulation is thermodynamic in origin and the distribution

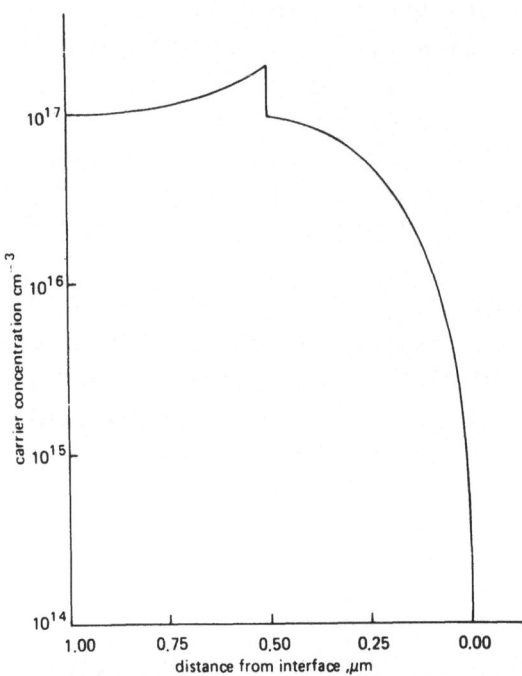

Figure 7 - Expected carrier profile for tin doping with either an incorporation rate constant or a bulk/surface distribution coefficient of 10^{-3}, for no predeposition and $F_{Ga} = 2.2.10^5 \, F_{Sn}$. The effect of changing growth conditions to reduce surface accumulation by a factor of 2 is shown at 0.5μm.

coefficient between bulk and surface $C_b/C_s \simeq 10^{-3}$. The expression describing the situation is exactly that shown in (37). K is in this case the distribution coefficient, C_b/C_s, and again is expected to vary with the growth temperature and arsenic coverage. Several features favor the thermodynamic origin of the surface accumulation. The high crystal quality suggests kinetically facile reaction steps and in fact tin is incorporated efficiently at lower temperatures ($\simeq 500°$C). Other dopants are also incorporated without kinetic hindrance at similar concentrations of V_{Ga} in surface and bulk. Tin is also known from recent work[46] to diffuse very quickly on the Ga (001) surface, leading to the formation of discrete islands. This effect occurs at bulk concentrations of $\simeq 10^{19}$ cm^{-3}, consistent with the maximum concentration obtained in LPE, even when grown from liquid tin.[47]

The suggestion that $C_b/C_s \simeq 10^{-3}$ indicates that Sn_{Ga} in the bulk is energetically unfavorable versus the surface site, and it is known that the distribution coefficient between GaAs solid and Ga liquid is even lower for tin.[48] These effects could be due to tin atoms being larger than gallium. They also lead to a further possible explanation of the results. C_b/C_s would be $\simeq 10^{-3}$ throughout the range 500-600°C. The tin would fit reasonably and diffuse effectively within the surface layer. However as this layer is covered and becomes bulk, the larger size of tin will lead to an increased free energy and a driving force to diffuse out to the surface layer again. It is reasonable to suggest that at 500°C this diffusion is kinetically blocked and only becomes fast enough to allow equilibrium to be reached at higher temperatures. Increased arsenic surface coverage would be expected to hinder this reaction step, in line with the observed results. At this stage none of these possibilities appears to be either proved or disproved. Careful analysis of the consequences of each model would need to be explored, and then further experiments carried out to distinguish between them.

5. CONCLUSION

Thermodynamics and kinetics are inextricably linked and should be regarded as complementary. Thermodynamics is particularly useful in delineating boundaries between the two regimes where forward or reverse reaction dominate. It is also useful in that the data required can be obtained from a wide base, so that guidelines can be obtained before the MBE experiments are carried out. For the normal situation, where only thermal sources of energy are involved, several firm statements can be made. If the thermodynamic conclusion is that a desired reaction should not proceed, i.e., the free energy of the products is higher than that of the reactants, then it will not. If according to thermodynamics the reaction could proceed, then the method allowed comparison between the desired reaction and other competing reactions - involving either the same reactants or contaminants. When the desired reaction is the thermodynamically preferred one, then comparison with experimental results allows clearer identification of any kinetically hindered reaction steps and hence a better understanding of the reaction mechanism.

In GaAs, where extensive experimental results are available for examination, it is clear that under conditions where layers most suitable for device application are grown there are very few effects due to kinetic limitations. Straightforward excess crystal defects can be limited to the part per billion range, and non stoichiometry effects have been identified on the part per million range if As_4 is used instead of As_2. Against this background, most doping processes can be successfully described by the thermodynamic approach. The two most clearly identified kinetic limitations involve not straightforward incorporation but surface desorption (S doping) and the fact that incorporation is facile only via the surface layer, not by direct diffusion into the bulk (Mg doping).

It is suggested that the determination of the boundaries of regimes where reactions are thermodynamically allows, will be most useful as MBE is extended to a larger range of compounds and alloys. In more detailed cases, where these boundaries are already known, calculation of the predictions of both thermodynamic and kinetic models will allow the design of more definitive experiments to increase understanding.

6. ACKNOWLEDGEMENTS

I would like to thank particularly G.J. Davies, D.A. Andrews, and K.A. Prior for helpful inputs and interesting discussions. Acknowledgement is also made to the Director of Research, British Telecom for permission to publish this paper.

REFERENCES

1. J.R. Arthur, T.R. Brown, J. Vac. Sci. Technol. 12, 200 (1975).
2. S. Nagata, T. Tanaka, J. Appl. Phys. 48, 940 (1977).
3. P.M. Petroff, Second IUPAP/UNESCO Semiconductor Symp, Trieste (1982).
4. B. Goldstein, Phys. Rev. 121, 1305 (1961).
5. J.R. Arthur, J. Phys. Chem Solids 28, 2257 (1967).
6. N. Matsumoto, K. Kumabe, Jap. J. Appl. Phys. 19, 1583 (1980).
7. J.H. Neave, B.A. Joyce, J. Cryst. Growth 43, 204 (1978).
8. S.R. McAfee, D.V. Lang, W.T. Tsang, 161st Electrochem Soc. meeting, Montreal (1982).
9. W.T. Tsang, Appl. Phys. Lett. 39, 786 (1981).

10. J.H. Neave, P. Blood, B.A. Joyce, Appl. Phys. Lett. 36, 311 (1980).

11. H. Kunzel, J. Knecht, H. Jung, K. Wunstel, K. Ploog, Appl. Phys. A 28, 167 (1982).

12. C.T. Foxon, B.A. Joyce, Surf. Sci. 50, 434 (1975).

13. D.T.J. Hurle, J. Phys. Chem. Solids 40, 613 (1979).

14. P.D. Kircher, J.M. Woodall, J.L. Freeouf, D.J. Wolford, G.D. Pettit, J. Vac. Sci. Technol. 19, 604 (1981).

15. A.Y. Cho, M.B. Panish, J. Appl. Phys. 43, 5118 (1972).

16. C.E.C. Wood, D. DeSimone, K. Singer, G.W. Wicks, J. Appl. Phys. 53, 4230 (1982).

17. B.A. Joyce, C.T. Foxon, Jap. J. Appl. Phys. 16, Suppl. 16-1, 17 (1977).

18. R. Heckingbottom, C.J. Todd, G.J. Davies, J. Electrochem. Soc. 127, 444 (1980).

19. O. Kubaschewski, C.B. Alcock, Metallurgical Thermochemistry, 5th Ed., Pergamon Press, London (1979).

20. M. Naganuma, K. Takahashi, Phys. Status Solidi A 31, 187 (1975).

21. C.E.C. Wood, Appl. Phys. Lett. 33, 770 (1978).

22. G.J. Davies, D.A. Andrews, R. Heckingbottom, J. Appl. Phys. 52, 7214 (1981).

23. M. Heyen, H. Bruch, K.H. Bachem, P. Balk, J. Cryst. Growth 42, 127 (1977).

24. H.C. Casey, M.B. Panish, J. Cryst. Growth 13/14, 818 (1972).

25. D.T.J. Hurle, Proc 6th Int. Symp. on GaAs and Related Compounds, Inst. Phys. Conf. Series 33a, 113 (1977).

26. J.B. Mullin, J. Cryst. Growth 42, 77 (1977).

27. E. Venhoff, M. Maier, K.H. Bachem, P. Balk, J. Cryst. Growth. 53, 598 (1981).

28. M.A. Savva, J. Electrochem. Soc. 123, 1498 (1976).

29. D.J. Ashen, P.J. Dean, D.T.J. Hurle, J.B. Mullin, A.M. White, P.D. Greene, J. Phys. Chem. Solids 36, 1041 (1975).

30. G.J. Davies, D.A. Andrews, R. Heckingbottom, 2nd Int. Symp. on MBE, Tokyo (1982).

31. O.M. Uy, D.W. Muenow, P.J. Ficalora, J.L. Margrave, Trans. Faraday Soc., 64, 2998 (1968).

32. M.B. Panish, J. Electrochem Soc. 127, 2729 (1980).

33. D.M. Collins, J.N. Miller, J. Appl. Phys. 53, 3010 (1982).

34. R. Heckingbottom, G.J. Davies, K.A. Prior, Surf. Sci., 132, 375 (1983).

35. D.T.J. Hurle, J. Phys. Chem. Solids 40, 647 (1979).
36. R. Heckingbottom, G.J. Davies, J. Cryst. Growth. 50, 644 (1980).
37. A.Y. Cho, I. Hayashi, J. Appl. Phys. 42, 4422 (1971).
38. A. Munoz-Yague, S. Baceiredo, J. Electrochem Soc. 129, 2108 (1982).
39. J.C.M. Hwang, H. Temkin, T.M. Brennan, R.E. Frahm, Appl. Phys. Lett. 42, 66 (1983).
40. Y.G. Chai, R. Chow, C.E.C. Wood, Appl. Phys. Lett. 39, 800 (1981).
41. J.M. Balliingall, C.E.C. Wood, Appl. Phys. Lett. 41, 947 (1982).
42. D.T.J. Hurle, private communication.
43. A.Y. Cho, J. Appl. Phys. 46, 1733 (1975).
44. C.E.C. Wood, B.A. Joyce, J. Appl. Phys. 49, 4854 (1978).
45. F. Alexandre, C. Raisin, M.I. Abdalla, A. Brenac, J.M. Masson, J. Appl. Phys. 51, 4296 (1980).
46. J.J. Harris, B.A. Joyce, J.P. Gowers, J.H. Neave, Appl. Phys. A. 28, 63 (1982).
47. M.B. Panish, J. Appl. Phys. 44, 2659 (1973).
48. H. Kressel, J.K. Butler, "Semiconductor Lasers and Heterojunction LED's", Academic Press, New York (1977).

SOME ASPECTS OF SURFACE SCIENCE RELATED TO MBE

Winfried Mönch

Laboratorium für Festkörperphysik, Universität Duisburg
D-4100 Duisburg, Germany

ABSTRACT: Crystal growth by the technique of molecular beam epitaxy occurs on clean surfaces under well controlled ultra-high vacuum conditions. The quality of the growing film is usually monitored in situ by the observation of surface-sensitive electron diffraction. Therefore, knowledge about the properties of clean surfaces is essential for MBE. This contribution will consider GaAs only, since with surfaces of this semiconductor a wealth of experimental data is available. In particular, we will discuss the crystallography of (111) and (001) surfaces, which are used in MBE, and the correlation between different reconstructions and the respective chemical surface compositions. Another issue is occupied and empty intrinsic surface states as well as extrinsic electronic surface properties. Finally, results on the chemisorption of oxygen, of hydrogen, and of H_2S are examined. With all the different topics the properties of cleaved surfaces are looked at for reasons of comparison. Studies with such surfaces have recently questioned the bulk homogeneity of the substrates used in MBE since segregation of anion atoms was detected on surfaces cleaved from GaP, GaAs, GaSb, and InP.

1. INTRODUCTION

As a physical process, molecular beam epitaxy is a surface phenomenon. Beams of different atoms and/or molecules evaporated from
Knudsen cells under well controlled ultra-high vacuum conditions meet
on the surface of a clean and well oriented substrate. Primarily, the
atoms and molecules arriving at the surface are adsorbed and desorbed,
and molecules may be dissociated. Proper adjustment of the beams
incident and the temperature of the substrate cause the atoms adsorbed
on the surface to form a continuous and single-crystalline film which is
growing in a layer-by-layer mode and with a desired chemical composition. Epitaxy is achieved if the unit meshes of the substrate surface
and the planes of the overgrowth are matching. The growth process of
semiconductor films and the simultaneous incorporation of doping
atoms are discussed in the contributions by others to this volume.

The present contribution will concentrate on some selected
aspects of surface science which are

- the atomic arrangement at the low-indexed (111), (001), and (110)
  surfaces,
- the chemical composition correlated with different atomic reconstructions on (111) and (001) oriented surfaces,
- the dispersion of electronic surface states,
- the chemisorption of oxygen, of hydrogen, and of hydric sulfide,
- the influence of surface imperfections on electronic and chemical
  surface properties, and
- the segregation of substrate atoms on clean surfaces.

Only results obtained with GaAs will be considered since for this
semiconductor a wealth of experimental data and theoretical studies has
been published that tends to give a coherent picture.

In all of the following chapters the (110) surface and its properties will be discussed in detail – although this orientation is *not* used in
MBE. The reasons for this are twofold. First of all, this surface can be
prepared very easily and reproducibly by cleavage in ultra-high vacuum,
and therefore the many experimental results published by different
groups may be reproduced and thus compared quite well. Some examples will be given. This surface and studies done with it may be and are
used as a reference. The second reason is a more personal one: the

author and his coworkers have studied cleaved III-V surfaces during recent years and feel somewhat "familiar" with its properties.

2. SURFACE GEOMETRY AND CHEMICAL COMPOSITION OF CLEAN GaAs SURFACES

2.1 Cleaved GaAs(110) Surfaces

In the zincblende lattice the unit mesh of a (110) plane contains one anion and one cation atom each. Such planes are thus electrically neutral, and they are cleavage planes. In a bulk layer the atoms are arranged in zigzag chains running along a [$\bar{1}$10] direction as shown in Fig. 1.

In their systematic studies of low-energy electron diffraction (LEED) with surfaces cleaved from III-V crystals MacRae and Gobeli[1] observed diffraction patterns containing only spots that could be indexed with integral indices. This implies that the surface unit mesh equals the one of an equivalent bulk plane. However, MacRae and Gobeli noticed two factors that suggest the surface arrangement of the Ga and the As atoms to differ from that found in bulk plane. They are an extreme asymmetry in the intensities of (hk) and (h$\bar{k}$) beams and

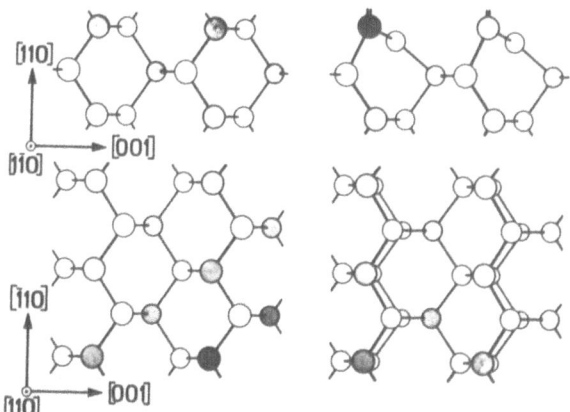

Figure 1 - Models of unreconstructed and reconstructed GaAs(110) surfaces.

strong intensities of the (10) and ($\bar{1}$0) beams. Since the scattering factors of gallium and arsenic are expected to be quite similar the (10) and the (10) beams should be quite weak because the scattering from the rows of Ga and As atoms would be out of phase for a bulk-like atomic arrangement in the top layer. MacRae and Gobeli concluded that the (10) beam must be mainly due to a change of the structure of the top layer while the intensities of the (01) and (0$\bar{1}$) spots result from both surface and subsurface distortions.

These general conclusions drawn by MacRae and Gobeli were confirmed in every respect by Duke, Mark, and their coworkers.[2] They have measured intensity-versus-voltage curves for 11 different LEED beams and have analyzed these data by applying a dynamical calculation. The resulting atomic geometry, which is sketched in Fig. 1, agrees well with the one determined by Chadi[3] from a minimization of the total energy. The surface structure of GaAs(110) is characterized by a rotation of the atoms in the top layer by 27° and a contraction of the first interlayer distance by 0.05 Å with the As atoms moved outward and the Ga atoms moved inward. The LEED analysis also established movements in the second layer, which are not shown in Fig. 1, while the energy-minimization calculations gave additional third-layer distortions.

The atomic arrangement in cleaved GaAs(110) surfaces indicates a dehybridization of the sp^3 tetrahedral bonds. The trivalent Ga approaches planar sp^2 bonds while the group V atom As tends to a pyramidal $AsGa_3$ configuration with bond angles close to 90° of pure p states. Such dehybridizations of sp^3 bonds at surfaces of covalently bonded semiconductors with diamond and zincblende structure, respectively, were already proposed by Haneman.[4]

The different bonding configuration of the surface atoms are accompanied by a charge redistribution. This shall be discussed in a very simple picture. Each sp^3 dangling bond should contain 3/4 of an electron on a Ga and 5/4 of an electron on an As surface atom. The dehybridization now transfers the dangling-bond charge from the Ga atoms to the As atoms leaving the latter ones with 5/4 + 3/4 = 2 electrons per dangling bond. This charge transfer from Ga to As atoms changes the electrostatic potential at the cores of those atoms, and thus shifts of the core-level binding-energies are expected with the surface atoms. Such core-level shifts were indeed observed with cleaved GaAs, GaSb, and InSb (110) surfaces[5,6] by using soft x-ray photoemission spectroscopy (SXPS). Experimental results of Eastman et al.[5] are displayed in Fig. 2.

The escape depth of the photoemitted electrons and thereby the sensitivity to bulk or surface atoms can be adjusted by proper choosing of the energy of the incident photons (see Fig. 4). With GaAs(110) the (3d) core-level binding energies were found to be by 0.28 eV larger at the Ga and by 0.34 eV smaller at the As surface-atoms with respect to the bulk values. These experimental results support the expectations based on the above outlined simple considerations.

2.2 GaAs (111) and ($\overline{1}\overline{1}\overline{1}$) Surfaces

In the (111) direction AB crystals with zincblende structure are ideally terminated by either an (111) A on a ($\overline{1}\overline{1}\overline{1}$) B polar face. Such

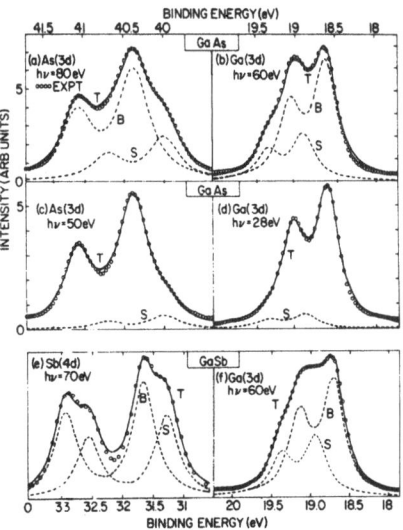

Figure 2 - Photoemission spectra for (3d) and (4d) core levels in GaAs(110) and GaSb(110). Spectra for lower photon energies show mainly bulk emission (B) while spectra for higher photon energies show additional surface core-level emission (S). (After Ref. 5).

surfaces cannot be obtained by cleavage, and oriented wafers have to be cut from larger ingots. The two different faces can be identified by the difference in etching behavior: Using, for example, a Br_2 + CH_3OH etch, the A face becomes rich in etch pits, the B face smooth and shiny.[7] Clean surfaces were prepared by ion-bombardment and subsequent annealing in UHV or by molecular beam epitaxy (MBE). LEED patterns observed with these surfaces exhibit extra-spots in addition to those normal-spots characterized by integral indices, i.e., these surfaces are reconstructed. Following the conventions described by Wood,[8] a m×n reconstruction means that the two surface primitive vectors are m and n times, respectively, larger than the bulk ones, i.e., in the surface the unit mesh is by m×n larger compared with an equivalent underlying bulk layer.

With the (111) A or Ga face a 2x2 reconstruction was found.[1] Several observations give evidence that this GaAs(111)-(2×2) surface contains a complete or almost complete Ga layer:

- The intensity ratio of the high-energy Ga(1070 eV) and As(1228 eV) Auger lines amounts to 2.3[9] which value is larger than the one observed with the Ga-stabilized ($\overline{111}$) surface[10] (see table 1).
- The sticking coefficient of oxygen on this surface exceeds those observed with cleaved GaAs(110) and Ga-stabilized GaAs($\overline{111}$) surfaces which both contain half a monolayer of gallium and arsenic atoms each. This behavior will be discussed later in section 5.1.

The atomic arrangement in the GaAs(111) surface leading to the (2×2) LEED pattern is not known.

The ($\overline{111}$) B or As surface may be prepared with either a 2×2 or a $\sqrt{19}×\sqrt{19}$ R 23.4° LEED pattern.[1,11,12] Those different reconstructions are correlated with different atomic compositions at the surface. The bulk stoichiometry is thus not necessarily conserved at the surface.

The 2×2-structure contains approximately half a monolayer of As atoms more than the $\sqrt{19}$ structure. This was first shown by Arthur[12] in his experiments using a modulated molecular-beam technique. Figure 3 shows the time dependence of the As_2 flux being

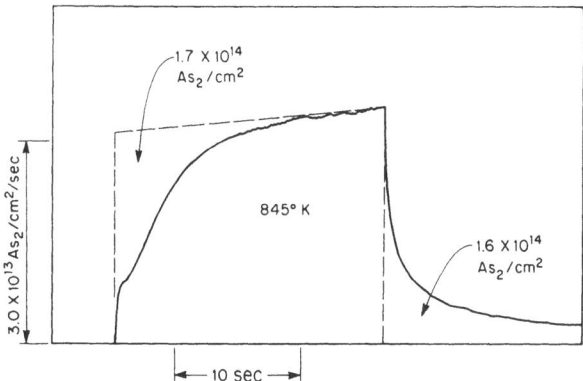

Figure 3 - Time dependence of As_2 desorption flux from a GaAs($\overline{111}$) surface exposed to an As_2 beam of intensity 3×10^{13} molecules/cm²s. Crystal temperature 845K. Dashed vertical lines indicate time when beam was turned on and off. (After Ref. 12.)

reflected and desorbed from a GaAs($\overline{111}$) surface being held at 845 K and exposed to an As_2 beam. After each pulse of As_2 molecules approximately half a monolayer ($\sigma_{111} = 7.22 \times 10^{14}$ sites per cm²) of arsenic desorbs which then is again taken up from the next pulse. When a ($\overline{111}$) surface is heated to above 800K in the absence of any As_2 or As_4 flux always the $\sqrt{19} \times \sqrt{19}$ R 23.4° LEED pattern is observed. Exposure of a surface exhibiting this structure to an arsenic beam produces a 2×2 pattern.

By using Auger-electron spectroscopy and a layer-model, to be discussed in the following section, Ranke and Jacobi[10] have determined the surface compositions correlated with both structures. They found, as shown in Table I, surfaces exhibiting the 2×2- and the √19 structure to contain 87% and 47%, respectively, of a monolayer of arsenic. The difference between both compositions amounts to 0.4 of a monolayer of arsenic and is in good agreement with Arthur's before mentioned result. As the cleaved (110) surface and the ($\overline{111}$)−√19 structure thus exhibits the same density of arsenic and gallium atoms.

TABLE 1. Surface structure and chemical composition of some GaAs surfaces

Surface structure	Preparation	AES intensity ratio		Θ_{As}	Reference
		$\dfrac{\text{Ga(55 eV)}}{\text{As(31 eV)}}$	$\dfrac{\text{Ga(1070 eV)}}{\text{As(1228 eV)}}$		
(110)-1 × 1	Cleavage	0.48 ± 0.02	1.5 ± 0.08	≡ 0.5	(9,13,17)
(111)-2 × 2	IBA	--	2.3 ± 0.1		(9)
$(\bar{1}\bar{1}\bar{1})$-2 × 2	MBE	0.24 ± 0.01	1.23 ± 0.09	0.87	(10)
$(\bar{1}\bar{1}\bar{1})$-√19	MBE	0.32 ± 0.02	1.30 ± 0.05	0.47	(10)
(001)-c(2 × 8)		--	1.18		(15)
		0.44 ± 0.02	0.8		(18)
		0.33		0.61	(17)
		0.35 ± 0.06		1.05	(16)
		0.52		0.82	(19)
c(6 × 4)		0.37		0.37	(17)
				0.9	(16)
c(8 × 2)		--	1.49		(15)
		--	1.1		(18)
		0.57 ± 0.03		0.22	(17)
		0.5		0.4	(16)

These differences in the surface composition also manifest in the distinct uptake of oxygen by both structures. The oxygen uptake-versus-exposure curves are almost the same for GaAs($\overline{1}\overline{1}\overline{1}$) √19×√19 R 23.4°[10] and cleaved GaAs(110) surfaces[13] as shown in Fig. 19. The ($\overline{1}\overline{1}\overline{1}$)−2×2 surface, on the other hand, takes up less oxygen at the same exposures. This behavior is to be understood since ($\overline{1}\overline{1}\overline{1}$)−√19 and (110) surfaces exhibit the same Ga:As ratio at the surface while the 2×2 structure only or almost only contains As atoms at the surface. This will be discussed in more detail in section 5.1.

To now, no detailed LEED analysis was carried out for either the ($\overline{1}\overline{1}\overline{1}$) 2×2 or the ($\overline{1}\overline{1}\overline{1}$)−√19 structures. Ranke and Jacobi[10] have speculated on possible atomic arrangements based on the ratios of arsenic to gallium atoms as given by them. However, the precision of these values seems to be not sufficient. The escape depths of the Auger-electrons Ranke and Jacobi have used are by 10% smaller than those determined more recently[14] and displayed in Fig. 4. Utilizing these newer values for an analysis of the Auger-electron spectroscopy data reported by Ranke and Jacobi increases the arsenic contents by approximately 10%.

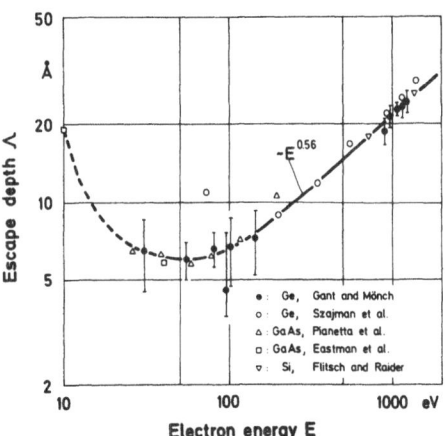

Figure 4 - Electron escape depth as function of electron energy for GaAs, Ge, and Si. (After Ref. 14.)

2.3 Evaluation of Chemical Surface Composition from AES

With GaAs(111) and also with GaAs(001) surfaces, the latter ones will be considered in the following section, more than one recon-struction pattern is observed by LEED or RHEED. The different atomic arrangements observed for the same orientation are distin-guished by markedly varying intensity ratios I(Ga)/I(As) of the low - as well as the high - energy Auger-electron lines as shown in Table I. This means that the reconstructions originate in different chemical surface compositions.

Bulk(001) as well as (111) layers contain either Ga or As atoms. To account for the variation of the intensity-ratios of the Auger-electron lines a layer model is used which is explained in the following. Since structural transitions are accompanied by desorption of As_2 molecules the different surface compositions are characterized by their particular "arsenic coverages" Θ.

At normal incidence an electron beam is exponentially attenuated in the sample as

$$I_p(z) = I_{po} \exp(-z/\lambda_p), \qquad (1)$$

where λ_p is the energy-dependent penetration depth of the primary electrons. At depth z below the surface Auger-electrons of energy E_i are excited at atoms of type i according to the energy-dependent cross section A_i. On the way to the surface the Auger-electron current is also exponentially attenuated but with the escape length λ_i. The cylin-drical mirror analyzer used to measure the energy distribution of the electrons emitted from the surface only accepts such electrons that are leaving the surface at a mean angle α, measured from the surface normal, into the entrance aperture of the CMA. The contribution of the Auger-electrons generated at depth z below the surface amounts to

$$I_i(z) = I_{po}A_i c_i(z) \exp[-z(1/\lambda_p + 1/\lambda_i < \cos \alpha >)], \qquad (2)$$

where $c_i(z)$ is the density of type i atoms at depth z. Along a specific crystal direction layers of the same composition are stacked up with an interplanar spacing d_{hkl}. The total current of a particular Auger trans-ition then reads

$$I_i = \sum_n I_i(n \cdot d_{hkl}) =$$

$$I_{po}A_i\sum_n c_i(nd_{hkl}) \exp\left[-nd_{hkl}(1/\lambda_p + 1/\lambda_i < \cos\alpha >)\right],$$

$$= I_{po}A_i\sum_i^{hkl}. \tag{3}$$

In an AB compound the ratio of two Auger-electron intensities results as

$$\frac{I_A}{I_B} = \frac{A_A}{A_B} \cdot \frac{\sum_A^{hkl}}{\sum_B^{hkl}}. \tag{4}$$

As mentioned in the preceding paragraph different surface reconstructions are characterized by varying coverages Θ_B of B atoms on an A layer. For the Auger-electron currents this means that a fraction Θ_B of the surface provides an additional attenuation of the signal coming from the layers underneath. Furthermore, the total Auger-electron current from B atoms is increased by the additional surface coverage Θ_B. The intensity ratio is thus given by

$$\frac{I_A}{I_B} = \frac{A_A}{A_B} \cdot$$

$$\frac{\sum_A^{hkl}[(1-\Theta_B) + \Theta_B \exp\{-d(\frac{1}{\lambda_p} + \frac{1}{\lambda_A < \cos\alpha >})\}]}{\Theta_B + \sum_B^{hkl}[(1-\Theta_B) + \Theta_B \exp\{-d(\frac{1}{\lambda_p} + \frac{1}{\lambda_B < \cos\alpha >})\}]}. \tag{5}$$

The ratio of the cross-sections of the Auger-electron excitations is determined from the intensity ratio measured with (110) surfaces which are expected to have the bulk 1:1 stoichiometry. In this respect, however, bulk non-stoichiometry may cause additional problems. These are discussed in section 3. For an energy of primary electrons of 3 keV the ratio of the cross-sections was determined as:[13]

$$\frac{A_{Ga(55eV)}}{A_{As(31eV)}} = 0.505, \qquad (6a)$$

$$\frac{A_{Ga(1070eV)}}{A_{As(1228eV)}} = 1.58. \qquad (6b)$$

The evaluation of coverages with chemisorbed atoms from the intensity ratios of the Auger-electron lines originating from the add-atoms and one of the substrate atoms follows exactly the same model. The oxygen coverages, to be discussed in section 5.1, were determined in this way.

For the surface analysis the intensity-ratios of the low-energy Auger-electron lines are used. These AES signals are very surface sensitive since electrons of around 50 eV exhibit escape depths of only 6 Å. Experimentally determined escape depths are displayed in Fig. 4.[14] For an energy of 2 keV of the primary electrons approximately 40% of the AES signal strength originates from the first layer.

2.4 GaAs(001) Surfaces

In an AB-compound crystallizing in the zincblende lattice, (001) surfaces are polar, too. In the bulk the distances between the alternate A and B layers are equal while along a (111) direction A-B double layers are stacked.

Among the low-indexed GaAs surfaces, the (001) oriented surface can be prepared with the largest variety of reconstructions. In his MBE experiments Cho[20] observed a sequence of different surface structures by using RHEED, the appearance of which depends on the substrate temperature and the Ga-to-As$_2$ arrival rate in a systematic way. This is explained in Fig. 5. For a given Ga-to-As$_2$ flux ratio he has observed a transition from an "As-stabilized" c(2×8) to a "Ga-stabilized" c(8×2) structure by increasing the temperature of the substrate. In two independent investigations, van Bommel and Crombeen[7] and Cho[15] determined the orientations of these reconstructions with respect to the crystal axis. For the indication of a (001)-m x n reconstruction the convention is such that the surface primitive vectors being m- and n-times larger than analog vectors in a bulk plane are pointing in [1̄10] and [1̄1̄0] directions, respectively. The "c" denotes, as usual,[8] a centered unit mesh. This indication means that

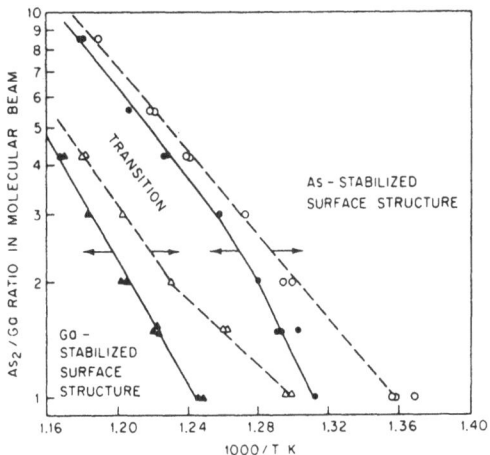

Figure 5 - Arsenic-to-gallium ratio in the molecular beam as a function of the substrate temperature when the transition of an As-stabilized surface structure and a Ga-stabilized surface structure takes place. Since there is hysterisis of the transitions as the substrate temperature is varied, two sets of curves, one for increase in temperature and the other for decrease in temperature, are presented. (After Ref. 21.)

for the As-stabilized $c(2\times 8)$ and the Ga-stabilized $c(8\times 2)$ structures the double surface periodicity is in the direction of the As and the Ga dangling bonds, respectively. These findings suggested to Cho[15] that the twofold axis periodicity might be due to a pairing of the dangling bonds or, in other words, a dimer formation, which had been proposed by Schlier and Farnworth[22] to be the mechanism causing the 2×1 reconstruction they had observed on Si(001) surfaces. We shall return to this at the end of this section.

Later on, more intermediate reconstructions were observed. Their surface compositions have been studied by Drathen et al.[17] using Auger-electron spectroscopy and by Bachrach et al.[23] using soft x-ray photoelectron spectroscopy. The results of both groups are displayed in Fig. 6. Additional data are given in Table I. First of all, both groups agree on the *ordering* of the various structures observed, however, with the exception of the c(8×2) Ga-stabilized structure. Bachrach et al. reported that structure to contain an equal density of Ga and As surface atoms while Drathen et al. concluded that it exhibits an excess

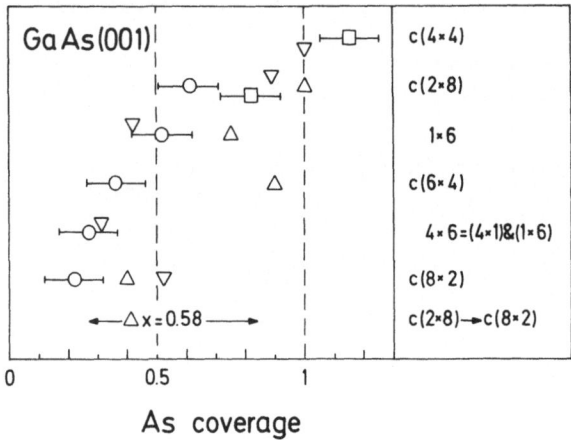

As coverage

Figure 6 - Correlation between atomic arrangement and chemical composition on GaAs(001) surfaces. (Δ from Ref. 16, o from Ref. 17, □ from Ref. 19, ∇ from Ref. 23, Δx from Ref. 12.)

of Ga atoms. The AES data of Drathen et al. were confirmed by measurements of van Bommel et al.[16] Our own reevaluation of the AES data using our more recent values of the escape depths of the Auger electrons verified the compositions as assigned to the various surface structures by Drathen et al., to within ± 0.02. There is additional evidence that the ordering given by Drathen et al. might be the correct one. On ion-bombarded GaAs(100) surfaces which were annealed at temperatures not higher than 650°C van Bommel et al.[16] have observed the following reversible structural transitions:

$$c(2\times 8) \quad \begin{matrix} 450°C \\ \rightarrow \\ \leftarrow \\ 350°C \end{matrix} \quad c(6\times 4) \quad \begin{matrix} 500\text{-}600°C \\ \rightarrow \\ \leftarrow \\ 450°C \end{matrix} \quad c(8\times 2).$$

Massier et al.[19] reported the following sequence of LEED structures which they observed during heating of their samples initially exhibiting a c(2×8) As-stabilized surface:

$$c(2 \times 8) \rightarrow (1 \times 6) \rightarrow c(6 \times 4) \rightarrow (3 \times 1) \rightarrow (4 \times 1) \rightarrow c(8 \times 2).$$
$$450 \div 500°C \qquad\qquad 550 \div 600°C \qquad\qquad 650°C$$

These results prove the $c(6 \times 4)$ structure to be intermediate between the $c(2 \times 8)$ and the $c(8 \times 2)$ structures in agreement with the ordering given by the analysis of the AES data reported by Drathen et al.[17]

Many authors (see for example Refs. 19,24,25) have pointed out that they have observed variations of the intensity ratio of the Ga-to-As Auger-electron signals over a wide range while the LEED and RHEED patterns, respectively, remained the same. Unfortunately, no LEED I/V curves were measured as function of these intensity ratios. Such studies could possibly clarify the correlation between atomic arrangement and chemical composition on these surfaces.

The structural transition from the As-stabilized $c(2 \times 8)$ to the Ga-stabilized $c(8 \times 2)$ surface is accompanied by a desorption of As_2 molecules while no desorption of Ga is observed. Between 500 and 875K (226 and 600°C) Arthur[12] measured a total loss of 1.8×10^{14} As_2-molecules per cm^2 from the As-stabilized structure which value equals 0.58 of a monolayer of arsenic ($\sigma_{001} = 6.26 \times 10^{14}$ sites per cm^2).

In more recent studies, 2×4 As-stabilized and 4×2 Ga-stabilized surfaces were prepared. Those structures were first observed by Cho[15] when he was annealing his samples in a vacuum containing a very low partial pressure of arsenic. Larsen et al.[26] prepared the 2×4 structure during MBE growth of GaAs at 800K with an As_4/Ga flux ratio of 10/1. The structure is very stable, and it is maintained on cooling to room temperature.

Subjecting this 2×4 structure to an As_4 flux in the temperature range of 670 to 720K produces a $c(4 \times 4)$. From his photoemission studies, to be discussed later in section 4.1.2, Larsen et al.[27] concluded that the $c(4 \times 4)$ structure contains extra, chemisorbed As_2 molecules that break some of the As-As dimer bonds. This proposal is consistent with those AES results by Massies et al.[19] plotted in Fig. 6 that result in an arsenic coverage larger than one monolayer for the $c(4 \times 4)$ structure (see Fig. 6). These results then also support those findings that assign a complete top-layer of arsenic to the $c(2 \times 8)$ structure.

Heating the As-stabilized 2×4 surface to temperatures above 800K in a vacuum of 10^{-10} Torr produces the Ga-stabilized 4×2 structure.[24] On cooling below 770K this structure decays to a mixed $(4 \times 6) + (3 \times 6)$ structure[24] that contains *more* arsenic. The source of the extra arsenic will be discussed in section 3. The 4×2 structure can also be produced during MBE growth with nearly equal Ga and As_4 fluxes impinging on the surface.[28]

Since the 2×4 As-stabilized structure is thought to contain a complete top-layer of arsenic structural models were proposed for this reconstruction. From RHEED patterns Larsen et al.[26] have observed, they concluded that this structure is considerably more ordered along the [$\bar{1}$10] direction than along the orthogonal [110] direction. The [$\bar{1}$10] direction is the direction of the arsenic dangling bonds.[7,15] Therefore, Larsen et al., as did previously Cho,[15] concluded that a dimerization of adjacent surface atoms is the primary reconstruction mechanism of this surface. The desorption of only As_2 molecules during the transformation of the As- to the Ga-stabilized structure[29] is thought to support this conclusion. Energy-minimization calculations by Chadi et al.[30] have indeed proven the dimerization to be energetically very favorable. As with the Si(001) surface[31] the asymmetric dimer was found to be the optimized geometry. Fig. 7 shows a possible arrangement of asymmetric and symmetric dimers forming a 2×4 reconstructed surface. The model combines configurations proposed by Dobson et al.[28] and by Chadi et al.[30] As on Si(001), the formation of asymmetric dimers is also supported by electronic considerations. At

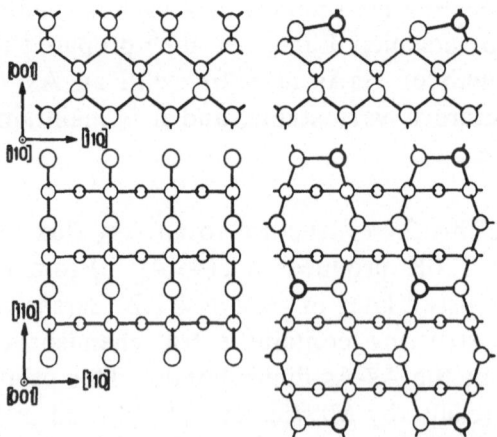

Figure 7 - Models of unreconstructed and reconstructed GaAs(001) surfaces.

an (001) surface every atom possesses two dangling bonds. An arsenic surface atom thus contains $2 \times 5/4 = 5/2$ electrons in its dangling bonds. With a complete As top-layer semiconducting surface bands are obtained by reconstructions with a unit mesh having an area 4n times larger (n integer) than the unit mesh of a bulk (001) layer.[32] A 2×2 mesh is the smallest one to meet this condition. This would result in 5 filled surface bands accommodating 10 electrons. A 2×4 as well as a $c(2 \times 8)$ would fulfill the condition, too.

So far, no geometrical models for reconstructions observed with GaAs(001) surfaces other than the 2×4 and the $c(2 \times 8)$ ones have been proposed.

3. EXCESS GROUP V ATOMS IN III-V SINGLE CRYSTALS

Recent experiments with (110) surfaces cleaved from GaAs, GaP, GaSb, and InP single crystals[33-35] revealed the samples used to exhibit spatially nonuniform chemical compositions. Two types of segregates formed by group V atoms were detected by spatially re-solved studies of the electron current absorbed by the samples and identified by spatially resolved Auger-electron as well as low-energy electron energy-loss spectroscopy. On GaAs,[33] for example, type A segregates were observed immediately after cleavage while type B ones form and grow during storage at room temperature or elevated tempera-tures up to 170°C. Increasing the temperatures above 200°C causes the segregates gradually to disappear. After cooling the sample to below 170°C segregates are observed again. The diameter of the arsenic segregates observed varies between 30 and 300 μm, where the lower limit was set by the diameter of the electron beam used. Such segregates were found on surfaces cleaved from n- as well as p-type crystals grown by the Horizontal Bridgman Technique, the Traveling Heater Method, and the Liquid Encapsulation Czochralski growth method.

A line-scan of the Ga(1070eV) and the As(1226eV) AES-signal across a typical segregation is shown in Fig. 8. Across the spot the Ga-intensity decreases to zero and the As-peak increases by a factor of two compared to the surrounding GaAs. Most of the As zones exhibit-ing a diameter larger than 70 μm show a fine-structure: a core contain-ing Ga and As is surrounded by pure arsenic.

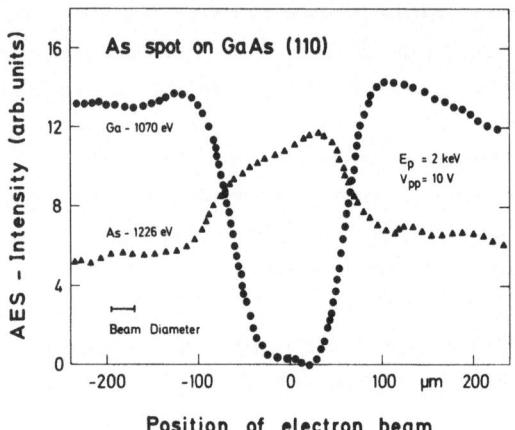

Figure 8 - AES line scan across an As-zone on a cleaved GaAs(110) surface. (After Ref. 33.)

With some of the larger segregates EELS could be studied, too. The 100 eV electron beam could only be focussed to approximately 100 μm. The energy-loss spectra measured with the beam on and besides the spot are displayed in Fig. 9. The lower trace shows all the energy-loss peaks usually found on GaAs surfaces, due to inter-band transitions (bb), to bulk and surface plasmon excitations (bp, sp) as well as to transitions from Ga(3d) and As(3d), respectively, core levels. However, no Ga(3d) signals are present in the spectrum recorded from the arsenic spot, and the bulk plasmon is shifted by 1.25 eV to higher energies. This shift is to be explained by the fact that GaAs and arsenic have four and five valence electrons per atom, respectively.

These experiments with GaAs(110) surfaces cleaved in UHV from single crystals demonstrate that those ingots have an excess of arsenic. There have been published some early indications in MBE studies which are supporting these observations.

Already in 1974, Arthur[12] "postulated the existence of a diffusion process supplying As from the bulk to the surface in order to explain the time-independent evaporation of As_2 at high temperatures without any corresponding change in surface composition".

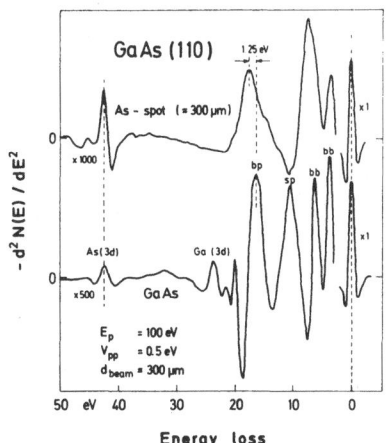

Figure 9 - EELS spectra of an As-spot on a cleaved GaAs(110) surface and of the surrounding, clean GaAs surface. (After Ref. 33.)

Some years later, Neave and Joyce[24] reported: "The concept of diffusion of As from the bulk is supported by the observation that, in agreement with previous results, heating an As-stable surface to > 800K in a vacuum of $< 10^{-10}$ Torr produced a Ga-stable (4×2) structure but on cooling from this state to < 770K, it immediately formed a mixed (4×6) + (3×6) reconstruction. This represents a surface with a relatively higher proportion of As, which could presumably only arise from As diffusing from the bulk and forming a sufficient surface population at the lower temperature to affect the diffraction pattern. Certainly it results from a process involving a *limited supply of As*, since it can be only repeated four to five times."

At the same time, van Bommel et al.[16] published similar observations: "Heat treatments at different temperatures revealed several reversible structure transitions, which were divided in two cycles

$$c(2 \times 8) \;\rightleftharpoons\; c(6 \times 4) \;\rightleftharpoons\; c(8 \times 2) .$$

This cycle was repeated two or three times and then going out from the c(8×2) structure changed into the cycle:

$$(1 \times 6) \quad \underset{\longrightarrow}{\longleftarrow} \quad c(8 \times 2) .$$

This cycle was also repeated a few times, after which the GaAs(001)-c(8×2) was stable at all temperatures up to 600°C."

Furthermore, Cullis et al.[36] have identified small precipitates in GaAs single crystals to consist of elemental, hexagonal arsenic. They have employed high-resolution diffraction utilizing an analytical electron-microscope. The precipitates exhibit rounded outlines with diameters of typically 200 to 1000Å. They are attached to line dislocations. This study made apparent that the presence of impurity species is not of primary importance for the formation of the precipitates. This is in agreement with the findings of Bartels et al.[33] who detected the much larger arsenic segregates in p- as well as in n-doped GaAs prepared by different techniques. Cullis et al. also pointed out that the existence of elemental As inclusions in GaAs strongly suggests that As exists as an important mobile point defect species in GaAs. Since Mönch and coworkers[33-35] found group V inclusions not only with GaAs but also with GaP, GaSb, and InP this statement of Cullis' et al. holds also for the other III-V compounds.

4. ELECTRONIC SURFACE STATES

The rearrangement of the surface atoms is obviously accompanied by a rehybridization of the surface bonds. In the top layer this includes, for example, dangling bonds, back bonds towards the second layer and dimer bonds. Of special interest are surface states with energies within the bulk band gap since, depending on whether they are of acceptor- or donor-character, they cause a band bending at the surface. However, although such space charge layers at a clean surface need the presence of charged surface states this does not necessarily imply that intrinsic rather than extrinsic surface states are present. As will be discussed later, a low density of surface states, amounting to approximately 1/100 of a monolayer, is sufficient to cause considerable band bendings. Such low densities of defects introduced, for example, by cleavage or left after ion-bombardment followed by annealing cannot be detected by LEED.

4.1 Occupied, Intrinsic Surface States and Angular-Resolved Photoemission

The occupied states of a surface band structure $E_s(\underline{k}_{"})$ can be mapped by using angle-resolved photoemission (ARPS). A photon of energy $\hbar\omega$ excites an electron from an initial state of E_{si} to a final state at E_{sf}. With the final state above the vacuum level E_{vac} the emitted electron carries a kinetic energy

$$E_{kin} = \frac{\hbar^2}{2m_o} \cdot k_{vac}^2. \tag{7}$$

The transitions in the semiconductor surface are direct, and the conservation of k-vectors gives

$$\underline{k}_{"}^{vac} = \underline{k}_{"} \pm \underline{G}_{"} \tag{8}$$

where $\underline{k}_{"}^{vac}$ is the component parallel to the surface of the wave vector of the emitted electrons and $\underline{G}_{"}$ is a vector of the surface reciprocal lattice. The emitted electron is detected at a polar angle Θ, and it holds

$$k_{"}^{vac} = k_{vac} \sin \Theta. \tag{9}$$

The conservation of energy gives

$$\hbar\omega = E_{sf} - E_{si},$$

$$= E_{kin} + I - E_{si}^{B}, \tag{10a}$$

$$= E_{kin} + \Phi + (E_F - E_{vs}) - E_{si}^{B}, \tag{10b}$$

where $I = E_{vac} - E_{vs}$ is the ionization energy, $\Phi = E_{vac} - E_F$ is the work function and $E_{si}^{B} = E_{si} - E_{vs}$ is the binding energy of the initially occupied surface state with respect to the top of the valence band. The parallel component of the wave vector of the emitted electrons then amounts to

$$k_{"}^{vac} = \sin \Theta \cdot \sqrt{\frac{2m_v}{\hbar^2}(\hbar\omega - I + E_{si}^{B})}. \tag{11}$$

One of the main problems in the mapping of band structures is how to distinguish between bulk and surface contributions. One procedure frequently used is to identify surface states by their sensitivity to the adsorption of foreign atoms. An other feature of the emission from surface states is that such peaks should not depend on changes of the perpendicular component of the wave vector for fixed parallel component $k_{\shortparallel}$. This behavior can be tested by keeping $\underline{k}_{\shortparallel}$ fixed while $\underline{k}_{\perp}$ is varied by changing the photon energy as well as the polar angle according to Eq. (11).

The energy dispersions of the occupied surface bands were experimentally determined for cleaved (110) surfaces,[37-39] ($\overline{1}\overline{1}\overline{1}$) surfaces exhibiting the 2×2 As-rich structure,[40] and (001) surfaces showing the 2×4 As-stabilized structure.[41] In the present paper the results for the cleaved and the (001)-2×4 surfaces shall be discussed.

The dispersion of the occupied surface states along the main symmetry lines of the surface Brillouin zone was measured by Lapeyre and coworkers[37,38] and by Huijser et al.[39] Their data are plotted in Fig. 10 together with the results of a recent calculation by Beres et al.[42] This paper references all of the earlier theoretical approaches. The shaded areas in Fig. 10 give the projected bulk band structure.·

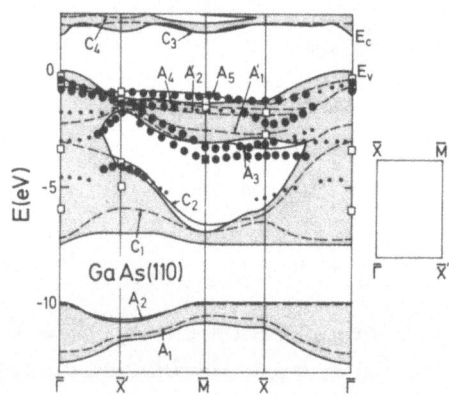

Figure 10 - Measured and calculated dispersion curves for surface bound states (solid lines) and surface resonances (dashed lines) at cleaved GaAs(110) surfaces along the symmetry lines of the surface Brillouin zone. The shaded areas give the projected bulk bands. (After Ref. 42.) □ After Ref. 38, o After Ref. 39.

The good agreement between experimental and theoretical results is clearly discernible. The labeling A_i and C_i of the calculated dispersion curves denotes states mainly associated with anion and cation atoms, respectively. A_5 and C_3 are the occupied and the empty dangling bond states, respectively. They are overlapping the bulk valence and conduction states, and the bulk band gap is free from any intrinsic surface states. Photoemission can only detect occupied surface states. For experimental studies of empty surface states other techniques have to be used. They will be discussed in the following section 3.2.

Larsen et al.[27] have mapped the dispersion of occupied surface-state bands on GaAs(001)-2×4 surfaces. As mentioned earlier, this As-stabilized surface structure is prepared at around 800K and is maintained during cooling the sample to room temperature. The angular-resolved electron energy-distribution curves showed four peaks that were identified as emissions from surface states. The dispersions of these peaks along the main symmetry lines of the surface Brillouin zone are plotted in Fig. 11. The energy dispersion of the surface state S_2 clearly exhibits the expected two-fold periodicity of the 2×4 surface Brillouin zone along the [$\bar{1}$10] direction. As was shown previously by RHEED studies,[26] this is the main ordering direction of the 2×4 structure. However, no four-fold periodicity along the [110] azimuth was observed with the surface-state emissions. Therefore, Larsen et al. defined the surface wave vectors with respect to a 2×1 rather than a 2×4 surface Brillouin zone. This is explained in the scheme of the surface Brillouin zone given in Fig. 11. Due to this lack of four-fold periodicity along [110] in the surface-state emission it is difficult to compare the experimental dispersion curves with calculated surface band structures. However, the surface state S_4 could be identified as the bonding As-As dimer band.[43] In the theoretical calculations asymmetric As-As dimers forming a 2×1 structure were considered. This result thus confirms asymmetric As-As dimers to be the building blocks of, at least, the As-stabilized GaAs (001) structures.

4.2 Empty Intrinsic Surface States Detected by Isochromat Spectroscopy

For the detection of empty surface states surface photoconductivity (SPCS) and surface photovoltage spectroscopy (SPVS)[44-47] and isochromat spectroscopy[48] have been applied successfully. In surface photovoltage spectroscopy, for example, which can be used only with semiconductors, electrons are excited from the bulk valence band into empty surface states or from occupied surface states into the bulk conduction band. The threshold energy then gives the energetic distance of the lowest (highest) empty (occupied) surface state above the valence band (below the conduction band) while the spectral shape of the photoresponse indicates whether the transitions are direct or indirect.

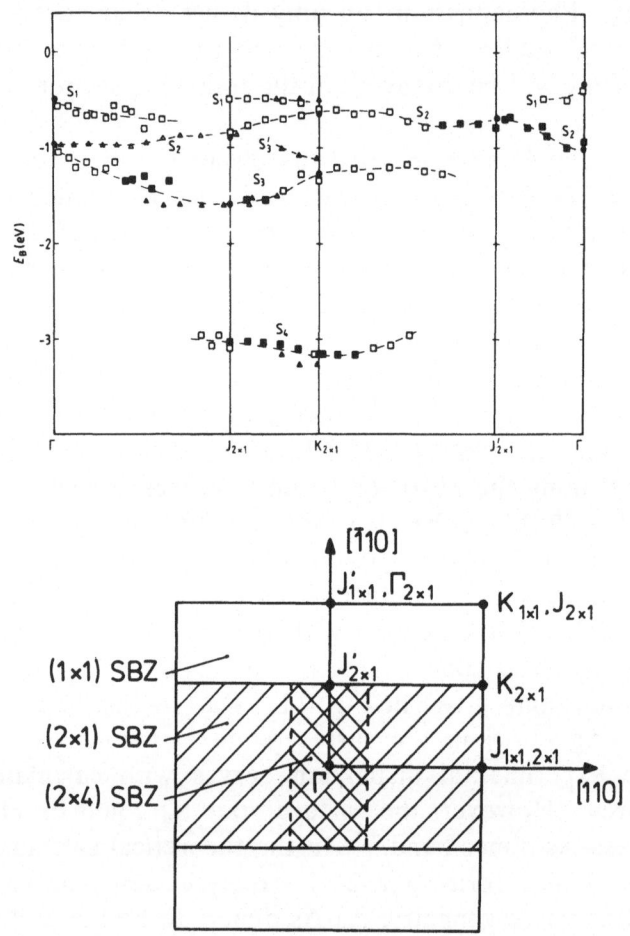

Figure 11 - Measured dispersion curves for surface bound states at GaAs(001)-2 × 4 surfaces along symmetry lines. (After Ref. 27.)

The most appropriate method for determining the density of unoccupied bulk and surface states is isochromat spectroscopy the potential of which for the evaluation of surface states has been recently demonstrated by Dose et al.[48] with GaAs(110) surfaces. The experiment is as follows: The sample to be investigated is bombarded with monoenergetic electrons, and from the Bremsstrahlung emitted the intensity at one fixed photon energy $\hbar\omega_0$ is recorded as function of the energy of the incident electrons. The curve resulting is an isochromat spectrum. The physics involved may be most simply described as an inverse photoeffect.[49,50] Those electrons which had an initial energy E above the Fermi level of the sample and undergo a radiative transition by emitting a photon of energy $\hbar\omega_0$ end up in final states at energy E $-$ $\hbar\omega_0$ above the Fermi level. Neglecting a possible dependence of the transition matrix element involved, the transition rate and thus the intensity of the photons emitted is proportional to the density of empty final states. To first approximation, an isochromat spectrum thus represents the density of unoccupied states as function of energy above the Fermi level. Fig. 12 gives the isochromat spectrum measured by Dose et al.[46] with GaAs(110) surfaces prepared by ion bombardment and subsequent annealing in UHV. Due to the photon energy used the experiment was very surface-sensitive since the electron escape depth is

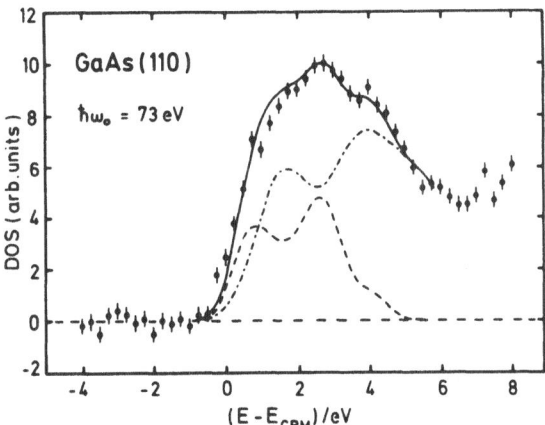

Figure 12 - Isochromat spectrum measured with a GaAs(110) surface cleaned by ion bombardment and annealing in UHV. The solid line is the total density of states above the conduction band minimum (cbm) at the surface. The surface and the bulk contributions are shown as dashed and dash-dotted lines, respectively. (After Ref. 48.)

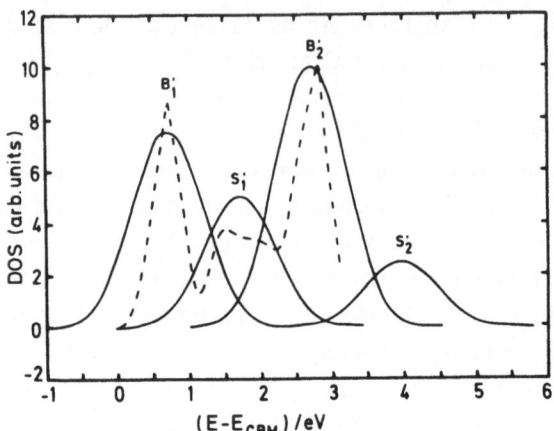

Figure 13 - The surface contribution to the isochromat spectrum of Figure 12 is decomposed into four individual surface states. The dashed line is the unbroadened result of a theoretical calculation (52). (After Ref. 48.)

minimal between 50 to 80 eV (see Fig. 4). From the measured isochromat spectrum Dose et al. subtracted the "bulk density of states" taken from an empirical non-local pseudo-potential calculation.[51] The resulting surface contribution, given by the dashed line in Fig. 12, can be decomposed into four surface state peaks. As shown in Fig. 13, three of them agree in energetic position and relative intensity with the density of surface states as calculated by Chadi[52] for a bond relaxation model, the parameters of which are compatible with the LEED data[2] discussed in section 2.1.

Chadi[52] has also computed the density of surface states for a *rigid* rotation of the surface atoms by 27° which is another surface model the geometry of which is also compatible with the LEED analysis. The densities of the occupied surface states are quite similar to each other while they are markedly different for the empty states. Since there is good agreement of the data calculated using the geometry of the bond-relaxation model and the measured isochromat spectrum Dose et al. conclude that the atomic arrangement of this model is more likely. However, in this comparison it has to be kept in mind that in his calculation Chadi employed a tight-binding scheme which is expected to describe the unoccupied surface states less well than the occupied ones.

The data on the atomic and the electronic structure of GaAs(110) surfaces form a coherent picture: the dispersion of the occupied and the density of empty surface states as calculated by using the atomic coordinates evaluated from LEED I/V curves as well as from minimizing the total energy agree with the experimental data of angular-resolved photoemission and isochromat spectroscopy, respectively. This makes the clean GaAs(110) surface to be one of all the semiconductor surfaces presently understood best.

4.3 Extrinsic Surface States on Clean GaAs Surfaces

The dangling bond states at the GaAs(110) surface are driven out of the fundamental band gap by the rearrangement of the surface atoms.[53] Thus, GaAs(110) surfaces exhibit no intrinsic surface states in the gap, and for a clean and perfectly cleaved GaAs surface the bands are flat up to the surface. Any band bending observed is thus caused by extrinsic surface states related to imperfections.

By now, it is well established that cleavage may introduce extrinsic acceptor-type surface states the density of which is related to the quality of the cleave.[54-58] This was most convincingly demonstrated by the work function measurements displayed in Fig. 14 and 15. The cleavages are characterized by $< \tan \alpha >$ where α is the optically determined angle of misorientation between the cleaved surface and a flat (110) surface.[59,60] The work function of a semiconductor may be written as

$$\Phi = E_{vac} - E_F = I - e_o V_s - W_p, \tag{12}$$

where $e_o V_s = E_{vb} - E_{vs}$ is the band bending at the surface and $W_p = E_F - E_{vb}$ gives the energetic position of the Fermi level above the top of the valence band in the bulk.

For a p-type semiconductor the hole concentration amounts to

$$p = N_v \exp \left(\frac{E_{vb} - E_F}{k_B T} \right). \tag{13}$$

With all acceptors exhausted, the hole concentration is independent of temperature, and the position of the Fermi level is given by

$$W_p = E_F - E_{vb} = k_B T \ \ln \frac{N_v}{p}. \tag{14}$$

For GaAs(110) surfaces the ionization energy was found to be independent of temperature.[61] The work function thus decreases linearly with increasing temperature as

$$\Phi = I - k_B T \ \ln \ \frac{N_v}{p} \quad \text{if} \quad e_o V_s = 0, \tag{15}$$

i.e for a perfect cleave.

The experimental results obtained by using a Kelvin probe are plotted in Fig. 14 and reveal the behavior expected from Eq. 15 up to 350°C but above this temperature the Fermi level becomes "pinned".[58] No donor states, characterized by charging states 0/+ depending on the position of the Fermi level, are present on the cleaved surface. However, above 350°C the now more broadened tail of the Fermi-Dirac distribution function reaches into acceptor-type surface states, exhibiting a charging characteristics −/0, and these states become partly populated and then cause band bending.

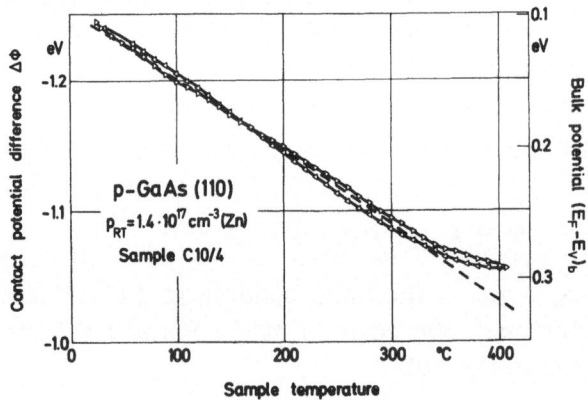

Figure 14 - Temperature dependence of the work function for a (110) surface cleaved from a p-type GaAs crystal. Data points ▷ and ◁ were measured during heating and cooling, respectively; --- ($E_F - E_{vb}$) as calculated from the bulk doping. (From Ref. 58.)

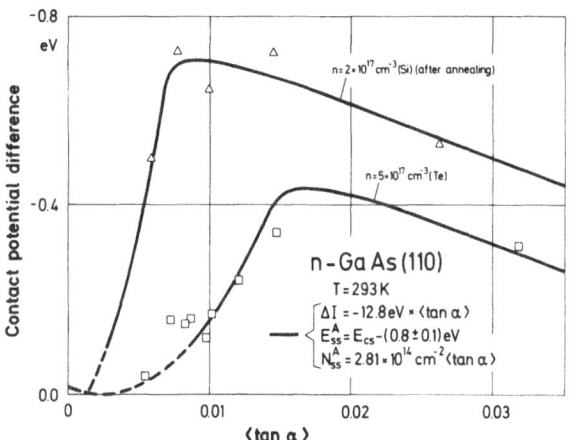

Figure 15 - Work function of (110) surfaces cleaved from n-type GaAs as function of the cleavage perfection. The solid lines are calculated by using Eq. (12) and (18) with the parameters in the insert. (After Ref. 62.)

Acceptor-type surface states are more easily detected on n-type samples. In Fig. 15 the work function measured with a number of surfaces cleaved from n-type crystals is plotted versus the $\langle \tan \alpha \rangle$ of these surfaces.[62] The work function of these surfaces is increased by cleavage imperfections and thus depletion layers have formed.

The concepts of band bending at semiconductor surfaces and of the space-charge region associated with it are discussed in detail in Ref. (63) and (64). On an n-type semiconductor the space charge in a depletion layer is given by

$$Q_{sc}^n = \{2\varepsilon\varepsilon_o N_D[e_o | V_s | - k_B T + \exp\left(-\frac{e_o | V_s |}{k_B T}\right)]\}^{1/2}. \quad (16)$$

The charge density in discrete, acceptor-type surface states of density N_{ss}^A at energy E_{ss}^A amounts to

$$Q_{ss}^A = -e_o N_{ss}^A \{g_A \exp \frac{E_{ss}^A - E_F}{k_B T} + 1\}^{-1}, \quad (17)$$

where g_A is the degeneracy factor. The condition of charge neutrality at the surface requires

$$Q_{sc} + Q_{ss} = 0. \tag{18}$$

This equation is explained in Fig. 16 where the open circles mark those band bendings that are established for particular densities of surface states. Those band bendings are plotted over the density of surface states in Fig. 17.

The main trend of the work function as function of $\langle \tan \alpha \rangle$ as plotted in Fig. 15, is reproduced by the dependence of the band bending as function of the density of surface states. This finding suggests that the density of the cleavage-induced surface states is proportional to $\langle \tan \alpha \rangle$. The decrease of the work function for larger $\langle \tan \alpha \rangle$, remaining to be explained, indicates a decrease of the ionization energy with increasing $\langle \tan \alpha \rangle$. The full lines drawn in Fig. 15 are least square fits to the experimental data considering both cleavage-induced surface states and changes of the ionization energy. The fit parameters are given in Fig. 15.

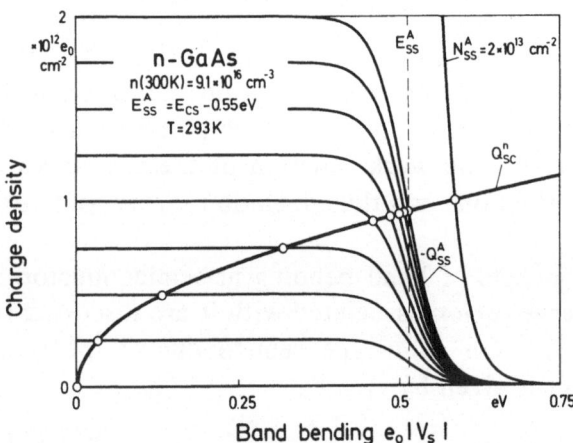

Figure 16 - Space charge in a depletion layer on n-GaAs and occupancy of acceptor-type surface states both as function of band bending. (After Ref. 65.)

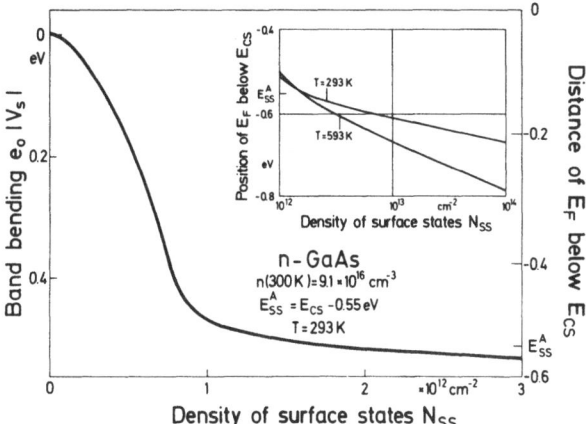

Figure 17 - Surface band bending in a depletion layer on n-GaAs as function of the density of acceptor-type surface states. (After Ref. 65.)

Cleavage-induced imperfections lower the work function of p-type samples by a reduction of the ionization energy, too.[58] The proportionality between ΔI and $<\tan \alpha>$ was found to be the same on both p- and n-type crystals. The reduction of the ionization energy means that cleavage induces additional surface dipoles. The linear correlation with $<\tan \alpha>$ indicates that the dipoles are too far apart to interact. The same behavior was found earlier with stepped metal[66,67] and silicon (111) surfaces.[68] Therefore, the extra surface dipoles and surface states introduced by cleavage are most probably correlated with cleavage steps.

By using a Kelvin probe Massies et al.[19] have investigated the variation of the work function as function of the surface composition of GaAs(001) surfaces. Their results are shown in Fig. 18. The authors have used the As coverage scale given by Drathen et al.[17] Independent of the particular surface structure, no significant differences of the work function were reported for p- and n-type substrates. The Fermi level is thus pinned by surface states in the gap, and acceptor- as well as donor-type surface states have to be present. These surface states

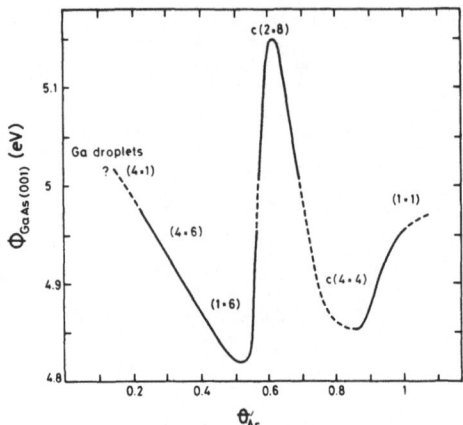

Figure 18 - Work function of GaAs(001) surfaces exhibiting different surface structures as function of arsenic surface coverage. Coverage scale from Ref. 17. (After Ref. 19.)

are most probably associated with surface defects. Unfortunately, no additional measurements are available to determine the position of the Fermi level with respect to the band edges at the surface and thus to evaluate the energetic positions of these surface states pinning the Fermi level.

5. THREE EXAMPLES FOR CHEMISORPTION ON GaAs SURFACES

5.1 Chemisorption of Oxygen

During the last 20 years the interaction of oxygen with GaAs surfaces has been studied very extensively. However, until recently the picture was rather incomplete since the experimental data reported for the (110) surface did not fit the results obtained with other surfaces. In Fig. 19 those data which are, to the opinion of the author, the most reliable ones, are compiled.[10,13,69-71] In these studies the oxygen coverage was determined from the intensity ratio of the O(510 eV) and the Ga(1070 ev) Auger-lines. Only such experiments were considered in which care was taken to avoid any stimulation of the oxygen uptake by an "excitation" of the impinging oxygen molecules at hot filaments or in ion-pumps,[72] by electrons hitting the surface[69] or by photons creating electron-hole pairs in the semiconductor surface[13] during the exposure.

The GaAs(110) surface shall be discussed first. The two sets of data shown in Fig. 19 for this surface were measured with cleaved[13,71] and with ion-bombarded and annealed surfaces.[70] On the cleaved surface, three different adsorption mechanisms are to be distinguished. This was done by a combination of Auger-electron and low-energy electron energy-loss spectroscopy.[13,71] Up to an exposure of approximately 10^5 L the oxygen uptake is determined by cleavage-introduced imperfections which are most probably cleavage steps. As Fig. 20 shows, the intensity of the 20 eV energy-loss, which was attributed to an intraatomic transition from Ga(3d) core-levels,[73] is reduced and has vanished after a total exposure of approximately 10^5 L. Exposures above this value cause a steep increase of the oxygen uptake. Above a coverage of approximately half a monolayer the intensity of the 42.5 eV energy-loss, which originates from transitions from As(3d) core-levels to conduction band states, decreases while new losses at 45 eV and 7 eV are built up, the intensity of which increases proportional to the oxygen coverage. Shifted As(3d) core-levels and a correlated

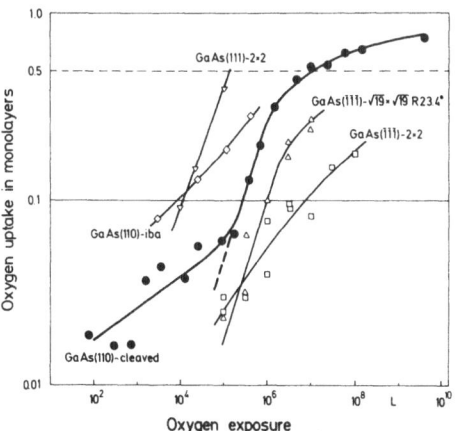

Figure 19 - Oxygen uptake in monolayers as function of exposure to "unexcited" molecular oxygen on cleaved (13,71) and ion-bombarded and annealed (70) GaAs(110), GaAs(111)-2x2 (69), GaAs($\bar{1}\bar{1}\bar{1}$)−2×2 and √19 x √19R23.4° (10) surfaces.

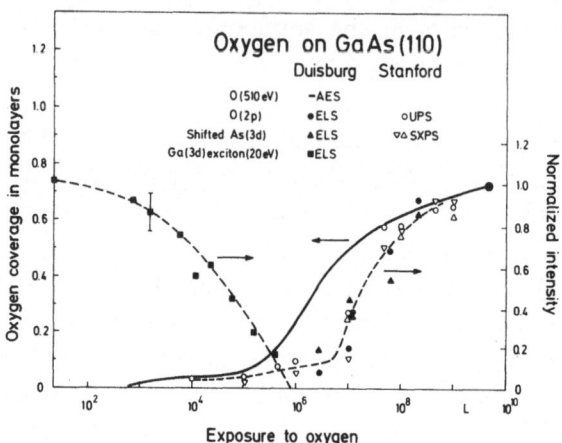

Figure 20 - Oxygen coverage and relative intensity of Ga(3d)-, As(3d)-, and O(2p)-levels involving transitions as observed with AES (71), EELS (71), and SXPS as well as UPS (72). (After Ref. 71.)

O(2p) level have been observed earlier by Spicer and co-workers[72] in their photoemission experiments but they had not extablished a coverage scale. As shown in Fig. 20 these data of the Stanford group follow exactly the behavior observed by the EELS studies. We conclude that on GaAs(110) surfaces the initial oxygen uptake preferentially occurs at the surface Ga atoms. Ludeke and Koma[74] had earlier reached the same conclusion. It was based on their studies with oxygen-exposed GaAs(110), (100), and (110) surfaces which, however, were all prepared by heavy ion-bombardment followed by annealing in UHV. The study by Ranke et al.[70] with (110) surfaces more carefully prepared by ion-bombardment and annealing is supporting this conclusion. As shown in Fig. 19, the slope of the uptake-versus-exposure curve of the ion-bombarded and annealed surface is the same as the one observed with the cleaved surface in the range below 10^5 L where it is determined by cleavage imperfections. It is well known that an iba treatment leaves the surface highly stepped[75] and, due to a preferential sputtering of As,[76,77] Ga-rich. For higher coverages, the uptake-versus-exposure curves of the iba and the cleaved surfaces are finally expected to merge. Then, adsorption on the terraces should dominate even on an iba surface.

The coverage-versus-exposure curve of the GaAs($\overline{1}\overline{1}\overline{1}$)$-$2x2 surface[10] shows almost the same slope as observed with iba(110) surfaces and cleaved (110) surfaces below 10^5 L. The ($\overline{1}\overline{1}\overline{1}$)$-$2x2 surface was prepared by ion-bombardment and annealing, too, and it may be suspected that the oxygen uptake is determined by residual steps and/or Ga atoms. At this surface at least 87% of the sites are occupied by arsenic atoms,[10] and, taking into account the results observed with the cleaved surface, the oxygen uptake is expected to rise steeply above exposures above 10^6 L. Additional studies with this surface are desireable.

The shapes of the uptake-versus-exposure curves are very similar for the Ga-stabilized (111)-2×2,[69] the ($\overline{1}\overline{1}\overline{1}$)$-\sqrt{19}$ (10), and above 10^5 L the cleaved (110) surfaces.[13,71] The ordering of these surfaces along the exposure scale may be explained. On both the cleaved surface and the ($\overline{1}\overline{1}\overline{1}$)$-\sqrt{19}$ surface[10] half of the surface sites are occupied by arsenic and gallium atoms, respectively. Therefore, it is small wonder that both curves in Fig. 19, are that close to each other. On the other hand, the (111)-2×2 structure is extremely Ga-rich as indicated by the large intensity ratio of the Ga (1070 eV) and As (1228 eV) Auger-lines[9] (see Table I). This might explain the increased initial sticking coefficient of oxygen. In any case, the chemisorption mechanisms seem to be the same on the flat terraces of these three surfaces, and they differ from that mechanism correlated with steps and other imperfections produced either by cleavage or left after ion-bombardment and annealing and resulting in a high initial sticking coefficient.

5.2 Chemisorption of Hydrogen

The interaction of hydrogen with GaAs surfaces has recently attracted some interest. Bringans and Bachrach[78] have started with two different (100) and ($\overline{1}\overline{1}\overline{1}$) surface structures each while Surkamp and Mönch[79] have used cleaved (110) surfaces.

The chemisorption of hydrogen on semiconductors causes some experimental difficulties. First of all, molecular hydrogen was found not to adsorb on GaAs,[80] and thus atomic hydrogen has to be used. Customarily, it is produced by thermal dissociation of molecular hydrogen at a hot filament placed in line of sight with the surface under study. Thus, as observed with the oxidation of GaAs(110) surfaces[13,71] possible stimulation of chemisorption by the light emitted from the filament has to be considered. The temperature of the filaments and the distance from the filaments to the samples vary from one group to another. Therefore the experimental results may not easily be compared. Secondly, the hydrogen uptake cannot be determined by those techniques, as XPS and AES, usually employed for surface chemical analysis. An appropriate experiment would be flash-desorption which has not been done till now. Therefore the coverages can only be estimated indirectly.

Surkamp and Mönch[79] have followed the uptake of atomic hydrogen on cleaved GaAs surfaces by measuring the changes of the work function with a Kelvin probe, in low-energy electron energy-loss spectra and in Auger-electron spectra. Part of their results are given in Fig. 21. The work function initially decreases on both p- and n-type samples. This indicates that the chemisorption of hydrogen causes depletion layers on both types of substrates via the generation of acceptor- and donor-type surfaces. The same behavior was reported in a series of papers by Mönch and Gant[81-83] for Ge and by Spicer and

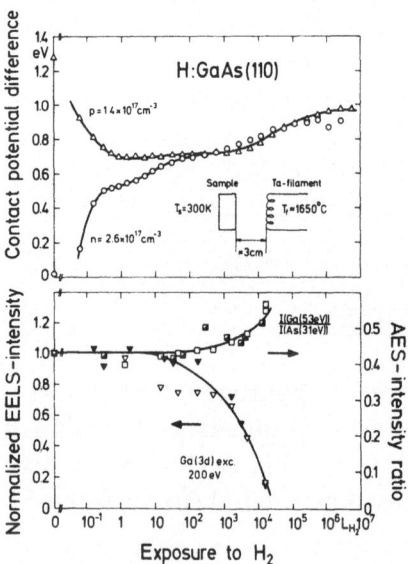

Figure 21 - Chemisorption of atomic hydrogen at cleaved GaAs(110) surfaces as followed by changes in the work function, the Ga(3d) 20 eV-loss, and the Ga(55 eV)/As(31 eV) AES-intensity ratio. (After Ref. 71.)

co-workers for metals and oxygen on GaAs(110) surfaces.[84,85] The analysis of the H-on-GaAs data reveals that chemisorption induced dipoles are changing the ionization energy by less than 100 meV and the change of work function is mainly due to band bending. At an exposure of 1L the coverage is estimated to be approximately 0.3 of a monolayer.

Up to this exposure the intensity of the prominent 20 eV loss, involving intraatomic transitions from Ga(3d) core levels, is not altered. Comparison with the previously discussed chemisorption of oxygen leads to the conclusion that hydrogen first bonds to the surface arsenic atoms. However, no shifted As(3d) line is observed. Above an exposure of 10 L the 20 eV loss begins to decay, and the line can no longer be observed after exposures above some 10^4 L.

In the same range of exposures the surfaces lose arsenic. The intensity ratio of the surface-sensitive As(31 eV) and Ga(55 eV) lines increases while the Ga(55 eV) signal itself remains unchanged. At the same time, AsH_3 was detected in the residual gas of the chamber by using a sensitive quadrupole mass spectrometer. Similar observations were reported by Tu et al.[86] with InP. They have exposed their samples to a hydrogen plasma and have found that at higher power densities phosphorus is preferentially removed by the hydrogen atoms, leaving the surface rich in In.

By the larger exposures the GaAs surfaces thus become enriched in Ga content, and they now are extremely sensible to traces of oxygen and/or water molecules in the residual gas. After hydrogen exposures above 10^5 L the uptake of oxygen can hardly be avoided while recording an Auger-electron spectrum. These experiments and the results obtained clearly demonstrate that atomic hydrogen is able to drastically change the surface properties of GaAs surfaces. This has to be taken into account with MBE experiments employing AsH_3 sources.

5.3 Chemisorption of H_2S

Massies et al.[87] have studied the interaction of H_2S with GaAs(001) surfaces at room temperature and at 700K. Their results with samples exhibiting the c(2×8) structure and being held at 700K during the exposure to H_2S are shown in Fig. 22. Massies et al. argued that at this temperature the H_2S molecules are dissociated at the GaAs

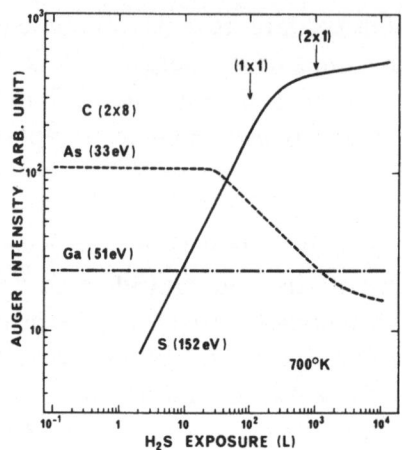

Figure 22 - Behavior of As, Ga, and S Auger-electron lines with increasing H_2S exposure of a GaAs(001)-c(2 × 8) surface at 700K. (After Ref. 87.)

surface. This conclusion is strongly supported by the experiments on the interaction of atomic hydrogen with cleaved GaAs surfaces by Surkamp and Mönch.[19] They found that higher exposures to atomic hydrogen causes cleaved surfaces to lose arsenic. The same was observed by Massies et al. with (001) surfaces exposed to more than some twenty L of H_2S. Surkamp and Mönch found the As-deficient and thus Ga-enriched cleaved surfaces to be highly reactive to oxygen- and even sulfur-containing species, which could not be identified till now, in the residual gas of the UHV system.

Such reactions might be essential for the *in situ* doping of MBE grown films.

6. SOME FINAL REMARKS

The present contribution did and could not attempt to give a critical review of all surface science aspects which are relevant for molecular beam epitaxy. Only a few examples were selected. During the process of preparing the manuscript it became evident to the author that many basic data not only for cleaved but especially for the (001) surface, used so successfully in MBE, are still missing. As an example, surface diffusion shall be mentioned. No further investigations have been published after Arthur's[88] early study of field emission from GaAs cathodes.

We did not discuss the chemisorption of gallium and arsenic on GaAs surfaces since these topics will be dealt with in other contributions. Detailed experimental results have been published by Arthur[12,89,90] and, later on, by Foxon et al.[91-93] However, for a quantitative description of part of the data, Jewsbury and Holloway[94] *adjusted* the registered As_4 fluxes.

In the present paper only the low-indexed (111), (001) and (110) surfaces were considered. However, other surface orientations, as for example (211), should be also studied since they have been discussed and also successfully used to overcome the problem of site allocation in the growth of polar semiconductors on nonpolar semiconductors.[95]

Although in many aspects of surface science which are relevant for molecular beam epitaxy the experimental data meanwhile give a coherent picture further experimental and theoretical studies are needed.

ACKNOWLEDGEMENT

The author should like to thank his co-workers for many discussions during the preparation of the manuscript. The typing of the manuscript was done by A. Kohs, and the figures were drawn by J. Krusenbaum.

REFERENCES

1.   A.U. MacRae and G.W. Gobeli, in Semiconductors and Semimetals, Vol. 2, Eds. R.K. Williams and A.C. Beer (Academic Press, New York, 1966) p. 115.

2.   R.J. Meyer, C.B. Duke, A. Paton, A. Kahn, E. So, J.L. Yeh, and P. Mark, Phys. Rev. B19. 5194 (1979) and references given therein.

3.   D.J. Chadi, Phys. Rev. Lett. 41, 1062 (1978).

4.   D. Haneman, Phys. Rev. 121, 1093 (1961) and 170, 705 (1968).

5.   D.E. Eastman, T.-C. Chiang, P. Heimann, and F.J. Himpsel, Phys. Rev. Lett. 45, 656 (1980).

6.   M. Taniguchi, S. Suga, M. Seki, S. Shin, K. Kobayashi and H. Kanzaki, J. Phys. C: Solid State Phys. 16, L45 (1983).

7.   A.J. van Bommel and J.E. Crombeen, Surf. Sci. 57, 437 (1976).

8.   E.A. Wood, J. Appl. Phys. 35, 1306 (1964).

9.   J.M. Chen, Surf. Sci. 25, 305 (1971).

10.   W. Ranke and K. Jacobi, Surf. Sci. 63, 33 (1977).

11.   A.Y. Cho, J. Appl. Phys. 41, 2780 (1970).

12.   J.R. Arthur, Surf. Sci. 43, 449 (1974).

13.   F. Bartels, H.J. Clemens, and W. Mönch, to be published.

14.   H. Gant and W. Mönch, Surf. Sci. 105, 217 (1981).

15.   A.Y. Cho, J. Appl. Phys. 47, 2841 (1976).

16.   A.J. van Bommel, J.E. Crombeen, and T.G.J. van Dirschot, Surf. Sci. 72, 95 (1978).

17.   P. Drathen, W. Ranke, and K. Jacobi, Surf. Sci. 77, L 162 (1978).

18.   K. Ploog, J. Vac. Sci. Technol. 16, 838 (1979).

19.   J. Massies, P. Etienne, F. Dezaly, and N.T. Linh, Surf. Sci. 99, 121 (1980).

20.   A.Y. Cho, J. Appl. Phys. 42, 2074 (1971).

21.   A.Y. Cho, J. Vac. Sci. Technol. 8, S 31 (1971).

22.   R.E. Schlier and H.E. Farnsworth, J. Chem. Phys. 30, 917 (1959).

23.   R.Z. Bachrach, R.S. Bauer, P. Chiaradia, and G.V. Hansson, J. Vac. Sci. Technol. 19, 335 (1981).

24.   J.H. Neave and B.A. Joyce, J. Crystal Growth 44, 387 (1978).

25.   T.G. Andersson and S.P. Svensson, Surf. Sci. 110, L578 (1981).

26.   P.K. Larsen, J.H. Neave, and B.A. Joyce, J. Phys. C : Solid State Phys. 14, 167 (1981).

27.   P.K. Larsen and J.D. van der Veen, J. Phys. C : Solid State Phys. 15, L431 (1982).

28.   P.J. Dobson, J.H. Neave, and B.A. Joyce, Surf. Sci. 119, L 339 (1982).

29.   C.T. Foxon, J.A. Harrey, and B.A. Joyce, J. Phys. Chem. Solids 34, 1693 (1973).

30. D.J. Chadi, C. Tanner, and J. Ihm, Surf. Sci. 120, L425 (1982).

31. D.J. Chadi, J. Vac. Sci. Technol. 16, 1290 (1979).

32. J.A. Appelbaum, G.A. Baraff, and D.R. Hamann, Phys. Rev. B14, 1623 (1976).

33. F. Bartels, H.J. Clemens, and W. Mönch, Physica 117B and 118B, 801 (1983) and Vac. Sci. Technol. B1, 149 (1983).

34. H. Gant, L. Koenders, F. Bartels, and W. Mönch, Appl. Phys. Lett. 42, (1983), in press.

35. L. Koenders, H. Gant, and W. Mönch, to be published.

36. A.G. Cullis, P.D. Augustus, and D.J. Stirland, J. Appl. Phys. 51, 2556 (1980).

37. J.A. Knapp and G.J. Lapeyre, J. Vac. Sci. Technol. 13, 757 (1976).

38. G.P. Williams, R.J. Smith, and G.J. Lapeyre, J. Vac. Sci. Technol. 15, 1249 (1978).

39. A. Huijser, J. van Laar, and T.L. van Rooy, Phys. Lett. 65A, 337 (1978).

40. K. Jacobi, C. v. Muschwitz, and W. Ranke, Surf. Sci. 82, 270 (1979).

41. P.K. Larsen and J.D. van der Veen, J. Phys. C : Solid State Phys. 15, L431 (1982).

42. R.P. Beres, R.E. Allen, and J.D. Dow, Solid State Commun. (1982) in press.

43. P.K. Larsen, J.F. van der Veen, A. Mazur, J. Pollmann, J.H. Neave, and B.A. Joyce, Phys. Rev. B26, 3222 (1982).

44. W. Müller and W. Mönch, Phys. Rev. Letters 27, 250 (1971).

45. J. Assmann and W. Mönch, Surf. Sci. 99, 34 (1980).

46. J. Clabes and M. Henzler, Phys. Rev. B21, 625 (1980).

47. W. Mönch, P. Koke, and S. Krueger, J. Vac. Sci. Technol. 19, 313 (1981).

48. V. Dose, H-J. Gossmann, and D. Straub, Phys. Rev. Lett. 47, 608 (1981); Surf. Sci. 117, 387 (1982).

49. V. Dose, Appl. Phys. 14, 117 (1977).

50. J.B. Pendry, Phys. Rev. Lett. 45, 1356 (1980).

51. J.R. Chelikowsky and M.L. Cohen, Phys. Rev. B14, 556 (1976).

52. D.J. Chadi, J. Vac. Sci. Technol. 15, 631 (1978).

53. D.J. Chadi, Phys. Rev. B18, 1800 (1978), and references therein.

54. J. van Laar and J.J. Scheer, Surf. Sci. 8, 343 (1967).

55. A. Huijser and J. van Laar, Surf. Sci. 52, 202 (1975).

56. W. Gudat, D.E. Eastman, and J.L. Freeouf, J. Vac. Sci. Technol. 13, 250 (1976).

57. W.E. Spicer, I. Lindau, P.E. Gregory, C.M. Garner, P. Pianetta, and P.W. Chye, J. Vac. Sci. Technol. 13, 780 (1976).
58. W. Mönch and H.J. Clemens, J. Vac. Sci. Technol. 16, 1238 (1979).
59. M. Henzler, Surf. Sci. 19, 159 (1970).
60. H. Ibach, K. Horn, R. Dorn, and H. Lüth, Surf. Sci. 38, 433 (1973).
61. W. Mönch, R. Enninghorst, and H.J. Clemens, Surf. Sci. 102, L54 (1981).
62. H.J. Clemens and W. Mönch, to be published.
63. A. Many, Y. Goldstein and N.B. Grover, Semiconductor Surfaces, (North-Holland Publ. Comp., Amsterdam 1965).
64. D.R. Frankl, Electrical Properties of Semiconductor Surfaces, (Pergamon Press, Oxford 1967).
65. W. Mönch, Proc. 2nd Trieste Semiconductor Symposium, 1982; Surf. Sci. (1983) in press.
66. K. Besocke and H. Wagner, Surf. Sci. 53, 351 (1975).
67. K. Besocke, B. Krahl-Urban, and H. Wagner, Surf. Sci. 68, 39 (1977).
68. S. Krueger and W. Mönch, Surf. Sci. 99, 157 (1980).
69. W. Ranke and K. Jacobi, Surf. Sci. 47, 525 (1975).
70. W. Ranke, Y.R. Xing, and G.D. Shen, Surf. Sci. 120, 67 (1982).
71. F. Bartels, L. Surkamp, H.J. Clemens, and W. Mönch, J. Vac. Sci. Technol. B1, 756 (1983).
72. P. Pianetta, I. Lindau, C.M. Garner, and W.E. Spicer, Phys. Rev. Lett. 35, 1356 (1976) and 37, 1166 (1976).
73. R. Murschall, H. Gant, and W. Mönch, Solid State Commun. 42, 787 (1982).
74. R. Ludeke and A. Koma, J. Vac. Sci. Technol. 13, 241 (1976).
75. D.G. Welkie and M.G. Lagally, J. Vac. Sci. Technol. 16, 784 (1979).
76. G.D. Davis, D.E. Savage, and M.G. Lagally, J. Electron Spectrosc. 23, 25 (1981).
77. I.L. Singer, J.S. Murday, and L.R. Cooper, Surf. Sci. 108, 7 (1981).
78. R.D. Bringans and R.Z. Bachrach, J. Vac. Sci. Technol. B1, 142 (1983).
79. L. Surkamp and W. Mönch, to be published.
80. D.D. Pretzer and H.D. Hagstrum, Surf. Sci. 4, 265 (1966).
81. W. Mönch and H. Gant, Phys. Rev. Lett. 48, 512 (1982).
82. H. Gant and W. Mönch, Applic. Surface Sci. 11/12, 332 (1982).

83. W. Mönch, R.S. Bauer, and H. Gant, J. Vac. Sci. Technol. 21, 498 (1982).

84. W.E. Spicer, P.W. Chye, P.R. Skeath, C.Y. Su, and I. Lindau, J. Vac. Sci. Technol. 16, 1422 (1979) and 17, 1019 (1980).

85. P. Skeath, I. Lindau, P.W. Chye, C.Y. Su, and W.E. Spicer, J. Vac. Sci. Technol. 16, 1143 (1979).

86. C.W. Tu, R.P.H. Chang, and A.R. Schlier, Appl. Phys. Lett. 41, 80 (1982).

87. J. Massies, J. Chaplast, M. Laviron, and N.T. Link, Appl. Phys. Lett. 38, 693 (1981).

88. J.R. Arthur, J. Appl. Phys. 37, 3057 (1966).

89. J.R. Arthur, J. Appl. Phys. 39, 4032 (1968).

90. J.R. Arthur, in Structure and Chemistry of Solid Surfaces, Ed. G.A. Somorjai, (J. Wiley & Sons, 1969).

91. C.T. Foxon, M.R. Bondry, and B.A. Joyce, Surf. Sci. 44, 69 (1974).

92. C.T. Foxon and B.A. Joyce, Surf. Sci. 50 434 (1975).

93. C.T. Foxon and B.A. Joyce, Surf. Sci. 64, 293 (1977).

94. P. Jeswbury and S. Holloway, J. Phys. C : Solid State Phys. 9, 3205 (1976).

95. S.L. Wright, M. Inada, and H. Kroemer, J. Vac. Sci. Technol. 21, 534 (1982).

MOLECULAR BEAM EPITAXIAL III-V COMPOUNDS: DOPANT INCORPORATION, CHARACTERISTICS AND BEHAVIOR*

Colin E.C. Wood[†]

School of Electrical Engineering, Cornell University
Ithaca, NY 14853 USA

ABSTRACT: In this lecture, an attempt is made to present all the available knowledge on the incorporation and behavior of impurities into III-V Semiconductors grown by molecular beam epitaxy. Initially, attention is paid to unintentional impurities with regard to their possible origin and mechanisms of incorporation. Intentional donors and acceptors are then systematically treated from the viewpoints of their incorporation mechanism, their behaviors, solubilities and limitations as donors species for practical applications. Wherever possible the incorporation kinetics are considered algebraically and used to give some insight to mechanism of incorporation.

1. INTRODUCTION

The first years of molecular beam epitaxy (MBE) research were devoted to understanding the preparation of stoichiometric III-V compounds; initially GaAs and subsequently, more difficult binary compounds, followed by ternary and quaternary alloys.[1,2,3] Having established the basic parameters for stoichiometric growth, the next task was to determine the origin and find methods to minimize unintentional impurities, and define growth conditions such that their densities are significantly low.

In parallel with this, much effort has been spent in intentional doping studies. *A priori* it was naively assumed that atoms intentionally incident on growing MBE films would be incorporated immediately. It was soon apparent that this was not so.[4] Further, for dopants that do incorporate immediately, their electrical activities are subject to growth conditions.

This chapter overviews the developments in understanding of dopant incorporation and electrical behavior mechanisms. The reader can thus decide the most appropriate dopant for his particular purpose and determine growth parameters for the best compromise between efficient doping characteristics, structural, electrical and optical properties.

No apology is made for the emphasis on GaAs as it is an archetypal III-V compound, and lessons learned in the doping of GaAs are instrumental in predicting the behavior of other III-V compounds.

2. UNINTENTIONAL IMPURITIES

Carrier concentrations can only be controlled by intentionally adding impurities when the identities and sources of unintentional or background impurities are understood and their concentrations minimized. Below, the current understanding of these troublesome species and their deleterious effects is explained.

2.1 Unintentional Shallow Acceptors

Shallow impurities are those that can be measured quantitatively at 300K or 77K by standard Hall measurements. The chemical identities of shallow donors are difficult to determine by simple electrical or optical methods and resort has to be made to far-infrared photothermal ionization[5] and very high resolution Zeeman photoluminescence (PL) measurements.[6] Shallow acceptors however can be readily identified by 4K PL measurements.

2.1.1 Carbon - At the time of writing, GaAs grown in "state of the art" MBE machines typically shows residual p-type conductivity.[7] The consensus is that the residual acceptor(s) responsible is substitutional carbon[8] which has a binding energy of 25 meV. Figure 1 shows a

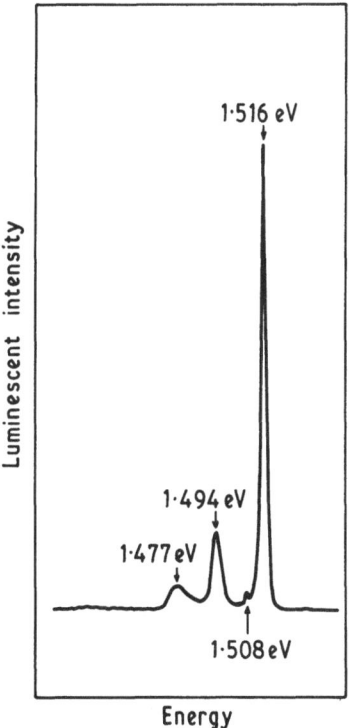

Figure 1 - 4K photoluminescence spectrum of 4×10^{13} cm^{-3} Si doped GaAs showing substitutional carbon-associated luminescence peak. Also shown are the carbon related defect excitation band and replica at 1.494 and 1.477 eV respectively.

representative PL scan of a "high purity" (4×10^{13} cm^{-3} Si doped) MBE film in which the carbon peak is apparent. Another emission band $\sim$ 30 mV below the exciton peak has been attributed to germanium,[9] and more recently to a defect exciton recombination.[10,11] High resolution 4K PL (Fig. 2) has shown this band to be comprised of some 9 (or more) other peaks which have replicas 26 millivolts lower in energy,[11] suggesting that carbon is associated at least with the latter set of recombinations. The source of carbon remains unclear, however reaction of carbon monoxide or dioxide with either surface arsenic or gallium, liberating oxygen, either as a volatile arsenic oxide or gallium suboxide (Ga_2O), and a free carbon atom is most probable.[12] A model involving Ga_2O from the Ga effusion cell was proposed earlier[13] however, the molecularity of this model is not consistent with the source to substrate dilution factor ($\sim 10^{-3}$) density of species present on growing surfaces. In support of a direct surface reaction, Stringfellow et al.[14] reported a relationship between carbon monoxide partial pressure and carbon related PL peak intensity. The concentration of acceptors from 300K Hall measurements is $\lesssim 10^{14}$ cm^{-3},[15] however, such low concentrations are difficult to quantify because of surface- and interface-depletion effects[16] (see Fig. 3). Quoted values can therefore be subject to large errors; however, a recent 26 μm film grown by the author demonstrated 8×10^{13} acceptors with a 77K mobility over 8,400 cm^2 v^{-1} s^{-1}. A more reliable method is to overcompensate with donors and measure (from liquid nitrogen mobilities μ_{77} and room-temperature free carrier densities n_{300}) the compensating acceptor concentrations using the well-known Brooks-Herring relation[17] and Wolf and Stillman[18] theoretical treatments.

In order to reduce the incorporation of background-gas related impurities, the density and the excitation state of background molecules such as carbon monoxide etc. should be reduced to a minimum. To help achieve this all hot filaments not necessary during epitaxy, i.e., ion-gauges, quadrupole mass-spectrometers, and most importantly reflection electron diffraction (RED) and Auger electron spectrometer (AES) guns should be turned off.[19] Reduction of ambient and quiescent cell temperatures which are not being used in the particular epitaxial run also helps. In this context it is also imperative that temporal or spatial variations in temperature of LN_2 cryo-pumping areas do not fluctuate during the epitaxial growth as this would result in variations in liberation and absorption of background gas species. This can be

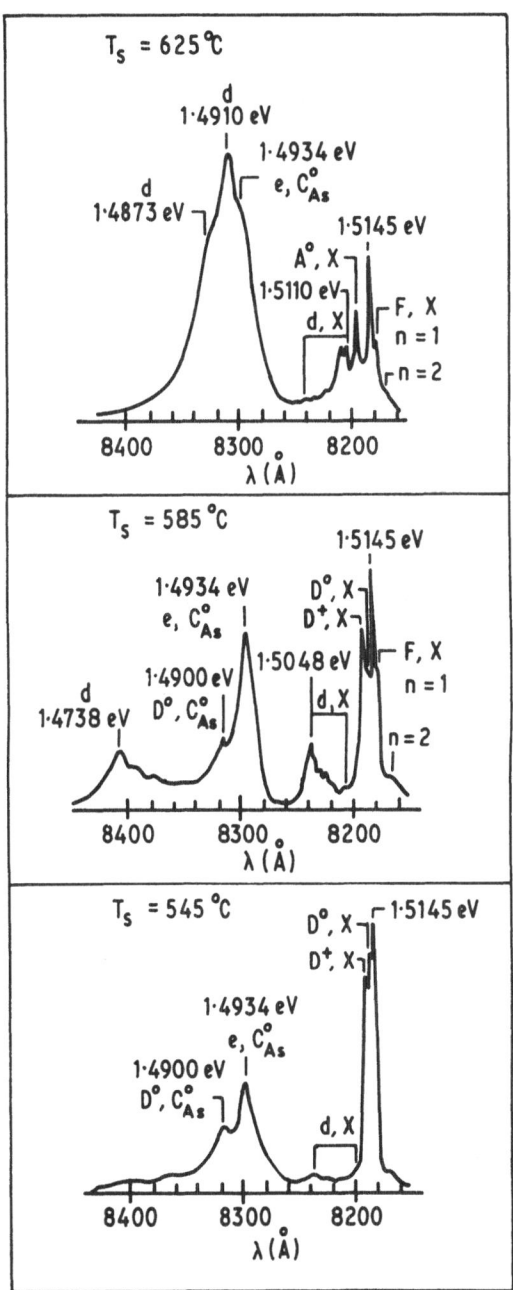

Figure 2 - High resolution 4K PL spectra of MBE GaAs grown at different temperatures showing the carbon associated defect excitation recombinations (After Ref. 11).

achieved by constant LN$_2$ flow controls and reducing the total power dissipation inside the growth chamber.

Compensating acceptor densities are falling as a function of time and experience and at the present state-of-the-art are in the mid 10^{13} cm^{-3} region. Happily there are few devices which require epitaxial films with doping control down in the 10^{14} cm^{-3} concentration range. Therefore above ~ 10^{15} cm^{-3} intentional doping residual acceptors can usually be ignored. Deep levels at this concentration however, cannot and their effects are discussed later.

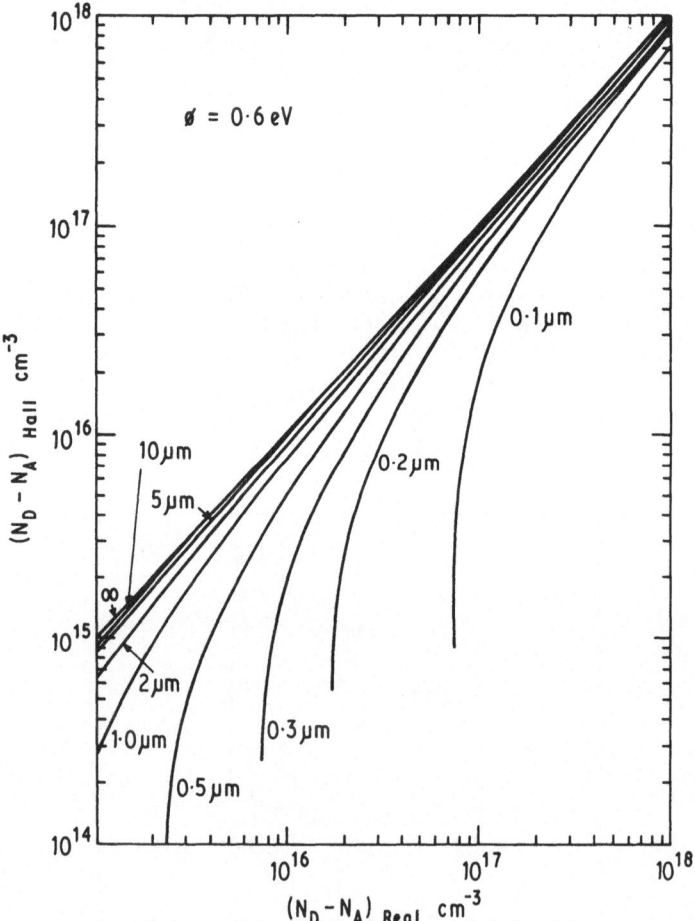

Figure 3 - Curves for correcting n$_{Hall}$ for surface and interface depletion effects (After Ref. 16).

2.1.2 Silicon - Other shallow acceptors that have been identified from their PL fingerprints are silicon and manganese.[20] There is no source of silicon in the MBE process *per se*. However if a bulk GaAs charge is used as an As_2 source then cell temperatures are sufficiently high that residual silicon in the charge can dope epitaxial films.[7] The advantage of using a GaAs source is that it produces a dimeric species[21] (rather that the tetrameric As_4 which is in thermodynamic equilibrium with elemental arsenic at ~ 300-350°C). The introduction of thermal cracking furnaces in various laboratories[22,23,24] has alleviated the need for GaAs sources of As_2 and thus silicon as a residual acceptor is typically no longer observed. As_2 can produce lower deep-level densities[22] and is currently under study[25] to help stabilize alloy surfaces during high temperature growth, and as a means to reduce the auto-compensating behavior of Si and Ge at high T_s values.[26,27]

2.1.3 Manganese - Early reports of manganese in MBE GaAs as a residual acceptor indicated that hot stainless steel components liberated Mn which subsequently incorporated in growing films.[20] However, concentrations found and the temperatures required to liberate such concentrations[28] are not self-consistent. Heat-treatment of bulk chromium-doped semi-insulating GaAs showed conducting (or "converted") surfaces.[29] The species responsible was identified from PL spectra as surface concentrations of Mn (p-type) and in certain cases Si (n-type). Secondary-ion mass-spectrometry (SIMS) measurements of impurity redistributions, in such heat-treated GaAs, confirmed large surface Mn concentrations.[29] Mn is known to have a high diffusion coefficient in GaAs;[30] however, recent studies in the author's laboratory have shown Mn diffusion to be insignificant during MBE growth below ~ 620°C.[31] Thus Mn in MBE films is believed to arise from surface-accumulations which are formed during substrate heat-treatment prior to epitaxy. The driving force for the accumulation, as with chromium, is probably the extensive supersaturation at the heat treatment temperature which is quenched in during bulk growth at the melting point. From current understanding of background-impurity incorporation in GaAs, methods have been developed for bulk growth of high resistivity (> 10^8 Ω cm) GaAs without chromium (and the associated manganese).[32] Thus surface accumulated Mn concentrations are now not usually significant and MBE layers grown on such substrates show little or no Mn PL peaks.[33]

2.2 Unintentional Shallow Donors

2.2.1 Silicon - Silicon behaves predominantly as a donor in GaAs although amphoteric character allows its identification by PL as a substitutional shallow acceptor, which explains why early GaAs grown with As_2 (GaAs) sources was n-type (Si_{Ga}) despite the fact that PL indicates a Si_{As} species.[9]

2.2.2 Sulphur and lead - Recently far infrared photothermal-ionization techniques have demonstrated both lead and sulphur in addition to silicon as residual donor species[34] (see Fig. 4). However, Pb has only been "seen" in samples grown using AsH_3 sources. It is not easy to give quantitative estimates but high associated μ_{77} values indicate that concentrations are exceedingly small, (certainly below 10^{14} cm^{-3}), as residual carriers are normally holes.

2.3 Residual Deep Centres (Deep Levels)

It is more convenient to refer to deep centres as electron- or hole-traps. Methods for their identification and quantitative estimation form a well established family under the generic title of "deep level transient spectroscopy". The transient parameter can be either capacitance or conductance. More indirect evidence for the presence of deep-levels comes from reduced PL intensity, as deep levels act as efficient parasitic non-radiative recombination centers.[33] Exciton recombination peak intensities are most affected (reduced intensity) and in certain alloys (e.g., AlInAs and GaInP) where deep-level densities are very high, they are absent completely.

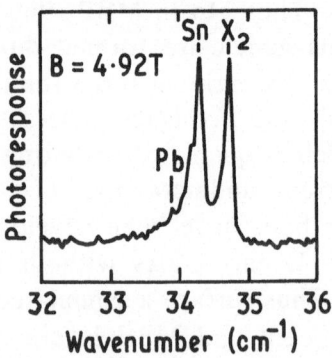

Figure 4 - Photothermal ionization spectrum of MBE GaAs samples showing, Sn and Pb and unknown donor peaks (After Ref. 5).

As in the case of shallow levels, there are two easily identifiable deep level origins. The first, and most consistent sources are substrates. Circumstantial evidence has indicated that chromium, copper and in certain cases, iron are present in films grown in semi-insulating GaAs.[35] In addition to these elements (which appear as hole traps) there are four electron traps which are consistently seen in spectra from many laboratories. In Figure 5 a representative DLTS spectrum of a very lightly Si back-doped n-type MBE GaAs which shows these preponderant electron traps. They have been arbitrarily named M1 through M4 by Lang et al.[36] and are present in GaAs grown under As stabilized MBE conditions. The two dominant levels M1 and M3 are often present in large concentrations (10^{11} - 10^{12} cm^{-3}); however, under certain conditions (usually high substrate temperatures and low arsenic excess fluxes) the densities of M2 and M4 are very low ($\leq 10^{10}$ cm^3).[37] The dependence of both concentrations (N_T) and type of deep

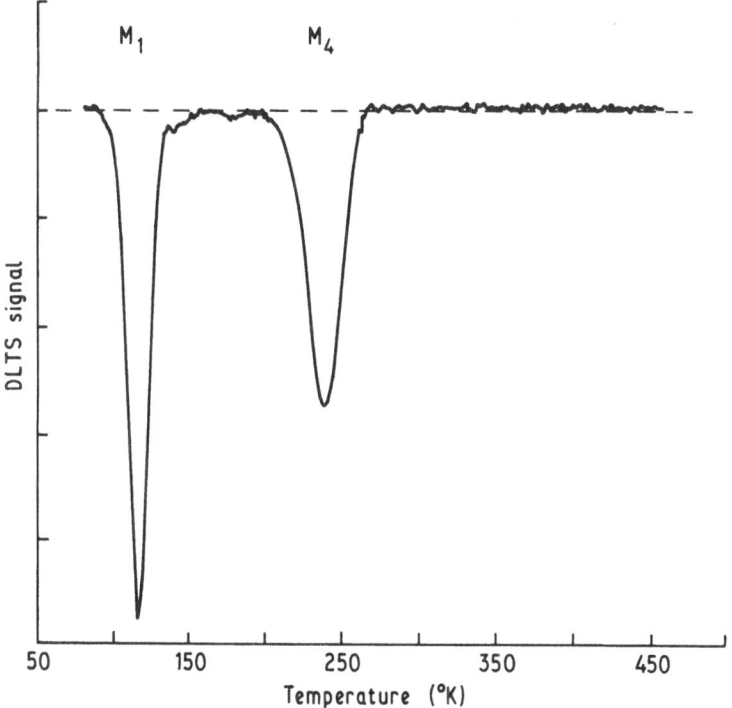

Figure 5 - Deep level transient capacitance spectrum of lightly Si doped (2×10^{14} cm^{-3}) MBE GaAs grown under As$_4$ stabilized conditions at 580°C.

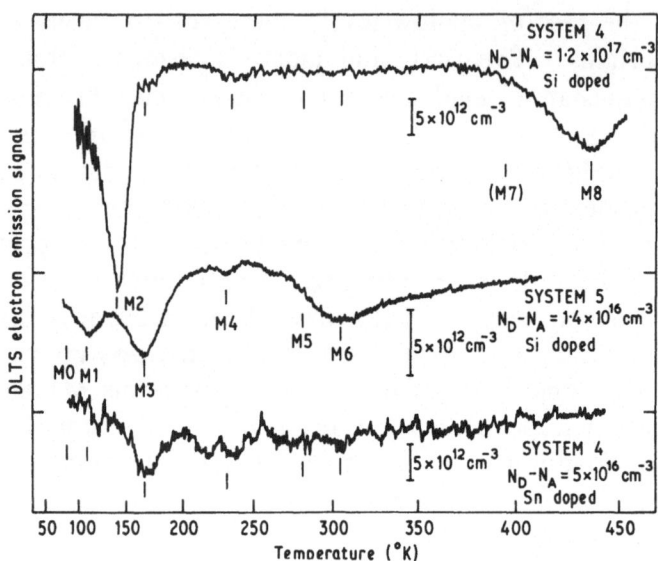

Figure 6 - Deep level transient capacitance spectrum of MBE GaAs grown under Ga stabilized conditions compared with similar layers grown under As stabilized conditions (After Ref. 36).

levels on growth parameters have been studied by several authors.[35-39] The consensus is that N_T increases with reducing T_s and with increasing J_{As_4}/J_{Ga} ratios leading to the conclusion that stoichiometric centers, (either simple or complexed with impurities) are responsible. Further corroboration arises from the completely different DLTS "fingerprint" of layers grown under Ga stabilized surface growth conditions[36] shown in Fig. 6. Such spectra are dominated by V_{As} related species as opposed to V_{Ga} related species under conventional As_4 stabilized conditions. Despite the fact that at low T_s ($\sim 300°C$) (see Fig. 7) simple V_{Ga} centers have been identified in MBE material,[39] (by reference to electron beam irradiation studies[40] no absolute identification of levels M1 through M4 has been possible to date. It is suggested that this problem should be studied more fully so that deleterious deep-levels can be further understood and N_T values minimized.

The density of hole-traps generally exceeds that of electron-traps by approximately two orders of magnitude.[7] In the author's laboratory good correlation between substrate impurities and most of these hole

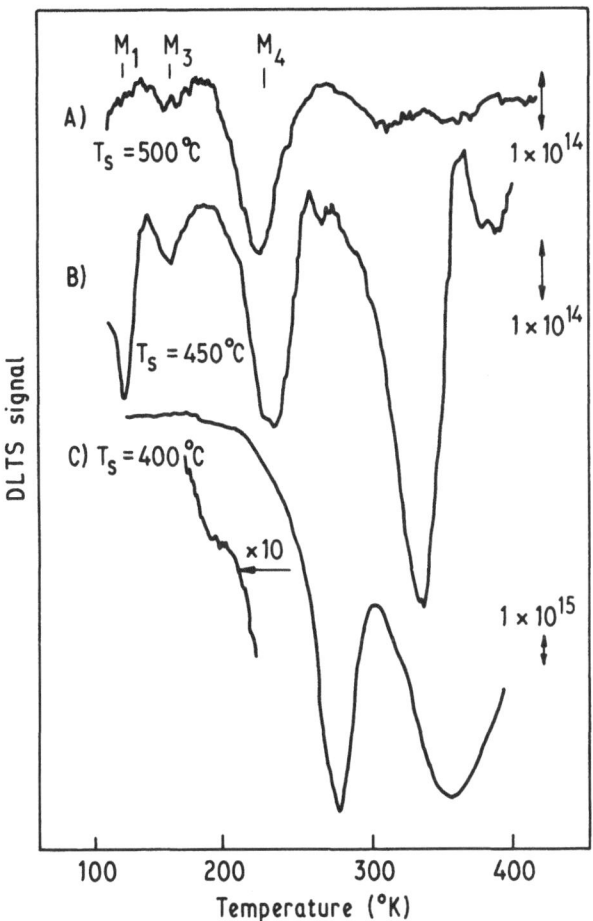

Figure 7 - Deep level spectra of MBE GaAs grown at different T_s values (from Ref. 64).

traps have been found, specifically Cu and Fe[39] (see Fig. 8). It should also be mentioned here that even MBE layers grown on undoped semi-insulating bulk GaAs can demonstrate substrate-related trap densities, specifically EL2 when grown at high temperatures ($\sim 700°C$). These can be reduced by a 24 hour $750°C$ treatment of polished substrates in a hydrogen atmosphere followed by ~ 10 micron etch to remove surface accumulated impurities.[41] In summary then, the understanding of origins and methods for the reduction of deep-levels in MBE GaAs is at present satisfactory but could be improved.

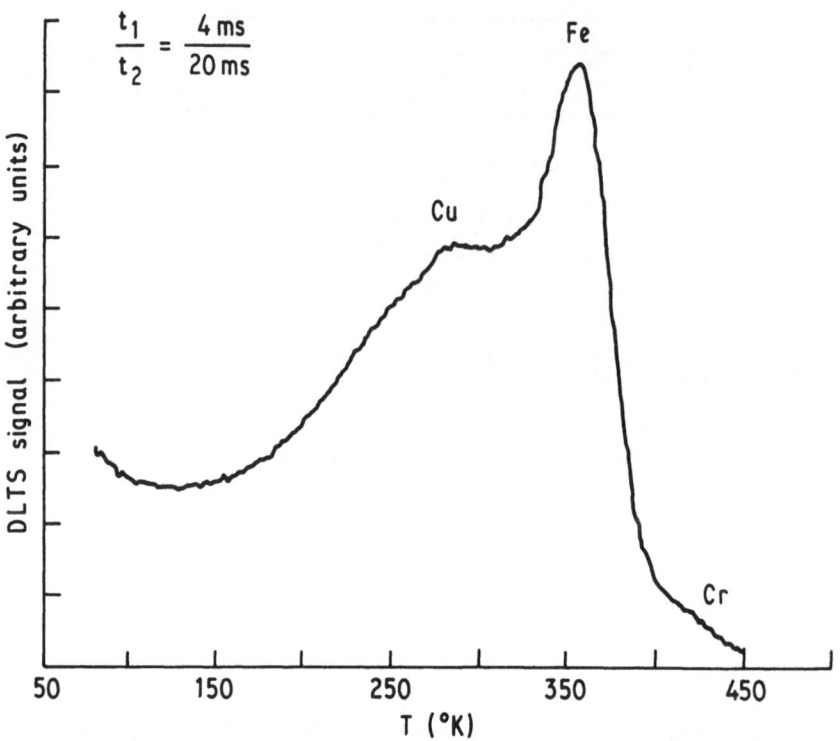

Figure 8 - DLTS hole-trap spectra of Cr, Cu and Fe levels originating from substrate (from Ref. 64).

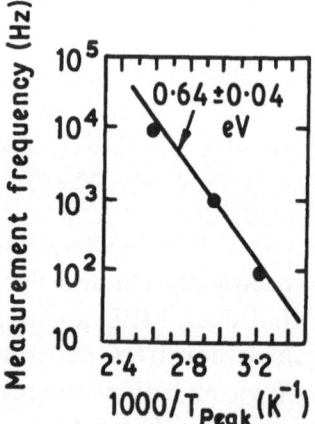

Figure 9 - Frequency versus temperature dependence of the conduction peak of oxygen doped AlGaAs (After Ref. 42).

2.4 Deep Levels in MBE Alloy Semiconductors

AlGaAs is probably the most extensively studied MBE material apart from the binary compounds. In contrast to GaAs, deep-level spectra are dominated by one electron trap (Fig. 9) which is fairly well documented as an oxygen related deep donor.[42] Oxygen doped $Al_xGa_{1-x}As$ as an insulator has been studied by Casey, et al.[42] and Tsang[43] for possible MIS application; however, little recent interest suggests that there are some relatively insurmountable problems. The intriguing question of the identity of X in the DX center[44] in AlGaAs alloys still remains and is the bête-noire of many DLTS studies. It is clear that raising T_s substantially above the values used for GaAs, significantly reduces N_T as evidenced by direct DLTS measurements[45,46] and by extremely good PL line shapes (Fig. 10) and intensities.[47] Further evidence for their reduction at high T_s and low J_{As_4}/J_{Ga} values comes from lowering of threshold-current densities of GaAs, GaAlAs DH lasers.[48]

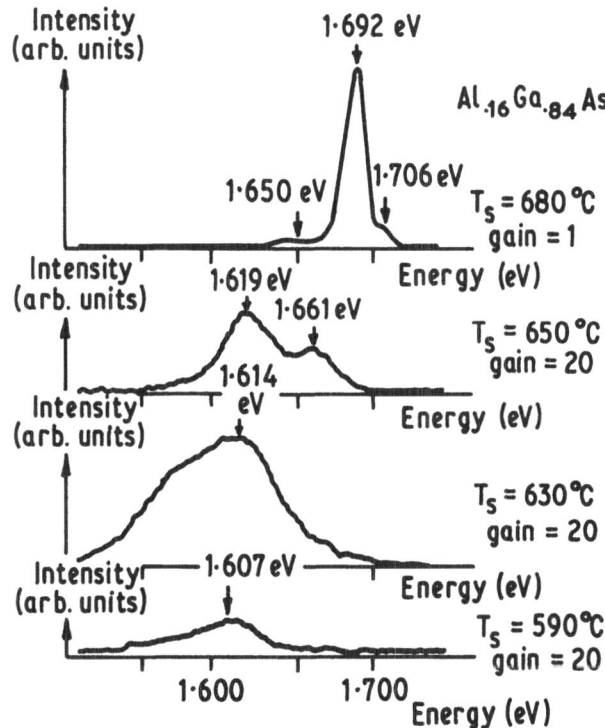

Figure 10 - PL spectra of AlGaAs grown at different substrate temperatures (After Ref. 47).

Some of the deep levels produced during growth at low T_s are extremely mobile and GaAs layers grown at 580°C on either GaAs buffers or AlGaAs buffers grown below 480°C or 600°C respectively, show severe compensation of free electron densities.[49] The identity of diffusing deep-levels is currently under study in the author's laboratory; however, the use of very high resistance GaAs or AlGaAs grown at low temperature for buffer layers is precluded by out-diffusion of compensating deep centers. The electrical degradation of GaAs layers grown on such films is mirrored by a much reduced 4K PL intensity as shown in Fig. 11. Similar degraded PL spectra are found for the ternary alloys AlInAs and GaInAs[50] and the quaternary alloy GaAlInAs[51] grown on InP (Fig. 12). Here the problem arises from an incompatibility in the optimum growth conditions for the constituent binary arsenides. More explicitly, it is necessary to grow AlAs containing alloys above 650°C to avoid oxygen related deep levels, but it is not possible without extremely large excess As_4 fluxes to grow InAs-containing alloys much above 550°C without decomposition.

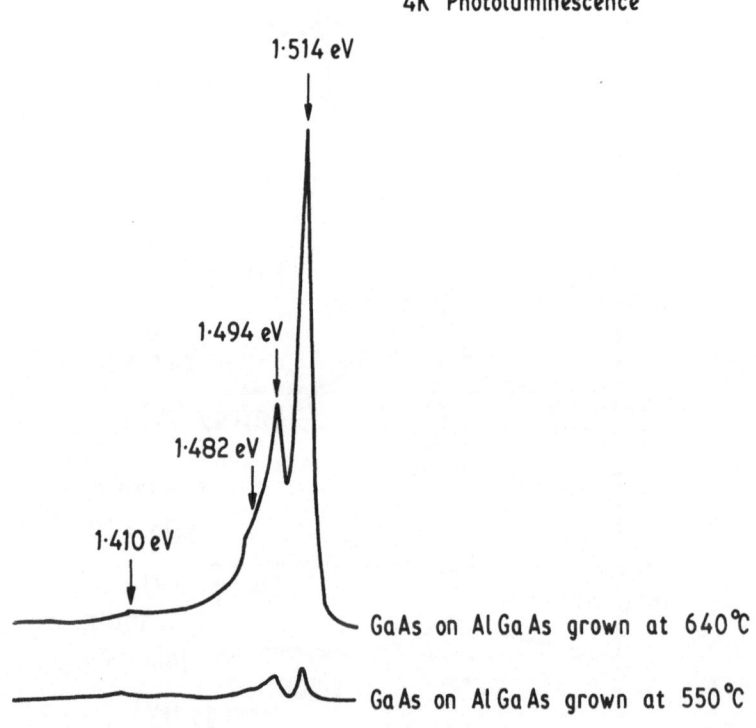

Figure 11 - 4K PL of MBE GaAs films AlGaAs grown at 640°C upper and 550°C lower trace.

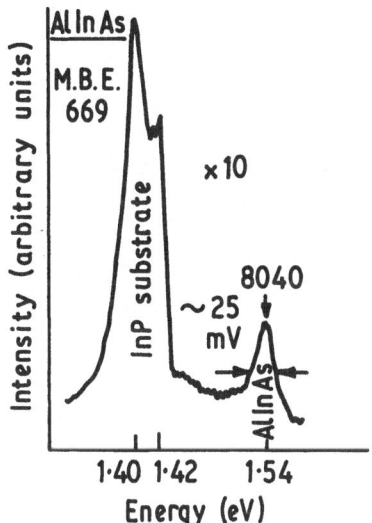

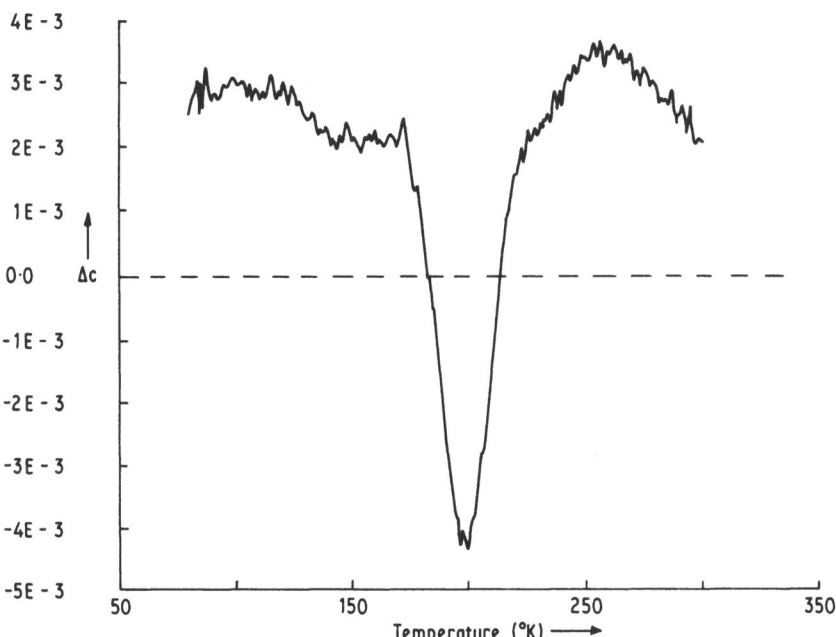

Figure 12 - (a) 4K PL of 10^{17} cm^{-3} Si doped AlInAs film and (b) deep level spectra showing dominant trap level.

3. INTENTIONAL SHALLOW DONOR IMPURITIES

3.1 Group IV Elements

3.1.1 Carbon - As mentioned under the heading "Unintentional Impurities", carbon is a ubiquitous companion of MBE GaAs and other III-V compounds, in which it behaves predominantly as an acceptor. Attempts to produce a clean atomic carbon dopant source have not been successful[52] and carbon-ion implantation/annealing have obtained only $\sim 8 \times 10^{15}$ cm^{-3} C_{As} related holes.[53] Most attention has therefore been paid to the remaining group IV elements Ge, Si, Sn and Pb and the group VI elements S, Se and Te. The substantial literature on these elements is discussed below.

3.1.2 Silicon - Silicon was early shown to be a useful donor. However, attempts to produce high quality back-doped GaAs initially produced heavily compensated layers.[4]   This was ascribed to volatile furnace components,[54,55] from reactions such as:

$$Al_2O_3 + Ta, Mo, W \text{ etc.} \iff Al_2O \uparrow + Mo_2O_3 \uparrow \text{ etc.}$$

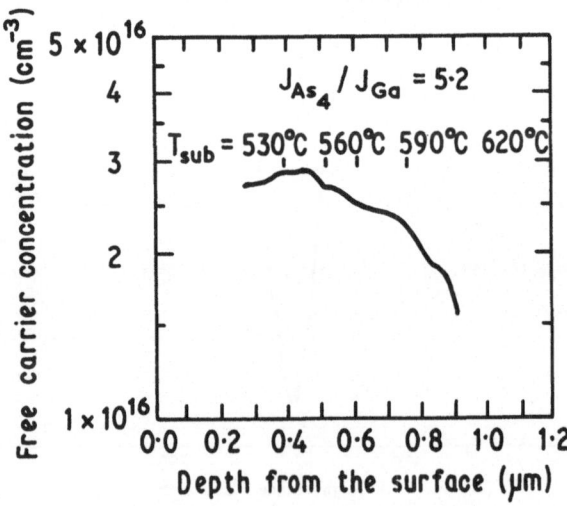

Figure 13 - C-V profiles of a Si doped MBE GaAs layer showing the T_s dependence of free electron density (After Ref. 58).

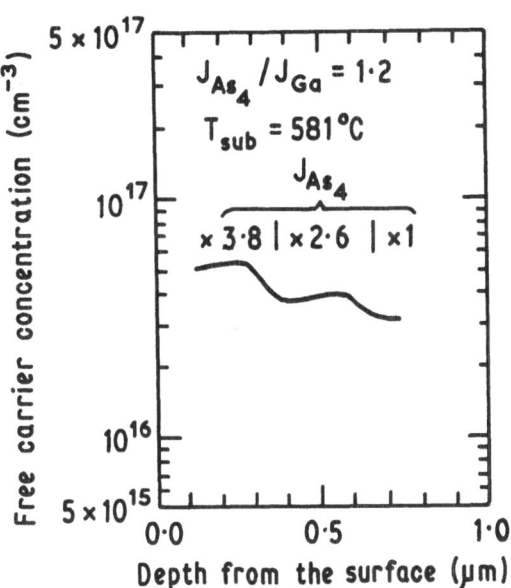

Figure 14 - C-V profile of Si doped GaAs layer (grown at 581°C) showing the effect of arsenic to gallium flux ratio J_{As_4}/J_{Ge} (After Ref. 58).

and to amphoteric (Si) behavior; both of which would lead to carrier compensation and low mobilities, for corresponding n_e values. More recently contamination from cell/furnace component reaction products has been effectively eliminated by substitution of pyrolitic boron nitride for alumina insulators and the exclusive use of vacuum melted tantalum for resistive windings and heat shields etc. in commercial MBE furnace design.

Melting the silicon (as for Beryllium) in the BN crucible[56] is now a standard practice to reduce radiation-cooling of the Si (Be) charge (and consequent need for increased power input).

The most attractive feature of Si as an n-type dopant is its apparent unity incorporation coefficient K_i, its "non"-amphoteric behavior, and its non-diffusivity, etc. The electrical solubility limit of silicon in GaAs grown by MBE has been variously reported to be 1.3×10^{19} cm^{-3},[57] (see Fig. 15) and $5\text{-}6 \times 10^{18}$ cm^{-3},[58] above which precipitation of second phases occurs.[59]

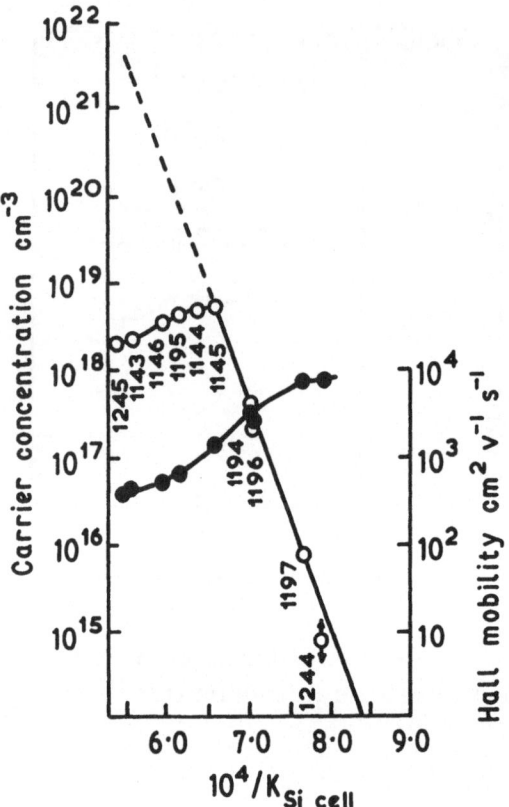

Figure 15 - Free electron concentration versus reciprocal Si cell temperature for MBE GaAs (After Ref. 58).

Recently we have seen evidence for significant T_s and J_{As_4}/J_{Ga} dependent Si amphoteric behavior (see Figures 13, 14 and the section on germanium below). These two parameters are not mutually independent, as increasing T_s affects the thermodynamics of substitutional-site occupancy.

$$Si_{Ga} \overset{T_s}{\Longleftrightarrow} Si_{As}$$

and reduces the surface lifetime (T_{As_4}) of incident arsenic, hence the dynamic surface coverage C_{As_4}.

There is evidence from attempts to grow very sharp profiles (changes in [Si] by 3-4 orders of magnitude in < 10-15 Å) that silicon <u>does</u> surface accumulate for a finite (albeit very short) time[60] and therefore Si doping levels may not be changed on a monolayer depth scale. PL spectra of Si doped MBE GaAs often shows D-Si$_{As}$ recombination peaks[9] however under optimum conditions ($\sim 600°$ with $J_{As_4}/J_{Ga} = 1$) auto-compensation is not significantly large.

Increasingly, higher T_s values are found to improve the quality of GaAs and its alloys with AlAs. We believe that Si diffusion is not negligible at these elevated temperatures in some of the very thin device-layer structures such as quantum-well devices, selectively-doped heterojunctions and planar-doped-barrier-transistors and are currently studying this problem. Very recently evidence of Si and/or Be diffusion has been found indirectly from I.V. characteristics of MBE grown tunnel diode junctions.[57]

To date the highest MBE GaAs electron mobility has been demonstrated with silicon doping. Because of its limited surface-accumulation behavior it is also the favored donor for modulation doping structures. Silicon in association with beryllium is also the donor of choice for doping superlattices (nipi structures).[61] By repetitively varying the separation and concentration of p and n dopant regions, artificial bandgap behavior can be simulated. Obviously dopant diffusion, surface accumulation, amphoteric or interstitial behavior would interfere with growth of these structures. Be and Si are therefore the obvious acceptor and donor choices respectively.

The amphoteric nature of Si is most exaggerated in growth on (110) and (111) oriented GaAs.[62] In the former it behaves predominantly as an acceptor above $\sim 550°C$ and as a donor below $\sim 550°C$ in analagous fashion to Ge in (100) MBE GaAs. On (111) oriented crystals, self-compensated incorporation has been observed yielding high resistivity films in the author's laboratory although early papers on[1] Si doped GaAs on (111)b GaAs did report some electrical activity.

3.1.3 Germanium - Germanium doping of MBE GaAs has received much attention because of its interesting site-occupancy-dependence on J_{As_4}/J_{Ga}[63-67] and T_s.[64]

Cho et al.[63] demonstrated p-n junctions in GaAs with Ge doping alone by simply changing from As$_4$- to Ga-stabilized growth surfaces. Under Ga-stabilized conditions however, three-dimensional growth results. Results of a systematic study of the effects of T_s and the

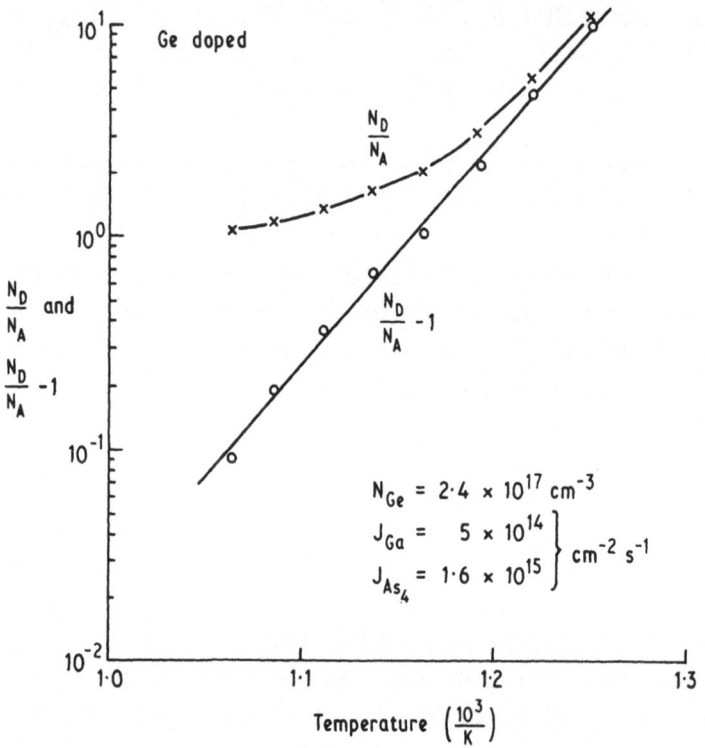

Figure 16 - Effects of (a) substrate temperature and (b) the arsenic to gallium flux ratio on the auto-compensation ratio of Ge in MBE GaAs. (Note the deviation from linearity at higher temperatures) (After Ref. 64).

$[Ge_{As}]/[Ge_{Ga}]$ ratio are shown in Figs. 16 and 17. Subtle deviations from the predicted thermodynamic behavior:[64]

$$\log_e \frac{[Ge_{Ga}]}{[Ge_{As}]} = \log_e f \frac{J_{As_4}}{J_{Ga}} - \frac{\Delta G}{RT}$$

(where ΔG = free energy difference between Ge_{As} and Ge_{Ga}) were observed and explained by a T_s-dependent unintentional C_{As} acceptor incorporation.[65] Germanium amphoteric behavior is now very well documented; however, the attention to T_s and J_{As_4}/J_{Ga} which is necessary to achieve a required carrier density is restrictive. There is therefore little advantage of germanium over silicon for n-type doping of

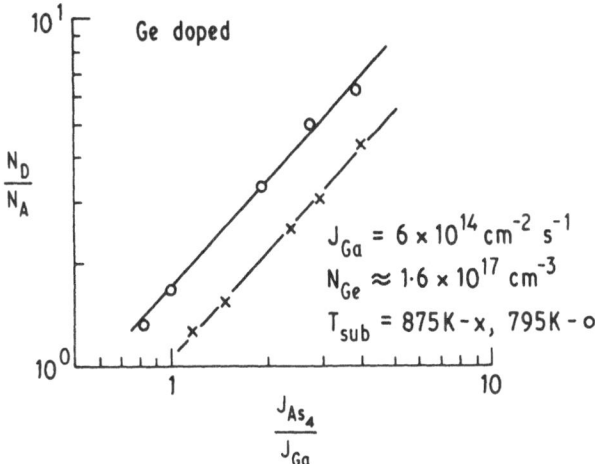

Figure 17 - Effect of germanium concentration on the free carrier concentration and type in MBE GaAs (After Ref. 64).

MBE GaAs despite the demonstration of 80,000 cm^2 v^{-1} s^{-1} 77K electron mobility in MBE germanium doped GaAs.[67]   Germanium is used as a donor in indium containing III-V compounds and alloys[68] in which the growth temperature is typically lower than GaAs and its alloys with AlAs.   In these materials Ge is predominantly donor-like which may be related to the larger size of the indium vacancy compared to that of gallium.   In MBE GaSb$_x$As$_{1-x}$, Ge, Si and Sn (see Fig. 18) are predominantly acceptors[69] for x > 0.2.[69]

Early attempts to produce abrupt doping-level changes by varying dopant cell temperature is limited by the thermal mass and thus the thermal time-constants of dopant-cell/furnace assemblies.   Indeed commercial MBE furnaces and cells are relatively large and do not respond rapidly to changes in set temperature.   For this reason it is now usual to have two or more n-type dopant sources in an MBE system such that carrier concentrations can be abruptly changed by modulating dopant-source shutters.   An alternative to multiple cells for a single carrier type was shown possible by use of atomic-plane doping using germanium.[70] (Although any dopant that does not diffuse or surface accumulate significantly can equally well be employed). In this technique growth of a nominally undoped layer is suspended, calibrated concentrations of dopant are deposited and growth resumed. By varying the separation of such doped planes and/or the concentration of germanium on each plane, a doping profile can thus be digitally syn-

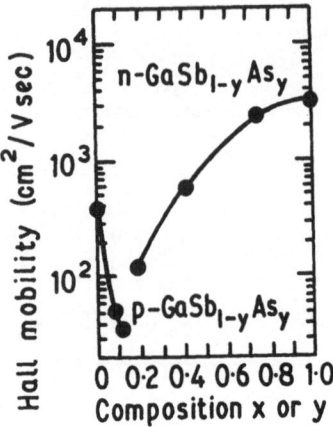

Figure 18 - Free electron mobility for ($\sim 10^{17}$ cm^{-3}) tin doped GaSb$_x$As$_{1-x}$ alloys as a function of x (After Ref. 69).

thesized (see Figs. 19a and b). The tendency for electrons to diffuse away from a dopant-concentration-gradient allows free electron concentrations to be effectively "smoothed" providing that neither the plane-to-plane separation nor the concentration of dopant on each plane is too large. In one variation of profile synthesis an attempt was made to use a single atomic doping plane as the source of donors for field effect transistor active layers.[71] It was found however, that high deep-level densities were associated with such heavily doped planes when the area doping concentration exceeded about 10^{11} cm^{-2}. This deep-level problem is not simply associated with germanium or donor species as it has also dominated the current voltage characteristics in planar doped barriers[72] in which the acceptor atoms are confined to a single atomic plane (see Fig. 20).

Although atomic-plane free electron profile synthesis was found easy under high As$_4$ fluxes and low T_s (conditions for low compensation n-type material), attempts to produce p-type atomic plane profiles with germanium on Ga stabilized surfaces during the growth suspension were not successful. In fact heavily compensated n-type or high-resistance layers resulted. This can be most readily explained by a non-unity germanium incorporation coefficient under Ga stabilized conditions (i.e. surface segregation and accumulation).

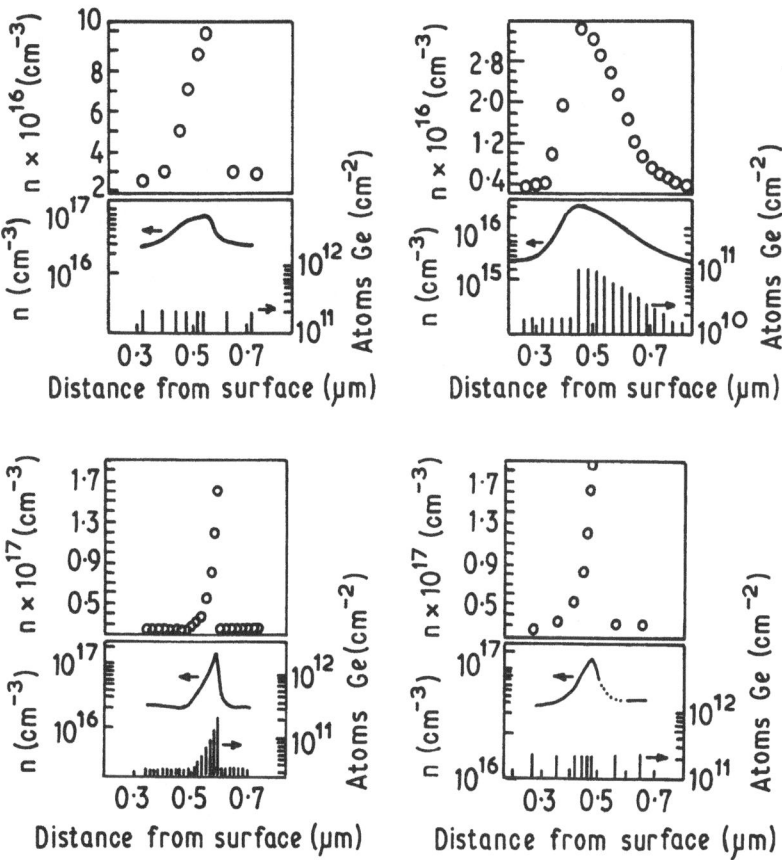

Figure 19 - (a) Linear graded and (b) power 10 free-electron profile synthesized by Ge lamella doping (After Ref. 76).

At high concentrations, Ge increasingly occupies arsenic sites until $\geq 10^{19}$ cm^{-3} the dominant carriers are holes.[73] From $\sim 1\%$ Ge across the Ge/GaAs alloy system, free-hole concentrations are practically invariant around 2×10^{20} cm^{-3} (see Fig. 21).

3.1.4 Tin - Tin is a widely used donor in MBE growth of III-V compounds because of its non-amphoteric behavior, (simple Sn$_{As}$ recombination peaks have not been found in tin doped MBE GaAs). In fact, Sn is still the favored donor in DH laser diode layer growth for this reason. Tin is however, predominantly acceptor-like in GaSb.[69]

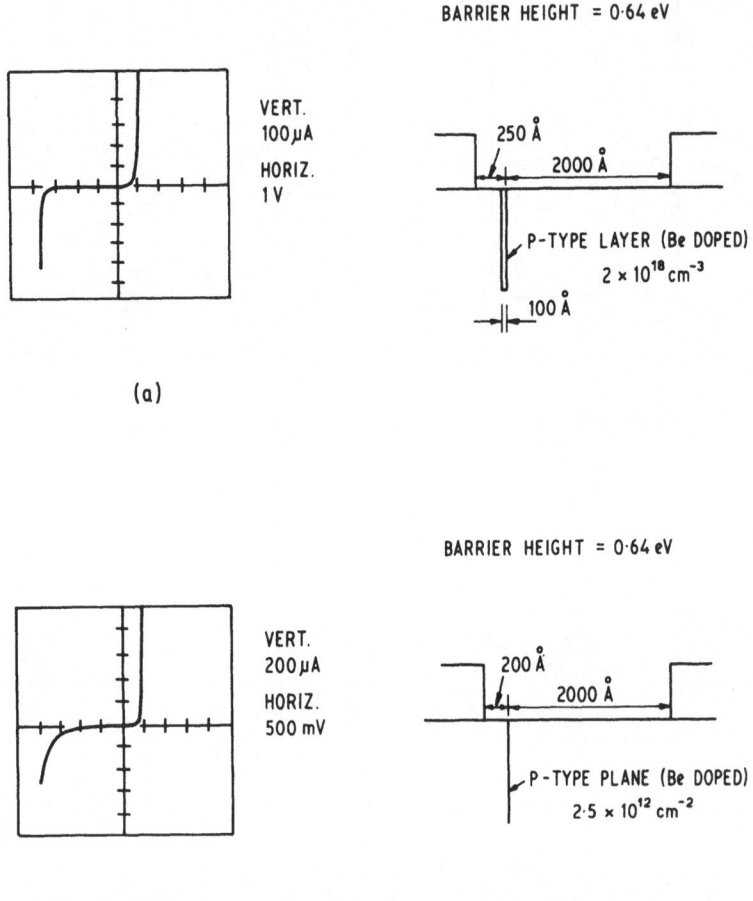

Figure 20 - I.V. characteristics of two planar-doped-barrier structures (a) with the acceptor atoms (Be) on a single plane, and (b) distributed over ~ 100 Å. Note the premature "break-down" due to deep-level-induces leakage (After Ref. 72).

Threshold currents of GaAs/GaAlAs DH lasers are reduced by increasing T_s.[48] As later experience demonstrated the need to grow GaAs above ~ 550°C to avoid deep traps, tin surface-accumulation behavior became apparent.[76] Sn incorporation has been extensively studied[4,74-78] and its behavior modeled[74,75,78] as an archetypal surface-accumulating species for epitaxial film growth. The energy difference for

$$V_{Ga} + Sn \Longleftrightarrow Sn^+Ga + e^-$$

was found[75] by the temperature dependence of its incorporation rate-constant to be ~ 1.3 eV (see Fig. 22). Accumulation and subsequent incorporation is dependent on the surface V_{Ga} concentration, which is inversely proportional to the dynamic surface arsenic coverage, $[As_s]$ which is in turn, exponentially dependent upon $1/T_s$. Tin incorporation data can be explained by rate-limited kinetics or by thermodynamic distribution-coefficient type of arguments.[79]   Increasing K_i values at lower T_s are not consistent with the distribution coefficient behavior of Sn (found in GaAs LPE studies).[80]

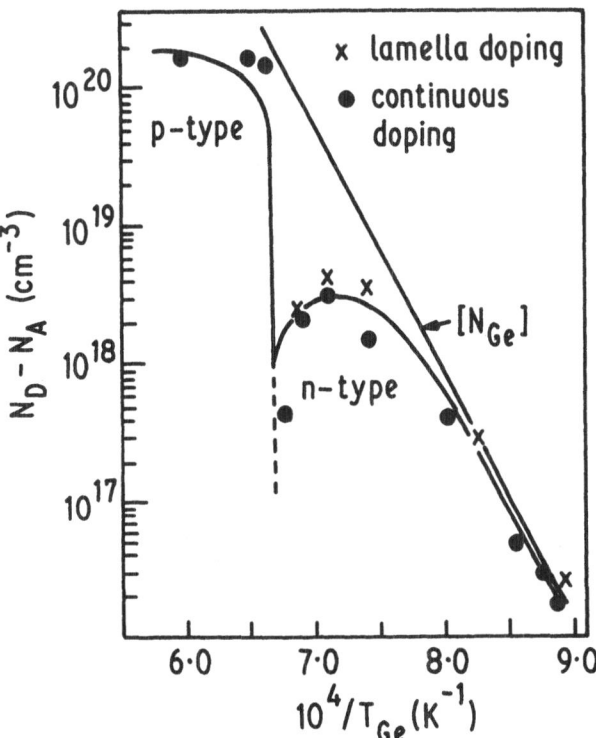

Figure 21 - Electrical characteristics of heavily Ge doped GaAs showing the type change at Ge concentrations much above 10^{19} cm^{-3} (After Ref. 73).

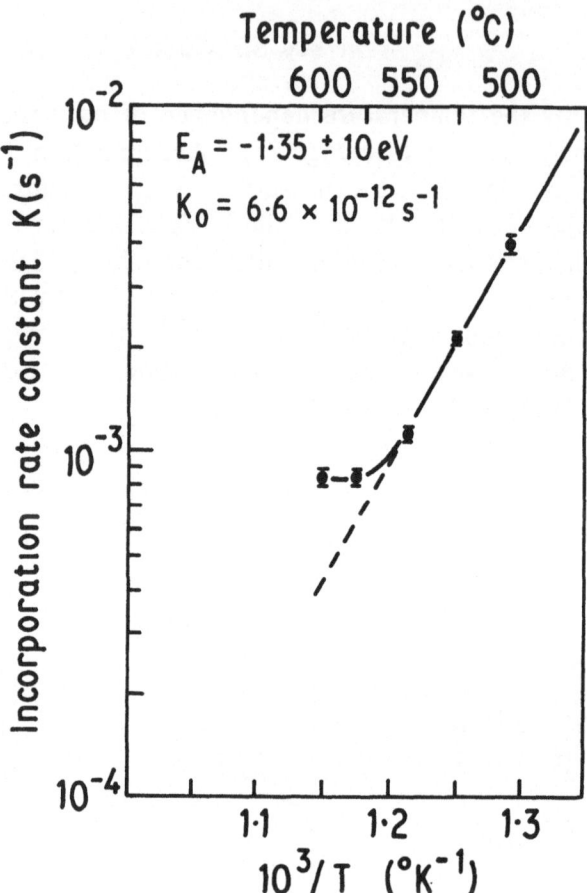

Figure 22 - Tin incorporation rate constant activation energy from slope (After Ref. 75).

Using a surface-tin conservation argument kinetically limited incorporation behavior can be described by the following relation:[75,78]

$$dC_{Sn}/dt = J_{Sn} - KC_{Sn} - DC_{Sn}$$

where C_{Sn} is the surface concentration of tin, K_{Sn} is the incorporation rate-constant, J_{Sn} is the incident Sn flux and D is the desorption rate-constant. Below $\sim 600°C$ desorption is not significant and was ignored in earlier treatments.[75] Later work[78] included the desorption term to explain the reduction in steady-state free-electron concentrations for layers grown with $T_s > 600°C$.

Predictions of the models,[75,78] later supported by experimental results, were twofold. Firstly, abrupt increases in free-electron concentration are not possible by changing J_{Sn} alone and can only be achieved by stopping growth and depositing tin atoms ("predeposition" or in a transient form by rapidly lowering T_s during growth). This is especially important when n (tin-doped) layers are grown on n-type substrates. In this case free-electron dips are formed at the interfacial region of the layer. Secondly, abrupt reductions of carrier concentrations can only be achieved by rapid increases in T_s and in association with stopping the tin flux J_{Sn}. The former process is now extensively used for FET active-layer growth and the latter for mixer-diode layers. This behavior is similar to that observed in Ga doping of MBE silicon[81] in which the desorption process is stimulated by elevating T_s in a process termed "flash off". These authors proposed a model very similar in derivation to that above.

The maximum free-electron density obtained in tin-doped MBE GaAs is approximately 3×10^{19} cm^{-3} for uniformly doped samples;[82] however, by shock-cooling, transient concentrations up to mid 10^{19} cm^{-3} have been reported.[83] Above $\sim 1 \times 10^{19}$ cm^{-3} however, the surface density of tin is such that 2nd-phase separation occurs and droplets of free tin (probably containing Ga and As in solution) are observed[82] on grown films.

3.1.5 Lead - All attempts to produce free carriers in GaAs using lead as a dopant have proven unsuccessful;[84] however, there is evidence from far infrared photoconductivity measurements from certain laboratories that Pb can be a trace impurity in MBE GaAs.

3.2 Group VI Elements

3.2.1 Oxygen - The role of oxygen is still an enigma in GaAs and all attempts to introduce oxygen as a shallow impurity by any method have been either inconclusive or unsuccessful.

3.2.2 Sulfur, selenium and tellurium - No attempts to use S or Se as elemental doping sources have been reported to the author's knowledge. Tellurium has been used[68] where the group IV elements are predominantly acceptor-like (e.g. in MBE GaSb); however, doping level control was not found possible below $\sim 10^{19}$ cm^{-3}. Arthur[7] reported a very strong Ga/Te surface interaction, compound formation and "floating" on the growing surface leaving uncontrollably high ($>10^{19}$ cm^{-3}) free-electron densities.

The "non-incorporation" behavior of Pb in GaAs was used in exchange doping from PbS, PbSe[85] and very recently PbTe[86] as sources of S, Se, and Te dopants. PbS, PbSe and PbTe evaporate as molecules with equilibrium vapor-pressures which are far lower than those of the group VI elements they contain.[87] Thus good doping concentration control below 10^{16} cm^{-3} was obtained with these molecular sources. No surface accumulation behavior was found (notably with PbTe) and doping spikes could easily be incorporated.

The maximum free-electron concentration achievable with lead selenide was $\sim 8 \times 10^{18}$ cm^{-3}, above which surface topography was degraded. Above $T_s \sim 550°C$ significant desorption of the chalcogenides occurred.[85]

Recently H_2S has been used as a gas doping source for MBE GaSb.[88] Doping efficiencies were low presumably because the H_2S was not thermally dissociated before or on impact with the substrate surface.

A real-time sulphur-beam generation source has demonstrated high controllability.[89] In this technique sulphur is electrochemically liberated from a solid silver sulfide source by an applied voltage. Again at substrate temperatures much above $\sim 550°C$ desorption became problematical.

The novel combination of a group IV (Sn) and group VI (Te) element in a molecular form as a double dopant has proved successful in recent years.[90,91] The incorporation mechanism is predominantly molecular. However, with increasing T_s, dissociation, desorption and surface accumulation of Sn and Te occur, although much attenuated with respect to tin. An elaborate model which fits the observed behavior has been developed in Ref. 91.

4. INTENTIONAL DEEP ACCEPTORS

4.1 Chromium

Apart from substrate surface segregation (and subsequent redistribution in growing films) of substrate chromium and manganese, there is only one report of intentional chromium doping of gallium arsenide by MBE.[92] The solubility as a function of temperature was correlated with intentional donor densities and the ability to create semi-insulating GaAs. The chemical solubility is low but growth of semi-insulating layers was possible. Above $\sim 10^{15}$ cm^{-3} surface topographical degradation occurs.

High resistance (undoped or Cr doped) GaAs for buffer layers has now been superceded by GaAlAs which can be used as a much better electron-confining buffer-layer for devices such as FETs,[50] and provides excellent isolation for discrete devices and integrated circuits. There is therefore little continuing interest in Cr doping of MBE compounds.

4.2 Iron Doping

There have been two reports of data, of iron doping of MBE GaAs.[93,94] the solubility of iron was found to be higher than that of chromium.[93] High resistance GaAs was not demonstrated because of unintentional (sulphur) donors which presumably accompanied the impure iron sources. Iron has been identified[33] as a hole-trap diffusing into n-type layers from semi-insulating substrates. Again we believe that the chromium sources used for semi-insulating GaAs bulk growth has traces of iron and manganese which find their way into the crystal and are subsequently distributed into epitaxial films.

5. INTENTIONAL SHALLOW ACCEPTORS

5.1 Group IIa Elements

5.1.1 Beryllium - Beryllium is a well behaved acceptor in MBE GaAs producing a shallow level $\sim$ 19 meV above the valence band. It has a unity K_i and electrical incorporation coefficient K_e, an electrically active solubility up to 1.3×10^{20} cm^{-3} before surface topographical degradation, a convenient-cell temperature/effusion-flux relationship and a high resistance to diffusion.[95,96] Be is toxic and is carcinogenic and is very efficient at gettering oxygen containing gas species, forming electrically inactive centers such as BeO in growing films. Providing adequate precautions *ex-situ* are taken and operating pressures *in situ* are sufficiently low then the lower limit of controllable free-hole densities with beryllium is ∼ 10^{15} cm^{-3}. Alexandre[97] has given evidence for increasing interstitial incorporation above ∼ 1×10^{19} cm^{-3} and McLevige[98] has shown the anomalously low diffusing properties of Be doped MBE GaAs. Be is now the most widely used controllable MBE acceptor despite the fact that commercially available Be is not better than 4.9's purity.

5.1.2 Magnesium - Magnesium was first reported[99] as an acceptor in MBE GaAs with exceedingly low incorporation coefficient K_i and then[100] with a very high K_i but with low electrically activity Ke. More recently very high K_i with high electrical activity has been reported.[101]

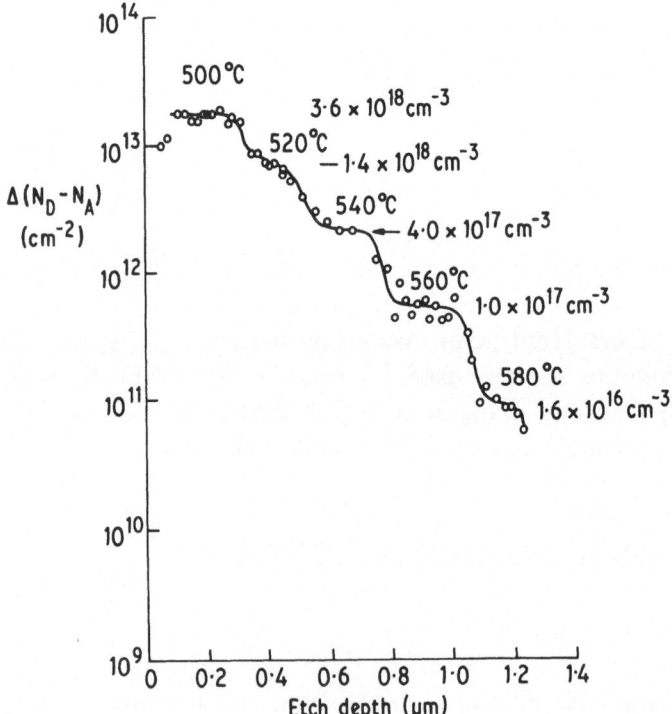

Figure 23 - Free-hole depth-profile of Mg doped GaAs layer grown at different substrate temperatures as shown (After Ref. 101).

This conflicting series was clarified by attention to the T_s dependence of K_e (see Fig. 23). It was found[102] that below $\sim 500°C$ K_i rapidly approaches unity. K_i decreases exponentially with increasing T_s above $500°C$ until at $\sim 600°C$ it is less than 10^{-3}.

This behavior precludes Mg as a viable acceptor for GaAs, in the light of the rapidly increasing deep-level incorporation behavior below $\sim 550°C$. For materials that can be successfully grown below $\sim 500°C$ however, Mg is, and should be used as a very successful non-toxic, non-carcenogenic acceptor.

The competition between incorporation in the growing crystal and desorption is successfully modeled by the same conservation relation described for tin where the desorption rate constant D is assumed exponential with temperature and K_e, the electrical incorporation rate constant is a simple function of the growth rate.

Consider an instantaneous surface concentration of Mg atoms, C_{Mg}. The rate of change in C_{Mg} will be given by three terms; namely the arrival rate (J_{Mg}), the desorption rate (DC_{Mg}) and the incorporation rate ($K\,C_{Mg}$). Thus

$$\frac{dC_{Mg}}{dt} = J_{Mg} - DC_{Mg} - KC_{Mg} \tag{1}$$

Now the desorption-rate constant can be written

$$D = D_o \exp - \frac{E_d}{kT_s} \tag{2}$$

where E_d is the activation energy for desorption.

The rate-constant of incorporation is a function of growth rate and can be written:

$$K = K' J_{Ga} \tag{3}$$

Now under equilibrium growth conditions

$$\frac{dC_{Mg}}{dt} = 0 \tag{4}$$

Substituting and solving for C_{Mg} at equilibrium,

$$C_{Mg} = J_{Mg}/(D_o \exp -\frac{E_d}{kT_s} + K'J_{Ga}) \tag{5}$$

Hence, the rate of incorporation is given by

$$K'C_{Mg}J_{Ga} = \frac{K' J_{Ga} J_{Mg}}{(K'J_{Ga} + D_o \exp -\frac{E_d}{kT_s})} \tag{6}$$

Assuming complete ionization, the resultant acceptor density is then equal to the rate of incorporation divided by the growth rate; i.e.,

$$p = \frac{K' J_{Ga} J_{Mg}}{(K'J_{Ga} + D_o \exp -\frac{E_d}{kT_s})} E \frac{J_{Ga}}{N_{Ga}} \tag{7}$$

where N_{Ga} is the Ga atom density of GaAs. Equation simplifies to

$$p = \frac{\hat{p}}{1 + \frac{D_o}{K'J_{Ga}} \exp -\frac{E_d}{kT_s}} \tag{8}$$

where $\hat{p}$ ($= N_{Ga}J_{Mg}/J_{Ga}$) is the maximum hole concentration corresponding to unity incorporation; i.e., low values of T_s and insignificant desorption.

The growth rate dependence of p is then seen from Eq. 8 to be

$$p = \frac{N_{Ga} J_{Mg}}{(J_{Ga} + \frac{D_o}{K'} \exp -\frac{E_d}{kT_s})} \tag{9}$$

Thus for low T_s, $p \propto \dfrac{1}{J_{Ga}}$ in accordance with conventional doping behavior, but for high T_s, p becomes independent of growth rate as the competition from desorption increases.

5.1.3 Calcium, strontium, and barium - These elements all have exceptionally large atomic volumes and would not be expected to be very soluble on the Ga (or arsenic) sub-lattice. Only one attempt[102] has been made (to the author's knowledge) to investigate any of these elements as alternative acceptors in MBE. Ca from a Ca_3As_2 source incident on growing MBE GaAs did not produce a measurable hole density, despite the fact that the surface structure was modified to that resembling the magnesium (2 × 2) structure, and very high J_{Ca} was used ($J_{Ca}/J_{Ga} \sim 1/10$).

5.2 Group IIb Elements

5.2.1 Zinc, cadmium and mercury - Hg is a very large atom and has a very high vapor pressure even at room temperature. It is therefore not even considered as a viable dopant. The vapor pressures of Zn and Cd are much lower that Hg but still several orders of magnitude too high at T_s values used for growth of low deep-level density GaAs, and attempts at their incorporation in GaAs have been unsuccessful.[103] Recently high incorporation efficiency of Zn in MBE InP has been reported but free hole densities were very low.

The volatilities of other possible acceptor species specifically zinc, cadmium and magnesium are very high.[28] Thus, not surprisingly, other residual acceptor species are typically not seen.

5.2.2 Manganese - Manganese has recently been found[31] to generate a Mn stabilized (4 × 2) surface-reconstruction pattern at surface coverages ~ 0.01 monolayer. Growth with [Mn] in the 10^{18} cm^{-3} range consistently produces wavey surfaces. Above $\sim 10^{19}$ cm^{-3} atomic concentrations, precipitation of Mn_2Ga as a second phase occurs.[59] By SIMS it was shown that Mn, like tin, surface accumulates, like Mg competitively desorbs and in addition forms a complex with As_4 on the surface (see Fig. 24). These problems make Mn a virtually unusable dopant for GaAs. In GaInAs, Mn has shown well behaved incorporation and electrically-active acceptor characteristics,[104] presumably because of the low T_s for growth of this alloy. However, there were apparently two different site preferences depending upon the excess As_4 flux employed. (See Fig. 25).

5.3 Ionized Impurity Atom Beam Doping

Relatively volatile species such as zinc and cadmium atoms
incident on III-V surfaces immediately evaporate at typical substrate
temperatures before they can be incorporated. This problem is also

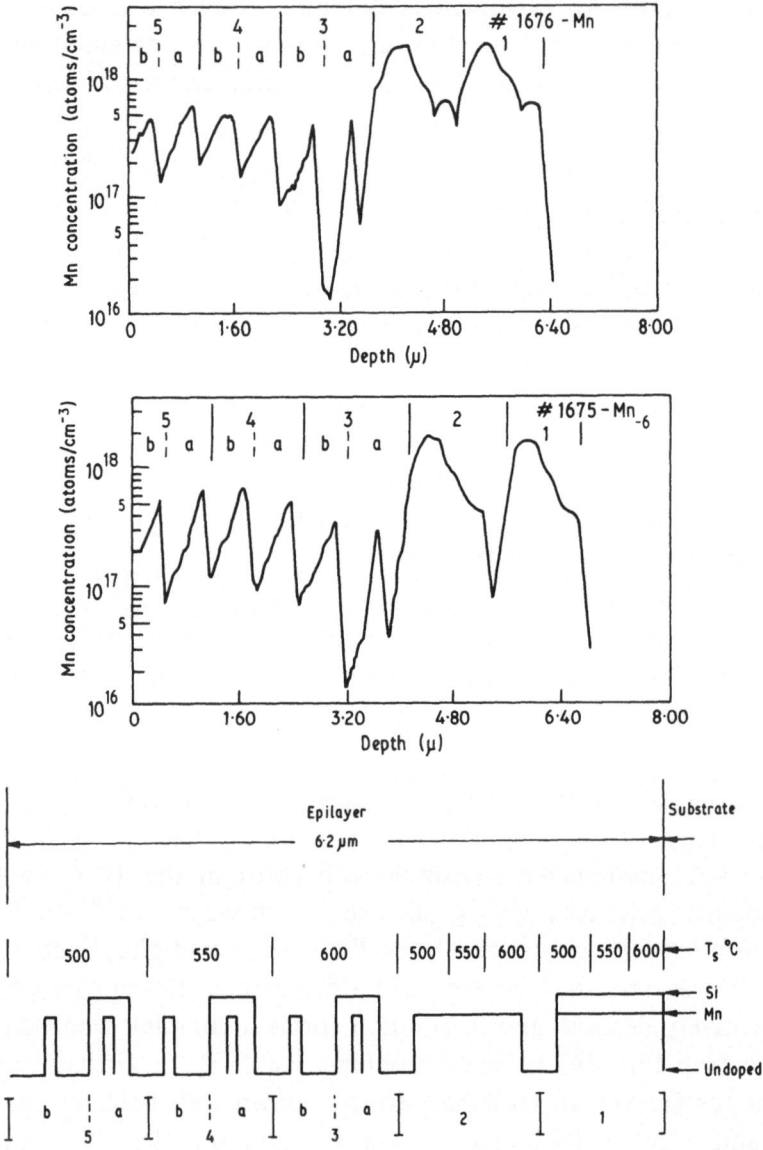

Figure 24 - SIMS profiles of two Mn doped layers grown at different
substrate temperatures under (a) high arsenic flux (b) low arsenic flux
with the Mb cell open and shuttered as shown in (c) (After Ref. 31).

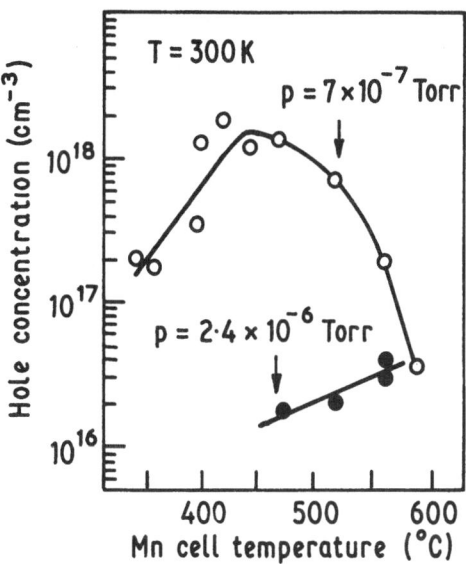

Figure 25 - Mn doping characteristics of MBE GaInAs (After Ref. 104).

experienced although to a lesser extent (as a reduction in expected hole concentrations for Mn and Mg above ~ 500°C) with the use of S, Se and Te above ~ 550°C and with tin above T_s ~ 620°C. Indeed only Si and Be do not appear to suffer from competitive desorption.

A technique, that has been demonstrated to improve the incorporation efficiency of Zn could probably help in the case of Cd and improve those of S, Se and Te, involves ionization of the dopant atom species which is then attracted to and implanted just below the growing III-V layer surface by a potential of 100-500 volts applied between the surface and the ionizer.[104-109] There has been much speculation about residual damage induced during this sub-surface implantation.

The effective electrical incorporation coefficient is at best 0.03 for Zn ions using this technique. In general mobilities are inferior for similar doping concentrations to bulk p-type GaAs samples which infers that all the damage is not annealed out or at best, much of the Zn is incorporated in interstitial sites. Photoluminescence spectra from later, more careful studies confirm this suspicion. Because of the extra complications of ionizers, the non-unity incorporation coefficient and the existence of Be as a very well behaved neutral acceptor species during MBE, no practical use is currently being made of ionized dopant implantation during MBE.

6. CONCLUSION

The characteristics displayed by the various dopant elements in the study of their incorporation behaviors are many and varied. Surface accumulation, desorption, surface association, complex formation, and auto-compensation are prevalent problems with most of the elements studied.

There is little doubt now that silicon and beryllium are the most convenient and controllable donor and acceptor species respectively for GaAs, and thus are most widely used. High cell temperatures for silicon are no longer problematical because of the substitution of boron nitride crucibles for alumina ceramic furnace components.

REFERENCES

*   Work to appear in "Technology and Physics of Molecular Beam Epitaxy", edited by E.H.C. Parker, Plenum Publishing Corporation, New York.

†   Currently affiliated with GEC Hirst Research Centre, East Lane, Wembly, Middlesex, U.K.

1. A.Y. Cho, M.B. Panish and I. Hiyashi, Inst. of Phys. Conf. Series #9, 18-29, (1971).
2. S. Gonda and Y. Matsushima, J. Appl. Phys. $\underline{47}$, 4198-4200 (1976).
3. K. Tateishi, M. Naganuma and K. Takahashi, Jap. J. Appl. Phys. $\underline{15}$, 785-789 (1976).
4. A.Y. Cho and I. Hayashi, J. Appl. Phys. $\underline{42}$, 4422-4425 (1971).
5. T.S. Low, G.E. Stillman, A.Y. Cho, H. Morkoc and A.R. Calawa, Appl. Phys. Lett. $\underline{40}$, 611-613 (1982).
6. R.J. Almassy, D.C. Reynolds, C.W. Litton, K.K. Bajaj and D.C. Look, J. Electronic Materials $\underline{7}$, 263-277 (1978).

7.   A.Y. Cho and J.R. Arthur, Prog. Solid State Chem. 10, 157-191 (1975).

8.   M. Ilegems and R. Dingle, Int. Phys. Conf. Series 24, 1-9 (1975).

9.   A.Y. Cho, J. Appl. Phys. 46, 1733 (1975).

10. H. Kunzel and K. Ploog, Appl. Phys. Lett. 37, 416-418 (1980).

11. F. Briones and D.M. Collins, J. Electronic Materials *(to be published)*.

12. C.E.C. Wood, "III-V Alloy Growth by MBE" Chapter 4 in GaInAsP Alloy Semiconductors, Ed. T. Pearsall, John Wiley & Sons, N.Y., 1982, p. 87-106.

13. P.D. Kirchner, J.M. Woodall, J.L. Freeouf and G.D. Pettit, Appl. Phys. Lett. 38, 427-429 (1981).

14. G.B. Stringfellow, R.A. Stall and W. Koschel, Appl. Phys. Lett. 38, 156-157 (1981).

15. J. Hwang and A.R. Calawa, *private communication*, 1982.

16. A. Chandra, C.E.C. Wood, D.W. Woodard and L.F. Eastman, Solid State Elec. 22, 645-650 (1979).

17. H. Brooks, "Self compensation of donors in high purity GaAs", in Advances in Electron Physics, (Ed. L. Manton, Academic, N.Y. 1958).

18. C.M. Wolfe and G.E. Stillman, Appl. Phys. Lett. 27, 564-566 (1975).

19. C.E.C. Wood, "Progress, Problems and Applications of MBE", 'Physics of Thin Films', Ed. G. Hass, M. Francombe (Academic Press, NY), Vol. II, 35-103 (1980).

20. M. Ilegems, R. Dingle and L.W. Rupp, Jr., J. Appl. Phys. 46, 3059-3065 (1975).

21. J.R. Arthur, J. Phys. Chem. Solids 28, 2257-2267 (1967).

22. J.S. Roberts and C.E.C. Wood, *unpublished* 1977 and Ref. 38.

23. A.R. Calawa, Appl. Phys. Lett. 39, 701-703 (1981).

24. M.B. Panish, J. Electrochem. Soc. 127, 2729-2733 (1980).

25. C.E.C. Wood and C.R. Stanley, Appl. Phys. Lett. *accepted*.

26. C. Stanley and C.E.C. Wood, Appl. Phys. Lett. *accepted*.

27. H. Kunzel, J. Knecht, H. Jung, W. Wunstel and K. Ploog, *submitted to Applied Physics (1982)*.

28. R.E. Honig and D.A. Kramer, RCA Review 30, 285-305 (1969).

29. A. Mircea-Roussel, G. Jacob and J.P. Hallais, "Influence of annealing on the electrical properties of semi-insulating GaAs", Proc. Conf. on Semi-insulating III-V Materials, pp. 133-141, Nottingham, U.K., Ed. G.J. Rees (Pub. Shiva, 1980).

30. S.M. Sze, Physics of Semiconductor Devices, 2nd Ed. (New York: Wiley-Interscience, 1981).

31. D. DeSimone, C.E.C. Wood and C.A. Evans, Jr., J. Appl. Phys. 53, 4938-4942 (1982).

32. R.N. Thomas, H.M. Hobgood, D.L. Barrett and G.N. Eldridge, "Large diameter, undoped semi-insulating GaAs for high mobility direct implanted FET technology", Proc. Conf. on Semi-insulating III-V Materials, Nottingham, U.K. Ed. G.J. Rees (Pub. Shiva, 1980).

33. W. Schaff and G.W. Wicks, *Private Communications* (1982).

34. T.S. Low, G.E. Stillman and C.M. Wolfe, I.O.P. Series #63, Ch. 4, p. 143-148 (1982).

35. R. Stall, "Growth, Characterization, and Applications of Gallium Arsenide and Germanium Layers Grown by Molecular Beam Epitaxy", Ph.D. Thesis, Cornell Univerysity, August 1980.

36. D.V. Lang, A.Y. Cho, A.C. Gossard, M. Ilegems and W. Wiegmann, J. Appl. Phys. 47, 2558-2564 (1976).

37. W. Schaff and C.E.C. Wood, *unpublished*.

38. J.H. Neave, P. Blood and B.A. Joyce, Appl. Phys. Lett. 36, 311-312 (1980).

39. R.A. Stall, C.E.C. Wood, P.D. Kirchner, L.F. Eastman, Electronic Lett. 16(5), 171-172 (1980).

40. D.V. Lang, R.A. Logan and L.C. Kimerling, Phys. Rev. B15, 4874-4882 (1977).

41. S.C. Palmateer, W.J. Schaff, A. Galuska, J.D. Berry and L.F. Eastman, Appl. Phys. Lett. (January 1983).

42. H.C. Casey, Jr., A.Y. Cho, and E.H. Nicollian, Appl. Phys. Lett. 32, 678-679 (1978).

43. W.T. Tsang, Appl. Phys. Lett. 33, 426-428 (1978).

44. D.V. Lang, R.A. Logan, D.M. Jaros, Phys. Rev. B 19, 1015 (1979).

45. W. Schaff and W.I. Wang, *unpublished*.

46. S.R. McAfee, W.T. Tsang and D.V. Lang, J. Appl. Phys. 52, 6165-6167 (1981).

47. G.W. Wicks, W.I. Wang, C.E.C. Wood, L.F. Eastman and L. Rathbun, J. Appl. Phys. 52, 5792-5796 (September 1981).

48. W.T. Tsang, Appl. Phys. Lett. 34, 473-475 (1979).

49. W.I. Wang, S. Judaprawira, C.E.C. Wood and L.F. Eastman, Appl. Phys. Lett. 38, 708-710 (May 1981).

50. H. Ohno, C.E.C. Wood, L. Rathbun, D.V. Morgan, G.W. Wicks and L.F. Eastman, J. Appl. Phys. 52, 4033-4037 (1981).

51.   J. Barnard, C.E.C. Wood and L.F. Eastman, Inst. of Phys. Conf. Series #63, 461-466 (1982).
52.   J.R. Arthur, *Private Communication* (1979).
53.   W.M. Theis, C.W. Litton, W.G. Spitzer and K.K. Bajaj, Appl. Phys. Lett. 41, 70-72 (19??).
54.   R.F.C. Farrow and G.M. Williams, This Solid Films 55, 303-315 (1978).
55.   C.E.C. Wood and B. Clegg, *unpublished.*
56.   D. Collins, J. Vac. Sci. Technol. 20, 250-251 (1982).
57.   D.L. Miller, S.W. Zehr and J.S. Harris, Jr., J. Appl. Phys. 53, 744-748 (1982).
58.   Y.G. Chai, R. Chow and C.E.C. Wood, Appl. Phys. Lett. 39, 800-803 (November 1981).
59.   C.B. Carter, D.M. DeSimone, T. Griem and C.E.C. Wood, "Defects in Heavily-doped MBE GaAs", presented at 40th Annual Electron Microscopy Society of America, Washington, D.C. August 9-13, 1982.
60.   M. Hollis, S.C. Palmateer, L.F. Eastman, C.E.C. Wood, P. Maki and A. Brown, "Fabrication and Performance of GaAs Planar-Doped Barrier Diodes and Transistors", Workshop on Compound Semiconductors and Materials for Microwave Active Devices, February 23, 1982.
61.   G.H. Döhler and K. Ploog, "Periodic doping structures in GaAs", Progress in Crystal Growth and Characterization, Ed. B.R. Pamplin, (Pergamon Press, Oxford), 1981.
62.   J. Ballingall and C.E.C. Wood, Appl. Phys. Lett. *accepted,* (1982).
63.   A.Y. Cho and I. Hayashi, J. Appl. Phys. 42, 4422-4425 (1971).
64.   C.E.C. Wood, J. Woodcock and J.J. Harris, Inst. of Phys. Conf. Series #45, 28-37 (1979).
65.   H. Kunzel, A. Fischer and K. Ploog, Appl. Phys. 22, 23-30 (1980).
66.   R. Heckingbottom and G.J. Davies, J. Crystal Growth 50, 644-647 (1980).
67.   R.J. Malik, *Private Communication*, (1982).
68.   G. Wicks, A. Brown, K. Hsieh and D. Welch, *Private Communication.*
69.   C.A. Chang, R. Ludeke, L.L. Chang and L. Esaki, Appl. Phys. Lett. 31, 759 (1977).
70.   C.E.C. Wood, G. Metze, J. Berry and L.F. Eastman, J. Appl. Phys. 51, 383-387 (1980).

71.   C.E.C. Wood, S. Judaprawira and L.F. Eastman, "Hyperthin Channel MBE GaAs Power FETs by Single Atomic Plane Doping", Proc. Int. Electron Device Meeting, Washington, DC, 388-389 (1979).

72.   R.J. Malik, K. Board, L.F. Eastman, C.E.C. Wood, T.R. AuCoin and R.L. Ross, Inst. of Phys. Conf. Series #56, 697-710 (1981).

73.   G.M. Metze, R.A. Stall, C.E.C. Wood and L.F. Eastman, Appl. Phys. Lett 37, 165-167 (1980).

74.   C.E.C. Wood and B.A. Joyce, J. Appl. Phys. 49, 4854-4861 (1978).

75.   C.E.C. Wood, D. DeSimone and S. Judaprawira, J. Appl. Phys. 51, 2074-2078 (1980).

76.   A.Y. Cho, J. Appl. Phys. 46, 1733-1735 (1975).

77.   K. Ploog and A. Fischer, J. Vac. Sci. Technol. 15, 255-259 (1978).

78.   F. Alexandre, C. Raisin, M.I. Abdalla, A. Brenac and J.M. Mason, J. Appl. Phys. 51, 4296-4304 (1980).

79.   R. Heckingbottom, C.J. Todd and G.J. Davies, J. Electrochem. Soc. 127, 444-450 (1980).

80.   M.B. Panish, J. Appl. Phys. 44, 2659-2666 (1973).

81.   S.S. Iyer, R.A. Metzger and F.G. Allen, J. Appl. Phys. 52, 5608-5613 (1981).

82.   R.A. Stall, C.E.C. Wood, K. Board, N. Dandekar, L.F. Eastman and J. Devlin, J. Appl. Phys. 52(6) 4062-4069 (1981).

83.   P.A. Barnes and A.Y. Cho, Appl. Phys. Lett. 33, 651-653 (1978).

84.   C.E.C. Wood, *unpublished*.

85.   C.E.C. Wood, Appl. Phys. Lett. 33, 770-772 (1978).

86.   D. Siang, Y. Makita, K. Ploog and H.J. Queisser, J. Appl. Phys. 53, 999-1006 (1982).

87.   V. Hirama, J. Chem. Eng. Data 9, 65-68 (1964).

88.   H. Gotoh, K. Sasamoto, S. Kuroda and M. Kimata, "S-doping from H_2S gas of M.B.E. GaSb", Abstract in Proc. 3rd Int. Workshop on Molecular Beam Epitaxy, Santa Barbara, CA (1981).

89.   G.J. Davies, D.A. Andrews and R. Heckingbottom, J. Appl. Phys. 52, 7214-7218 (1981).

90.   D.M. Collins, Appl. Phys. Lett. 35, 67-69 (1979).

91.   D.M. Collins, J.M. Miller, Y.C. Chai and R. Chow, J. Appl. Phys. 53, 3010-3018 (1982).

92.   H. Morkoc and A.Y. Cho, J. Appl. Phys. 50, 6413-6415 (1979).

93. D. Covington, J. Comas and P.W. Yu, Appl. Phys. Lett. 37, 1094-1096 (1980).

94. M. Nakaya, T. Shimae and A. Nara, "Fe doped - GaAs by MBE", Proc. Jap. Phys. Soc. Meeting, Tokyo, paper 1p-D-11, (1980).

95. M. Ilegems, J. Appl. Phys. 48, 1278-1287 (1977).

96. J.S. Roberts and C.E.C. Wood, *unpublished* (1976).

97. N. Duhamel, P. Henoc, J.P. Alexandre and E.V.K. Rao, Appl. Phys. Lett. 39, 49-51 (1981).

98. W.V. McLevige, K.V. Vaidyanathan, B.G. Streetman, M. Ilegems, J. Comas and L.P. Lew, Appl. Phys. Lett. 33, 127-129 (1978).

99. A.Y. Cho and M.B. Panish, J. Appl. Phys. 43, 5118-5123 (1972).

100. B.A. Joyce and C.T. Boxon, Jap. J. Appl. Phys. 16, 17-23 (1977).

101. P.D. Kirchner, J.M. Woodall, J.L. Freeouf, D.J. Wolford and G.D. Pettit, J. Vac. Sci. Technol. 19(3), 604-606 (1981).

102. C.E.C. Wood, D. DeSimone, K. Singer and G.W. Wicks, J. Appl. Phys. 53, 4230-4235 (1982).

103. J.R. Arthur, Surface Science 38, 394-412 (1973).

104. H. Asahi, Y. Kawamura, M. Ideka and H. Okamoto, Jap. J. Appl. Phys. 20, L87-L90 (1981).

105. N. Matsunaga and K. Kakahashi, "Precise doping control for molecular beam epitaxial film with ionized beam", Proc. World Electrochem. Congress, Moscow (June 1977).

106. M. Naganuma and K. Kakahashi, Appl. Phys. Lett. 27, 3420344 (1975).

107. N. Matsunaga, T. Susuki and K. Takahashi, J. Appl. Phys. 49, 5110-5715 (1978).

108. L. Esaki and J.C. McGroddy, IBM Tech. Disclosure Bull. A, 3108-3109 (1975).

109. J.C. Bean and R. Dingle, Appl. Phys. Lett. 35, 925-927 (1979).

GROWTH AND PROPERTIES OF III-V SEMICONDUCTORS BY MOLECULAR BEAM EPITAXY*

A.Y. Cho

Bell Laboratories
Murray Hill, NJ 07974

ABSTRACT: Advances in solid-state device technology in the sixties established III-V materials as a new class of semiconductors for high-speed microwave and highly efficient optical devices. Molecular beam epitaxy (MBE) is an extremely versatile thin film technique which can produce single crystal layers with atomic dimensional controls and thus permit the preparation of novel structures and devices tailored to meet specific needs. Important factors to achieve high quality MBE growth such as in-situ analysis, substrate preparations, growth conditions, and layer properties are discussed.

1. INTRODUCTION

Molecular beam epitaxy (MBE) is a versatile technique for epitaxial growth of semiconductor, metal and insulator thin films.[1-8] It distinguishes itself from previous vacuum evaporation techniques with its significantly more precise control of the beam fluxes and deposition conditions. MBE is an epitaxial growth process involving the reaction of one or more thermal beams of atoms or molecules with a crystalline surface under ultra-high vacuum conditions. The knowledge of surface physics and the observation of surface atom rearrangements resulting from the relations between the atoms arrival rate (beam flux) and the substrate temperature allow considerable understanding of how to prepare high quality thin films with compilation of atomic layer upon atomic layer.[9-12]

Advances in solid-state device technology in the sixties established III-V materials as a new class of semiconductors for high speed microwave and highly efficient optical devices. These compound semiconductors usually consist of the group III elements, Ga, Al, and In, and the group V elements, As, P. and Sb. Several compounds such as $GaAs$[9,10], GaP[13], $Al_xGa_{1-x}As$[13], $GaAs_xSb_{1-x}$[14], and $Ga_xIn_{1-x}As_yP_{1-y}$[2] were first studied. A list of the semiconductor materials which have been grown with MBE is shown in Figure 1. The potential for excellent dimensional control of MBE was first demonstrated by the growth of $GaAs/Al_xGa_{1-x}As$ periodic structures.[15] High quality microwave and optical devices requiring precise layer thickness were then fabricated.[1] In the process of evaluating device performance, it was found that the photoluminescent intensity increased more than an order of magnitude when the substrate temperature was increased from 540° to 650°C during growth.[16] Excellent results with double-heterostructure lasers[17-21] and microwave field effect transistors,[22-26] coupled with the high throughput and highly uniform growth with rotating sample holders[27-28] made MBE an important thin film technology.

2. IN SITU ANALYSIS

In the initial development of MBE, surface analysis performed during deposition played a major role in the understanding of the growth process. As the MBE technology matured with time and for a modern MBE system to be used for supplying device fabrications, only

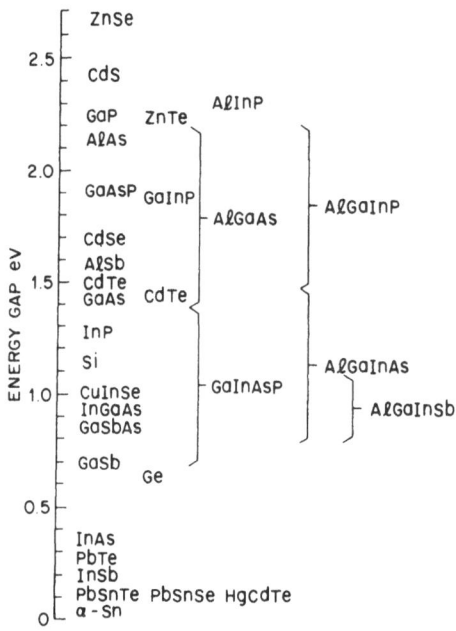

Figure 1 - A list of semiconducting materials which have been grown by MBE.

a high energy electron diffraction (HEED) apparatus and an ion gauge in the growth chamber are essential. A mass spectrometer is convenient to have for detecting a leak in the vacuum system or to measure the water vapor background in the residual gas.

2.1 Mass-Spectrometry

The mass-spectrometer was first used to study the adsorption-desorption kinetics of atoms on solid surfaces.[29-32] One can determine the mean adsorption life-time, sticking coefficient, and the activation energy of certain elements on the substrate. As to the determination of the ratio of the species desorbed from the substrate, care must be taken to interpret the results. This is because the intensity of the detected peak is a strong function of the resolution setting, ionization cross-section, ion accelerated voltage, ionization energy, and the geometry of the spectrometer. Details of the mass spectrometry study are discussed in the chapter on Kinetic Aspects and will not be repeated here.

2.2 High Energy Electron Diffraction (HEED)

It is most essential to have a HEED apparatus in the growth chamber because it gives the information about substrate cleanliness and proper growth conditions. It is used in the first five minutes of growth for every run. The HEED gun should not be left on throughout the growth because it may provide excess water vapor and other contaminants. Too high an electron beam current may even polymerize the residual hydrocarbon gases resulting in carbon contamination on the substrate.

In analyzing the HEED pattern, we can imagine a crystal made up of sets of parallel net planes in which the atoms are located. The well-known Bragg's law relates wavelength to the angle through which the electron beam is diffracted and it can be written

$$2d_h \cdot Sin\theta = \lambda$$

where d_h is the plane spacing, θ is the incident angle, and λ is the electron beam wavelength (deBroglie wavelength).

$$\lambda = \sqrt{\frac{150}{V(1+10^{-6})}} V\mathring{A}$$

An electron accelerating voltage range from 3 to 100 keV would correspond to wavelengths from 0.22Å to 0.04Å.

The net plane spacing d_h in cubic structures can be expressed in terms of Miller indices h_1, h_2, h_3 stated as

$$d_h = \sqrt{\frac{1}{(h_1/a)^2 + (h_2/a)^2 + (h_3/a)^2}}$$

The different crystallographic spacings d_{110} and d_{111} are therefore equal to $a\sqrt{2}$ and $a\sqrt{3}$ respectively, where a is the lattice constant.

The relation between the atom arrangements on the substrate and the observed diffraction pattern on the flourescent screen can best be described in Figure 2 and may be expressed as

$$d_h = \frac{2\lambda L}{D}$$

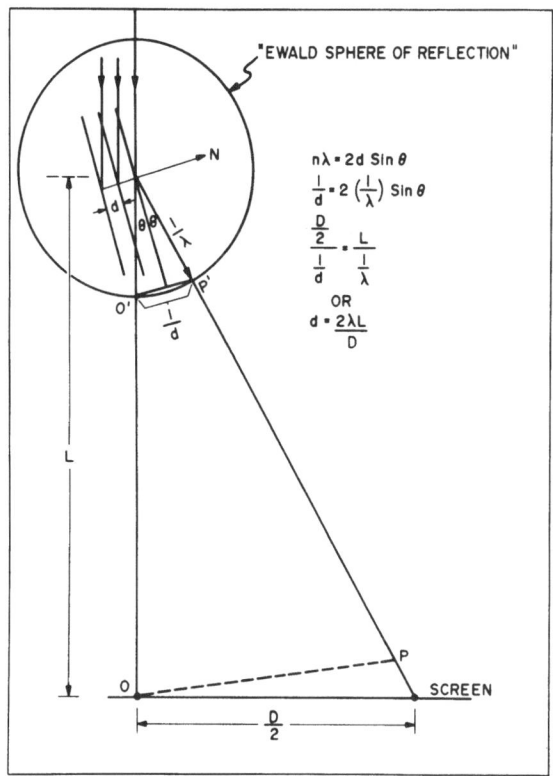

Figure 2 - Interpreting high energy electron diffraction patterns.

where L is the distance between the substrate and the flourescent screen (camera constant) and D is twice the distance between the diffraction spots measured on the flourescent screen. One can therefore deduce the atom periodicities by measuring spot spacings.

For example, the electron wavelength at 5keV is 0.173Å. For a camera constant of 12 in and (110) plane spacing for GaAs being 4Å, the diffracted patterns should have a spacing of 0.53 in on the flourescent screen.

2.3 Auger Electron Spectroscopy

An Auger electron spectrometer is useful for surface studies where the impurity concentration is more than 1 percent of the surface atoms. It is not sensitive enough to study the dopant concentration used in the semiconductor devices which is on the order of one part per million. In the initial development of MBE, Auger was used to study the procedure for obtaining an oxygen and carbon free substrate surface.[33] It may also be useful to some extent for observing the relative surface compositional changes[34,35] as shown in Figure 3.

3. SUBSTRATE PREPARATION

The success of MBE growth relies most importantly on the proper substrate preparation. This is unlike liquid phase epitaxy (LPE) or chemical vapor deposition (CVD) where in-situ melt-back or etch-back may be carried out before growth. For a simplified MBE growth process, ion sputtering cleaning of the III-V compound substrate is generally not used. An oxide protection and in-situ oxide desorption process[33] is utilized as described in the following.

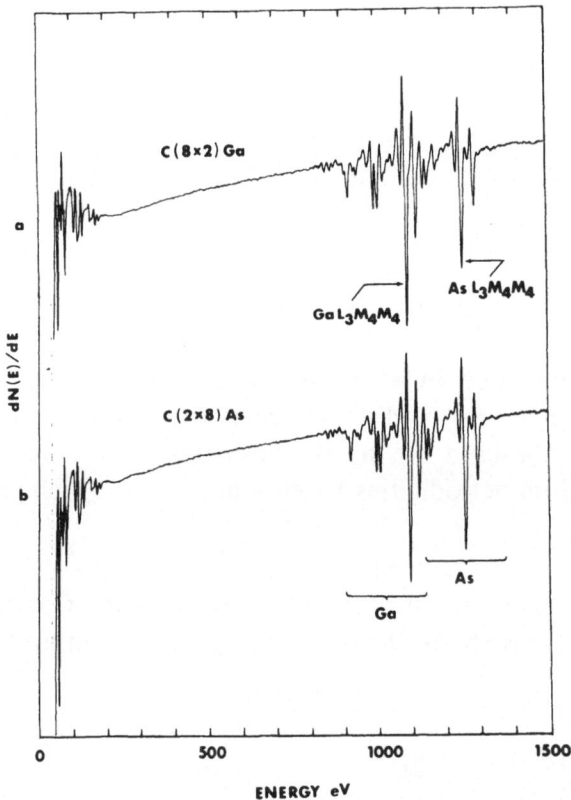

Figure 3 - Auger spectra of a C(2x8) As-stabilized, and a C(8x2) Ga-stabilized (001) GaAs surface. Notice the increase in Ga peak intensity and decrease in As peak intensity of the latter (After Ref. 34).

3.1 GaAs Substrate

The substrate is first boiled in trichloroethylene, rinsed in acetone and methanol, and then etch-polished with Br_2-methanol soaked lens paper. After rinsing in methanol and deionized water, the substrate is then etched in a stagnant solution of $H_2SO_4:H_2O_2: H_2O$ (4:1:1 at 60°C) for ten minutes, rinsed again in deionized water, and then etched in HCl for five minutes to remove any oxide and organic material on the surface. Finally, it is rinsed in deionized water for passivation[33,36] and blow dried with filtered nitrogen gas. The substrate is then mounted on a preheated (160°C) Mo sample holder with In solder and loaded into the MBE system immediately.

This substrate preparation procedure was a result of earlier Auger spectrometric studies.[33] It was found that Auger peak heights for chemically etched GaAs depend strongly on the exact etching procedures. In particular, it is possible to enhance the oxygen content and reduce the carbon content by flushing the etching solution away from the crystal with methanol and then water before the surface is exposed to air. GaAs surfaces passivated in this manner are relatively stable to laboratory air and do not adsorb carbon containing gases readily. This is important because oxygen can be removed by heating GaAs substrate in vacuum to about 580°C (measured with an infrared pyrometer with emissivity setting at 0.6) while carbon is hardly affected by heating. Figure 4 shows Auger spectra of an oxygen-passivated surface before and after heating.

3.2 InP Substrate

In many respects the preparation of InP substrate is similar to that of GaAs. Since InP is much "softer" than GaAs, a slight pressure on the substrate may generate dislocation lines. Extra care must be taken during polishing and handling. In the case of GaAs, HCl is used to remove the oxide on the surface. Since HCl will etch InP, other solutions such as KOH or dilute Br_2-methanol are used to remove the oxide on InP. The procedure[37] is to first degrease by boiling in trichloroethylene and acetone, then rinse in methanol and etch in KOH. The saw-cut damage is then removed by polishing the substrate to a mirror-like finish on lens paper soaked with 0.5% Br_2-methanol. The substrate is then etched in $H_2SO_4:H_2O_2: H_2O$ = 4:1:1 solution for ten minutes followed by a three minute etching in 0.3% Br_2-methanol.

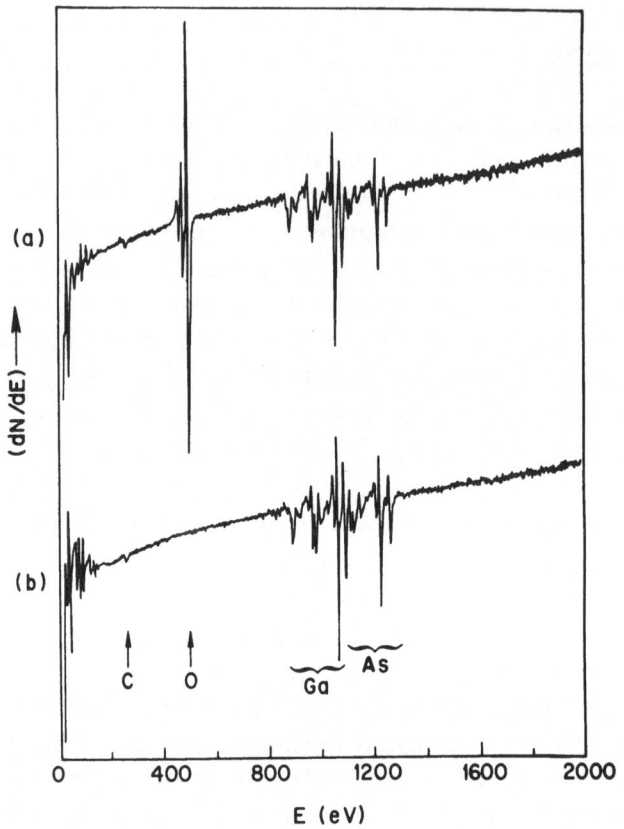

Figure 4 - Auger spectra of (001) GaAs (a) etched in Br$_2$-methanol and rinsed in deionized water; (b) after heating to 580°C (After Ref. 33).

Again the substrate is passivated in water for one minute before being blown dry with filtered nitrogen.

A substrate cleaning study was carried out in an MBE system described earlier[37] where an Auger spectrum can be taken within 10 sec after deposition. The Auger electron gun and cylindrical analyzer compartment are bellows mounted on the chamber and can be rolled in and out rapidly to provide fast data acquisition. This is a desirable feature for studying the cleanliness of the InP substrate surface which is heated in an As molecular beam because it is a rate process.

The Auger spectrum of an oxygen passivated (001) InP surface is shown in Figure 5(a). The surface shows no trace of chemical reagents but a passivated thin oxide layer. It was found that in contrast to that for GaAs substrates, the oxide on InP substrates does not evaporate even up to a temperature where the substrate decomposes. The substrate cleaning process for InP therefore involves an additional As or P beam directed toward the InP substrate during heating to prevent the decomposition of the substrate at elevated temperatures.

Figure 6 shows the variation in intensity of Auger peaks of In, O, P and As as a function of substrate temperature. The As beam used in the cleaning process gives an arrival rate of about $10^{17}As/cm^2$ sec. It is seen here that the In and P Auger peaks increase in intensity while that of O decreases with increasing substrate temperature. At about 400°C, the P peak intensity reaches a maximum and the As peak starts

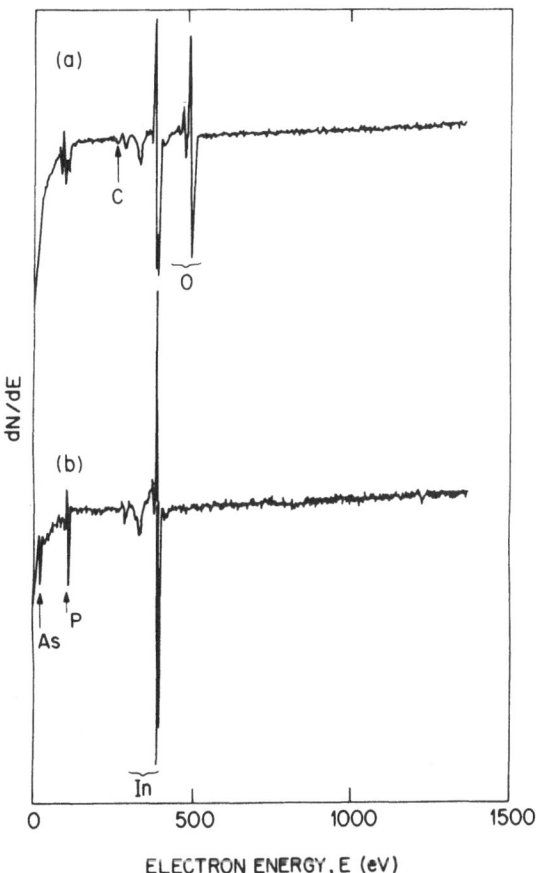

Figure 5 - Auger spectra of (001) InP (a) etched in Br$_2$-methanol and rinsed in deionized water; (b) after heating to 500°C under an arsenic molecular beam (After Ref. 37).

to appear and increase rapidly. This is because the P_2O_5 formed on the surface during the passivation process starts to decompose and desorb. Under the exposure of the As_4 molecular beam, the vacant P site is replaced by As, and therefore P peak intensity starts to decrease as the As peak intensity starts to increase. When the substrate temperature approaches 450°C, the free energy of formation of arsenic oxide is comparable to that of indium oxide and even becomes smaller at around 500°C.[38] This means that the indium oxide formed during the oxide passivation process will be transformed into arsenic oxide. Since the As_2O_3 will be desorbed at 460°C,[38] a large increase in In peak intensity above that temperature is expected as shown in Figure 6. The InP surface is completely cleaned when the substrate temperature reaches 500°C, where all the indium oxide is replaced by As_2O_3 and desorbed. The Auger spectrum of an InP surface cleaned as such is shown in Figure 5(b).

4. GROWTH OF III-V COMPOUNDS

Epitaxial growth of III-V compounds by MBE involves a series of events: (1) adsorption of the constituent atoms and molecules; (2) surface migration and dissociation of the adsorbed molecules; (3) incorporation of the atoms to the substrate resulting in nucleation and growth. The thin film grown in this manner has a crystallographic structure related to that of the substrate. An important understanding of the epitaxial growth process for GaAs was obtained by the kinetic studies of Ga and As atoms on GaAs surfaces.[31,39] Briefly, As has a very low sticking coefficient above 500°C unless it is combined with a Ga atom to form GaAs. Stoichiometric GaAs is formed as long as an excess As is supplied at the growing surface. The As which does not form a bond to Ga will simply reevaporate from the surface. However, in the case of growing ternary and quaternary compounds such as $Ga_xIn_{1-x}As$ and $Al_xIn_{1-x}As_yP_{1-y}$, precise ratios of the beam fluxes are required to grow the compounds with desired mole fractions.

4.1 Growth Apparatus

A typical MBE system for III-V compounds is shown in Figure 7. The sample exchange load-lock permits the maintenance of ultra-high vacuum while changing substrates. A liquid-nitrogen-cooled shroud is used to enclose the entire growth area in order to minimize the residual water vapor and carbon-containing gases in the vacuum chamber during

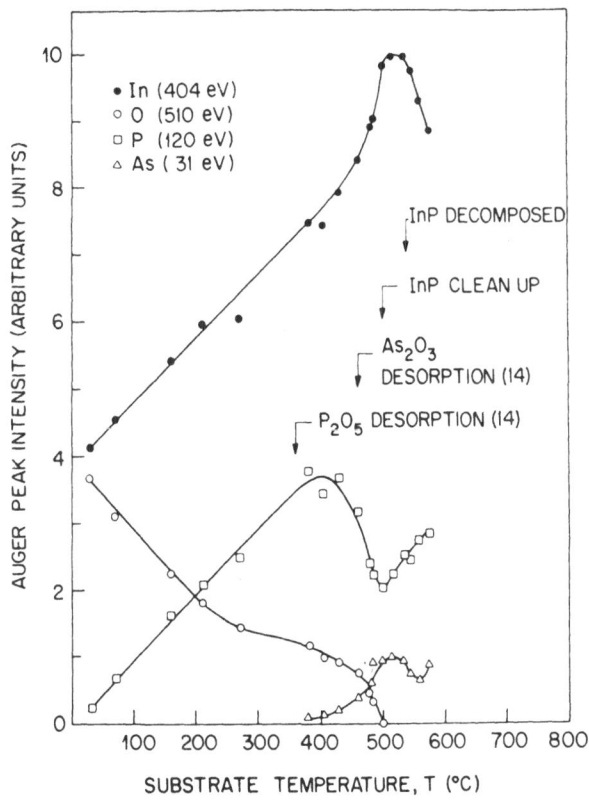

Figure 6 - Auger peak intensities of indium, oxygen, phosphorus, and arsenic on an oxidized (001) InP substrate surface under the exposure of arsenic molecular beam (10^{17} As/cm^2 sec) as a function of the substrate temperature (After Ref. 37).

epitaxy. The substrate holder can accommodate wafers 5 cm in diameter, and it can rotate continuously at a speed of from 0.1 to 5 rpm to achieve extremely uniform epitaxial layers.[40,41] The effusion cells are generally 2.5 cm in diameter and 7.5 to 12 cm in length, and they are made out of pyrolytic BN. A new approach is the introduction of gas sources such as AsH$_3$ and PH$_3$ with a high temperature dissociation baffle.[42,43] The ability to produce As$_2$ and P$_2$ beams made possible the growth of InP and Ga$_x$In$_{1-x}$As$_y$P$_{1-y}$ with MBE.

4.2 Calculation of the Arrival Rate From Vapor Pressure Data

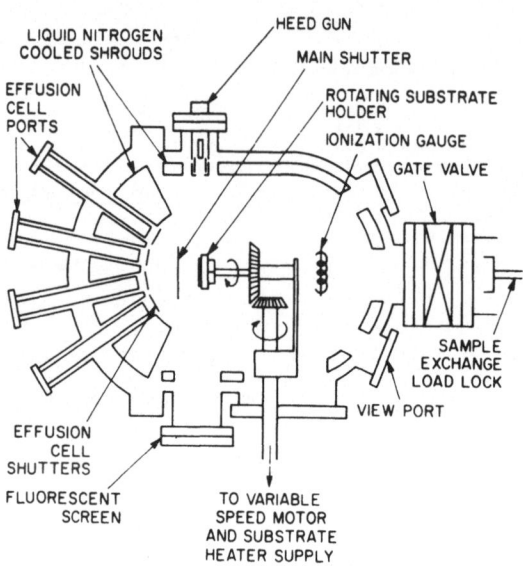

Figure 7 - Schematic of the MEB system viewed from the top. The rotating sample holder has a variable speed from 0.1 to 5 rpm (After Ref. 27).

Both the film constituent and the dopant atom arrival rate at the substrate may be calculated from the vapor pressure data. Dopants having a lower vapor pressure than the film materials generally have unity sticking coefficients. If we assume the vapor in the effusion cell is near equilibrium condition and the aperture of the cell has an area A, the total number of atoms escaping through the aperture per second is

$$\Gamma = \frac{pAN}{\sqrt{2\pi MRT}}$$

where p is the pressure in the cell, N in Avogardro's number, M is the molecular weight, R is the gas constant, and T (°K) is the temperature of the cell. If p is expressed in Torr the effusion rate is then

$$\Gamma = 3.51 \times 10^{22} \frac{pA}{\sqrt{MT}} \text{ molecules/sec}$$

If the substrate is positioned at a distance ℓ from the aperture and is directly in line with the aperture, the expression for the number of molecules per second striking the substrate of unit area is

$$G = 1.118 \times 10^{22} \ \frac{pA}{\ell^2 \sqrt{MT}} \ \text{molecules/cm}^2 \ \text{sec}$$

Taking Ga as a typical case, for T = 970°C (or 1243°K) the vapor pressure is 2.2 x 10⁻³ Torr. If we substitute M = 70, A = 5 cm² and ℓ = 12cm, the arrival rate on the substrate is 2.94 x 10¹⁵/cm² sec.

In calibrating the growth rate, an estimated layer thickness of 5 to 10μm is first grown on the substrate. The cleaved cross-section of the layer is then examined under a phase-contrast microscope after staining or etching to delineate the interface. In the case of a GaAs layer on a GaAs substrate, $HNO_3:H_2O$ = 1:3 may be used for the delineation of the interface. For $Al_xGa_{1-x}As$ layers, a solution of H_2O_2 mixed with NH_4OH giving a pH value of 7.02 may be used. The actual growth rate is then determined by the measured layer thickness divided by the growth time. The AlAs mole fraction, x, in $Al_xGa_{1-x}As$ may be determined by the relation

$$x = \frac{G(Al_xGa_{1-x}As) - G(GaAs)}{G(Al_xGa_{1-x}As)}$$

where $G(Al_xGa_{1-x}As)$ and $G(GaAs)$ are the growth rates of $Al_xGa_{1-x}As$ and GaAs, respectively. Similar procedures may be used for the $Ga_{0.47}In_{0.53}As$ and $Al_{0.48}In_{0.52}As$ systems.[37]

The commonly used n-type dopants for III-V compounds are Sn, Si and Ge.[45] The best p-type dopant is Be,[46] however, Mg,[47] Mn,[48] and Ge[49] can also be used. A universal chart for the III-V compound doping concentration as a function of dopant effusion cell temperatures when the group III element arrival rate is 3 x 10¹⁵/cm² sec, is shown in Figure 8.

4.3 Growth Conditions

For III-V compounds, the passivated oxide layer serves as a protection for the freshly chemical-etched substrate from atmospheric contamination before epitaxial growth. After the MBE system is pumped down, the liquid nitrogen shroud cooled, and the effusion cells brought up to the desired temperatures, one begins to heat the sub-

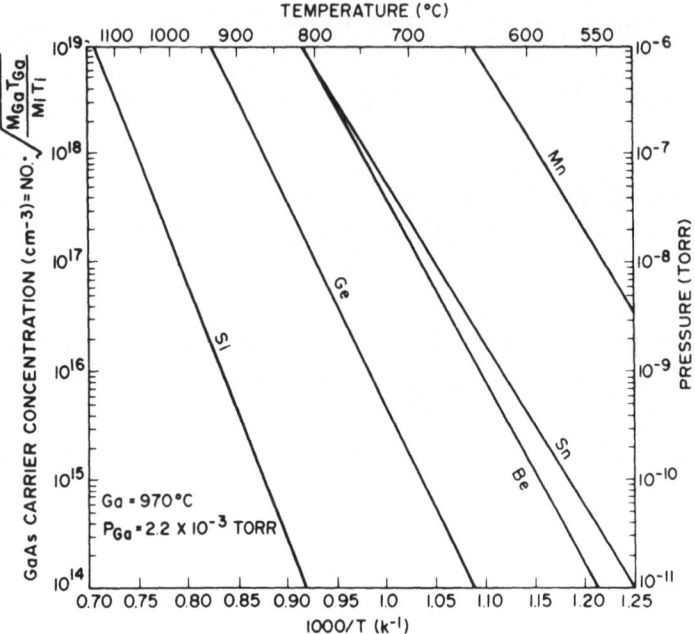

Figure 8 - III-V compound doping concentration as a function of dopant effusion cell temperatures.

strate. In the case of GaAs, the oxide on the substrate desorbed between 580° - 600°C, and for InP, the oxide desorbed at about 520°C.[37] At this point, the substrate is nearly atomically clean and ready for epitaxial growth. Assuming the substrate is properly prepared and atomically clean, the epitaxial layer will be mirror shiny if the group V to group III ratio in the molecular beam is above a certain value, giving an As-stabilized surface structure.[11] This value is also a function of the substrate temperature. An approximate relationship which is sometimes referred to as an "MBE phase diagram" is shown in Figure 9. In commercial MBE systems, a GaAs growth rate of 10μm/hr. may be achieved with a substrate temperature at 620°C.

The construction of this phase diagram is made possible from the knowledge gained about surface atom structures by the use of high energy reflection electron diffraction (HEED).[10,11] In the case of GaAs in the <100> and <111> directions, the crystal is formed with alternate layers of Ga and As atoms. The terms Ga-rich and As-rich surface structures are used to describe the growth conditions where the top layer is terminated with Ga or As respectively.[10,11] On the (100)

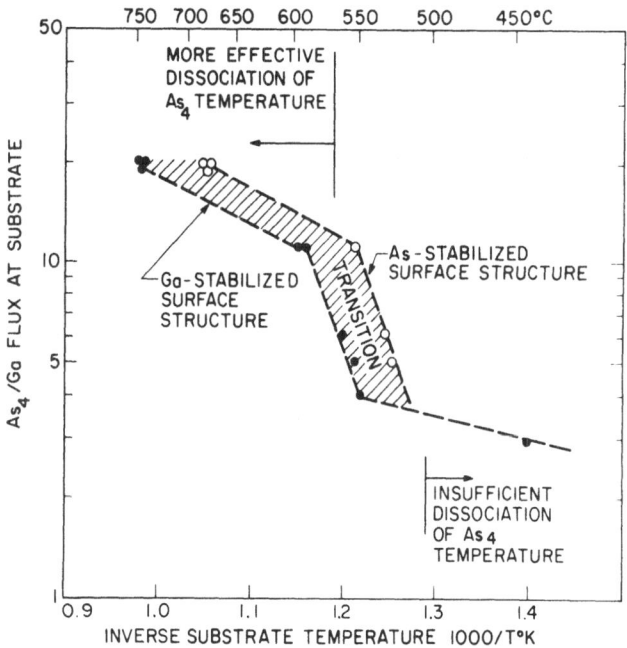

Figure 9 - As$_4$/Ga molecular beam flux ratio as a function of substrate temperature when the transition between As-stabilized and Ga-stabilized structures occurs on the (001) GaAs surface. The beam flux was measured by an ion gauge at the substrate position with Ga flux equal to 8 x 10^{-7} Torr giving a growth rate of about 1μm/hr.

surface, the Ga-stabilized surface structure is C(8x2) and the As-stabilized surface structure is C(2x8). These results were later confirmed with excellent mass spectrometry[12,44] and Auger[34,35] studies. There are many more surface structures reported on (100)[34] and (111)[10] surfaces. The ratio of As/Ga in the molecular beam to form an As-stabilized surface on a (100) surface is different from that on a (111) surface for a given substrate temperature; the latter requires a higher As/Ga ratio.

High energy electron diffraction (HEED) is sensitive to surface morphology with characteristic diffraction features indicating the presence of microscopic roughness, twinning, or oriented inclined facets on an otherwise flat surface. Information of this nature is extremely useful in studies of the early stages of crystal growth. Figure 10 shows HEED patterns (electron beam along the (110) azimuth) from a (001) GaAs

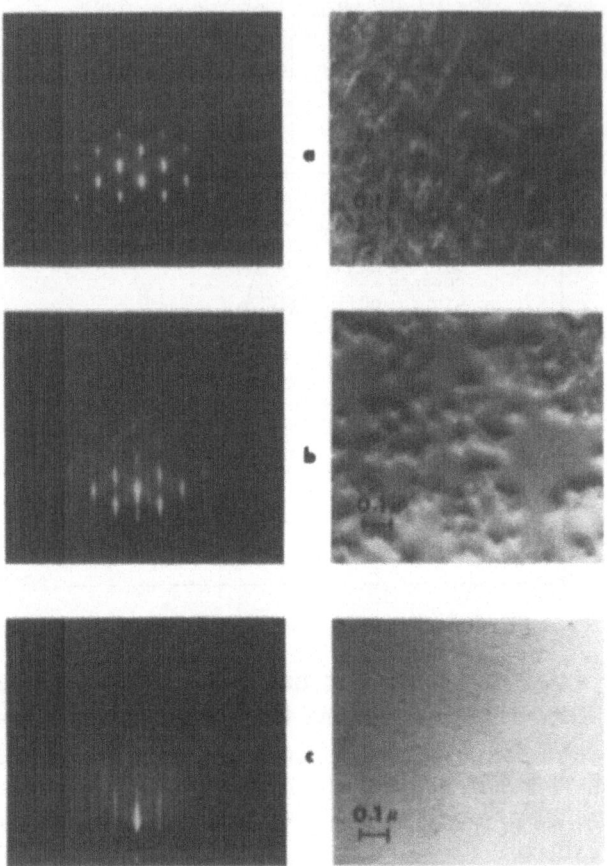

Figure 10 - HEED pattern (40 keV, $\overline{1}10$ azimuth) and the corresponding electron micrographs (38,400x) of Pt-c replica of the
same surface.   (a) Br_2-methanol polish-etched GaAs heated in
vacuum to 580°C; (b) 150Å layer of GaAs deposited on surface
of (a); (c) 1μm GaAs deposited on surface of (a) (After Ref.
50).

surface and corresponding electron micrographs of pt-shadowed carbon
replicas of the surface.[50] It can be seen that the initial thermal cleaning
of a chemically etched surface is rough, which produces a diffraction
pattern of spots; the electron beam can penetrate the surface asperities
and produce a transmission diffraction pattern. After the deposition of
150Å of GaAs by MBE, the surface has become flatter and the electron
diffraction has become elongated in the direction normal to the surface.
At the same time, additional diffraction features have appeared halfway

between the original columns of diffraction spots. Further deposition produces a surface appearing completely flat and featureless, while the diffraction pattern is uniformly streaked normal to the crystal surface. This is due to a relaxation of the Laue condition in the direction along the surface normal because the high attenuation of the electron beam into the crystal means that the diffracted beams are sampling effectively only a two-dimensional crystal. Streaking is also enhanced by refraction of the electron beam as it crosses the interface at low angles. Both of these effects depend on the angle at which the electron beam strikes the crystal, and thus are a measure of the microscopic roughness when the macroscopic angle of incidence is fixed at $\approx 1°$ as in Figure 7. If additional surfaces inclined to the macroscopic surface were to form by thermal facetting or three-dimensional nucleation, diffraction streaking perpendicular to these surfaces would indicate their presence.[10]

Higher growth temperature resulted in higher quality epitaxial layers and it was related to the efficiency of photoluminescence in earlier studies.[16] It has also been experienced that this luminescence result is directly related to the performance of double-heterostructure (DH) lasers.[16] In Figure 11, room temperature photoluminescence

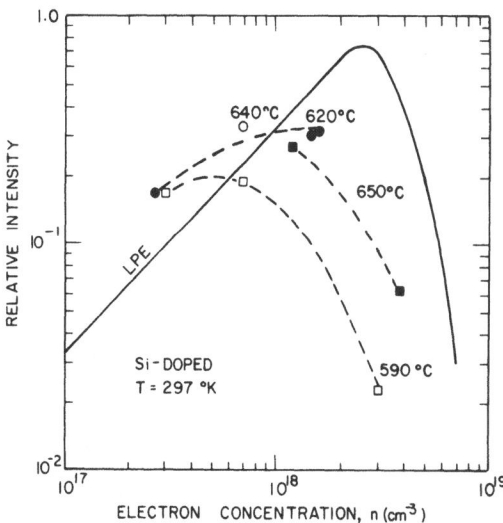

Figure 11 - Relative photoluminescent (PL) intensity. The LPE sample is Sn-doped. The numbers on the figure were the substrate temperatures during MBE growth of the Si-dope GaAs layer (After Ref. 16).

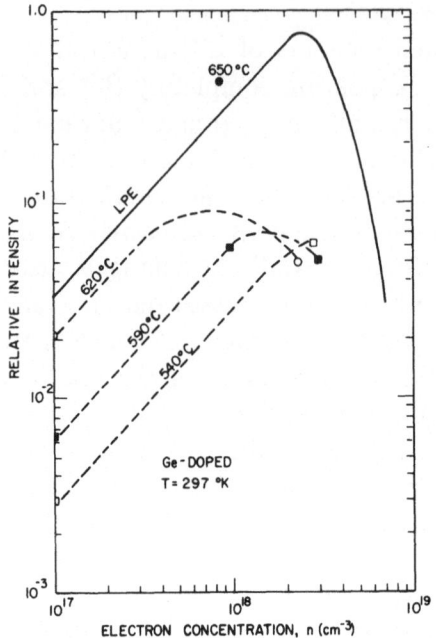

Figure 12 - Relative peak intensity. The LPE sample is Sn-doped. The numbers on the figure were the substrate temperatures during growth of the Ge-doped GaAs MBE layers (After Ref. 16).

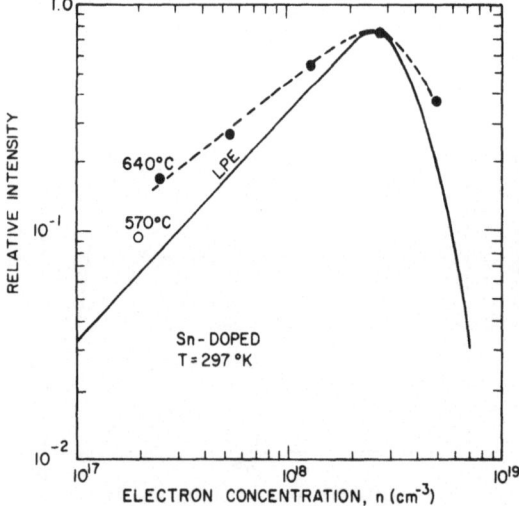

Figure 13 - Relative PL peak intensity. The LPE sample is Sn-doped. The number on the figure were the substrate temperatures during MBE growth of the Sn-doped GaAs layers (After Ref. 16).

intensities of Si-doped GaAs grown by MBE at various temperatures are compared with those of Sn-doped GaAs grown by liquid phase epitaxy (LPE). The carrier concentrations were determined by Hall measurements. Results from Ge- and Sn-doped MBE GaAs are shown in Figures 12 and 13. It is seen here that a higher substrate temperature during growth improves the photoluminescence intensities particularly for layers with carrier concentrations below $10^{18}/cm^3$.[16] Similar results were reported for optical pumping of GaAs-Al_xGa_{1-x}As double heterostructures.[51]

The upper limit of the growth temperature is controlled by the availability of group V over pressure, or the group V arrival rate that prevents the noncongruent evaporation of the compound. A higher growth temperature therefore requires a larger As consumption in the case of growing GaAs or Al_xGa_{1-x}As. Furthermore, at above 640°C, the Ga adsorption lifetime becomes sufficiently short to affect the growth rate.[52] The growth rates of GaAs, AlAs, and the GaAs fraction in Al_xGa_{1-x}As as a function of the substrate temperature are shown in Figure 14. Below 640°C the growth rates are nearly independent of the substrate temperature, implying that the sticking coefficient of Ga is nearly unity. This is in agreement with the earlier mass-spectrometry studies.[31] It is interesting to note that only a small percent of Al in the beam can significantly increase the Ga sticking coefficient at high temperatures.[52]

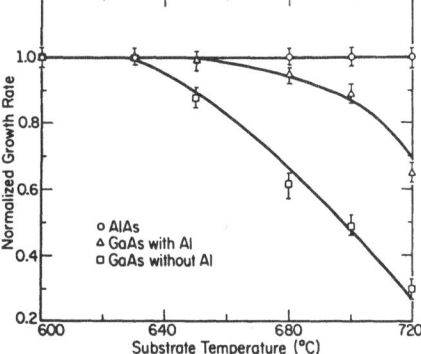

Figure 14 - Growth rates normalized to low temperature ($\leq$ 620°C) values as a function of substrate temperature. The curve labeled "GaAs with Al" represents the Ga fraction of the growth rate (After Ref. 52).

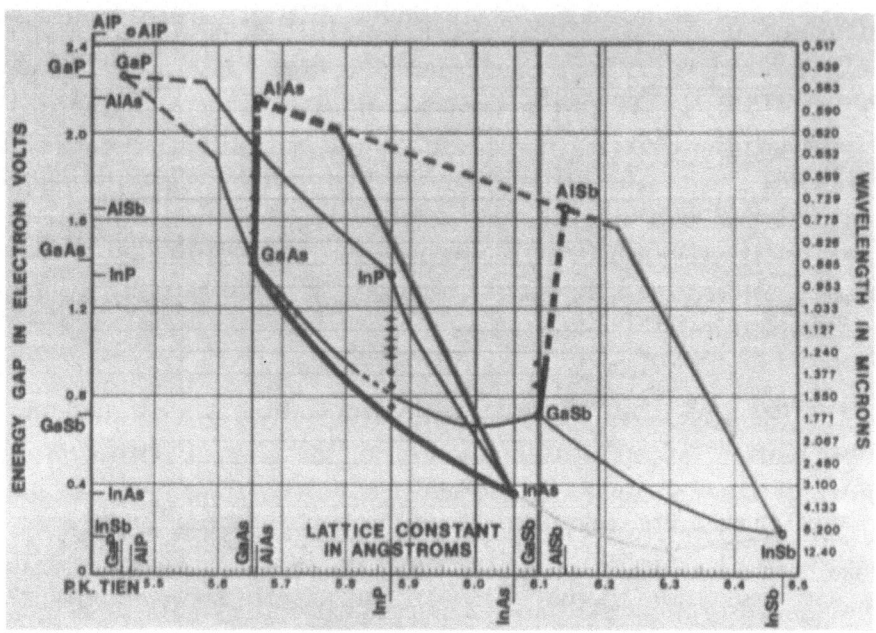

Figure 15 - Energy gap and lattice constant for several III-V compounds. The boundaries joining the binary compounds give the ternary energy gap and lattice constant.

4.4 III-V Lattice Matching Systems

Many microwave and optoelectronic devices require specific carrier transport properties, energy gap and refractive index discontinuities at the heterojunctions that provide carrier and optical confinement. For high quality layers that can be used for device fabrications, the heterojunctions must be formed by materials that are lattice matched. Many III-V systems fulfill these requirements and some examples are illustrated in Figure 15.[53] This diagram shows the variation of energy gap as a function of lattice constant for several ternary and quaternary solid solution systems. The lines joining the binary compounds give energy gaps and lattice constants of the ternary compounds with various mole fractions.

One of the most recently studied ternaries which requires precise control in mole fractions in order to lattice match to the binary substrate is $In_xGa_{1-x}As$ on InP.[28,37,54-58] In contrast to the growth of $Al_xGa_{1-x}As$ on GaAs substrates where the ternary is closely lattice-

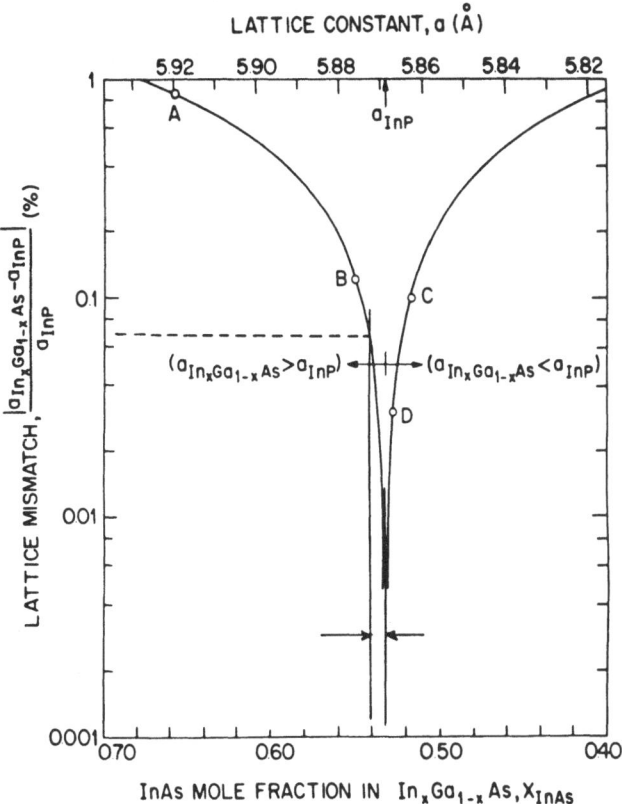

Figure 16 - Lattice mismatch between the $In_xGa_{1-x}As$ epitaxial layer and the InP substrate as a function of the InAs mole fraction in the ternary epitaxial layer. Lattice mismatch is expressed in absolute value to accommodate both positive and negative values on the same coordinate. Notice that a variation of 1% in InAs mole fraction from x = 0.53 will cause a lattice mismatch of 7 x 10^{-4} (After Ref. 37).

matched to the binary substrate for all values of x, the growth of $In_xGa_{1-x}As$ on InP is lattice matched only when x = 0.53. Figure 16 shows the lattice mismatch between the $In_xGa_{1-x}As$ epitaxial layer and the InP substrate as a function of the layer composition x_{InAs} as well as its lattice constant.[37] The lattice mismatch was calculated as the ratio of the difference in lattice constant between the ternary layer and the InP substrate, i.e., $A_{In_xGa_{1-x}As} - A_{InP}/A_{InP}$ or $\frac{\Delta A}{A}$. It is shown here that a variation of 1% in InAs mole fraction from x = 0.53 will cause a lattice mismatch of $\frac{\Delta A}{A} \approx 7$ x 10^{-4}. For any layers intended for device fabrication, $\frac{\Delta A}{A}$ should be less than 10^{-3}. Therefore, for the MBE growth of this ternary systems, one has to control the mole fraction to better than 1%. Results on high electron mobilities in modulation doped $Ga_{0.47}In_{0.53}As$ / $Al_{0.48}In_{0.52}As$ heterojunctions,[55,59] cw operation of $Ga_{0.47}In_{0.53}As$ / InP buried heterostructure lasers[57] and high transconductance FET's[25,26] have all been reported.

There is a unique quaternary alloy which is most suited for MBE growth and will play an important role in the preparation of optoelectronic devices. This alloy is $(Al_xGa_{0.47-x})In_{0.53}As$. It is possible to choose different values of x to cover the entire wavelength region between $\lambda \approx 1.65\mu m$ and $\lambda \approx 0.8\mu m$ and at the same time be lattice matched to the InP substrate.[60-62] Figure 17 shows the variation of energy gap as a function of the AlAs mole fraction in this quaternary. The reason that this alloy system is suited for MBE growth is that it contains three group III elements, Ga, Al and In, whose sticking coefficients are nearly unity at the epitaxial temperature, and thus the mole fractions can be controlled precisely for lattice matching.

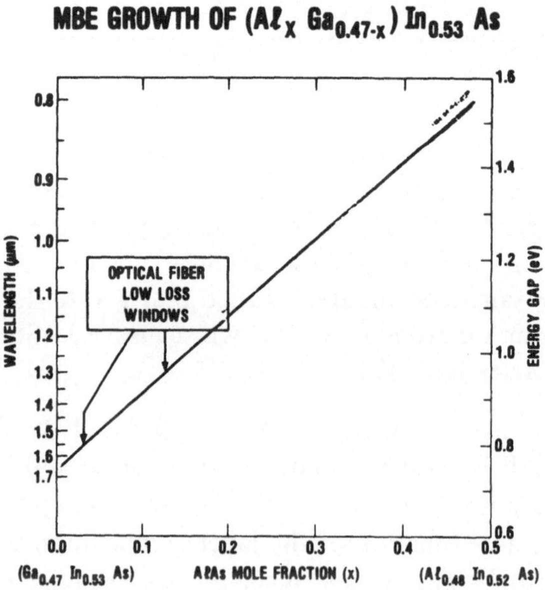

Figure 17 - Energy gap as a function of the AlAs mole fraction in $(Al_xGa_{0.47})In_{0.53}As$.

Another well studied quaternary system is $Ga_xIn_{1-x}As_yP_{1-y}$.[2,63,64] Major effort is focused on the control of the high vapor pressure group V elements. Unlike the growth of GaAs or $Al_xGa_{1-x}As$ where stoichiometry can be achieved by an excess supply of As_4, since the sticking coefficient of As_4 depends upon the Ga population, the amount in excess of As_4 is not critical. In the case of $Ga_xIn_{1-x}As_yP_{1-y}$ precise ratios of these elements are required. The arsenic and phosphorus beams have to be dissociated into As_2 and P_2, respectively, by a high temperature zone in order to increase their sticking coefficients and the control of the As to P ratio in the alloy.[64] It is equally important to design the MBE system for such growth so that there is no memory effect to facilitate rapid switching from one composition to another. There are many other III-V lattice matched systems and an extensive summary is given in Ref. 65.

5. GROWTH WITH MASK

Patterned structures may be grown with masks by MBE. One approach is to evaporate an oxide mask directly onto the substrate. This could be SiO_2, anodic oxide, or any amorphous layer such as Si_3N_4 which can withstand the growth temperature of 600°C. The other approach utilizes the fact that the atoms and molecules in the beam have sufficiently long mean free path so that they do not suffer collisions between the effusion cell and substrate. This unidirectional beam permits the growth of three-dimensional structures with shadow masks, which may again be divided into two categories: one is called the self-aligned channel mask where preferentially etched channels are formed on the substrate to serve as the shadow mask, and the other is the removable mechanical mask which may be made of W, Ta, Mo, and Si.

5.1 Oxide Mask

For the oxide mask, a thin SiO_2 film (100-1000Å), for instance, is first grown on the substrate and windows are opened by the conventional photolithographic technique. These window areas define the locations of the active devices. When GaAs is deposited onto this wafer, single crystal GaAs is formed in the window and polycrystalline GaAs is formed on the SiO_2 covered areas. The polycrystalline GaAs is semi-insulating even if the single crystal is doped to have a carrier concentration of $10^{18}/cm^3$.[66] Figure 18 shows photomicrographs of the substrate before and after the GaAs deposition as viewed with a Nomarski phase contrast microscope. Figure 18(a) shows a GaAs substrate covered with SiO_2 where windows have been opened to expose the GaAs substrate for growth of the epitaxial single crystal layers where active devices are to be located. Figure 18(b) resulted after depositing $6\mu m$ on GaAs. It is clear from the micrograph that the growth in the window is single crystal and that on the SiO_2 is polycrystalline with an extremely fine grained structure. A (110)-cleaved cross-section reveals the overgrowth layer in the window area with featureless interface and that on the SiO_2 covered area as shown in Figure 18(c). It is also shown here that the upper surface grown in the window is approximately level with the surrounding area covered with polycrystalline GaAs; the two levels differ only by the thickness of the oxide used as the mask. It is therefore called MBE planar technology. Fine stripe geometries[2] and device fabrications[67,68] with this oxide masking have also been demonstrated.

5.2 Self-Aligned Shadow Mask

An etch-and-fill technique was used to study the growth characteristics of single and multilayer structures of $GaAs-Al_xGa_{1-x}As$ over preferentially etched channels aligned along the $(\overline{1}10)$ and (110) directions on a (001) GaAs substrate.[69] These etched channels were formed by using a preferential chemical etchant such as $H_2SO_4:H_2O_2:H_2O$ (1:8:10, 24°C) and employing Shipley AZ-1350 J as an etching mask. Two different channel profiles were obtained: one under-cut and one over-cut. After removal of the photoresist and substrate cleaning, the substrate was mounted in the MBE system for epitaxial growth. The cross sectional views of the channel and the overgrown layer in the $(\overline{1}10)$ and $(1\overline{1}0)$ channels are shown in Figure 19(a) and (b), respectively.

5.3 Mechanical Shadow Mask

Mechanical masks made of refractory materials have been used for MBE to grow optical waveguides.[70] More recently, thin (001) Si

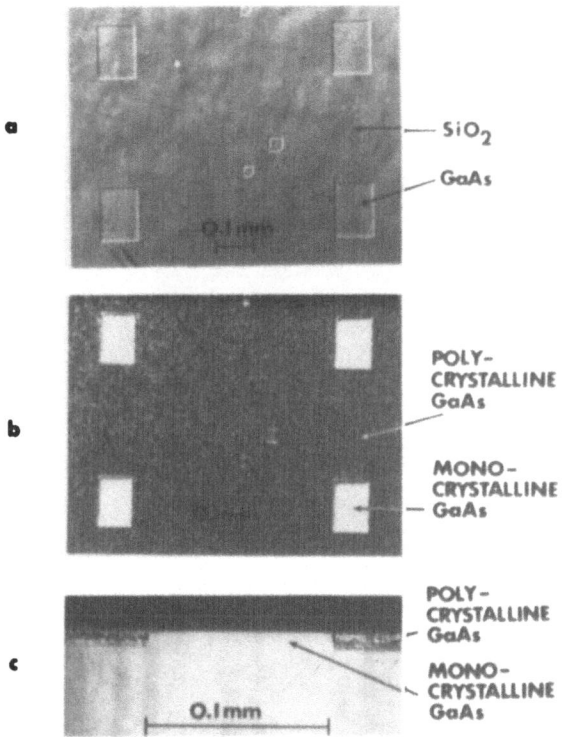

Figure 18 - Photomicrographs of (a) SiO_2 patterns on a GaAs substrate; (b) after depositing 6μm of GaAs by MBE; (c) an enlarged view of the (110)-cleaved cross section of the wafer, showing single crystal featureless growth interface between the epitaxial layer and the substrate, and polycrystalline growth on SiO_2 mask (After Ref 66).

wafers (50μm thick) were used to make masks for MBE growth.[71] Many anisotropic etchants can be used for opening the windows in the Si mask.[72-74] One of them in KOH:H_2O:isopropyl alcohol = 50 g:100 ml:100 ml, at 80°C.[72] A problem with the use of Si masks is the dimensional control of the window openings over a large mask because the dimension of the hole opening is a strong function of the mask thickness. However, Si masks are noncontaminating and mechanically strong. Besides static masking, molecular beam writing[75] of patterned structures may be accomplished by a relative motion between the mask and the substrate during deposition as shown in Figure 20. This technique makes possible the variation of chemical composition in the

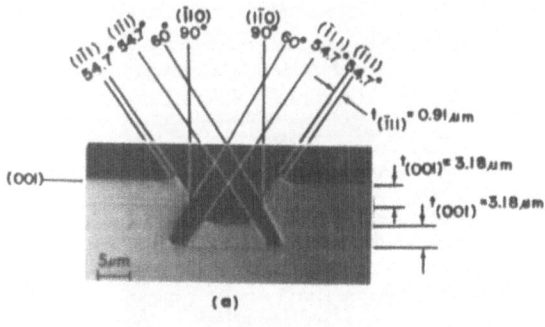

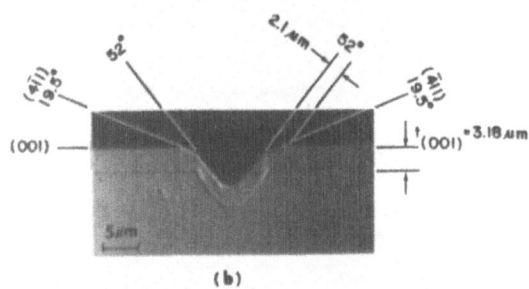

Figure 19 - (a) Shows the cross-sectional view of MBE layer (3.18μm thick) over a channel aligned the ($\bar{1}$10) direction on a (001) GaAs substrate; (b) shows a similar photomicrograph for a channel aligned along the ($1\bar{1}$0) direction (After Ref. 69).

epitaxial layer structure in both the lateral and vertical directions. MBE mask-growth is not only useful for applications to integrated optics but also to many novel microwave devices and as contacts to embedded layers.

6. TRANSPORT AND OPTICAL PROPERTIES

The ability to grow high quality epitaxial layers requires meticulous care in substrate preparation, precise substrate heating schedules, and deposition rates. Most importantly, also, the MBE apparatus has to be leak-free and all the heated parts (substrate holder and effusion cells) have to be made of extremely high quality refractory materials.

6.1 GaAs

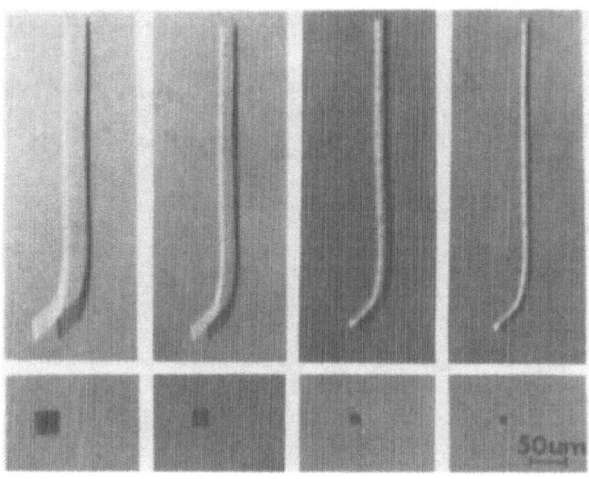

Figure 20 - GaAs epitaxial beam writing through a series of square mask openings with different sizes. The smallest width is $10\mu m$ (After Ref. 75).

The unintentionally doped GaAs is generally p-type having net hole concentrations at 300 and 78°K of 7.8 x 10^{13} cm^{-3} and 2 x 10^{13} cm^{-3}, respectively. The first MBE GaAs layer having an electron mobility over 100,000 cm^2/Vsec at 78°K is shown in Figure 21;[76] the epitaxial layer was doped with Sn to a carrier concentration of 4 x 10^{14} cm^{-3}.[77] More recently, high purity GaAs layers grown with AsH$_3$ source gave electron mobility as high as 133,000 cm^2/Vsec at 55°K with an unintentionally doped electron concentration of 2 x 10^{14} cm^{-3}.[77] Most recently, high purity GaAs layers grown with solid As sources in a commercial MBE system resulted in semi-insulating layers when they were not doped. The layer doped with Si to a concentration 10^{14} cm^{-3} gave a Hall mobility of 140,000 cm^2/Vsec at 55°K.[78]

For a more complete characterization, photoluminescence studies were carried out using a krypton-ion laser-pumped dye laser with the samples held at cryogenic temperatures.[79] The results are displayed in Figure 22. Considering first the low-excitation nonresonant spectrum (b), all sharp features seen in the best-purity vapor phase epitaxy or liquid phase epitaxy samples are observed.[80] A well separated free-exciton (polariton) luminescence (x), a sharp line at 1.514 eV due to excitons bound to neutral donors (D°-x), a band at 1.5132 eV originat-

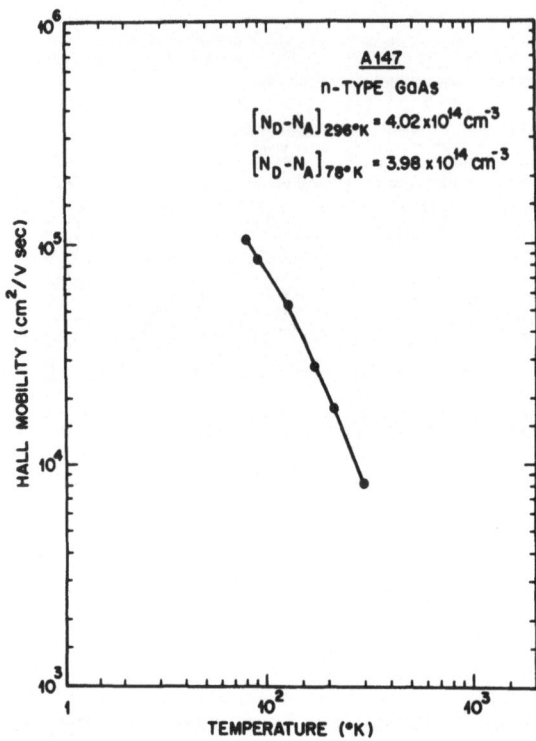

Figure 21 - Hall mobility of GaAs epitaxial layer as a function of temperature ($\mu = 105,000$ cm^2/Vsec at 78°k) (After Ref. 76).

ing from the recombination of holes on neutral donors (D°-h), a doublet structure at 1.5123 eV due to the recombination on neutral donors (D°-h), and a doublet structure at 1.5124 eV due to the recombination of neutral acceptor-bound excitons are illustrated. Similar spectra were observed in MBE samples measured in other laboratories.[81-83] The spectrum observed under higher pumping intensity (Figure 22a) shows the usual high-intensity behavior also similar to that observed in other crystal growth materials.[84] A new feature, apparently quite common in MBE material, is the line 1.5112 eV. At lower energies, the main feature is a broad band at ≈ 1.492 eV due to the recombination of donor-acceptor pairs (D°-A). The acceptor is identified as carbon which is the principal impurity. For the purpose of comparison, the spectrum of an undoped p-type sample ($\mu_{300} = 450$ cm^2/Vsec, $\mu_{77K} = 8440$ cm^2/Vsec) is shown in Figure 22(d). Other spectroscopy studies of these high purity GaAs materials have also been reported.[85]

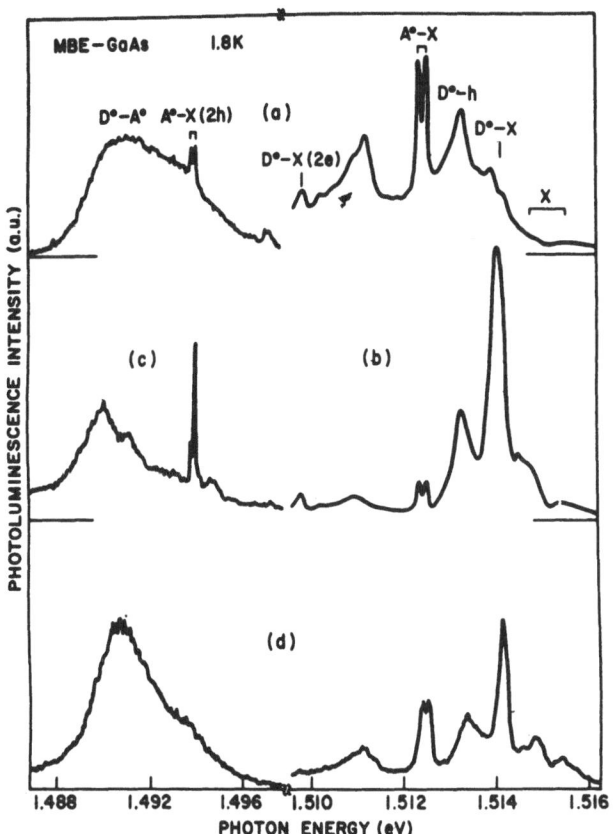

Figure 22 - Low temperature photoluminescence spectra of MBE-grown GaAs under dye laser excitations. (a) n-type sample under focused ($\approx$ 5W/cm^2) nonresonant excitation at 1.63 eV. The various luminescence lines are labeled by their mechanism (see text); (b) n-type sample under unfocused (10 mW/cm^2) nonresonant excitation at 1.63 eV (Ref. 79). (Only the high energy part of the spectrum is shown.); (c) Two-hole replica spectrum under unfocused (10 mW/cm^2) excitation resonant with the upper component of A$^\circ$-X$^\circ$; (d) p-type sample under unfocused ($\approx$ 10 mW/cm^2) nonresonant excitation at 1.63 eV.

6.2 $Al_xGa_{1-x}As$

The transport properties of $Al_xGa_{1-x}As$ for layers doped with Sn[86] and Si[87] were studied. It was found that the mobility has a strong dependence on the substrate temperature during growth. The lower substrate temperatures resulted in lower electron mobility.[86] It was also found that the electron concentration decreases with increasing the

AlAs mole fraction for the same Sn dopant flux. Donor levels increased from less than 3 to 40 meV as the value of the AlAs mole fraction increased from 0.17 to 0.375. The reduction of the free carrier concentration in $Al_xGa_{1-x}As$ as a function of x can be explained by the increase in the donor activation energy.[87] $Al_{0.22}Ga_{0.78}As$ layers having a net electron concentration of about 1.7 x 10^{18} cm^{-3} and grown at 630°C exhibited a mobility of 868 and 1095 cm^2/Vsec at 300 and 78°K, respectively.[86] Further increase in growth temperature or growth under a condition where the As/Ga ratio in the molecular beam gives a near Ga-stabilized structure may further improve the mobility.[86] This is because, as reported earlier,[88] the growth near the Ga-stabilized structure resulted in less trap concentration when studied with deep-level transient capacitance spectroscopy (DLTS).

Similar to the case of $GaAs^{16}$, the photoluminescence (PL) intensity of $Al_xGa_{1-x}As$ increases as a function of the epitaxial growth temperature.[89] The PL intensity of Si-doped $Al_{0.28}Ga_{0.72}As$ increases by more than an order of magnitude as the growth temperature increases from 665°C to 720°C. The $Al_xGa_{1-x}As$ layers grown at 720°C have PL intensity comparible to those grown by liquid phase epitaxy.[89]

6.3 $\underline{Ga_{0.47}In_{0.53}As}$ $\underline{on}$ $\underline{InP}$

$Ga_xIn_{1-x}As$ has high mobility and large drift velocity which give this material potential for application to high frequency field effect transistors and high-speed logics. Both Sn and Si doped $Ga_{0.47}In_{0.53}As$ have electron mobilities in good agreement with the theoretically calculated values involving the alloy scattering mechanism at 77 and 300°K[90-92] as shown in Figure 23. The undoped layers were n-type 2.4 x 10^{15} cm^{-3}, and the mobility measured in the dark was found to vary from 7800 to 8800 cm^2/Vsec at 300°K and from 35,000 to 40,000 cm^2/Vsec at 77°K.[90,92] Under illumination, this maximum increases to 45,000 cm^2/Vsec at liquid nitrogen temperature.[92]

7. SUMMARY

Molecular beam epitaxy can be used to achieve extreme dimensional control in both chemical compositions and doping profiles. Single crystal multi-layered structures with dimensions of only a few atomic layers started a new branch of experimental quantum physics.[94,95] Utilizing the ability to produce abrupt interfaces and

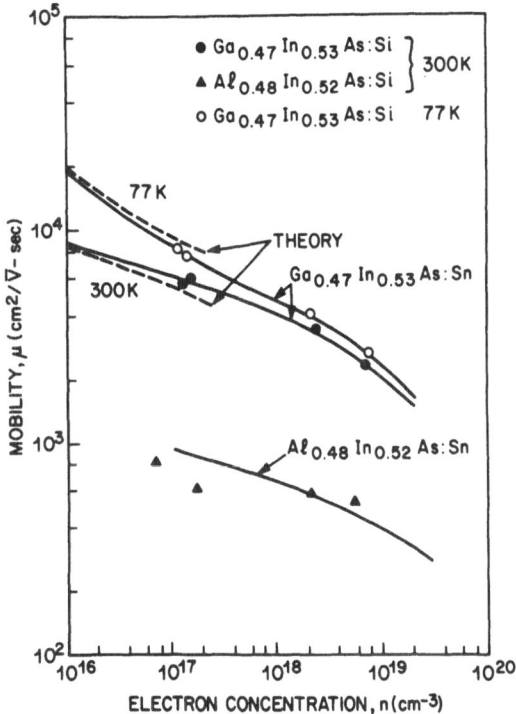

Figure 23 - Hall mobilities as a function of carrier concentration for Si-doped $Ga_{0.47}In_{0.53}As$ and $Al_{0.48}In_{0.52}As$ layers[91]. The solid lines are the experimental mobility results of Sn-doped $Ga_{0.47}In_{0.53}As$ and $Al_{0.48}In_{0.52}As$ layers (After Ref. 90). The dashed curves are theoretical calculations on electron Hall mobilities with alloy scattering (After Ref. 93).

doping variations, devices with desirable and novel transport and optical properties may be fabricated. In situ masking and molecular beam writing allow the formation of three dimensional structures monolithically integrated on the wafer. Besides the growth of semiconducting materials discussed here, MBE can also be used to grow metal and insulating layers.[2] This extraordinarily versatile epitaxy technique coupled with the recent addition of a load-lock for fast sample exchange, and a rotating sample holder for large area uniform growth, make it an important device fabrication technique.

REFERENCES

* Work appeared in "Thin Solid Films", Vol. 100, p. 291-317, Feb. 1983, published by Elsevier Sequoia SA, Lausanne.

1. A.Y. Cho, J.R. Arthur, Progress in Solid-State Chemistry, edited by G. Somorjai and J. McCaldin (Pergammon, New York, 1975) Vol. 10, p. 157.
2. A.Y. Cho, J. Vac. Sci. Technol. 16, 275 (1979).
3. K. Ploog, Crystal Growth, Properties, and Applications, edited by H.C. Freyhardt (Spring-Verlag, Berlin Heidelberg, 1980) Vol. 3, p. 73.
4. K. Ploog, Am. Rev. Mater. Sci. 11, 171 (1981).
5. L.L. Chang, R. Ludeke, Epitaxial Growth, ed J.W. Mathews (Academic New York, 1975) p. 37.
6. C.T. Foxon, B.A. Joyce, Current Topics in Materials Science, edited E. Kaldis (North Holland, Amsterdam/New York, 1980) Vol. 7.
7. R.F.C. Farrow, Crystal Growth and Materials, edited E. Kaldis and H.J. Schul (North Holland, Amsterdam/New York 1977), Vol. 1, p. 237.
8. J.C. Bean, Growth of Doped Silicon by Molecular Beam Epitaxy, ed. F.F.Y. Wang (North Holland, Amsterdam/New York 1977) Vol. 1, p.237.
9. A.Y. Cho, Surface Sci. 17, 494 (1969).
10. A.Y. Cho, J. Appl. Phys. 41, 2780 (1970).
11. A.Y. Cho, J. Appl. Phys. 42, 2074 (1971).
12. J.R. Arthur, Surface Sci. 43, 449 (1974).
13. A.Y. Cho, M.B. Panish, I. Hayashi, Proc. Third International Symposium on GaAs, (The Inst. of Phys., London, 1970) p. 18.
14. A.Y. Cho, H.C. Casey, Jr., P.W. Foy, Appl. Phys. Lett. 30, 397 (1977).
15. A.Y. Cho, Appl. Phys. Lett. 19, 467 (1971).
16. H.C. Casey, Jr., A.Y. Cho, P.A. Barnes, IEEE J. Quantum Elec. QE-11, 467 (1975).
17. W.T. Tsang, Appl. Phys. Lett. 34, 473 (1979).
18. W.T. Tsang, C. Weisbuch, R.C. Miller, Appl. Phys. Lett. 35, 673 (1979).
19. W.T. Tsang, R.A. Logan, IEEE Quantum Elect. QE-15, 451 (1979).

20. D.M. Collins, 1982 MBE Workshop, Urbana, IL, October 21-22, 1982.
21. S. Yamakoshi, O. Wada, T. Fujii, S. Hiyamizu, T. Sakurai, International Electron Devices Meeting, San Francisco, California, December 13-15, 1982, Technical Digest, p. 342.
22. H. Morkoc, T.J. Drummond, M. Omori, IEEE Trans. Elect. Dev. ED-29, 222 (1982).
23. M. Feng, V.K. Eu, I.J. D'Haenens, M. Braunstein, Appl. Phys. Lett. 41, 633 (1982).
24. J.C.M. Hwang, D.G. Flahive, S.H. Wemple, Elect. Dev. Lett. EDL-3, 320 (1982).
25. C.Y. Chen, A.Y. Cho, K.Y. Cheng, T.P. Pearsall, P.O'Connor, IEEE Elec. Dev. Lett. EDL-3, 152 (1982).
26. P. O'Connor, T.P. Pearsall, K.Y. Cheng, A.Y. Cho, J.C.M. Hwang, K. Alavi, IEEE Elec. Dev. Lett. EDL-3, 64 (1982).
27. A.Y. Cho, K.Y. Cheng, Appl. Phys. Lett. 38, 360 (1981).
28. K.Y. Cheng, A.Y. Cho, W.R. Wagner, Appl. Phys. Lett. 39, 607 (1981).
29. H. Shelton, A.Y. Cho, J. Appl. Phys. 37, 3544 (1966).
30. A.Y. Cho, C.D. Hendrick, J. Appl. Phys. 40, 3339 (1969).
31. J.R. Arthur, J. Appl. Phys. 39, 4032 (1968).
32. C.T. Foxon, J.A. Harvey, B.A. Joyce, J. Phys. Chem. Solids 34, 1603 (1973).
33. A.Y. Cho, J.C. Tracy, Jr., U.S. Patent 3969164.
34. A.Y. Cho, J. Appl. Phys. 47, 2841 (1976).
35. K. Ploog, A. Fischer, Appl. Phys. 13, 111 (1977).
36. A.Y. Cho, H.C. Casey, Jr., C. Radice, P.W. Foy, Electronics Lett. 16, 72 (1980).
37. K.Y. Cheng, A.Y. Cho, W.R. Wagner, W.A. Bonner, J. Appl. Phys. 52, 1015 (1981).
38. HANDBOOK OF CHEMISTRY AND PHYSICS, 56th ed., ed. R.C. West (Chemical Rubber Company, Cleveland, Ohio, 1975) p. D-61.
39. J.R. Arthur, Proc. Int. Mater. Symp. Struct. Chem. Solid Surfaces, Berkeley, p. 46 (1968).
40. A.Y. Cho, K.Y. Cheng, Appl. Phys. Lett. 38, 360 (1981).
41. K.Y. Cheng, A.Y. Cho, W.R. Wagner, Appl. Phys. Lett. 39, 607 (1981).
42. M.B. Panish, J. Electrochem. Soc. 127, 2729 (1980).
43. A.R. Calawa, Appl. Phys. Lett. 38, 701 (1981).
44. J.H. Neave, B.A. Joyce, J. Cryst. Growth. 44, 387 (1978).

45. A.Y. Cho, J. Appl. Phys. 46, 1722 (1975).
46. M. Ilegems, J. Appl. Phys. 48, 1278 (1977).
47. A.Y. Cho, M.B. Panish, J. Appl. Phys. 43, 5118 (1972).
48. M. Ilegems, R. Dingle, L.W. Rupp, Jr., J. Appl. Phys. 46, 3659 (1975).
49. A.Y. Cho, I. Hayashi, J. Appl. Phys. 42, 4422 (1971).
50. A.Y. Cho, J. Vac. Sci. Technol. 8, S31 (1971).
51. W.T. Tsang, I.K. Reinhart, J.A. Ditzenberger, Appl. Phys. Lett. 36, 118 (1980).
52. R. Fischer, J. Klem, T.J. Drummond, R.E. Thorne, W. Kopp, H. Morkoc, A.Y. Cho, J. Appl. Phys. 54, 2508 (1983).
53. P.K. Tien, International Summer School, Hong Kong, (1982).
54. B.I. Miller, J.H. McFee, J. Electrochem. Soc. 125, 1310 (1978).
55. K.Y. Cheng, A.Y. Cho, T.J. Drummond, H. Morkoc, Appl. Phys. Lett. 40, 147 (1982).
56. J. Massies, J. Rochette, P. Delecscluse, P. Etienne, J. Chevrier, N.T. Linh, Elect. Lett. 18, 758 (1982).
57. Y. Kawamura, Y. Noguchi, H. Asahi, H. Hagai, Elect. Lett 18, 91 (1982).
58. J. Barnard, C.E.C. Wood, L.F. Eastman, Elect. Dev. Lett. EDL-2, 8 (1981).
59. A. Kastalsky, R. Dingle, K.Y. Cheng, A.Y. Cho, Appl. Phys. Lett. 41, 274 (1982).
60. D. Olego, T.Y. Chang, E. Silberg, E.A. Caridi, A. Pinczuk, Appl. Phys. Lett. 41, 476 (1982).
61. J.A. Barnard, C.E.C. Wood, L.F. Eastman, Elect. Dev. Lett. EDL-3, 318 (1982).
62. K. Alavi, H. Temkin, W.R.Wagner, A.Y. Cho, Appl. Phys. Lett. February 1, 1983.
63. G.D. Holah, F.L. Eisele, E.L. Meeks, N.W. Cox, Appl. Phys. Lett. 41, 1073 (1982).
64. W.T. Tsang, F.K. Reinhart, J.A. Ditzenberger, Appl. Phys. Lett. 41, 1094 (1982).
65. H.C. Casey, Jr., M.B. Panish, HETEROSTRUCTURE LASERS, PART B, Academic Press, New York, 1978.
66. A.Y. Cho, W.C. Ballamy, J. Appl. Phys. 46, 783 (1975).
67. W.B. Ballamy, A.Y. Cho, IEEE Trans. Elect. Dev. ED-23, 481 (1976).
68. T.P. Lee, A.Y. Cho, Appl. Phys. Lett. 29, 164 (1976).
69. W.T. Tsang, A.Y. Cho, Appl. Phys. Lett. 30, 293 (1977).
70. A.Y. Cho, F.K. Reinhart, Appl. Phys. Lett. 21, 355 (1972).

71. W.T. Tsang, M. Ilegems, Appl. Phys. Lett. 31, 301 (1977).
72. Y. Tami, Y. Komiya, Y. Harada, J. Electrochem. Soc. 118, 118 (1971).
73. D.B. Lee, J. Appl. Phys. 40, 4569 (1969).
74. R.M. Finne, D.L. Klien, J. Electrochem. Soc. 114, 965 (1967).
75. W.T. Tsang, A.Y. Cho, Appl. Phys. Lett. 32, 491 (1978).
76. H. Morkoc, A.Y. Cho, J. Appl. Phys. 50, 6413 (1979).
77. A.R. Calawa, Appl. Phys. Lett. 38, 701 (1981).
78. J.C.M. Hwang, H. Temkin, T.M. Brenan, R.E. Frahm, Appl. Phys. Lett. 42, 66 (1983).
79. R. Dingle, C. Weisbuch, H.L. Stormer, H. Morkoc, H. Morkoc, A.Y. Cho, Appl. Phys. Lett. 40, 507 (1982).
80. H.B. Bebb, E.W. Williams, SEMICONDUCTORS AND SEM-IMETALS, edited by R.K. Williardson, A.C. Beer (Academic Press, New York, 1972), Vol. 8, p. 182 and 321.
81. D.W. Covington, C.W. Litton, D.C. Reynolds, R.J. Almassy, G.L. McCoy, in Gallium Arsenide and Related Compounds 1978 (Inst. Conf. Series 45), p. 71.
82. G.B. Scott, J.S. Roberts, in Gallium Arsenide and Related Compounds 1978 (Inst. Phys. Conf. Ser. 45), p. 181.
83. H. Kunzel, K. Ploog, in Gallium Arsenide and Related Compounds 1980 (Inst. Phys. Conf. Ser. 56), p. 519.
84. D.J. Ashen, P.J. Dean, D.T.J. Thurle, J.B. Mullis, A.M. White, J. Phys. Chem. Solids 36, 1041 (1975).
85. T.S. Low, G.E. Stillman, A.Y. Cho, H. Morkoc, A.R. Calawa, Appl. Phys. Lett. 40, 611 (1982).
86. H. Morkoc, A.Y. Cho, C. Radice, Jr., J. Appl. Phys. 51, 4882 (1980).
87. T. Ishibashi, S. Tarucha, H. Okamoto, Japan J. Appl. Phys. 21, L476 (1982).
88. D.V. Lang, A.Y. Cho, A.C. Gossard, M. Ilegems, W. Wiegmann, J. Appl. Phys. 47, 2558 (1976).
89. T. Fujii, S. Hiyamizu, O. Wada, T. Sugahara, S. Yamakoshi, T. Ishikawa, T. Sakurai, H. Hashimoto, Second International Symposium on Molecular Beam Epitaxy, August 27-30, 1982, Tokyo, Japan (p. 85).
90. K.Y.Cheng, A.Y. Cho, W.R. Wagner, J. Appl. Phys. 52, 6328 (1981).
91. K.Y. Cheng, A.Y. Cho, J. Appl. Phys. 53, 4411 (1982).
92. J. Massies, J. Rochette, P. Delescluse, P. Etienne, J. Chevriet, N.T. Linh, Elect. Lett. 18, 758 (1982).

93. Y. Takeda, A. Sasaki, Japan J. Appl. Phys. <u>19</u>, 383 (1980).
94. R. Dingle, W. Wiegmann, C.H. Henry, Phys. Rev. Lett. <u>33</u>, 827 (1974).
95. L. Esaki, L.L. Chang, Phys. Rev. Lett. <u>33</u>, 495 (1974).

MBE GROWTH OF II-VI AND IV-VI COMPOUNDS AND ALLOYS

R.F.C. Farrow

Westinghouse Electric Corporation
Research and Development Center
Pittsburgh, PA   15235

ABSTRACT: MBE growth of II-VI and IV-VI compounds and alloys is driven both by device requirements and by the ability of the technique to generate quantum well structures which can probe areas of narrow band gap semiconductor physics. The present state of knowledge of growth mechanisms and MBE growth techniques of these materials is reviewed. Recent advances in growth of device quality films are described and remaining problems and future trends in the field are discussed.

1. INTRODUCTION

In this review the application of molecular beam epitaxial (MBE) growth techniques to the preparation of two classes of technologically significant compound semiconductor materials is considered. The first class (II-VI compounds) comprises compounds or alloys formed from group IIB and VIA elements. In particular, materials in the (Zn, Cd, Hg: S, Se, Te) system form zinc-blende semiconductors with a variety of band gaps. The second class (IV-VI compounds) comprises compounds or alloys formed from group IVA and VIA elements, in particular, the chalcogenides of Pb and Sn. These materials and some rare earth chalcogenides form rock salt-structure semiconductors with a variety of band gaps. The application of MBE and MBE-related techniques to the growth and characterization of II-VI compounds and alloys is a rapidly expanding field. The main driving forces for this expansion are listed in Table 1. Currently, a major topical area of research is the growth and characterization of the narrow-gap semiconductor alloy $Hg_{1-x}Cd_xTe$ which has[1] near-ideal electrical properties as a detector of infrared radiation. The thermal instability[2] of this alloy and the need for extreme spatial uniformity of alloy composition over large substrate areas demands the development of a low temperature film growth technique with a capability of handling large-area substrates. MBE is a promising candidate for such a technique. In addition, the built-in surface diagnostic techniques in MBE systems provide the capability for in situ studies of critical device processes such as alloy contacting, passivation and ion milling. In parallel with this device-related driving force the ability of the technique to generate quantum well structures is attractive since such structures have been proposed[3-7] as a new class of materials with novel optical and electrical properties. However, although there is plenty of theoretical modelling of such structures these models have not included the effects of uniaxial strain which are present in lattice-matched structures. Uniaxial strain has now been observed[7,8] in several lattice-matched, MBE-grown, narrow gap semiconductor structures and is a symmetry breaking effect which can significantly modify the optical and transport properties of such structures.

More generally, the wide gap II-VI semiconductors (Zn, Cd: S, Se Te) are of potential importance in optoelectronic devices such as short wavelength (visible) light emitting diodes, electroluminescent

TABLE 1. Driving Forces for MBE Research on II-VI and IV-VI
Compounds and Alloys

(A) Device Applications

1. $Hg_{1-x}Cd_xTe$ photoconductive and photovoltaic detectors
    for thermal imaging arrays. $CdTe/Hg_{1-x}Cd_xTe$ hetero-
    junction diodes

2. IV-VI diode lasers for tunable IR sources

3. Electroluminescent devices - AC/DC panels ZnS, ZnSe,
    $ZnTe_{1-x}Se_x$

4. II-VI solar cells: CdTe, CdS

(B) Narrow-Gap Semiconductor Physics

1. 2D - subband quantum well structures in zero gap semi-
    conductors of α-Sn type:

    $$\alpha - Sn, HgTe, HgSe$$

2. Symmetry breaking effect of uniaxial strain in lattice
    matched structures:

    $$CdTe/InSb \qquad Hg_{1-x}Cd_xTe/CdTe$$
    $$\alpha-Sn/CdTe \qquad \alpha-Sn/InSb$$

3. HgTe-CdTe superlattice

panels, optical waveguides and photovoltaic solar cells. However, our control and understanding of intrinsic and extrinsic doping in such thin film structures and in bulk II-VI compounds is far from complete. For example, the degree of control of conductivity has been hampered by the self-compensation effect[9,10] in these materials. While the interpretation of this effect is still contentious it is clear that low temperature, film growth techniques, in which the diffusion and segregation of native defects is suppressed and in which the level of impurities is reduced, can alleviate this problem. MBE is a particularly promising growth technique in this respect.

Although the naroow-gap semiconductor alloy $Hg_{1-x}Cd_xTe$ has replaced the IV-VI alloy $Pb_{1-x}Sn_xTe$ and related Pb-salts in infrared detector array applications, the latter materials remain of importance for temperature-tunable infrared laser sources for spectroscopy and pollution monitoring[11] applications.

Following earlier reviews of MBE growth of II-VI[12,13] and IV-VI[14] compounds considerable advances have been made in the MBE growth of device quality films of II-VI and IV-VI compounds and alloys. These advances, the underlying growth mechanisms, remaining problems and future trends in the field will be reviewed.

2. GROWTH MECHANISMS

In gaining an understanding of the mechanisms and critical parameters in MBE growth a knowledge of the growth species and dissociation behavior of the compound, or alloy, to be grown is essential. For example, the dissociation rate of the film sets the ultimate upper limit to the film growth temperature at given primary beam impingement conditions. In addition, if film growth occurs from separate elemental species, one of which is a dimer (or tetramer) then the necessary surface dissociation of the dimer could, in principle, set a lower temperature limit to stoichiometric film growth. Fluctuations in the arrival rate of primary growth species can also lead to non-stoichiometric film growth if, as is the case at very low substrate temperature, both species have a unity sticking coefficient. In addition, states of congruent vaporization of the compounds, under Knudsen effusion conditions, may exist which can be utilized to overcome this latter problem.

2.1 II-VI Compounds

Apart from HgSe (which dissociates[15] into Hg and dominant high order Se_m polymeric molecules) all of the II-VI compounds, MX, where M is the metal and X the chalcogen dissociate according[15] to

$$MX(s) \rightarrow M(g) + 1/2 \, X_2(g)$$

However, the II-VI compounds can be subdivided into two classes depending on their dissociation behavior under Knudsen effusion conditions:

I. congruent dissociation with emergent flux ratio $J_{X_2}/J_M = 1/2$ and time independent.

II. incongruent dissociation with emergent flux ratio $J_{X_2}/J_M < 1/2$ and time dependent.

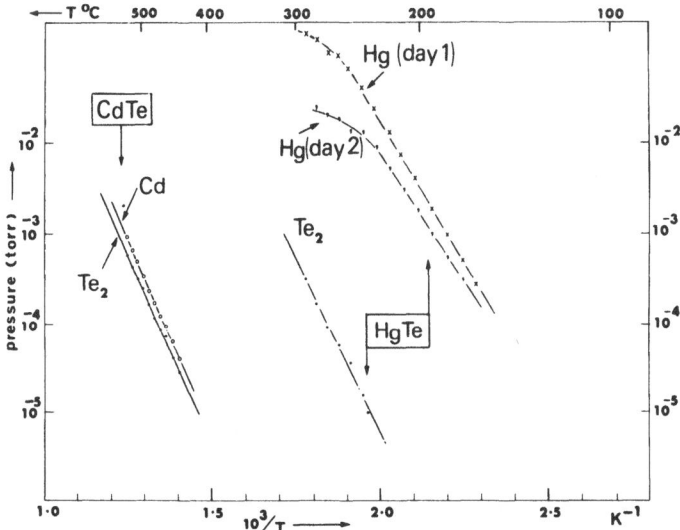

Figure 1 - Arrhenius plots of the dissociation vapor pressures developed over HgTe and CdTe under Knudsen effusion conditions. These measurements were made using the technique of modulated beam mass spectrometry. Note that HgTe dissociates at very low temperatures, and behaves as a pure Hg source at T < 200°C. However, this incongruent dissociation leads to Hg depletion from the charge and to a progressive decrease in P_{Hg}. In contrast CdTe dissociates congruently and at much higher temperatures.

The II-VI compounds in the (Zn, Cd: S, Se, Te) system fall into class I as a result of the strong dependence of dissociation vapor pressures P_M, P_{X_2} on crystalline charge, MX(c), composition within the solidus field. If the charge composition is on the X-rich side of the solidus field then $P_{X_2} > P_M$ and $J_{X_2} > 1/2 J_M$. The reverse is true if the charge is rich in M. As a result the beam and charge composition both approach a steady state in which $J_{X_2}/J_M = 1/2$. This effect was described, in the case of CdTe, by Brebrick and Strauss[16] and was confirmed by early mass spectrometric data[15] for II-VI compounds in the (Zn, Cd: S, Se, Te) system and recently for CdTe in particular.[8] On the other hand, the II-VI compounds in the (Hg: S, Te) system vaporize[15] incongruently at low (< 200°C) temperatures with $J_M >> J_{X_2}$ as a result of the very weak Hg-X bond and the volatility of Hg. This contrasting behavior is well illustrated by the dissociation vapor pressure developed over the compounds CdTe and HgTe under Knudsen effusion conditions. Figure 1 shows the results of modulated molecular beam mass spectrometry (MBMS) studies of these compounds by the author.[17] The steady state sublimation of CdTe at temperatures above 400°C contrasts strongly with the incongruent dissociation of HgTe at much lower temperatures. The instability and drift of P_{Hg} over HgTe is due to the formation of a Hg-depleted surface layer of Te(s) with consequent diffusion rate limitations to further loss of Hg. This instability was a likely contributory cause of the poor structural and electrical quality of the early MBE-grown[18,37] $Hg_{1-x}Cd_xTe$ films in which HgTe was used as a Hg source and separate Cd and Te sources were used as sources of Cd and Te_2. Even if the HgTe effusion oven is controlled by a feedback system incorporating a mass spectrometer monitoring J_{Hg}, the significant and constantly changing value of J_{Te_2} (arising from HgTe dissociation) can cause compositional problems (e.g. Te precipitates) in the film. Short time-constant pulses of both Hg and Te_2 also occur as a result of the unstable structure of the Te-rich surface boundary layer.[17] In contrast to this behavior, the steady state of congruent sublimation for CdTe has been used[8] to advantage in growth of high structural perfection films of CdTe and in $CdTe/Hg_{1-x}Cd_xTe$ photodiode[19] preparation by MBE. The grossly incongruent dissociation[2] of the narrow gap alloy $Hg_{1-x}Cd_xTe$ reflects this great difference in dissociation behavior of the two end members of the alloys. It also shows the need for low temperature growth.

Separate sources of the elements have been used for growth of II-VI compounds and alloys. However, because of the inevitable fluctuations in primary beam impingement flux ratio in this case, reasonable quality films are obtained only at elevated growth temperatures where the self-regulating regime applies. This is a regime of growth temperature, first identified by Smith and Pickhardt[12], in which both elements are volatile and in which the only mechanism for sticking is surface reaction to form the II-VI compound. These authors used the technique of II-VI compound film growth on a heated quartz crystal resonator to study the temperature dependence of film growth rate as a function of beam impingement rates from separate sources of the elements in the (Zn, Cd: S, Se, Te) system. In particular, with the Zn flux and substrate temperature (300°C) held constant, the film growth rate increased linearly with Te_2 flux up to the point where $2J_{Te_2} \approx J_{Zn}$. At higher Te_2 impingement rates excess Te_2 was desorbed and the film growth rate levelled off. The analogous effect was observed when the Te_2 flux was held constant and Zn flux increased. Other II-VI compounds in the system exhibited the same behavior. This confirmed that the minority beam flux controlled the growth rate and that, as in growth of III-V compounds, large variations in impingement flux ratios could be tolerated in achieving compound film growth. Since these studies were restricted to polycrystalline film growth some doubt remained as to the validity of Smith and Pickhardt's data for epitaxial growth conditions. Yao, *et al*[20] later provided qualitative support for the self-regulating growth mode for ZnSe and ZnTe grown epitaxially onto GaAs substrates but numerical discrepancies persist in the data. These authors proposed a highly simplified model of II-VI compound growth which provided qualitative agreement with the measured growth rate - impingement flux curve. In this model all collisions between group II (VI) beam atoms and group VI (II) surface atoms were assumed to result in II-VI compound bonds. Collisions between group II (VI) beam atoms and group II (VI) surface atoms were assumed to result in instant desorption of impinging atoms. Assuming steady state coverages θ_{II}, θ_{VI} for group II and VI adatoms, respectively, the film growth rate (in units of II-VI bonds/sec) is given by

$$G = 1/2 \, (J_{II} \, \theta_{VI} + J_{VI} \, \theta_{II}) - R \qquad (1)$$

where R is the congruent dissociation rate of the compound at the growth temperature. For steady state coverages:

$$J_{II} \, \theta_{VI} = J_{VI} \, \theta_{II} = J_{VI}(1 - \theta_{VI}) \qquad (2)$$

and so

$$\theta_{VI} = \frac{J_{VI}}{J_{II} + J_{VI}} \qquad (3)$$

$$G = \frac{J_{II} J_{VI}}{J_{II} + J_{VI}} - R \qquad (4)$$

Equation 4 gives a curve of the same shape as the experimental one, however, the large measured increases in growth rate at $J_{Se_2} \gg J_{Zn}$ and $J_{Te_2} \gg J_{Zn}$ are not accounted for by the theory and may arise from systematic errors in measured impingement rates. The theory also neglects the possibility of adsorption of beam species into weakly physisorbed precursor states with the possibility of desorption from these states prior to chemisorption. Also, the model clearly breaks down at growth temperatures sufficiently low for elemental condensation without formation of II-VI bonds. This is certainly possible for Te at temperatures below 200°C where its elemental vapor pressure is only 10^{-7} Torr. Growth of single phase II-VI compound films (especially tellurides) at very low ($\lesssim$ 200°C) temperatures therefore requires the use of the single (compound) source technique. It is interesting that despite the dissociative mode of vaporization of the II-VI compounds, stoichiometric epitaxial films (for example of CdTe) can be grown[21] at room temperature. This shows that the surface reactions of group VI dimer dissociation and II-VI atom recombination do not provide kinetic barriers to growth at least at normal growth rates $\sim$ 1μm h^{-1}.

2.2 IV-VI Compounds

The IV-VI compounds and alloys vaporize, predominantly as molecules:

$$MX(s) \rightarrow MX(g)$$

This tendency towards strong intramolecular IV-VI bonding arises from the "inert electron pair" effect[22] in which outer S-shell electrons in the group IV metal are precluded from participating in the intramolecular bond. This induces greater overlap of the outer shell p-electrons of both group IV and VI atoms. The same effect lies behind the weak II-VI intramolecular bond. (In fact, no II-VI molecules have been observed[15] over II-VI compounds). Minority dissociation vapor species

M, X, X$_2$, M$_2$X, M$_2$X$_2$ are also observed in small quantities. This is illustrated by the MBMS spectra[17], shown in Figures 2(a) and 2(b), of the beam generated by Knudsen effusion of PbTe.

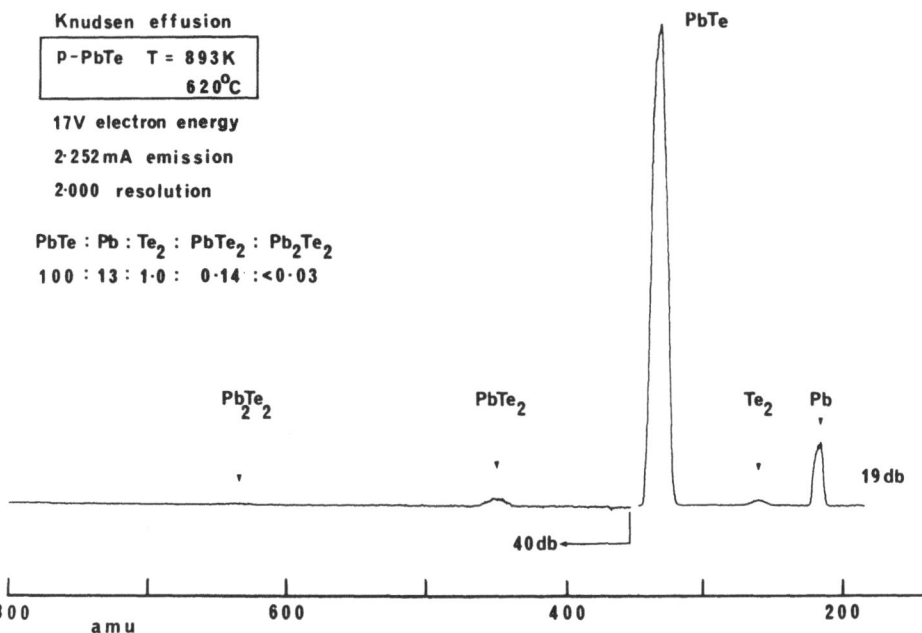

Figure 2(a) - Modulated beam mass spectra of beam generated by Knudsen effusion of PbTe. Ionizing electron energy 17 eV. Low resolution setting to minimize mass discrimination in quadrupole filter. Beam modulation frequency 70 Hz. Knudsen effusion oven temperature 620°C. Spectrum recorded with main peaks PbTe$^+$, Te$_2$$^+$, Pb$^+$ at same amplifier gain (19 dB). Note that the main beam species is PbTe but that higher order units are just detectable at higher gain.

Even at the low electron energy of 17 eV the Pb$^+$ signal is still largely due to fragmentation of the PbTe molecules. The small but significant PbTe$_2$$^+$, Pb$_2Te_2$$^+$ ion species have significantly different temperature dependences than PbTe$^+$ and can, respectively, be attributed to PbTe$_2$ and Pb$_2$Te$_2$ parent molecules. The existence of analogous molecules in the vapor over SnSe and SnTe has been established by Colin and Drowart[52]. The small but significant quantities of Pb$_2$Te$_2$ and PbTe$_2$ molecules represent large building blocks of the solid and may play a significant role as crystal building units during nucleation and growth. For example, such molecules may have large cross sections for surface adatom capture.

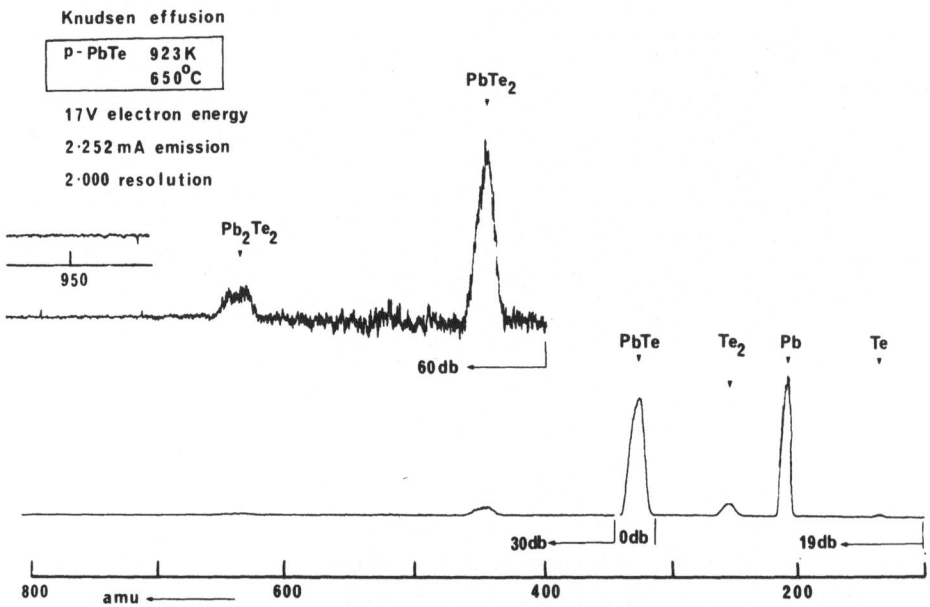

Figure 2(b) - Modulated beam mass spectra of beam generated by Knudsen effusion of PbTe. Knudsen effusion oven temperature 650°C. At this temperature and at high amplifier gain (60 dB) the large molecular units at high mass are well resolved.

This predominant molecular vaporization of the II-VI compounds means that these materials grow by a simple process of sublimation in both MBE and vapor transport techniques. The use of separate elemental sources in IV-VI film growth has no advantages over growth by molecular sublimation. The rare earth chalcogenides (Eu, Yb: S, Te) also exhibit congruent molecular sublimation and YbTe forms solid solutions, $Pb_{1-x}Yb_xTe$ with PbTe which have been grown[49] by MBE.

In the case of the IV-VI alloy $Pb_{1-x}Sn_xTe$, the near coincidence of the sublimation enthalpies of the two end members of the alloy system (49.80 ± 2 kcal mole[-1] for PbTe, 50.7 ± 2 kcal mole[-1] for SnTe) leads to quasi-congruent sublimation of the alloy from a single Knudsen effusion oven. Figure 3 shows the MBMS spectrum[17] of the beam generated by Knudsen effusion of $Pb_{0.804}Sn_{0.196}Te$ at T = 650°C.

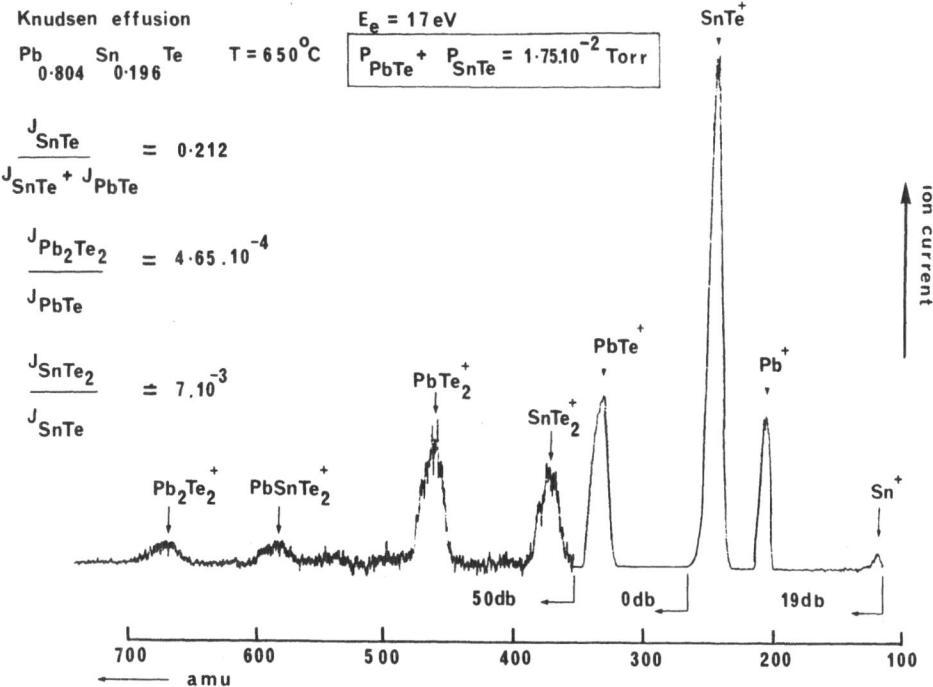

Figure 3 - Modulated beam mass spectrum of beam generated by Knudsen effusion of $Pb_{0.804}Sn_{0.196}Te$ at T = 650°C. Modulation frequency 70 Hz. Note the changes in amplifier gain for different segments of the mass spectrum. The dominant vapor species are PbTe and SnTe with smaller amounts of polymeric units extending to Pb_2Te_2. There is a slight perferential loss of SnTe which leads to a slow but significant drift, with time, of the alloy composition towards smaller x values.

The dominant vapor species are PbTe and SnTe with smaller amounts of polymeric units extending to Pb_2Te_2. There is a slight preferential loss of SnTe which leads to a slow but significant drift ($\Delta x = -0.08$ for 50% weight loss of charge) in charge composition towards smaller x values. Single source growth of alloy films has been reported[14] but the optimum MBE growth technique for IV-VI alloys such as $Pb_{1-x}Sn_xTe$, $PbTe_{1-x}Se_x$ or $Pb_{1-x}Sn_xSe$ is to use[14] an isothermal double oven containing the binary end members of the alloy. By fixing the orifice area ratio for the two compartments of the oven, to give the appropriate x-value of the film, very uniform composition films for array applications[14] can be prepared. In addition, temperature fluctuations in the oven cause minimal changes in the x-value of the film since the effusion rate of both compounds changes in the same direction during the fluctuations.

3. MBE GROWTH OF II-VI COMPOUNDS AND ALLOYS

3.1 (Zn, Cd: Se, Te) System

The heteroepitaxial growth of II-VI compounds (Zn, Cd: Se, Te) on a variety of substrate crystals including GaAs, CdS, CdSe, CaF_2 and BaF_2 was first reported by Smith and Pickhardt[12] in 1975. However, these studies were purely morphological and no electrical or optical data were reported. More recently, high structural quality, lattice-matched n-type CdTe films grown onto InSb substrates have been prepared[8] and the photoluminescence properties of In-doped films explored.[23] MBE growth of Ga-doped ZnSe on GaAs has been studied by Niina *et al*[24] and strong blue photoluminescence observed under optimized growth conditions.

3.1.1 CdTe. MBE growth of CdTe films on (111) and (001) orientation CdTe substrates was reported in 1981 by Faurie and Million.[21] A single CdTe source was used and growth was studied over the substrate temperature range of room temperature to 250°C. Faurie and Million did not state which polar face of the (111) orientation CdTe substrates was used but acheived epitaxial growth on only one of the polar faces over this complete temperature range. In situ RHEED (reflection high energy electron diffraction) studies showed that following growth on Ar ion bombarded and (300°C) annealed (111) orientation substrates film growth proceeded with a (2 x 2) reconstruction. Scanning electron microscopy revealed smooth featureless surfaces on films grown at 200°C. Films grown at higher temperatures exhibited a rough surface texture and degraded RHEED patterns. Following growth of a 3000Å thick buffer layer at 200°C it was found that epitaxial growth was maintained as the growth temperature was reduced to room temperature. On the (001) orientation epitaxial growth was observed over the temperature range 80-200°C. Below 80°C the quality of epitaxy degraded to polycrystalline growth, often indicated by the appearance of twin diffraction spots in the RHEED pattern. However, no indication of the film growth rate at which this transition occurred was given and no data on the electrical or photoluminescence properties of the films was reported. In subsequent studies of HgTe-CdTe multilayer structures Faurie *et al*[25] found that in growth from a single CdTe source the film perfection, as judged from the RHEED patterns, was significantly higher at 200°C than 120°C. Nevertheless x-ray rocking curves for the CdTe films grown at 200°C were very wide (30 minutes of arc) compared with rocking curves for the CdTe substrate (3-4 minutes of arc). This implies that substrate defects such as dislocations and low-angle grain boundaries propagate into the film and multiply. It

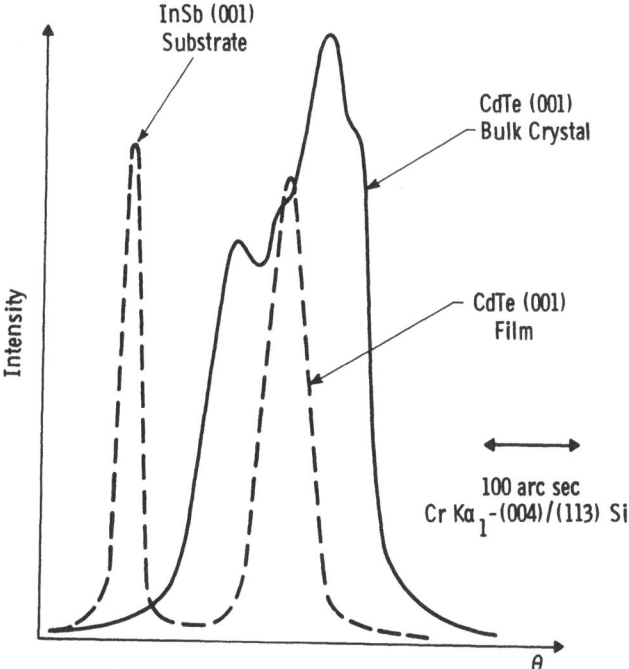

Figure 4 - Double crystal rocking curves for bulk CdTe and for a 0.8μm thick CdTe film grown at 200°C on an InSb (001) orientation substrate (dashed curve). The rocking curves are for (004) Bragg diffractions. Note that the epitaxial film has a much sharper rocking curve than the bulk sample and does not exhibit the multiple peaking which is clearly present for the bulk sample. This confirms that the film is free of low angle grain boundaries.

highlights a long-standing problem in homo- and heteroepitaxial growth on CdTe substrates, namely the generally poor structural perfection of bulk CdTe crystals and consequently of wafers cut from them. This problem arises from the pronounced tendency for twinning and low angle boundary formation in seeded melt growth. CdTe is particularly vulnerable to such effects since it is particularly soft (Knoop hardness 45 at 300K) and has a very low thermal conductivity (0.07 W cm^{-1} K^{-1} s^{-1} at 300K). Double crystal dispersive x-ray topography[26] studies of CdTe wafers from a wide variety of sources invariably reveal a mosaic structure of grains, a few hundred microns in size, differing in orientation by 10-20 arc sec. X-ray rocking curves in double-crystal x-ray diffraction from such wafers are typically 180-200 arc sec wide with multiple peaking arising from the low angle boundaries intercepted by the primary beam. A double crystal rocking curve of this type is shown in Figure 4. One approach to this problem pioneered by the author and coworkers is to use the excellent lattice match ($\Delta d/d \simeq 500$ ppm) between InSb (a = 6.4798A, 25°C) and CdTe (a = 6.4829A, 25°C) to generate heteroepitaxial, high structural perfection, lattice-matched

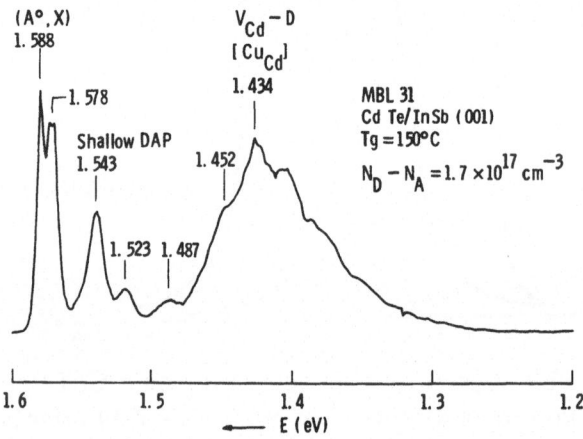

Figure 5 - Photoluminescence (recorded at 4K) of a 1.5μm thick CdTe film grown on an InSb (001) orientation substrate by MBE at T_g - 150°C. The intensity of the luminescence was strong which is remarkable for a heteroepitaxial film grown at such a low temperature (see text).

films of CdTe. InSb substrates are free of low angle boundaries, have
low (< 500 cm^{-2}) dislocation densities and are available as large (2
inch diameter) area wafers. CdTe films grown[8] onto ion-bombarded,
annealed InSb surfaces are exactly lattice-matched and have very
narrow rocking curves compared with bulk CdTe. Figure 4 shows a
double-crystal rocking curve for a 0.8μm CdTe film grown at 200°C
superimposed on a double-crystal rocking curve for a bulk CdTe wafer.
Films of this perfection show strong photoluminescence and exhibit
n-type conductivity. Figure 5 shows a photoluminescence spectrum[27]
from a lattice-matched, 1.5μm thick film of CdTe grown at 150°C.
The spectrum shows a strong peak at 1.588 eV which can be assigned[28]
to a neutral acceptor (probably Ag)-bound exciton (A°, X) recombination. The origin of the shoulder at 1.578 eV is not clear at this time.
The peak at 1.543 eV can be assigned[28] to donor-acceptor pair recombination associated with a Li impurity in the film. The broad emission
centered at 1.434 eV has been assigned by Barnes and Zanio[29] to
radiative recombination at a deep acceptor complex: $V_{Cd}In_{Cd}$. This
peak was observed by both Barnes and Zanio[29] and Sugiyama[23] to
scale with In concentration in CdTe. It is likely[28] that there is also a
contribution to this peak from Cu donor-acceptor pair luminescence at
1.469 eV. The origin of the high (n_D-n_A ≈ 10^{17} cm^{-3}) free electron
concentration in this particular film (run number MBL 31) is not clear
at present. No evidence of deep In diffusion into the CdTe was observed in secondary ion mass spectrometry studies of this or other films
and the free electron concentration was found to be constant through
the film. Moreover, subsequent films grown (by G.M. Williams of
RSRE, Malvern) from the same CdTe source material showed a progressive reduction in free electron concentration to a background plateau
level of ≤ 10^{15} cm^{-3}. This could possibly result from source purifica-
tion by desorption of a voltaile residual impurity from the charge.
Minority carrier (hole) diffusion length measurements[30] for the film
MBL 31 showed this parameter has a value greater than the film thickness. This is consistent with the high degree of structural perfection of
the film. Films grown at higher growth temperatures ~ 200°C also had
free electron concentrations ≤ 10^{15} cm^{-3} and exhibited a small but
significant dilation of lattice parameter with increasing growth temperature. Although much work remains to be done to examine this effect in
detail it seems likely that since Cd is much more volatile at the growth
temperature than Te it may desorb preferentially from its precursor
state at high (≳ 200°C) growth temperatures leading to incorporation
of cadmium vacancies and possibly excess Te in the film. Cross-section

transmission electron microscopy studies of the films and interfac region are presently under way to examine this latter possibility.

Sugiyama[23] has studied the MBE growth of CdTe on (001) orientation InSb substrates on which a buffer layer of InSb was deposited immediately prior to CdTe film growth. The reconstruction of the InSb surface prior to growth was the Sb-stabilized "pseudo (1x3) state"[31]. A (2x1) surface reconstruction was observed during CdTe film growth at 200°C. Unintentionally doped films were n-type with free carrier concentrations in the range 10^{15} - 10^{16} cm^{-3}. Controlled In doping over the range 10^{16} - 10^{18} cm^{-3} was achieved using an In Knudsen effusion oven. The broad photoluminescent peak centered at 1.43 eV increased with In doping level reaching a saturation intensity at ~ 5 x 10^{17} cm^{-3}. At higher doping levels the near band edge emission at 1.58 eV decreased, possibly as a result of decreasing film quality related to precipitation of In. The weak, broad, defect-dominated spectra observed by Fugiyama *et al* at In doping levels below 10^{16} cm^{-3} may be indicative of poor structural quality films. This is likely since the use by Sugiyama of In solder to mount the substrate wafers is known to introduce strain and inhomogeneous distortion into both substrates and film. The use of a carbon based suspension (e.g. "aquadag") to mount substrates (as Noreika *et al*[31] and Farrow *et al*[8] have done) avoids this problem.

MBE growth of CdTe on (111)B InSb surfaces has recently been reported by Myers *et al.*[32]. However, these authors used heat cleaning of InSb substrates at temperatures well above that (230°C)[33] at which Sb is preferentially desorbed from InSb (111)B surfaces. This would have resulted in formation of In microprecipitates at the InSb surface which could well be responsible for the poor crystallographic quality of the films indicated by the arcs in the oscillation x-ray diffraction patterns. It is significant that the CdTe films prepared by Myers *et al* were semi-insulating as were the CdTe films evaporated onto InSb substrates by Varlamov *et al.*[34]. The author and co-workers have found that poor crystallographic quality (i.e. broad x-ray rocking curve) CdTe films were invariably semi-insulating probably as a result of carrier traps at grain boundaries and other defect sites.

3.1.2 ZnSe. Recently, Niina *et al.*[24] have prepared Ga-doped ZnSe films, heteroepitaxially by MBE onto GaAs substrates. The bulk lattice mismatch in this case is: $\Delta d / <d> \sim 5$ x 10^{-4} at 25°C which is comparable to the value for CdTe/InSb. In addition, semi-insulating GaAs wafers are very convenient from the point of view of availability and suitability for post-growth electrical characterization. Following in situ heat cleaning of semi-insulating GaAs wafers, ZnSe films were grown at $T_g = 360°C$ using separate Zn and Se sources and a Ga source for doping. The beam equivalent pressure ratio of Zn and Se species was held near unity. The film growth rate was $0.8\mu m$ h^{-1} and films were grown over a range of Ga oven temperatures from 300-600°C to explore the electrical and optical properties of Ga-doped films. Hall

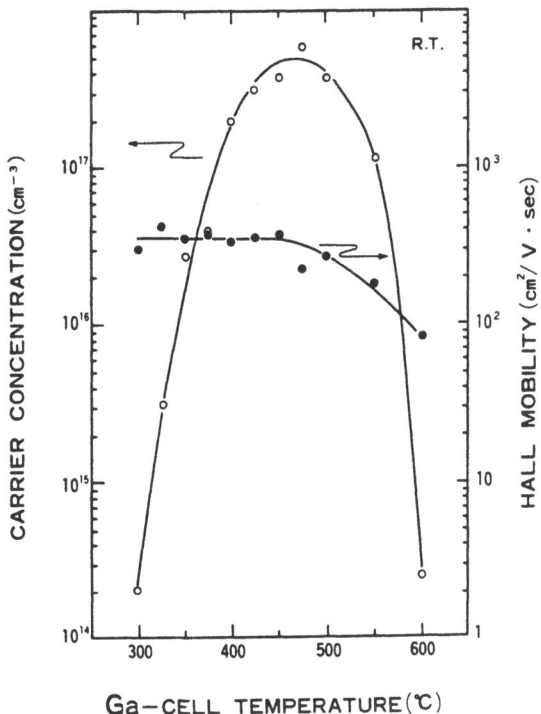

Figure 6 - Carrier concentration (open circles) and Hall mobility (solid circles) for MBE-grown, Ga-doped ZnSe layers measured at room temperature as a function of Ga-cell temperature. (After Ref. 24)

mobility studies of the films at room temperature showed (see Figure 6) that the free electron concentration increased smoothly with Ga-oven temperature (i.e. also with J_{Ga}) up to a maximum of n_D-$n_A \approx 5.10^{17}$ cm^{-3}. Carrier concentrations covering the range 2.10^{14} cm^{-3} (at T_{Ga} =

300°C) to 5.10^{17} cm^{-3} (at T_{Ga} = 475°C) were achieved. At higher temperatures (T_{Ga} = 475-600°C) the free carrier concentration decreased. This decrease occurred in the same temperature range as a broadening of the x-ray rocking curve and a dilation of the film. This suggests that above T_{Ga} = 475 °C the solubility limit of substitutional Ga was exceeded with consequent formation of Ga precipitates and other lattice defects including V_{Zn} - Ga complexes which would act as compensating acceptors. Room temperature photoluminescence spectra recorded from films grown over the T_{Ga} range 300-600°C showed that the near band edge emission at λ = 4610Å reached a maximum intensity at T_{Ga} = 425°C. At higher values of T_{Ga} the intensity of a broad emission due to a Zn vacancy-donor complex, increased at the expense of the near band edge emission.

The promising aspect of this study is the low (0.05 Ωcm) film resistivity achieved at the optimized Ga oven temperature (475°C). This is lower than that achieved by closed-ampoule growth[34] of ZnSe platelets from Ga solution. In addition, the low background carrier concentration ($\lesssim 2.10^{14}$ cm^{-3}) of the films suggests that interdiffusion between film and substrate is not a serious problem at the growth temperature of 360°C.

3.2 Hg_xCd_xTe Ternary Alloy Films and HgTe-CdTe Mulitlayer Structures

3.2.1. $Hg_{1-x}Cd_xTe$ Ternary Alloy Films. The early vacuum evaporation experiments of Hohnke *et al.*[36] showed that single phase $Hg_{1-x}Cd_xTe$ films could be prepared from separate Hg, Te$_2$ and Cd beams provided that the Hg impingement rate at the sampe was equivalent to a beam pressure at least as high as the equilibrium Hg pressure at the Te-rich phase boundary of the alloy homogeneity region. The poor background vacuum (p>10^{-7} Torr) and lack of clean, ordered, lattice-matched substrates in this work led to growth of essentially polycrystalline films. Independent confirmation was obtained by the author and co-workers of the ability of Hg to react with Te$_2$ at low (room temperature to 200°C) temperatures. A near-stoichiometric beam of Hg + Te$_2$ (i.e. $J_{Hg} \approx 2J_{Te_2}$) derived[2] from a Knudsen effusion oven containing $Hg_{0.8}Cd_{0.2}Te$ was impinged on a room temperature glass slide. Since the Hg is volatile (P_{Hg} over Hg $\approx 10^{-3}$ Torr) at room temperature it would desorb from the substrate surface if it did not undergo a reaction with Te adatoms to form HgTe bonds. EDAX and x-ray diffraction analysis of the condensate on the glass side confirmed that HgTe was present thus showing that surface dissociation of Te_2 and reaction with Hg was possible even at room temperature. In a similar way, the $Cd + 1/2\ Te_2 \to CdTe$ reaction was also found to proceed at room temperature. $Hg_{1-x}Cd_xTe$ alloy films grown from the alloy source onto glass slides were, however, microcrystalline and clearly the next step was to explore growth of the alloy on clean, ordered substrates with a good lattice match to the alloy. In 1981, Faurie et al.[18] showed for the first time that epitaxial films of the alloy could be grown on CdTe substrates. HgTe was used as a Hg source since it provided a source of Hg with a small $(J_{Te_2}/J_{Hg} \lesssim 1\%)$ Te_2 contribution at temperatures below 220°C. This permitted the MBE system, including the HgTe Knudsen effusion oven, to be baked at $\sim 150°C$ prior to film growth. Alloy films grown[38] at a substrate temperature of 100°C exhibited low mobilities and unusually low carrier concentration characteristics[39] of poor structural quality films. A major contributory reason for this was probably the drift and short term fluctuations in both Hg and Te_2 fluxes generated from the HgTe source (see section 2.1). The low growth temperature may also have been a contributory factor. To achieve alloy growth at higher temperatures, which is desirable for an improvement in film crystallinity, very high fluxes of Hg are required. Even at 120°C the minimum Hg impingement flux required for single phase, epitaxial alloy film growth is[40] 5×10^{16} cm^{-2} s^{-1}. In order to provide the stable, high impingement fluxes required for alloy film growth Faurie et al.[38,39,40] used a liquid Hg source. In these studies, it was found that the Hg sticking coefficient was particularly low and strongly temperature dependent. Table 2 summarizes the minimum Hg impingement flux (J_{Hg}) required[40] for single phase alloy growth of epitaxial p-type films at a Te_2 flux corresponding to a film growth rate (on the (111) orientation) of 2 μm h^{-1}. Since the film growth rate is controlled by the non-volatile species Te_2, the effective sticking coefficient of Hg is

$$\sigma_{Hg} = (2J_{Te_2} - J_{Cd})/J_{Hg} = 1.6 J_{Te_2}/J_{Hg} \text{ for } x = 0.2 \qquad (5)$$

and is listed in Table 2. The Hg impingement fluxes, J_{Hg}', corresponding to the equilibrium pressure of Hg at the Te-rich boundary of $Hg_{0.8}Cd_{0.2}Te$ are also listed. The high values of J_{Hg} required at growth temperatures of 160°C and above correspond to considerable mass depletion rates of the Hg charge. For a Hg oven orifice of 1 cm^2 in area at a distance of 15 cm from the substrate surface the calculated

TABLE 2. Hg impingement fluxes and source conditions for MBE-growth of $Hg_{0.8}Cd_{0.2}Te$ films

T_s °C	J_{Hg} (cm^{-2}/s)	$J_{Hg'}$ (cm^{-2}/s)	σ_{Hg}	P_{Hg} (Torr)	$\overset{*}{m}$ gm/h
120	5×10^6	1.1×10^{15}	1×10^{-2}	0.25	42
160	1.8×10^{17}	1.6×10^{16}	3×10^{-3}	0.89	152
180	5×10^{17}	8.6×10^{16}	1×10^{-3}	2.47	423

Notes:

1. J_{Hg} values are those required for growth of p-type films, data from Faurie, et al. (Ref. 40)

2. $J_{Hg'}$ values are the calculated impingement fluxes needed to compensate the Hg loss rate from the alloy at the Te-rich phase boundary. Data from Ref. 36.

3. σ_{Hg} is the effective sticking coefficient for Hg, i.e., 1.6 J_{Te_2} / J_{Hg} where J_{Te_2} = 3.34 x 10^{14} cm^{-1} s^{-1} and the film growth rate is $\sim$ 2µm h^{-1}.

4. P_{Hg} is the Hg pressure in the effusion oven needed to sustain the Hg impingement flux J_{Hg}. $\overset{*}{m}$ is the mass depletion rate of the Hg source. These values assume molecular beam effusion conditions and an orifice of 1 cm^2 at 15 cm from the sample. However, at these pressures, collisions in and in front of the orifice may result in lower values of $J_{Hg'}$ then those calculated for the values of P_{Hg} listed. Conversely, higher values of P_{Hg} may be required to sustain the listed fluxes. The values of $\overset{*}{m}$ will remain similar to those listed.

source pressure (P_{Hg}) and mass depletion rate ($\dot{m}$) are listed in the final columns of Table 2. At $T_s = 160°C$, the Hg source pressure required is 0.89 Torr which corresponds to a source depletion rate of 152 gm h^{-1}. Thus a 40 cc charge of Hg, which weighs 544 grams contains only sufficient material for growth of a single $7\mu m$ thick film of the alloy. Since films of at least this thickness are required for photoconductive and photovoltaic detector applications, this problem and the high room temperature Hg vapor pressure sets several technological problems. Firstly, the Hg charge requires replenishing after each growth run. Faurie *et al.* achieved this [39,40] by transferring a filled Hg effusion oven through a vacuum lock prior to each run. However, this is not an ideal solution since some oxidation of Hg might occur in loading. Also baking of the filled oven in UHV to desorb atmospheric gases is not possible because of the volatility of Hg. These problems may be responsible for the run to run variability of structural and electrical properties of the alloy films observed by Faurie *et al.*[40]. An alternative approach[41] is a continuous feed system in which Hg can be introduced from an external reservoir through a feeder tube which passes through the main system wall into the orifice of the Hg oven. During alloy film growth most of the Hg effusing from the source will condense onto cryopanels around the ovens and substrate. At the end of a sequence of film depositions the Hg can be transferred, by mild $\sim$ 130°C bakeout, into a separate Hg storage chamber which can be isolated after the transfer.

The growth of $Hg_{1-x}Cd_xTe$ films for infrared detector arrays places severe requirements on alloy uniformity both spatially and through the depth of the film. In addition, control of both n- and p-type carrier concentrations down to background levels of $\sim 10^{14}$ cm^{-3} is required to satisfy the requirements of a variety of types of detector structure. For example, for the newly emerging "SPRITE" photoconductive detector arrays,[42] based on scanning of the infrared image in synchronism with the drift of minority carriers (holes), x = 0.2 material with good ($\gtrsim 5\mu S$) minority carrier lifetime and low ($n_D - n_A \lesssim 10^{15}$ cm^{-3}) carrier concentration is required. Similarly, for the fabrication of photodiode arrays by ion implantation in p-type alloy films carrier concentrations $\lesssim 10^{16}$ cm^{-3} are needed. For advanced monolithic detector structures[43] in which signal processing on the focal plane is carried out using charge coupled devices in CdTe or wide-gap $Hg_{1-x}Cd_xTe$ films, arbitrary control of carrier concentration in both materials is required. At the present time Faurie and co-workers have demonstrated[39,40] that high purity alloy films, both n- and p-type can be prepared by MBE in the alloy composition range (x = 0.2 → 0.3) of interest for detector array applications in the wavelength range 8-12 μm and 3-5 μm. Table 3 lists the properties of some of the more promising films prepared to date.

TABLE 3. Electrical properties of some $Hg_{1-x}Cd_xTe$ MBE-grown films (After Faurie, et al.[40])

T_g (°C)	Film Thickness (m)	x	Carrier Conc.* (cm^{-3})	Mobility* ($cm^2V^{-1}s^{-1}$)
180	6	0.23	p 2.0 10^{15}	660
180	7	0.20	p 3.0 10^{17}	300
180	3.8	0.21	n 3.0 10^{15}	2.2 x 10^4
180	7	0.18	n 1.2 10^{15}	1.85 x 10^5

* From Hall mobility measurements at 40K for p-type films and 77K for n-type films.

The highest mobility p- and n-type films are of device quality and are comparable in electrical quality with films grown by liquid phase epitaxy and bulk samples. The lateral uniformity of alloy composition achieved in these samples is Δx = 0.01 across areas of ~ 1 cm², but this could be significantly improved and extended to larger areas by sample rotation in a second generation MBE system. The surface morphology of the alloy films, grown at 180°C was similar to that (surface roughness ~ 0.1μm) of films grown at 100°C.

3.2.2 HgTe-CdTe Multilayer Structures. It has been suggested, on the basis of a variety of calculations[3-6] that quantum well and superlattice structures comprising the symmetry-induced, zero-gap semiconductors,

HgTe or α-Sn, confined by lattice-matched CdTe films should show novel and potentially useful properties. In particular these suggestions have provided a driving force for studies of HgTe-CdTe multilayer structures. Faurie *et al.*[25] have presented preliminary information on the preparation of such structures by MBE techniques. The growth was carried out with three effusion ovens. CdTe for growth of CdTe; Te and Hg for growth of HgTe. The substrates were (111) orientation CdTe wafers which were heat-cleaned at 300°C prior to film growth. Alternating HgTe-CdTe multilayer structures were grown at three different growth temperatures as indicated in Table 4.

TABLE 4.   Characteristics of multilayers grown at different temperatures; (n is the number of layers and e is the layer thickness)

Sample	T_s (°C)	CdTe		HgTe	
		n	e (Å)	n	e (Å)
1	120	7	600	7	1600
2	160	13	150	14	400
3	200	100	44	100	180

From RHEED observations during growth it was clear that although the structures were epitaxial at 120°C the film perfection was much improved at 200°C. The CdTe film quality was particularly sensitive to growth temperature. At 120°C and 160°C film quality deteriorated (indicated by increasing broadening and diffuseness of RHEED spots) with increasing CdTe film thickness. Auger and SIMS depth profiles for the multilayer grown at 160°C indicate an upper limit to interdiffusion of about 40Å No data for interdiffusion in the multilayers grown at

200°C were presented and no optical or electrical data for the structures were presented.*

4. MBE GROWTH OF IV-VI COMPOUNDS AND ALLOYS

As pointed out in the introduction, interest in MBE growth of IV-VI compounds and related alloys has continued, largely because of their suitability as materials for temperature tunable laser diodes for pollution monitoring applications. In 1979, Holloway and Walpole[14] reviewed MBE growth techniques for growth of IV-VI compounds and alloys on BaF_2 and SrF_2 substrates for applications in infrared photodiode detector arrays. The detectivities achieved in linear arrays of $PbSe_{0.8}Te_{0.2}$ (cut-off wavelength 6μm at 150K) and $Pb_{0.93}Sn_{0.07}Se$ (cut-off wavelength 12μm at 80K) reached background limited values. However, problems remained in amplifier matching at high frequencies because of the high dielectric constants of the alloys. Furthermore, as Charlton[44] has pointed out, the large difference in thermal expansion match between the IV-VI alloys and silicon was a major reason for the choice of the alloy $Hg_{1-x}Cd_xTe$ as a more versatile detector material which could be interfaced to silicon circuitry in hybrid schemes for focal plane signal processing.

The use of MBE techniques for growth of double heterostructure Pb-salt injection laser structures was also reviewed by Holloway and Walpole[14]. Although the highest operating temperatures for such structures were achieved by MBE techniques, the laser threshold current densities were large ($\gtrsim 10^3$ A cm^{-2}) even at low ($\sim$ 20K) heat sink temperatures. These high threshold currents have been attributed[45] to a high surface recombination velocity which is caused by lattice mismatch at the heterostructure interfaces. In this context it is worth noting that compared with the GaAs-AlAs materials system the lattice mismatch is often large in IV-VI heterostructures. For example, the numerical value of the mismatch for $PbTe$-$Pb_{0.8}Sn_{0.2}Te$ is $\Delta d/<d>$ = 0.43% which inevitably results in appreciable misfit elastic strain and misfit dislocation formation in the active layer of the structure. Partin and Lo[46] have subsequently shown that MBE-grown homojunction lasers, in which thick buffer layers were used and carrier confinement was achieved by tailoring the carrier concentration profile in the structure had much lower threshold current densities that DH laser structures. In addition to developing MBE techniques[47] for such structures Partin has also explored growth of $Pb_{1-x}Ge_xTe$[48] and $Pb_{1-x}Yb_xTe$[49] in

an attempt to develop viable structures for short (<6μm) wavelength
lasers.

4.1. PbTe

In a series of experiments leading up to growth of low threshold
PbTe homojunction lasers, Partin[47] studied MBE growth and extrinsic
doping of PbTe films. In these experiments, PbTe films were grown

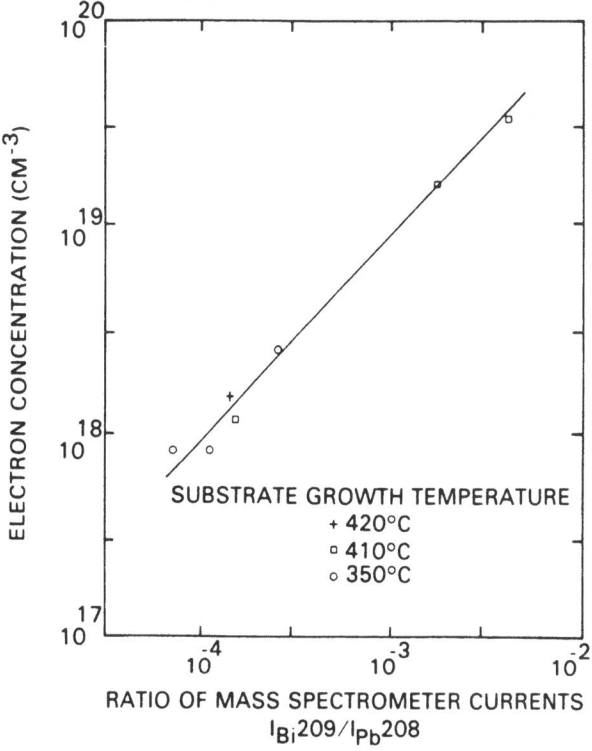

Figure 7 - Free electron concentration in PbTe films as a function of
fractional BiTe flux for MBE growth of Bi-doped PbTe from a single
PbTe effusion source and a separate Bi_2Te_3 effusion source. Note the
straight line fit and absence of any significant dependence of free
electron concentration on growth temperature. These features are
consistent with a unity incorporation coefficient of Bi into the film
(After Ref. 47).

using a single PbTe compound source. Compound sources of Bi_2Te_3
and Tl_2Te were used as sources of Bi and Tl for n- and p-type doping
respectively. As Strauss has shown[50] Bi (group V) and Tl (group III)
occupy Pb lattice sites in PbTe and their doping action is relatively

unaffected by the presence of excess Pb or Te in the crystal lattice. Smith[13] had shown earlier that Bi and Tl were suitable dopants in PbTe when supplied as atoms from separate elemental dopant sources. However, to achieve p-type doping levels above 10^{18} cm^{-3} using the Tl source it was found to be necessary to supply an additional Te$_2$ flux to generate Pb vacancies. Partin found that if Bi$_2$Te$_3$ and Tl$_2$Te were used as compound sources for doping, levels well into the 10^{19} cm^{-3} range could be achieved without an auxiliary Te$_2$ flux. This is not surprising since dissociation of the compound provides an automatic source of Te$_2$. For example, Uy and Drowart[51] have shown that Bi$_2$Te$_3$ dissociates according to:

$$Bi_2Te_3(s) \rightarrow 2\, Bi\, Te(g) + 1/2\, Te(g)$$

and that BiTe molecules have a similar dissociation energy (228.5 $\pm$ 11.3 k J mol^{-1}) to PbTe (246.1 $\pm$ k J mol^{-1}). This means that not only are the Bi atoms supplied to the growth surface in a stable molecular unit with Pb substituted by Bi but also the small excess flux of Te$_2$ provided by compound dissociation provides the necessary tendency towards Pb vacancies in the film. As a result Bi is incorporated into PbTe with unity incorporation efficiency. Figure 7 shows the measured free electron concentration (n_D - n_A) in Bi-doped PbTe as a function of fraction BiTe flux in the beam measured mass spectrometrically. (Clearly, Partin used a high, > 35 eV value of electron energy in these measurements so that the Pb$^+$ ion current was the dominant ion species and proportional to the PbTe beam flux). The linear dependence of n_D - n_A on [Bi$^+$]/[Pb$^+$] is consistent with a unity incorporation coefficient, i.e.

$$\frac{n_D - n_A}{N_t} = \alpha_o \frac{W_d}{G_t}$$

where N_t is the total density of moelcules cm^{-3} in the grown films, W_d is the dopant flux on the substrate, G_t is the total number of PbTe molecules cm^{-2} s^{-1} incorporated into the growing film and α_o is the number of charge carriers per dopant molecule (BiTe).

Partin[47] found that the mobility of homoepitaxial PbTe increased by a factor of >3 as the film thickness was increased from 1 to 7μm, possibly as a result of the decreasing influence of substrate-film interfacial dislocations. Another possibility might be the decreasing contribution of surface scattering of carriers resulting from increasing film surface smoothness with increasing film thickness.

Subsequent fabrication[46] of broad area homojunction laser diode structures, consisting of an MBE-grown p-n junction on a $6\mu m$ thick buffer layer of PbTe, resulted in devices with CW threshold current densities as low as 40A cm^{-2}, high temperature operation (85K CW, 150K pulsed) and high CW output power (3.5 mW).

4.2 Pb$_{1-x}$Ge$_x$Te, Pb$_{1-x}$Yb$_x$Te

In order to extend the emission wavelengths of injection lasers to shorter wavelengths than $6\mu m$, new adjustable gap IV-VI alloys are required. While Pb$_{1-x}$Cd$_x$Te, Pb$_{1-x}$Mn$_x$Te and Pb$_{1-x}$Ge$_x$Te were originally thought to be promising candidates they all have properties which make them unsuitable in device applications. For example at MBE growth temperatures ($\sim 400°C$ for optimum quality films of IV-VI compounds) the solubilities of Cd (2%), Mn (2%) and Ge (7%) are small.[53-55] In addition, as Partin[49] has pointed out, Cd is a fast diffusing n-type dopant in PbTe, Mn atoms tend to cluster in PbTe and Ge tends to re-evaporate from the film surface as GeTe during MBE growth of Pb$_{1-x}$Ge$_x$Te[48]. The limited solubilities are a result of the different crystal structures of CdTe, MnTe and GeTe from the rock salt structure of PbTe. The compounds YbTe and EuTe, on the other hand, have the rock salt structure and are expected to have extended solid solubilities in both PbTe and Pb$_{1-x}$Sn$_x$Te. YbTe (Eg = 1.8eV) and EuTe (Eg = 2.0eV) are wide gap semiconductors and so offer the possibility of visible emission wavelengths. The lattice parameter of YbTe (6.375Å) is smaller than that of PbTe (6.460Å) whereas EuTe is larger (6.585Å) so that Pb$_{1-x}$Yb$_x$Te is a potentially more useful alloy since it can be lattice-matched to Pb$_{1-x}$Sn$_x$Te. Partin has studied the MBE growth of this alloy on BaF$_2$ and PbTe substrates using PbTe, Yb, Te, Bi$_2$Te$_3$ (n-dopant) and Tl$_2$Te (p-dopant) sources. Although epitaxial films of the alloy with x-values up to 0.2 could readily be grown it was found that p-doping of the films with Tl was not possible due to the presence of an intrinsic compensating donor level associated with the Yb^{+2} = Yb^{+3} + e$^-$ transition. More promising p-type doping characteristics were obtained for the quaternary alloy Pb$_{1-x-y}$Yb$_x$Sn$_y$Te, for particular y values, as a result of the valence band edge moving above the intrinsic donor level with increasing y. The practicability of such an alloy system seems questionable, however, in view of the extreme reactivity of Yb to atmospheric oxygen and even the hydrogen background in the growth system. Also, the possibility of fast oxygen diffusion down dislocations or low angle boundaries deep into the laser structure is a real possibility.

5. DEVICE APPLICATIONS

5.1 II-VI Compounds

5.1.1 Infrared Detectors. The first reported use of an MBE technique for preparation of $Hg_{1-x}Cd_xTe$ photodiodes was recently reported by Migliorato *et al.*.[19] In this technique CdTe was deposited from a single source onto a p-type $Hg_{1-x}Cd_xTe$ substrate with x = 0.295. As a result of the low (80°C) deposition temperature the films were esentially polycrystalline and were semi-insulating (>10^7 Ωcm). However, following evaporation of In contact pads onto the top of the CdTe film and annealing the structure in a reducing atomsphere at temperatures in the 80-160°C range the CdTe film and a region of the substrate under the pad exhibited n-type conductivity due to In-diffusion. This is schematically illustrated in Figure 8. The structure is essentially a top-contacted, passivated homojunction and acts as an efficient lateral collection photodiode. The lateral collection effect was used to meas-

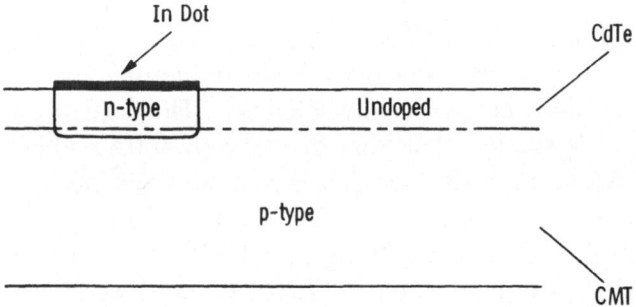

Figure 8 - Schematic diagram of CdTe/$Hgd_{1-x}Cd_xTe$ In-diffused diodes. (After Ref. 19).

ure the minority carrier diffusion length by photocurrent measurements during He-Ne laser ($\lambda = 1.15$ μm) spot scans. Figure 9 shows the measured photocurrents as a function of spot distance (y) from the junction edge. For negligible surface recombination and for distances from the junction large compared with the spot diameter (10μm) the photocurrent decayed as $\exp(-y/L_n)$ where L_n is the electron diffusion length in the substrate. As indicated in Figure 9, the values of L_n were 15, 25 and 45 μm at 77K, 195 and 245K respectively. The diode saturation current as a function of $1/T$ was in good agreement with theory suggesting that recombination at the $CdTe/Hg_{1-x}Cd_xTe$ interface was negligible. This behavior is consistent with a slightly (compositionally) graded $CdTe/n$-type $Hg_{1-x}Cd_xTe$ interface in which electrons are repelled from the surface by a "bump" in the conduction band.[56] The technique has potential application in array fabrication and suggests that CdTe may be useful in passivating more convential photodiode arrays.

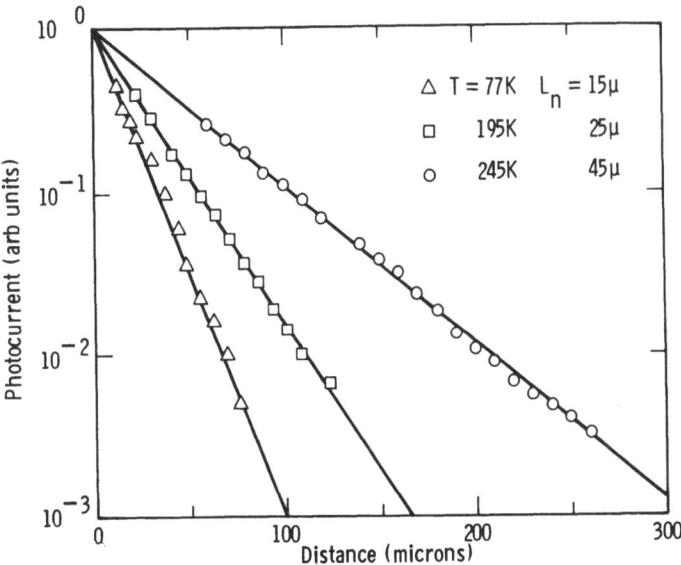

Figure 9 - Photocurrent (arbitrary units) versus spot distance (y) from the junction edge. For lateral collection homojunction in $Hg_{.705}Cd_{.295}Te$. Temperatures: 77K (Δ), 195K ($\square$), 245K (o). The three sets of data have been normalized at y = 0. The values of L_n deduced from the slope of the best-fit straight lines are also reported in the figure. (After Ref. 19)

Currently there is considerable interest in epitaxial techniques for preparation of thin films of $Hg_{1-x}Cd_xTe$ for application in photoconductive and photovoltaic devices. This interest stems directly from the major difficulties in achieving compositional uniformity and homogeneity, in addition to carrier concentration control, in large bulk crystals of the alloy. Since only thin (5-10μm thick) films of the alloy are required for device applications the attractions of epitaxial growth of the alloy onto large-area lattice matched substrates such as CdTe are obvious. Faurie *et al.*[39,40] have shown that device quality films of the alloy can be prepared by MBE on CdTe substrates and the first device result for an MBE grown photovoltaic device structure has recently[57] been reported. The device comprised a 5μm thick n-type alloy film grown onto a p-type alloy film on a CdTe buffer layer. The detectivity was reported to be $D^* = 3.5 \times 10^{10}$ cm $Hz^{1/2}$ W^{-1} with a quantum efficiency of 50% and cut-off wavelength of 9.4μm. This is a promising first result. The lateral alloy uniformity is clearly not yet adequate for device array fabrication but the use of a second generation MBE machine with wafer rotation can be expected to result in lateral x variations $\lesssim 0.5\%$ (i.e., comparable with the uniformity of MBE-grown films of $Ga_{1-x}In_xAs$ and $Al_xGa_{1-x}As$). The achievement of a similar uniformity in alloy composition through the depth of the film will be more difficult and techniques for stabilizing the $[Cd]/[Te_2]$ beam intensity ratio during growth will need to be developed.

5.1.2. Multilayer and Quantum Well Structures. The ability of MBE to generate multilayer quantum well structures of CdTe-HgTe has been convincingly demonstrated.[25] In the calculation of Broerman[6] a HgTe film of 128.6Å sandwiched between CdTe electron confining layers is needed to produce an effective optical gap of 0.08 eV. However, these calculations did not take into account the overlap of the degenerate valence bands in the HgTe band structure. The numerical value of this overlap is still in question but it is clear that for useful (for infrared device applications) bandgaps (0.08 - 0.30 eV) the HgTe well thickness should be $\lesssim$ 128.6Å. Also in the calculations of Schulman and McGill[3-5] for HgTe-CdTe superlattices the HgTe thickness required for bandgaps in the useful range are less than 10 monolayers ($\sim$ 65Å). These dimensions have not yet been achieved in the multilayer structures grown to date. In fact, Faurie has recently reported[57] for multilayer 3 (see Table 4), for which each period consists of 44Å CdTe, 180Å HgTe, the effective optical gap is < 5 meV. This is clearly a topical area which will attract considerable experimental effort in the next few years. One reason for this is that for quantum well and superlattice structures the requirement of bandgap uniformity is achievable by accurat control of film thicknesses rather than by depth and lateral uniformity of alloy composition. In addition to the prospect of a sharper band edge, the different occupancy statistics for a two dimensional quantum well structure may lead to a lower intrinsic carrier concentration in the well at a given temperature compared with the random alloy. This may have advantages in improved dark current in photoconductive detectors. However, perhaps a major materials advantage of a multilayer structure would be the greater thermal stability of such a structure. CdTe is much more thermally stable than HgTe (see Figure 1) and will encapsulate the HgTe. However, the significant lattice-mismatch ($\sim 0.30\%$) between these materials will inevitably lead to elastic strain (lateral tension, i.e., uniaxial compression) in the HgTe wells which will break the cubic symmetry and may significantly modify the effective gap. The growth and exploration of such structures is a promising new field in which MBE grown structures will permit critical areas of narrow-gap semiconductor physics to be probed in a direct way. Also, the ability of MBE to prepare relatively thick multilayer structures will allow, for the first time, fabrication of a number of novel structures (e.g., Ref. 58) for signal processing on the focal plane.

5.1.3 Light Emitting Devices. Recent progress in achieving high conductivity n-type films of CdTe and ZnSe by MBE (see Section 3.1.2) has been encouraging. However, for injection LED's high conductivity p-type material is also required. Progress towards p-doping has been made using a variety of techniques. For example high conductivity ($\lesssim 1$ Ωcm) Tl-doped ZnSe has recently been reported by Nakau et al.[60] who used Tl diffusion in a Zn-rich atmosphere. Other attempts[61] to dope ZnSe with Li were less successful. As pointed out in the Introduction (Section 1) these difficulties may[10] be related to the combined effects of compositional inhomogeneities, such as Te precipitates, and impurities such as Cu and Li in bulk crystals. Since the preparation of high perfection, high purity CdTe films has now been achieved[8] by MBE techniques extrinsic doping experiments with both n- and p-dopants are now timely. Recent studies[8,62] indicate that the stoichiometry of the MBE grown CdTe can be adjusted within the solidus region without corresponding changes in film conductivity. This suggests that native defects are not electrically active (at least in CdTe) in the absence of foreign impurities.

5.2 IV-VI Compounds

The fabrication of high performance MBE-grown PbTe homo-junction lasers by Partin and Lo[46] is a major advance in this area. However, the prospects for shorter wavelength laser structures based on new IV-VI alloys such as $Pb_{1-x}Yb_xTe$ do not seem bright in view of difficulties in controlling p-type doping levels. Whether the quaternary system (PbSnYbTe) or an alternative ternary system (PbSrS) presently under investigation[63] prove viable material systems for device applications is an open question at present.

6. CONCLUSIONS

Molecular beam epitaxy is a particularly powerful and versatile technique for the growth and exploration of II-VI and IV-VI compounds and alloys. Modulated beam mass spectrometry studies have revealed that II-VI compounds and alloys containing Hg dissociate readily at very low ($\lesssim 200°C$) temperatures. Low growth temperatures for HgTe and the important narrow gap alloy $Hg_{1-x}Cd_xTe$ are therefore called for and can be readily achieved by MBE techniques. Device quality epitaxial films of $Hg_{1-x}Cd_xTe$ have been prepared and although alloy uniformity problems remain, the use of sample rotation and stable beam sources in a second generation beam system provides a potential solution. Multilayer CdTe-HgTe structures have been prepared by MBE and the interdiffusion distances ($\lesssim 40A$) appear to be sufficiently small for quantum well and superlattice structures to be prepared and investigated. The low ($\lesssim 200°C$) growth temperatures which are possible by MBE using the single Knudsen oven approach has enabled high purity, high structural perfection films of CdTe to be grown on InSb. This is a substrate which is available in large areas and with high structural perfection. Since it transmits beyond $\lambda = 5.7\mu m$ it is likely to be a useful substrate for infrared detectors (grown by MBE) operating in the $8-13\mu m$ atmospheric window.

On a more fundamental materials science level the availability of high purity, high structural perfection CdTe films, grown by MBE on InSb substrates, has opened up the prospects for extrinsic doping studies on well-characterized material. In addition, double crystal x-ray diffraction techniques are seen to be sufficiently sensitive to permit a study of the relationship between film stoichiometry and growth parameters. A similar approach may be possible for other II-VI compounds which sublime congruently and have a near lattice-matching III-V compound substrate of high structural perfection, e.g. ZnSe-GaAs.

Developments in MBE-growth of wide gap II-VI compounds with high conductivity, n-type doping have been encouraging in the cases of CdTe and ZnSe. Attempts at p-type doping can be expected.

In the area of MBE-growth of IV-VI compounds, significant progress in the growth and controlled extrinsic doping of PbTe homo-junction laser structures has been made but major problems remain in preparing devices quality films of IV-VI compounds and alloys suitable for shorter wavelength ($< 6\mu$m) lasers.

REFERENCES

* Magneto-optical investigations of a HgTe-CdTe superlattice hae now been reported by Y. Guldner, G. Bastard, J. Vieren, M. Voos, J.P. Faurie, A. Million, Phys. Rev. Lett. 51, 907 (1983).

1. P.W. Kruse in "Semiconductors and Semimetals", Vol. 18, page 1, Chapter 1, edited by R.K. Williardson and A.C. Beer, Academic Press, 1981.
2. R.F.C. Farrow, G.R. Jones, G.M. Williams, P.W. Sullivan, W.J.O. Boyle, J.T.M. Wotherspoon, J. Phys. D: Appl. Phys. 12, L117 (1979).
3. J.N. Schulman, T.C. McGill, Appl. Phys. Lett. 34, 663 (1979).
4. J.N. Schulman, T.C. McGill, J. Vac. Sci. Technol. 16, 1513 (1979).
5. J.N. Schulman, T.C. McGill, Solid State Comm. 34, 29 (1979).
6. J.G. Broerman, Phys. Rev. Lett 45, 747 (1980).
7. R.F.C. Farrow, D.S. Robertson, G.M. Williams, A.G. Cullis, G.R. Jones, I.M. Young, P.N.J. Dennis, J. Crystl. Growth 54, 507 (1981).
8. R.F.C. Farrow, G.R. Jones, G.M. Williams, I.M. Young, Appl. Phys. Lett. 39, 954 (1981).
9. Y. Marfaing, Prog. Crystal Growth Characterization 4, 317 (1981).

10. J.L. Pautrat, N. Magnea, J.P. Faurie, J. Appl. Phys. 53, 8668 (1982).

11. See for example: H. Sano, R. Koga, M. Kosaks, K. Shinohara, Japan J. Appl. Phys. 20, 2145 (1982).

12. D.L. Smith, V.Y. Pickhardt, J. Appl. Phys. 46, 2366 (1975).

13. D.L. Smith, Prog. Cryst. Growth Charact. 2, 33 (1979).

14. H. Holloway, J.N. Walpole, Prog. Crystl. Growth Charact. 2, 49 (1979).

15. P. Goldfinger, M. Jeunehomme, Trans. Faraday Soc. 59, 2851 (1963).

16. R.F. Brebrick, A.J. Strauss, J. Phys. Chem. Solids 25, 1441 (1964).

17. R.F.C. Farrow, to be published.

18. J.P. Faurie, A. Million, J. Cryst. Growth 54, 582 (1981).

19. P. Migliorato, R.F.C. Farrow, A.W. Dean, G.M. Williams, A.M. White, Infrared Physics 22, 331 (1982).

20. T. Yao, Y. Miyoshi, Y. Makita, S. Maekawa, Japan J. Appl. Phys. 16, 369 (1977).

21. J.P. Faurie, A. Million, J. Crystl. Growth 54, 577 (1981).

22. See for example: "Introduction to Advanced Inorganic Chemistry", P.J. Durrant, B. Durrant, Wiley 1962, p. 487.

23. K. Sugiyama, Japan J. Appl. Phys. 21, 665 (1982).

24. T. Niina, T. Minato, K. Yoneda, Japan J. Appl. Phys. 21, L387 (1982).

25. J.P. Faurie, A. Million, J. Piaguet, Appl. Phys. Lett. 41, 713 (1982).

26. G.R. Jones, I.M. Young, B. Cockayne, G.T. Brown, Inst. Phys. Conf. Series No. 60: Section 5, 265 (1981).

27. Spectrum recorded by P.J. Dean (RSRE Malvern U.K.) of a film grown by G.M. Williams, RSRE, Malvern, U.K.

28. Personal communication with P.J. Dean.

29. C.E. Barnes, K. Zanio, J. Appl. Phy. 46, 3959 (1975).

30. D.R. Wight, D. Bradley, G. Williams, M. Astles, S.J.C. Irvine, C.A. Jones, J. Cryst. Growth 59, 323 (1982).

31. A.J. Noreika, M.H. Francombe, C.E.C. Wood, J. Appl. Phys. 52, 7416 (1981).

32. T.H. Myers, Yaw-cheng Lo, J.F. Schetzina, S.R. Jost, J. Appl. Phys. 53, 9232 (1982).

33. B. Goldstein, "III-V Surface Studies" Final Report for period Nov. 1, 1974 to June 30, 1975., U.S. Army, Night Vision Lab.,

Contract DAAK02-74-C-0081, RCA Laboratories, Princeton, NJ.

34.   I.V. Varlamov, L.A. V'yukov, A.M. Gulyaev, L.P. Lazarenko, A.N. Solyakov, Sov. Phys. Semicond. 14, 1213 (1980).

35.   B.J. Fitzpatrick, B.N. Bhargava, S.P. Herko, P.M. Harnack, J. Electrochem. Soc. 126, 341 (1971).

36.   D.K. Hohnke, H. Holloway, E.M. Logothetis, R.C. Crawley, J. Appl. Phys. 42, 2487 (1971).

37.   R.F.C. Farrow, Review paper on "MBE Growth of II-VI and IV-VI Compounds", presented at First European Conference on Molecular Beam Epitaxy, March 30-April 1, 1981, Stuttgart, Germany.

38.   J.P. Faurie, A. Million, G. Jacquier, Thin Solid Films 90, 107 (1982).

39.   J.P. Faurie, A. Million, Appl. Phys. Lett. 41. 264 (1982).

40.   J.P. Faurie, A. Million, J. Piaguet, J. Cryst. Growth 59, 10 (1982).

41.   D.J. Williams, (VG Scientific) personal communication.

42.   C.T. Elliot, Electron Lett 17, 312 (1982).

43.   W.A. Gutierrez, J.H. Pollard, U.S. Patent 4,228,365 (1980).

44.   D.E. Charlton, J. Cryst. Growth 59, 98 (1982).

45.   D. Kasemset, C.G. Fonstad, Appl. Phys. Lett. 34, 432 (1979).

46.   D.L. Partin, W. Lo, J. Appl. Phys. 52, 1579 (1981).

47.   D.L. Partin, J. Electronic. Mat. 10, 313 (1981).

48.   D.L. Partin, J. Vac. Sci. Technol. 21, 1 (1982).

49.   D.L. Partin, in press, J. Vac. Sci. Technol. (1982).

50.   A.J. Strauss, J. Electron. Mater. 2 553, (1973).

51.   O.M. Uy, J. Drowart, Trans. Farad, Soc. 65, 3221 (1969).

52.   R. Colin, J. Drowart, Trans. Farad. Soc. 60, 673 (1964).

53.   A.J. Rosenberg, R. Grierson, J.C. Wooley, P. Nikolic, Trans. Metal. Soc., AIME 230, 342 (1964).

54.   V.G. Vanyarkho, V.P. Zlomanov, A.V. Novoselova, Inorg. Mater. 6, 1352 (1970).

55.   D.K. Hohnke, H. Holloway, S. Kaiser, J. Phys. Chem. Solids 33, 2053 (1972).

56.   P. Migliorato, A.M. White, Solid State Elect. 26, 65 (1983).

57.   J.P. Faurié, "Récent Developments in MBE Growth of $Hg_{1-x}Cd_xTe$" paper presented at 1983 U.S. Workshop on the Physics and Chemistry of Mercury Cadmium Telluride, Dallas, TX, February 8-10, 1983.

59. W.A. Gutierrez, J.H. Pollard, U.S. Patent 4,228,365, Oct. 14, 1980.
60. T. Nakau, T. Fujiwara, S. Yoshitake, H. Takenoshita, N. Itoh, M. Okuda, J. Cryst. Growth 59, 196 (1982).
61. G.E. Newmark, S.P. Herko, J. Cryst. Growth 59, 189 (1982).
62. G.M. Williams, Private communication.
63. H. Holloway, G. Jesion, Phys. Rev. B. 26, 5617 (1982).

SILICON MOLECULAR BEAM EPITAXY

F. W. Saris and T. de Jong

FOM-Institute for Atomic and Molecular Physics, Kruislaan 407, 1098 SJ Amsterdam, The Netherlands

ABSTRACT: This paper gives a review of our experience in the past four years with Silicon Molecular Beam Epitaxy. First, our equipment for Si-MBE is discussed along with two new methods for surface preparation, then comes the determination of the lowest epitaxial temperatures on Si(111) and Si(100) surfaces; results of doping by off-line ion implantation and of Si-MBE on GaP are given. Where appropriate a comparison is made to other work and the bibliography may serve as a guide through other literature.

1. INTRODUCTION

Surface engineering of silicon in high vacuum started in the sixties with great optimism, but soon the physicists lost a race and today integrated circuits are fabricated on silicon wafers via chemical processing techniques. The demand for very large scale integration, however, has become a major driving force for reducing process temp-eratures and gradually physical processing techniques are penetrating the silicon integrated circuit manufacturing facilities. In this respect the growth under ultrahigh vacuum conditions of "device-quality" epitaxial silicon films has meant a major break-through. This is reflected in the steep increase since then of the number of publications on silicon molecular beam epitaxy (see Fig. 1 and Ref. 1). Today many groups are active in this field which has led to significant results regarding: UHV chamber designs, sample preparation, growth conditions, doping param-eters, heteroepitaxy capabilities and last but not least novel device structures.

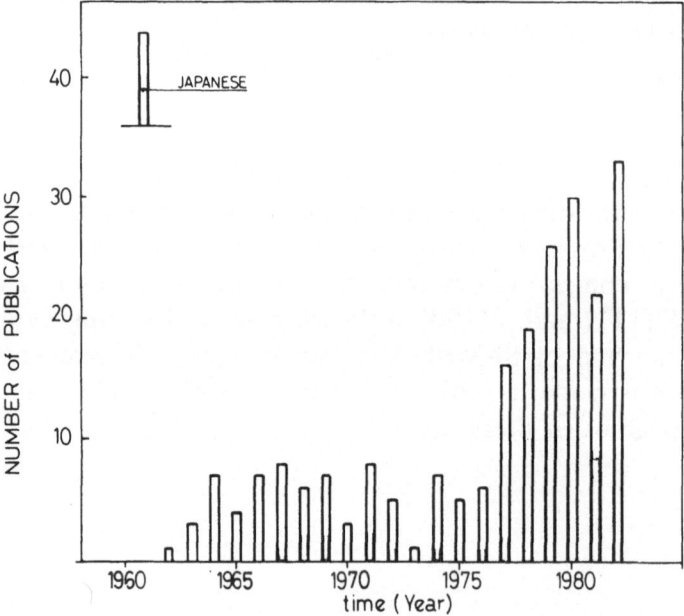

Figure 1 - Number of Si:MBE publications as a function of time.

Architecture of silicon structures with atomic dimensions using beams of atoms, molecules, ions, photons and electrons has become a challenging objective of our laboratory for atomic and molecular physics. In the past four years we have gained experience on which we will report at here.[2] We will discuss: (1) our equipment for Si-MBE and surface preparation by pulsed laser irradiation, (2) determination of the lowest epitaxial temperatures on Si(111) and Si(100) surfaces, (3) doping by off-line ion implantation, (4) silicon molecular beam epitaxy on gallium phosphide and finally (5) some conclusions that can be drawn from this work, especially with regard to surface diffusion and epitaxial growth mechanisms. Where appropriate a comparison is made to other work and the bibliography may serve as a guide through the literature.

2. EXPERIMENTAL

2.1 Apparatus

Our MBE apparatus consists of three UHV, stainless steel chambers, separated by manually operated valves, see Fig. 2. The upper chamber contains the sample and analysis equipment, the lower chamber houses a UHV adapted 10 kW electron-gun evaporator. Samples are introduced into a storage chamber also used as fast load lock. All chambers are equipped with turbo-molecular pumps, liquid nitrogen cooled titanium sublimation units and ion pumps. Background pressure in all chambers is less than 1×10^{-8} Pa. A liquid nitrogen cooled copper panel envelops the evaporator. During evaporation the pressure rises to 5×10^{-7} Pa. which consists mainly of hydrogen, partial pressures of O_2, CO, NO, and H_2O being in the 10^{-9} Pa. range. With the low sticking coefficient of contaminating gases in the order of 10^{-3} and with a typical deposition rate of 0.1 nm/s, we calculate the concentration of oxygen and other contaminants in deposited layers to be $<10^{-7}$ ($<10^{15}$ atoms/cm^3). In the experiments a Q-switched ruby laser (λ = 694.3 nm), delivering pulses of 20 ns duration at energies of 2.0 J/cm^2, was used. Surface sensitive analysis consists of low energy electron diffraction (LEED) and Auger electron spectroscopy (AES). For this purpose a four-grid LEED system and a single-pass cylindrical mirror analyser (CMA) are present in the upper chamber. The above apparatus has been derived from the design of surface analysis chambers. It cannot handle the standard silicon sample formats of 3" or 4" wafers, neither is in situ doping facilitated. Clearly, this has limited our program to more fundamental growth problems of Si auto- and hetero-epitaxy. On the other hand we will show also results of combining the MBE equipment with a conventional ion implantation machine.

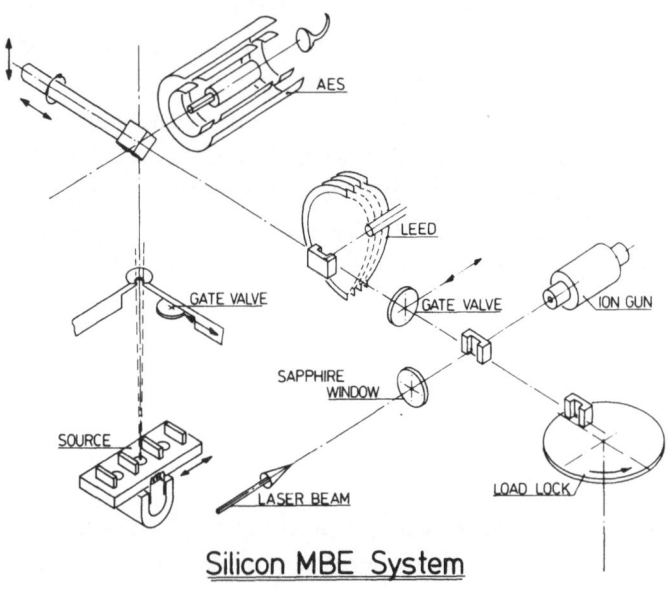

Figure 2 - Schematic view of Si:MBE system, not shown are thickness monitor and quadrupole mass spectrometer.

2.2 Surface Preparation

One of the key problems in Si-MBE is the cleaning of silicon surfaces by a preferably low temperature process. It has been reported that a 1 keV Ar+ beam can be used to sputter-clean silicon samples, followed by 1120 K anneal to reorder the surface and expel the argon.[3] In our group, however, we have the experience that sputtering through the native oxide causes damage to the silicon surface which cannot be restored to the quality required for surface analysis experiments. High quality clean Si surfaces are obtained reproducibly[4] by first heating to 870 K for 5 minutes then to 1170 K for 2 minutes, followed by mild sputtering (1 keV Ar ions, $20\mu C. cm^{-2}$), angle of incidence 70° with the surface normal and annealing at 1370-1520 K for 3 minutes, followed by cooling down at a rate of 100 K/min. A major drawback in this procedure is the rather high annealing temperature which may have adverse effects on the electrical properties of the wafers and which will lead to smearing out of possible doping profiles already present. Therefore we have investigated two new substrate preparation techniques.

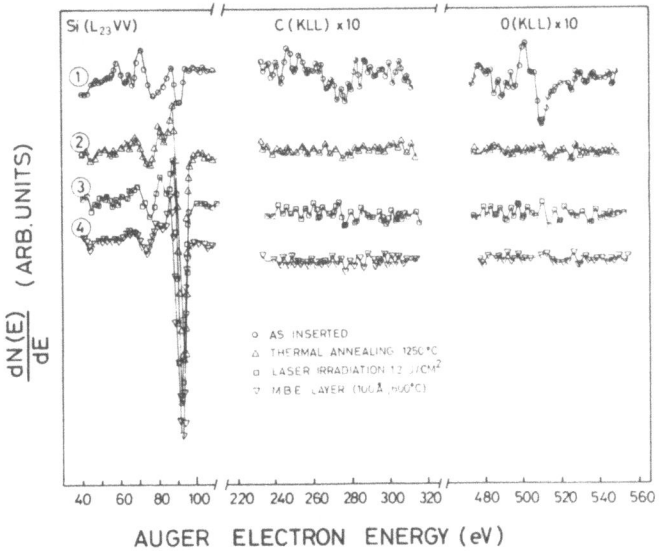

Figure 3 - Derivative Auger spectra of Si (91 eV), C (270 eV) and O (504 eV): 1) as inserted, 2) after thermal annealing at 1520 K, 3) after laser irradiation with 1.2 J/cm², 4) after epitaxial growth of 10 nm at 870 K.

First, in order to obtain well ordered, atomically clean surfaces, repeated pulsed laser irradiation may be used.[5,6,7] In Figure 3 our AES data on samples after surface preparation are summarized. We show the derivative Auger spectra of silicon, carbon and oxygen. No other elements could be detected on our samples at any stage of the experiments. The as inserted samples (curve 1) show a Si Auger line typical for a native oxide on the surface. Also some C was detected. After mild sputtering and thermal annealing the ratio of the peak heights of both C and O compared to that of Si were less than 5×10^{-3}, being the detection limit. At the same time a Si-LVV line at 91 eV was observed typical for a clean surface (curve 2). Similar clean surface results were obtained after 5-8 laser pulses of 1.5 J/cm^2 (curve 3). For comparison, curve 4 shows an AES spectrum taken after a 10 nm thick film has been grown by Si-MBE at 870 K.

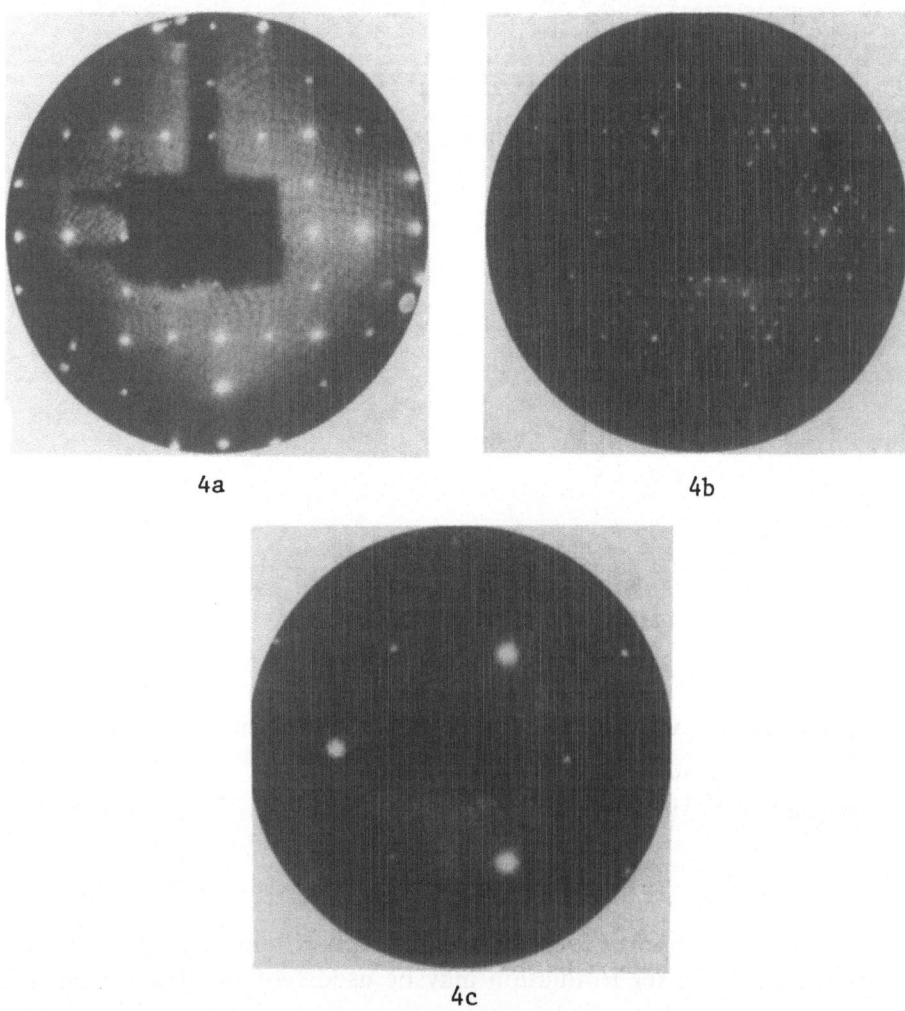

4a   4b

4c

Figure 4 - LEED patterns (80 eV) taken from clean surfaces: a) Si(100) thermally annealed, (2×1), b) Si(111) thermally ann-ealed, (7×7), c) Si(111) laser irradiated, apparent (1×1).

On all clean surfaces low-background LEED patterns were observed. Some examples, taken at 80 eV are reproduced below. On

Si(100) a (2×1) pattern is seen while on the thermally annealed Si(111) the (7×7) pattern is observed (Figs. 4a and 4b). In case of Si(100) a laser pulse does not change the (2×1) pattern but on Si(111) the (7×7) pattern is no longer seen: instead a (1×1) pattern appears (Fig. 4c). Laser-irradiated Si(111) surfaces have not only been studied by LEED but more recently also by photoelectron spectroscopy and medium energy ion scattering.[8] From the photoelectron spectroscopy and ion scattering work it was concluded that the atomic structure of the laser-irradiated Si(111) is similar to that of the (7×7) surface but does not have its long range order. More evidence regarding this laser irradiated surface structure can be derived from molecular beam epitaxy as will be discussed below.

The second low temperature cleaning process, which we have employed is the reduction of the surface oxide through the reaction between the SiO_2 with Si from the molecular beam at a temperature of 1120 K. This reaction is known[9,10,11] to result in oxygen-free surfaces through the mechanism:

$$SiO_2 + Si_{beam} \rightarrow SiO_{volatile}$$

In Fig. 5 the Auger data are summarized at four stages of the process. The first set (marked 1) were taken from an As^+ implanted sample. Around 80 eV a typical SiO_2 Auger spectrum is recorded along with C and O peaks at 270 and 504 eV. No As peak was observed since the surface concentration of As in the Si wafer escapes detection. An almost identical set of curves is recorded (marked 2) after a 60 minutes thermal anneal, of the ion implantation damage, at 870 K. We have tried to remove C and O contamination by heating the substrate to 1120 K and exposing it to small amounts of Si. During 10 minutes about 3 nm Si was "deposited" on the sample by 10 exposures of 1.6×10^{15} at/cm². The Si does not stick to the surface but reduces the native SiO_2 and desorbs as SiO. The AES data confirm this, as can be seen from the third set of curves in Fig. 5. Hardly any O can be detected and a strong Si (LVV) transition at 91 eV is recorded. Carbon however is still present. We estimate the surface coverage with C and O to be 0.14 and 0.05 of a Si monolayer respectively. At this stage a (2×1) LEED pattern is observed of which an example is reproduced in Fig. 6a. After characterizing the substrate with LEED and AES a 60 nm thick Si layer was deposited at 870 K substrate temperature. This resulted in epitaxial growth, which was demonstrated by the observation of a LEED (2×1) pattern of which we reproduce an example in Fig. 6b. From the Auger data obtained after epitaxial growth (Fig. 5, set 4) it is seen that a clean Si surface results in C and O coverages of less than 0.02 of one Si(100) monolayer, which is the detection limit.

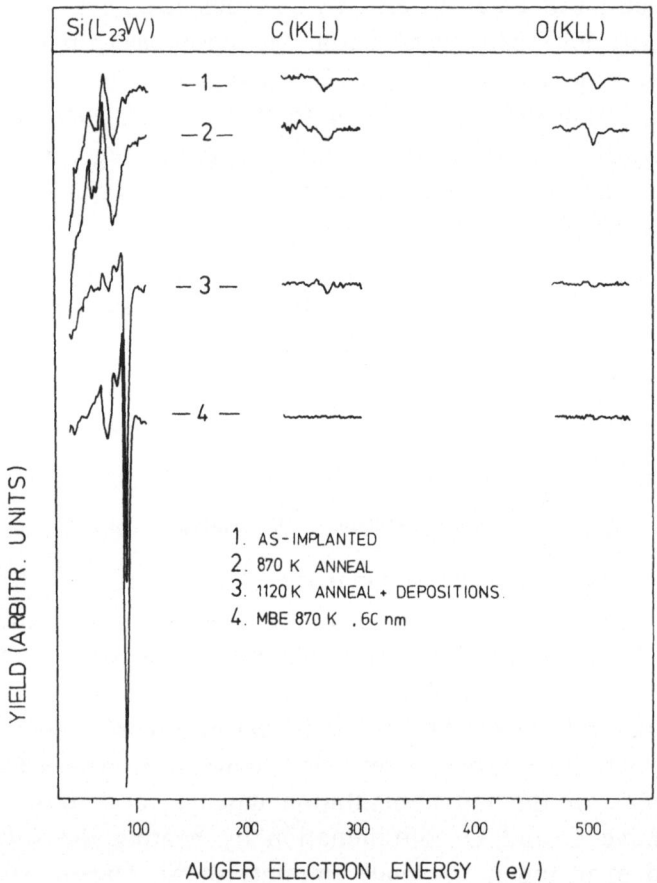

Figure 5 - Auger spectra of buried layer at four stages of processing: 1) As-inserted in MBE chamber, 2) after thermal annealing at 870 K for 60 minutes, 3) after thermal annealing at 1120 K plus small depositions: O disappears, 4) after epitaxial growth of 60 nm Si at 870 K: clean surface.

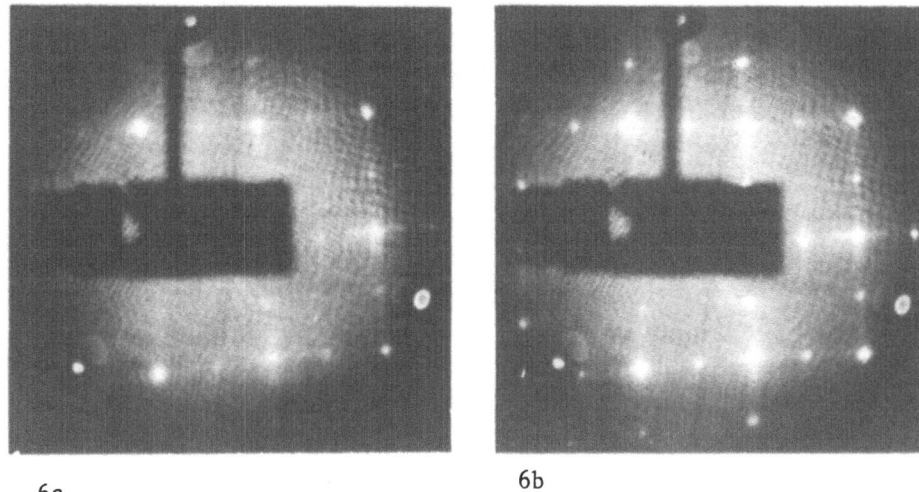

6a   6b

Figure 6 - LEED patterns taken from buried layers at two stages of processing: a) after thermal annealing plus small depositions, b) after epitaxial growth of 60 nm Si at 870 K.

Finally we like to point out that the results discussed above, especially AES curves[4] in Figs. 3 and 5, have shown that Si-MBE is itself a very good low temperature method of producing clean and well-ordered Si surfaces.

3. EPITAXIAL TEMPERATURES

Provided that ultra-high vacuum conditions are maintained during growth, Si-MBE on clean Si surfaces requires growth temperatures of 850-1100 K, considerably lower than temperatures necessary in chemical vapour deposition (CVD) (1250-1450 K). Not only the growth temperatures of CVD and MBE differ, but also entirely different growth characteristics are reported. While in CVD the growth rate decreases with decreasing temperature, one finds an almost constant growth rate in MBE for a very large temperature range.

Epitaxial growth of Si on Si(111) was studied with evaporated Si under UHV conditions by Abbink et al.[12] and the growth mechanism was identified to be step-flow of double-layer steps, as described theoretically by Burton, Cabrera and Frank (BCF).[13] The growth process can be separated into two parts. First a single Si atom impinges on the Si surface, sticks there and thermalizes. The second step is diffusion over the surface to a growth site, for instance a step. From the step-to-step distance Abbink et al. calculated an activation energy for surface diffusion of 0.2 eV. Recently, Kasper[14] determined the condensation coefficient, defined as the probability that a Si atom impinging on the surface will not desorb later on, for a large temperature range to be near unity. In the framework of the BCF theory an upper limit was calculated for the activation energy of surface diffusion: $E^{sd} < 1.1$ eV. On the basis of desorption experiments[2] we arrive at a value of E^{sd} of 1.0 eV. We conclude that in Si-MBE under UHV conditions the surface diffusivity of Si is higher than found in CVD type growth, which explains why epitaxial growth in MBE takes place at temperatures much lower than in CVD.[15]

In early work on Si-MBE[9] it was recognized that vacuum quality, particularly during growth, in combination with substrate surface condition, influenced the growth mode of the process. Residual surface contaminants remaining after in-situ surface preparation (notably carbon) or contaminating species sticking on the surface during growth can cause the growth mode to become three dimensional island type growth. In this case high growth temperatures are required to lower the sticking coefficients on the growing surface as well as to provide an annealing treatment for possible defects, entrapped in the growing overlayer. In today's UHV system these problems no longer exist. Therefore, we set-out to measure with Low Energy Electron Diffraction the epitaxial temperature on Si(111) and Si(100) substrates.

3.1 Si(111) Substrates

On thermally annealed and cleaned Si(111) substrates series of 10 nm Si depositions were made at a rate of 0.1 nm/s and substrate temperatures of 870, 770, 670 and 580 K. Between depositions the substrate was "regenerated" by high temperature annealing (4 min. at 1270 K, 1 min. at 1520 K and cooling at a rate of 100 K/min). In Figs. 7a through 7d we reproduce the LEED patterns, taken at 65 eV for all four depositions. Using this series one can define the epitaxial temperature as that particular substrate temperature at which an epitaxial overlayer, grown on the clean substrate, exhibits the same diffraction pattern as the substrate itself. It can be seen from Figs. 7a through 7d that the epitaxial temperature for thermally annealed Si(111) lies at 870 K. For lower growth temperatures we find an enhanced background with a gradual disappearance of the $1/7^{th}$ fractional order spots. Also, the intensity of the integral-order spots decreases with decreasing growth temperatures. At a growth temperature of 580 K the pattern should be indexed as (1×1).

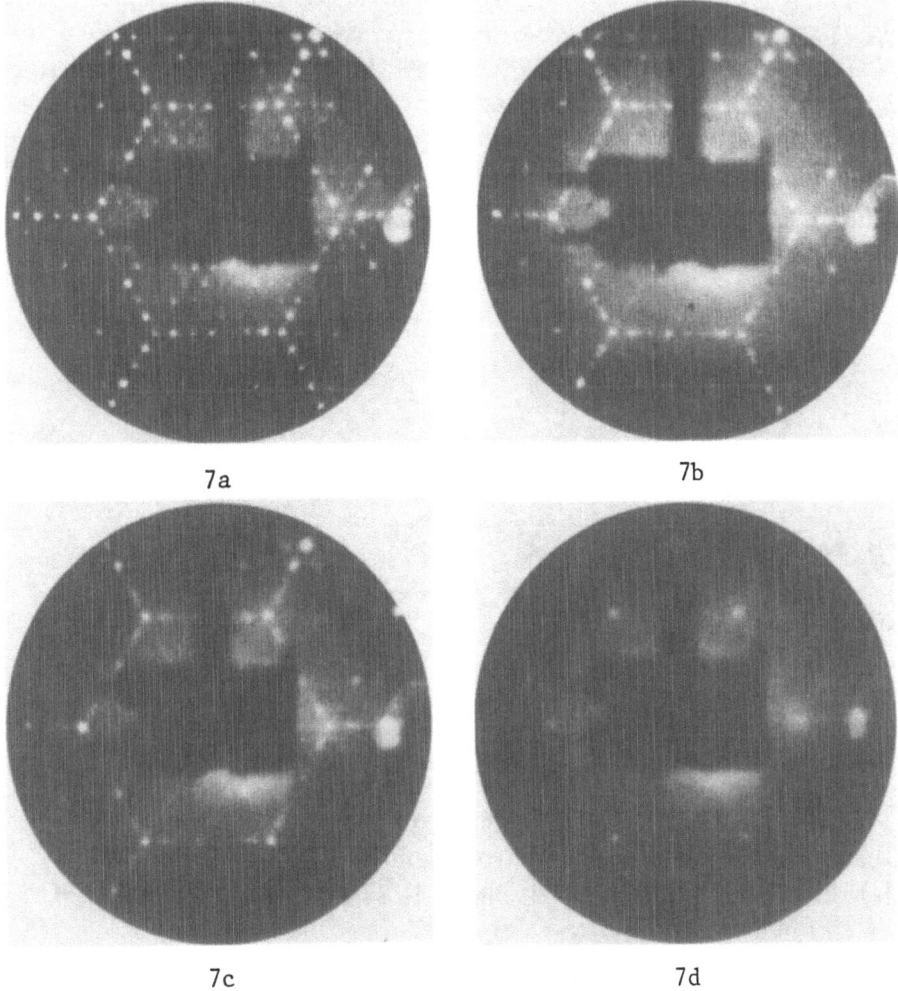

7a   7b

7c   7d

Figure 7 - LEED patterns (65 eV) taken from Si(111) after deposition of 10 nm Si. From this series the epitaxial temperature is determined. a) at 870 K, b) at 770 K, c) at 670 K, d) at 580 K.

In order to determine the influence of steps on the growth process we have used vicinal Si(111) surfaces, tilted $4.0\pm0.5°$ from the <111> surface normal towards the nearest (110) plane. Provided only double layer steps are present this will result in the formation of [112] oriented steps of a density of $2\times10^6 cm^{-1}$. The LEED pattern seen on thermally annealed, clean, vicinal (111) silicon is a (7×7) pattern. We have carried out a similar set of 10 nm depositions of Si on vicinal Si(111). From the LEED patterns the epitaxial temperature was determined to be 770 K, significantly lower than for the (111) substrate. The LEED pattern obtained after growth at 770 K is given in Fig. 8a. The lowest growth temperatures at which LEED patterns were observed from these 10 nm depositions on the vicinal (111) surface was found to be 410 K, see Fig. 8b, again lower than for (111) surfaces (compare Fig. 7d).

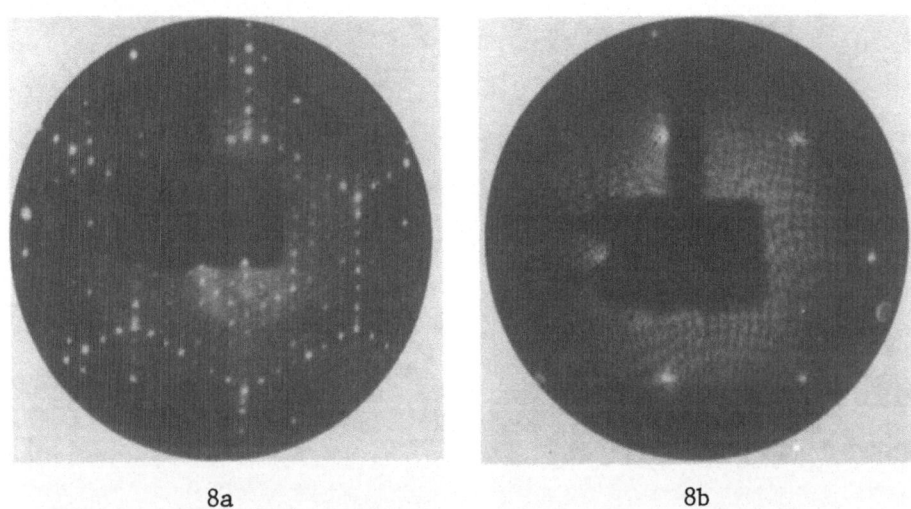

8a   8b

Figure 8 - LEED patterns taken from 4° vicinal Si(111) (note that patterns were obtained from samples, positioned differently): a) after deposition of 10 nm Si at 770 K, b) after deposition of 10 nm Si at 410 K.

The very first stages of epitaxial growth on laser-irradiated (111) surfaces have been investigated.[16] Apart from the observation that epitaxy occurs on laser-irradiated clean Si, which is in itself important, it is possible to make use of the fact that the laser-irradiated and thermally annealed Si(111) surfaces have different but related structures. The difference in LEED patterns enables us to compare the annealing

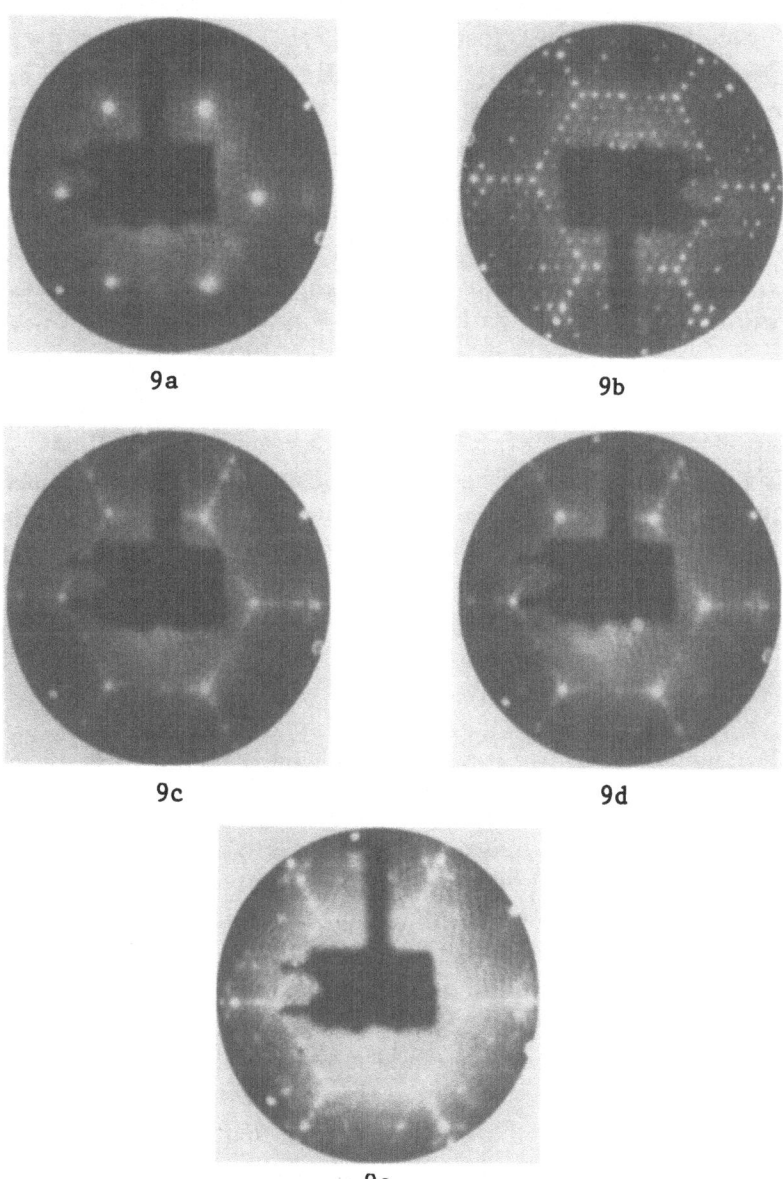

9a

9b

9c

9d

9e

Figure 9 - LEED patterns taken from Si(111) showing the transition of (1×1) to (7×7) and initial stages of epitaxial growth. a) laser irradiated surface; b) thermal anneal at 870 K, 21 sec, of the laser irradiated surface complete transition seen; c) thermal anneal at 770 K, 16 sec, of the laser irradiated surface transition incomplete; d) same as c, but with 0.3 nm Si deposited during anneal transition incomplete; e) same as c, but with 1.0 nm Si deposited during anneal (7×7) has developed.

behaviour with the initial stages of epitaxial growth. By single pulse irradiating with 1.2 J/cm^2, clean (111)- (7×7) surfaces were prepared in the metestable phase (1×1). They exhibit the LEED pattern reproduced in Fig. 9a. A 21 sec. anneal at 870 K restored the (7×7) pattern which is shown in Fig. 9b. Identical (7×7) patterns are obtained by depositing silicon layers from 0.3 to 100 nm thickness on the 870 K heated substrate (no pattern shown). Thermal annealing at 770 K as such, does not bring about the (1×1) to (7×7) transition (see Fig. 9c), but small depositions during the anneal do lead to overlayers exhibiting (7×7) patterns (see Figs. 9d and e). Depositions at 670 K show the same trend as for the 770 K experiment but at neither temperature are the grown layers perfect.

The above results can be understood in a step flow mechanism. The Si(111) surface can be terminated by two possible monoatomic layers, one of which has three and the other only one dangling bond per surface atom. The latter syrface is the more stable configuration. Due to this difference double layer steps are present on diamond type (111) surfaces. These steps will be separated by (111) terraces. On the (111) terraces an adatom can form only one bond with the substrate (i.e. the dangling surface atom bond) whereas at a [112] step two bonds can be formed at once. Therefore step flow will be an important growth mechanism. As the growth temperature is lowered, however, the surface diffusivity is reduced and it takes longer for adatoms to reach a step, hence the concentration of diffusing adatoms increases and two-dimensional nucleation becomes important. As long as adatoms and nucleated islands are able to arrange themselves in the energetically most favourable position epitaxial growth should occur. On Si(111) there is always the possibility that adatoms grow on a (111) terrace not at a regular crystal configuration but rather at a twin position. If this happens the crystal structure will be disturbed, resulting in an imperfect layer. If new atoms arrive before proper reordering of the previously arrived atoms has occured, the not yet fully ordered overlayer is buried. As a result an imperfect overlayer will be formed. This is what we suggest to occur below T_{epi}. So we believe in two stable growth regimes in Si-MBE under UHV: single crystalline growth occurs at temperatures above T_{epi} through step flow, whereas two-dimensional nucleation becomes a competitive growth mode below T_{epi} and long range order cannot be maintained, explaining the disappearance of fractional order spots, high background and isotropically diffuse integral order spots. Due to the larger concentration of steps on vicinal (111) surfaces, adatoms will be able to reach a step at lower temperatures than in the case of (111) surfaces. To picture the laser irradiated (1×1) surface as a disordered (7×7) would be consistent with the ion-scattering analysis.[8] Adatoms arriving on this laser irradiated surface will diffuse to a step which then grows. The step is part of the locally reconstructed surface and adatoms are likely to settle in phase with the already reconstructed atoms. Thus the outgrowing step is capable of introducing long-range order and the transition from (1×1) to (7×7) is explained.

3.2 Si(100) Substrates

In order to determine the epitaxial temperature on Si(100) a series of Si depositions were made on both laser-irradiated and thermally annealed Si(100) substrates, all showing the familiar (2×1) LEED pattern. We did not find any evidence for differences in epitaxial temperatures on either surface. The LEED patterns shown below were all obtained on thermally annealed samples, but the results hold for both types of surfaces. On the clean (2×1) reconstructed Si(100) surface 10 nm depositions were performed at 670, 570, 470 and 335 K substrate temperatures. The corresponding LEED patterns, all taken at 35 eV, are shown in Figs. 10a-d. No change in the patterns is observed until growth temperatures of 470 K and below. Depositions of 10 nm layers at room temperature on thermally annealed Si(100) substrates did not result in layers exhibiting (2×1) patterns. Room temperature depositions of 1.0 nm Si layers did result in partly ordered overlayers, as can be seen in Fig. 10e. On Si(100) surfaces prepared by Si-MBE, overlayers of 2.5 nm Si could be grown, see Fig. 10 f.

On the (100) surface it seems very probable that only monoatomic steps are present since each monolayer can terminate with only two dangling bonds per surface atom. According to BCF, growth on a low-index plane should take place at monoatomic steps of low index orientation, such as [100] or [110]. On the Si(100) surface, however, growth of adatoms at such mono-atomic steps seems just as probable as direct growth at the low-index oriented terraces (100) between the steps. Seen from a structural point of view the adatom can form two bonds whether it grows on the (100) terrace or at any low-index monoatomic step. In our opinion the fact that on Si(100) adatoms can grow at every site rather than predominantly at steps, as is the case for Si(111), may very well be the explanation for the much lower epitaxial temperature of Si(100).

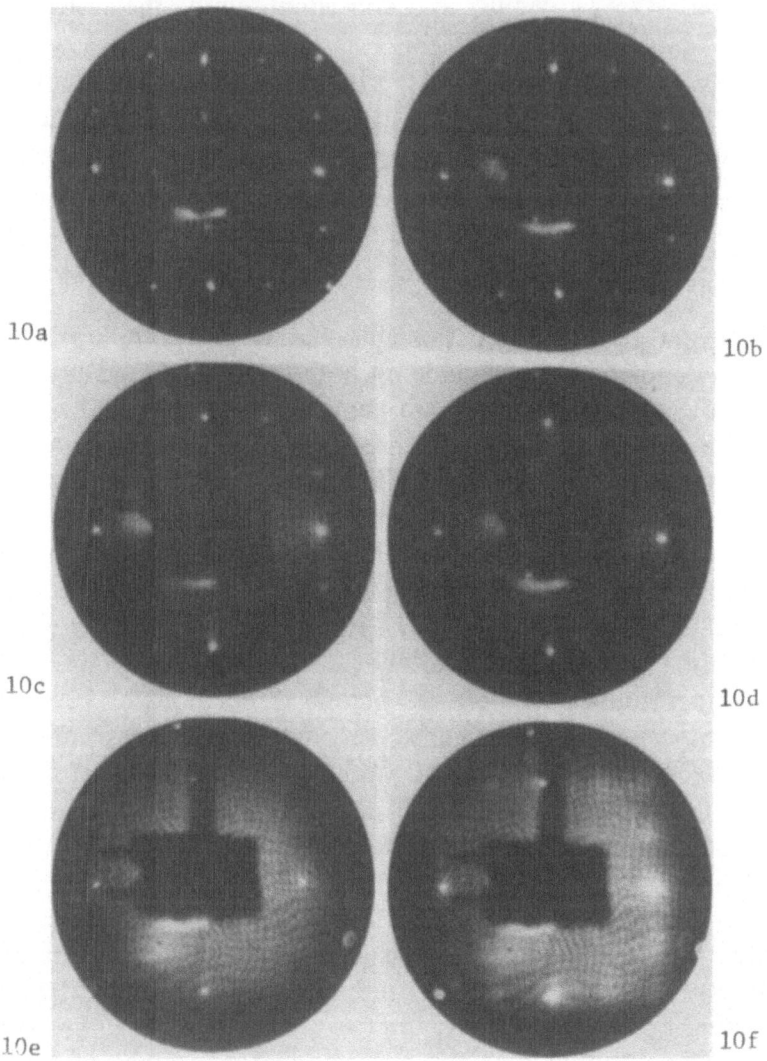

Figure 10 - LEED patterns (35 eV) taken from Si(100) from this series the epitaxial temperature is determined. a) after deposition of 10 nm Si at 670 K, b) after deposition of 10 nm Si at 570 K, c) after deposition of 10 nm Si at 470 K, d) after deposition of 10 nm Si at 360 K, e) after deposition of 1.0 nm Si at room temperature on thermally annealed substrate, f) after deposition of 2.5 nm Si at room temperature on MBE epilayer.

4. DOPING

Differently doped silicon layers with specified dopant concentration profiles and well-defined interfaces are an integral part of most semiconductor devices. In the quest of higher packing densities and faster devices considerable research effort is now being devoted to the development of techniques that will give better control over the depth profile of dopant atoms.

Most techniques that are currently employed for introducing dopant atoms into silicon require high temperatures, either during the process or afterwards. At these high temperatures dopant diffusion is no longer negligible and will therefore broaden the dopant concentration profile, setting a lower limit to the thickness of the layers. Molecular beam epitaxy offers a way for making ultra thin layers with arbitrary dopant concentration profiles at any depth.

At present two schemes are used to introduce the dopant species into the growing layer. Dopants can be added during growth either by coevaporation or by ion implantation. Coevaporation of dopant species during growth is achieved by Knudsen cell evaporation during epitaxial growth of Si.[17] A complication arises since some dopant species evaporate as clusters (e.g. As_4, P_8 and Sb_4) which may have its effect on Si epitaxial growth. Another complication lies in the fact that the sticking coefficient of evaporated dopant particles on the growing Si substrate varies rapidly with temperature. Becker and Bean[18] reported Ga sticking coefficients verying from 10^{-1} to 10^{-5} in the temperature range of 720 to 1120 K.

The second possibility for dopant introduction into the epitaxially growing Si crystal is by means of combined ion implantation and MBE. By adding an ion implanter to his ultra-high vacuum Si-MBE chamber Ota[19] produced abrupt doping profiles in Si. Sticking coefficients of ions on the substrate surface are near unity and do not vary as much with temperature. Beam handling, such as scanning and current monitoring, are straight forward. The energy of the implanted species must be sufficiently low as not to sputter the grown overlayer[20]. This requires sophisticated ion optics to transport a low energy ion beam properly to the growing substrate. It is also possible to use simplified schemes for ion doping in Si-MBE as was shown by Sugiura.[21] Recently Bean et al.[22] reported on a specialized ion accelerator for Si-MBE with which As and B can be implanted during deposition. A complication arises since ion implantation of heated substrates is known to yield extended defects that are difficult to anneal. Apparently for low energy low dose implants this problem is not so serious because Ota[19,23], Sugiura[21,24] and Bean[3,22] all report low dislocation densities and minority carrier lifetimes of $\sim 5~\mu$ sec . in both N and P type doped films.

We have investigated an alternative combination of ion implantation and Si-MBE.[25] We have used a conventional ion implanter, transported the samples in air to the Si-MBE apparatus, cleaned the substrates in-situ and grown epitaxial layers thus achieving buried layers and modulation doping structures.

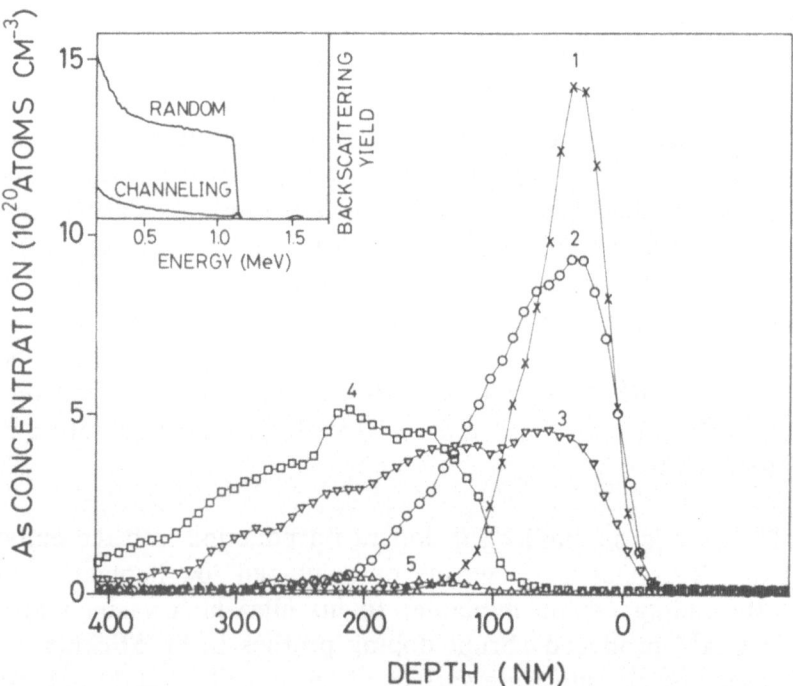

Figure 11 - Arsenic signal of RBS spectrum, with the energy of the backscattered particles converted to depth and the yield converted to As concentration. Spectrum 1: As implanted; 2: after 1 laser pulse; 3: after 5 laser pulses; 4: after 5 laser pulses and 100 nm deposition, random direction; 5: ibid, channeling direction. The insert shows a complete RBS spectrum.

4.1 Buried Layers

Recently we have shown[26] that in-situ laser irradiation of off-line ion implantated Si(100) removes the implantation damage and results in atomically clean and ordered substrates. Overgrowth at 870 K with Si leads to buried As doped Si layers with an epitaxial overlayer free of autodoping. As a result of repeated laser irradiation, however, indiffusion of the dopant occurs over several hundred nanometers below the original implant depth, see Fig. 11.

Ideally one should like to anneal the implantation damage and clean the substrate surface without disturbing the doping profile. Therefore, we have studied the possibility of thermal annealing combined with reduction of the oxide layer through the reaction between the SiO_2 with Si from the molecular beam, discussed in Section 2b. Rutherford backscattering analysis of the overgrown implantations is shown in Fig. 12, where the As profile of the as-implanted Si is also given for comparison. The As profile after surface oxide reduction and epitaxial

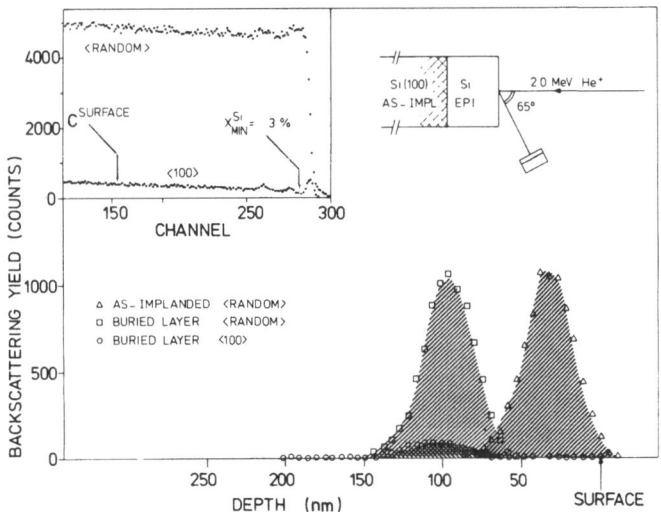

Figure 12 - RBS results of buried layer, As profile shown, Si spectra in insert. Profiles shown of as-implanted (random) and overgrown (random and channeling) samples.

growth of 60 nm Si is shifted to greater depth. This profile was taken in a random direction, whereas under <100> channeling conditions a much reduced yield is found showing good substitutionality of the dopant. The As profile of the as-implanted and the overgrown sample have the same width and height. It may be concluded that during the annealing and growth stages no profile broadening has occured. This was expected since the calculated diffusion of As in Si, for the process described in Section 2b, is less than 7 nm. It is remarkable that although growth has taken place on a partly contaminated and disordered surface, as was shown in Figs. 5 and 6, the epitaxial layer improves in quality upon growth and the minimum channeling yield near the very surface is better than at the interface between the epi-layer and the original substrate.

4.2 Modulation Doping

By repeating the implantation/low T annealing and cleaning/growth cycle several times a modulation doping structure in silicon has been fabricated. We have implanted series of (100) oriented Si samples with 10 keV As$^+$ ions to a dose of 10^{15} cm^{-2}. After transportation and loading in the MBE chamber of a batch, the same UHV annealing process (as described in Section 2b) was carried out. In addition to the one hour 870 K anneal plus 1120 K anneal and depositions, we have flashed the samples for 20 sec. at 1370 K to remove C contamination probably originating from the ion-implanter. Unfortunately, this step was not found to be very effective and reproducible results could not be obtained. After annealing and cleaning a 60 nm MBE layer was grown at 870 K. Epitaxial growth was demonstrated by LEED patterns seen from the overlayers.After processing the batch was loaded again in the implantation machine, where the second implantation with 10^{15} As$^+$/cm^2 at 10 keV was carried out. This implantation energy is so low that the second implantation does not overlap with the previous one, the projected range of 10 keV As$^+$ is ~30 nm. After implantation and transport a second MBE layer was grown. In the second epitaxial layer the final implantation with 10keV As$^+$ to a dose of 10^{15} cm^{-2} was done and annealing took place in the MBE chamber.

The samples were analyzed by RBS in random and channeling conditions, the results of which are shown in Fig. 13. The As profile obtained in random incidence shows three distinct peaks at depths of 15, 64 and 122 nm. The peak widths of ~30 nm are slightly larger than our depth resolution of 20 nm and may well be caused by dopant diffusion during the three annealing steps. Although the peaks are clearly separated the dopant concentration between peaks is not negligible. In all three As doped layers more than 80% of the dopants were found to be on substitutional sites. The enhanced dechanneling seen in the Si spectrum just below the surface peak may well be caused by imperfect epitaxial growth on the carbon contaminated substrates.

It is to be expected that more sophisticated protective and cleaning methods should result in higher quality structures. The work described above demonstrates that a combination of Si-MBE and low energy ion implantation provides the depth control required for making ultra-fast switching devices such as hot electron transistors.[27] It is anticipated that these devices will soon be added to the long list of the diodes, transistors and detectors already fabricated by Si-MBE.[3]

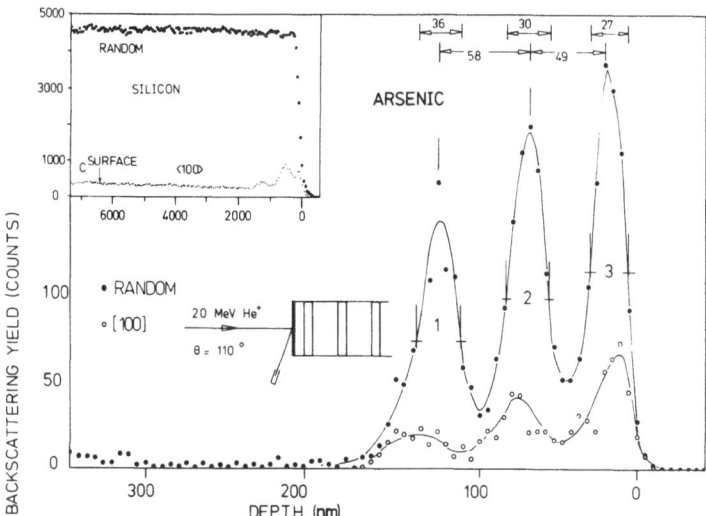

Figure 13 - RBS results of modulated doping structure, As profile shown, Si spectra in profiles shown of completed structure in random and channeling conditions.

5. HETEROEPITAXY OF Si ON GaP

To this point only silicon on silicon growth has been discussed, but the low temperatures inherent in the MBE process also introduce many heteroepitaxial growth possibilities. In heteroepitaxy metallic, semiconducting and insulating layers can be grown epitaxially with MBE on Si. The fact that these layers are epitaxial implies that they can in turn be overgrown with single crystalline Si. This has been demonstrated for the Si-CoSi$_2$-Si (matallic)[28] and Si-CaF$_2$-Si (insulating) [29] double heterostructure. Thus buried metallic and insulating layers have been made which can be applied as contacts or insulation between parts of a three-dimensional device. Epitaxial growth of one semiconductor on another also offers many interesting possibilities as is discussed extensively in this volume. Here we shall present the results of an investigation of Si-MBE on GaP(100), to our knowledge not reported previously in the literature.

A GaP-Si heterostructure built-up from a GaP substrate and an epitaxial Si overlayer with a thickness in the order of 1 μm is in fact a thin Si crystal transparent for light in the visible region. Applications of this type of "transparent Si" are to be found in the integrated opto-electronic devices, e.g. Si-diodes realized in such an epitaxial layer as a light transmitter/detector to be implented in the Lightpen of the optical pickup of a Compact Disk or Laser Vision Player.[30]

GaP crystals (as-inserted) were measured in the UHV system by means of AES. Ga (55 eV) and P (120 eV) were detected as well as trace amounts of C (270 eV) and O (504 eV). No other elements were found. Clean surfaces were prepared[31] by ion bombardment with 800 eV Ar$^+$ ions after which neither C nor O could be detected. The disorder introduced by the ion sputtering was removed by thermal annealing for 60 to 90 minutes at 550°C. The topmost spectrum of Fig. 14 is a derivative AES spectrum of an ion bombarded and annealed (IBA) surface of GaP(100). From the peak heights of Ga at 55 eV and P at 120 eV we estimate the surfaces, prepared by IBA to be rich in Ga. This is caused by preferential sputtering of P and by evaporation of P during the thermal anneal. One of the LEED patterns observed after IBA is reproduced in Fig. 15a. Assuming that the surface is Ga-rich we index the pattern as (4×2). After deposition of a few monolayers Si on GaP(100) the LEED pattern has changed from the (4×2) to the (2×1) pattern typical for a clean reconstructed Si(100) surface. In Fig. 15b this (2 × 1) pattern is reproduced, obtained by growing a Si layer of 0.8 nm thickness at 870 K.

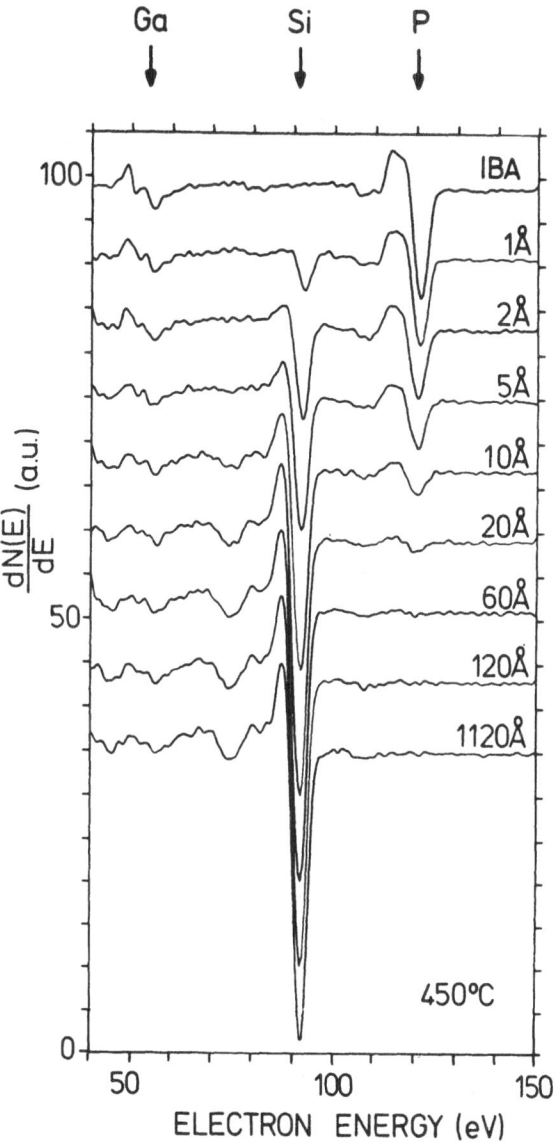

Figure 14 - Derivative Auger spectra, each taken after growth of an epitaxial Si layer on GaP(100). Growth was interrupted in order to take a spectrum: full Si-coverage is indicated in nm. The peaks correspond to the Ga 55 eV, Si 91 eV and P 120 eV Auger transition.

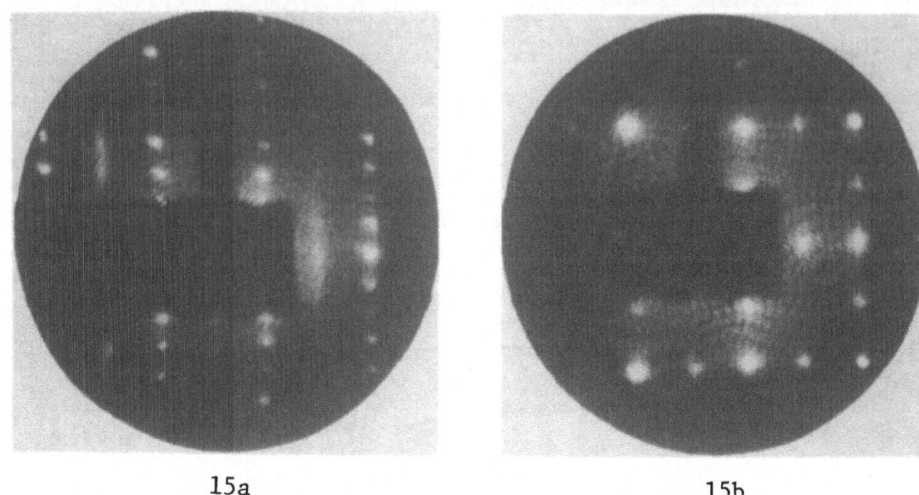

15a   15b

Figure 15 - Low Energy Electron Diffraction patterns, each takeb at 60 eV primary energy, of a GaP(100) clean substrate, and a GaP:Si(100) heterostructure. a) (4×2) pattern of clean GaP, prepared by ion bombardment and annealing, b) (2×1) pattern of GaP(100) with 0.8 nm Si grown at 870 K by MBE.

In order to determine the surface composition of the Si films we took AES spectra at a number of Si coverages. In Fig. 14 Auger derivative spectra are given, taken from epitaxial Si layers grown at 450°C. The deposited thicknesses of the epitaxial Si films are indicated in the figure. The intensities of the Ga peaks drop exponentially for Si overlayer thicknesses of $0 - 10\text{Å}$ (0-4 momolayers). At higher coverages the Auger spectra still indicate the presence of Ga surface coverage of 0.4 monolayer even at a Si overlayer thickness of 1120Å. Surface segregation of P to the Si surface during growth also takes place, but no P could be detected at Si overlayer thicknesses of 1120Å. This difference with Ga segregation can be understood in that segregation of Ga and P seems to occur immediately after Si growth takes place. The segregated elements are driven out in front of the growing Si layer. Since the solid solubility of Ga is lower by at least one order of magnitude than that of P,[12] the P will dissolve more easily in Si than Ga leading to a quasi saturation of Ga, and no detectable coverage with P. At the growth temperatures used, bulk diffusion of P and Ga through Si can be excluded. Preliminary electrical measurements revealed n-type conductance of the Si films in support of the above mentioned explanation.

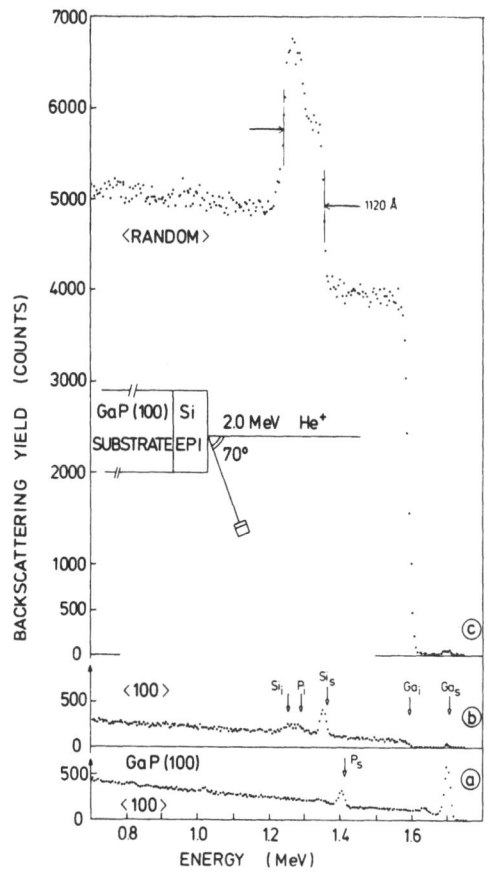

Figure 16 - Rutherford backscattering spectra taken from a GaP: Si(100) heterostructure grown at 720 K to a Si thickness of 112 nm; spectra are taken in random and <100> channeling conditions using 2.0 MeV He scattering. Arrows mark the position of Ga, Si and P at the surface and at the interface. Spectrum 1: GaP(100) in <100> channeling conditions. Spectrum 2: GaP:Si(100) heterostructure in <100> channeling conditions. Spectrum 3: GaP:Si(100) heterostructure now in random alignment; Si thickness was calculated from this spectrum.

Further characterization of the samples was performed by Rutherford backscattering and ion channeling. In Fig. 16 we show spectra of bare GaP(100) and the heterostructure grown at 720 K on GaP(100). On bare GaP(100) spectrum 1 was taken in a <100> aligned configuration. The corresponding random spectrum is not shown. A Ga minimum channeling yield of 3% was found. Surface peaks of Ga and P may be noted at 1.70 and 1.40 MeV. After growth of an epitaxial Si layer, spectra 2 (channeling <100>) and 3 (random incidence) were taken. In the random spectrum the backscattering yield from Si is seen from 1.36 to 1.25 MeV, coinciding with the leading edge of P in the substrate. From the energy difference between the Si leading and trailing edges we calculate the overlayer thickness to be 1120Å. Due to the stopping in the overlayer the initial Ga and P contributions (see spectrum 1) are shifted towards lower energies by the same energy difference.

At the initial Ga surface position (around 1.70 MeV) a small peak may be noted. This is attributed to segregated Ga, which could be confirmed by analysing in different scattering geometries. The amount of Ga segregated was calculated to be 0.63×10^{15} at/cm^2, corresponding to about one monolayer coverage. This value was also estimated from the Auger peak height (Fig. 1) and found to be 0.4 monolayer. With RBS we found spot-to-spot differences of 30% in the amount of segregated Ga. The discrepancy between the Auger data (0.4 ml) and RBS results (1.0 ± 0.3ml) indicates that the Auger data are only accurate within a factor of two.

In the channeling spectrum (2) the Si contribution has dropped considerably, which is indicative of good crystal order. A Si surface peak is seen with enhanced yield near the position of the GaP-Si interface (1.25 MeV). The Ga and P leading edges seen in the bare GaP(100) channeling spectrum (1) as surface peaks (Ga$_s$ and P$_s$) have now shifted towards lower energies, but in the spectrum of the heterostructure no Ga nor P interface peaks (Ga$_i$ and P$_i$) are seen. By subtracting the GaP contribution from the (100) channeling spectrum, which is done by extrapolation from the 1.45-1.55 MeV part of the spectrum, the Si dechanneling just below the Si surface peak can be calculated. In the random spectrum the GaP contribution can also be subtracted. From these data the Si minimum channeling yield was calculated to be 3%. This low minimum yield was reproduced several times and corresponds to the yield values obtained for monocrystalline (100) Si of high quality.

Although surface segregation during heterojunction formation is in itself highly interesting, it is of course an undesirable effect which will lead to Si overlayers which will be doped with Ga and P. One way to eliminate the segregation of Ga and P during growth of Si is by using a Si layer as-grown, i.e. a layer which contains segregated Ga and P. Ga and P are subsequently removed from the surface by ion bombardment followed by thermal annealing to remove ion bombardment damage. On such a clean surface, exhibiting a (2×1) LEED pattern, MBE growth has been continued, resulting in Si overlayers with no detectable surface segregation (no Ga and P lines seen in the AES spectrum).

Another successful method to remove Ga and probably P too, is thermal annealing at 1070 K of the substrate with overlayer.[32] Si overlayers, grown at 870 K were analyzed with RBS and found to have a Ga-coverage of 10^{15} at/cm^2. If a post-anneal of 60 seconds at 1070 K was done, no Ga could be detected with RBS at the substrate surface, nor in depth ($<0.1\%$ concentration of Ga in top 50 nm Si). The bulk diffusion of Ga and P through the Si overlayer as a result of this post-anneal can be estimated to be less than 2.5 nm. On the cleaned Si surface MBE growth could be continued leading to overlayers free of Ga.

6. CONCLUDING REMARKS

After a slow start the field of Si-MBE is maturing rapidly and today the manufacturing of three-dimensional devices of silicon made by physical processing with beams of molecules, ions, photons and electrons under ultra-high vacuum seems much less a fantasy than a fact. Yet, before Si-MBE will be introduced into fully integrated device structures a number of problems has to be studied further, some of which we have dealt with in this paper.

We have discussed two new substrate preparation methods, using the Si molecular beam and using pulsed laser irradiation. Both methods yield good templates for epitaxial growth without excessive heating of the wafer. The SiO$_2$ reduction is only applicable where surfaces are already free of carbon. Pulsed laser cleaning may well find wide application upon the advent of powerful excimer lasers that will process 4" wafers in a few seconds.

For the two technologically most important Si surfaces the epitaxial temperature has been determined for deposition rates of 0.1 nm/sec. On (100) oriented wafers we have found T_{epi} to lie around 470 K whereas on (111) oriented surfaces a much higher value of 870 K was found. Introducing a large number of misorientation steps (by taking vicinal (111) surfaces) T_{epi} could be lowered to 770 K. We propose that on the (111) surface the step flow mechanism, which is the generally accepted growth mechanism at high temperatures, gradually is replaced by two dimensional nucleation as the substrate temperature is lowered below T_{epi}. On the Si (100) surface we propose that it is not very favourable for an adatom to grow at a step, other sites are energetically rather similar, which might explain the much lower T_{epi} at Si(100). A morphological study of epitaxial Si layers on Si(100) and Si(111) grown above and just below T_{epi} should help to verify these suggestions. In addition, we have recently observed that the Si molecular beam from e-gun evaporators predominantly consists of Si atoms but a few percent Si_2, Si_3 and Si_4 are present also[2]. It is as yet not known whether these molecules adsorb on the Si surface dissociatively or whether thay stay in tact and act as nucleation centers. Further, in solid phase epitaxial regrowth of amorphously deposited or amorphized silicon doping has been observed to affect the growth rate strongly.[33] Similar effects are to be expected in Si-MBE but to our knowledge this has not been studied yet.

By combining conventional ion implantation with Si-MBE we have manufactured buried layers and modulating doping structures in Si. Ishiwara et al.[34] have applied ion implantation into amorphously deposited Si layers followed by low temperature annealing to induce solid phase epitaxial growth. The latter method avoids channeling effects during implantation thus improving the depth control and more importantly no damage is done to the substrate from where solid phase epitaxial growth starts. Clearly both methods of making hyperabrupt junctions and modulation doping structures in Si should be characterized and compared further especially by electrical measurements.

The list of hetero epitaxy possibilities involving silicon has been increased further by our observation that Si can be grown on GaP(100). Hetero epitaxy of Si on GaP can lead to new applications in optoelectronics and advanced heterostructures, so far realizable only by the low temperature process of molecular beam epitaxy.

ACKNOWLEDGEMENTS

This work is part of the research program of the Stichting voor Fundamenteel Onderzoek der Materie (Foundation for Fundamental Research on Matter) and was made possible by financial support from the Nederlandse Organisatie voor Zuiver-Wetenschappelijk Onderzoek (Netherlands Organization for the Advancement of Pure Research).

REFERENCES

1. J. C. Bean and S. R. McAfee, Proceedings Int. Meeting on Relationship Between Epitaxial Growth Conditions and Properties of Semiconducting Epitaxial Layers, Perpignan, France 1982, J. de Physique, Colloque C5-153 (1982).
2. T. de Jong, Thesis Univ. of Amsterdam, 1983.
3. J. Bean, Proc. 29th AVS Meeting, Baltimore 1982, (to be published in J. Vac. Sci. Technol.)
4. R. Tromp, Thesis Univ. of Utrecht, 1982.
5. S. M. Beldair and H. P. Smith, J. Appl. Phys. $\underline{40}$, 4776 (1969) and S. M. Beldair, Surface Sci. $\underline{42}$, 595 (1974).
6. D. M. Zehner, C. W. White and G. W. Ownby, Appl. Phys. Lett. $\underline{36}$, 56 (1980).
7. Z. L Wang, H. Westendorp and F. W. Saris, Nucl. Instr. & Meth., Nucl. Instr. & Methods $\underline{211}$, 193 (1983).
8. R. Tromp, E. J. van Loenen, M. Iwami and F. W. Saris, Solid State Communications $\underline{44}$, 971 (1982).
9. B. A. Joyce, Rep. Prog. Phys. $\underline{37}$, 363 (1974).
10. H. Siguira and Yamaguchi, Jap. J. Appl. Phys. $\underline{19}$, 583 (1980).
11. M. Tabe, K. Arai and H. Nakamura, Jap. J. Appl. Phys. $\underline{20}$, 703 (1981).
12. H. Abbink, R. M. Broudy and G. P. McCarthy, J. Appl. Phys. $\underline{39}$, 4673 (1968).
13. W. K. Burton, N. Cabrera and F. C. Frank, Phil. Trans. Roc. Soc. (London) $\underline{243}$, 299 (1951).
14. E. Kasper, Appl. Phys. $\underline{A28}$, 129 (1982).
15. J. Bloem and L. J. Giling, in Current Topics in Materials Science, Vol. 1, ed. by E. Kaldis, North-Holland Publ. Co. 1978.

16. T. de Jong, L. Smit, V. V. Korablev and F. W. Saris, in Laser and Electron-Beam Interactions with Solids, ed. by B. R. Appleton and G. K. Celler, North-Holland Publ. Co. 1982, p. 215.

17. U. Koenig, H. J. Herzog, H. Jorke, E. Kasper and H. Kibbel, in Proc. 2nd Int. Symp. MBE and Related Clean Surface Techniques, Tokyo, 1982.

18. G. E. Becker and J. C. Bean, J. Appl. Phys. 48, 3395 (1977).

19. Y. Ota, J. Appl. Phys. 51, 1102 (1980).

20. P. C. Zalm and L. J. Beckers, Appl. Phys. Lett. 41, 167 (1982).

21. H. Siguira, J. Appl. Phys. 51, 2630 (1980).

22. J. C. Bean, in Impurity Doping Processes in Silicon, ed. by F. F. Y. Wang, North-Holland Publ. Co. 1981.

23. Y. Ota, J. Electrochem. Soc. 126, 1761 (1979).

24. H. Siguira, Jap. J. Appl. Phys. 19, 583 (1980).

25. T. de Jong, W. A. S. Douma and F. W. Saris, Materials Lett. 1, 157 (1983).

26. L. Smit, T. de Jong, D. Hoonhout and F. W. Saris, Appl. Phys. Lett. 40, 64 (1982).

27. J. Shannon, Nucl. Instr. & Meth. 182/183, 545 (1981).

28. S. Saitoh, H. Ishiwara and S. Furukawa, Appl. Phys. Lett. 37, 203 (1980).

29. H. Ishiwara and T. Asano, Appl. Phys. Lett. 40, 66 (1982).

30. Philips Techn. Rev. 40, 150 (1982).

31. B. W. Lee, R. K. Ni, N. Masud. X. R. Wang, D. C. Wang and M. Rowe, J. Vac. Sci. Technol. 19, 294 (1981).

32. S. L. Wright and H. Kroemer, Appl. Phys. Lett. 36, 210 (1980).

33. I. Suni, G.Goltz, M. G. Grimaldi, M. A. Nicolet and S. S. Lau, Appl. Phys. Lett. 40, 269 (1982) and ref. therein.

34. H. Ishiwara et al., (to be published in Proc. Int. Conf. Ion Beam Modification of Materials, Grenoble, 1982).

293

MOLECULAR BEAM EPITAXIAL GROWTH AND CHARACTERIZATION OF III-V/III-V AND IV/III-V HETEROSTRUCTURES

Chin-An Chang

IBM Thomas J. Watson Research Center
Yorktown Heights, New York 10598 U.S.A.

ABSTRACT: Semiconductor heterostructures, both single interfaces and superlattices, have been made by molecular beam epitaxy (MBE) for many different materials. In this paper we describe the MBE growth of a variety of III-V/III-V systems including $GaAs-Ga_{1-x}Al_xAs$, $InAs-GaAs$, $In_{1-x}Ga_xAs-In_{1-y}Ga_yAs$, $GaSb-GaAs$, $InAs-GaSb$, $In_{1-x}Ga_xAs-GaSb_{1-y}As_y$, $AlSb-GaAs$, $AlSb-GaSb$ and $AlSb-InAs$, and two IV/III-V systems of $Ge-GaAs$ and $Si-GaP$. The various properties of these systems are also described including metallurgical, structural, optical and electrical characteristics.

1. INTRODUCTION

Heteroepitaxy of thin film semiconductors has been of increasing interest and importance in recent years. A variety of growth techniques has been employed including liquid phase epitaxy,[1] vapor-phase epitaxy,[2] metal-organic chemical vapor deposition,[3] sputtering,[4] and molecular beam epitaxy (MBE).[5]

In this paper we describe the MBE growth of heteroepitaxial junctions and superlattices for a variety of systems including both III-V/III-V and IV/III-V semiconductors. The III-V/III-V systems include GaAs-GaAlAs, InAs-GaAs, $In_{1-x}Ga_xAs-In_{1-y}Ga_yAs$, GaSb-GaAs, InAs-GaSb, InGaAs-GaSbAs, AlSb-GaAs, AlSb-GaSb and AlSb-InAs. We also describe two IV/III-V systems, Ge-GaAs and Si-GaP. Characterizations of these systems include the use of reflection high energy electron diffraction (RHEED), Auger electron spectroscopy (AES), Rutherford backscattering spectrometry (RBS) and channeling, X-ray diffraction, transmission electron microscopy (TEM), cross-sectional transmission electron microscopy (XTEM), optical and electrical measurements.

The most important parameter in choosing a heteroepitaxial system is the lattice matching between the two materials. Study of the mismatch effect is illustrated using the InAs-GaAs and $In_{1-x}Ga_xAs-In_{1-y}Ga_yAs$ systems. The electron mobilities measured are related to the defect densities observed in these films. The effect of thermal stress on the lattice constants of the heteroepitaxial films grown is illustrated using the GaSb-GaAs system where the large effect of thermal stress is shown to exert an appreciable change on the lattice constant of the GaSb films grown. The InAs-GaSb system is interesting both for its electrical and metallurgical properties. The interface structures are probed by He^+-ion channeling; the electrical transport measurements show interesting junction and superlattice properties. AlSb is interesting in the possibility of making an InAs-AlSb-GaSb three-component stucture; the interface properties of AlSb-GaAs, AlSb-GaSb and AlSb-InAs, important to such a structure, are described. Ge-GaAs is an ideal IV/III-V system which contains both elemental and compound semiconductors. The interface properties of this system have been studied by various techniques and will be described in detail. A brief description of the Si-GaP system is also given.

2. III-V/III-V HETEROSTRUCTURES

2.1 GaAs-Ga$_{1-x}$Al$_x$As

GaAs-Ga$_{1-x}$Al$_x$As, with a lattice mismatch $\leq 0.15\%$, is the most widely used III-V/III-V system for various structures and devices. Structural and device studies of this system are the subjects of several papers in this Symposium and will not be discussed in detail in this paper. Here we describe some experimental studies important to the MBE growth of GaAs, which is the key material in most III-V/III-V and IV/III-V heterojunctions, and to those of GaAlAs and GaAs-GaAlAs superlattices. Structural studies of GaAs-Ga$_{1-x}$Al$_x$As superlattices will also be briefly described.

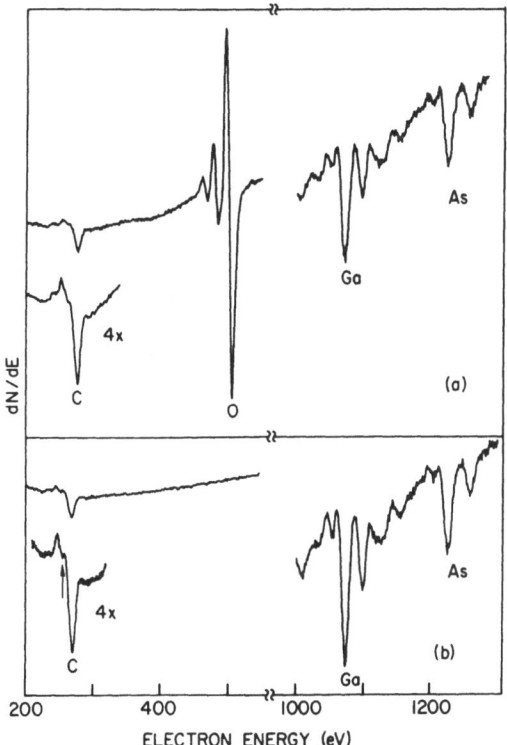

Figure 1 - Auger spectra of a GaAs substrate with an extensive exposure to the Auger beam (see text): (a) before heating, (b) after heating to 560°C.

In the MBE growth of GaAs and other materials, the substrate surface treatment has a profound effect on both the quality of the films grown and on the electronic transport across the film-substrate interface. A freshly prepared GaAs substrate, by degreasing and chemical etching, upon loading into the MBE chamber, always contains oxygen and carbon on its surface. Oxygen can be thermally removed by heating the substrate to ~550°C. Removal of carbon, however, depends sensitively on the surface conditions. For example, it is observed that an extensive exposure of the substrate surface to an Auger beam, especially during substrate heating below ~350°C, makes it very difficult to remove the surface carbon thermally.[6] Figure 1 shows the Auger spectra of such a GaAs surface where carbon remains on the surface after oxygen has been thermally removed at ~550°C. The remaining carbon can only be sputtered off using energetic ions and the substrate is subsequently annealed to remove the damage from sputtering. It has been observed, however, that such a sputter-annealed substrate gives rise to a high resistance region under the surface which is detrimental to the electronic transport across the film-substrate interface. An I-V characteristic is shown in Fig. 2 for a mesa-etched GaAs film grown on such a sputter-annealed GaAs substrate.[7] Similar I-V characteristics have also been observed for the nonsputtered GaAs substrates which have a high content of surface carbon as that shown in Fig. 1b.

The above problems can be solved by either (1) depositing a mono- or submonolayer of Sn on the sputter-annealed GaAs substrate surface prior to the growth of GaAs,[7] or (2) minimizing the exposure of the substrate to an irradiating beam and thermally removing the surface carbon above ~ 350°C.[6] The Auger spectra of a GaAs substrate treated in the latter way are shown in Fig. 3 which shows no trace of carbon after removing oxygen above 550°C. Growth of GaAs films on a substrate treated in either of the above ways gives an ohmic I-V characteristic across the film-substrate interface, as shown in Fig. 4.

Another useful means of improving the qualities of GaAs as well as GaAlAs and GaAs-GaAlAs superlattices is the use of hydrogen during MBE growth which we observed in 1975.[8] Impinging a hydrogen beam toward the substrate during growth improved the electron mobilities and photoluminescence yield of the Sn-doped GaAs, GaAlAs and GaAs-GaAlAs superlattices, the surface finish of GaAlAs and GaAs-

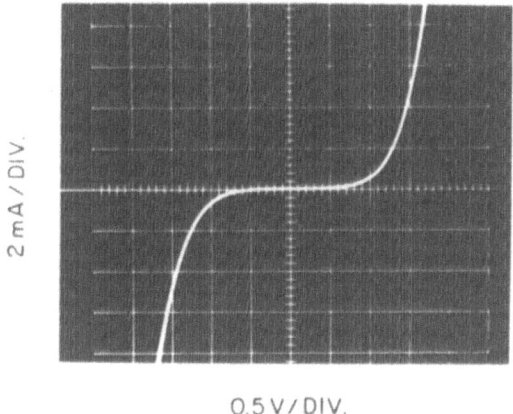

0.5 V / DIV.

Figure 2 - I-V characteristic of a 2μm Sn-doped GaAs film (n ~ 1×10^{19}cm^{-3}) deposited on a sputter-annealed GaAs substrate. Mesa structure size: 100 x 100 μm^2.

GaAlAs superlattices, and the incorporation efficiency of Sn in GaAlAs.[8] These have contributed to our successful resonant Raman studies[9] and the first observation of Shubnikov-de Haas oscillations in the GaAs-GaAlAs superlattices.[10] Improvements using hydrogren have later been reported by other workers for GaAs[11] and very recently for GaAlAs.[12] The advancement of modulation doping and improved design of MBE chambers have in recent years increased dramatically the electron mobilities of the GaAs-GaAlAs structures.[13] The use of hydrogen, however, is expected to provide further improvement for the films grown. In this respect, it is worth mentioning the reported in-crease in electron mobility in GaAs using arsine as an arsenic source, which was attributed to a superior growth using As$_1$ cracked from arsine.[14] In this author's opinion, the presence of atomic hydrogen on the film surface generated by the thermal cracking of arsine can be very effective in removing or combining with the oxygen-containing species during growth to significantly reduce the deterious effect of these species on the films grown.

One unique feature of MBE is the capability of growing atomi-cally smooth and abrupt interfaces which are essential to a successful

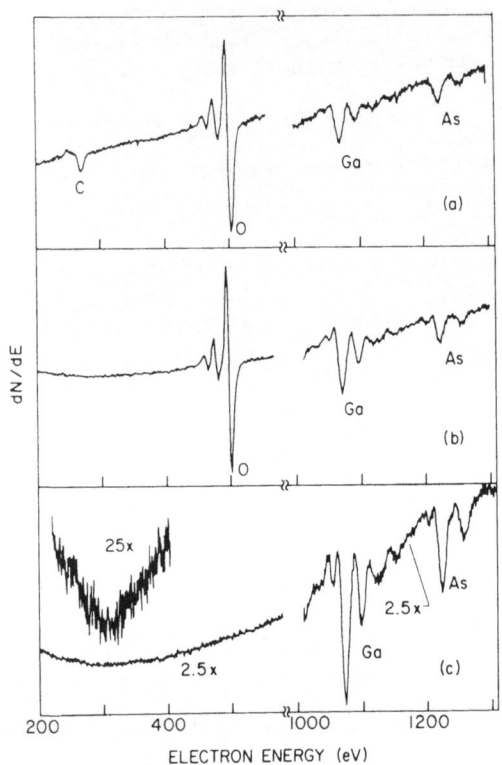

Figure 3 - Auger spectra of a GaAs substrate with only ~2 min. exposure to the Auger beam at room temperature (see text):   (a) before
heating, (b) at 400°C, (c) at 560°C.

fabrication of ultrathin-layer structures such as superlattices.   The
GaAs-AlAs superlattices, for example, have been analyzed by x-ray
interference measurement and showed smooth interfaces on the scale of
atomic dimensions over the thickness range of 15 to 200Å for the
individual GaAs and AlAs layers.[15] The periodic structures of the
superlattices can also be resolved using Auger depth profiling, as demonstrated for a $GaAs(50Å)$-$Ga_{0.75}Al_{0.25}As(50Å)$ superlattice.[16] In
addition, $GaAs$-$Ga_{1-x}Al_xAs$ superlattices in the nearly monolayer
thickness range have also been fabricated and analyzed.[17]

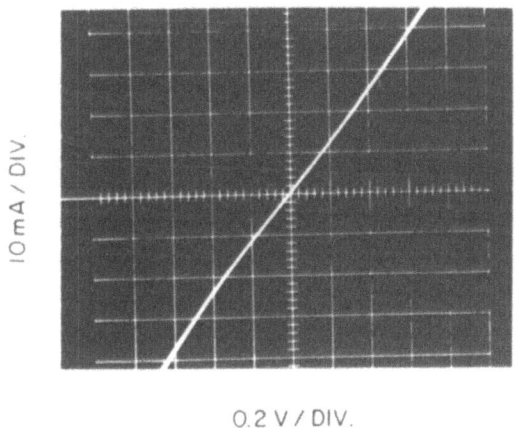

0.2 V / DIV.

Figure 4 - I-V characteristic of a 2μm Sn-doped GaAs film (n ~ 1 x 10^{19}cm^{-3}) deposited on a sputter-annealed GaAs substrate with a predeposition of monolayer Sn. Mesa structure size: 100 x 100μm^2.

2.2 InAs-GaAs and $In_{1-x}Ga_xAs$-$In_{1-y}Ga_yAs$

The lattice mismatch between InAs and GaAs is 7%. The InAs-GaAs and $In_{1-x}Ga_xAs$-$In_{1-y}Ga_yAs$ systems have several interesting properties suitable for the studies of the effect of lattice mismatch: (a) InAs has a high electron mobility which can be sensitive to the defects present in the films. (b) Growth of InAs and GaAs involves the same group V element, As, which allows an easier control of the alloy composition for the InGaAs growth than those involving different group V elements. (c) The above advantages allow a convenient test of new growth techniques to reduce the effect of lattice mismatch, which can be applied to other heteroepitaxial systems with both scientific and technological importance.

InAs and InGaAs were grown at 400-600°C on (100)GaAs substrates. For the growth of InGaAs, both the growth rate and alloy composition are determined by the individual fluxes of In and Ga. Undoped InAs films are n-type with carrier concentrations of

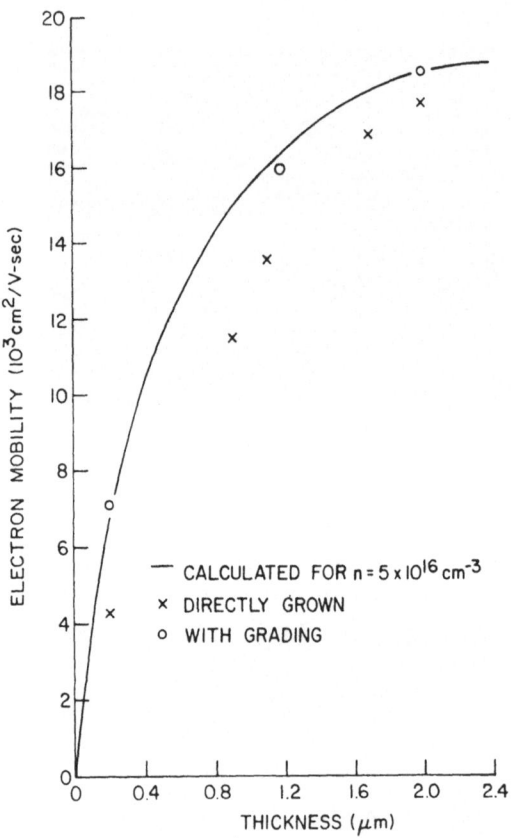

Figure 5 - Observed and calculated electron mobilities of InAs films grown on GaAs with different film thickness.

~$1 \times 10^{16} cm^{-3}$. The (100)InAs has a surface reconstruction pattern of c(2x8) and c(8x2) for the As- and In-stabilized surfaces, respectively,[18] similar to the (100)GaAs ones.

The electron mobilities of the InAs films depend very sensitively on the film thickness as shown in Fig. 5, which decreases rapidly for the thinner films. For comparison, we have calculated the dependence of electron mobility on the film thickness for an InAs crystal using the bulk mobility data and the equation $\mu_F = \mu_B(1 + 2\ell/t)$ assuming diffused surface scattering.[19] Here, μ_F and μ_B are the film and bulk

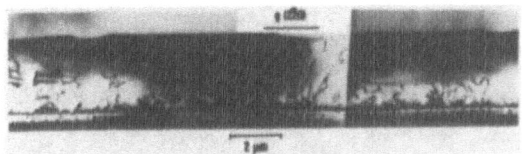

Figure 6 - Cross-sectional transmission electron micrograph of a 2-μm InAs film grown directly on GaAs.

mobilites, respectively, ℓ is the mean free path, and t the film thickness. Using literature mobilities for bulk InAs,[20] and estimating ℓ by $\ell = m\mu v/e$, m being the effective mass, and v the unilateral mean velocity, $(3KT/\pi m)^{1/2}$, the calculated μ_F's are also shown in Fig. 5. It is seen that the calculated μ_F's also decrease for the thinner layers, but not as rapidly as the observed values. We attribute this difference to the increasingly high density of defects in the thinner layers of InAs, as revealed by the TEM studies.[21] Figure 6 shows a cross-section TEM (XTEM) micrograph of a 2μm InAs films grown directly on GaAs. The density of dislocations is $\sim 10^8$cm^{-2} near the film surface, but rapidly increases to $\geq 10^{12}$cm^{-2} near the InAs-GaAs interface. Such a high density of dislocations within a region of ~ 2000Å from the interface is believed to be responsible for the rapid decrease in electron mobilities in the thinner InAs films. To find a remedy to reduce the effect of lattice mismatch, we have developed a step-grading growth technique using MBE which employs several grading layers of In$_{1-x}$Ga$_x$As with different alloy compositions, each several thousand angstroms thick. The XTEM micrograph of one such structure is shown in Fig. 7 for the growth of a 2000Å InAs layer on GaAs.[22] The density of dislocations remain high in the three In$_{1-x}$Ga$_x$As grading layers, $\sim 10^{11}$cm^{-2}, but decreases to $\sim 10^9$cm^{-2} in the last InAs layer, a reduction of three orders of magnitude from the first 2000Å region of an InAs film grown directly on GaAs as shown in Fig. 6. The InAs layers thus grown using the three In$_{1-x}$Ga$_x$As step-grading layers invariably show improved electron mobilities, also shown in Fig. 5. Using a similar grading technique, improved electron mobilities have been obtained for the InGaAs alloy layers, InAs-GaSb superlattices and InGaAs-GaSbAs superlattices grown on GaAs substrates.[21] Application of this growth technique to other systems with similarly large lattice mismatch should also be useful in both improving the film qualities and allowing for growth of a variety of materials on a largely lattice mismatched substrate.

The effectiveness of the step-grading technique on the reduction of defects due to lattice mismatch has been further studied by varying the magnitude of lattice mismatch across the interface of a heteroepitaxial system. Figures 8 and 9 show the XTEM micrographs of two examples, with the lattice mismatches ranging from 1.75 to 0.35% which correspond to an In concentration change of 25 to 5%, respectively. For the cases with an In concentration change of $\geq 20\%$, a high concentration of dislocations is observed, similar to those in the grading layers of Fig. 7 which have an In concentration change of 25%. The density of dislocations is drastically reduced for smaller In concentration changes, being the least at 15% as shown in Fig. 8 for the $In_{0.15}Ga_{0.85}As$ layer grown on GaAs, which corresponds to a misfit stress of 1%. Nearly all the misfit dislocations are confined in the film plane at the $In_{0.15}Ga_{0.85}As$-GaAs interface with few defects in the bulk layer. The confinement becomes less effective at lower In concentration changes as shown in Fig. 9. It seems therefore that a certain

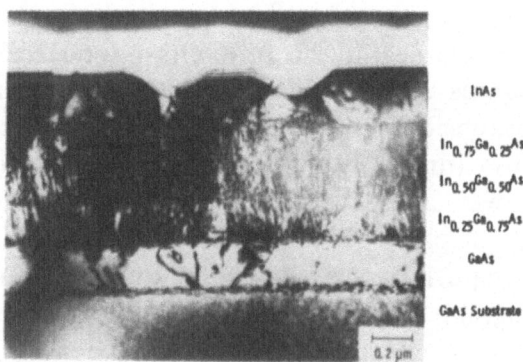

Figure 7 - Cross-sectional transmission electron micrograph of a 2000Å-InAs film grown by the step-grading technique.

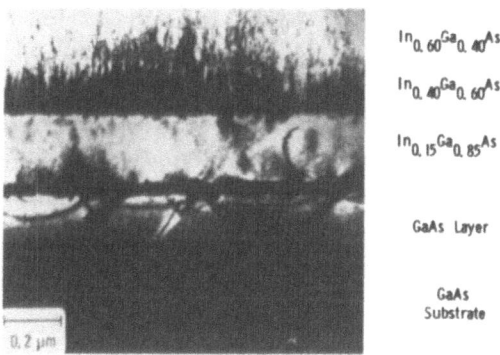

Figure 8 - Cross-sectional transmission electron micrograph of step-graded grown InGaAs layers.

amount of misfit stress is needed to prevent the propagation of misfit dislocations from the interface into the grown layer, the optimal stress being ~1% in the InGaAs case.   It is also interesting to notice that different mechanisms[23] can be involved in relieving the misfit stress as shown at different interfaces of Fig. 9.

InAs films grown on GaAs by MBE have been analyzed by He[+]-ion channeling measurement.[24]   A high density of defects was

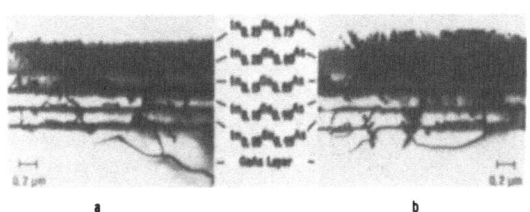

Figure 9 - Cross-sectional transmission electron micrographs of step-graded grown InGaAs layers: (a) and (b) for different regions.

observed near the first 2000Å of the interface, consistent with the XTEM work shown above. MBE growth of $In_{0.53}Ga_{0.47}$ As and $In_{0.53}Ga_{0.47}As$-$In_{0.52}Al_{0.48}As$ heterojunctions lattice matched to InP substrates have recently been reported[25] and are discussed elsewhere in this Symposium. The InGaAs-GaAs and $In_{1-x}Ga_xAs$-$In_{1-y}Ga_yAs$ systems have also been used for strained-layer superlattices where a certain degree of lattice mismatch is tolerated in making ultra-thin layer structures[26]. The studies described above on the misfit stress effect can be very useful to the understanding of such strained structures also.

2.3 GaSb-GaAs

The GaSb-GaAs system has a mismatch of 7.5%. GaSb has been grown on both GaAs and GaSb substrates by MBE at 450-600°C.[18,27] Undoped GaSb is p-type with a hole concentration of $\sim 1x10^{16}$ and $\sim 3x10^{15} cm^{-3}$ at 300 and 77°K, respectively; the corresponding hole mobilities are 710 and 4700 cm^2/V-sec for a $2\mu m$ film grown on GaAs. Doping with Sn and Ge both give p-type GaSb to a maximum doping level of 3-$5x10^{18} cm^{-3}$.

Table 1. Observed and calculated lattice constants of GaSb and InAs grown on GaAs substrates.

	Bulk	Observed	Misfit Stress	Thermal Stress
	$a_\perp$	$a_\perp$	$a_\perp$	$a_\perp$
GaSb	6.095Å	6.079Å	~6.52Å	6.083Å
InAs	6.058Å	6.061Å	~6.48Å	6.060Å

GaSb films grown on GaAs have been analyzed by x-ray diffraction, and showed smaller lattice constant than the bulk value, opposite to that of the InAs films grown on GaAs. The results are shown in Table I. It is pointed out that these values are the lattice constants perpendicular to the film plane as measured by x-ray diffraction. From the consideration of lattice mismatch between the film and the GaAs substrate, one expects a reduced lattice constant in the film plane, a_{11}, but an expanded one perpendicular to the film plane, $a_\perp$. The calculated $a_\perp$ for InAs and GaSb using full misfit stress are shown in Table I which are both larger than the corresponding bulk values. Relaxation of this misfit stress should lead to high density of defects such as those observed in InAs by XTEM shown earlier. The residual misfit stress, if any, should leave an $a_\perp$ larger than the bulk value, consistent with the observed one for InAs. The fact that a reduced $a_\perp$ is observed for GaSb indicates the presence of other stresses in addition to the misfit one. One likely candidate is thermal stress which arises from a difference in thermal expansion coefficients between the film and the substrate. We have calculated $a_\perp$ due to thermal stress for both GaSb and InAs films grown on GaAs and shown in Table I.[28] The value for GaSb is indeed smaller than the bulk one and agrees well with the observed value in magnitude. For InAs, thermal stress has only a small effect, leading to a slight increase in its $a_\perp$. It is thus concluded that the misfit stress has been relaxed into defects for both GaSb and InAs films grown on GaAs; the large thermal stress in GaSb becomes dominant and is responsible for its reduced $a_\perp$ observed. Such an understanding of the thermal stress effect also helps studies of other systems involving GaSb. The alloy compositions needed for the lattice matching between $GaSb_{1-y}As_y$ and $In_{1-x}Ga_xAs$, for example, are modified by such a consideration.[28]

2.4 InAs-GaSb and $In_{1-x}Ga_xAs$-$GaSb_{1-y}As_y$

InAs-GaSb is an interesting system due to the bandedge relation between the two semiconductors: the conduction bandedge of InAs lies below the valence bandedge of GaSb.[29] Upon alloying with GaAs, the similar bandedge relation between $In_{1-x}Ga_xAs$ and $GaSb_{1-y}As_y$ persists to $x \sim y \sim 0.2$, beyond which the conduction bandedge of $In_{1-x}Ga_xAs$ lies above the valence bandedge of $GaSb_{1-y}As_y$ resembling ordinary p-n junctions of most materials. The transport, optical, metallurgical and structural studies of these systems will be described after the MBE growth of GaSbAs.

The MBE growth of GaSbAs differs from that of InGaAs: the growth rate is determined by the Ga flux, but the alloy composition by a competing incorporation mechanism between As and Sb. Figure 10 illustrates the growth of GaSbAs at different fluxes of Sb and As and at different substrate temperatures. At 470°C, for example, and a Sb/Ga

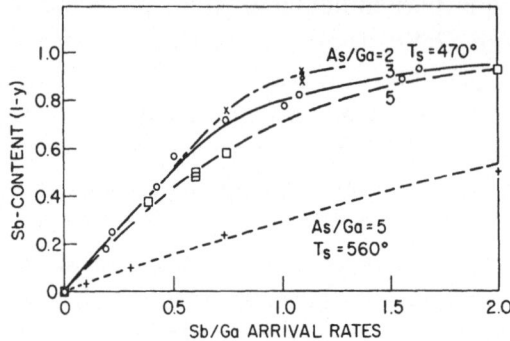

Figure 10 - Dependence of Sb content in the GaSb$_{1-y}$As$_y$ films on the Sb arrival rate for various As arrival rates, which are normalized with respect to that of Ga.

flux ratio of unity and that of As/Ga = 3, the GaSbAs film grown contains 80% Sb and 20% As, indicating a much higher incorporation efficiency of Sb than As. The Sb incorporation rate, however, decreases rapidly at higher temperatures, which can be related to the thermodynamics of the evaporation of Sb and As.[18]

RHEED studies show that the surface reconstruction patterns of GaSb$_{1-y}$As$_y$ resemble the GaSb one for y < 0.2, ie., c(2x6) for the Sb-stabilized surface, and the GaAs one for y > 0.5, i.e., c(2x8) for the As-stabilized surface. Doping of GaSb$_{1-y}$As$_y$ alloys with Sn shows a p-to-n transition, since Sn is an acceptor in GaSb but a donor in GaAs. The transition takes place at y ~0.2 as shown in Fig. 11, which also shows the doping behavior of Sn in In$_{1-x}$Ga$_x$As, giving n-type materials over the whole composition range. We have made a series of n In$_{1-x}$Ga$_x$As-p GaSb$_{1-y}$As$_y$ diodes and measured their I-V characteristics.[30] The results are shown in Fig. 12. The (x,y) compositions and the bandedge overlap E$_s$ in eV of the four diodes measured, (x,y,E$_s$), are: A: (0.62, 0.64, 0.47eV); B: (0.52, 0.56, 0.33eV); C: (0.50, 0.28, 0.14eV) and D: (0.16, 0.10, -0.09eV). Except for sample D which has a negative E$_s$, the others all have positive E$_s$'s and show different degrees of rectification, being most rectified for sample A and least for sample C. Sample D, on the other hand, shows an ohmic I-V characteristic; a similar ohmic behavior is also seen for the n InAs-p

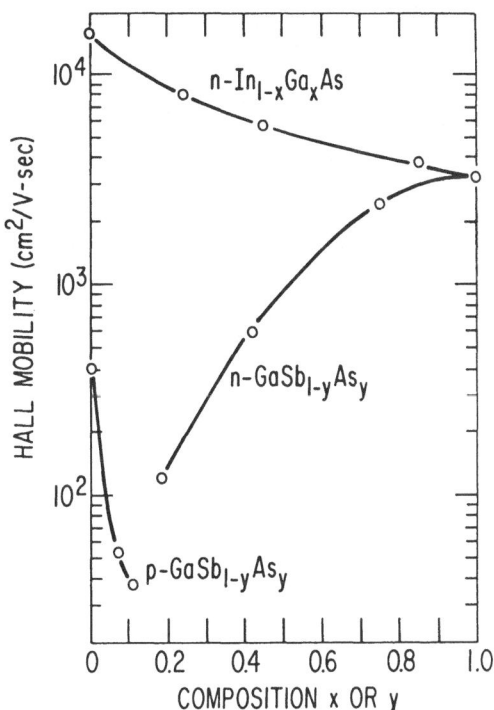

Figure 11 - Doping behavior of Sn in $GaSb_{1-y}As_y$ and in $In_{1-x}Ga_xAs$. The room temperature mobilities are for films with carrier concentration of $3 \times 10^{17} cm^{-3}$.

GaSb diode which has an E_s of $\sim-0.15eV$. The ohmic behavior for the p-n junctions with negative E_s can be understood by the insertion of band energy diagram of Fig. 12 for sample D. Accumulation of electrons on the $In_{1-x}Ga_xAs$ side and of holes on the $GaSb_{1-y}As_y$ side of the interface is believed to be responsible for the ohmic behavior observed. Samples A, B, and C, on the other hand, behave similarly to ordinary p-n junctions and therefore show the rectifying I-V characteristics.

We have also studied $In_{1-x}Ga_xAs$-$GaSb_{1-y}As_y$ superlattices using both optical absorption and electroreflectance measurements. The absorption measurements give band gap energies in good agreement with calculated values.[31] The electroreflectance measurement provides

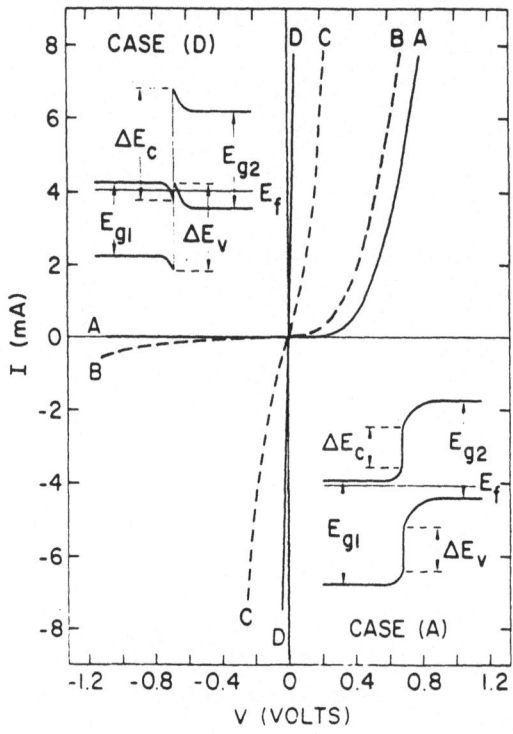

Figure 12 - Room temperature I-V characteristics of n-In$_{1-x}$Ga$_x$As-p-GaSb$_{1-y}$As$_y$ heterojunctions. The inserts are the energy band diagrams for Samples A and D whose junction area is 8 x 10^{-5}cm^2.

information on the effect of superlattice potential away from the Brillouin zone center. The E_1 and $E_1 + \Delta_1$, from spin-splitting, of In$_{1-x}$Ga$_x$As-GaSb$_{1-y}$As$_y$ superlattices are shown in Fig. 13 along with those of the In$_{1-x}$Ga$_x$As and GaSb$_{1-y}$As$_y$ alloys. It is seen that, above x~y > 0.2, the measured E_1 and $E_1 + \Delta_1$ of the superlattices deviate appreciably from those of the alloys. The physics involved in such an observation is still unclear.[3] It is interesting to notice the close resemblance in transition region, at x~y≈0.2, between the electroreflectance observations of Fig. 13 and that of the energy bandedge overlapping of the In$_{1-x}$Ga$_x$As and GaSb$_{1-y}$As$_y$ alloys.

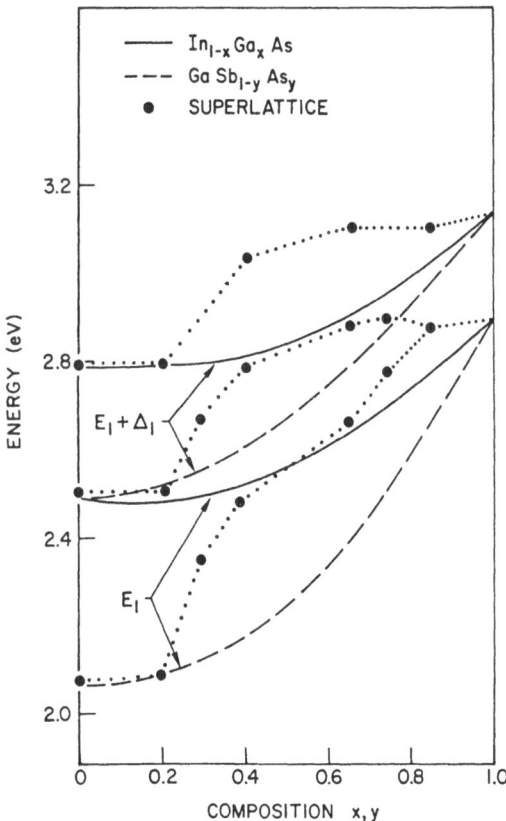

Figure 13 - Variation of the E_1 and $E_1 + \Delta_1$ transitions with composition for $In_{1-x}Ga_xAs$-$GaSb_{1-y}As_y$ superlattices and for $In_{1-x}Ga_xAs$ and
$GaSb_{1-y}As_y$ systems.

Magnetic transport measurements of the InAs-GaSb superlattices
have revealed interesting properties of this system involving
semiconductor-to-semimetal transitions,[33] the magnetic field effect,[34]
etc. Details of these studies are described elsewhere in this Symposium.

The structural studies of InAs-GaSb superlattices using He+-ion
channeling have revealed interesting information of the lattice structures involved. The channeling measurement provides valuable information on the crystallinity, defect nature and distribution, and interface
structures. For the zinc-blende and diamond structures, channeling

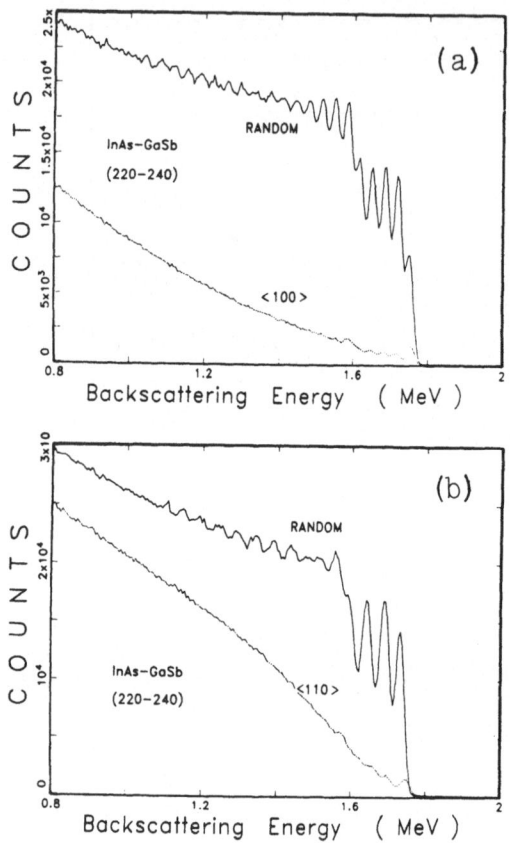

Figure 14 - Backscattering and channeling spectra of InAs(220Å)-GaSb(240Å) superlattice: (a) [100]; (b) <110>.

along <110> axis gives a lower backscattering yield, or dechanneling yield, than that along <100> axis. Channeling of the InAs-GaSb superlattices grown on (100)GaSb substrates, however, all show much enhanced <110> dechanneling yields with low dechanneling yields along the [100] growth direction. Figure 14 shows the RBS spectra at a random incidence angle and the channeling spectra along [100] and <110> axes of an InAs(220Å)-GaSb(240Å) superlattice. The oscillatory random spectra reveal the periodic structure of the superlattice, and the channeling spectra show a much enhanced dechanneling yield along <110> axis. The low dechanneling yield along [100] indicates good epitaxy and low density of defects in this superlattice. The high <110> dechanneling yield therefore indicates a certain lattice structure deviated from normal InAs and GaSb ones, which can only be detected along <110> but not along the [100] axis. The presently adopted model has to do with the interface bonding relaxation in these structures. Although the mismatch between InAs and GaSb is small, 0.7%, at the interfaces there can be alternating Ga-As and In-Sb bonds whose bond lengths are shorter and longer, respectively, than those of Ga-Sb and In-As by 7%. The corresponding contraction and expansion in interfacial layer spacings take place along the [100] growth direction and cannot be detected by [100] channeling. It can, however, be detected by <110> channeling and give rise to enhanced <110> dechanneling yields. A model of this interface relaxation is shown in Fig. 15. Details of this work have been described elsewhere.[35,36] Studies are presently in progress to understand quantitatively the channeling results observed, and to consider other contributions to the changes in lattice structures, e.g., the lattice distortion due to the misfit stress between GaSb and InAs.

2.5 AlSb-GaAs, AlSb-GaSb and AlSb-InAs

AlSb and AlGaSb have been materials of optoelectronic interest and fabricated by various techniques.[37] AlSb is of special interest to the InAs-GaSb system due to its large bandgap of 1.6eV and small lattice mismatch of 0.6 and 1.3% with GaSb and InAs, respectively.

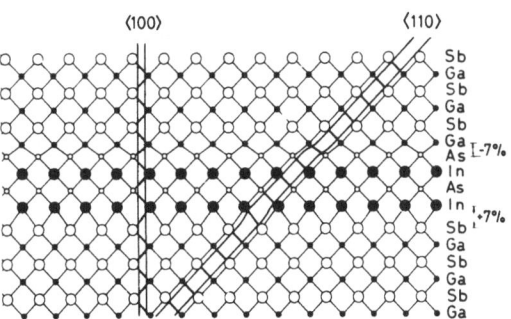

Figure 15 - A bond-relaxation model of InAs-GaSb superlattice which gives enhanced dechanneling yield along <110> but not along [100].

The possibility of forming InAs-AlSb-GaSb three- component junctions and polytype superlattices has been proposed.[38]

AlSb has been grown by MBE between 400 and 550°C. Undoped AlSb films are p-type with a carrier concentration of $\sim 10^{14}$cm^{-3} and a resistivity of $\sim 200\Omega$cm. Doping with Ge usually gives p-type AlSb films; at very high Sb fluxes, e.g., Sb/Al > 10, n-AlSb has been observed.[39] In the latter case the high Sb flux is believed to push Ge into the cation site, behaving as a donor, similar to the behavior of Ge in GaAs.[40] (100) AlSb exhibits a c(2x6) or (2x3) surface reconstruction pattern for the Sb-stabilized surface, but a (4x2) one for the Al-stabilized surface. The RHEED patterns are shown in Fig. 16.

Growth of AlSb on GaAs has been monitored with RHEED for the effect of lattice mismatch, 7%, on the interface morphologies. Figure 17 shows the RHEED patterns at different stages of growth of AlSb on (100) GaAs. The spotty diffraction pattern of Fig. 17c indicates the formation of a three-dimensional nucleation stage similar to those observed for the growth of GaSb on GaAs.[18] The diffraction pattern becomes streaking at a thicker growth of AlSb, an indication of smoothing out of the surface which persists for the growth of AlSb to a thickness beyond 1μm. For the AlSb-GaSb, AlSb-InAs and GaSb-AlSb-InAs systems, smooth and abrupt interfaces are essential to the successful fabrication of heterojunctions and superlattices of these materials. Figures 18 and 19 show the RHEED patterns for the growth of the AlSb-GaSb and AlSb-InAs systems, respectively. It is seen that both the growths of GaSb on AlSb and of AlSb on GaSb show smooth and abrupt interfaces from the RHEED patterns. For the AlSb-InAs case, the RHEED patterns are somewhat diffused for both the growths of InAs on AlSb and of AlSb on InAs, which can be due to either a larger mismatch effect or a transition between the (2x3) and (2x4) patterns, or both. However, no nucleation stage similar to that of Fig. 17 is observed for either InAs-AlSb interface. We have also grown InSb on (100)AlSb at 300°C with a lattice mismatch of 8%. The RHEED patterns shown in Fig. 20 also indicate an initial nucleation stage; the surface is later smoothed out by subsequent growth of thicker InSb films. The (100)InSb film shows a (2x3) and a (3x2) surface reconstruction pattern for the Sb- and In-stabilized surfaces, respectively.

[1Ī0]   [100]   [110]

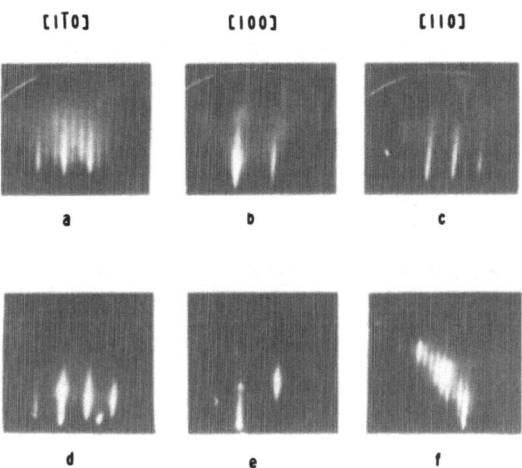

Figure 16 - RHEED patterns of (100)AlSb at 10KV and different azimuths.

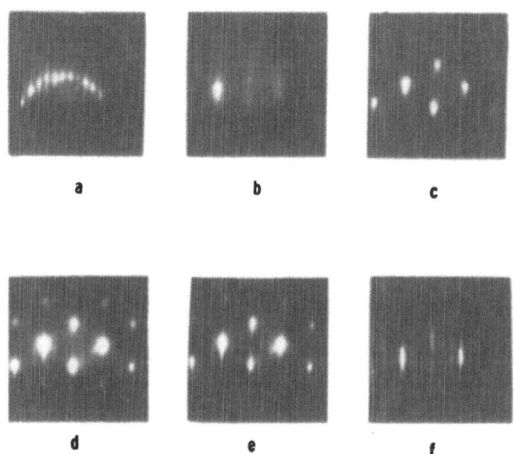

Figure 17 - RHEED patterns at 10KV and [1$\bar{1}$0] azimuth for the growth of AlSb on (100)GaAs: (a) GaAs film; after a total growth of (b) 2Å, (c) 6Å, (d) 10Å, (e) 20Å, and (f) 250Å of AlSb at 420°C.

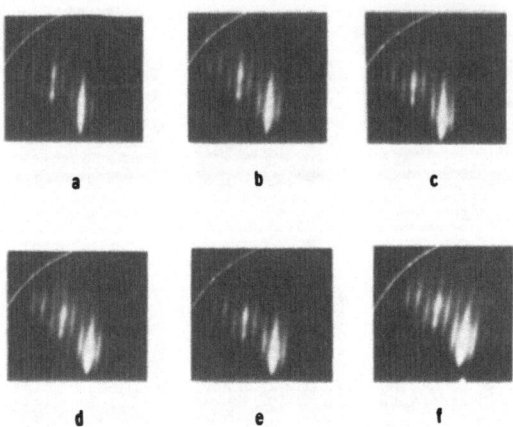

Figure 18 - RHEED patterns at 10KV and [1$\bar{1}$0] azimuth for the growth of (100) GaSb-AlSb system at 420°C: (a) AlSb film; after a total growth of (b) 2Å, (c) 6Å, and (d) 20Å of GaSb on AlSb; and after a total growth of (e) 2Å, and (f) 10Å of AlSb on GaSb.

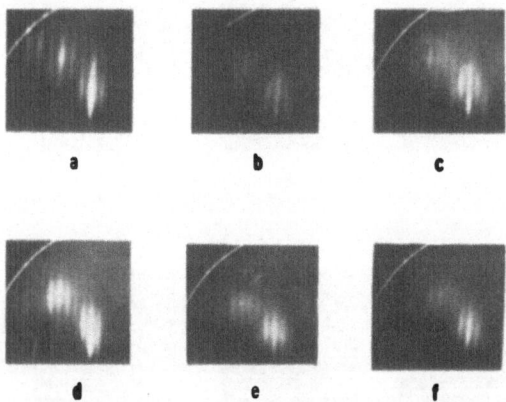

Figure 19 - RHEED patterns at 10KV and [1$\bar{1}$0] azimuth for the growth of (100) InAs-AlSb system at 420°C: (a) AlSb film; after a total grown of (b) 2Å, (c) 5 Å, and (d) 30Å of InAs on AlSb; and after a total growth of (e) 2Å and (f) 5Å of AlSb on InAs.

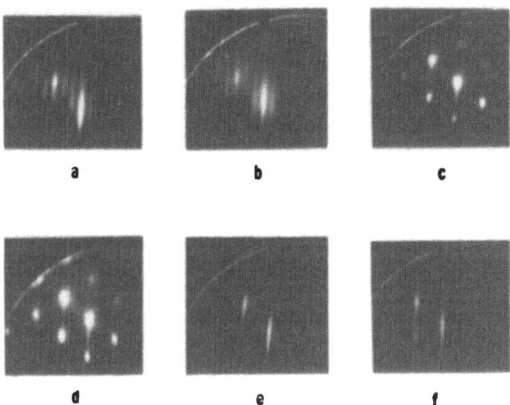

Figure 20 - RHEED patterns at 10KV and [1$\bar{1}$0] azimuth for the growth of InSb on (100) AlSb: (a) AlSb film; after a total growth of (b) 2Å, (c) 10Å, (d) 30Å and (e) and (f) 250Å of InSb. (e) for Sb-rich surface and (f) for In-rich surface of InSb.

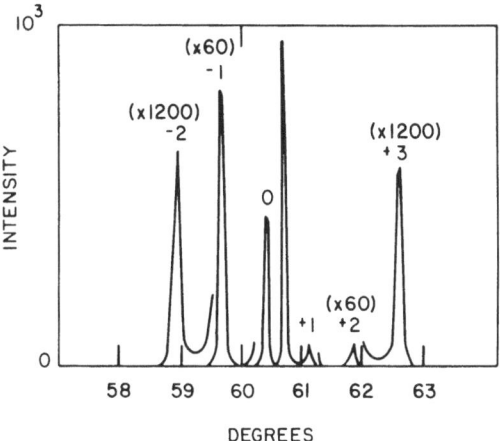

Figure 21 - X-ray diffraction pattern of the GaSb(80Å)-AlSb(60Å) superlattice.

Both the AlSb-GaSb and AlSb-InAs superlattices have been grown and their properties studied. Figure 21 shows an x-ray diffraction pattern of the AlSb-GaSb superlattice with the satellite peaks revealing its periodic structure. The periodicity estimated from the x-ray analysis agrees well with the layer thickness intended for growth. The formation of superlattice band structure of the AlSb-GaSb system has also been studied and confirmed by optical absorption,[41] electroreflectance[42] and photoluminescence studies.[42,43]

Preliminary measurements of an n InAs-AlSb-p GaSb structure with a 100Å AlSb barrier layer showed nonlinear I-V characteristics.[44] Detailed studies of this three-component structure are in progress to understand the mechanism involved.

3. IV/III-V HETEROSTRUCTURES

3.1 Ge-GaAs

The Ge-GaAs system has been of extensive interest and studied from both theoretical and experimental point of views. The lattice mismatch is nearly zero, 0.07%. Both elemental and compound semiconductors are involved, giving rise to the orientation-dependent bondings at the interfaces: the GaAs layer is nonpolar for the (110) orientation and polar for the (100) and (111) orientations, while Ge remains nonpolar for all orientations. The two materials have different energy gaps and GaAs is direct but Ge indirect. All of these make Ge-GaAs an interesting system, with both theoretical[45-50] and experimental[51-57] studies reported.

MBE growth of the Ge-GaAs system has been made at 350-550°C, with most films grown at 400°C to maintain good epitaxy while reducing interdiffusion between Ge and GAs.[54] Undoped Ge films grown on GaAs are n-type with a carrier concentration of $3 \times 10^{18} cm^{-3}$.[54,57] The high carrier concentration of donors in Ge has been related to the outdiffusion of As from the GaAs substrate, as evidenced by the observation of $GeAs_x$ on the Ge surface.[51, 52] The outdiffusion of As is also expected from the consideration of charge neutrality at the Ge-GaAs interface when the last layer of GaAs is As-rich from the MBE growth conditions used.[58] The carrier concentrations in Ge can be reduced to the high $10^{17} cm^{-3}$ range by reducing the As background pressure, from 10^{-7} to $< 10^{-9}$ Torr, during the Ge growth.[59] This indicates that part of the donors observed in Ge described above comes from the background As vapor. Further reduction in carrier concentration to $\sim 3 \times 10^{17} \text{cm}^{-3}$ has been achieved by epitaxially growing Ge on GaAs at 350°C with a background As pressure $\leq 10^{-9}$ Torr.[59]

Interdiffusion between Ge and GaAs has been observed by several groups of workers[46,47,49] Figure 22 shows the Auger spectra of both 10Å Ge-on-GaAs and 10ÅGaAs-on-Ge, grown at 400°C on (100)GaAs substrates and maintained at 400°C for 4 hrs. The Auger spectra clearly show interdiffusion in both cases, ~ 10Å after heating at 400°C for 4 hrs..[49] A similar interdiffusion rate has also been observed by others.[47]

RHEED studies of the Ge-GaAs system showed interesting growth morphologies which depend on orientations of the substrates used. The (100)Ge surface has a (2x2) reconstructing pattern, those of

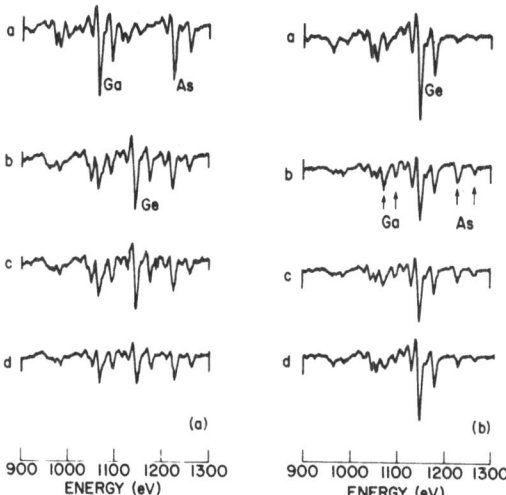

Figure 22 - Auger spectra of 10-Å Ge deposited on GaAs (left) and of 10-Å GaAs deposited on Ge (right). The four traces in each case correspond to (a) before deposition, (b) immediately after deposition at 400°C, (c) after 20 min. heating, and (d) after 4 hrs. heating at 400°C.

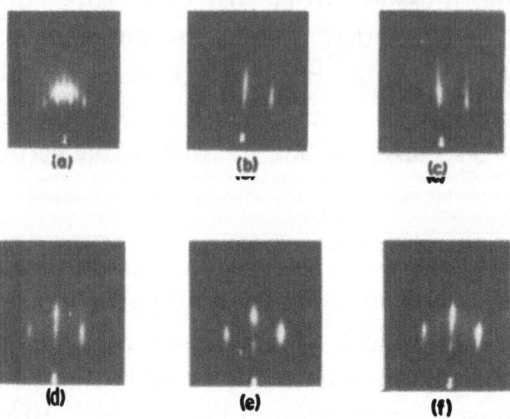

Figure 23 - RHEED patterns at 10KV and $[1\bar{1}0]$ azimuth for the growth of (100)Ge-GaAs: (a) GaAs film grown at 580°C; (b) GaAs film grown at 400°C; (c) Ge film grown on GaAs of (b); a total of (d) 2Å and (e) 5Å of GaAs grown on Ge of (c); (f) 5Å of Ge grown on 10Å of GaAs which has a RHEED pattern of (e). (b)-(f) all grown at 400°C.

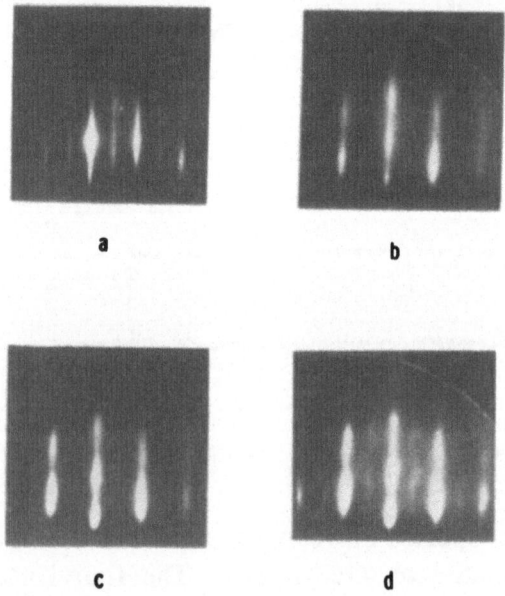

Figure 24 - RHEED patterns at 10KV and [111] azimuth for the growth sequence of (111) Ge-GaAs at 400°C: (a) GaAs, (b) 10Å of Ge, and a total growth of (c) 10Å and (d) 40Å of GaAs on Ge.

(110)Ge include (1x1), (2x3) and (4x4), and a (5x5) pattern is observed for (111)Ge.[52]   Figure 23 shows the RHEED patterns of (100)Ge-GaAs at different stages of epitaxy. The growth of Ge on GaAs shows a smooth surface; that of GaAs on Ge, however, shows a rougher surface on an atomic scale, which can always be smoothed out by a further growth of ~3-5Å of Ge.[55] The surface roughness of the GaAs layers grown on Ge is thus estimated to be ~3-5Å. Similar growth morphologies have also been observed for the (111) orientation,[56] as shown in Fig. 24. A different growth morphology of GaAs grown on Ge is observed for the (110) orientation, shown in Fig. 25, which indicates a smooth surface for both the GaAs layers grown on Ge and the Ge layers grown on GaAs. The rough surfaces observed for the (100) and (111)GaAs on Ge are attributed to the formation of antiphase domains in the GaAs layers. This will be discussed later along with our measurements on the (100)Ge-GaAs superlattices.

A number of (100) Ge-GaAs superlattices have been grown at 400°C with the individual layer thickness ranging from 25 to 1000Å. X-ray diffraction measurements, shown in Fig. 26, show that the periodic structures can be resolved to a layer thickness of 35Å. For thinner layers, e.g., 25Å, a broadened x-ray diffraction peak is observed with no satellite peaks resolved.

Channeling measurements of the (100)Ge-GaAs superlattices show an increasingly enhanced dechanneling yield along the [100] growth direction with decreasing layer thickness; those along the <100> axis remain very low.[58, 60] Figures 27 and 28 show the channeling spectra of two superlattices: Ge(100Å)-GaAs(80Å) and Ge(25Å)-GaAs(25Å). The very low <110> dechanneling yields in both structures indicate very good epitaxy and low density of defects. The much enhanced [100] dechanneling yield for the thinner layer superlattice is believed to be related to the different growth morphologies of GaAs on Ge: both surface roughness and grooving at the antiphase boundaries can be the major causes.[58,61] The same causes can also contribute, along with interdiffusion, to the broadened x-ray diffraction observed for the Ge(25Å)-GaAs(25Å) superlattice described above.

We have also used high-resolution XTEM to study the structure of a Ge(100Å)-GaAs(80Å) superlattice. Figure 29 shows a bright-field image of this superlattice. The periodic structure of the superlattice is

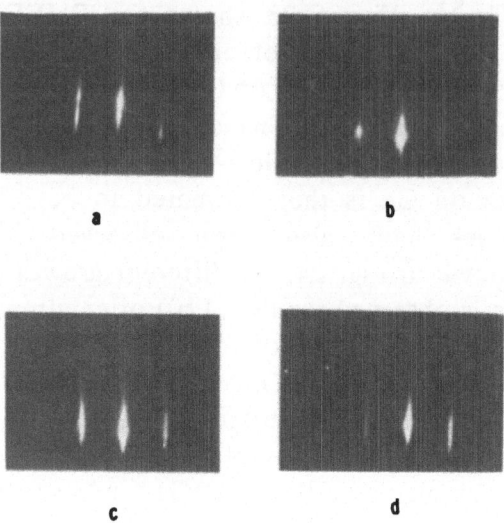

Figure 25 - RHEED patterns at 10KV and [110] azimuth for the growth sequence of (110)Ge-GaAs at 400°C: (a) GaAs, (b) 5Å of Ge, and a total growth of (c) 6Å and (d) 20Å of GaAs on Ge.

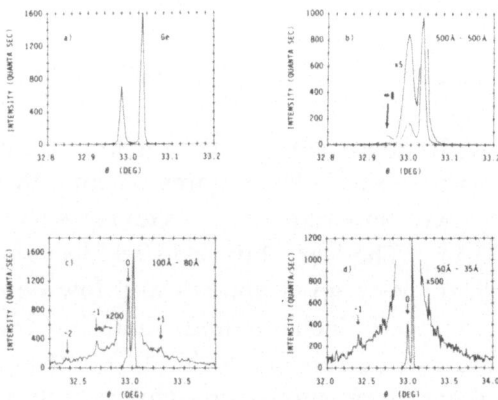

Figure 26 - X-ray diffraction patterns of (a)Ge and (b), (c), (d)Ge-GaAs superlattices grown on GaAs substrates.

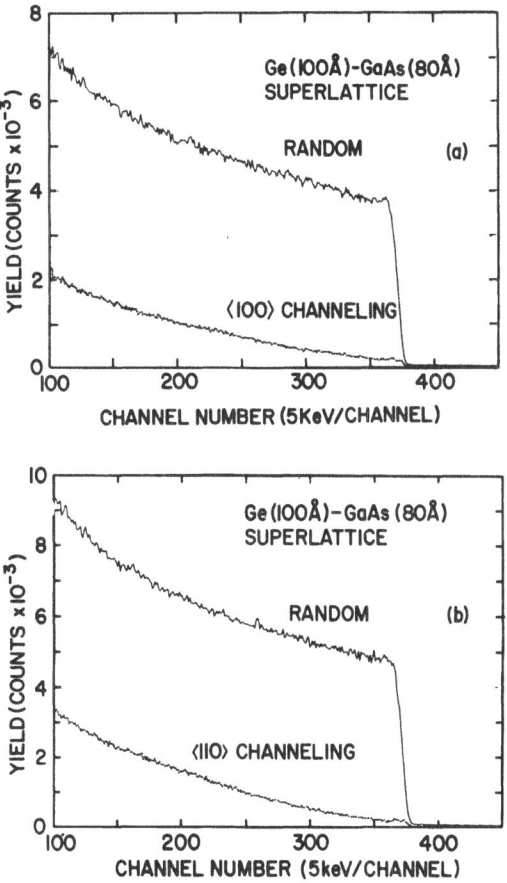

Figure 27 - He+-ion channeling spectra for the Ge(100Å)-GaAs(80Å) superlattice: (a) <100> channeling, (b) <110> channeling.

clearly resolved, with the layer thickness of Ge and GaAs in good agreement with those intended for growth. It also shows contrasting features indicating the presence of planar defects in the GaAs layers but not in the Ge layers. However, a 13-beam lattice image, shown in Fig. 30, indicates no lattice defects in the GaAs layers. In addition, perfect lattice structures in both GaAs and Ge layers and a perfect lattice match at the Ge-GaAs interfaces are observed. The planar structures observed in the GaAs layers in Fig. 29 are interpreted to be

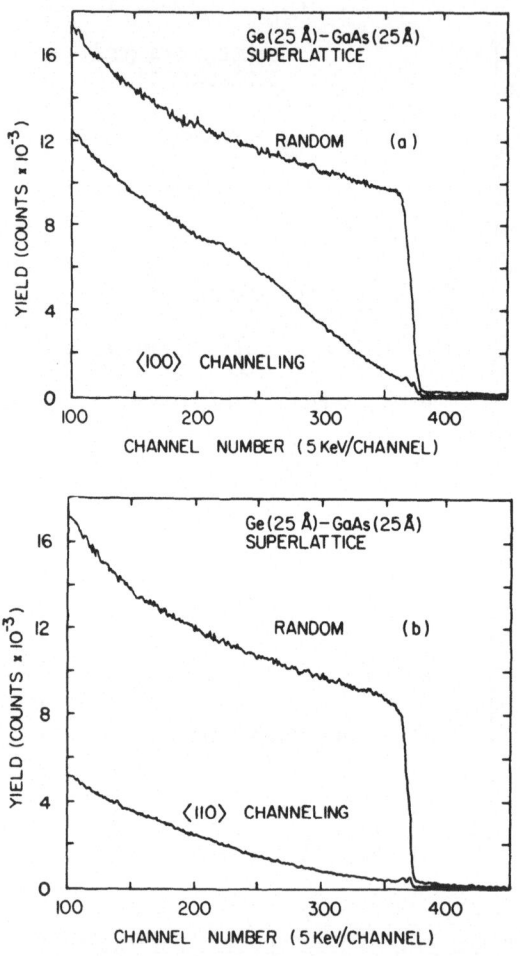

Figure 28 - He⁺-ion channeling spectra for the Ge(25Å)-GaAs(25Å) superlattice: (a) <100> channeling, (b) <110> channeling.

antiphase boundaries which support the other observations described above using various techniques. Details of the structural studies of Ge-GaAs superlattices using XTEM have been described elsewhere.[61,62]

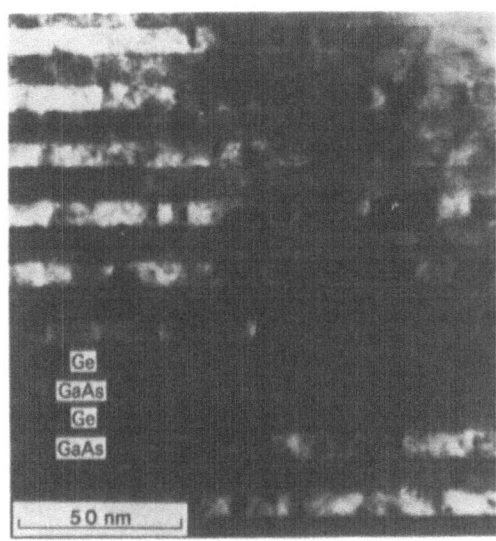

Figure 29 - Cross-sectional TEM bright-field image from a Ge(100Å)-GaAs(80Å) superlattice.

Optical studies of the Ge-GaAs superlattices using electroreflectance have shown two sets of transitions for both the Ge(50Å)-GaAs(50Å) and Ge(100Å)-GaAs(100Å) superlattices which correspond to the respective E_1 and $E_1 + \Delta_1$ transitions of bulk Ge and GaAs but shift to higher energies, an indication of the effect of superlattice potential. A thicker sample, Ge(500Å)-GaAs(500Å), whose last layer is GaAs, shows only one set of transitions corresponding to the E_1 and $E_1 + \Delta_1$ of bulk GaAs, consistent with the penetration depth of the radiation. The thinner sample, Ge(25Å)-GaAs(25Å), on the other hand, shows only a broad reflection spectrum whose origin is difficult to assess, but can be correlated with the various structural studies of this superlattice described above.

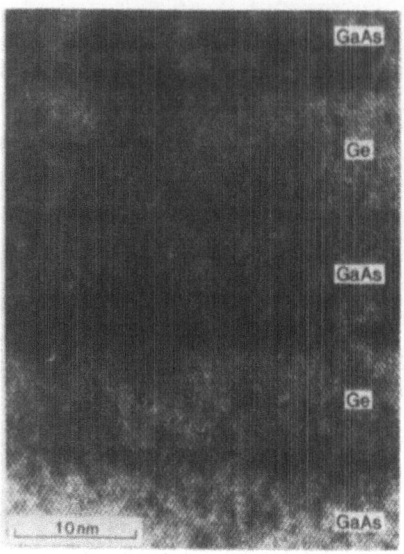

Figure 30 - A 13-beam cross-sectional lattice image of the same super-lattice of Figure 29.

Band discontinuity between Ge and GaAs has been of great interest and studied by many workers.[63-66] Details of these studies are described elsewhere in this Symposium.

3.2. Si-GaP

Si-GaP, with a lattice mismatch of 0.4%, consists of another IV/III-V system which has recently attracted some studies.[67-70] Both MBE[67,68] and vapor growth[69] techniques have been used to fabricate this system. The growths have, however, been limited to that of GaP on Si.

MBE growth of GaP on Si substrates was carried out at ~580-650°C using either elemental P[67] or GaP[68] as source materials of phosphorus. Si substrates used included (100), (110), (111) and (211).

The Si surface treatment was found to have profound effects on the GaP films grown, leading to polycrystalline GaP in some cases which was attributed to the residual Si oxide on the substrate surface.[67] Using a sputter-annealing procedure for the substrate cleaning, epitaxial GaP was achieved at 580°C with fairly good crystalline quality from x-ray diffraction analysis.[67] The autodoping of Si in the GaP films from the Si substrate was found to be limited to the first 1μm of the GaP film grown with a doping concentration of $\leq 10^{17}$cm^{-3}.[63] The doping level reaches ~2x10^{18}cm^{-3} for the GaP films grown at 650°C.[68] Studies of the orientation dependence indicated a better growth of GaP on (211)Si under certain growth conditions than those of other orientations.[68]

4. CONCLUSIONS

We have described the MBE growth, characterization and interface properties of a variety of semiconductor heterostructures. Different interface properties are seen to be present for these systems. It is shown that the studies described have provided better understandings of both the MBE growth mechanisms involved and the effects of various interface properties on the qualities of the films grown. Improved growths and film qualities have also been illustrated for certain systems.

REFERENCES

1.   See, for example, K-W Benz and E. Bauser, in "Crystals, Growth, Properties and Applications," Vol. 3, III-V Semiconductors, Springer-Verlag, Berlin (1980), p. 1.

2.   See, for example, Proceedings of the 5th International Conference on Vapor Growth and Epitaxy, San Diego, Calif., July 1981.

3.   See, for example, Proceedings of the International Conference on Vapor Phase Epitaxy, Ajaccio, France, May 1981.

4.   See, for example, M. H. Francome, in Epitaxial Growth, Ed. J. W. Matthews, Vol. A. Academic Press, New York, 1975, p. 109.

5.   See, for example, Collected Papers of 2nd Intern. Symp. on Molecular Beam Epitaxy and Related Clean Surface Techniques, Tokyo (1982).

6.   C-A Chang, J. Vac. Sci. Technol. 21, 663 (1982).

7. C-A Chang, M. Heiblum, R. Ludeke and M. Nathan, Appl. Phys. Lett. 39, 229 (1981).

8. C-A Chang, L. L. Chang and L. Esaki, U. S. Patent No. 4,239,584 (1980).

9. P. Manuel, G. A. Sai-Halasz, L. L. Chang, C-A Chang and L. Esaki, Phys. Rev. Lett. 37, 1701 (1976).

10. L. L. Chang, H. Sakaki, C-A Chang and L. Esaki, Phys. Rev. Lett. 38, 1489 (1977).

11. A. R. Calawa, Appl. Phys. Lett. 33, 1020 (1978).

12. K. Kondo, S. Muto, K. Nanbu, T. Ishikawa, S. Hiyamizu and H. Hashimoto, Ref. 5, p. 173.

13. R. Dingle, H. L. Stormer, A. C. Gossard and W. Wiegmann, Appl. Phys. Lett. 33, 665 (1978); S. Hiyamizu in Ref. 5, p. 113.

14. A. R. Calawa, Appl. Phys. Lett. 38, 701 (1981).

15. L. L. Chang, A. Segmüller and L. Esaki, Appl. Phys. Lett. 28, 39 (1976).

16. R. Ludeke, L. Esaki and L. L. Chang, Appl. Phys. Lett. 24, 417 (1974).

17. P. M. Petroff, A. C. Gossard, W. Wiegmann and A. Savage, J. Crystal Growth 44, 5 (1978).

18. C-A Chang, R. Ludeke, L. L. Chang and L. Esaki, Appl. Phys. Lett. 31, 759 (1977).

19. A. Many, Y. Goldstein and N. B. Grover, Semiconductor Surfaces (North-Holland, Amsterdam, 1965), p. 308.

20. T. C. Harman, H. L. Goering and A. C. Beer, Phys. Rev. 104, 1562 (1956).

21. C-A Chang, C. M. Serrano, L. L. Chang and L. Esaki, J. Vac. Sci. Technol. 17, 603 (1980).

22. C-A Chang, C. M. Serrano, L. L. Chang and L. Esaki, Appl. Phys. Lett. 37, 538 (1980).

23. C. M. Serrano and C-A Chang, Appl. Phys. Lett. 39, 808 (1981).

24. R. S. Williams, B. Paine, W. Schaffer and S. Kawalczyk, J. Vac. Sci. Technol. 21, (1982).

25. K. Y. Cheng and A. Y. Cho, in Ref. 5, p. 103.

26. S. T. Picraux, R. M. Biefeld, L. R. Dawson, G. L. Osbourn and W. K. Chu, J. Vac. Sci. Technol., to be published.

27. T. Waho, S. Ogawa and S. Maruyama, Jpn. J. Appl. Phys. 16, 1875 (19 77).

28. C-A Chang and A. Segmüller, J. Vac. Sci. Technol. 16, 285 (1979).

29. G. A. Sai-Halasz, R. Tsu and L. Esaki, Appl. Phys. Lett. 30, 651 (1977).

30. H. Sakaki, L. L. Chang, R. Ludeke, C-A Chang, G. A. Sai-Halasz and L. Esaki, Appl. Phys. Lett. 31, 211 (1977).

31. G. A. Sai-Halasz, L. L. Chang, J-M Welter, C-A Chang and L. Esaki, Solid State Commun. 27, 935 (1978).

32. E. E. Mendez, C-A Chang, L. L. Chang and L. Esaki, J. Phys. Soc. Japan 49 (1980) Suppl. A, p. 1009.

33. L. L. Chang, N. J. Kawai, E. E. Mendez, C-A Chang and L. Esaki, Appl. Phys. Lett. 38, 30 (1981), and references therein.

34. N. J. Kawai, L. L. Chang, G. A. Sai-Halasz, C-A Chang and L. Esaki, Appl. Phys. Lett. 36, 369 (1980).

35. W. K. Chu, F. W. Saris, C-A Chang, R. Ludeke and L. Esaki, Phys. Rev. B 26, 1999 (1982).

36. F. W. Saris, W. K. Chu, C-A Chang, R. Ludeke and L. Esaki, Appl. Phys. Lett. 37, 931 (1980).

37. R. Druilhe, J. Cryst. Growth 54, 330 (1981); J. E. Johnson, J. Appl. Phys. 36, 3193 (1965); M. Leroux, A. Tromson-Carli, P. Gibart, C. Vérié, C. Bernard and M. Schouler, J. Cryst. Growth 48, 367 (1980); P. Gautier, A. Joulle, G. Bougnot and C. H. Champness, J. Cryst. Growth 51, 336 (1981); J. P. David, L. Capela, L. Lande and S. Martinuzzi, Revue de Physique Applique 1, 172 (1966).

38. L. Esaki, L. L. Chang and E. E. Mendez, Jpn. J. Appl. Phys. 20, L529 (1981).

39. C-A Chang, H. Takoka, L. L. Chang and L. Esaki, Appl. Phys. Lett. 40, 983 (1982).

40. A. Y. Cho and I. Hayashi, J. Appl. Phys. 42, 4422 (1971).

41. P. Voisin, G. Bastard, M. Voos, E. E. Mendez, C-A Chang, L. L. Chang, and L. Esaki, J. Vac. Sci. Technol., B1, 409 (1983).

42. E. E. Mendez, C-A Chang, H. Takaoka, L. L. Chang and L. Esaki, J. Vac. Sci. Technol., B1, 152 (1983).

43. M. Naganuma, Y. Suzuki and H. Okamoto, Int. Symp. on GaAs and Related Compounds, Oiso, Japan, 1981, Inst. Phys. Conf. Ser. 63, p. 125.

44. H. Takaoka, C-A Chang, E. E. Mendez, L. L. Chang and L. Esaki, Proc. 16th Int. Conf. Physics of Semicond., Montpellier, France, 1982.

45. G. A. Baraff, J. A. Appelbaum and D. R. Herman, Phys. Rev. Lett. 38, 237 (1977).

46. W. E. Picket, S. G. Louis and M. L. Cohen, Phys. Rev. Lett. 39, 109 (1977).

47. W. A. Harrison, E. A. Kraut, J. R. Waldrop and R. W. Grant, Phys. Rev. B18, 4402 (1978).

48. J. Pollman and S. T. Pantelides, Phys. Rev. B21, 709 (1980).

49. R. Martin, J. Vac. Sci. Technol. 17, 978 (1980).

50. H. Kroemer, K. J. Polasko and S. C. Wright, Appl. Phys. Lett. 36, 763 (1980).

51. R. Z. Bachrach and R. S. Bauer, J. Vac. Sci. Technol. 16, 1149 (1979); R. S. Bauer, J. Vac. Sci. Technol. in press (1983).

52. W. Mönch and H. Grant, J. Vac. Sci. Technol. 17, 1094 (1980).

53. P. M. Petroff, A. C. Gossard, A. Savage and W. Wiegmann, J. Cryst. Growth 46, 172 (1979).

54. C-A Chang, A. Segmüller, L. L. Chang and L. Esaki, Appl. Phys. Lett. 38, 912 (1981); C-A Chang, W. K. Chu, E. E. Mendez, L. L. Chang and L. Esaki, J. Vac. Sci. Technol. 19, 567 (1981).

55. C-A Chang, J. Appl. Phys. 53, 1253 (1982).

56. C-A Chang, Appl. Phys. Lett. 40, 1037 (1982).

57. R. A. Stall, C. E. C. Wood, K. Board, N. Dandekar, L. F. Eastman and J. Devlin, J. Appl. Phys. 52, 4062 (1981).

58. C-A Chang, J. Vac. Sci. Technol. B1, 346 (1983).

59. C-A Chang, unpublished

60. C-A Chang and W. K. Chu, Appl. Phys. Lett. 42, 463 (1983).

61. C-A Chang and T. S. Kuan, J. Vac. Sci. Technol. B1, 315 (1983).

62. T. S. Kuan and C-A Chang, Proc. of 10th Intn. Congress on Electron Microscopy, Hamburg (1980), p. 401; T. S. Kuan and C-A Chang, J. Appl. Phys. 54, 4408 (1983).

63. R. L. Anderson, Solid-State Electron. 5, 341 (1962).

64. A. G. Milnes and D. G. Feucht, Heterojunctions and Metal-Semiconductor Junctions (Academic, New York 1972).

65. F. F. Fang and W. E. Howard, J. Appl. Phys. 35, 612 (1964).

66. W. Mönch, R. S. Bauer, H. Gant and R. Murschall, J. Vac. Sci. Technol. 21, 498 (1982), and references therein.

67. I. Gonda, Y. Matsushima, S. Mukai, Y. Makita and O. Igarasin, Jpn. J. Appl. Phys. 17, 1043 (1978).

68. S. L. Wright, M. Inada and H. Kroemer, J. Vac. Sci. Technol. 21, 534 (1982).

69. T. Katoda and M. Kishi, J. Electron. Materials 9, 783 (1980); and references therein.

70.   A. Madhuka and J. Delgado, Solid State Commun. <u>37</u>, 199 (1981).

THEORY OF HETEROJUNCTIONS: A CRITICAL REVIEW

Herbert Kroemer

Department of Electrical and Computer Engineering
University of California, Santa Barbara, CA 93106

ABSTRACT: The central aspect of an abrupt heterojunction, and the point of departure for all its device properties, is the exact lineup of the bands of the two semiconductors at the interface. Band lineups vary over a wide range. Lineup data from numerous heterosystems have been reported in the literature, but only a few can be considered truly reliable. Of several theories that have been proposed to explain and/or predict band lineups, the Harrison Atomic Orbital (HAO) Theory has been by far the most successful; it agrees with those experimental data that are considered most trustworthy to within ± 0.13 eV (standard deviation). Also reviewed are the Frensley-Kroemer Pseudopotential (FKP) theory, the Electron Affinity Rule (EAR), and Self-Consistent Interface Potential (SCIP) calculations of band lineups.

Although heterojunctions between two III/V semiconductors grown by high-performance technologies appear to be well-understood, the origins of observed technology-dependent variations remain obscure. Heterojunctions involving different columns in the periodic table on the two sides are prone to various severe technology-sensitive complications that lie outside existing theories.

1. INTRODUCTION

1.1 The Scope of this Paper

This paper attempts to describe the present status of those aspects of the theory of heterojunctions that are of concern to the applied physicist ultimately interested in using these heterojunctions in actual devices or in device-like research structures. The emphasis will be on the static aspects of heterojunctions, that is, on the energy band structure and the associated wave functions, rather than on the electron transport within that energy band structure.

Within the energy level aspects, the emphasis will be on the intrinsic properties of "good" interfaces made by a "good" technology such as MBE, rather than on heterojunctions whose properties are dominated by defects. Although defects are never absent, in order for a heterojunction to be of interest for devices, the defect densities must be sufficiently low that their effect can be treated as a perturbation on the physics of the intrinsic interface, rather than dominating that physics.

From a device point-of-view, the central property of an abrupt semiconductor junction, and the point of departure for everything else, is the exact lineup of the bands at the interface. We assume that the transition from one semiconductor to the other takes place over at most a few lattice constants, and that the band structure changes, to the first order, abruptly at the interface. The theory of band lineups at such a quasi-abrupt transition is the central topic of this paper.

The present paper complements, and is a sequel to, a recent paper[1] under the title "Heterostructure Devices: A Device Physicist looks at Interfaces." In that earlier paper, I addressed myself to some of the utilitarian and empirical aspects of band lineups, including specifically a detailed critique of the reliability (or lack thereof) of the various methods employed to determine band lineups experimentally. But I had deliberately said very little about the problems of understanding and predicting band lineups theoretically, the central topic of the present paper. Throughout the present paper, we will often need to draw on material discussed in detail in the earlier work, but I have shunned the practice of using this as an excuse to re-publish once more what is published already. Instead, the reader will frequently be referred directly to the earlier material, with only as much repetition as is necessary to maintain the continuity of the presentation.

1.2 Some General Comments

Band lineups between two different semiconductors vary over a wide range. The most common lineup is that in which the two band edges of the wider-gap semiconductor straddle those of the narrower-gap semiconductor (Fig. 1a). Several of the most widely studied heterojunction pairs, notably GaAs/(Al,Ga)As, are of this kind.[2-5] The most extreme possibility is the broken-gap lineup (Fig. 1c), known to occur in the InAs/GaSb system.[6] In-between is the staggered lineup, in which both band edges of one semiconductor are shifted in the same direction relative to those of the other (Fig. 1b). It occurs over a wide composition range in the (Ga,In)As/Ga(As,Sb)

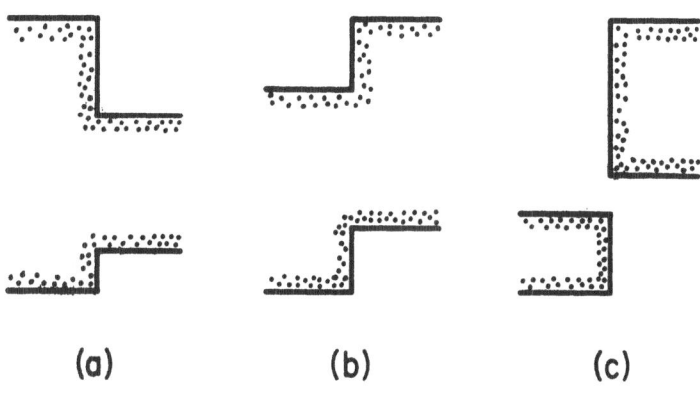

(a)   (b)   (c)

Figure 1 - Types of energy band lineups: (a) straddling, (b) staggered, and (c) broken-gap lineup.

system,[6] and it has also been reported for the CdS/InP system.[7] This staggered lineup has drawn far less attention than either the straggling or the broken-gap lineup - probably quite unjustly so.

The idea of an abrupt change in band structure is not without conceptual difficulties. Band edges have a rigorous meaning only when they are constant. A step in a band edge is similar to a potential step in quantum mechanics. Although the potential may change discontinuously, the electron distribution does not, due to tunneling into the step. To give a well-defined meaning to the concept of a band edge discontinuity, especially in the presence of fields, we must view it as an extrapolation (Fig. 2a,b).

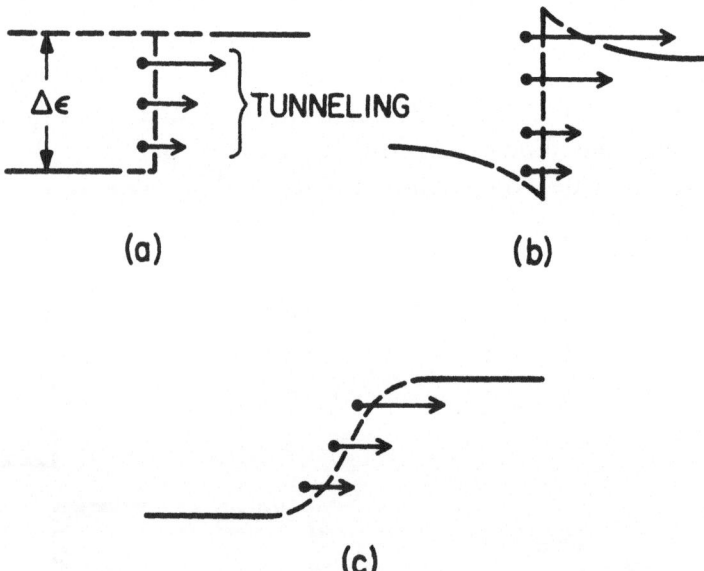

Figure 2 - Band discontinuities as an extrapolated concept: (a) in the absence of <u>macroscopic</u> fields, (b) in the presence of band bending due to macroscopic space charges. (c) shows the effects of composition grading. Band edges of abrupt heterojunctions can be viewed as being discontinuous only on a semi-macroscopic scale large compared to interatomic distances and tunneling distances.

Some distance away from the interface, the band edge gradually becomes well-defined. If macroscopic electric fields are present (weak compared to interatomic fields), the band edges have a finite slope. The band "discontinuity," in the sense in which we are using the term here, is the discontinuity that is obtained by extrapolating the sloping bulk band edges to the interface.

When the heterojunction is itself compositionally graded (Fig. 2c) over a sufficiently side distance, we regain a continuously varying band structure. We ignore graded heterojunctions here.

Theories of band lineups attempt to serve two quite different purposes:

(a) A predictive purpose, to forecast band lineups of systems for which no reliable experimental lineups are known. This use is of particular interest to the device physicist: If a particular hypothetical device structure requires a very specific band lineup not exhibited (at least not quantitatively) by any semiconductor pair whose lineups are known, then a reliable predictive theory is vastly preferable over the tedious experimental evaluation of the band lineups of random semiconductor pairs, which may require very extensive prior technology development, followed by rather difficult experimentation to determine the band lineups themselves. Reference [1] gives several examples illustrating this point.

(b) A retrodictive purpose, to test our understanding of the physics of heterojunctions, by calculating what the band lineup should be on a given physical model, and comparing the calculation with accurately known lineups.

The two purposes are not completely separable, and most theories address themselves, to various degrees, to both purposes. But the distinction is nevertheless important conceptually, especially in assessing a theory: A theory may reveal a great deal of the underlying physics but be too inaccurate for device-type predictions, or it may be an excellent cookbook theory without giving any insight. We shall often need to make this distinction.

Because the energy gaps of the semiconductors may be considered as being known with a high accuracy, band lineups may be specified in terms of either the conduction band offsets or the valence band offsets. Once one is known, the other is readily obtained from the known energy gaps. For device purposes, the conduction band offset is usually of greater interest. Theoretically, however, the valence band offsets are usually less difficult to estimate, and tend to be more accurate. We will express here the band lineups in terms of their valence band offsets.

2. THE EMPIRICAL BASIS

2.1 Uncertainties of Empirical Band Lineups

To assess the validity of any theory of band lineups, it is necessary to compare its predictions with band offsets that are already known experimentally with a degree of reliability sufficient to permit a meaningful test. This qualifier turns out to be surprisingly restrictive. Although the literature contains a very large number of lineup data for many different semiconductor pairs, few can be considered as reliable enough to permit a meaningful test of lineup theories.[1] Often the scatter of data signals their unreliability: For the widely-studied Ge/GaAs system, conduction band offsets ranging from 0.09 eV to 0.54 eV have been claimed in the literature, a range corresponding to 68% of the energy gap of Ge. For GaP/GaAs, the range of claims goes from 0.0 eV to 0.65 eV. Many of those values must be wrong, and this automatically makes all experimental data suspect.

Ignoring ordinary measurement inaccuracies, one can identify three problem areas.

2.1.1 **Measurement problems.** Many techniques that have been employed to extract band offsets determine those offsets only very indirectly, by projecting the results of whatever measurement is employed upon a preconceived model of the heterojunction. If the model is not valid, the resulting offset values may be invalid, too. In particular, small residual interface charges tend to grossly distort the results of some measurement techniques. I have given a critical assessment of various measurement techniques in reference [1], the results of which may be summarized as follows.

Probably the most reliable data are those obtained from UPS or XPS photoemission experiments[8] on very thin heterojunctions, <u>provided</u> the heterojunction itself was prepared by a technology that satisfies the criteria under 2.1.2 below. A close second to UPS/XPS measurements are optical <u>absorption</u> (not emission!) measurements on multi-quantum well structures (superlattices).[10] Capacitance-voltage (C-V) measurements on heterojunctions may or may not be reliable, depending on the exact nature of the measurements.[1] Least reliable are I-V measurements, and of these, band offsets extracted from I-V measurements on isotype heterojunctions (usually n-n junctions) are essentially worthless.[1]

2.1.2 Technology problems.   Heterojunction band offset data invariably tend to depend on technological details of how the heterojunction is prepared. Exactly why this is so, is in itself an unsatisfactorily-solved problem. But whatever the reason, it is clear that one cannot fully trust data that were taken on structures prepared under conditions significantly different from those employed for those device-quality structures whose lineups are the real object of the theory.

2.1.3 Chemically-induced interface dipoles.   Many heterojunctions that have been studied involve two semiconductors from different columns or column-pairs of the periodic table, such as Ge/GaAs, Ge/ZnSe, GaAs/ZnSe, InP/CdS, and many others. In all such systems, any interchanges of atoms across the interface will introduce atomic dipole moments that change the band offsets. Such atom interchange effects can, in general, not be prevented. In fact, it has been shown[11] that for most crystallographic orientations atom interchanges across the interface are <u>necessary</u> to prevent the accumulation of a huge interface-destabilizing net interface charge. The final result will be an interface with both a residual interface dipole and interface charge, the magnitudes of which depend sensitively on technology. These effects can be minimized by working with the electrically-neutral (110) cleavage planes of the compounds, and by growing the junction at a low temperature. But the latter is only a compromise, because the low-temperature growth tends to lead to poor bulk properties, not representative of a device-quality semiconductor. We will return to this topic in Sec. 9.

2.2 Selected "Iso-Columnar" Reference Systems for Theory Testing

2.2.1 Lattice-matched III/V systems.   When all these problems are taken into consideration, only two heterosystems remain that can truly serve as standards of comparison for lineup theories: The $Al_xGa_{1-x}As/GaAs$ system and the InAs/GaSb system.

For the first of these, both superlattice absorption data[10] and XPS data[12] for (100)-oriented abrupt heterojunctions show that the valence band offset is 15% ± 3% of the direct energy gap at k = 0, both for x in the range 0.2 to 0.3, and for x ≅ 1. If one assumes a linear relation with x, the data can be described by

$$\Delta\varepsilon_v[Al_xGa_{1-x}As/GaAs] = (0.19 \pm 0.04)x \quad eV.$$

To specify the sign of $\Delta\varepsilon_v$, we adopt the convention that $\Delta\varepsilon_v[A/B]$ shall be counted positive if semiconductor A has the lower valence band.

For InAs/GaSb, various data[6] show beyond the shadow of a doubt that this system is of the broken-gap variety (Fig. 1c), with a break in the gap of about 150 ± 50 meV. Combined with the 300K energy gap of InAs (0.36 eV), this yields

$$\Delta\varepsilon_v[InAs/GaSb] = 0.51 \ eV \ .$$

Although the accuracy of this value is not as high as one might wish, the very unusual nature of this broken-gap lineup more than makes up for its limited accuracy, and the system serves as a severe test of any theory of heterojunction lineups. Any theory that comes any-where near an accurate prediction of this unusual lineup without "fudging" is unlikely to be far from the truth, especially if it also succeeds in predicting the noticeably unsymmetric lineup of the (Al,Ga)As system. Conversely, any theory that is inherently incapable of yielding such well-established unsymmetrical lineups cannot possibly be correct, no matter how well it may agree with less extreme lineup data, especially if the latter are less-well established. On such experi-mental grounds alone both the Adams-Nussbaum theory[13] and the von Roos theory[14] are ruled out, even if they had not already been demonstrated[15] to be without theoretical foundation, and based on conceptual errors. The recent attempt by Nussbaum[16] to justify his theory by comparing its predictions to a set of UPS lineup data by Margaritondo et al.[17] is spurious: The comparison omits all III/V-

III/V systems with their disagreeing but more reliable lineups, and to the extent that the remaining data fit the Adams-Nussbaum theory, they fit the Harrison Atomic Orbital (HAO) theory[18-20] (to be discussed below) much better. But Nussbaum never even mentions the existence of that theory, even though Margaritondo et al.[17] specifically point out the surprisingly good overall fit of their data to HAO! Instead, Nussbaum compares his predictions only with the (poorer) fit of the data to the von Roos theory and to the Electron Affinity Rule (EAR),[21,2] also discussed later. Such a biased selection only discredits the whole comparison.

2.2.2   Lattice-mismatched   systems.   Compared   to   the (Al,Ga)As/GaAs and InAs/GaSb data, all other lineup data suffer from one uncertainty or another. Most likely to be reliable are probably the XPS/UPS data for InAs-on-GaAs[22] and for Ge-on-Si,[17] for which the following lineups have been reported:

$$\Delta\varepsilon_v[\text{InAs/GaAs}] = -0.17 \text{ eV} ,$$

$$\Delta\varepsilon_v[\text{Ge/Si}] = -0.2 \text{ eV} .$$

The trouble with both systems is that they are badly mismatched (7% and 4%). One must expect that the exact lineups depend on how exactly this mismatch is accommodated at the interface; hence they should be technology-dependent. In the Ge/Si case this criticism is aggravated by the fact that the Ge was grown at an unrepresentatively low temperature. As a result, it is not clear to what extent the band offsets for both systems should agree with any theory, which necessarily must make idealizing assumptions (even if implicitly) about the atomic structure of the interface and of the crystal itself. Any comparison with theory for such systems is therefore only partially a test of the theory. Given a theory that is already credible, the comparison is at least as much a test of the ideality of the interfaces.

Of the two interfaces, InAs/GaAs is the more interesting one, precisely because of the highly unsymmetric lineup of that system, which is probably more significant for the purpose of theory-testing than the exact numerical value quoted.

2.3 The Anion Correlation Rule

There is strong independent evidence that in systems such as (Al,Ga)As/GaAs and InAs/GaAs, in which the anion atom species (As) on both sides of the heterojunction is the same, the valence band offsets should be much smaller than the conduction band offsets. It was discovered by McCaldin et al.[23] that gold Schottky barriers on various zincblende-type semiconductors exhibit very predictable band lineups: The valence band energy at the interface, measured relative to the Fermi level, correlates very strongly with the electronegativity of the anion species and is almost independent of the cation species. McCaldin et al. explained this Equal Anions Rule in terms of the theoretically well-established fact[24] that the valence band wave functions derive largely from the anion atom wave function. This, together with the fact that the valence band wave functions tend to be more localized than the conduction band wave functions, yields valence band energies that correlate strongly with the anion species. It was pointed out by Frensley and Kroemer[25] that this rule should remain at least approximately applicable to heterojunction lineups. The InAs/GaAs data are the first data confirming this expectation. This is significant beyond its value as an individual data point, because it suggests that one should expect any credible theory of band lineups to yield valence band offsets that correlate at least approximately with the anion electronegativities. For semiconductor pairs with equal anions, the valence band offsets should always be small compared to the conduction band offsets. For semiconductor pairs with equal cations "X", the valence band energies at the interface should correlate with the different anion electronegativities. For the III/V compounds this implies

$$\varepsilon_v(XP) < \varepsilon_v(XAs) < \varepsilon_v(XSb) \ . \tag{1}$$

If the correlation is strong enough, one might expect (1) to persist even if both anion and cation are different. As we shall see, the HAO theory[18,19] exhibits this expected correlation.

The equal-anions rule is of particular interest for the important system $Cd_xHg_{1-x}Te/CdTe$, which is closely lattice-matched. To my knowledge, experimental band offsets for this system have not been determined, and it will be interesting to see to what extent such data, once they become available, will confirm the first-order expectation of a very small band offset.

2.4 Selected Mixed-Column Reference Systems for Theory Testing

There exist large numbers of lineup data on heterojunctions in which the two semiconductors come from different columns of the periodic table. As was explained earlier, all such systems are prone to exhibit technology-dependent interface charges and interface dipoles. These nuisance effects will depend very strongly on the crystallographic orientation of the interface; the possibility for their being weak exists only for those interface orientations for which one of the <111> bond directions lies in the interface plane.[26,1] The two simplest such interface orientations are (110) and (112). The widely-used (001) and (111) orientations are highly non-ideal for such systems, no matter how ideal they may be for III/V-only heterojunctions.

To the limited extent that these problems are discussed at all in the literature,[11,27,26] this discussion has invariably been in the context of heterojunctions between one of the III/V compounds such as GaAs or GaP and one of the column-IV elemental semiconductors, Ge or Si. but they apply even more to heterojunctions between one of the II/VI semiconductors and Ge or Si, and most of the complications occur also for II/VI-III/V systems.

2.4.1 ZnSe-on-GaAs (110).   Of all mixed-column lineup data in the literature the ones I consider least likely to suffer from any of these complications are the XPS data of Kowalczyk et al.[28] for heterojunctions of ZnSe grown on GaAs (110) at 300°C (not their 23°C-growth data):

$$\Delta\varepsilon_v[\text{ZnSe} - \text{on} - \text{GaAs}(110)] = 0.96 \pm 0.03 \text{ eV} \ .$$

The reasons for the credibility of these data are these: (a) The (110) orientation is one of the nonpolar orientations least subject to interface charges and dipoles. (b) The growth temperature was high enough to suggest reasonably good crystallographic ordering of the ZnSe deposit, yet low enough to minimize atomic interchanges between the ZnSe and the GaAs, a circumstance further encouraged by the low tendency of elemental Zn and Se to react with GaAs. (c) Last but not least, the XPS technique is one of the most reliable techniques for lineup determinations, and these measurements were performed with exemplary care and attention to detail.

These ZnSe/GaAs data are, in my judgment, very much preferable as a representative example of II/VI-on-III/V heterojunctions over the CdS/InP data of Shay et al.[7]: $\Delta\varepsilon_v[\text{CdS/InP}] = 1.63$ eV. In that case the crystallographic orientation was the highly suspect (111) orientation,[29] known to be prone to interface charge and dipole effects,[11,26,27] and the technique to determine the band lineups was the C-V intercept technique[5], known to be highly sensitive to such disturbances, as discussed elsewhere.[1]

2.4.2 Ge-on-ZnSe (110). In heterojunctions between one of the column-IV elements and a III/V or a II/VI compound the problem of interface charges and dipoles is greatly compounded by the problem of severe antiphase disorder that is likely to occur if the compound semiconductor is grown on the elemental semiconductor substrate, rather than in the opposite order. There probably does not exist a heterosystem more ill-suited to a test of band lineup theories than GaAs grown on (001)-oriented Ge; yet this combination has been widely studied - with predictably irreproducible results. In my judgment, such compound-on-element systems should be expected not to satisfy any simple lineup theory. Only systems in which the element was grown upon the compound should be considered for testing such theories.

Probably the least-suspect data on any element-on-compound system are the XPS data, again of Kowalczyk et al.,[28] for Ge grown on ZnSe (110):

$$\Delta\varepsilon_v[\text{Ge} - \text{on} - \text{ZnSe}(110)] = 1.52 \pm 0.03 \text{ eV} .$$

My reasons for this preference are the same as for the selection of the same authors' ZnSe-on-GaAs data; my only reservation is that the Ge growth was performed at an undesirably low temperature (23°C) which, even though followed by a 300°C anneal, is not likely to provide the same crystalline quality as, say, a 300°C growth would have.

2.4.3 Ge-on-GaAs (110) and beyond. It is only with considerable reluctance that I include amongst the reference systems what is one of the most widely studied heterosystem, Ge-on-GaAs. The lineup data on this system scatter so widely[30] that it appears difficult to decide which of the data are least unreliable, and the strong chemical interaction of Ge with As makes the system prone to chemical interface reactions.[31] However, recent data on MBE-grown Ge-on-GaAs (100) heterojunctions have tended to converge towards what appears to be the most carefully determined value, that of the Rockwell group,[32,9]

$$\Delta\varepsilon_v[\text{Ge} - \text{on} - \text{GaAs}(110)] = 0.53 \pm 0.03 \text{ eV} \ .$$

obtained again by XPS, on junctions grown at 425°C. However, with this system we have clearly reached the border between systems for which an agreement with a lineup theory represents a meaningful test of that theory, and systems for which such an agreement would represent a test of the internal atomic structure of the heterojunction itself.

Considering the problems inherent in mixed-column heterojunctions, I consider most remaining experimental lineup data on such junctions as being of very limited value for the testing of lineup theories. The value of such data lies in something different: Given a credible theory of band lineups, such data have a large retrodictive value. The extent to which such data do or do not fit such a theory contains valuable quasi-empirical information about the extent to which the atomic arrangement at the interface does or does not satisfy the idealizing assumptions under which these theories should be valid. If, as appears to be the case surprisingly frequently, the data on any single heterojunction seem to fit a particular theory, this might be highly gratifying, but it should be regarded as hindsight, not as a test of that theory, so long as there are numerous other heterojunctions that should agree just as well but don't, for no ostensible a-priori reason.

It is only when large numbers of such data on different heterojunctions tend to fall into an overall pattern of agreement, that this pattern itself tends to support the theory even in the face of individual data that disagrees. But the recognition of a pattern of agreement requires that the theory be formulated first. It is to this task that we turn next.

3. A FIRST-ORDER MODEL: LINEAR SUPERPOSITION OF ATOMIC-LIKE POTENTIALS (LSALP)

I pointed out in 1975[33] that the problem of theoretically understanding and predicting band lineups is the problem of determining the relative alignment of the two periodic potentials in the two participating semiconductors. We may view the energy band structure of each semiconductor itself as known. Modern self-consistent band structure calculation techniques have also provided us with increasingly reliable self-consistent periodic potentials. They always include automatically the knowledge of the alignment of the bands relative to those potentials, even though this lineup is often not stated in the papers on band structure calculations. The problems of the band lineups then reduces to the problems of the potential lineups.

It is useful to discuss the problem of the potential lineups in terms of a simple first-order reference model; the model of a Linear Superposition of Atomic-Like Potentials (LSALP).

Given the two periodic potentials within the two individual bulk semiconductors, we may always view them as a linear superposition of overlapping atomic-like potentials (Fig. 3). Near the atomic nuclei, the atomic-like potentials resemble the potentials inside the free atoms. But in the regions between the atoms, especially within the interstices in the diamond and zincblende structure, these potentials will be different from free-atom potentials. Hence our designation atomic-like potentials.

For any given periodic potential such atomic-like potentials can always be defined, although their actual extraction appears to have been performed rarely. But conceptually at least, the bulk band structure may be viewed as being known relative to the atomic-like potentials of the crystals.

The simplest possible atomic theory of band lineups would then be one in which the potential throughout the entire structure is viewed as a superposition of unmodified overlapping atomic-like potentials. In the immediate vicinity of the interface itself the potential would contain contributions from atoms on both sides of the interface, but with each atomic-like potential still being the same as deep inside the bulk of the particular semiconductor. In such a model, the relative lineup of the two bulk potentials is well defined. The band lineups are then also well-defined, and the only problems are those of calculational technique. In fact, the most successful one of the lineup theories, the Harrison theory,[18,19] is explicitly of this kind, and the Frensley-Kroemer theory[34,25] (prior to applying certain dipole corrections) is implicitly of this kind, even though no atomic-like potentials are explicitly introduced in either theory.

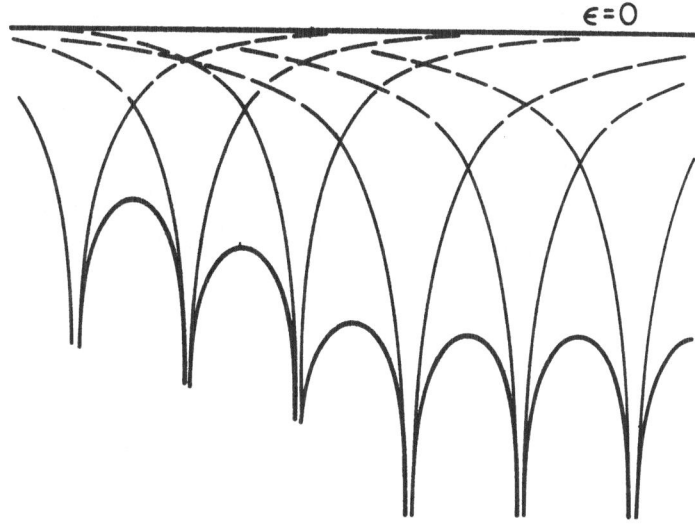

Figure 3 - Simple Model of the potential energy within a few atoms of a heterointerface, as a linear superposition of overlapping atomic-like potentials. Within each semiconductor the individual atomic-like potential for each atomic species is the same for all atoms of that species. Near the interface the potentials from the two sides overlap.

It should be noted that in the strict LSALP model there is no place for any crystallographic orientation dependence or technology-dependence of the band lineups. All those must be due to deviations from the model.

In the vicinity of the interface one must expect charge and potential re-adjustments relative to the predictions of a simple linear superposition of atomic-like bulk charges and potentials. The charges will re-adjust in response to various forces, such as image forces, quantum-mechanical exchange forces, tunneling, etc. The overall result of these re-adjustments is an <u>electronic</u> interface re-adjustment dipole, which shifts the bands relative to one another, compared to the linear superposition model.

In addition to these electronic dipoles we must expect <u>atomic</u> (or ionic) interface dipoles to occur if, for whatever reason, atoms from one semiconductor cross over the ideal interface into the other semiconductor. At least in mixed-column systems this can produce large net dipoles.

These dipoles - of either origin - do not appear to be large (usually at most a few tenths of 1 eV) and hence are probably not of great concern for the retrodictive use of the theory. But they are the major bottleneck in the use of lineup theories for accurate predictions on the level of accuracy desired for device applications. For example, for practically all heterosystems, variations of band lineups with technology and with crystallographic orientation have been reported. Such variations are inherently outside the possibility of the simple LSALP, and are therefore necessarily consequences of such dipole shifts, presumably atomic ones.

In my judgment, a better understanding of these dipole shifts, and particularly of their variability, is one of the most urgent (and as yet unfulfilled) needs of heterojunction lineup theory.

4. THE HARRISON ATOMIC-LIKE ORBITAL (HAO) THEORY OF BAND LINEUPS

4.1 The Idea

If one ignores those theories that are in hopeless disagreement with experimentally observed band lineups of the conceptually and technologically simplest heterojunctions, between two nearly lattice-matched III/V-compounds, the simplest remaining theory is Harrison's Atomic(-like) Orbital (HAO) theory.[18,19] It also appears to give the best agreement with experiment, a combination that makes it the standard of comparison against which all others must be measured.

Conceptually, HAO comes closest to the simple LSALP model of a heterojunction introduced above. However, Harrison never explicitly introduces an atomic-like potential, but instead proceeds as follows. He sets up a LCAO (Linear Combination of Atomic Orbitals) band structure calculation formalism (for nearest neighbors only) that would be used <u>if</u> true atomic-like potentials and energy eigenfunctions were known. This formalism does not require the knowledge of the full atomic-like potential nor of the full wave functions: All that is needed are selected unperturbed atomic energy eigenvalues and four different kinds of matrix elements coupling the relevant atomic states between nearest neighbors. Harrison simply takes as unperturbed atomic energy values the accurate theoretical values calculated by Herman and Skillman[35] for free atoms. This choice specifies the energies of all bands relative to the energy at infinity of the free atoms. To the extent that the free-atom potentials differ from the atomic-like potentials inside the solid, this choice is only an approximation, but it is what makes the problem tractable. Harrison's approximation for the matrix elements is just as daring. In principle, one would need a different set of matrix elements for each semiconductor. But Harrison argues that the principal reason for differences between matrix elements for different compounds is the difference in interatomic distance d, and he asserts that each of the four kinds of matrix elements can be approximated well for all compounds by writing it in the form $\eta \hbar^2/md^2$, where η is a constant that is the same for all zincblende-type semiconductors, and d is the nearest-neighbor distance in the particular semiconductor. There remains then only a set of four η-parameters, and these are determined by adjusting them in such a way that certain experimentally accurately known bulk properties of selected III/V compounds, including selected band structure data, are fitted to the theory to give the best possible overall agreement.

Up to this point, HAO is simply a (surprisingly successful) attempt to match the energy band structures of the III/V compounds to a simple LCAO model with a minimum number of parameters. It becomes a theory of band lineups by postulating that the zero of the energy scale, which is established by the use of the Herman-Skillman free-atom energy levels, may be used as the energy at infinity for the atomic-like potentials introduced in Sec. 3 as the conceptual basis of the lineup theory, but never calculated in HAO. Note the wording: "may be used as...". The point is that even a large error in this zero is irrelevant if the error is the same for each III/V compound. It is only any difference in errors that would falsify the final lineups.

Because the valence band structure of the zincblende semiconductors is much simpler than the conduction band structure, and because band structure calculations tend to reproduce this structure better, Harrison expresses the band lineup in terms of the valence band offsets, which are expected to be more accurate than the conduction band offsets. The latter are obtained more accurately indirectly, from the predicted valence band offsets, by adding the difference between the accurately-known <u>experimental</u> energy gaps.

4.2 <u>Lineup</u> Tables

We express here the results of the Harrison theory in a form somewhat different from Harrison's, as a set of simple tables, one each for the III/V compounds, the II/VI compounds, the I/VII compounds, and the elemental semiconductors. For convenience, we have re-expressed all energies relative to the top of the valence band of GaAs rather than using Harrison's pseudo-vacuum level. The columns in Table 1a and 1b represent equal anions, the rows equal cations.

The bottom entry in each box is the energy of the valence band edge, the top entry the conduction band edge, obtained by adding the experimental 300K energy gaps. For the III/V compounds the gaps were taken from the excellent compilation by Casey and Panish,[3] which are believed to be more accurate than the values used by Harrison. In those cases where the conduction band edge is not at k = 0, the notation X, Δ, or L has been added to indicate the location of the band edge, and the value at the Γ-point has been given in parentheses. In his own work, Harrison gives energies for several additional materials, which have been omitted here because they are of little interest to the applied semiconductor physicist.

In my judgment, these tables represent the best theoretical estimates that can be given <u>at this time</u>. As we shall see presently, for heterojunctions between two III/V compounds, even lattice-mismatched ones, the agreement tends to be better than $\pm.2$ eV. For heterojunctions involving semiconductors from different columns of the periodic table, the agreement is somewhat poorer, as <u>should</u> be expected on theoretical grounds.

Tables 1a - 1d   Band edge energies in eV of various semiconductors relative to the top of the valence band of GaAs, from the Harrison Atomic-Like Orbital (HAO) theory. The bottom entry in each box represents the valence band edge, the top the conduction band edge. The designations X, Δ or L preceding the conduction band value indicates an indirect gap; the direct gap is then given in parentheses.

Table 1a: III/V Compounds		
	P	As	Sb

	P	As	Sb
Al	X: 1.95 (3.1)	X: 2.17 (2.52)	X: 2.44 (3.08)
	-0.50	-0.04	+0.86
Ga	X: 1.79 (2.31)	+1.42	+1.57
	-0.47	0.00	+0.84
In	+1.24	+0.68	+1.29
	-0.11	+0.32	+1.12

Table 1b: II/VI Compounds

	S	Se	Te
Zn	+1.93	+1.82	+2.42
	-1.87	-1.05	+0.03
Cd	+0.97	+1.02	+1.81
	-1.59	-0.82	+0.21

Table 1c: The Cuprous Halides. Only the valence band energies are given.

	CuCl	CuBr	CuI
	-3.58	-2.37	-1.09

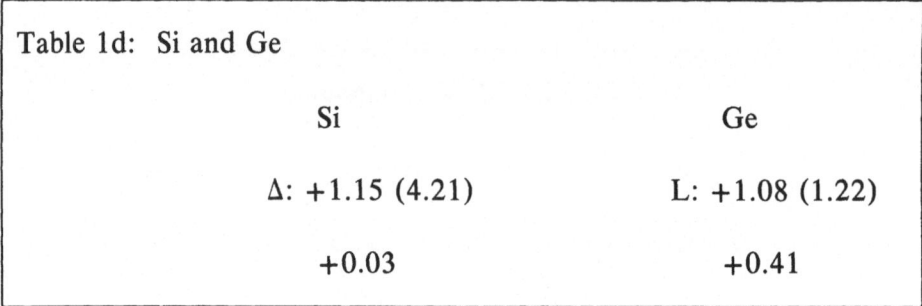

Table 1d: Si and Ge

	Si	Ge
	Δ: +1.15 (4.21)	L: +1.08 (1.22)
	+0.03	+0.41

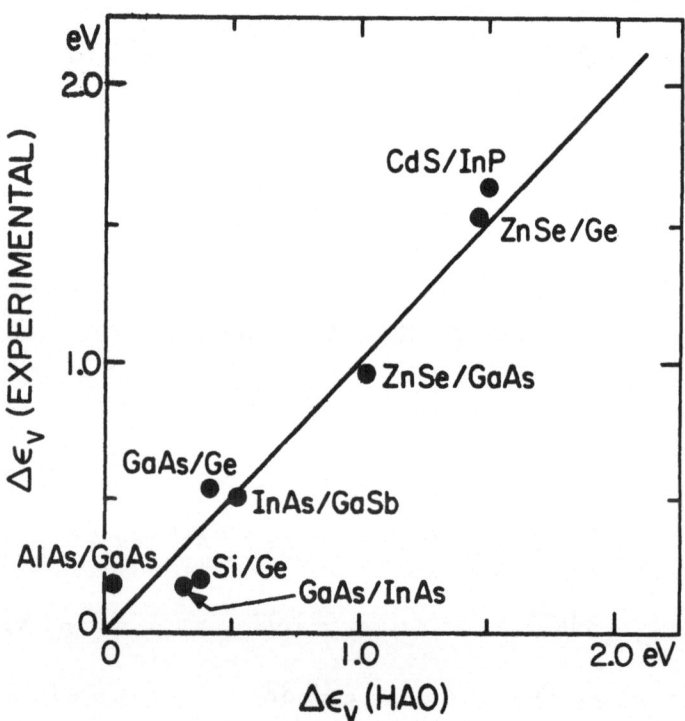

Figure 4 - The experimental valence band offsets in eV for the seven reference systems selected in Sec. 2, plotted as a function of the theoretical valence band offsets as predicted by the Harrison theory. The values for CdS/InP are also shown. The AlAs/GaAs value is extrapolated from $Al_xGa_{1-x}As$/GaAs for $x \cong 0.3$. A numerical tabulation, with references, is given in Table 2.

4.3 Comparison with Experiment

Figure 4 shows a comparison of the experimental data for the seven reference systems discussed in Sec. 2 with the theoretical predictions of the HAO theory. The fit speaks for itself. Table 2 gives the actual data. If one weighs all seven data points equally, one finds a mean error of only -0.016 eV, with a standard deviation of 0.13 eV. I have also added the CdS/InP data point of Shay et al.,[7] which I consider as not fully trustworthy, but which is in no poorer agreement than AlAs/GaAs and InAs/GaAs. Considering that our reference systems deliberately stressed heterojunctions with unsymmetric lineups, the agreement must be considered excellent.[36]

Table 2   Comparison of the valence band discontinuities (in eV) predicted by the Harrison Atomic Orbital (HAO) theory for selected reference systems, with experimental values. The values for CdS/InP are also given. The AlAs/GaAs value is the value extrapolated from $Al_xGa_{1-x}As/GaAs$ value with $x \cong 0.3$.

Heterojunction	HAO	Experimental	Error
AlAs/GaAs	0.04	0.19 [10]	+0.15
InAs/GaSb	0.52	0.51 [6]	-0.01
GaAs/InAs	0.32	0.17 [22]	-0.15
Si/Ge	0.38	0.20 [17]	-0.18
ZnSe/GaAs	1.05	0.96 [28]	-0.10
ZnSe/Ge	1.46	1.52 [28]	+0.06
GaAs/Ge	0.41	0.53 [9]	+0.12
CdS/InP	1.48	1.63 [7]	+0.15

Furthermore, by comparing the valence band edge energies within each equal-anions column in Tables 1a and 1b, one readily sees that the range for a given anion species, although far from zero, is sufficiently small that it does not overlap with the range of predictions for the other columns. Evidently, the HAO theory contains within itself the anion correlation rule discussed earlier.

How well do some of the remaining heterojunction pairs fit? Fig. 5 shows the data of Margaritondo et al.,[17] for the various Ge-on-compound systems investigated by that group. The fit for the wider-gap compounds is almost as good as in Fig. 4, but for the narrower-gap

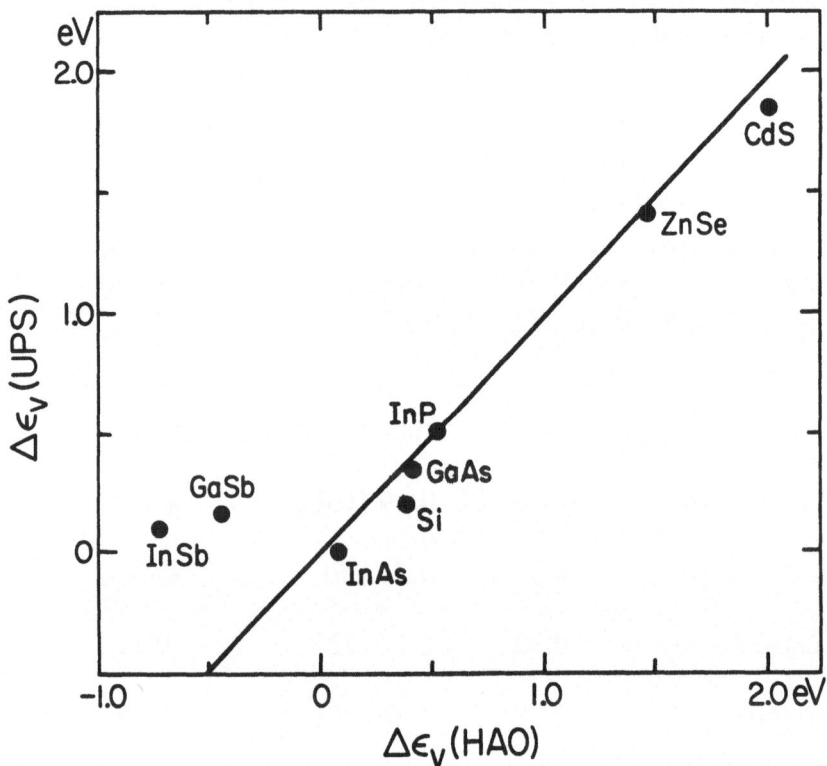

Figure 5 - Experimental valence band offsets $\Delta\varepsilon_v[X/Ge]$ for several Ge-on-compound semiconductor systems, as measured by Margaritondo et al.[17], plotted against theoretical valence band offsets as predicted by the Harrison theory. The Si/Ge point is the same as in Figure 4, but the GaAs/Ge point is that of Margaritondo et al.[17], not that of ref. [9]. The highly uncertain GaP data point has been omitted.

compounds the deviations are larger, up to 0.8 eV for Ge-on-InSb. The authors attempt to explain the deviations as a lattice mismatch effect, and they apply a bond length correction. It improves the fit for some materials, but degrades it for others. Inasmuch as these polar/nonpolar systems can be expected to show deviations from the simple HAO theory for various reasons, and inasmuch as the heterojunctions in that work were apparently all grown at room temperature without annealing, which shold introduce additional uncertainties, I am not convinced of the validity of such a correction. I have therefore omitted it from Fig. 5 and not applied such a correction to the data of Fig. 4.

4.4 Critique of the HAO Theory

It is clear that the HAO theory is only an approximation. This calls for a critique of whatever approximation steps were made. The need for doing so is by no means eliminated by the observation that the theory appears to fit the data better than one might have expected. This merely changes the nature of the critique from asking "what is wrong?" to "why is it right - or is it?".

In assessing any theory of heterojunction it is always useful to distinguish between the qualitative conceptual aspects underlying the theory, and the quantitative numerical detail. This is particularly important in the case of the HAO theory, whose originator has deliberately and specifically stressed the simplicity of the theory over high numerical accuracy. His aim was principally to show that an excellent overall physical understanding of band lineups can be achieved with a very simple physical model. Still, there are areas that appear to call either for refinements or for a better understanding of why certain approximations work as well as they appear to do. I see principally the three areas of concern, listed in the following.

4.4.1 Atomic potentials.   Why is it apparently allowed to calculate the lineup of the periodic potentials as if those potentials consisted of the linear superposition of free-atom potentials? Charge transfer along the bands from cation to anion atoms shifts the cation and anion potentials relative to one another, and the potentials outside the cores should bear little resemblance to free-atom potentials. It would be highly desirable to investigate to what extent it is indeed valid to use free-atom potentials, and why. The issue is not the use of free-atom potentials for the individual band structure calculations but only for the relative lineup of those band structures. I suspect that a two-fold cancellation of errors is at work here: First, the core shifts due to charge transfer from cation to anion do not affect the average potential of each crystal, and besides, the matrix elements are adjusted to force the band structure of each semiconductor to come out correctly anyway. Second, the deviations from free-atom potentials in the space farthest away from the atoms may be sufficiently similar from semiconductor to semiconductor that the error in the potential at infinity, while possibly large, cancels out of the band lineups. Whatever the explanation, this is the weakest point of the HAO theory. A detailed study would certainly be in order.

4.4.2   Electronic dipole shifts.   Harrison neglects any electronic dipole shifts associated with charge transfer across the heterointerface, causing deviations from a simple linear superposition of unmodified atomic potentials. The neglect is based on a rough estimate of the magnitude of such shifts, which caused Harrison to conclude that the shifts are indeed small (below 0.1 eV), in striking disagreement with Frensley and Kroemer,[25] who came to a rather different conclusion. The difference appears to lie at least partially in diametrically different assumptions about screening. We will return to this point later. Empirically, the good agreement between HAO and experimental does not appear to call for the addition of interface dipole shifts. Be that as it may, this problem also calls for an investigation. One possibility that should be explored is that the error in neglecting interface dipole shifts might tend to cancel the error involved in using neutral free-atom potentials rather than more realistic non-neutral atomic-like potentials to estimate the mutual lineup of the two periodic potentials.

4.4.3   Quantitative refinements.   Despite its simplicity, the HAO theory agrees sufficiently well with observations on a semi-quantitative level, that the hope appears not unreasonable that it could be refined (at some loss of simplicity) to the point that it can serve as a quantitative predictive tool that truly serves the needs of the device physicist. To do this, it would probably be necessary, in addition to clarifying the questions raised above, to refine the band structure calculation portion of the theory beyond the simplest nearest-neighbor LCAO calculation that uses only four different kinds of off-diagonal matrix elements which, moreover, differ from compound to compound by only a scaling factor allowing for bond length variations. The spin-orbit splitting of the valence bands should also be taken into account.

5. THE FRENSLEY-KROEMER PSEUDOPOTENTIAL (FKP) THEORY

5.1 The Idea

The HAO theory was not the first attempt to explain heterojunction band lineups in terms of (a) the periodic potentials actually present inside the two semiconductors, (b) the alignment of the bands relative to each potential, and (c) the alignment of the two potentials relative to each other. It was preceded by the Frensley-Kroemer pseudopotential (FKP) theory,[34,25] which was not quite as successful, however. The differences between the two theories are more in methodology than in concept. Impressed by the spectacular advances in band structure calculations by self-consistent pseudopotentials, we re-calculated the band structures of most III/V semiconductors self-consistently with the associated periodic potentials. Such calculations automatically yield the relative alignment of bands and potentials, giving potentials that, in contrast to LCAO potentials, are most accurate in the important space between the atoms. Because such calculations are essentially perturbation calculations that start from free electrons rather than isolated atoms as the unperturbed state, the resulting periodic potentials are not in any obvious way related to a linear superposition of atomic-like potentials. However, because of the charge shifts from cation to anion that take place in actual crystals, the potential should in any event not be a superposition of neutral-atom potentials, but of ion potentials. Outside the ion itself, any ion potental is a simple Coulomb potential. Now, there exist, inside the zincblende structure, two kinds of large interstitial spaces, which are largely free of charge, and inside which the electrostatic potentials are well-defined. A simple electrostatics argument shows[25] that, under certain conditions, the average of the two interstitial potentials is just the potential at infinity! In the spirit of a linear-superposition-of-ion-potentials model the band offsets are therefore postulated to be governed by the requirement that the mean interstitial potentials on the two sides of the heterojunction are the same.

The resulting "absolute" valence band energy predictions are given in Table 3, following a format similar to Table 1. Table 3a is somewhat less complete than 1a, because band structures for AlP and InSb were not calculated, nor are there any predictions for the cuprous halides. The predictions for lattice-matched pairs are similar to those

Tables 3a - 3c: Valence band edge energies in eV of various semi-conductors relative to the top of the valenceband of GaAs, from the Frensley-Kroemer (FKP) theory.

Table 3a:	III/V Compounds		
	P	As	Sb
Al	unknown	0.00	+0.02
Ga	-0.16	0.00	+0.07
In unknown	-0.62	-0.42	

Table 3b:	II/VI Compounds		
	S	Se	Te
Zn	-1.38	-1.11	-0.78
Cd	-1.46	-1.33	-0.94

Table 3c:	Si and Ge	
	Si	Ge
	+0.8	+0.71

by the HAO theory, and are in similar agreement with experiment. However, for mismatched pairs, the predictions differ significantly, with the HAO predictions being in distinctly better agreement with observation.

5.2 Critique of FKP, and Comparison with HAO

5.2.1 Quality of pseudopotentials.   Like HAO, FKP is an approximation, subject to critique on both qualitative and quantitative grounds.
In contrast to HAO, I believe that the most important criticism of our
work pertains less to the conceptual core of the theory, than to its
quantitative detail. Self-consistent pseudopotential calculations require
a rather formidable calculational effort, going far beyond the pocket
calculator level that was Harrison's explicitly-stated aim. To keep this
computational effort low, all our calculations were done with the simplest (local) pseudopotentials, involving only two adjustable parameters
per atomic species, the parameters furthermore being independent of
the compound in which the atomic species occurs. Such oversimplified
pseudopotentials permit only a crude fit to actual band structures, and
they hence yield only crude self-consistent periodic potentials. We
believe that this purely calculational constraint is the principal limitation of the FKP theory in its present form. We would expect that, in
order to obtain lineup predictions on the level of accuracy of interest to
the device physicist, the addition of a third, non-local term to the
pseudopotential is necessary, but probably also sufficient. However, no
such attempt has been undertaken. In contrast to the HAO theory,
corrections for the spin-orbit splitting of the valence band have been
included in the FKP theory.

5.2.2 Potential matching scheme.   A second critique addresses itself
to the matching scheme via the mean interstitial potential. As stated
above, the mean interstitial potential equals the potential at infinity
only under certain conditions. These are: (a) It must be possible to
describe the total charge distribution as a superposition of spherical
charge distributions centered on the atoms. (b) Although different
spheres may overlap, they must not extend to the center of the interstices, that is, beyond the interatomic distance. Our calculations showed
that the second condition is well satisfied, but the first one is definitely
not, due to the strong tetragonal distortion of the charge distribution by
the bond charges. A more realistic model is to split off the bond
charges and to treat the overall charge distribution as a sum of three
spherical charge distributions:[37] One spherical charge centered on each
cation, one centered on each anion, and a spherical bond charge of two
electrons somewhere along each bond, nearer to the anion. It is not
difficult to determine all necessary parameters from the band structure

calculations themselves, and to adjust the potential matching scheme, but the details have not been worked out.

5.2.3 Interface dipoles.   In the form in which we have described it so far, the FKP theory suffers from the neglect of interface dipoles in exactly the same way as HAO.   However, in our work we did attempt to estimate the interface charge readjustment dipole theoretically.   We assumed a charge transfer along all bonds crossing the mathematical interface.   The magnitude of the transferred charge was assumed proportional to the difference of the electronegativities of the two atoms connected by the bond.   This attempt was not a success: The calculated dipole shifts appear to be too large and in those cases where accurate experimental lineups are known, their agreement with theory is better without the dipole corrections than with them.

The principal difference between our treatment of the electronic interface dipole and Harrison's, appears to lie in diametrically opposite ways to include (or ignore) screening in translating any electronic charge adjustments near the interface into an electrostatic dipole.   We were treating the charges as if they resided in vacuum, on the (unstated) grounds that most charges that would contribute to the screening of any bond polarizaiton were located far outside that bond, and were thus not able to screen the bond polarization very much. Harrison, by contrast, simply used the bulk dielectric constant to arrive at his conclusion that electronic interface dipoles are negligible. Both points-of-view are extreme, and the truth almost certainly is somewhere in between, but it remains to be worked out.

6. THE ELECTRON AFFINITY RULE (EAR)

6.1 The EAR as an Attempt to Relate Band Offsets to Free-Surface Data

No discussion of band lineup theories is complete without a discussion of the Electron Affinity Rule (EAR).[21,2] It is the oldest of the rules for the prediction of heterojunction band lineups, predating the developments of efficient band structure calculation techniques. It is not an atomistic theory of band lineups, but simply an attempt to relate the band lineups to other (presumably known) experimental properties, specifically to the electron affinities of the two semiconductor. Electron affinities are the energies required to remove an electron

from the bottom of the conduction band to the outside of the semiconductor, just beyond the range of any dipole and image forces (Fig. 6).
If band bending is present near the surface, the electron affinity is the
removal work from directly beneath the surface, not from the deep
bulk.

   Actual measurements of the electron affinity are invariably
indirect. They almost always involve measuring the photo-ionization
threshold energy Φ of the semiconductor, that is, the work required to

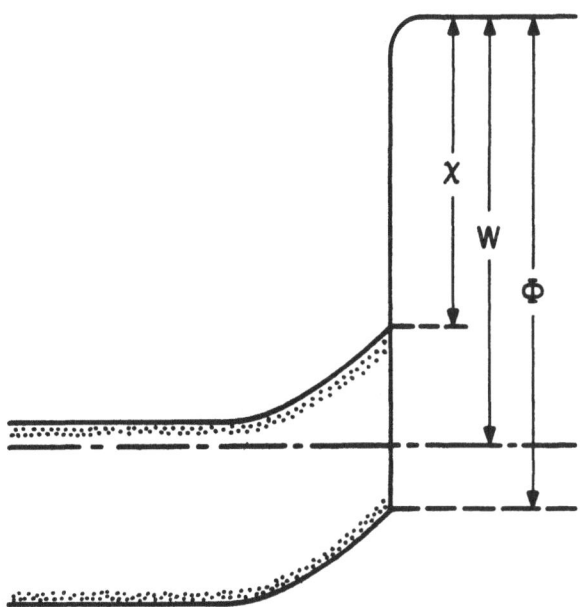

Figure 6 - The electron affinity χ, the work function W, and the
ionization energy ϕ all represent the work required to remove an electron to the outside of the semiconductor, just beyond the range of
dipole and image forces, but from different starting levels. For the
electron affinity, the initial energy is the bottom of the conduction
band, for the ionization energy (or photothreshold) it is the top of the
valence band. In both cases, the band energies are taken at the surface, not deep in the bulk. For the work function the starting energy is
the Fermi level.

remove an electron from the top of the <u>valence</u> band. The electron affinity is obtained by subtracting the known gap energy from that energy.

The electron affinity rule asserts that the conduction band offset at an abrupt heterojunction is simply equal to the difference in electron affinities between the two semiconductors, with signs chosen such that the semiconductor with the lower electron affinity has the higher conduction band:

$$\Delta\varepsilon_c \equiv \varepsilon_{c2} - \varepsilon_{c1} = \chi_1 - \chi_2 \ . \tag{2}$$

Inasmuch as the actually measured quantities are the ionization energies, and inasmuch as the valence band offset appears to be a more reliable measure of the band lineup, it would be preferable to express this rule as an <u>ionization energy</u> rule,

$$\Delta\varepsilon_v = \varepsilon_{v2} - \varepsilon_{v1} = \Phi_1 - \Phi_2 \ , \tag{3}$$

but the term electron affinity rule is too deeply entrenched to be displaced. We shall therefore continue to use it, but understand it to mean (3) as well as (2).

I pointed out already in 1975[33] that the EAR is at best an approximation that should hold quantitatively only if the surface dipoles present during the electron affinity measurements cancel any dipole present at the interface. There was no reason at the time to expect that there should be an even approximate cancellation. It now appears that the various dipoles might be sufficiently small to begin with[7,38] that their non-cancellation introduces only a relatively small error into the EAR, small enough to make the EAR much better than nothing at all, although still quite inferior to the HAO theory. We will return to this point later.

But even if the EAR might be a practical "cookbook recipe" with some utility, from the fundamental point-of-view of "understanding" the physics of the band lineups, the EAR continues to be, as I expressed it in 1975, "no answer at all, but rather a rephrasing of the original question in terms of a quantity....that is even less well understood." Furthermore, "the electron affinity is a true surface property that contains many complications...at the free surface. ...There is something inherently unsatisfactory about a procedure that, in principle, first includes all of these free-surface complications, which in the end should cancel out..." Harrison[18] says flatly "... this replaces one simple problem by two very difficult problems."

Despite those criticisms, the EAR continues to be used, and it has in fact found vocal defenders,[7,38] probably largely because better predictive theories, meaning nowadays the HAO theory, have in the past appeared not so overwhelmingly superior to have caused its abandonment.

6.2 Critique of the EAR

In the face of this persistence of the EAR, it is probably appropriate at this point to re-state exactly what are the conditions under which the EAR should be valid.

A detailed analysis of the problem was presented in 1975 by Kroemer,[33] of which we give here an updated outline, re-worded in terms of ionization energies and valence band offsets rather than electron affinities and conduction band offsets. An essential ingredient of the argument is that it is crucially important to recognize that the work required to move an electron from one position to another is NOT simply the quantity $-q\Delta\phi$, where $\Delta\phi$ is the difference in electrostatic potential between the two points. Electrostatic potential differences represent the work per unit charge only in the limit of a fictitious infinitesimally small test charge, when the test charge itself does not influence the electrons in the surrounding medium. An electron itself influences the distribution of the other electrons, not only through Coulomb repulsion by the finite electron charge, but through quantum-mechanicsl exchange effects (essentially, the Pauli exclusion principle). In particular, the energies required to remove an electron from the top of the valence band to "infinity" (the ionization energy) must consist of an electrostatic part and a polarization or correlation part:

$$\Phi = \Phi^{(e)} + \Phi^{(p)} . \tag{4}$$

An example of a polarization contribution to the work required to remove an electron from a substance is the work required against the image force near the surface of the substance.

Only in the space outside both semiconductors, beyond the range of both image forces, does the work required to move an electron coincide with -q$\Delta\phi$. There is, in fact, a finite electrostatic potential difference between the two semiconductor surfaces, equal to the work function difference.

The overall work required to move an electron along a closed path must vanish. This condition then leads to the relation

$$\Delta\varepsilon_v = \Phi_1^{(p)} - \Phi_2^{(p)} + \Phi_i^{(e)} = \Phi_1 - \Phi_2 - [\Phi_1^{(e)} - \Phi_2^{(e)} - \Phi_i^{(e)}]. \quad (5)$$

Here the so far unspecified term $\Phi_i^{(e)}$ represents electrostatic work across any atomic-scale interface dipole that may - and in general will - be present. It does not contain any dipole associated with conventional band bending, nor do the free-surface dipoles in (5) contain band bending dipoles.

The electron affinity rule (3) evidently results only if the three electrostatic dipole terms in (5) cancel,

$$\Phi_i^{(e)} \stackrel{?}{=} \Phi_1^{(e)} - \Phi_2^{(e)} . \quad (6)$$

If the interface dipoles were somehow the same as a simple linear superposition of the two (oppositely directed) free-surface dipoles, the EAR would indeed be valid. But considering the completely different physical origins of the various dipoles, there is no basis for such an assumption, and an accidental cancellation of three non-negligible terms of three unrelated origins would be highly improbable. Hence, the EAR cannot be expected to be generally valid.

Such an improbable cancellation would of course be unnecessary if each of the electrostatic dipole terms were in itself small. As we saw, the band lineups in lattice-matched III/V-only systems appear to be described quite closely by both the HAO and the FKP theory, which implies the empirical statement that the electrostatic interface dipoles are indeed small, at least in such systems. The same can definitely not be said for surface dipoles in general: It is well known that work functions (and hence electron affinities) depend strongly both on crystallographic orientation and on the chemical status of the surface, both of which imply large and varying dipoles. Such a dependence itself rules out a general validity of the EAR, because it is preposterous to assume that the chemical status of the free surface should have an influence on the band lineups at a heterojunction far away. However, the few electron affinity values for semiconductors for which reliable

experimental data actually exist, were not measured on arbitrary surfaces, but on the natural cleavage planes of the crystals, under UHV
conditions under which surface dipoles due to absorbed atoms were
specifically minimized. The natural cleavage surfaces are surfaces
where each atomic plane contains an equal number of cation and anion
atoms. This eliminates the largest source of dipoles on clean free
surfaces, surface dipoles due to the polar nature of most compound
semiconductor surfaces. Even the cleavage planes cannot be completely dipole-free, however. It is well-known that the (110) surfaces of the
zincblende-type semiconductors all reconstruct, with the anion atoms in
the top atomic plane pushed outward, the cation atoms inward. The
resulting reconstruction dipole is probably the dominant free-surface
dipole on the cleavage planes, and it is by no means negligible. If the
strengths of these dipoles are similar on different III/V compounds, the
free-surface dipoles would at least partially cancel, to an extent that the
residual dipole difference might indeed be small. Together with the
apparent smallness of the interface dipole, this would lead to an at least
approximate validity of the EAR.

6.3 Comparison with Experiment

If one accepts the restriction to experimental electron affinity
values taken on vacuum-cleaved surfaces, how well does the EAR
work? Probably its most impressive prediction is that of a broken-gap
lineup for InAs/GaSb, contained in the electron affinity data of Gobeli
and Allen,[39] χ(InAs) = 4.90 eV, χ(GaSb) = 4.06 eV, which imply

$$\Delta\varepsilon_c = 0.83\text{eV} = \varepsilon_g(\text{GaSb}) + 0.11\text{eV} \quad , \quad \Delta\varepsilon_v = 0.47\text{eV} \quad ,$$

which agrees very well with the experimental data (see Table 2). The
higly unsymmetric lineup on InAs/GaAs heterojunctions is also predicted well: With χ = 4.07 eV, one obtains $\Delta\varepsilon_c$[GaAs/InAs] = -0.83 eV
or $\Delta\varepsilon_v$ = +0.23 eV. Unfortunately, this almost spectacular agreement
collapses when one uses the more recent electron affinity value,
χ(InAs) = 4.54 eV, by the same group of investigators,[40] and based on
more detailed data. We must leave it up to the reader which data to
prefer. Maybe the dilemma speaks for itself.

Of the other reference systems, the Ge/Si prediction, $\Delta\varepsilon_c$ = 0.12
eV, $\Delta\varepsilon_v \cong$ 0.33 eV is in reasonable agreement with the data. For the
two ZnSe systems, the agreement is distinctly inferior to HAO, as

Kowalczyk et al.[28] have already pointed out. Swank[41] gives $\Phi(ZnSe)$ = 6.82 eV, which, with $\Phi(GaAs)$ = 5.47 eV and $\Phi(Ge)$ = 4.80,[39] yields $\Delta\varepsilon_v$ = 1.35 eV and 2.02 eV, in rather poor agreement with experiment. Overall, the HAO theory agrees definitely better with the reference system data than the EAR.

In our above enumeration the most important heterojunction of them all, (Al,Ga)As/GaAs, is conspicuous by its absence. It cannot be compared with the EAR prediction because, to my knowledge, no electron affinity values for AlAs or (Al,Ga)As were ever determined. The values sometimes found in the literature are fictitious ad-hoc values, determined from the known band lineup by applying the EAR in reverse to extract an electron affinity value from the band lineups - hardly suited for testing the theory itself.

For cases such as this, it has been argued by Shay et al.[7] and by Phillips[38] that the EAR remains a useful predictive tool if one simply replaces the missing electron affinity data by van Vechten's theoretical values.[42] Such a procedure might conceivably be justifiable if there existed absolutely no alternatives, but even in that case it would be subject to Harrison's comment that the electron affinity rule replaces one simple problem by two very difficult ones. But there is in fact no necessity for such a conceptual detour: (a) The HAO theory has shown itself capable of giving predictions better than the EAR even in those cases in which reliable electron affinity data are available, and it offers predictions with a much better theoretical justification than the EAR in the rest of cases. (b) In those cases where accurate experimental electron affinities are available, the van Vechten predictions sometimes disagree substantially. For example, it was pointed out recently[1] that for GaP the disagreement is as large as 0.6 eV, hardly negligible. This last comment should not be construed as a claim that the experimental electron affinity values should lead to better predictions than the theoretical ones. Considering the poor theoretical basis of the EAR, one should not be surprised if the opposite were the case. In fact, for one of the prime examples of Shay et al.,[7] CdS/InP, electron affinities have been published $\chi(CdS)$ = 4.79 eV,[41] $\chi(InP)$ = 4.40 eV,[43,39] which, although not far different from the van Vechten values (4.87 eV, 4.35 eV) deviate in a direction of poorer agreement of the EAR with experiment.

Inasmuch as the use of theoretical rather than experimental photo-thresholds is conceptually not very far from the HAO theory, all this does is lead us back to the HAO theory.

7.   LINEARITY, TRANSITIVITY AND TECHNOLOGY VARIA-TIONS

7.1 The Concept of Transitivity

All the theories we have discussed assumed that there is a specific absolute energy associated with the various band edges of every individual semiconductor, and that the band offsets are simply the differences between the respective absolute band energies of the two semiconductors. With our sign convention,

$$\Delta\varepsilon_v[A/B] = \varepsilon_v(B) - \varepsilon_v(A) \ . \tag{7}$$

Such theories may be called linear theories. Presumably, linearity is only an approximation, and one should really write

$$\Delta\varepsilon_v[A/B] = \varepsilon_v(B) - \varepsilon_v(A) + \delta(A/B) \ , \tag{8}$$

where $\delta(A/B)$ is a correction that cannot be written as a linear difference between two individual properties of semiconductor A and B, but is a true property of the specific combination A/B. The excellent agreement of experimental band lineups with the HAO theory, which is manifestly linear, sugests that $\delta(A/B)$ is in fact small, below 0.2 eV.

Note that the linearity property (7) is a property independent of exactly what theory is employed to calculate $\varepsilon_v(A)$ and $\varepsilon_v(B)$ individually. Even an arbitrary large error in such a calculation would not matter if the error is the same for all semiconductors and hence cancels out. For example, the (unsuccessful) dipole correction of the FKP theory is itself of the form $V_D[A/B] = V_D(B) - V_D(A)$, and hence it does not change the character of FKP as a linear theory. The property (7) is a property of an entire class of theories. It is possible to test experimentally whether or not the experimental data can possibly satisfy a linear theory, without actually invoking a specific theory. One characteristic feature of all such theories is the following: Given the band lineups of two different semiconductor pairs A/B and B/C, having one semiconductor (B) in common, the lineup for the third possible pair A/C follow by simple addition: With our sign convention,

$$\Delta\varepsilon_v[A/C] = \Delta\varepsilon_v[A/B] + \Delta\varepsilon_v[B/C] \tag{9}$$

or

$$\Delta\varepsilon_v[A/B] + \Delta\varepsilon_v[B/C] + \Delta\varepsilon_v[C/A] = 0 \ . \tag{10}$$

This property has been referred to as transitivity by Frensley and Kroemer.[25] For example, from the well-established lineups for InAs/GaSb (0.51 eV,[6]) and GaAs/InAs (0.17 eV,[22]) transitivity predicts that the band lineup for GaAs/GaSb should exhibit a valence band discontinuity

$$\Delta\varepsilon_v[GaAs/GaSb] = 0.51eV + 0.17eV = 0.68eV \ .$$

with the GaSb valence band being the higher one. This prediction is independent of whatever theory one might wish to invoke for the band lineups, so long as it is a theory of this transitive class. Because transitivity should be a common feature of entire classes of theories, tests for transitivity are useful tests of great generality, which do not involve fitting experimental data to any particular lineup theory, but which may rule out entire classes of theories if unsuccessful.

In 1979, Waldrop and Grant[44] studied, by XPS, the lineup for the two highly unconventional heterojunctions CuBr/Ge and CuBr/GaAs. They found that the lineups, taken together with the already-known lineup for Ge/GaAs, did not satisfy the transitivity sum rule condition (10). Instead, the sum of the three valence band lineups, taken with proper signs, added up to the value 0.64 eV. It was subsequently argued by Phillips[38] that this large sum is more likely to be due to antiphase disorder in the CuBr-on-Ge growth than to a true non-transitivity. Phillips' point is that CuBr-on-Ge represents the growth of a zincblende-type semiconductor (CuBr) with two different atoms per primitive cell, on a diamond structure substrate (Ge) with two identical atoms. We ourselves have repeatedly and emphatically pointed out in the similar context of GaP-on-Si and GaAs-on-Ge growth,[27,26,1] that such growth should be highly prone to antiphase disorder. We therefore agree with Phillips' criticism: One shold not invoke as speculative an explanation as a true non-transitivity when it can in fact be guaranteed that one of the heterojunctions under test suffers from defects as drastic as heavy antiphase disorder.

7.2 Ge/GaAs/ZnSe: A Positive Test for Transitivity

In their more recent ZnSe-work already mentioned,[28] the Rockwell group has studied the systems Ge/ZnSe and ZnSe/Ge which, together with the old standby Ge/GaAs, form another nearly lattice-matched triplet suitable for testing transitivity under difficult conditions.   In an important departure from the CuBr/Ge work, the ZnSe/Ge heterojunctions were evaluated both as ZnSe-on-Ge junctions and a Ge-on-ZnSe junctions, yielding significantly different lineups:

$$\Delta\varepsilon_v[Ge-on-ZnSe(110)] = 1.52 \pm 0.03 eV \quad ,$$
$$\Delta\varepsilon_v[ZnSe-on-Ge(110)] = 1.29 \pm 0.03 eV \quad .$$

The large difference speaks for itself; it must somehow reflect a different atomic arrangement for the two growth sequences.   There can be little doubt that the Ge-on-ZnSe sequence is far more likely to represent a structurally simple junction than ZnSe-on-Ge.   In fact, as we stated in Sec. 2, we consider these data on Ge-on-ZnSe as being the most trustworthy of all mixed-column systems, and we had specifically included them amongst the experimental reference data for theory testing.

For the ZnSe-on-GaAs heterojunctions, the authors also give two different lineups: $\Delta\varepsilon_v \cong 1.10 \pm 0.03$ eV for ZnSe layers deposited at 23°C and then annealed at 300°C, and $\Delta\varepsilon_v \cong 0.96 \pm 0.03$ eV for ZnSe deposited at 300°C.   The difference is not likely to be due to different degrees of interdiffusion (both junctions "see" the same 300°C temperature eventually) but is almost certainly due to either a poorer crystalline perfection or - worse - a poorer stoichiometry of the low-temperature deposit.   We therefore consider the 0.96 eV-value the more trustworthy one.   It, too, was included amongst the reference data in Sec. 2.   When the two preferred lineups are combined with the Ge-on-GaAs lineup $\Delta\varepsilon_v \cong 0.53$ eV, one finds the closure sum

$$\Delta\varepsilon_v[Ge-on-ZnSe] + \Delta\varepsilon_v[ZnSe-on-GaAs]$$
$$-\Delta\varepsilon_v[Ge-on-GaAs] \quad ,$$
$$= -1.52eV + 0.96eV + 0.53eV = -0.03eV \quad ,$$

an extremely small remainder, below the accuracy of the measurement themselves. We conclude that, within the experimental accuracy of the data, the transitivity sum rule is very well satisfied, in contradiction to Waldrop and Grant's earlier conclusion.[44]

7.3 More on Technology Variations

This critical review of data necessary to test the transitivity postulate leads us naturally into the question of the dependence of the band offsets on growth technology, crystallographic orientation, or even growth sequence.

The difference between the two different ZnSe-on-GaAs lineup values shows that lineup data taken on heterojunctions grown at low temperature, are not necessarily representative of lineups for junctions grown at more "device-like" temperatures. Quite possibly, some of the larger discrepancies of the HAO predictions with the experimental data of Margaritondo et al.[17] are due to the fact that all junctions in the latter work were prepared at low temperatures. Certainly, these data must be expected to differ somewhat from (nonexistent) data obtained by "better" technologies, but in the absence of actual data nothing beyond this can be said.

We attributed the growth-sequence dependence of the ZnSe/Ge lineups to heavy antiphase disorder in ZnSe grown on Ge. Differences in any atomic exchange between ZnSe and Ge might conceivably also play a role. None of these mechanisms, however, can explain the most embarrassing deviation from the idea that band offsets are simply differences between absolute bulk energies: By comparing GaAs-on-AlAs (110) lineups with AlAs-on-GaAs (110) lineups, Waldrop et al.[12] found that in the latter case the valence band offsets agreed reasonably well with AlAs-on-GaAs (001) offsets, while in the GaAs-on-AlAs case the valence band offset was 0.25 eV larger! If this result is in fact real, it must somehow be due to a rather large growth sequence- and orientation-dependent dipole shift.

Somehow such different dipole shifts must in return reflect different atomic arrangements in the vicinity of the interface, for the two growth sequences, and for different orientations in general. As we shall see in Sec. 9, in the case of mixed-column heterojunctions there are natural mechanisms that yield major atomic dipole shifts as a result of minor variations in the exact microscopic interface geometry, or of minor diffusion effects across the interface. But it has proven difficult to come up with an atomistic model for such large and variable dipole shifts at any III/V-only heterojunction, unless one assumes that either a significant number of column-III sites are not occupied by column III atoms, and/or column-V sites not by column-V atoms. Some widely-observed growth sequence dependences of the electrical properties of (Al,Ga)As/GaAs (001) heterojunctions can apparently be explained by the Miller-Tsang-Munteanu carbon accumulation model.[45] It postulates the accumulation of background carbon on a growing (Al,Ga)As surface, but not on a growing GaAs surface. The result is a deterioration of the (Al,Ga)As morphology, followed by p-type doping of the subsequent GaAs growth by the accumulated carbon. This model does not require actual band offset variations, which would in any event conflict with Dingle's superb fit[10] of his superlattice absorption data to a symmetric quantum well model. But it is not easy to see how this mechanism could explain the data of Waldrop et al..[12] The least implausible (ad-hoc) explanation appears to be the hypothesis that the dipole shift is due to the formation and incorporation of Ga-As or Al-As antisite pairs at a GaAs-on-AlAs (110) interface.

8.   SELF-CONSISTENT INTERFACE POTENTIAL (SCIP) THEORIES

As we pointed out, all the theories discussed so far achieve their successes - whatever they may be - by ignoring the electronic interface dipole shift problem rather than solving it. To the extent that the theories, especially the HAO theory, are able to account for the experimentally observed band lineups, this disregard may be <u>empirically</u> justified, but conceptually it hardly represents a satisfactory state of affairs: <u>Why</u> are the interface dipole shifts negligible - or are they indeed? Are the residual deviations between theory and experiment due to the neglected electronic interface dipoles?

One possible approach towards the problem of electronic interface dipoles consists of a self-consistent quantum-mechanical treatment of the interface region itself. The idea is this: The electron wave functions and hence the charge distribution in the interface region depend on the exact potential in that region. But this potential is in turn generated, in part, by the charge distribution, and hence by the wave functions in the potential. Evidently, a knowledge of the potential requires a knowledge of the wave functions, and vice versa. It is possible to obtain both kinds of quantities self-consistently by an iterative process: Given an approximate potential one can, from the wave functions in that potential, determine a new and presumably improved potential. The process may evidently be repeated, and if it converges, it results in a self-consistent set of potentials, wave functions, and charge distributions, which automatically contain whatever electronic dipole shifts may be present. Convergence of the iteration is not automatic, but can be enforced by systematically applying suitable corrections at each iteration step.

The calculational effort required is formidable, but powerful techniques have been developed for handling such self-consistency problems on large high-speed compters at a tolerable cost. Following the pioneering work of Baraff et al.,[46] several efforts have been made to determine heterojunction band lineups in this fashion, especially by Cohen and his co-workers.[47-50]

Unfortunately, these calculations have so far not fulfilled the hope that they would provide a reliable predictive tool for accurate values of the band lineups, free of the questionable simplifications of the simpler theories, such as HAO, FKP and EAR: Except for the simple (Al,Ga)As/GaAs heterojunction, the band lineups predicted by these self-consistent interface potential (SCIP) calculations have been in much poorer agreement than HAO with reliable observations. Table 4 gives the comparison, which speaks for itself. The discrepancy is especially strong for ZnSe/GaAs (110), where Ihm and Cohen predict a staggered lineup quite incompatible with the experimental data.

Evidently, the SCIP theories in their present form are not yet satisfactory tools for the prediction of any but the mildest interface discontinuities, and it is not clear that they offer much retrodictive insight beyond AlAs/GaAs either.

Considering the outstanding success of pseudopotential theories in simpler problems, I suspect that the reasons for this unsatisfactory state of affairs are not fundamental, but are purely matters of calculational accuracy. Presumably, this could be improved, and I suspect that it will eventually be improved, at least to the point that the most glaring discrepancies with experiment (as for ZnSe/GaAs) have been removed, and the empirical band lineups in such systems have been reproduced

Table 4:   Comparison of Self-Consistent Interface Potential (SCIP) predictions with best experimental data and with HAO predictions. All data are valence band energies in eV, followed by references to their origins. The HAO data are from Table 2.

	SCIP	Experimental	HAO
AlAs/GaAs	0.25 [47]	0.19 [10]	0.04
ZnSe/GaAs	2.0 [49]	0.96 [28]	1.05
ZnSe/Ge	2.0 [48]	1.62 [28]	1.46
GaAs/Ge	0.35 [47]	0.53 [9]	0.41

sufficiently well to satisfy one's curiosity about retrodictively understanding the self-consistency problem at such interfaces.

However, even if this is achieved I perceive one major continuing problem: The SCIP theories appear to be, by their very nature, "single-shot theories" in which each semiconductor pair is a case unto itself. I do not perceive anything in those theories that makes the results of such calculations naturally appear in the quasi-linear form (8), with a small correction $\delta(A/B)$. If these corrections were large, this would of course be the principal strength of the SCIP theories, because it offers a hope to obtain - sometime in the future - reliable quantitative estimates and perhaps even a physical understanding of exactly what these nonlinear correction terms are. But to the extent that these correction terms appear to be small, the single-shot nature of present SCIP calculations makes those theories miss an essential point: The highly-systematic interrelation between band lineups of different heterojunctions, as expressed in the transitivity sum rule (10). At the present state of these theories there appears to exist no way to express the SCIP results in the form (8) except by brute-force "hindsight fitting" of large numbers of individually calculated (accurate) band lineups. Considering the formidable calculational effort required for SCIP calculations, which will probably have to be increased considerably to improve the accuracy to acceptable levels, this is hardly an attractive proposition. Unless and until this situation can be changed, the SCIP theories cannot in practice compete with such simpler if less-well founded approaches as HAO.

I suspect that, in order to express SCIP results naturally in the form (8), it will be necessary to introduce into the theory explicitly some natural common reference potential similar to the mean interstitial potential in FKP or (less likely) to the free atom potential at infinity in HAO. The task of the theory should then be to determine the deviations of the band offsets from the first-order lineup in which this natural reference potential is simply the same in both semiconductors.

9. THE MIXED COLUMN PROBLEM

9.1 General Comments

Throughout this paper I have repeatedly pointed out that there are numerous fundamental complications that can occur at heterojunctions in which the two semiconductors are not from the same column, or pair of columns, in the periodic table. Inasmuch as these complications do not appear to have been fully appreciated yet, sometimes not even by those actively working on such heterojunctions, at least a brief review is in order. Because most of the relevant points have been discussed in considerable detail in the recent literature,[11,27,26,1] this review will concentrate on simply directing the reader to that literature, going into detail only where new and additional points must be brought up.

9.2 Interface Neutrality Problem

The point of departure for any rational understanding of mixed-column heterojunctions is a classical 1978 paper by Harrison, Kraut, Waldrop and Grant (=HKWG, [11]), who investigate the electrostatics of "ideal" Ge/GaAs interfaces for several low-index orientations. "Ideal" here means interfaces in which the two semiconductors are separated by a mathematical plane, with no atoms intermixed across that plane. The authors showed that for most interface orientations such an ideal interface would contain a huge net interface charge, which would make the interface energetically unstable. They postulate that the interface would reconstruct during growth in such a way as to minimize both the interface charge and any electrostatic interface dipole. A notable exception was the ideal (110) interface, which was shown to be naturally free both of interface charges and of atomic interface dipoles.

Although the HKWG argument was worded for Ge/GaAs, it applied to all mixed-column heterojunctions.

It was subsequently argued by Kroemer et al.[27] that the HKWG reconstruction was very unlikely to go to completion during growth, and that the result would be a complex non-ideal interface whose exact atomic configuration was likely to be a frozen-in, highly technology-dependent state, with technology-dependent residual interface charges and interface dipoles. The band lineups of such heterojunctions should then also be technology-dependent rather than obeying any simple theory that does not take such atomic rearrangements into account. Hence the observed very wide fluctuations in the band lineups reported for Ge/GaAs are what one should expect, rather than signalling something obscure that calls for an explanation. Analogous arguments apply to all other mixed-column heterojunctions.

These pessimistic predictions do not automatically apply to the (110) orientation, for which HKWG do not require an interface reconstruction to achieve either neutrality or freedom from atomic interface dipoles. Indeed, the experimental band lineups reported for Ge-on-GaAs (110) grown by MBE are less inconsistent with each other than other data on Ge/GaAs heterojunction.

It was subsequently discovered by Wright et al.[26] that the (110) orientation is not the only orientation that offers at least the possibility of being charge- and dipole-free: There is an entire set of higher-index orientations with this property, of which the (112) orientation is the most interesting one. For a detailed discussion see reference [1]. There are no lineup data for the (112) orientation available yet.

9.3 Chemical Interface Reactions

Even for the (110) and (112) orientations, freedom from technology-dependent atomic dipoles cannot be guaranteed. For those orientations, atomic interchanges are not necessary to annihilate net interface charges and interface dipoles. Instead, interchanges that do take place, driven by chemical bonding forces, will create net interface charges and interface dipoles. Such chemical effects are quite likely.

Especially in the Ge/GaAs system, the chemical interaction between Ge and As is strong. Extensive work by Bauer et al.[31] and by Mönch et al.,[30] has shown that these interactions have a strong influence on the Ge/GaAs heterojunction properties, especially at higher Ge deposition temperatures. It is for this reason that I expressed my extreme reluctance, in Sec. 2, to include Ge/GaAs amongst the reference systems. It remains to be seen how technologically stable the Rockwell value $\Delta\varepsilon_v \cong 0.53\mathrm{eV}$[9] really is, and to what extent its reasonable agreement with the HAO theory might be changed for better or for worse by future data.

If anything has been learned from the very extensive MBE work in the Ge/GaAs system in recent years, it is that this is not the simple model system it was once thought to be, but one of the most erratic systems, more so even than most other mixed-column systems.

9.4 Antiphase Disorder due to Site Allocation Ambiguity

In those mixed-column heterojunctions, in which a compound semiconductor is grown on an elementary (Si or Ge) substrate, an additional difficulty arises: An extreme sensitivity to antiphase disorder on the compound side. In the elementary semiconductors, the two sublattices that make up the diamond structure of these materials are occupied by the same atomic species. In a defect-free heterojunction one of the two sublattices, if extended across the interface, should everywhere be occupied by the compound cation atom while the other sublattice should everywhere be occupied by the anion atom. But if a specific mechanism forcing such a coherent occupation of the two sublattices is absent, the local nucleation must be expected to be random, leading to domains of alternating occupancy of the sublattices, that is, to antiphase disorder. Heterojunctions with such antiphase disorder can hardly be considered the kind of device-quality heterojunctions that are of interest here, nor can band offset measurements on such junctions be considered reliable, especially if they contradict other measurements. As was pointed out earlier, in one case in which both compound-on-element and element-on-compound lineup data are available (ZnSe-on-Ge and Ge-on-ZnSe [28]), the discrepancy is blatant. In such cases, the compound-on-element data are far more likely to be incorrect than the element-on-compound data. In fact, I believe that all compound-on- element lineup data should be flatly ignored, unless there is positive evidence that antiphase disorder is absent.

If compound-on-element heterojunctions of device quality are to be achieved, this calls for a suppression both of any built-in interface charges and of antiphase disorder. This is a formidable problem, but apparently not an unsolvable one: Wright et al.[26] have reported the successful growth of GaP apparently free from antiphase domains, on Si (112), one of the orientations that should also be free from natural interface charges. The interface quality was sufficient to permit the construction of (mediocre) GaP-on-Si n-p-n transistors. The quality of the junctions appears sufficient, though, that reliable band lineup data should be obrainable, and such data would indeed be very interesting because they would potentially add a system of an altogether different kind to the set of reference data.

10.   CONCLUSIONS

If we look back eight years to the time when Dingle presented the first truly reliable heterojunction lineup data,[10] and I myself called for something better than the Electron Affinity Rule to predict and understand such lineups,[33] it is clear that much progress has been made. Not one but three new theoretical approaches have emerged, and enough "good" data to decide between them. The Harrison Atomic Orbital theory has emerged sufficiently clearly as superior over the rivals as well as over the Electron Affinity Rule that there no longer appears to be any rational reason to use the latter even as an empirical cookbook recipe. Yet there remains a puzzle: We do not really understand WHY the HAO theory works as well as it does. Maybe it is time to go back to the theory and find out not what is wrong with HAO, but why it is right. Injecting some of the ideas of FKP into HAO might offer considerable insight.

As for FKP itself, it could probably be improved to give results every bit as excellent as HAO, but in doing so it would almost certainly merge into SCIP. In fact, this combination might be powerful: Understanding the systematic variations from one heterojunction to another appears to call for an injection of FKP-like ideas into SCIP anyway. This, together with the surprising failure of SCIP to give realistic band lineups for mixed-column systems, provides such a strong conceptual challenge to the very foundations of the self-consistent pseudopotential method, that I would expect the SCIP theorists not to rest until that has been cleared up.

What remains is the embarrassing puzzle of such effect like the growth sequence dependence of AlAs/GaAs heterojunctions, their anisotropy, and technology variations in general. The hope remains that this is truly not a fundamental phenomenon, but an ordinary atomic dipole caused by defects carrying the wrong charge, introduced by improper growth conditions. If so, they would be less a subject for profound theoretical study than one of down-to-earth technology, something for which the only purpose of understanding it is to make it go away.

And, of course, we need more and better data to test our theories, especially on as yet unexplored systems. But this lies outside the scope of a paper on Theory of Heterojunctions.

ACKNOWLEDGEMENT

I have benefitted from numerous discussions with many individuals but from no one more than E. A. Kraut (Rockwell) who, together with his colleagues J. R. Waldrop, R. N. Grant and S. P. Kowalczyk, and with R. S. Bauer (Xerox), has kept my interest in heterojunction lineups alive over many years.

REFERENCES

1. H. Kroemer, Surf. Sci., Vol. 132 (1983), in the press.
2. For an excellent recent review see Ch. 4 of ref.[3]. Still-useful older reviews are contained in refs. [4] and [5].
3. H. C. Casey, Jr. and M. B. Panish, Heterostructure Lasers, Academic Press, New York, 1978.
4. B. L. Sharma and R. K. Purohit, Semiconductor Heterojunctions, Pergamon Press, London, 1974. See especially Ch. 2.
5. A. G. Milnes and D. L. Feucht, Heterojunctions and Metal-Semiconductor Junctions, Academic Press, New York. 1972/
6. J. Sakaki, L. L. Chang, R. Ludeke, C.-A. Chang, G. A. Sai-Halasz, and L. Esaki, Appl. Phys. Lett. 31, 211 (1977). See also L. L. Chang and L. Esaki, Surf. Sci. 98, 70 (1980).

7. J. L. Shay, S. Wagner, and J. C. Phillips, Appl. Phys. Lett. $\underline{28}$, 31 (1976).

8. For an excellent review, see Kraut et al., ref.[9] below.

9. E. A. Kraut, R. W. Grant, J. R. Waldrop, and S. P. Kowalczyk, Phys. Rev. Lett. $\underline{44}$, 1620 (1980).

10. R. Dingle, in Festkörperprobleme/Advances in Solid State Physics, H. J. Queisser, editor, Vieweg, Braunschweig, 1975, Vol. 15, p. 21.

11. W. A. Harrison, E. A. Kraut, J. R. Waldrop, and R. W. Grant, Phys. Rev. B $\underline{18}$, 4402 (1978). See also ref.[1].

12. J. R. Waldrop, S. P. Kowalczyk, R. W. Grant, E. A. Kraut, and D. L. Miller, J. Vac. Sci. Technol. $\underline{19}$, 573 (1981).

13. M. J. Adams and A. Nussbaum, Solid-State Electron. $\underline{22}$, 783 (1979).

14. O. von Roos, Solid-State Electron, $\underline{23}$, 1069 (1980).

15. H. Kroemer, IEEE Electron Dev. Lett., $\underline{EDL}$-4, 28 (1983).

16. A. Nussbaum, Solid-State Electron. $\underline{25}$, 1201 (1982).

17. G. Margaritondo, A. D. Katnani, N. G. Stoffel, R. R. Daniels, and T.-X. Zhao, Solid-State Commun. $\underline{43}$, 163 (1982).

18. W. A. Harrison, J. Vac. Sci. Technol. $\underline{14}$, 1016 (1977).

19. See Sec. 10F of ref. [20].

20. W. A. Harrison, Electronic Structure and the Properties of Solids: The Physics of the Chemical Bond, Freeman, San Francisco, 1980.

21. R. L. Anderson, Solid-State Electron. $\underline{5}$, 341 (1962).

22. S. P. Kowalczyk, W. J. Schaffer, E. A. Kraut, and R. W. Grant, J. Vac. Sci. Technol. $\underline{20}$, 705 (1982).

23. J. O. McCaldin, T. C. McGill, and C. A. Mead, Phys. Rev. Lett. $\underline{36}$, 56 (1976).

24. For a very "physical" discussion see Chs. 1-3 and Ch. 6 of Harrison, ref. [20].

25. W. R. Frensley and H. Kroemer, Phys. Rev. B $\underline{16}$, 2642 (1977).

26. S. L. Wright, M. Inada, and H. Kroemer, J. Vac. Sci. Technol. $\underline{21}$, 534 (1982).

27. H. Kroemer, K. J. Polasko, and S. L. Wright, Appl. Phys. Lett. $\underline{36}$, 763 (1980).

28. S. P. Kowalczyk, E. A. Kraut, J. R. Waldrop, and R. W. Grant, J. Vac. Sci. Technol. $\underline{21}$, 482 (1982).

29. S. Wagner, J. L. Shay, K. J. Bachmann, and E. Buehler, Appl. Phys. Lett. $\underline{26}$, 229 (1975).

30. See, for example, W. Mönch, R. S. Bauer, H. Gant, and R. Murschall, J. Vac. Sci. Technol. 21, 498 (1982), and the references given there.

31. See, for example, R. S. Bauer and J. C. Mikkelsen, J. Vac. Sci. Technol. 21, 491 (1982), which contains extensive references to earlier work.

32. R. W. Grant, J. R. Waldrop, and E. A. Kraut, Phys. Rev. Lett. 40, 656 (1978); J. Vac. Sci. Technol. 15, 1451 (1978).

33. H. Kroemer, CRC Crit. Revs. Solid State Sciences 5, 555 (1975).

34. W. R. Frensley and H. Kroemer, J. Vac. Sci. Technol. 13, 810 (1976).

35. F. Herman and S. Skillman, Atomic Structure Calculations, Prentice Hall, Englewood Cliffs, N. J. 1963.

36. It should perhaps be stated explicitly that the seven reference systems were in no way chosen to aid in a good or bad fit to one theory or another, but strictly on their inherent credibility.

37. W. R. Frensley, unpublished.

38. J. C. Phillips, J. Vac. Sci. Technol. 19, 545 (1981).

39. G. W. Gobeli and F. G. Allen, Phys. Rev. 137, A245 (1965); also in Semiconductors and Semimetals, R. K. Willardson and A. C. Beer, editors, Academic Press, New York, 1966. See pp. 263ff.

40 T. E. Fischer, F. G. Allen, and G. W. Gobeli, Phys. Rev. 163, 703 (1967).

41. R. K. Swank, Phys. Rev. 142, 519 (1966).

42. J. A. van Vechten, Phys. Rev. 187, 1007 (1969). This paper gives an extensive table of theoretical ionization energies, from which the electron affinities are easily obtained by subtracting the gap energies.

43. T. Fischer, F. G. Allen, and G. W. Gobeli, Phys. Rev. 163, 701 (1967).

44. J. W. Waldrop and R. W. Grant, Phys. Rev. Lett. 43, 1686 (1979).

45. R. C. Miller, W. T. Tsang, and O. Munteanu, Appl. Phys. Lett. 41, 372 (1982).

46. G. A. Baraff, J. A. Applebaum, and D. R. Hamann, Phys. Rev. Lett. 38, 237 (1977); J. Vac. Sci. Technol. 14, 999 (1977).

47. W. E. Pickett, S. G. Louie, and M. L. Cohen, Phys. Rev. B 17, 815 (1978).

48. W. E. Pickett and M. L. Cohen, Phys. Rev. B 18, 939 (1978).

49. J. Ihm and M. L. Cohen, Phys. Rev. B 20, 720 (1979).

50.   For an excellent review see M. J. Cohen, Adv. Electronics and Electron Physics 51, 1 (1980).

ENVELOPE FUNCTION APPROACH TO THE SUPERLATTICES BAND STRUCTURE

G. Bastard

Groupe de Physique des Solides de l'Ecole Normale Superieure 24, rue Lhomond, 75231 Paris Cedex 05, France

ABSTRACT: The superlattices are artificial semiconductors obtained by stacking alternatively layers of two host semiconductors A and B. In this lecture we incorporate the superlattice effect into a Kane type analysis of the hosts band structure assuming flat band condition. Various limits are analyzed and specific examples are presented. We also discuss the in-plane electron motion within an approximate scheme. Optical selection rules for interband transitions are derived and we show that in type II structures (e.g. InAs-GaSb) the optical matrix element strongly depends on the superlattice wavevector. Impurities and excitons effects in isolated quantum wells are briefly outlined.

1. INTRODUCTION

In 1970 Esaki and Tsu,[1] in search of new devices exhibiting a negative differential resistance, proposed a new and revolutionary concept: the semiconducting superlattices. They suggest that if it were possible to grow periodic semiconducting structures by alternating two closely lattice-matched semiconductors A and B, the quasicontinuum of the electron motion along the growth axis will split into several mini-bands characterized by one-dimensional dispersion relations ϵ_n (q) such as

$$\epsilon_n(q) = \epsilon_o - 2t_n \cos qd \qquad (1.1)$$

where q is the electron wave vector along the superlattice axis (i.e. the growth direction), d the superlattice period and $4t_n$ the bandwidth. If these new, man-made materials were subjected to external electric field F and provided the electron mean free path were large enough, the electrons being accelerated according to

$$\hbar\frac{dq}{dt} = -eF \qquad (1.2)$$

will move along the miniband (1.1) in the q space and before being scattered will reach a point in the mini-Brillouin zone $(-\pi/d, +\pi/d)$ where their effective masses will become negative. Hence the drift velocity will decrease with increasing F and negative resistance will occur. Moreover if the scattering time were long enough (eF zd > 2π) the whole miniband (1.1) will be travelled along by the electrons which, being Bragg-reflected back and forth at the zone edges, will exhibit a periodic motion at the frequency ν = eFd/h (Bloch oscillator). For modest field (F = 10^3 V/cm) and d = 100Å, ν = 250 GHz. Such a high frequency oscillator will enormously enhance the capability of microwaves devices.

To realize the required artificial periodicity, Esaki and Tsu proposed two ideas: a) To use a single material (e.g. GaAs) but to dope it alternatively n or p type (Fig. 1a). This idea has been extensively developed by Dölher, Ploog and collaborators[2] and is now referred to as "ni-pi superlattices". This subject is reviewed in this volume by Ploog.[3]; b) To alternate layers of two lattice-matched materials (e.g. GaAs and GaAlAs) where the bandgap of one material (GaAlAs) overlaps the bandgap of the other material (GaAs). One is left with a periodic variation of the conduction and valence band edges (Fig. 1b).

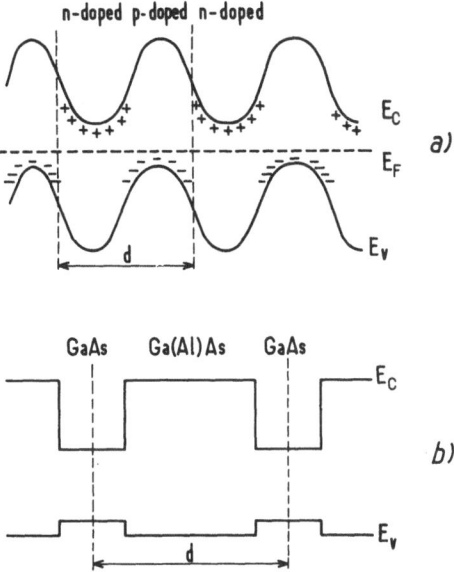

Figure 1 - (a) Spatial dependence of the band edge energies along the growth axis of a doping superlattice; (b) Spatial dependence of the band edges energies along the growth axis of a type I superlattice with flat band (e.g. undoped GaAs-Ga(Al)As superlattice).

It is clear that the mobility of superlattices (b) is a priori larger than the mobility of doping superlattices (a). This mobility can be tremendously enhanced if one incorporates the modulation doping idea due to Dingle et al.[4] to the superlattice (b). One ends up with (c) the modulation doping superlattices in which the barrier (e.g. GaAlAs) only is doped (Fig. 2a). The electrons provided by the donors fall down in the GaAs potential wells where they are confined. Hence one solves the old dilemma of the semiconductors physics by achieving a high carrier concentration (typically 10^{12} cm^{-2}) with a high mobility because they are spatially separated from their parent donors. At low temperatures mobility as high as 10^6 cm^2/Vs has been achieved in modulation-doped heterojunctions. The modulation doping superlattices are reviewed by Gossard in this volume.[5]

The superlattices 1b, 2a are so-called type I materials in which the bandgap of the barrier-like components completely overlaps the bandgap of the other component. Hence both electrons and holes are mostly confined in the same layers, those corresponding to the material with the smaller bandgap. The IBM group[6] has developed a second class of superlattices, the type II materials, in which, due to a peculiar band edges ordering of the two host materials, the electrons are mostly confined inside one kind of layers and the holes in the other kind of layers (Fig. 2b). The prototype of type II superlattices is the InAs-GaSb system: the top of the GaSb valence band is located in energy above the bottom of the InAs conduction band. As will be shown hereafter, the InAs-GaSb superlattices can be either semiconducting or semimetallic despite the fact that both host materials are semiconductors.

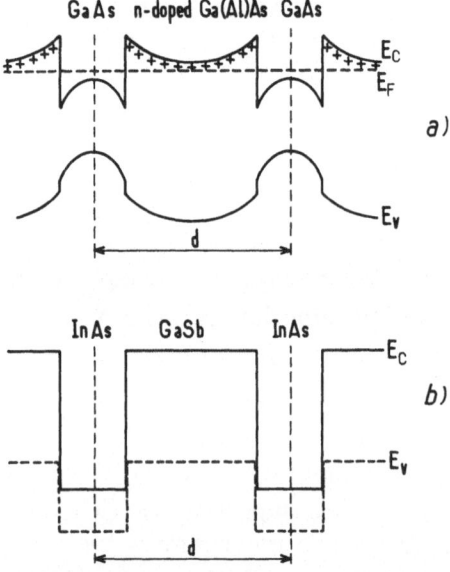

Figure 2 - (a) Spatial dependence of the band edges energies along the growth axis of a modulation-doped superlattice (the dopant is a donor); (b) Spatial dependence of the band edges energies along the growth axis of the InAs-GaSb superlattices. (semiconducting configuration)

To experimentally achieve the growth of a 10^2 Å period superlattice is a formidable task. The molecular beam epitaxy (see Cho's[7] and C.A. Chang's[8] lectures in this volume) has proved capable of producing extremely sharp interfaces (especially in the GaAs-GaAlAs materials). This success has generated an impressive blow-up of the superlattice studies both in the fields of fundamental and applied researches. The present lecture is by no means exhaustive and the reader is urged to consult both the references listed at the end of this article and the other courses of the volume to get a more complete overview of the superlattice field. The aim of the present lecture is to present a simple and analytical approach of the superlattice band structure. The focus will be placed on the effects of the periodic variations of the band edges on the slowly varying envelope functions which characterize the host materials A and B.

The section 2 will recall the essential ingredients of the band structure of the direct gap III-V and II-VI compounds (Kane model) which are of relevance for our purposes. In section 3 we will derive the superlattice band structure in the envelope function approximation for a zero electron wave vector in the layer plane. The section 4 will be devoted to a presentation of specific examples: undoped GaAs-GaAlAs, InAs-GaSb materials and the exotic HgTe-CdTe superlattices. In section 5 we will discuss the effective mass which characterizes the carrier motion in the layer plane and derive the density of states in a superlattice. In section 6 the band-to-band absorption will be calculated for type I and II materials. The section 7 will be devoted to a brief discussion of the Coulombic bound states in quantum wells which are the limiting cases of superlattices with wide barriers.

Most of the experimental results obtained so far in superlattices have involved magnetotransport or magnetooptics experiments. These magnetoproperties are the subject of L.L. Chang's lecture[29] and therefore will not be included in the present course. As indicated before doping superlattices (a) or modulation doping superlattices (c) are lectured by Ploog[3] and Gossard[5]. We will therefore only discuss the superlattice band structure under flat band conditions.

2. DISPERSION RELATIONS IN DIRECT GAP III-V AND II-VI COMPOUNDS

All the materials considered hereafter crystallize in the zinc-blende lattice. This lattice consists of two interpenetrating f-c-c lattices on which are located the atoms of the column III or II and the atoms of the columns V or VI respectively. The primitive cell contains two different atoms. The bonds are mostly covalent but also display some small ionic character. We are interested in obtaining the dispersion

TABLE 1 - Periodic parts of the Bloch functions u_l for the eight ban-dedges under consideration

u_i	$\lvert j\ m_j \rangle$	ψ_{jm}	$E(k=0)$
u_1	$\lvert \frac{1}{2},\frac{1}{2} \rangle$	$\lvert S\uparrow \rangle$	0
u_3	$\lvert \frac{3}{2},\frac{1}{2} \rangle$	$-\sqrt{\frac{2}{3}}\lvert Z\uparrow \rangle + \frac{1}{\sqrt{6}}\lvert (X+iY)\downarrow \rangle$	$-\epsilon_0$
u_5	$\lvert \frac{3}{2},\frac{3}{2} \rangle$	$\frac{1}{\sqrt{2}}\lvert (X+iY)\uparrow \rangle$	$-\epsilon_0$
u_7	$\lvert \frac{1}{2},\frac{1}{2} \rangle$	$\frac{1}{\sqrt{3}}\lvert (X+iY)\downarrow \rangle + \frac{1}{\sqrt{3}}\lvert Z\uparrow \rangle$	$-\epsilon_0-\Delta$
u_2	$\lvert \frac{1}{2}-\frac{1}{2} \rangle$	$\lvert S\downarrow \rangle$	0
u_4	$\lvert \frac{3}{2}-\frac{1}{2} \rangle$	$-\frac{1}{\sqrt{6}}\lvert (X-iY)\uparrow \rangle -\sqrt{\frac{2}{3}}\lvert Z\downarrow \rangle$	$-\epsilon_0$
u_6	$\lvert \frac{3}{2}-\frac{3}{2} \rangle$	$\frac{1}{\sqrt{2}}\lvert (X-iY)\downarrow \rangle$	$-\epsilon_0$
u_8	$\lvert \frac{1}{2}-\frac{1}{2} \rangle$	$-\frac{1}{\sqrt{3}}\lvert (X-iY)\uparrow \rangle + \frac{1}{\sqrt{3}}\lvert Z\downarrow \rangle$	$-\epsilon_0-\Delta$

relations in the vicinity of the center of the Brillouin zone (the
Γ-point). The crystal hamiltonian is:

$$\mathcal{H} = P^2/2m_0 + V(\vec{r}) + \frac{\hbar}{4m_0^2c^2}\, \vec{\sigma}\cdot(\nabla V \times \vec{p}) \qquad (2.1)$$

where m_0 is the bare electron mass and $\vec{\sigma}$ the electron spin. The third
term is the split-orbit coupling which is quite large since the elements
involved in the bonds are heavy. We assume we know the eigenfunctions and the eigenstates of (2.1) at $k = 0$. The band edges of interest
contains 8 states which eigenfunctions are listed in Table 1.   The
energy origin has been taken at the bottom of S-type band (usually the
conduction band).   The bandgap separating the conduction and first
valence band is denoted by ϵ_0. The quantity Δ is the spin-orbit coupling

$$\Delta = -\frac{3i\hbar}{4m_0^2c^2}\, < X \,|\, \vec{\nabla}V \times \vec{p} \,|\, Z > \qquad (2.2)$$

It is of the order of 1 eV in the materials of our interest. The function
S, X, Y, Z are periodic functions which transform like atomic s, x, y, z
functions under the symmetry operation of the T_d group.   At finite k
we expand the Bloch functions solutions of (2.1) in terms of the Kohn-
Luttinger basis functions:

$$\Psi_{\vec{k}} = \sum_n c_n(\vec{k})\chi_{n0}(\vec{r}) \qquad (2.3)$$

$$\chi_{n0}(\vec{r}) = \Omega^{-1/2}\, u_{n0}(\vec{r})\, \exp\, i\vec{k}\cdot\vec{r} \qquad (2.4)$$

u_{n0} are the periodic part of the Bloch function at $k = 0$ which for the
band edges of interest have been listed on Table 1. We have:

$$\sum_m \left\{ (\epsilon_{n0} + \frac{h^2k^2}{2m_0} - \epsilon)\delta_{nm} + \frac{\hbar}{m_0}\vec{k}\cdot\vec{p}_{nm} \right\} c_m(\vec{k}) = 0 \quad (2.5)$$

In practice, the band edges of relevance are closer to one another than
the other $k = 0$ states. We therefore subdivide[9] the labels n, m into: a)
the closely spaced bands (hereafter labelled as ℓ, $1 \leq \ell \leq 8$); b) the
remote bands (hereafter labelled as v). The off-diagonal terms $p_{\ell\ell'}$ are
taken exactly into account within the closely spaced set whereas the
coupling $\vec{p}_{\ell v}$ is retained only up to the second order. The matrix D of

the $\vec{k} \cdot \vec{p}$ coupling is consequently written as the sum of two 8x8 matrices D^0 and $\vec{\delta} \cdot D^0_{\ell \ell'}$ includes all the $\vec{p}_{\ell \ell'}$ terms and all the elements of $\delta_{\ell \ell'}$ arise from $p_{\ell \nu}$ coupling. In a first approximation, we can diagonalize only $D^0_{\ell \ell'}$. If necessary $\delta_{\ell \ell'}$ terms will be taken into account. The matrix $D^0_{\ell \ell'}$ has been first obtained by Kane[9,10] and is written in Table 2. In this table

$$k_{\pm} = 2^{-1/2} (k_x \pm ik_y) \qquad (2.6)$$

$$P = \frac{1}{m_0} < S \, | \, p_x \, | \, X > \qquad (2.7)$$

The dispersion relations are isotropic and we can therefore quantized the angular momentum along k. Then the even-odd off-diagonal terms of $D^0_{\ell \ell'}$ disappear and we obtain:

$$\epsilon (\epsilon + \epsilon_0)(\epsilon + \epsilon_0 + \Delta) = \hbar^2 k^2 P^2 (\epsilon + \epsilon_0 + 2\Delta/3) \qquad (2.8)$$

$$\epsilon = -\epsilon_0 \qquad (2.9)$$

Each of the solutions (2.8, 2.9) are twice degenerate. The solutions (2.9) do not exhibit any dispersion. They correspond to the heavy hole band whose k-curvature can only be obtained through the inclusion of $\delta_{\ell \ell'}$. Hence the heavy hole mass will be much larger than the effective masses deduced from (2.8). We note that for k//z the heavy hole states correspond to $m_j = \pm 3/2$ whereas the solutions of (2.8) correspond either to $m_j = + 1/2$ (u_{2n+1}) or $m_j = -1/2$ (u_{2n}). In all what follows the eigenstates corresponding to (2.8) will be referred to as light particles and the solutions of (2.9) (modified by a partial inclusion of $\delta_{\ell \ell'}$) will be referred to as heavy particles.

In the vicinity of k = 0 we can find the effective masses of the S and P, $m_j = \pm 1/2$ type bands. We have

$$\frac{1}{m_c} = \frac{2P^2(\epsilon_0 + 2\Delta/3)}{\epsilon_0(\epsilon_0 + \Delta)} \qquad (2.10)$$

$$\frac{1}{m_v} = -4P^2/3\epsilon_0 \qquad (2.11)$$

where c and v stand for conduction and valence (light) effective masses. Note that both light particles exhibit non-parabolicity effects in their dispersion relations, i.e., m_c, m_v increase when the energy states of interest move away from the band edges. Even though it is not a good quantitative approximation in the relatively wide gap materials as GaAs, one usually uses in superlattices calculations a two-band model in which the split-off valence band is entirely neglected. This simplified scheme is obtained by setting $\Delta = \infty$ in (2.8, 2.10). According to (2.8) the energy segments $[-\epsilon_0, 0]$ and $[-\epsilon_0 -\Delta, -\epsilon_0 - 2\Delta/3]$ are forbidden gaps as k^2 becomes negative. No electrons can propagate in these energy ranges for bulk materials (because of the Bloch theorem). Evanescent states are however possible in a finite layer. Their dispersion relations are obtained by setting $k = i\kappa$ in (2.8). Figure 3 shows a sketch of the real and evanescent states dispersion relations for a GaAs and AlAs.[11] In connection with superlattices the complex band structure of semiconductors has been extensively discussed by Schulman and Chang.[11] They have shown that the evanscent states within the fundamental gap $[-\epsilon_0, 0]$ of GaAs and AlAs are described as correctly by tight binding calculations as by the Kane model. When the quantization axis of the angular momentum is not taken parallel to k, the dispersion relations (2.8, 2.9) are still valid (with $k^2 = k_x^2 + k_y^2 + k_z^2$). The eigenfunctions of light particles however no longer correspond to pure $m_j = 1/2$ or $m_j = -1/2$ states (see Table 2). This k_+ - induced coupling of the $\pm 1/2$ states has been known for a long time and mani-

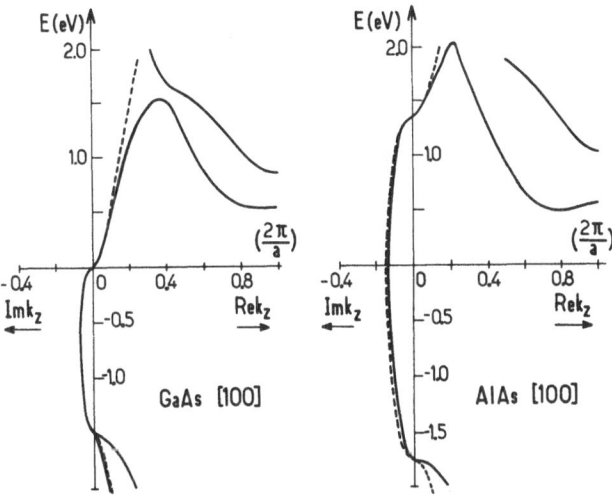

Figure 3 - Real and envanscent states dispersion relations of GaAs and AlAs. The solid lines correspond to tight-binding calculations and the dashed lines to the Kane (two-band) model (After Ref. 11).

TABLE 2 - $\vec{k}.\vec{p}$ matrix written in the subspace spanned by the functions u_ℓ.

	u_1	u_3	u_5	u_7	u_2	u_4	u_6	u_8
u_1	0	$-\sqrt{\dfrac{2}{3}}P\hbar k_z$	$P\hbar k_+$	$\dfrac{1}{\sqrt{3}}Pk_z$	0	$-\dfrac{1}{\sqrt{3}}P\hbar k_-$	0	$-\sqrt{\dfrac{2}{3}}P\hbar k_-$
u_3	$-\sqrt{\dfrac{2}{3}}P\hbar k_z$	$-\epsilon_0$	0	0	$\dfrac{1}{\sqrt{3}}P\hbar k_-$	0	0	0
u_5	$P\hbar k_-$	0	$-\epsilon_0$	0	0	0	0	0
u_7	$\dfrac{1}{\sqrt{3}}P\hbar k_z$	0	0	$-\epsilon_0-\Delta$	$\sqrt{\dfrac{2}{3}}P\hbar k_-$	0	0	0
u_2	0	$\dfrac{1}{\sqrt{3}}P\hbar k_+$	0	$\sqrt{\dfrac{2}{3}}P\hbar k_+$	0	$-\sqrt{\dfrac{2}{3}}P\hbar k_z$	$P\hbar k_-$	$\dfrac{1}{\sqrt{3}}P\hbar k_z$
u_4	$-\dfrac{1}{\sqrt{3}}P\hbar k_+$	0	0	0	$-\sqrt{\dfrac{2}{3}}P\hbar k_z$	$-\epsilon_0$	0	0
u_6	0	0	0	0	$P\hbar k_+$	0	$-\epsilon_0$	0
u_8	$\sqrt{\dfrac{2}{3}}P\hbar k_+$	0	0	0	$\dfrac{1}{\sqrt{3}}P\hbar k_z$	0	0	$-\epsilon_0-\Delta$

fests itself clearly in intraband magnetooptics[12] (e.g. by the existence of a "combined" transition where both the Landau level index and the spin states are changed through the absorption of a photon). This $k_\pm$ coupling arises from the non-zero value of the spin-orbit coupling.

The Kane model is insufficient to predict a correct effective mass for the heavy hole band. Hence we must include $\delta_{\ell\ell}$, to restore a finite heavy hole mass. If we want to keep spherical dispersion relations, the matrix δ depends, at zero magnetic field, on three small parameters called the Luttinger higher band parameters. They are i) $\gamma_1, \bar{\gamma}$ which take into account P-P interactions via virtual jumps to the remote bands and ii) F which describes S-S coupling via the same mechanism. The parameter F is usually neglected because the S band has already a finite curvature at zero δ. The matrix δ has no non-vanishing elements connecting S and P bands. Its expression, although known[13], is simple only at $k_\pm = 0$ where δ is diagonal. For our purpose we need only

$$< \frac{3}{2}, \pm\frac{3}{2}|\delta|\frac{3}{2}, \pm\frac{3}{2} > \; = \; -\frac{\hbar^2 k_z^2}{2m_{hh}} \; = \; (-\gamma_1 + 2\bar{\gamma})\frac{\hbar^2 k_z^2}{2m_o} \quad (2.12)$$

$$< \frac{3}{2}, \pm\frac{1}{2}|\delta|\frac{3}{2}, \pm\frac{1}{2} > \; = \; -\frac{\hbar^2 k_z^2}{2\mu} \; = \; (-\gamma_1 - 2\bar{\gamma})\frac{\hbar^2 k_z^2}{2m_o} \quad (2.13)$$

As the light hole band curvature is already finite without $\delta_{11'}$, the higher bands contribution to the light hole mass $(1/\mu)$ will be neglected in our qualitative approach. They heavy hole mass m_{hh} is on the other hand of primary importance to obtain reasonable description of the heavy hole superlattice subbands. Further information concerning the δ matrix as well as a complete discussions of the 8x8 effective Hamiltonian describing the carrier kinematiks in III-V and II-VI semiconductors can be found in Kane's and Weiler's review papers.[10,14]

With respect to superlattice calculations, the key simplifying feature of zinc-blende materials is the remarkable constancy of the Kane matrix element P when going from one material to the other one (see Table 3). Once ϵ_o, Δ are given the band structure of a given III-IV material depends only on a single parameter: P. The envelope function description of the superlattice will take a full advantage of this property. Namely P will be assumed to be the same in both host materials and fitted to reproduce the effective mass of one of the bands of our interest, say the conduction band mass of the A material. This

TABLE 3 - Low temperature band parameters of some III-V compounds. The value of $2m_0P^2$ is calculated from (2.10).

compound	InSb	InAs	InP	GaSb	GaAs
ϵ_0(eV)	0.237	0.42	1.423	0.813	1.519
Δ(eV)	0.81	0.38	0.11	0.752	0.341
$\dfrac{m_c}{m_0}$	0.0139	0.023	0.0803	0.041	0.0667
$2m_0P^2$(eV)	22.98	21.70	18.16	23.61	24.26

procedure is of course only valid for materials which are lattice-matched. Once P is determined the superlattice problem reduces to periodic variations of the band gap ϵ_0 and of the spin orbit energy Δ and the calculations become relatively simple.

3. SUPERLATTICE BAND STRUCTURE IN THE ENVELOPE FUNCTION APPROXIMATION

A binary superlattice AB is made by stacking alternatively layers of A material and of B material. The A and B materials are assumed to have the same lattice parameter and the same crystalline symmetry. Each layer of A(B) material has a length L_A (L_B). The superlattice system is invariant with respect to any translation along the superlattice axis z which is a multiple of the fundamental superlattice period $d = L_A + L_B$. The basic approximation which permits a simple description of the dispersion relations of a A-B superlattice is the assumption that the periodic part of the Bloch functions u_ℓ are the same in both A and B materials for the band edges of our interest. If this approximation is retained, inside the $\ell x \ell$ subspace which has been defined (see 2) the complete solution of the superlattice Hamiltonian

$\mathcal{H}(z)$	$= \mathcal{H}_A$	if z corresponds to A material	(3.1)
$\mathcal{H}(z)$	$= \mathcal{H}_B$	if z corresponds to B material	(3.2)
$\mathcal{H}(z+d)$	$= \mathcal{H}(z)$	for any z	(3.3)

can be written

$$\Psi(\vec{r}) = \sum_{\ell=1}^{8} F_\ell(\vec{r}) u_{\ell_0}(\vec{r}) \tag{3.4}$$

where $F_\ell(\vec{r})$ is a slowly varying function at the scale of the hosts unit cell. Inside the A layers $\mathcal{H}(z) = \mathcal{H}_A$, hence the functions $F_\ell(\vec{r})$ are linear combination of the ℓth component of the eigenstates of $\mathcal{H}_A$, where $\mathcal{H}_A$ is the 8x8 matrix given in Table 2 with ϵ_0 and Δ being equal to their actual values in the A material: ϵ_A, Δ_A. The fact that the superlattice hamiltonian matrix does not depend on $\vec{r} = (x,y)$ implies that

$$F_\ell(\vec{r}) = (L_x L_y)^{-1/2} \exp^{i\vec{k}_\perp \cdot \vec{r}_\perp} f_\ell(z) \tag{3.5}$$

where $L_x L_y$ is the area of the layers and $\vec{k}_\perp = (k_x, k_y)$ a two-dimensional wave vector: the transverse wave vector $k_\perp$ is a constant of motion. In this effective mass-like formalism the only difference between the A and B materials appear in the values of the band-edges energies which become periodic functions of z according to

$$\epsilon_\ell(z) = \epsilon_\ell^{(A)} \qquad \text{if z corresponds to A layers} \tag{3.6a}$$

$$\epsilon_\ell(z) = \epsilon_\ell^{(B)} \qquad \text{if z corresponds to B layers} \tag{3.6b}$$

$$\epsilon_\ell(z + d) = \epsilon_\ell(z) \qquad \text{for any z} \tag{3.7}$$

and $1 \leq \ell \leq 8$. Hereafter, we will denote by V_s the algebraic energy shift of the S type band edge ($\ell = 1,2$) when going from the A to the B material. Similarly V_p will denote the algebraic energy shift of the topmost valence band ($3 \leq \ell \leq 6$) when going from A to B layers and V_δ the corresponding quantity for the split-off valence band ($\ell = 7,8$). We have

$$V_s - \epsilon_B = V_p - \epsilon_A \tag{3.8}$$

$$V_s - \epsilon_B - \Delta_B = V_\delta - \epsilon_A - \Delta_A \qquad (3.9)$$

Hence the band edges variations (3.5-7) can be described by defining three periodic functions, V_s (z), V_p (z), V_δ(z) which are zero in A layers and equal to V_s, V_p V_δ in B layers (Fig. 4).

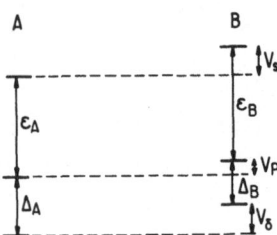

Figure 4 - Relations between V_s, V_p, V_δ and ϵ_A, ϵ_B, Δ_A, Δ_B.

As far as eigenvalues are concerned, we can forget about the u_ℓ functions and deal only with the envelopes F_ℓ. We have to solve an 8x8 differential system, for each k where the potential is described by (3.5-7) and where (because we deal with envelope functions) the wave vector k_z, which is no longer conserved, has to be replaced by the operator - $i\partial/\partial z$ like in usual effective mass calculations. In doping superlattices or modulation doping superlattices, the band edges are not constant versus z, even within a given layer. To the periodic variations (3.5-7) we need to add the diagonal (ℓ -independent) term -$e\phi_{sc}$ (z) which is also periodic in z. The quantity ϕ_{sc} (z) is the self consistent electrostatic potential which arises from charge transfer effects occurring inside a given layer or between different layers.[19] It is often calculated within the Hartree approximation by solving the Schrödinger equation together with the Poisson equation. A detailed discussion of these charge transfer effects will be found in Ploog's[3] and Gossard's[5] lecture in this volume.

The envelope function scheme should not be trusted blindly. Indeed the delicate problem of the exact nature of the interface is entirely neglected. Consider a A-B interface. It is clear that the last (first) atom of the A(B) layers has a different environment that the other atoms inside the A(B) layers. Even if A and B materials have the same lattice parameter, the interface atoms will not have the same bond length as in the bulk. In our approach we have neglected any phenomenon varying rapidly with z. We have assumed that only the slowly varying envelopes F_ℓ will experience the differences between A and B. Hence the strongly localized phenomena such as those occurring at the interfaces are beyond our formalism. To take them into account, we should include all the band edges of both host materials in our description. This is clearly too complicated.

There are two situations where our envelope function approximation will clearly fail. a) In very short period superlattices (few atomic planes per layers) most of the atoms are interface atoms. Thee is no clear cut distinction between envelope and u_ℓ functions. A L.C.A.O. calculation is here of relevance since it focuses the attention on atomic-like details. b) In any superlattices, if we are interested in energies which are such that they correspond to two different valleys (e.g. at the Γ and L points) in one (or both) of the host materials. Then the superlattice potential will have matrix elements connecting host functions corresponding to the two valleys. At least one of the involved wave vector will have the dimension of the host Brillouin zone. Hence at least, on of the envelope function will be as rapidly varying as the u_ℓ's invalidating the effective-mass treatment. We will need a complete description of the host band structure, valid over the whole Brillouin zone. Again this would require to extend the $\vec{k} \cdot \vec{p}$ treatment over a very large number of band edges. In all what follows, we shall discuss only the superlattice states which energies are close to the hose band edges.

We have quantized the angular momentum along the superlattice axis (z axis). This seems the most natural choice especially in view of the actual growth along the (001) directions of the host crystals. However, note that we have discarded any non spherical terms in the effective mass Hamiltonian (Table 2). Hence, the quantization of the angular momentum axis is somewhat arbitrary. It can also be taken in the layer plane.[18,20] Even though the host dispersion relations are taken as isotropic, the superlattice ones are not, due to the presence of a preferential axis. The anisotropy of the effective masses along and perpendicular to the superlattice axis can be very large. For instance in GaAs-GaAlAs superlattices we know that for a fixed GaAs thickness L_A the effective mass along the superlattice axis diverges when the GaAlAs layer thickness L_B becomes infinite. On the other hand, the transverse mass (i.e. in the layer planes) remains finite and in practice very close to the bulk GaAs mass. For very small L_B the longitudinal mass is finite and should approach the bulk GaAs mass when $L_B \to 0$ for a fixed L_A. If we let $k_\perp \neq 0$, we see from Table 2 that the $m_j = \pm 1/2$ states are coupled and moreover the matrix δ is fairly complex. As we are primarily interested in the subbands energy positions we will first let $k_\perp = 0$, postponing the $k_\perp \neq 0$ study to section 5.

3.1 Superlattice States at $\vec{k}_\perp = \vec{0}$

At $\vec{k}_\perp = \vec{0}$, the 8x8 matrix (Table 2) splits into two identical 3x3 matrices corresponding to $m_j = \pm 1/2$ respectively and into two identical equations corresponding to $m_j = \pm 3/2$. Like in bulk materials the light and heavy particles states are decoupled. The heavy particles states will be discussed separately (see below). To study the $m_j = \pm 1/2$ states we take the energy zero at the bottom of the S band of the A material. Since the various band-edges are periodic in z the Bloch theorem holds. Hence for any ℓ, $1 \leq \ell \leq 8$

$$f_\ell(z + d) = f_\ell(z) \exp iqd \qquad (3.10)$$

where q is the superlattice wave vector which can be restricted to the first superlattive Brillouin zone

$$-\frac{\pi}{d} \leq q \leq +\frac{\pi}{d} \qquad (3.11)$$

The Bloch theorem governs the long range behavior of the envelope functions $f_\ell(z)$. A superlattice unit cell contains two A-B interfaces. We need to know how the envelope functions and their derivatives behave at the interface. This is most easily discussed in terms of the effective hamiltonian governing the S-like envelopes ($\ell = 1,2$). It is obtained by projecting the 8x8 matrix onto the 2x2 subspace spanned by S↑ and S↓. We find two identical Schrödinger equations governing the behavior of the f_1 and f_2 functions:

$$\frac{P^2}{3} p_z \left[\frac{2}{\epsilon + \epsilon_A - V_P(z)} + \frac{1}{\epsilon_A + \Delta_A + \epsilon - V_\delta(z)} \right] p_z f_{1,2}$$

$$+ V_S(z) f_{1,2} = \epsilon f_{1,2} \qquad (3.12)$$

Note that p_z and the bracketed expression do not commute. Since $f_{1,2}$ must be continuous at the interfaces (because the u_ℓ's are linearly independent) the integration of (3.12) across an interface provides us with:

$$\left[\frac{2}{\epsilon+\epsilon_A-V_p(z)} + \frac{1}{\epsilon_A+\Delta_A+\epsilon-V_\delta(z)}\right] \frac{df_{1,2}}{dz}$$

(3.13)

continuous at the interface

It is now straightforward to use the Bloch and the boundary conditions (3.10, 3.12-13) to derive the superlattice dispersion relations. Inside each A or B layers of a superlattice unit cell f_1 or f_2 are the sums of an incoming and of an outcoming plane waves characterized by wave vectors k_A and k_B in each layer respectively. For both f's there are two unknown coefficients per layer. The four boundary conditions obtained at the two interfaces leaves us with a 4x4 homogeneous determinant. Non zero solutions exist if:[15-17]

(3.14)

$$\cos qd = \cos k_A L_A \cos k_B L_B - \frac{1}{2}\left(\eta + \frac{1}{\eta}\right)\sin k_A L_A \sin k_B L_B$$

where

$$\eta = \frac{\dfrac{k_B}{k_A}\left[\dfrac{2}{\epsilon+\epsilon_A-V_p} + \dfrac{1}{\epsilon+\epsilon_A+\Delta_A-V_\delta}\right]}{\left[\dfrac{2}{\epsilon+\epsilon_A} + \dfrac{1}{\epsilon+\epsilon_A+\Delta_A}\right]}$$

(3.15)

$$\epsilon(\epsilon+\epsilon_A)(\epsilon+\epsilon_A+\Delta_A) = \hbar^2 k_A^2 P^2(\epsilon+\epsilon_A+2\Delta_A/3)$$ (3.16)

$$(\epsilon-V_s)(\epsilon-V_s+\epsilon_B)(\epsilon-V_s+\epsilon_B+\Delta_B)$$
$$= \hbar^2 k_B^2 P^2(\epsilon-V_s+\epsilon_B+2\Delta_B/3)$$

(3.17)

We recognize in (3.14) the results of the familiar Kronig-Penney model in which however k_B/k_A is replaced by η. The dispersion relations (3.14) are therefore a generalization of the Kronig-Penney model to a multiband situation. Note that (3.14) is completely general and therefore applicable to any A-B superlattice provided the A and B band edges of relevance are well described by the Kane model. Specific examples will be given in section 4.

3.2 Miscellaneous Remarks

3.2.1 If the energy of interest ϵ is close enough from both host S type band edges we can neglect ϵ and ϵ - V_s with respect to ϵ_A and ϵ_B respectively. Hence making use of (2.10) and (3.8, 3.9) the condition (3.13) can be re-expressed as:

$$\frac{1}{m_A} \frac{df_{1,2}}{dz} = \frac{1}{m_B} \frac{df_{1,2}}{dz} \tag{3.18}$$

where m_A, m_B are the S band edge effective masses defined in (2.10). We see that for host materials with parabolic dispersion relations, a condition equivalent to $\epsilon \ll \epsilon_A$, ϵ - $V_s \ll \epsilon_B$, we recover Ben Daniel and Duke's results[21] for the continuity conditions of the envelope and $m^{-1}(z)$ times the derivative of the envelope at the interface. In the parabolic limit, the dispersion relations (3.14) are unchanged but η is modified into:

$$\eta_{para} = \frac{k_B \, m_A}{m_B \, k_A} \tag{3.19}$$

3.2.2 Quite frequently the split off valence band is discarded in actual calculations (two-band model). This limit is obtained by setting Δ_A, Δ_B infinite in (3.15-17). In this case η becomes:

$$\eta = \frac{k_B(\epsilon_A + \epsilon)}{k_A(\epsilon_A + \epsilon - V_P)} \tag{3.20}$$

3.2.3 The superlattice states often correspond to evanescent states in one of the host material (note that no superlattice state is allowed if ϵ is in both A and B bandgaps because of the Bloch theorem in a superlattice). Assume that ϵ is in the B bandgap. With the change $k_B = i$

κ_B we obtain $\eta = i\tilde{\eta}$ and (3.14) becomes:

$$\cos qd = \cos k_A L_A \cos h\kappa_B L_B + \frac{1}{2}\left[\tilde{\eta} - \frac{1}{\tilde{\eta}}\right]$$

$$x \sin k_A L_A \sin h\kappa_B L_B \qquad (3.21)$$

If the barriers become very thick ($L_B \rightarrow \infty$) we end up with isolated quantum wells and the allowed energy states become discrete:

$$\cos k_A L_A + \frac{1}{2}\left[\tilde{\eta} - \frac{1}{\tilde{\eta}}\right] \sin k_A L_A = 0 . \qquad (3.22)$$

This is a generalization of the well known finite quantum well bound states equation. If moreover the barrier become infinite κ_B diverges and we find $k_A L_A = n\pi$ as it should.

In the intermediate range of L_B where the coupling between wells is weak but non zero ($\kappa_B L_B$ finite) we can simplify (3.21). Let us denote by ϵ_n the discrete energies, solutions of (3.22) and by $\beta(\epsilon)$ the right hand side of (3.21). As $\kappa_B L_B$ is large $\beta(\epsilon)$ is close to one only in the vicinity of ϵ_n. Expanding $\beta(\epsilon)$ near ϵ_n, we obtain to the lowest order

$$\epsilon = \epsilon_n + S_n + 2t_n \cos qd \qquad (3.23)$$

where

$$S_n = -\beta(\epsilon_n)\frac{d\epsilon_n}{d\beta} \quad ; \quad t_n = \frac{1}{2}\frac{d\epsilon_n}{d\beta} \qquad (3.24)$$

The expressions (3.23-24) correspond to a tight-binding analysis of coupled quantum wells which if they were isolated would have bound states which are solutions of (3.22). The coupling between the wells through the evanscent states of the barrier manifest itself in two ways. First there is a shift S_n of the bound state ϵ_n. If N wells are non-interacting the level ϵ_n is N fold degenerate. The coupling lifts the degeneracy and gives rise to a band of finite width $4t_n$.

3.2.4 Tight binding description of weakly coupled quantum wells. Even though (3.21) is completely general and therefore permits a calculation of subband widths, subband positions, etc. it can be illustrative and useful to calculate directly the shifts and bandwidth from a knowledge of isolated quantum wells eigenfunctions. To do so we simplify the algebra by assuming that both host materials have parabolic bands. (To a reasonable approximation this is the case in the GaAs-GaAlAs system for $\epsilon \lesssim 100$ meV). Under this assumption we have

$$\Psi_{nqk_\perp}(\vec{r}) = (L_x L_y)^{-1/2} f_{nq}(z) \exp i\vec{k}_\perp \cdot \vec{r}_\perp \qquad (3.25)$$

where $f_{nq}(z)$ is the solution of

$$\left[\frac{1}{2}P_z \frac{1}{m(z)}P_z + V_s(z) - V_s\right] f_{nq}(z) = (\epsilon - V_s)f_{nq}(z) \quad (3.26)$$

where as usually done in the Kronig-Penney model the energy origin is now taken at the bottom of the $\underline{B}$ materials. $m(z)$ is m_A or m_B if z corresponds to A or B layers. The periodic potential is the sum of "atomic-like potentials":

$$V_s(z) - V_s = \sum_j V(z - jd) \qquad (3.27)$$

where

$$V(z - jd) = \begin{cases} -V_s \text{ if } |z - jd| \leq L_A/2 \\ \\ 0 \text{ elsewhere} \end{cases} \qquad (3.28)$$

For an isolated quantum well centered at $z_j = jd$ there exists one or several bound states with discrete energied

$$\epsilon_n = -V_s + \alpha_n \qquad (3.29)$$

α_n is the confinement energy which goes to $h^2\pi^2n^2/2m_A L_A^2$ when V_s becomes infinite. To the energies ϵ_n are associated square-integrable wave functions $f_n^0(z-z_j)$ centered at z_j. In addition to the bound states there exists a continuum corresponding to $\epsilon > 0$ which is associated eigen-functions which oscillate at large z.

The couplings between the wells broaden the discrete level ϵ_n. To write a superlattice states of wave vector q originating from ϵ_n, we expand the Bloch function f_q in terms of the "atomic-like" functions f_n^0 (z-z_j):

$$f_{nq}(z) = \frac{1}{\sqrt{N}} \sum_{j=-\infty}^{+\infty} e^{iqjd} f_n^0(z-jd) \qquad (3.30)$$

It has to be stressed that the expansion (3.30) actually means that we have neglected the couplings between ϵ_n and the other bound states ϵ_m of a given well and of other wells due to the tunelling across the barriers as well as the couplings between ϵ_n and the continuum $\epsilon > 0$ of the same origin. Hence (3.30) is appropriate for narrow bands as those originating from the deepest bound levels ϵ_n, $n = 1,2$. In any case it should be checked a posteriori that the bandwidth calculated from (3.30) remains smaller than the energy difference $|\epsilon_n - \epsilon_{n+1}|$. If not, more than a single atomic level ought to be included in (3.30). The normalization coefficient $N^{-1/2}$, where N is the (large) number of superlattice unit cells, is also an approximation. It amounts to neglect the overlap integral between "atomic" functions centered at different z_j. Again it is meaningful only if the barriers are thick enough. If necessary couplings between nearest neighbors can be retained provided we make the replacement:

$$N \rightarrow N[1 + 2O_d \cos qd] \qquad (3.31)$$

with

$$O_d = \int_{-\infty}^{+\infty} dz\, f_n^o(z)\, f_n^o(z - d)$$

which decreases exponentially with the barrier thickness. We now insert (3.30) into the Schrödinger equation (3.26), multiply by $f_n(z)$ and integrate to obtain:

$$(\epsilon_n - \epsilon)(1 + 2O_d \cos qd) + \sum_j e^{iqjd}$$
$$x \sum_{m \neq j} \int_{-\infty}^{+\infty} dz f_n^o(z) V(z - md) f_n^o(z - jd) = 0 \qquad (3.32)$$

To simplify further we assume that the three-centers integral in (3.32) vanishes unless it reduces to a two-centers integral. This happens if $j = 0$ or $m = 0$ and leaves us with

$$(\epsilon_n - \epsilon)(1 + 2O_d \cos qd) + S_n + 2\sum_{j > o} t_n^{(j)} \cos jqd = 0 \qquad (3.33)$$

where the shift S_n of the discrete state ϵ_n is given by

$$S_n = \sum_{m \neq o} \int_{-\infty}^{+\infty} |f_n^o(z)|^2 V(z - md) \qquad (3.34)$$

and the transfer integral $t_n^{(j)}$ by

$$t_n^{(j)} = \int_{-\infty}^{+\infty} dz \ f_n^o(z) \ V(z) \ f_n^o(z - jd) \qquad (3.35)$$

An extra simplification is obtained if we neglect the transfers beyond nearest neighbors and the dispersion relations simplify into

$$\epsilon = \epsilon_n + \frac{(S_n + 2t_n^{(j)} \cos qd)}{1 + 2O_d \cos qd} \qquad (3.36)$$

If futhermore O_d is neglected (3.23) is recovered. The integrals O_d, S_n $t_n^{(j)}$ can be readily evaluated. Let us restrict ourselves to the ground superlattice subband. In a single well centered at $z = 0$ the ground state wavefunction is:

$$\begin{array}{llll}
f_1(z) & = & A \ \cos k_A z & |z| < L_{A/2} \\
f_1(z) & = & B \ \exp \kappa_B \ (z + L_{A/2}) & z \leq -L_{A/2} \\
f_1(z) & = & B \ \exp -\kappa_B \ (z - L_{A/2} \ ) & z \geq L_{A/2}
\end{array} \qquad (3.37)$$

$$\frac{\hbar^2 \kappa_B^2}{2m_B} = -\epsilon \ ; \quad \frac{\hbar^2 k_A^2}{2m_A} = V_s - \epsilon \ ; \quad 0 \geq \epsilon \geq -V_s \quad (3.38)$$

The continuity of f_1 and $1/m(z) \ df_1/dz$ at $z = -L_A/2$ and the normalization condition for $f_1(z)$ imply that:

$$\cos \frac{1}{2} k_A L_A - \frac{m_B}{m_A} \frac{k_A}{k_B} \sin \frac{1}{2} k_A L_A = 0 \qquad (3.39)$$

$$\Phi_A = k_A L_A;$$

$$\frac{B}{A} = \cos \frac{1}{2} \Phi_A; \qquad (3.40)$$

$$1 = \frac{B^2}{\kappa_B} + \frac{1}{2} A^2 L_A \left(1 + \frac{\sin \Phi_A}{\Phi_A} \right)$$

Then, with $\Psi_A = k_B L_A$, we obtain:

$$t_n^{(j)} = -2V_s ABe^{\Psi_A/2} \frac{e^{-\kappa_B jd}}{\kappa_B^2 + k_A^2}$$

$$\times \left\{ \kappa_B \cos \frac{1}{2} \Phi_A \sin h \frac{1}{2} \Psi_A + \kappa_A \sin \frac{1}{2} \Phi_A \cos h \frac{1}{2} \Psi_A \right\} \qquad (3.41)$$

$$s_1 = -2V_sB^2 \frac{\sin h\Psi_A}{\kappa_B} e^{\Psi_A} e^{-2\kappa_B d} (1- \exp -2\kappa_B d)^{-1} \quad (3.42)$$

$$O_d = B^2[\kappa_B^{-1} + de_A^\Psi]e^{-\kappa_B d} + 4ABe^{(\Psi_{A/2} - \kappa_B d)} (\kappa_B^2 + k_A^2)_x^{-1}$$
$$\left\{\kappa_B \cos \frac{1}{2}\Phi_A \sin h\frac{1}{2}\Psi_A + k_A \sin \frac{1}{2}\Phi_A \cos h\frac{1}{2}\Psi_A\right\} \quad (3.43)$$

3.2.5   Symmetry properties of the eigenfunctions[22]. The superlattice potential is unchanged under a reflection with respect to a line passing either at the center of a A layer or at the center of a B layer. The consecutive actions of two such reflections for neighboring A,B layers is equivalent to a superlattice translation of d

$$R_B R_A = \tau_d \quad (3.44)$$

The eigenfunctions f_q of the superlattice hamiltonian are eigenfunctions of τ_d with eigenvalue e^{iqd}. Hence

$$R_B R_A f_q(z) = e^{iqd} f_q(z) \quad (3.45)$$

The relation (3.45) is of special interest at $q=0$ and $q = \pi/d$. For $q = 0$ the product of the two reflections leaves the eigenfunction unchanged. Suppose that we deal with the ground superlattice conduction subband. We know that it is derived from the ground bound state in an isolated well which eigenfunction is symmetric with respect to the center of the A layer. We can then infer that

$$R_A f_{q=0}(z) = + f_{q=0}(z)$$

and deduce from (3.45) that

$$R_B f_{q=0}(z) = + f_{q=0}(z) \quad (3.47)$$

the eigenfunction $f_q(z)$ has no mode. On the other hand at $q = \pi/d$ we have $R_A R_B = -1$. If there is only a small mixing between the bound states of the isolated well due to tunneling across the barrier R_A $f_{\pi/d}(z) = + f_{\pi/d}(z)$ will still hold and (3.45) will lead to

$$R_B f_{\pi/d}(z) = -f_{\pi/d}(z) \quad (3.48)$$

the wavefunction has a node at the center of B layer.

These considerations are of relevance when on discusses inter-band optical transitions in a superlattice (for an explicit calculation based on the simple tight binding model developed previously see section 6). The oscillator strength of an optical transition between valence and conduction subbands involves the square modulus of the overlap integral $I_{cv}(q)$ between valence and conduction superlattice state of the same wave vector q. In type I superlattices, electrons and holes are mostly confined within the same layers. If I_{cv} (q=0) is non-vanishing (because valence and conduction eigenfunctions have the same parity) and if the parity with respect to the center of the A layer is preserved when q increases from 0 to π/d then I_{cv} (q=π/d) will also be non-vanishing (Fig. 5 left pannel). Under the same assumption of parity conservation between q=0 and q=π/d consider a type II super-lattice (Fig. 5 right pannel). The electrons are mostly confined in A layers, the holes in B layers. Assume that both $f^{(c)}_{q=0}$ and $f^{(v)}_{q=0}$ are even with respect to the middles of the A(B) layers. Then from (3.45) they are also even with respect to the middles of the B(A) layers and there-fore I_{cv} (0) is nonzero. At $q = \pi/d$, $f^{(c)}_{q=\pi/d}$, $f^{(v)}_{q=\pi/d}$ are still even with respect to the middles of the B(A) layers. Hence I_{cv} (q=π/d) vanishes by symmetry. In a type II superlattice, if the eigenfunctions retain the same parity with respect to the center of the layer where they are mostly confined when q describes the superlattice Brilluoin zone, one of the two interband transitions I_{cv} (q=0) or I_{cv} (q=π/d) is forbidden. In the InAs-GaSb system, the rigidity of the parity within a given layer holds in the whole semiconducting regime.[22] Apparently, it becomes weakly violated in semimetallic compounds[23] presumably because of the energy proximity of several interacting host subbands.

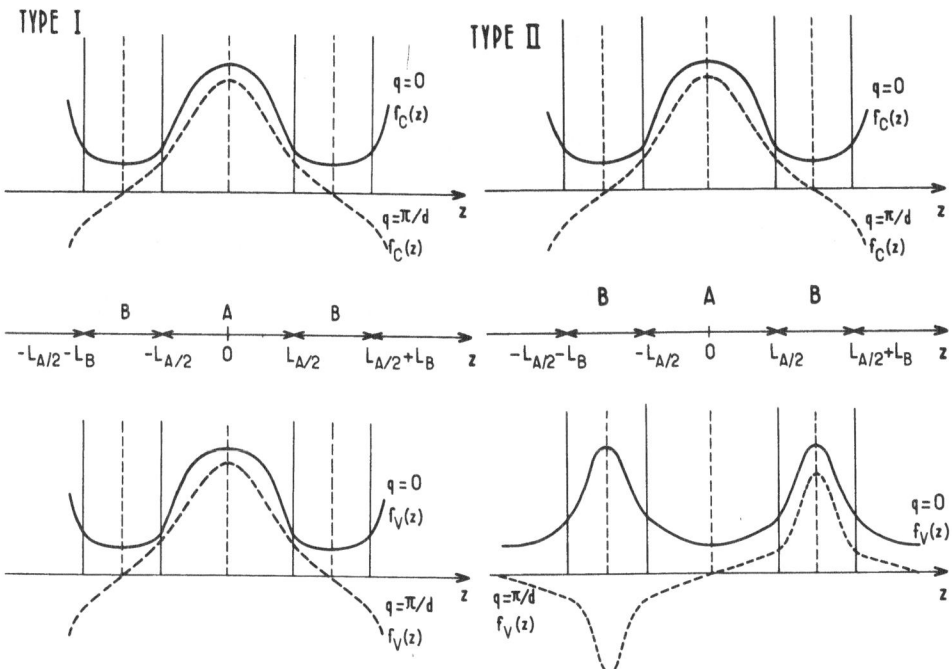

Figure 5 - Schematic electron wave functions at $q = 0$ and $q = \pi/d$ for valence and conduction subbands in a type I superlattice (left panel) and a type II superlattice (right panel).

3.2.6 Heavy hole superlattice states. So far we have only discussed the light particle ($m_J = \pm1/2$) dispersion relation. To obtain the heavy hole ($m_J = \pm3/2$) superlattice states we need to account for the δ matrix. At $k_\perp = 0$, it is diagonal with a matrix element between $(3/2, \pm 3/2)$ states given by (2.12). The same analysis as the one done for coupled $m_J = \pm1/2$ states can be performed for the heavy hole subbands. We find two fold degenerate heavy hole bands whose dispersion relations are again given by (3.14) with

$$\eta_{hh} = \frac{k_B}{M_B} \frac{M_A}{k_A} \qquad (3.49)$$

$$k_A = \left[\frac{2M_A}{\hbar^2} (-\epsilon - \epsilon_A) \right]^{1/2} \qquad (3.50)$$

$$k_B = \left[\frac{2M_B}{\hbar^2} (-\epsilon - \epsilon_A + V_P) \right]^{1/2} \qquad (3.51)$$

where M_A, M_B are the heavy hole effective masses in A and B materials respectively. The energy zero in (3.49-51) is still taken at the bottom of the S band of the A material. Depending on the energy range of interest k_A or k_B can become imaginary. Assume this is k_B. The heavy hole dispersion relations are then given by (3.21) with $\underset{\sim}{\eta}_{hh} = i\eta_{hh}$ and η_{hh} k_A, k_B given by (3.49-51).

4. SPECIFIC EXAMPLES

We will discuss in this paragraph the levels or subband positions of three different superlattice systems.

4.1 GaAs-Ga$_{1-x}$Al$_x$As

For this type I system, Dingle[24,25] derived the levels of isolated quantum wells assuming parabolic bands and equal effective masses in both host materials. Comparing the calculated energy levels with optical transmission data, he deduced the conduction band offset $V_s(x)$ between GaAs and Ga$_{1-x}$Al$_x$As. Dingle found:

$$V_s(x) = 0.85 \ x \ \left[\epsilon_0(x) - \epsilon_0(0) \right] \qquad (4.1)$$

where $\epsilon_0(x)$ is the Ga$_{1-x}$Al$_x$As band gap. Recently White and Sham[16] calculated the bound states of isolated GaAs quantum wells in the two-band model. They found an excellent agreement with Dingle's data[24,25]. With the same approximations as those made by Dingle, Palmier and Chomette[26] calculated the energy band of GaAs-Ga$_{1-x}$Al$_x$As superlattices. We present on Fig. (6) the results of

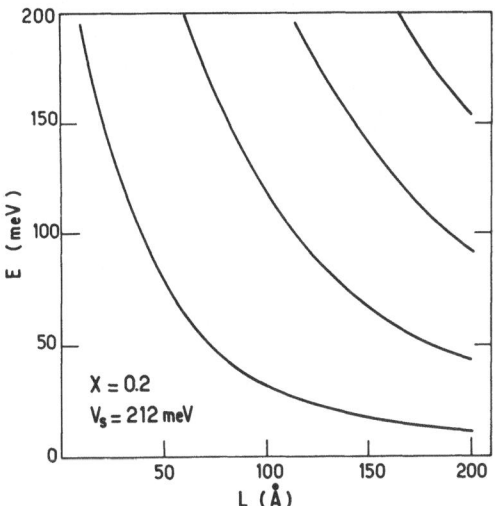

Figure 6 - Thickness dependence of the confinement energies of the bound states in a GaAs well clad between $Ga_{0.8}Al_{0.2}As$ barriers. The calculations are performed in the three-band model.

the three-band model applied to isolated GaAs quantum well. We have used equations (2.15-17. 22) taken $V_s(x)$ as given by (4.1) and used the room temperature values listed on Table 4. The Fig. (6) shows the results obtained for x = 20% (V_s = 212 meV) and 0 < L < 200Å. The dominant feature appearing on Fig. (6) is the effect of the finite barrier heights. For instance at L = 60Å we find confinement energy of $E_1 \simeq 63.8$ meV for the ground state which coresponds to $k_A L_A$ = 0.62π instead of π if V_s were infinite. For a given V_s and a given band-edge mass it is illustrative to compare the predictions of the three-band and of the parabolic models. We have plotted on Fig. (7) the L-dependence of the difference ΔE_1 = E_1^{para} - E_1^{3b} for the ground bound level. ΔE_1 vanishes at both L = 0 and L = ∞ limits and peaks near L = 30Å for x = 20%. The sign of ΔE_1 reflects the non-parabolicity effects which increases the apparent effective mass and therefore lower the confinement energy with respect to the parabolic band results. Both models are in rather close agreement (the maximum of ΔE_1 is ~ 4 meV). Obviously larger Al concentration allowing larger V_s and then larger E_1 will increase ΔE_1, as the non-parabolicity will become more important. However, the use of the three band model seems only required when detailed informations are needed; for instance when one wants to extract the exciton binding energy from optical absorption data. We have checked to the two-band and the three-band model give almost identical values: $|E_1^{3b} - E_1^{2b}| < 0.5$ meV for the previous example.

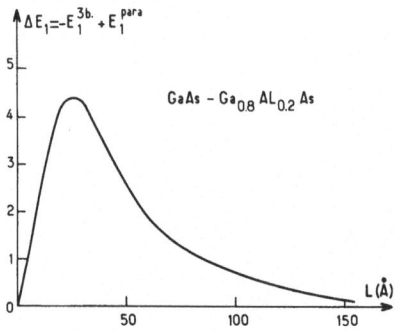

Figure 7 - Thickness dependence of the difference $E_1^{para} - E_1^{3b}$ between the confinement energies of the ground states of a GaAs quantum well clad bewteen $Ga_{0.8}Al_{0.2}As$ barriers calculated within the parabolic and the three-band models.

TABLE 4 - Room temperature band parameters of $Ga_{1-x}Al_xAs$ alloys used in section 4 (all energies are in meV).

$\epsilon_0(x)$	$= 1424 + 1247x$
$\Delta(x)$	$= 341 - 45x$
$V_S(x)$	$= 106x$
$m_c(GaAs)$	$= 0.064m_0$

4.2 The InAs-GaSb Superlattice

These superlattices are of special interest because of the peculiar band-edges ordering of InAs and GaSb. The InAs conduction edge is located in energy below the GaSb valence edge. The most commonly used value for the overlap between InAs (conduction) and GaSb (valence) bands is

$$\Lambda = E_v^{GaSb} - E_c^{InAs} = 0.15eV \qquad (4.2)$$

Hence, if InAs (GaSb) is the A(B) material of a A-B superlattice we have (at low temperature)

$$V_s = 0.96eV \ ; \ V_P = 0.57eV \ ; \ V_\delta = 0.198eV \qquad (4.3)$$

At first glance we expect to find the conduction (upper valence) electrons confined in the InAs (GaSb) layers. By increasing the InAs thickness, the conduction electron confinement energy will decrease. It will necessarily happen that the ground conduction bound state will pass below the top most valence electron level of the GaSb layers. This happens roughly when

$$\frac{\hbar^2 \pi^2}{2m_A^{(c)} L_B^2} = \Lambda - \frac{\hbar^2 \pi^2}{2M_B L_B^2} \qquad (4.4)$$

With (4.2), $m_A^{(c)} = 0.023 \ m_0$, $M_B = 0.33 \ m_0$ (GaSb heavy hole) and say $L_A = L_B$ we obtain $L_A \sim 108 \text{Å}$. For thicker InAs layers the GaSb valence electrons will not stay in the GaSb layers but will spontaneously flow in the neighbouring InAs layers leaving free holes in the GaSb slices. Two semiconductors when stacked together in a superlattice can then give rise to a <u>semimetal</u>.[6,27,28] The semimetallic configuration results from an intrinsic charge transfer as opposed to the extrinsic one found, e.g. in selectively doped GaAs-GaAlAs superlattices.[5] In both cases the charge transfer is accompanied by band bendings of opposite curvatures alternating from one layer to another layer. Note, however, that in the InAs-GaSb system the positive charges (holes) have a finite mobility, whereas, in the GaAs-Ga(Al)As system the ionized donors remain completely immobile.

Most of the experiments on InAs-GaSb superlattices have been performed in the presence of a strong magnetic field. This subject is covered in this volume by L.L. Chang[29] and will not be discussed here. For the purpose of subband calculations, we remark that the major difference between the InAs-GaSb and GaAs-GaAlAs materials lies in the necessity of hybridizing S and P electrons in the former system versus hybridizing S and S electrons in the latter system. A correct description of the hybridization requires then to use either a two-band or a three-band model. We show on Fig. (8) the evolution of the superlattice band structure of the InAs-GaSb superlattices, at $k_\perp = 0$, with increasing periodicity $d = L_A + L_B$, equal layer thickness: $L_A = L_B$, calculated in the two-band model.[15] Only the light particles are shown on this figure where the energy zero is taken at the bottom of the InAs conduction band. Heavy hole levels can be calculated separately according to (3.23, 49-51). The location of semiconductor $\rightarrow$ semimetal transition as defined by the onset of the charge transfers described previosuly occurs near $d \sim 170$Å. The band bending effects, effective for $d \gtrsim 170$Å, are not included in the calculations leading to Fig. (8). Noticeable in Fig. (8) is the anticrossing between the ground electrons (InAs) and light hole (GaSb) subbands which occurs near $d \sim 230$Å. Were the S-like and P-like bands treated as non-interacting this anticrossing would be replaced by a regular crossing.

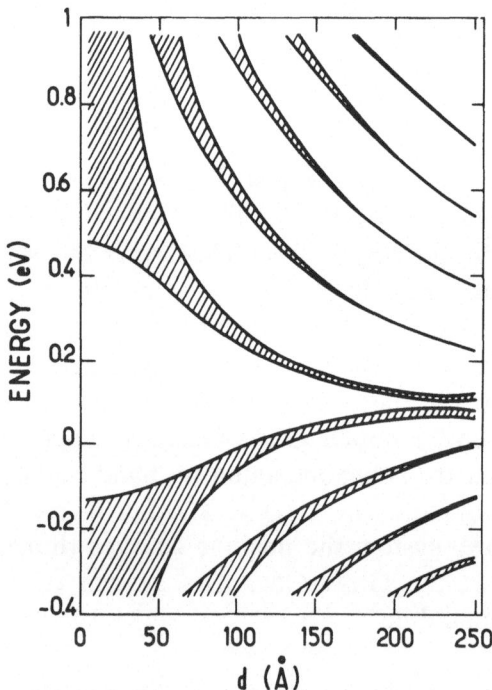

Figure 8 - Allowed energy bands (hatched areas) in InAs-GaSb super-lattices with equal layer-thicknesses and period d calculated in the two-band model. Only the light particle subbands are shown.

The two-band model assumes that the spin-orbit energies of the two host materials are much larger than, say, the valence energies of interest. The two-band model is then an acceptable approximation for states near the InAs conduction band edge but is unjustified for deep valence levels as, e.g. those located in energy near the InAs valence band edge, which corresponds in GaSb to valence states as deep as ~ 2/3 of the GaSb spin orbit energy. We therefore expect that the proximity of the GaSb split-off valence band will produce large perturbations to the superlattice states in this energy range. The three-band model has not yet been applied to the InAs-GaSb system.

4.3 The HgTe-CdTe System

This system starts to be fabricated[30] in view of applications as infrared sensors. HgTe is a zero-gap semiconductor with an inverted band structure, i.e. the S-like band lies below (by ~ 0.3 eV) the P-like (Γ_8) quadruplet. What was a light-hole band in regular semiconductor is now a conduction band, whereas the S-like band which is the conduction band in GaAs or CdTe has in HgTe a light-hole character (Fig. 9). The heavy hole band is as expected. CdTe has a standard, GaAs-like band structure. The valence band offset between HgTe and CdTe is unknown.

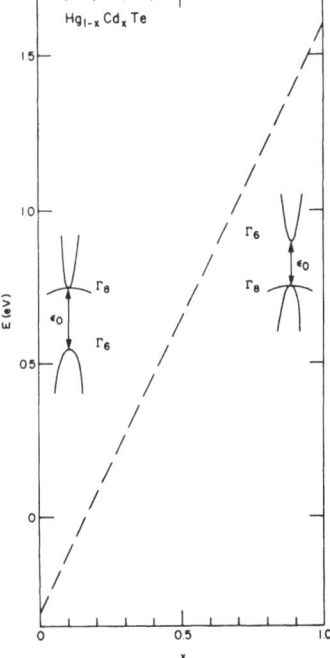

Figure 9 - Band structure of the ternary random alloys $Hg_{1-x}Cd_xTe$ (virtual crystal approximation).

The L.C.A.O. calculations would predict it is small because Te is a common anion and HgTe, CdTe have a very small lattice mismatch. The HgTe-CdTe superlattice band structure has first been calculated by Schulman and McGill[31], by tight-binding calculations, assuming a zero valence band offset. With such a zero valence band offset an interesting situation occurs for the light particle dispersion relations calculated in the envelope-function framework. Namely: we have to hybridize the states arising from two bands which have the same symmetry (P, $m_J = \pm 1/2$) and opposite curvature (electron-like in HgTe, light-hole -like in CdTe). A detailed analysis,[17] performed in the two-band model and including heavy holes, shows that the HgTe-CdTe superlattices should exactly behave as the ternary random alloys $Hg_{1-x}Cd_xTe$: for $\ell_{HgTe}/\ell_{CdTe} > 0.84./0.16$ the superlattices are zero-gap semiconductors like $Hg_{1-x}Cd_x$ alloys with $x \lesssim 0.16$. Conversely for $\ell_{HgTe}/\ell_{CdTe} < 0.84/0.16$ the HgTe-CdTe superlattices have a finite gap which increases with increasing CdTe layer thicknesses. The latter feature was also obtained by tight-binding calculations.[31]

5. SUPERLATTICE DISPERSION RELATIONS AT FINITE $k_\perp$

Once we have calculated the subband positions at $k_\perp = 0$ it is of primary importance to know the $k_\perp$ dispersion relations of the superlattice. Here, as always in this lecture, the subscript in $k_\perp$ refers to "perpendicular to the superlattice axis", that is to say the k - component parallel to the layer planes. Other authors used $k_{//}$ or k_t for our $k_\perp$. That the knowledge of $\epsilon_n (k_\perp)$ is important arises from the fact that most electrical or optical properties involve the in-plane effective mass. For instance, transport or magnetotransport practically takes place in the layer plane if the well coupling is too small; cyclotron resonance experiments probe the in-plane effective mass if the magnetic field is applied parallel to the superlattice axis etc... One immediately recognizes in Table 2 the messy situation brought about by non-zero $k_\perp$. This is further aggravated by the presence of δ, the higher band k.p matrix, which if $k_\perp \neq 0$ becomes very complicated. A finite $k_\perp$ induces coupling between light particles dispersion of different "spins". Furthermore, it leads to a coupling between light and heavy particles motions. The complexity is such that only numerical solutions are available. However, before resorting to these numerical solutions one should distinguish between: i) the semimetallic InAs-GaSb superlattices in which band non-parabolicity, heavy-hole-electron and heavy-hole-light hole degeneracies conspire to force us to a numerical calculation.[18]; ii) the superlattices which can display non-parabolicity effects but in which electron and hole subbands are well energy-separated (semiconducting InAs-GaSb, GaAs-GaAlAs superlattices). For the latter category an explicit, although tedious analytical calculations remains possible [17] because we may neglect δ when calculating conduction dispersion relations. This amounts to deal only with the Kane matrix (Table 2), which is a manageable task. Typically the error commited in such an approximation is $\sim$ (heavy hole confinement energy)/(energy separation between electron and heavy hole sub-bands). Within this approximate scheme, we need first to rederive the boundary conditions at finite $k_\perp$. This is again done by projection the 8x8 matrix (Table 2) onto the 2x2 subspace spanned by S↑ and S↓. The 2x2 effective hamiltonian is <u>not</u> diagonal, the off-diagonal terms reflecting the ↑, ↓ mixing arising from the combined actions of spin-orbit coupling and periodic variations of the band-edges. Let f_1, f_2 refer to the two envelope functions associated with each spin projection. We find that at the interfaces

$$f_1 , f_2 \tag{5.1}$$

$$\left[\epsilon_A + \epsilon - V_P(z)\right]^{-1} x \left[-\sqrt{2}\frac{df_1}{dz} - ik_+ f_2\right] \tag{5.2}$$

$$\left[\epsilon_A + \epsilon - V_P(z)\right]^{-1} x \left[-\sqrt{2}\frac{df_2}{dz} + ik_- f_1\right] \tag{5.3}$$

all are continuous. Moreover f_1, f_2 also fulfill the Bloch theorem. For a superlattice unit cell we have two interfaces, hence eight boundary conditions; but we have eight unknowns (two plane-wave amplitudes per layer and per spin). This is just right to provide us with a dispersion relations. In the two-band model it reads:

$$\tag{5.4}$$

$$\cos qd = \cos k_A L_A \cos k_B L_B - \frac{1}{2}\left[\frac{1}{\eta} + \frac{k_\perp^2}{4k_A k_B}(r + r^{-1} - 2)\right]$$

$$x \ \sin k_A L_A \sin k_B L_B$$

$$\eta = rk_A/k_B \ ; \ r = (\epsilon_A + \epsilon - V_P)(\epsilon_A + \epsilon)^{-1} \tag{5.5}$$

$$\frac{2P^2}{3}\hbar^2(k_A^2 + k_\perp^2) = \epsilon(\epsilon + \epsilon_A) \qquad (5.6)$$

$$\frac{2P^2}{3}\hbar^2(k_B^2 + k_\perp^2) = (\epsilon - V_s)(\epsilon + \epsilon_A - V_P) \qquad (5.7)$$

Two remarks should be made: i) the superlattice dispersion relations at finite $k_\perp$ are not only obtained by inserting $k_\perp^2$ in the host dispersion relations (5.6-7). There is an extra term in (5.4), explicitly $k_\perp$ - dependent which comes from the ↑, ↓ mixing in (5.2-3). ii) The super-lattice bands are not only shifted by $k_\perp$ as one expects but also deform with $k_\perp$ since $k_\perp$ affects k_A, k_B in a different way. This leads in partic-ular to a bandwidth dependence upon $k_\perp$. In the InAs-GaSb semicon-ducting superlattices the first electron subband width decreases with increasing $k_\perp$ (Fig. 10) Quite generally we may expand (5.4) in the vicinity of $k_\perp = 0$. To the lowest order we obtain

$$\epsilon_\nu(k_\perp, q) = \epsilon_{\nu o} + \frac{\hbar k_\perp^2}{2m_\nu} - 2t_{\nu o}(1 + \beta_\nu k_\perp^2)\cos qd, \quad k_\perp \to 0 \quad (5.8)$$

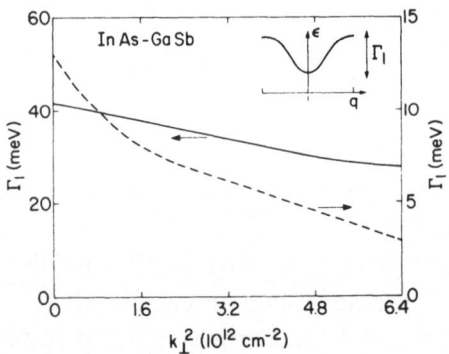

Figure 10 - Decrease of the E_1 bandwidth Γ_1 versus $k_\perp^2$ for two InAs-GaSb (A-B) superlattices: $(L_A, L_B) = 30\text{Å}, 50\text{Å}$ (left scale) and 65Å, 80 Å (right scale).

In (5.8) $\epsilon_{\nu o} - 2t_{\nu o}$ is the subband position at q, $k_\perp = 0$, m_ν will be the effective mass for carrier motion transverse to the superlattice axis, $4t_{\nu o}$ the $k_\perp = 0$ bandwidth. The term $(1 + \beta_\nu k_\perp^2)$ takes care of the band-width dependence with $k_\perp$, which in large gap materials like GaAs-GaAlAs can in a first approximation be neglected. The superlattice density of states $\rho_\nu(\epsilon)$ for the νth band is equal to

$$\rho_\nu(\epsilon) \;=\; \sum_{\sigma,\vec{k}_\perp,q} \delta(\epsilon - \epsilon_\nu(\vec{k}_\perp,q)) \tag{5.9}$$

σ is the "spin" index.   Assume for definiteness that m_ν, $t_{\nu0}$, β_ν are positive.   Then

$$\rho_\nu(\epsilon) \;=\; \begin{aligned} &\frac{\Omega}{4\pi^3} \int_{-\pi/d}^{+\pi/d} dq \int d^2k_\perp \; x \\[6pt] &\delta\!\left[\epsilon - \epsilon_0 - \frac{\hbar^2 k_\perp^2}{2m_\nu} + 2t_{\nu0}(1+\beta_\nu k_\perp^2)\cos qd \right] \end{aligned} \tag{5.10}$$

$$\rho_\nu(\epsilon) \;=\; \frac{\Omega}{\pi^2 d}\frac{m_\nu}{\hbar^2} \int_0^\pi dx \frac{Y(\epsilon - \epsilon_{\nu0} + 2t_{\nu0}\cos x)}{(1 - 4m_\nu t_{\nu0}\beta_\nu/\hbar^2 \cos x)} \tag{5.11}$$

where β_ν is assumed small enough to prevent the denominator in (5.11) to vanish.   For $\epsilon \le \epsilon_{\nu0} - 2t_{\nu0}$ $\rho(\epsilon)$ vanishes.   For $\epsilon \ge \epsilon_{\nu0} + 2t_{\nu0}$ $\rho(\epsilon)$ is a constant ρ_0 equal to

$$\rho_0 \;=\; \frac{\Omega m_\nu}{\pi d\hbar^2}\left[1 - 16m_\nu^2 t_{\nu0}^2\beta_\nu^2/\hbar^4\right]^{-1/2} \tag{5.12}$$

If we neglect β_ν we recover the familiar 2 dimensional density of states for a carrier with transverse effective mass m_ν, multiplied by L_z/d the number of superlattice unit cells in the crystal.   Finite β_ν only corrects the apparent transverse mass.   For $\epsilon_{\nu0} - 2t_{\nu0} \le \epsilon \le \epsilon_{\nu0} + 2t_{\nu0}$, $\rho(\epsilon)$ smoothly increases from 0 to ρ_0 displaying a three-dimensional behavior near the band edges:

$$\rho(\epsilon) \sim (\epsilon - \epsilon_{\nu0} + 2t_{\nu0})^{+1/2} \quad \text{near } q = 0 \tag{5.13}$$

$$\rho(\epsilon) \sim (\epsilon_{\nu0} + 2t_{\nu0} - \epsilon)^{+1/2} \quad \text{near } q = \pi/d \tag{5.14}$$

Note that for small $t_{\nu0}$ (i.e. quasi-isolated quantum wells) it may be difficult to experimentally distinguish between the smooth increase between 0 and ρ_0 due to finite bandwidth and the unavoidable band tailings of a purely two dimensional system (imperfect but isolated quantum wells).

6. BAND TO BAND ABSORPTION AND MAGNETOABSORPTION IN A SUPERLATTICE

Consider a binary superlattice AB and assume that it can be treated as weakly interacting quantum wells (see section 3.2.4). For convenience we write the ν'th and the νth valence subbands dispersion relations as

$$\epsilon^{(c)}_{\nu'k'_\perp q'} = \epsilon^{(c)}_{\nu'} + 2\lambda^{(c)}_{\nu'}(1- \cos q'd) + \frac{\hbar^2 k'^2_\perp}{2m_c} \tag{6.1}$$

$$\epsilon^{(v)}_{\nu k_\perp q} = - E - \epsilon^{(v)}_\nu - 2\lambda^{(v)}_\nu(1- \cos qd) - \frac{\hbar^2 k^2_\perp}{2m_v} \tag{6.2}$$

If a magnetic field H is applied parallel to the superlattice axis k^2 has to be replaced by $(2n+1)/ \ell^2$; n = 0, 1, 2 ... and $\ell^2 = \hbar c/e H$ is the magnetic length. The wave functions corresponding to (6.1, 2) are

$$<\vec{r}|c\nu'k'_\perp q'> = u_{co}(\vec{r})\frac{e^{i\vec{k}_\perp \cdot \vec{r}_\perp}}{\sqrt{L_x L_y}} \times \frac{1}{\sqrt{N}} \sum_{p=-\infty}^{+\infty}$$
$$e^{iq'pd}\, f^{(c)}_{\nu'}(z-pd) \tag{6.3}$$

$$<\vec{r}|v\nu k_\perp q> = u_{vo}(\vec{r})\frac{e^{i\vec{k}_\perp \cdot \vec{r}_\perp}}{\sqrt{L_x L_y}} \times \frac{1}{\sqrt{N}} \sum_{m=-\infty}^{+\infty}$$
$$e^{iqmd}\, f^{(v)}_\nu(z-md-\sigma\frac{d}{2}) \tag{6.4}$$

where u_{co}, u_{vo} are the periodic parts of the Bloch functions of the hosts, $L_x L_y$ the layers area and $f^{(c)}_{\nu'}, f^{(v)}_\nu$ the envelope functions which are solutions of the isolated well problems in both hands. The index σ is zero in type I superlattices (e.g. GaAs-GaAlAs) where electrons and holes are mostly confined in the same layers. σ is equal to 1 in type II superlattices (e.g. InAs-GaSb) where electrons and holes are mostly confined in adjacent layers. If a magnetic field is present, we replace in (6.3,4) exp $ik_x x$ by Ψ_n $(x+ \ell^2 k_y)$ where n = 0,1,2... and Ψ_n is the harmonic oscillator wavefunction of order n. An electromagnetic wave propagating along the superlattice axis induces interband transitions.

This leads to absorption or emission of light. The calculation of the transition rate $R_{inter}^{\nu\nu'}$ for interband transitions is straightforward. After some manipulations (which give the selection rules $q = q'$, $k'_y = k_y$, $k'_x = k_x$ (or $n' = n$ if $H \neq 0$)) $R_{inter}^{\nu\nu'}$ can be written

$$R_{inter} = A \sum_{\vec{k}_\perp, q} \delta . \left[\epsilon_{\nu k_\perp q}^{(c)} - \epsilon_{\nu k_\perp q}^{(v)} - \hbar\omega \right] |I_{\nu'\nu}^{cv}(q)|^2 \qquad (6.5)$$

$$I_{\nu'\nu}^{cv}(q) = \sum_{p=-\infty}^{\infty} e^{iqpd} \int_{-\infty}^{+\infty} dz f_{\nu'}^{(c)}(z) f_{\nu}^{(v)}(z - pd - \frac{\sigma d}{2}) \qquad (6.6)$$

where A is almost a constant in the energy range of interband transitions. We see at once in (6.5-6) that type I and type II superlattices behave differently. Retaining only nearest neighbour interactions in 6.6 we obtain $I_{\nu'\nu}^{cv}$ independent of q in type I superlattices ($\sigma = 0$ and only $p = 0$ contribute to 6.6). On the other hand in type II superlattices $\sigma = 1$; then $p = 0$ and $p = -1$ have to be retained in (6.6) and

$$I_{\nu'\nu}^{cv}(q) = \left[1 + (-1)^{\nu+\nu'} e^{-iqpd} \right] \qquad (6.7)$$

where the symmtery properties of the functions $f_\nu^{(c)}$ et $f_\nu^{(v)}$ have been exploited. The eq. (6.7) shows that if in a type II superlattice an interband transition is allowed at $q = 0$ it is forbidden at $q = \pi/d$ and vice versa. In type I structures, R_{inter} exactly reflects the joint density of states between the conduction and valence subbands, as $I_{\nu'\nu}^{cv}$ is a constant.

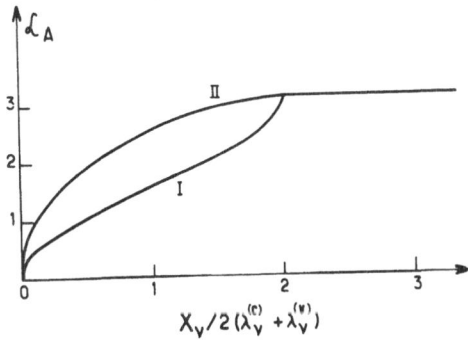

Figure 11 - Absorption lineshapes L_A (arbitrary units) in type I or type II superlattices. The conduction and valence subbands widths are $4\lambda_\nu^{(c)}$ and $4\lambda_\nu^{(v)}$ respectively. The quantity X_ν is equal to $\hbar\omega - E - \epsilon_\nu^{(c)} - \epsilon_\nu^{(v)}$.

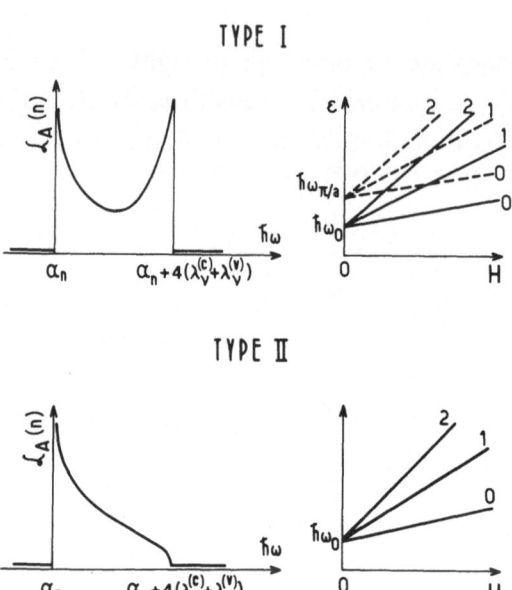

Figure 12 - Magneto-absorption lineshapes and related fan charts in type I and type II superlattices. The magnetic field is applied parallel to the growth axis. The conduction and valence subbands widths are $4\lambda^{(c)}$ and $4\lambda^{(v)}$ respectively. The quantity α_n is equal to E + $\epsilon_\nu^{(c)} + \epsilon_\nu^{(v)} + (n+1/2)\,\hbar eH\, m_c m_v/(m_c+m_v)$.

In type II superlattices this joint density of states is modulated by a factor $1 \pm \cos qd$, a feature which markedly affects the absorption (Fig. 11) and magnetoabsorption (Fig. 12) lineshapes. In magnetooptics two series of lines of comparable magnitude are expected in type I superlattices corresponding to $q = 0$ and $q = \pi/d$ resonances. The selection rules derived from (6.7) in type II structures makes the $q = \pi/d$ lines absent ($\nu+\nu'$ even). There has not been any detailed report of magnetoabsorption experiments in type I superlattices (but see (Ref. 32)). In semimetallic InAs-GaSb samples such experiments[23] have shown the existence of both the $q = 0$ and $q = \pi/d$ transitions, the latter being weaker than the former. The observation of a weak but finite $q = \pi/d$ transitions for $\nu' = \nu$ odd is not necessarily in contradiction with the previous statements. In semi-metallic InAs-GaSb superlattices there exist complicated interactions and mixings between electron-like, light hole-like, heavy hole-like bands of hosts. We conjecture that such band interactions invalidate the basic assumption underlaying the vanishing of the $q = \pi/d$ transitions; which is that both valence and conduction wavefunctions retain the same parity with respect to the center of the GaSb and InAs layers respectively when q increases from 0 to π/d. In semiconducting InAs-GaSb superlattices, the situation is more clear cut (subbands are well separated in energy) and we expect to see only either the $q = 0$ or the $q = \pi/d$ transitions.

7. COULOMBIC IMPURITY STATES AND EXCITIONS IN ISO-LATED QUANTUM WELLS

Once the eigenstates of a perfect superlattice have been calculated one should inquire about the effects of impurities, defects, etc.. Impurities are usually substutionnal, charged, coulombic centers. They act in two ways i) by deflecting the free carriers they limit their mobility (this matter will be extensively discussed by Gossard in this volume); ii) they can also produce bound electron states. The problem of coulombic bound states in a superlattice has not yet been solved, mostly because the usual effective-mass approximation fails badly as the impurity potential does not vary slowly at the scale of the superlattice unit cell. (For instance the effective Bohr radius for donors in GaAs is 100Å, i.e. comparable to a superlattice period). This significant spatial variation leads in the effective mass treatment to inter-subband terms which can be as large as intra-subband contributions. A second difficulty, also linked to the size of the super cell, is the large number of possible, non-equivalent, impurity sites inside the super cell. This contrasts with the bulk situation where impurities substitute only at one or two unequivalent lattice sites. This means that the impurity states are no longer degenerate with respect to the impurity position as they were in bulk materials. Instead, if there are N non-equivalent possible sites in the super cell, any reasonable calculation should produce N/2 distinct impurity levels (1/2 because of the mirror symmetry with respect to the center of the superlattice cell). In practice, N is usually so large that it is better to think of impurity bands instead of impurity levels. Note that these impurity bands are a purely geometrical effect and have nothing to do with the many-impurity effects found in highly doped bulk materials. Quite generally the impurity density of states will be defined as

$$\rho_{imp}(\epsilon) = 2\sum_{z_i} \delta\left[\epsilon - \epsilon(z_i)\right]\Psi(z_i); \quad 0 \le z_i \le \frac{d}{2} \qquad (7.1)$$

where $\epsilon(z_i)$ is the energy of the bound state (assumed to be unique) created by the impurity located at z_i and $\psi(z_i)$ an impurity doping profile function which accounts for the actual distribution of impurities inside a super cell. For a random distribution of impurities $\psi(z_i)$ is constant but one may also imagine $\psi(z_i)$ to peak or to vanish somewhere in the cell due to a selective doping of the superlattice unit cell. The coulombic bound state problem was discussed only in isolated quantum wells assuming first an infinite barrier[33] and more recently by taking into account finite barrier effects.[34] The photoluminescence associated with electron-neutral acceptor[35,36] or neutral donor-valence band recombination[36] has been recently studied in GaAs quantum wells clad between thick GaAlAs barriers. The Fig. (13) shows a comparison between Miller et al.[35] experimental determination of acceptor binding energy in GaAs wells of various thickness with calculations assuming infinite barrier height.[33] The agreement can be considered as acceptable owing to the crudeness of the calculations for acceptor levels.

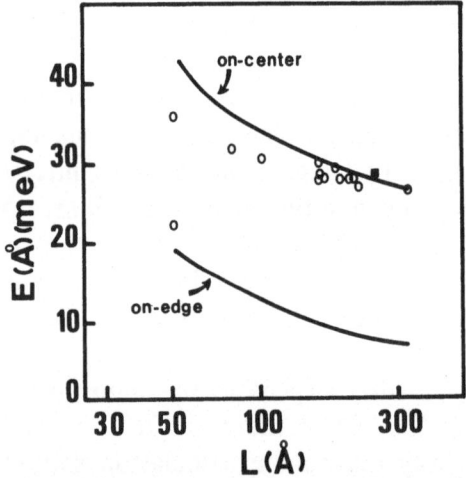

Figure 13 - Acceptor binding energy in GaAs quantum wells versus well thickness. The symbols correspond to experiments.[35] The solid line corresponds to a calculation assuming infinite barrier height.[33]

Similar experimental data were obtained by Lambert et al.[36] for acceptor or donor levels in GaAs quantum wells. The dominant feature is an increase of the binding energy with decreasing well thicknesses. This is not surprising since by decreasing L one evolves from a three dimensional (large L) to a two-dimensional (small L) coulombic problem and that is known that the two-dimensional hydrogen atom has a binding energy equal to four times the three-dimensional one. Obviously the two-dimensional limit is never really achieved because of the finiteness of the confining barrier. As pointed out by Mailhot et al.[34] the true limit at L = 0 is one time the <u>barrier</u> effective Rydberg. This implies that the coulombic binding energy has some maximum at a finite L which amplitude increases with increasing barrier height. In the specific case of $GaAs-Ga_{1-x}Al_xAs$, x ~ 40% L $\geq$ 50Å ensure that one may safely use the infinite barrier approximation. The exciton binding energy also increases markedly in type I quantum wells.[37,38] In type II materials the spatial separation between electrons and holes makes this increase a marginal one.[38] An exactly solvable model for exciton in quantum wells (contact exciton) has been reported by Satpathy and Altarelli.[39]

Besides coulombic impurities, interface defects (islands) have been reported to exist in GaAs quantum wells.[40] They are thought to ultimately control the exciton recombination linewidth in high quality materials.

8. CONCLUSION

This paper has been an attempt to describe the electronic band structure of superlattices in the envelope-functions approximation. We have not done justice to the alternative approach; the tight binding scheme, mostly because it does not allow analytical solutions of the superlattice problem. The reader should, however, remember that, despite its successes, the envelope function scheme has limitations. When it fails there is no other way but to handle the LCAO machinery. For superlattice and quantum well states not too far from the hosts Γ points the envelope function scheme works however well. It allows access to fine details, such as band non-parabolicity, which are of relevance in precise measurements. Its main virtue is its simplicity.

ACKNOWLEDGEMENTS

I am glad to thank the Ecole Normale Superieure (Y. Guldner, P. Voisin and M. Voos) and the IBM (C.A. Chang, L.L. Chang, L. Esaki and E.E. Mendez) groups for useful discussions on the envelope function scheme.

REFERENCES

A recent and complete review on the physics of two-dimensional electron gas can be found in: T. Ando, A.B. Fowler, F. Stern, Reviews of Modern Physics 54, 437 (1982).

1. L. Esaki and R. Tsu, IBM J. Res. Dev. 14, 61 (1970).
2. G.H. Döhler, H. Künzel, D. Olego, K. Ploog, P. Ruden, H.J. Stolz and G. Abstreiter, Phys. Rev. Lett. 47, 864 (1981)
3. K. Ploog, this volume.
4. R. Dingle, H. Störmer, A.C. Gossard and W. Wiegmann, Appl. Phys. Lett. 33, 665 (1978).
5. A.C. Gossard, this volume.
6. G.A. Sai-Halasz, L. Esaki and W. Harrisson, Phys. Rev. B18, 2812 (1978). For a recent review on InAs-GaSb see Y. Guldner, Proc. 16th Int. Conf. Phys. of Semiconductors, Montpellier 1982, to be published.
7. A.C. Cho, this volume.
8. C.A. Chang, this volume.
9. E.O. Kane, J. Phys. Chem. Solids 1, 249 (1957).
10. E.O. Kane in Semiconductors and Semimetals Vol. 1 edited by Willardson and Beer, Academic Press.
11. J.N. Schulman and Y.C. Chang, Phys. Rev. B24, 4445 (1981). A recent review on the theory of superlattices can be found in: J.N. Schulman and T.C. McGill, in "Synthetic Modulated Structure Materials", edited by L.L. Chang and B.C. Giessen (Academic Press, New York, in press).
12. C.R. Pidgeon in Handbook of Semiconductors, Vol. 2, edited by T.S. Moss, North Holland, 1980.
13. J.M. Luttinger, Phys. Rev. 102, 1030 (1956).
14. M.H. Weiler in Semiconductor and Semimetals, Vol. 16, edited by Willardson and Bee, Academic Press, 1981.
15. G. Bastard, Phys. Rev. B24, 5693 (1981).
16. S. White and L. Sham, Phys. Rev. Lett. 47, 879 (1981).
17. G. Bastard, Phys. Rev. B25, 7584 (1982).
18. M. Altarelli, Proc. 16th Int. Conf. Physics of Semiconductors, Montpellier, 1982, to be published.
19. T. Ando and S. Mori, J. Phys. Soc. Japan 47, 1518 (1979).
20. G.E. Marques and L.J. Sham, Surface Science 113, 131 (1982).

21. D.J. Ben Daniel and C.B. Duke, Phys. Rev. 152, 683 (1966) see also W.A. Harrisson, Phys. Rev. 123, 85 (1961).

22. P. Voisin, G. Bastard, M. Voos, L.L. Chang and L. Esaki, March Meeting APS (1982). See also P. Voisin, G. Bastard, M. Voos, Phys. Rev. B (1983) in the press.

23. J.C. Maan, Y. Guldner, J.P. Vieren, P. Voisin, M. Voos, L.L. Chang and L. Esaki, Solid State Commun. 39, 683 (1981).

24. R. Dingle, W. Wiegmann and C.H. Henry, Phys. Rev. Lett. 33, 827 (1974).

25. R. Dingle in Festkörperprobleme XV (Advance in Solid State Physics) edited by H.J. Queisser (Pergamon Viewag 1975).

26. J.F. Palmier, A. Chomette, J. Physique 43, 381 (1982).

27. L.L. Chang, N.J. Kawai, G.A. Sai-Halasz, R. Ludeke and L. Esaki, Appl. Phys. Lett. 35, 939 (1979).

28. Y. Guldner, J.P. Vieren, P. Voisin, M. Voos, L.L. Chang and L. Esaki, Phys. Rev. Lett. 45, 1719 (1980).

29. L.L. Chang, this volume.

30. J.P. Faurie and A. Million, Appl. Phys. Lett. 41, 264 (1982).

31. J.N. Schulman and T.C. McGill, Phys. Rev. 23, 4149 (1981).

32. R. Dingle, Surf. Sci. 73, 229 (1978).

33. G. Bastard, Phys. Rev. B24, 4714 (1981).

34. C. Mailhiot, Yia-Chung Chang and T.C. McGill, Phys. Rev. B26, 4449 (1982).

35. R.C. Miller, A.C. Gossard, W.T. Tsang and O. Munteanu, Phys. Rev. B25, 3871 (1982).

36. B. Lambert, B. Deveaud, A. Regreny and G. Talalaeff, Solid State Commun. 43, 463 (1982).

37. R.C. Miller, D.A. Kleinman, W.T. Tsang and A.C. Gossard, Phys. Rev. B24, 1134 (1981).

38. G. Bastard, E.E. Mendez, L.L. Chang and L. Esaki, Phys. Rev. B26, 1974 (1982).

39. S. Satpathy and M. Altarelli, Phys. Rev. B23, 2977 (1981).

40. C. Weisbuch, R. Dingle, A.C. Gossard and W. Wiegmann, Solid State Commun. 38, 709 (1981).

LIGHT SCATTERING IN SEMICONDUCTOR HETEROSTRUCTURES

G. Abstreiter

Physik-Department, Technische Universität München 8046 Garching, Fed. Rep. of Germany

ABSTRACT: We discuss the application of light scattering spectroscopy for the investigation of semiconductor thin films, heterostructures, and multilayer systems. Some aspects of the theory of phonon Raman scattering and of light scattering by free carrier excitations are presented. Special emphasis is made on the difference between two- and three-dimensional systems. The properties of the studied semiconductor systems are discussed together with the scattering geometries and the Raman spectrometer. The experimental results cover various different effects which allow the determination of electrical, structural or compositional properties of semiconductor heterojunctions. The quantization of carriers in multi-quantum well structures and at semiconductor surfaces and interfaces is investigated for several systems. The discussion of the tunable optical and electrical properties of doping superlattices closes this review.

1. INTRODUCTION

Light scattering in solids has been used extensively to study elementary excitations such as phonons, plasmons, magnons as well as interactions between them. From the measured spectra one can obtain information on the energy, the intensity, the lineshape, and the polarization properties of the excitations. The potential of Raman scattering as a tool for the investigation and characterization of various types of crystalline and amorphous solids became clear with the invention of visible lasers and especially when reliable cw dye lasers were commercially available. During the past ten years there was an enormous increase of interest in this type of work. The experimental and theoretical developments in this field have been collected and reviewed recently in four volumes of the series "Topics in Applied Physics" edited by M. Cardona and G. Güntherodt.[1-4] The present work presents in a comprehensive form special aspects of inelastic light scattering as an important and versatile tool to study semiconductor heterojunctions and multilayer systems as well as semiconductor surface and bulk properties. It is based on several extensive review articles.[5-8]

In Sec. 2 we discuss some basic concepts of light scattering. Phonon Raman scattering is presented both in a phenomenological way which describes the symmetry properties of the excitations and in a microscopic description which yields information on the coupling mechanisms. After the discussion of the phonon properties there follows a more extensive discussion of light scattering by free carriers. The nature of collective and single particle excitations, the coupling to LO-phonons, the resonance behavior, and the characteristic differences of two- and three-dimensional electron systems is treated theoretically. Section 3 is concerned with the specific optical and electronic properties of the samples studied. It also contains some information on scattering geometries and instrumentation for Raman spectroscopy.

In the main part of this lecture (Sec. 4) we discuss in detail experimental results which have been obtained for various different types of semiconductor structures. The usefulness of phonon Raman scattering for the characterization of single- and multilayer systems of, for example, GaAs and $Al_xGa_{1-x}As$ is demonstrated. The scattering of plasmon-like excitations leads to information on carrier concentration or effective mass and electron damping. The most exciting results, however, have been obtained for quasi-two-dimensional carrier systems during the past years. Two-dimensional electrons can exist at semiconductor-semiconductor interfaces, in metal-insulator-semiconductor (MIS) structures and in semiconductor multilayer systems. Their electronic properties have been reviewed by Ando, Fowler, and Stern.[9] Here we discuss the light scattering properties of such systems which were first proposed by Burstein et al.[10] and shortly after verified experimentally.[11] Electronic excitations have been studied during the past four years for carriers confined in GaAs-Al$_x$Ga$_{1-x}$As single heterojunctions and multilayer structures,[11-19] at Ge-GaAs interfaces,[20] in MIS-structures on InAs[21,22] InP,[23] and Si[24,25] and in GaAs doping superlattices.[26-28] Light scattering by two-dimensional carrier systems has been investigated theoretically.[29] We select a few examples of these experiments and discuss the informations which have been extracted from the measured single particle and collective carrier excitations. The section ends with the discussion of the optical and electrical properties of GaAs doping superlattices, which has been called recently an "exciting landmark in semiconductor physics."[30]

2. FUNDAMENTALS OF LIGHT SCATTERING IN SOLIDS

2.1. Phonon Raman Scattering

2.1.1. Phenomenological aspects. The interaction of light with a solid is described by the electric susceptibility χ. The electric field E of the incident light wave induces an oscillating dipole moment P which is given by

$$\vec{P} = \chi(\omega,\vec{k})\vec{E} \tag{1}$$

χ is a tensor which in general depends on the frequency ω and the wave vector k The electric susceptibility is closely related to the dielectric function

$$\varepsilon(\omega,\vec{k}) = 1 + 4\pi\chi(\omega,\vec{k}) \tag{2}$$

The emitted power of an oscillating dipole is classically proportional to the time average of the second derivative of the dipole moment P. If χ is stationary, then P oscillates with the frequency of the incident light wave ω_i. This describes transmission and reflection of light in a solid (Fig. 1). For inelastic light scattering one has to consider the modulation of χ due to fluctuations in space and time of, for example, phonons. The lattice oscillations are given by the time dependent displacement of the atoms.

$$u(\vec{r},t) = u_0 e^{i(\vec{q}\ \vec{r}+\Omega t)} \tag{3}$$

where u_0 is the unexcited position, q is the wave-vector of the phonon with frequency Ω. In the harmonic approximation the susceptibility χ can be developed with respect to the phonon amplitudes. The higher order terms of χ lead to inelastic light scattering (Brillouin- and Raman scattering). Scattering in first order occurs with frequency $\omega_s = \omega_i \pm \Omega$ and wave vector $k_s = k_i \pm q$, where $\omega_{i,s}$ and $k_{i,s}$ are the frequencies and wave vectors of the incident and scattered light, respectively. The minus sign stands for stokes scattering, the plus for antistokes. In opaque crystals one usually has to apply back scattering geometry (Fig. 1), such that $|\vec{q}| \approx |k_i| + |k_s|$, which in general is much smaller than the Brillouin zone boundary. Therefore in first order phonon Raman scattering one observes only excitations with wave vector $|q| \approx 0$.

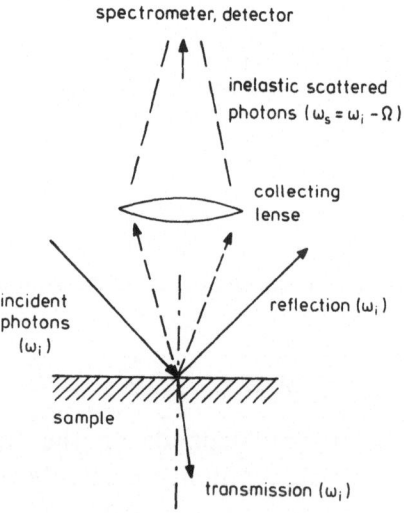

Figure 1 - Schematic diagram of the geometry for reflection, transmission, and inelastic light scattering.

a.)

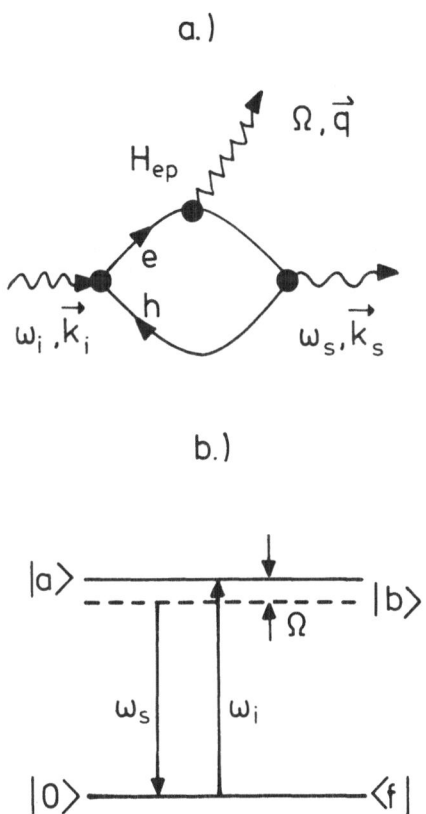

b.)

Figure 2 - Feynman diagram and energy level scheme for phonon Raman scattering.

2.1.2. Microscopic aspects. The Raman process for phonons is described by the Feynman diagram of Fig. 2a. The incident photon ω_i creates an electron-hole pair. In the excited state the electron (or the hole) emits a phonon Ω caused by the electron-phonon interaction H_{ep}. The scattered photon ω_s is created by the recombination of the electron-hole pair. The energy scheme for these processes is shown in Fig. 2b. If the groundstate and the excited state are real, one talks about resonant Raman scattering. For the transition susceptibility one can write

$$\chi \sim \; < f|\, \vec{E}_s \vec{p}|b > \; \frac{<b|H_{ep}|a> \; <\vec{p}|\, \vec{E}_i \,|0>}{(\omega_i - E_a)(\omega_i - \Omega - E_b)} \tag{4}$$

where $\vec{p}$ is the momentum operator, $\vec{E}_{i,s}$ are the electric field vectors of the incident and scattered photons, $E_{a,b}$ are the energies of the excited states, respectively. H_{ep} is the Hamilton operator for electron-phonon coupling, which for the deformation potential interaction describes the periodic modulation of the electronic states due to the phonon induced lattice distortion. There are, however, also other possible interaction mechanisms like electro-optic effect or Fröhlich interaction. For the latter the macroscopic electric field of the LO-phonons in polar semiconductors is important. It is also essential for the discussion of the so-called "forbidden" Raman scattering which has been used to study surface barrier heights.

2.1.3. Selection rules. The selection rules for allowed Raman scattering are determined by the structure and symmetry of the transition susceptibility χ, which is also called Raman tensor. These are listed for all crystal classes in (31). For crystals with diamond or zinc-blende type structure like GaAs there exist three irreducible components of the Raman tensor R. The optical phonons at $k \approx 0$ have Γ_{15}-symmetry. The Raman tensor component for this symmetry is

$$\Gamma_{15} : \begin{matrix} 0 & d & 0 \\ d & 0 & 0 \\ 0 & 0 & 0 \end{matrix} \ , \ \begin{matrix} 0 & 0 & d \\ 0 & 0 & 0 \\ d & 0 & 0 \end{matrix} \ , \ \begin{matrix} 0 & 0 & 0 \\ 0 & 0 & d \\ 0 & d & 0 \end{matrix}$$

The first order selection rules depend on the scattering configuration and can be evaluated by multiplying the polarization vectors of the incident and scattered light with the Raman tensor component of the corresponding phonon symmetry. In nonpolar crystals like Si or Ge the TO and LO-phonons are degenerate at $k \approx 0$. In polar crystals like GaAs on the other hand, the macroscopic electric field associated with the LO-phonons leads to an LO-TO splitting. From the selection rules it follows that in backscattering geometry from (100) surfaces scattering by LO-phonons is allowed. Scattering by TO-phonons on the other hand is symmetry-forbidden. This is opposite in backscattering from (110) surfaces. The symmetry properties of phonon Raman scattering have been used to obtain information on the crystal orientation of thin epitaxial films of GaAs.[32]

2.1.4. Forbidden LO-phonon scattering. The selection rules just discussed can be violated by different possible symmetry-breaking mechanisms.[33]

1. Intraband scattering of electrons by LO-phonons via the Frölich interaction, which becomes allowed for finite q-vectors of the LO-phonons.
2. Forbidden scattering induced by the presence of impurities in the sense that the electron is scattered elastically by an impurity to provide the necessary momentum change.
3. Electric-field induced Raman scattering which can be described with a Franz-Keldysh-type theory.

All three mechanisms are especially important for LO-phonon Raman scattering under resonance condition. Forbidden LO-phonon scattering has been used for example to study surface barrier heights on clean and oxygen covered (110) cleavage surfaces of GaAs.[34]

2.2 Light Scattering by Free Carriers (Three-Dimensional)

The theory of light scattering by electron plasmas in solids has been developed already in the early sixties.[35] The scattering cross section was found to be related to the spectrum of density fluctuations. At higher densities, however, the one-electron excitations are modified by dynamical screening effects with the longitudinal polarization of the plasma. It has also been recognized that the band structure influences light scattering in various ways. The resonance behavior of the scattering cross section close to optical interband energy gaps and the spin-orbit interaction of the valence bands have opened the possibility to observe excitations of single-particle character also at high electron densities. The first observation of laser light scattering by a solid state plasma was reported by Mooradian and Wright[36] in doped n-GaAs.

2.2.1. **Single-particle excitations.** The one-electron excitation in a solid state plasma is shown schematically in Fig. 3. The necessary momentum change $\vec{q}$ is provided by the wave vectors of the incident and scattered light k_i and k_s. In backscattering geometry one obtains

$$|\vec{q}| = |\vec{k}_1| + |\vec{k}_s| \approx 2\frac{2\pi n}{\lambda_L} \qquad (5)$$

where n is the refractive index and λ_L is the wavelength of the incident and scattered light which here is taken approximately as the laser wavelength. In GaAs $|\vec{q}|$ can be varied from $\simeq 0$ cm^{-1} to $\simeq 10^6$ cm^{-1} by using different laser excitation lines. For a three-dimensional elec-

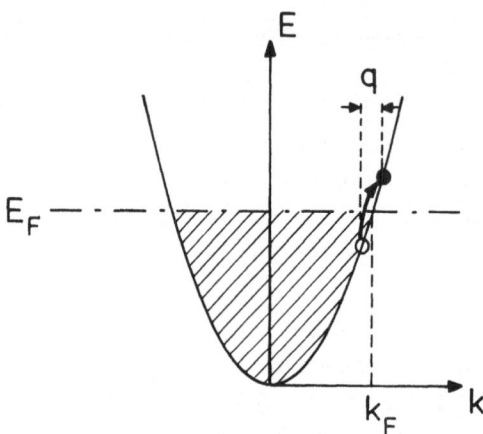

Figure 3 - Single-particle excitation process of electrons in a parabolic band.

tron plasma the integration over all possible excitations with given $\vec{q}$ leads to a triangular shaped spectrum with a cut-off around $q \cdot v_F$, where $v_F = \hbar k_F/m^*$ is the Fermi velocity. This is shown schematically in Fig. 4. The triangular line shape is smeared out due to finite damping. The single-particle excitation spectrum has been found to be proportional to the imaginary part of the dielectric function Im $\varepsilon(\omega, q)$ (see (8)). To include finite damping in the calculations the Lindhard-Mermin approximation was introduced to calculate the lineshape function Im $\varepsilon(q, \omega)$.[37]

There exists a class of excitations of single-, as well as multicomponent carrier systems which carry no net fluctuation in charge density and consequently have single-particle character. These are spin-density fluctuations in which the spin of the electrons is changed via the spin orbit interaction.[38] In GaAs these excitations become dominant for photon energies close to the $E_o + \Delta_o$ energy gap. It has also been predicted that light scattering by single-particle excitations without a spin-flip is possible in the case of nonparabolic electron bands.[39]

2.2.2. Collective excitations. Electron excitations which carry a charge density fluctuation are usually dynamically screened by the electron plasma. The scattering lineshape is closely related to Im $1/\varepsilon(q, \omega)$ which peaks around the plasma frequency $\omega_p(q)$. The dispersion of ω_p

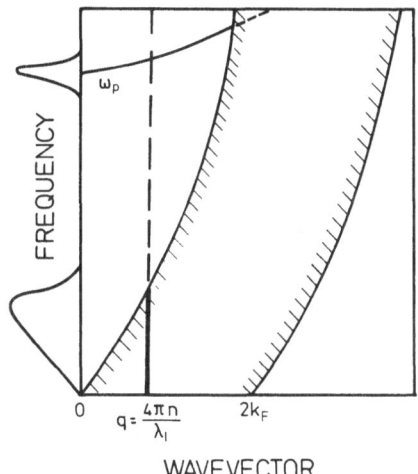

Figure 4 - Single-particle region and plasmon dispersion for electrons in a parabolic band.

is included in Fig. 4. In polar semiconductors the longitudinal plasma oscillations are coupled to the LO-phonons due to the macroscopic electric field. If we neglect damping and use a simple Drude expression for the dielectric function of the electron gas, one can write the total dielectric function:[35]

$$\varepsilon(q,\omega) = \varepsilon_\infty \left(\frac{\omega^2_{LO} - \omega^2}{\omega^2_{TO} - \omega^2} - \frac{\omega^2_p(q)}{\omega^2} \right) \qquad (6)$$

lattice   electrons

with $\omega^2_p(q) = \omega^2_p + 3/5(qv_F)^2$ and $\omega^2_p = 4\pi ne^2/\varepsilon_\infty m^*$.

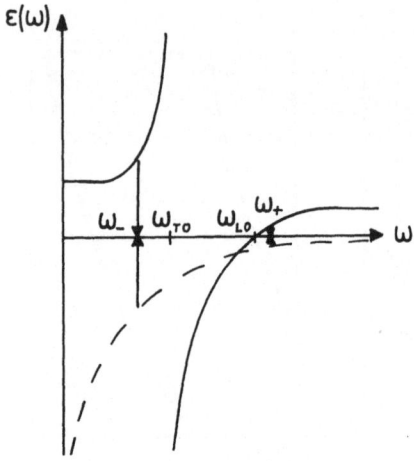

Figure 5 - Dielectric function $\varepsilon(\omega)$ separated for the electronic and lattice parts.

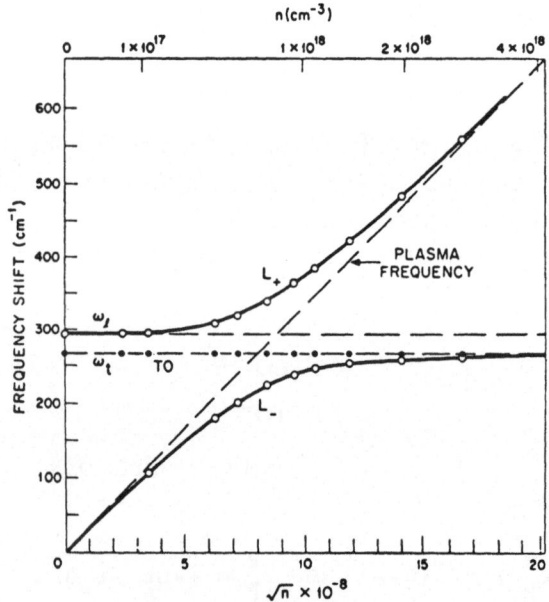

Figure 6 - Experimental and theoretical results for the coupled modes versus carrier concentration (After Ref. 36).

The frequencies of the resulting coupled phonon plasmon modes are determined by the zeros of Eq. 6. This is illustrated in Fig. 5. In Fig. 6 we show experimental and theoretical results of the coupled modes $\omega\pm$ for $\varepsilon(q\to o,\omega)$.[36] If one approaches the region of single-particle excitations and if damping is not negligible, one has to evaluate the peaks of Im $1/\varepsilon$. This has been done successfully using the Lindhard-Mermin dielectric function of the electron gas.[40]

2.2.3. *Resonance behavior.* Excitations by free carriers show a strong resonance enhancement close to optical gaps where carrier occupied states are involved in the transitions. In GaAs this has been studied extensively for laser excitation lines close to the $E_0 + \Delta_0$ energy gap.[8] In Fig. 7a we show the transitions which are important for the spin-flip single-particle excitations. In the first step an electron is excited from the spin orbit split-off valence band to the conduction band above the Fermi energy. An electron with opposite spin can recombine with the hole in the valence band whose wave function includes both spin directions due to spin-orbit interaction. Spin-flip excitations are found to be antisymmetric, the polarizations of the incident and scattered light are crossed. The scattering cross-section of these types of excitations is strongly enhanced at the $E_0 + \Delta_0$ energy gap of GaAs. As a consequence of the study of this resonance behavior it was suggested in Ref.

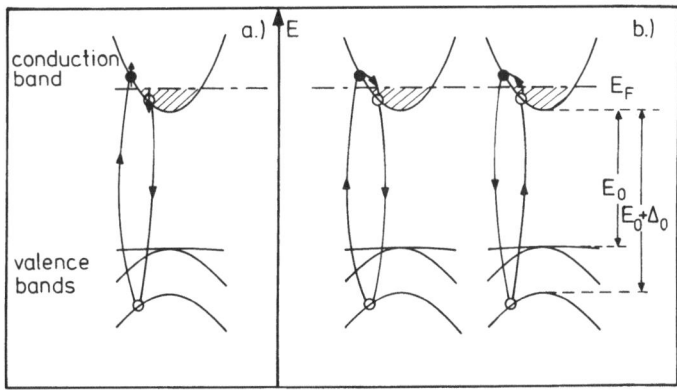

Figure 7 - Transitions involved in resonant electronic Raman scattering are shown schematically.

(37) that it should be possible to study also two-dimensional electron systems confined at semiconductor surface space charge layers.

In Fig. 7b we show transitions involved in scattering by collective excitations. The resonance condition is fulfilled for both incident and scattered photons. Therefore the resonance condition is not as sharp as for single-particle excitations. The creation of collective longitudinal excitations can be understood as the creation of single-particle electron-hole pairs by the incident photons which then emit plasmons or coupled phonon-plasmon modes. These excitations can also be observed at energy gaps which do not involve carrier occupied states via the phonon scattering mechanisms.

2.3 Light Scattering by Free Carriers (Two-Dimensional)

Two-dimensional carriers are characterized by the separation of motion perpendicular and parallel to the direction of quantization. While in the parallel direction the usual dispersion of the bands is maintained in the effective mass approximation, the carriers are bound in subbands with minimum energies E_0, E_1, E_2 ... in the direction perpendicular to the potential well. In Fig. 8a we show schematically a

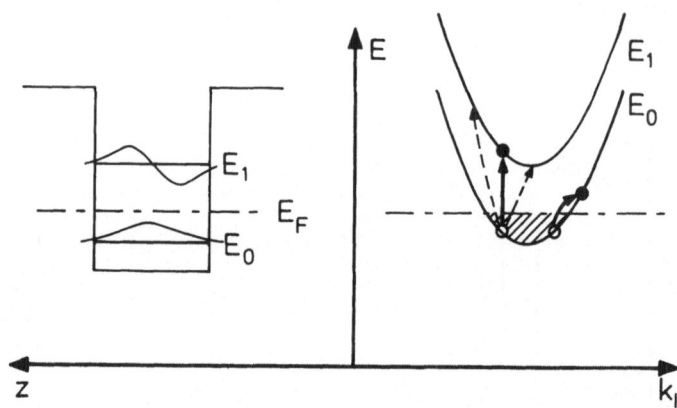

Figure 8 - One-dimensional rectangular potential well and dispersion of two subbands in the parallel direction. Also shown are single-particle excitation processes.

one-dimensional, rectangular potential well with finite height. E_0 and E_1 are subbands, E_F the Fermi energy. The more realistic nature of one-dimensional potential wells at semiconductor surfaces and hetero-junctions is discussed later. The dispersion of the subbands in $k_\parallel$ is shown in Fig. 8b. Depending on the scattering wavevector one can create both intra- and inter-subband excitations.

2.3.1. *Intrasubband excitations.* Excitations within one subband are only possible, if there exists a component of the scattering wavevector in the $q_\parallel$ direction. Similar to the three-dimensional case one can create both single-particle and collective excitations. Because of the two-dimensional nature of the electron system there exist, however, basic differences of the excitation spectra. The integration over all possible single-particle excitations for a given $q_\parallel$ differs from the lineshape found for the three-dimensional case. This is shown schemat-ically in Fig. 9. The collective intrasubband excitations are two-dimensional plasma oscillations of the electrons parallel to the surface or interface. The frequency of this plasmon tends to zero with decreas-ing $q_\parallel$. So far only collective intrasubband excitations have been observed in GaAs-Al_xGa$_{1-x}$As multilayer structures.[41]

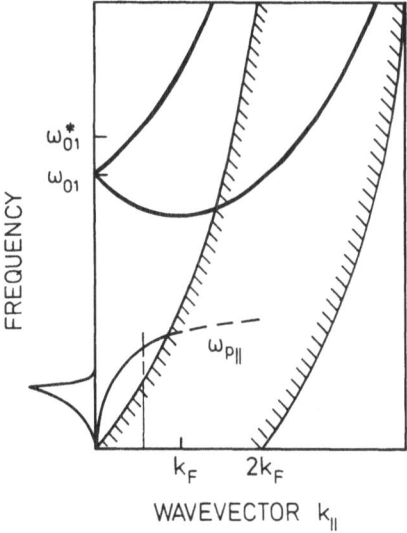

Figure 9- $k_\parallel$ -dispersion of electronic excitation in two-dimensional carrier systems.

2.3.2. Single-particle intersubband excitations. Single-particle inter-subband excitations are uncorrelated excitations of an electron below the Fermi energy in a lower subband to an empty state in the higher subband. Similar to the three-dimensional case, these unscreened excitations can be observed when the scattering occurs via spin-density fluctuations. The measured energies directly correspond to the subband splitting, if the scattering wavevector is perpendicular to the direction of quantization. This is usually the case in backscattering geometry. An additional component of the wavevector parallel to the layers leads to a strong broadening of the spin-flip single-particle intersubband excitations. This is shown schematically in Fig. 9. Subband energies have been determined with this method for various two-dimensional carrier systems which exist in semiconductor heterostructures like GaAs, InP, Si, Ge. In these materials the resonance condition for spin-flip excitations could be fulfilled with conventional Raman spectrometers.

2.3.3. Collective intersubband excitations. The collective intersubband excitations of a two-dimensional plasma reflect in a way the finite extension of the carrier system in the direction of quantization. The collective excitations involve charge density fluctuations and therefore are dynamically screened by Coulomb interactions. This screening causes an upward shift with respect to the single-particle excitations. The effect is called "depolarization shift" and has been treated theoretically.[42-44] It describes the dielectric response of the thin layer of carriers to the electron-hole excitation and can be written as an effective plasma frequency ω^*_p perpendicular to the layer. For a two-level model one finds

$$\omega_p^{*2} = \frac{8\pi n_s e^2}{\varepsilon \hbar} \omega_{01} f_{11} \tag{7}$$

where f_{11} is the Coulomb integral of the wavefunctions of the two subbands involved. The measured subband excitation is then given by

$$\omega_{01}^{*2} = \omega_{01}^2 + \omega_p^{*2} \tag{8}$$

The investigation of single-particle and collective inter-subband excitations consequently leads to direct information on Coulomb matrix elements in two-dimensional carrier systems.

In polar semiconductors the collective excitations are coupled to the LO-phonons. Similar to the three-dimensional case the coupled mode frequencies can be determined from the zeros of the total dielectric function

$$\varepsilon(q,\omega) = \varepsilon_\infty \left(\frac{\omega_{LO}^2 - \omega^2}{\omega_{TO}^2 - \omega^2} + \frac{\omega_p^{*2}}{\omega_{01}^2 - \omega^2} \right) \qquad (9)$$

Damping is neglected in Eq. 9. The dielectric function of the electrons is expressed in analogy to the three-dimensional case, discussed above, by a Drude-like expression where ω_p is replaced by ω_p^* and the denominator resonates at $\omega = \omega_{01}$. Equation 9 yields two coupled modes which are plotted in Fig. 10 (q = 0). For $\omega_{01} << \omega_{LO}$ one finds

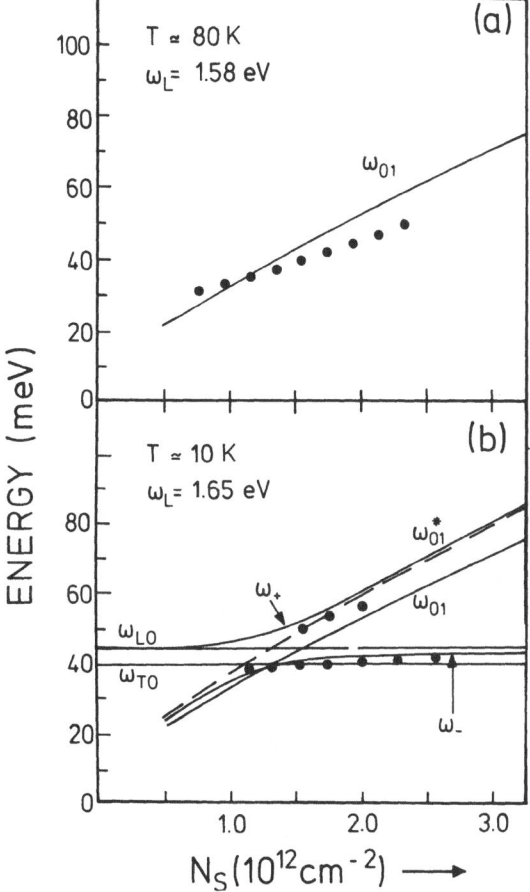

Figure 10 - Calculated and experimental results for single-particle and collective excitations of electron accumulation layers in InP (After Ref. 23).

$\omega_- \simeq \omega^*_{01}$, while for the opposite case $\omega_{01} >> \omega_{LO}$ the high frequency ω_+ -mode approaches ω^*_{01}. Contrary to the three-dimensional case, the ω_- -mode crosses the frequency of the transverse optical phonon ω_{TO} when $\omega_{01} = \omega_{TO}$ and falls in the reststrahlen region for higher subband splittings. In Fig. 10 we show both the bare subband splitting ω_{01} and the coupled modes versus number of carriers for an electron accumulation layer on InP as calculated self-consistently.[23]

3. EXPERIMENTAL ASPECTS

3.1. Sample Characteristics

In Section 2.2.3. we have seen that for "electronic Raman scattering" one usually has to work under resonant conditions. In Fig. 11 the band structure with the relevant optical energy gaps is shown for the example GaAs. Transitions which involve carrier occupied states are only possible for laser excitation energies close to E_0 and $E_0 + \Delta_0$.

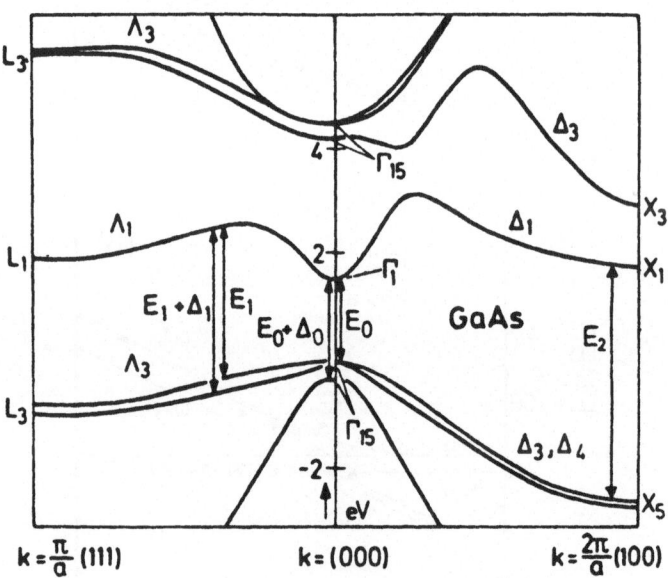

Figure 11 - Band structure of GaAs.

In Table 1 we have collected the relevant optical gaps for various semiconductors of interest.

Table 1 Optical energy gaps relevant for resonant inelastic light scattering via carrier density mechanisms

GaAs	electrons	$E_0 = 1.51 eV$, $E_0 + \Delta_0 = 1.85 eV$
GaAs	holes	$E_0 = 1.51 eV$
Ge	electrons	$E_1 = 2.2 eV$, $E_1 + \Delta_1 = 2.4 eV$
Si	electrons	$E_2 = 4.35 eV$
Si	holes	$E'_0 = 3.35 eV$
InP	electrons	$E_0 = 1.41 eV$, $E_0 + \Delta_0 = 1.52 eV$

Above the fundamental optical energy gap the semiconductors are opaque. The information depth of Raman spectroscopy is determined by the penetration depth of the laser light. Very often this is much less than $1 \mu m$. Consequently only properties of semiconductor heterojunctions very close to the surface can be investigated using resonant inelastic light scattering techniques. The design parameters and the electrical properties of semiconductor heterostructures studied are discussed for selected examples together with the experimental results in Sec. 4.

3.2. Scattering Geometries

In opaque semiconductors the most widely used scattering geometry is backscattering, where the wave vector of the incident light is normal to the surface and the scattered light is collected in backward direction. Because of the high refractive index ($n \sim 4$) of the semiconductors studied, nearly ideal backscattering conditions are fulfilled even when the incident laser beam is focussed under an angle of 45 degrees to the normal of the surface plane (see Fig. 1). In order to provide a larger component of the scattering wave vector parallel to the surface the incident light can be focussed on to the surface under glancing incidence. This opens the possibility to study also in-plane excitations of a two-dimensional electron gas.

3.3. Raman Set-Up

For inelastic light scattering in semiconductors conventional Raman spectrometers can be used. The incident light is provided by cw-ion gas lasers or by cw-dye lasers. To fulfill the resonance conditions for various semiconductors, photon energies covering the region from near ultraviolet to infrared have to be available. The laser beam is focussed on to the sample with a spherical or cylindrical lense. The sample is usually mounted in a temperature variable optical cyrostat. The backscattered light is collected and focussed to the entrance slit of a double grating spectrometer. A polarization analyser allows the separation of different scattering components. The scattered light is detected with a specially selected photmultiplier tube which is connected with special pulse counting electronics.

4. EXPERIMENTAL RESULTS

In this section we present selected experimental results where Raman scattering has been used to investigate special properties of semiconductor heterojunctions and multilayer systems. Because of limitations in space this collection is by no means complete. So we omit the excellent work which has been carried out in high magnetic fields,[45] the effect of Brillouin zone folding in superlattices phonons[46] and the surface sensitive Raman work on cleaved GaAs under ultrahigh vacuum conditions.[34]

4.1. Phonon Aspects

In section 2.1. we have learnt that under certain scattering configurations not all types of optical phonons can be observed in Raman scattering. This has been used in Ref. (32) to study the surface orientation of thin films of GaAs grown with MBE. In backscattering from (100) surfaces only scattering by LO-phonons is allowed. In the Raman spectra shown in Fig. 12 this is the case for the two upper spectra. The two lower spectra show a strong TO-mode even though obtained from thin MBE grown films with a nominally <001> orientation perpendicular to the surface. It could be shown that the strong "forbidden" TO-phonon is connected with a twinning of the growing films caused by carbon contamination of the substrate surface.

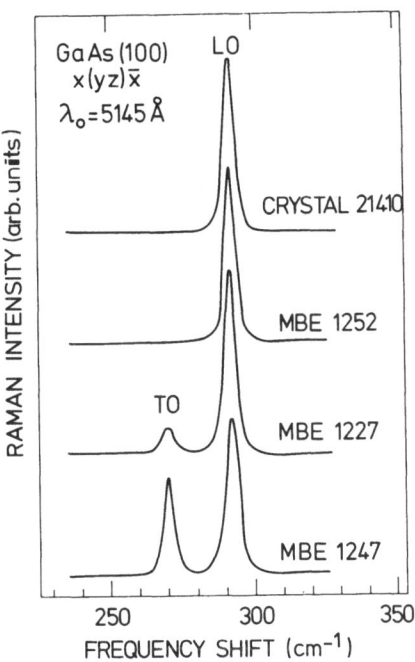

Figure 12 - Raman spectra of nominally <100> oriented thin films of GaAs (After Ref. 32).

In mixed crystals like $Al_xGa_{1-x}As$ phonon Raman scattering can be used to obtain information on the composition. The optical phonons in these crystals have a two-mode behavior with frequencies close to the modes of pure GaAs and pure AlAs. The frequency dependence of these modes on the molar fraction x of Al in GaAs is shown in Fig. 13. Raman spectroscopy has been used to study the depth profile of the Al-content in the $Al_xGa_{1-x}As$ films and the composition of unknown crystals and multilayer structures (see for example (32). We want to emphasize that Raman scattering experiments can be performed on very small spots on the sample surface. Therefore it is possible to study the properties with high spatial resolution which is of the order of the focussed laser beam (a few μm^2).

4.2. Three-Dimensional Carriers in Polar Semiconductors

In polar semiconductors plasmons are coupled to LO-phonons. The frequencies of the coupled phonon-plasmon modes ω_+ and ω_- depend on the carrier concentration, the linewidth on the electron

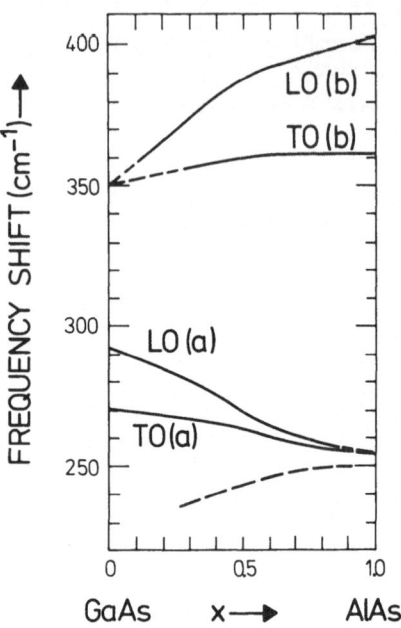

Figure 13 - Frequencies of the optical phonons in $Al_xGa_{1-x}As$ (After Ref. 32).

damping. Under resonance conditions one also can observe excitations with single particle character. In Fig. 14 we show resonance Raman spectra as obtained from a homogeneously doped single crystal of GaAs with $n = 7 \times 10^{17} cm^{-3}$. The spectrum $z(xy) \bar{z}$, where z and $\bar{z}$ are the propagation directions of incident and scattered light and x and y their polarization directions, respectively, exhibits a broad spin-flip single-particle excitation band sitting on top of the hot luminescence background around the $E_o + \Delta_o$ optical energy gap. The spectrum for parallel polarizations contains the two coupled modes at frequencies ω_- and ω_+ which depend on n and q. In Ref. 32 it has been demonstrated that these modes can be used to determine directly the carrier concentration and the scattering times with high spatial resolution. This method has been applied also for the characterization of other polar semiconductors like InP and InAs.

4.3. Semiconductor Single Heterojunctions

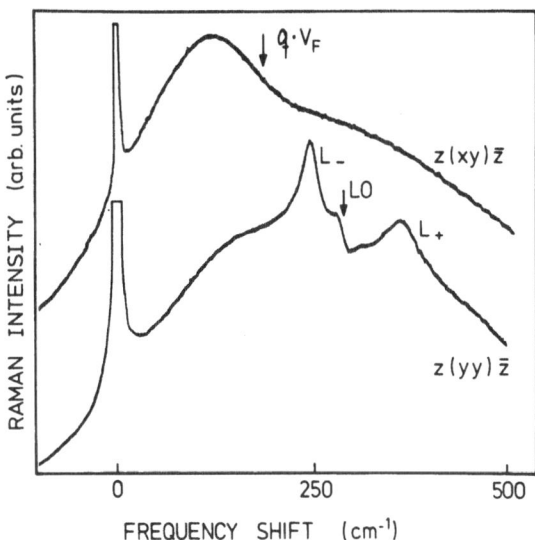

Figure 14 - Resonance Raman spectra of homogeneously doped n-GaAs
with n = 7 x 10^{17}cm^{-3}.

4.3.1.   The GaAs-Al$_x$Ga$_{1-x}$As system.   The extensive study of the
resonance behavior of single-particle and collective excitations by
three- dimensional carriers coincided with reports of the achievement of
high mobility two-dimensional systems in GaAs-Al$_x$Ga$_{1-x}$As heterostructures made by molecular beam epitaxy.[47]   It has been realized
immediately that these heterostructures are ideal candidates for electronic light scattering experiments.   Shortly afterwards the first observations of light scattering by electrons confined at the interface of
selectively   doped   GaAs-n-Al$_x$Ga$_{1-x}$As   heterostructures   were
reported.[11]   At these interfaces charge carriers are transferred from the
donors in the Al$_x$Ga$_{1-x}$As layers to the energetically lower conduction
band of GaAs forming a depletion layer on the Al$_x$Ga$_{1-x}$As side and an
accumulation or inversion layer on the GaAs side.   The energy band
diagram and the electric subbands are shown in Fig. 15.   Raman spectra
obtained from such a structure are plotted in Fig. 16.   Besides the
spin-flip single-particle excitation of the GaAs bulk carriers one can
identify single-particle intersubband excitation of the electrons confined
to the interface.   The energy has been compared with self-consistent
calculations in Ref. 13.   They are in good agreement with the subband
difference E$_{01}$.   In the work of (11) it has been demonstrated for the

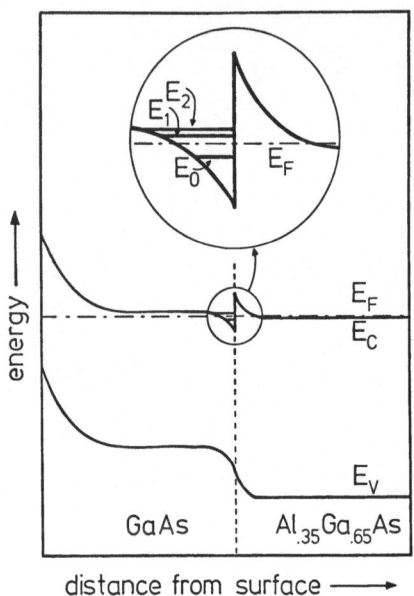

Figure 15 - Conduction- and valence bands in a GaAs-n-Al$_x$Ga$_{1-x}$As heterostructure.

first time that resonant inelastic light scattering is a sensitive tool to study electronic properties of two-dimensional systems. It was also shown that with a Schottky barrier arrangement one is able to control the carrier density. This opened the possibility of fabricating high mobility field effect transistors with semiconductor heterostructures where charge transfer occurs.

In the first experiments the relatively high carrier concentration in the individual layers yielded complicated spectra especially for the collective excitations. The experiments have been repeated in the work of (18) with samples of better quality. The results are shown in Fig. 17. These spectra exhibit both higher order transitions and collective intersubband transitions which are shifted to higher energies because of the depolarization field effect. The evaluation of this shift in terms of Coulomb matrix elements is discussed in more detail together with the results obtained in multiquantum well structures.

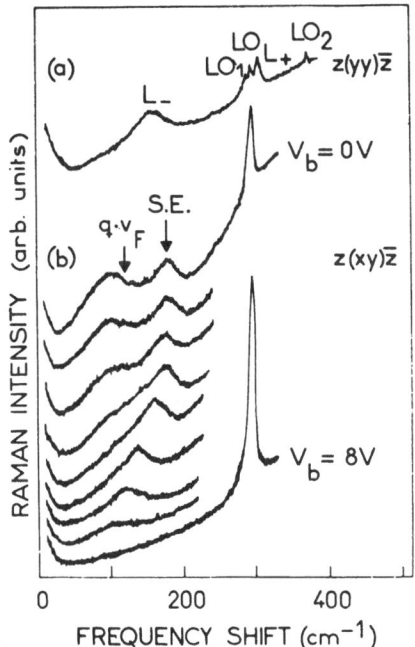

Figure 16 - Raman spectra of a GaAs-n-Al$_x$Ga$_{1-x}$As heterostructure (After Ref. 11).

4.3.2. Ge on GaAs. Recently the first results of light scattering studies of electrons confined to Ge-GaAs interfaces have been reported.[20] The electronic structure of this interface has received considerable attention because of the nearly perfect lattice matching. The band gap in GaAs is larger than in Ge. Therefore charge transfer can occur from the GaAs to the Ge forming a two-dimensional carrier system at the interface. The conduction band minima in Ge are at the Brillouin zone boundary along the <111> direction. The resonant optical gap for electronic Raman scattering therefore is E$_1$ = 2.22eV. In the work of Ref. (20) Ge-GaAs heterostructures grown by MBE have been investigated with laser excitation lines close to the E$_1$ gap of Ge. The samples consist of a thin Ge-layer ($\sim$ 300Å) on top of (100) GaAs layer. The Ge-film is highly doped with As. The spectra with crossed polarizations exhibit a broad asymmetric structure which peaks around 25meV. A theoretical fit of this spectrum indicates that it is due to interband transitions between a quasi-two-dimensional band and the continuum.

4.4. Metal-Insulator-Semiconductor Structures

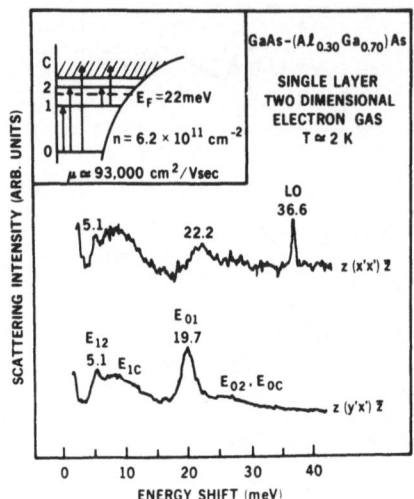

Figure 17 - Raman spectra of a high-mobility GaAs-n-Al$_x$Ga$_{1-x}$As heterostructure (After Ref. 18).

Electronic light scattering by two-dimensional carriers in MIS-structures has been observed so far for electrons in InAs[21,22] and InP[23] and for holes in Si.[24,25] The work in InAs was performed with laser lines close to the E_1 and $E_1 + \Delta_1$ energy gap where no carrier-occupied states are involved in the optical transitions. Consequently only collective intersubband excitations coupled to LO-phonons have been observed via the phonon scattering mechanisms. Only little information could be obtained from these spectra.

4.4.1. Electrons in InP. The band structure of InP is very similar to that of GaAs. The spin-orbit-split-off band gap $E_o + \Delta_o$ is about 1.6eV at low temperature. Recently good MIS-structures with voltage tunable surface carrier concentrations have been fabricated on InP.[48] Resonant light scattering experiments have been used to investigate spectroscopically the subband structure of electron accumulation layers in such samples. Both spin-flip single-particle intersubband- as well as collective excitations have been observed (Fig. 18). The measured energies of the coupled modes and the single-particle excitations can be satisfactorily compared with self-consistent calculations for this subband system (Fig. 10). However, the increase of the subband splittings with surface carrier concentration N_s is found to be smaller than theoretically predicted. Part of this discrepancy is due to the nonparabolici-

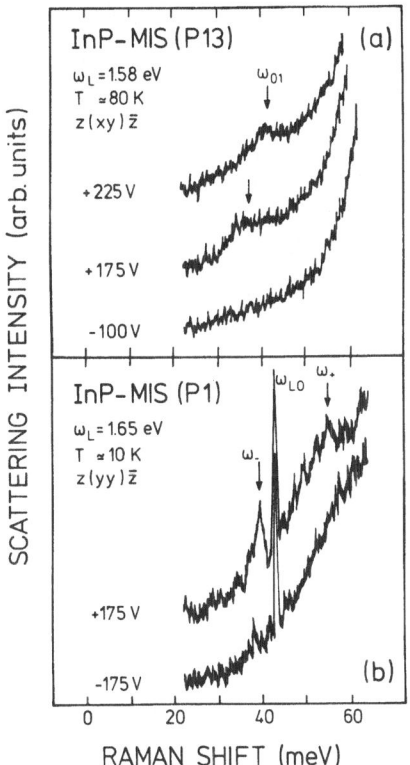

Figure 18 - Raman spectra of InP-MIS structures (After Ref. 23).

ty of the conduction band which has not been taken into account in the self-consistent theory.

4.4.2. Holes in Si. The most widely studied surface space-charge layers in semiconductors are accumulation and inversion layers on Si (see Ref. 9). Recently the first successful light scattering experiments on hole space-charge layers in Si (100) have been reported.[24] The resonant energy for holes is connected with the direct E'_0 gap at $k \simeq 0$, which is about 3.4eV. Both collective and spin-flip subband transitions have been observed.[25] Some of the spectra exhibit very broad and asymmetric bands due to the nonparabolic dispersion of the individual subbands. Different transitions between heavy and light hole subbands could be identified. The results are in good agreement with subband calculations from Ref. (49).

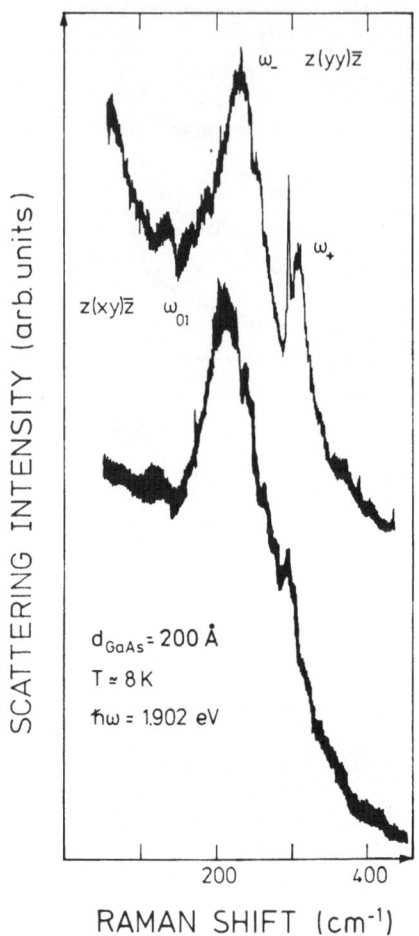

Figure 19 - Raman spectra of a GaAs-n-Al$_{1-x}$As quantum well structure.

4.5. Compositional Superlattices (GaAs-Al$_x$Ga$_{1-x}$As)

4.5.1. Subband energies and depolarization shifts. Electronic light
scattering has been used extensively for the investigation of
GaAs-Al$_x$Ga$_{1-x}$As quantum well structures. The first results obtained
with modulation doped multilayer structures[12] appeared shortly after
the successful experiments performed with single heterojunctions.[11]
The electrons in multilayer structures are quantized in potential wells
which are nearly rectangular due to the quantum size effect in the thin
layers which, however, are modified by the space-charge potential of

the transferred carriers. The subband structure for such systems was first calculated self-consistently by Mori and Ando.[50] Spin-flip single-particle excitations have been used extensively to study the subband energies in a number of samples with different parameters.[12,14,16,18] It was realized already in the first experiments that the positions of the collective excitations observed for parallel polarizations are shifted with respect to the single-particle peaks. The differences in energies reveal in a direct way the depolarization shifts which are related to the Coulomb matrix elements (see section 2.3.3.). In GaAs the collective excitations are coupled to the LO-phonons. The coupled modes have been observed for various samples and for different subband transitions having even and odd parity.[15,16] In Fig. 19 an example is shown for a sample with $d_{GaAs} = 200\text{Å}$ and a two-dimensional carrier concentration $n_s \simeq 4 \times 10^{11}\text{cm}^{-2}$. Such spectra have been used for quantitative determination of ω^*_p and the Coulomb matrix elements f_{nn}, which for the 0 - 1 transitions are of the order 10 to 20Å. The results are in excellent agreement with calculations using the numerical wavefuntions of the subbands.

4.5.2. Linewidth and mobility. It had been realized that the introduction of undoped $Al_xGa_{1-x}As$ spacers at the interfaces of these compositional superlattices lead to a stronger enhancement of the electron mobilities due to the further separation of electrons from their ionized parent impurities. Pinczuk et al.[18] have studied single-particle excitations in several multi-quantum well structures with different thicknesses of the undoped spacer layers but otherwise identical properties. The low temperature mobilities varied in these samples from $12,500\text{cm}^2/\text{Vs}$ to $93,000\text{cm}^2/\text{Vs}$. The Raman results are shown in Fig. 20. The subband excitations are much sharper for the samples with higher mobilities. A striking effect was found when the widths were studied for different laser photon energies. For the samples with lower mobilities the width of the E_{01} transitions has a peak for spectra obtained with photon energies close to the maximum in resonant enhancement. These results have been interpreted in terms of wave vector nonconservation due to scattering of electrons or holes by the Coulomb potential of the ionized impurities.

4.5.3. Photo-excited carriers. Resonant inelastic light scattering by photo-excited electrons in $GaAs-Al_xGa_{1-x}As$ multiple quantum well structures have been reported for both pure and modulation doped samples.[17,19] In undoped structures Pinczuk et al.[17] found no depend-

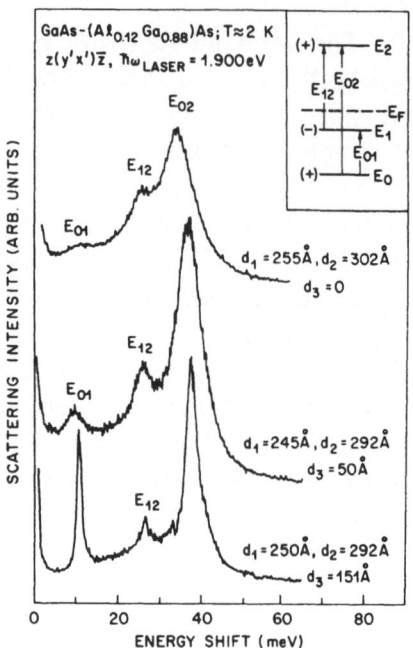

Figure 20 - Raman spectra of GaAs-n-Al$_x$Ga$_{1-x}$As multiquantum well structures with different mobilities (After Ref. 18).

ence of the subband splitting on the excitation intensity. This indicates that the creation of electrons and holes within the GaAs layers does not result in a distortion of the rectangular potential wells. No separation in space occurs for the photo-excited electron-hole pairs. Zeller et al.[19] studied modulation doped samples with high laser power densities. They found a shift of the subband splitting E_{01} to smaller energies with increasing excitation intensity. A decrease of the subband splitting is expected when the electron density in the GaAs layers is increased more strongly than the hole concentration. This might happen due to the presence of surface depletion layers or by hole traps in the Al$_x$Ga$_{1-x}$As layers.

4.5.4 In-plane excitations. Recently the plasma frequency dispersion in a layered electron gas has been measured by inelastic light scattering.[40] A modified backscattering geometry was applied to provide the necessary wave vector component parallel to the GaAs layers of modulation doped GaAs-Al$_x$Ga$_{1-x}$As multi-quantum well structures. The measured dispersion of the plasmon frequency was found to be linear in the in-plane component of the wave vector. This differs from the results obtained in pure two-dimensional and pure three-dimensional plasmas. It, however, confirms basic predictions of the plasmon behavior in layered carrier systems.

4.6. Doping Superlattices

4.6.1. Electrical and optical properties. One of the nicest examples where resonant inelastic light scattering acts as a tool for the investigation of semiconductor heterostructures is the recent work on periodic doping multilayer structures, so-called "nipi"-crystals. This type of semiconductor superlattices has been first proposed and analyzed by Döhler.[51] It is composed of a periodic sequence of ultrathin n- and p-doped layers of GaAs and exhibits various novel and exciting electrical and optical proprties which are caused by purely space-charge induced potential wells. A doping superlattice is simply an alternation of p-n and n-p junctions of an otherwise homogeneous semiconductor. When the concentration of the donors N_D times the thickness of the n-type layers is equal to the number of the acceptors N_A times the thickness of the p-type layers the nipi-crystal is compensated. The electrons from the donors are attracted by the acceptors in the p-type layers resulting in a periodic rise and fall of the conduction and valence band. Such structures exhibit the special feature of a semiconductor with an "indirect gap in real space." Excited electrons and holes are separated in space and may have recombination lifetimes orders of magnitude longer than in homogeneous bulk crystals. The reduced effective band gap which depends on the nonequilibrium electron and hole concentrations results in a strong tunability of the optical absorption and luminescence. In Fig. 21 the modulated conduction and valence band of a nipi-crystal is given in an excited situation. The transitions of photo-luminescence and resonant Raman scattering experiments are shown schematically. The energetic position of the luminescence depends on the excitation intensity and can be used to determine the nonequilibrium carrier concentration. Indirect photoluminescence spectra are shown in Fig. 22. These experiments directly demonstrate the tunability of the effective band gap.

4.6.2. Single-particle and collective excitations. To get direct information on the quantization of photo-excited carriers in doping superlattices, resonant inelastic light scattering experiments have been performed.[26-28] Spin-flip single-particle intersubband excitations have

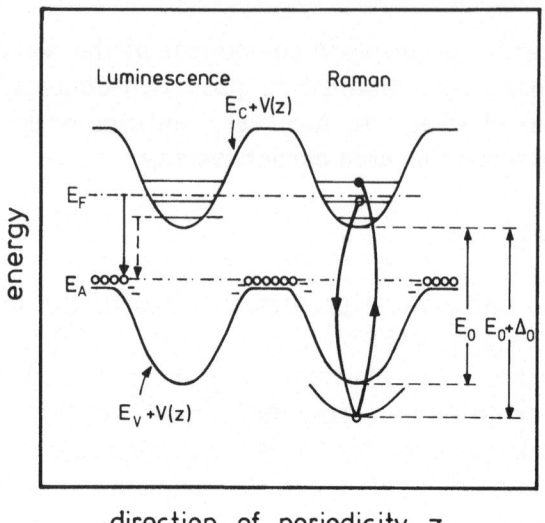

Figure 21 - Conduction and valence bands of an excited doping super-lattice.

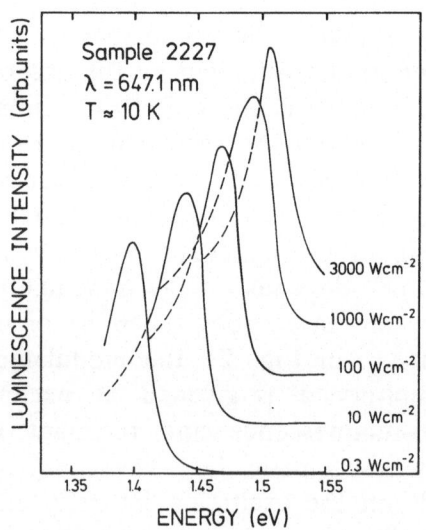

Figure 22 - Photoluminescence spectra of a doping superlattice (After Ref. 27).

been studied using different power densities of the incident laser (Fig. 23). At low excitation intensity several distinct peaks have been observed on top of a hot luminescence background (Fig. 23). These peaks could be identified as $\Delta = 1$, $\Delta = 2$, and $\Delta = 3$ intersubband transitions of photo-excited electrons in the conduction band of the doping superlattice. The subband energies obtained in this way are in excellent agreement with self-consistent calculations.

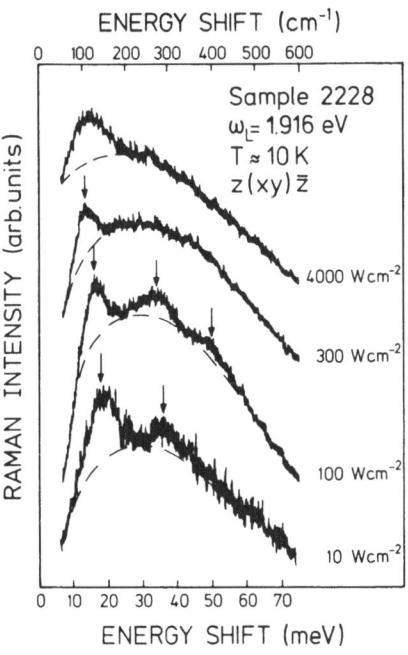

Figure 23 - Single-particle Raman spectra of a doping superlattice (After Ref. 27).

Raman spectra of collective excitations have also been observed in doping multilayer structures. Because of the occupation of several subbands already at relatively low power densities and the coupling to LO-phonons, the observed polarized spectra exhibit a complicated structure (Fig. 24). At low excitation intensities three peaks have been observed below the LO phonon mode which represent several ω_- -modes. A broad ω_+ -mode can be identified above the LO-phonon. The positions of these modes are in good agreement with recent calculations performed by Ruden and Döhler.[52]

4.6.3. Two- and three-dimensional effects. At high power densities the subband splittings get smaller and the higher occupied states show considerable dispersion in the direction perpendicular to the layers. This can be observed by studying both the single-particle and the collective excitations in the highly excited case. While at low laser power densities distinct subband transitions are observed, the individual peaks merge at high power densities into one broad single-particle excitation band which has a very similar lineshape as obtained for a homogeneously doped GaAs single crystal of comparable carrier concentration (see Figs. 23 and 14). The spectra of the coupled modes also change with increasing power density and finally they look very similar to coupled phonon plasmon modes of homogeneously doped n-GaAs (Figs. 24 and 14). The behavior with increasing power density is concomitant with the transition from a quasi-two-dimensional to a quasi-three-dimensional electron system.

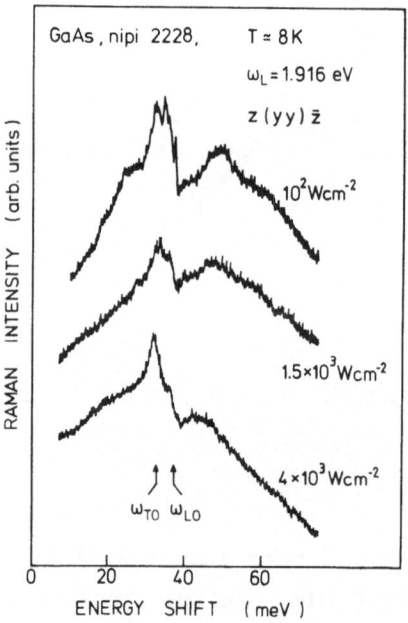

Figure 24 - Collective Raman spectra of a doping superlattice (After Ref. 28).

5. CONCLUDING REMARKS

We have demonstrated the usefulness of inelastic light scattering for the investigation of various properties of semiconductor thin layers, -heterojunctions and -multilayer systems. We want to emphasize that Raman scattering experiments can be performed on very small spots on the surface. Thus it is possible to study all the properties discussed in this work along the whole surface with high spatial resolution. Raman spectroscopy is a powerful technique not only for the investigation of light scattering properties by itself, but it also acts as a tool to characterize and analyse electrical and optical properties of new materials.

REFERENCES

1. M. Cardona (ed.): "Light Scattering in Solids", Topics in Applied Physics, vol. 8 (Springer, Berlin, Heidelberg, New York, 1975).
2. M. Cardona and G. Güntherodt (eds.): "Light Scattering in Solids II", Topics in Appl. Phys., vol. 50 (Springer, Berlin, Heidelberg, New York 1982).
3. M. Cardona and Güntherodt, G. (eds.): "Light Scattering in Solids III", Topics in Appl. Phys., vol. 51 (Springer, Berlin, Heidelberg, New York 1982).
4. M. Cardona and G. Güntherodt (eds.): "Light Scattering in Solids IV," Topics in Appl. Phys., Vol. 54 (Springer, Berlin, Heidelberg, New York 1983).
5. A. Pinczuk and E. Burstein, in Ref. 1, p. 23.
6. M. V. Klein, in Ref. 1, p. 147.
7. M. Cardona, in Ref. 2, p. 19.
8. G. Abstreiter, M. Cardona, and A. Pinczuk, in Ref. 4.
9. T. Ando, A. B. Fowler, and F. Stern, Reviews of Modern Physics, Vol. 54, 437 (1982).
10. E. Burstein, A. Pinczuk, and S. Buchner, Proc. of the 14th Int. Conf. on the Physics of Semiconductors, 1978, ed. B. L. H. Wilson (The Institute of Physics, London, 1979), p. 1231.
11. G. Abstreiter and K. Ploog, Phys. Rev. Lett. 42, 1308 (1979).
12. A. Pinczuk, H. L. Störmer, R. Dingle, J. M. Worlock, W. Wiegmann, and A. C. Gossard, Solid State Commun. 32, 1001 (1979).
13. G. Abstreiter, Surf. Sci. 98, 117 (1980).
14. A. Pinczuk, J. M. Worlock, H. L. Störmer, R. Dingle, W. Wiegmann, and A. C. Gossard, Surf. Sci. 98, 126 (1980).

15. A. Pinczuk, J. M. Worlock, H. L. Störmer, R. Dingle, W. Wiegmann, and A. C. Gossard, Solid State Commun. 36, 43 (1980).
16. G. Abstreiter, Ch. Zeller, and K. Ploog, Proc. 8th Int. Symp. on "GaAs and Related Compounds", Vienna, 1980, ed. H. W. Thim, Inst. Phys. Conf. Ser. 56 (Inst. Phys. London, 1981), p. 741.
17. A. Pinczuk, J. Shah, A. C. Gossard, and W. Wiegmann, Phys. Rev. Lett. 46, 1307 (1981).
18. A. Pinczuk, J. M. Worlock, Surf. Sci. 113, 69 (1982).
19. Ch. Zeller, G. Abstreiter, and K. Ploog, Surf. Sci. 113, 85 (1982).
20. R. Merlin, A. Pinczuk, W. T. Beard, C. E. E. Wood, J. Vac. Sci Technol. 21, 516 (1982).
21. L. Y. Ching, E. Burstein, S. Buchner, H. H. Wider, Proc. 15th Int. Conf. on Physics of Semiconductors 1980, eds. S. Tanaka and Y. Toyozawa (J. of Phys. Soc. Jpn. 49, Suppl. A, p. 951).
22. G. Tränkle and G. Abstreiter (unpublished), G. Tränkle, Diplom-thesis (1981), Tech. Univ. München.
23. G. Abstreiter, R. Huber, and G. Tränkle, Solid State Commun., 47, 651 (1983).
24. G. Abstreiter, U. Claessen, and G. Tränkle, Solid State Commun. 44, 673 (1982).
25. M. Baumgartner and G. Abstreiter, to be published in Phys. (C).
26. G. H. Döhler, H. Künzel, D. Olego, K. Ploog, P. Ruden, H. J. Stolz, and G. Abstreiter, Phys. Rev. Lett. 47, 864 (1981).
27. Ch. Zeller, B. Vinter, G. Abstreiter, and K. Ploog, Phys. Rev. B26, 2124 (1982).
28. Ch. Zeller, B. Vinter, G. Abstreiter, and K. Ploog, Proc 16th Int. Conf. on Physics of Semiconductors, Montpellier, 1982 (Physica 117B and 118B, p. 729).
29. E. Burstein, A. Pinczuk, and D. L. Mills, Surf. Sci. 98, 451 (1980).
30. J. M. Worlock, Nature 297, 360 (1982).
31. R. Loudon, Adv. Phys. 13, 423 (1964).
32. G. Abstreiter, E. Bauser, A. Fischer, and K. Ploog. Appl. Phys. 16, 345 (1978).
33. R. Trommer, G. Abstreiter, and M. Cardona, Proc. Int. Conf. Lattice Dynamics, ed. M. Balkanski (Flammarion Sciences, Paris, 1977), p. 189.
34. H. J. Stolz and G. Abstreiter, J. Vac. Sci. Technol. 19, 380 (1981).

35. See for example: D. Pines, "Elementary Excitations in Solids" (Benjamin, New York, 1963).
36. A. Mooradian, G. B. Wright, Phys. Rev. Lett. 16, 999 (1966).
37. A. Pinczuk, G. Abstreiter, R. Trommer, and M. Cardona, Solid State Commun. 30, 429 (1979).
38. D. C. Hamilton, A. L. McWhorter, "Light Scattering Spectra of Solids", ed. G. B. Wright (Springer, New York, Heidelberg, Berlin, 1969), p. 309.
39. P. A. Wolf in Ref. 38, p. 273.
40. G. Abstreiter, R. Trommer, M. Cardona, and A. Pinczuk, Solid State Commun. 30, 703 (1979).
41. D. Olego, A. Pinczuk, A. C. Gossard, W. Wiegmann, Phys. Rev. B25, 7867 (1982).
42. W. P. Chen, Y. J. Chen, and E. Burstein, Surf. Sci. 58, 263 (1976).
43. S. J. Allen Jr., D. C. Tsui, and B. Vinter, Solid State Commun. 21, 133 (1977).
44. D. Dahl and L. J. Sham, Phys. Rev. B16, 651 (1977).
45. Z. J. Tien, J. M. Worlock, C. H. Perry, A. Pinczuk, R. L. Aggarwal, H. L. Störmer, A. C. Gossard and W. Wiegmann, Surf. Sci. 113, 89 (1982).
46. C. Colvard, R. Merlin, M. V. Klein, and A. C. Gossard, Phys. Rev. Lett. 45, 298 (1980).
47. R. Dingle, H. L. Störmer, A. C. Gossard, and W. Wiegmann, Appl. Phys. Lett. 33, 665 (1978); H. L. Störmer, J. Phys. Soc. Jpn. 49, Suppl. A (1980), p. 1013.
48. H. C. Cheng and F. Koch, Phys. Rev. B26, 1989 (1982).
49. E. Bangert (unpublished) (1975).
50. T. Ando and S. Mori, J. Phys. Soc. Jpn. 47, 1518 (1979).
51. G. H. Döhler, Phys. Status Solidi B52, 79 and 533 (1972).
52. P. Ruden and G. H. Döhler, Phys. Rev. B 27, 3547 (1983).

ELECTRONIC PROPERTIES OF SEMICONDUCTOR HETEROS-TRUCTURES IN A MAGNETIC FIELD

L. L. Chang

IBM T. J. Watson Research Center
Yorktown Heights, New York 10598

ABSTRACT: This work is devoted to the electronic properties of semiconductor heterostructures under the application of a magnetic field. The focus is on the quantum regime where the heterostructure potential results in electric subbands and the field gives rise to Landau levels. Experiments are reviewed to demonstrate the unique capability of the magnetic field for the investigation of electron systems with reduced dimensionality.

1. INTRODUCTION

Semiconductor heterojunctions have long attracted the attention of researchers in the fields of solid state physics and electronics.[1] The additional degree of freedom achieved with two dissimilar semiconductors provides the opportunity to study new physical phenomena and device structures not possible with homogeneous materials.[2] With recent advancement in thin-film preparation techniques, notably molecular beam epitaxy and metalorganic chemical vapor deposition, heterostructures of atomically smooth and abrupt interfaces can be readily made.[3] The structure may contain a single interface as in a heterojunction; or two interfaces in a sandwich configuration, known as a single potential or quantum well; or a series of interfaces, usually referred to as a multiple-well structure or a superlattice if the layer spacings are periodic.[4] Investigations in heterostructures covering all these types have experienced tremendous progress in recent years, and are of intensive, current interest.

One of the important consequences of the high-quality interface is that it is possible to fabricate extremely thin layers with a thickness comparable to the electron wavelength. The electron motion, restricted in the direction perpendicular to the interface, becomes confined essentially to the plane in the potential well, leading to the creation of a quasi two-dimensional electron system. Space-charge effect is important and, in fact, dominates in defining the potential well at the interface of a single heterojunction. The situation in this case is similar to that in a surface inversion layer where two-dimensional electron behavior was first observed.[5] The narrow potential wells in heterostructures cause electric quantization with the formation of quantum states or subbands, which govern the electronic properties.

There are many experimental techniques to probe the two-dimensional electron system associated with the subband structure. The most powerful is perhaps that of applying a magnetic field. The field quantizes electron motion in the plane perpendicular to it. The density of states is profoundly modified, influencing significantly the behavior of the electron system. This work is devoted exclusively to the electronic properties in the quantum regime under a magnetic field, with special interest in those which are directly related to the subband structure. After considering some fundamental aspects, we review subsequently experimental results in magneto-resistance and magneto-absorption, the two areas where most extensive work of this kind has been performed. This is followed by brief descriptions of other effects under a magnetic effect. In terms of material, we focus on GaAs-GaAlAs and InAs-GaSb with emphasis on the superlattice structure. Other material systems will be touched upon only to the extent that they exhibit new features. We will not attempt to refer to original articles, but rely on recent available reviews which contain references to earlier work.

2. FUNDAMENTAL CONSIDERATION

2.1 Electric Quantization

The motion of electrons in a one-dimensional periodic potential is quite familiar, as exemplified by the Kronig-Penney model of a crystal lattice. In a semiconductor superlattice, such as that shown schematically in Fig. 1 for GaAs-GaAlAs, we consider electrons in the conduction band near the fundamental edge. The potential is defined by the magnitude ΔE_c, the band discontinuity at the interface between the two materials, and the period, $d = d_1 + d_2$, the sum of the two layer thicknesses. The system is described by the Hamiltonian,

$$H = -\frac{1}{2m}(p_x^2 + p_y^2) - \frac{1}{2m}p_z^2 + \Delta E_c(z). \tag{1}$$

Assuming separation of variables, the wave equation can be solved independently in the parallel direction (x, y) which gives free motion in the layer plane, and in the perpendicular direction (z) which results in electric quantization. The total energy can be written as

$$E = \frac{\hbar^2 k_\parallel^2}{2m} + E_z(k_z). \tag{2}$$

The dispersion, $E_z(k_z)$ which can be obtained readily by matching plane wavefunctions at the periodic boundaries, is given by

$$\cos kd = \cos k_1 d_1 \cosh k_2 d_2 - \gamma \sin k_1 d_1 \sinh k_2 d_2 \tag{3}$$

In this expression, $\gamma = 1/2(k_1/k_2 + k_2/k_1)$ and the wavevectors are

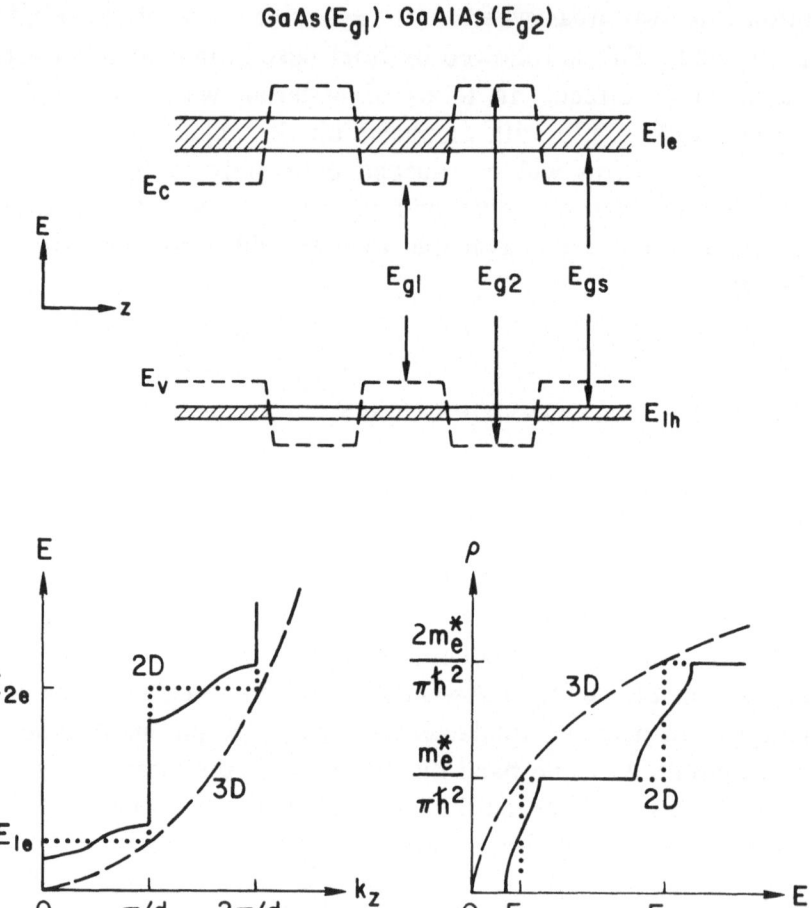

Figure 1 - Schematic diagrams of the GaAs-GaAlAs superlattice: the subband energy (upper), the dispersion relation (lower left) and the density of states (lower right). E_c and E_v represent the conduction and valence bandedges of the bulk at the Γ-point. E_{gs} denotes the energy gap of the superlattice, defined between the ground states of electrons, E_{1e}, and holes, E_{1h}. In the lower diagrams, electron systems of two-dimensional (dotted), three-dimensional (dashed) and intermediate (solid) cases are illustrated.

related to the energy by

$$k_1 = \sqrt{2mE}/\hbar$$

and

$$k_2 = \sqrt{2m(\Delta E_c - E)}/\hbar.$$

In Eq. (3), we have dropped the subscripts z to avoid complexity. By requiring $|\cos kd| \leq 1$, a series of allowed states or subband E_n are formed, which are separated by forbidden gaps at the zone boundaries $k_n = n\pi/d$. The density of states per unit volume can be obtained from the surface integral, $\rho(E) = (4\pi^3)^{-1} \int ds/|\nabla_k E(k)|$ where ds is the unit area on the constant energy surface. Using Eq. (2), this expression reduces to

$$\rho(E) = \frac{m}{\pi^2 \hbar^2} \int_o^E \frac{dE_z}{(dE_z/dk_z)}. \tag{4}$$

One of the most important features of the superlattice is its controllability of the dimensionality of the electron system by varying the potential profile. For a weak potential, characterized by a small ΔE_c and a thin d_2, the electrons in neighboring wells become coupled, leading to an increasing subband width ΔE_n. In the limit, Eq. (3) gives $E_z = \hbar^2 k_z^2/2m$ and Eq. (4), $\rho(E) = (2m^3 E)^{1/2}/\pi^2 \hbar^3$; the electron system is the familiar three-dimensional case. In the opposite extreme, with large ΔE_c and d_2, electrons are completely confined in the wells, capable of moving only in the two-dimensional layer plane (x,y). The dispersion relation becomes flattened. The energy is independent of k_z in each zone but takes discrete jumps at the boundaries, i.e., $E_n = n^2 \pi^2 \hbar^2/2md^2$. The density of states, now per unit area, is energy-independent, $\rho = m/\pi\hbar^2$, and changes abruptly at each E_n. Both the extremes and the intermediate dimensionality between two and three are illustrated in Fig.1 for the two lowest subbands, the ground state E_1 and the first-excited state E_2. For most superlattices investigated experimentally, the coupling between wells is usually small so that the electron system does not deviate much from the two-dimensional regime. For heterojunctions or single quantum well structures, such coupling in general does not exist.

In the above consideration, we have made a number of simplified assumptions to focus on the main features of electric quantization. In particular, we have treated the conduction band as if it were in vacuum and, in the same framework, would treat the valence band similarly. This approximation is not serious for GaAs-GaAlAs in which these two

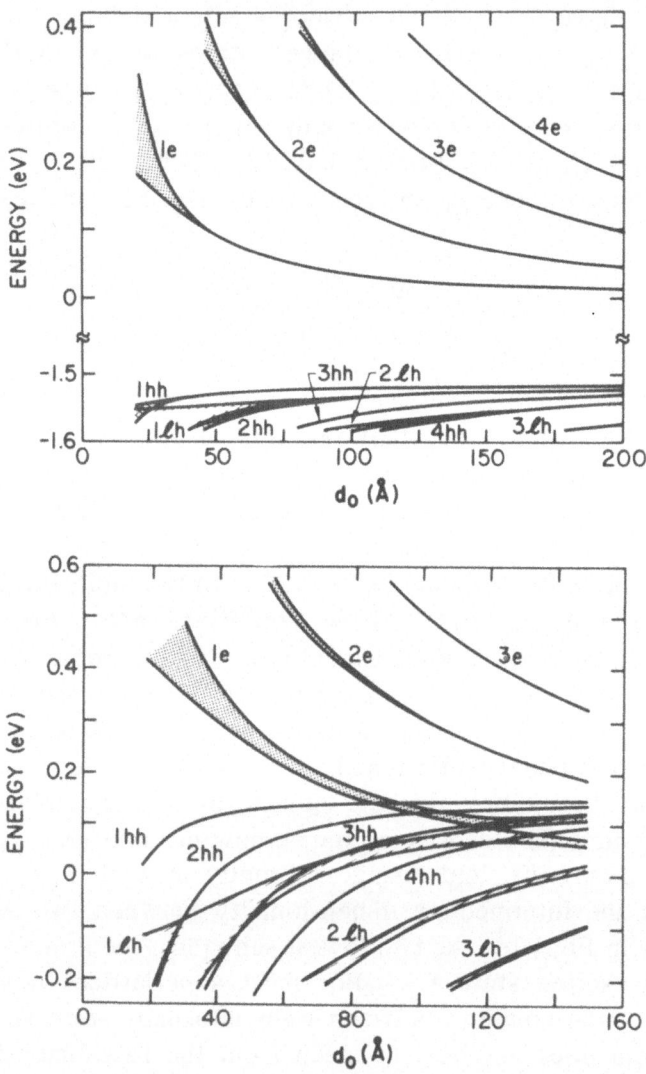

Figure 2 - Calculated subband energies as a function of (equal) layer thickness (d_0) by the envelope wavefunction method for GaAs-GaAlAs (upper) and InAs-GaSb(lower) superlattices. The subbands refer to electrons (e), heavy (hh) and light (lh) holes.

bands are far apart in energy. In other systems such as InAs-GaSb where E_c of the former component is close to and, in fact, below E_v of the latter, the situation is quite different.[6] Calculated subband energies for the two systems by the approach of envelope wavefunctions using the Kane model[7] are shown in Fig. 2 for equal layer thicknes $d_1 = d_2 = d_0$. The results for GaAs-GaAlAs are, as expected, essentially the same as those obtained from Eq. (3). This approach, treated in detail elsewhere in this volume, provides sufficient accuracy for the subbands in structures of experimental interest without involving excessive complexity in computation. Other sophisticated methods have been used, including the tight-binding[8] and the pseudopotential[9] methods. Strictly speaking, the introduction of layers of two different materials creates a new lattice, whose band structure can only be calculated in a fundamental way by treating the dissimilar atoms as a super cell.

2.2 Magnetic Quantization

The most significant effect by the application of a magnetic field is the creation of Landau levels with associated singularities in the density of states. In our present case, the presence of B in an arbitrary direction, $\overline{B} = \nabla \times \overline{A}$ where $\overline{A}$ is the vector potential, leads to the Hamiltonian

$$H = -\frac{1}{2m}(\overline{p} + e\overline{A})^2 + \Delta E_c(z), \qquad (5)$$

which cannot be separated in coordinates in general. Fortunately, in most cases of interest, the field is applied along z, perpendicular to the layer plane. Choosing the proper gauge $\overline{A} = A(0, Bx, 0)$, the Hamiltonian reduces to

$$H = -\frac{1}{2m}p_x^2 - \frac{1}{2m}(p_y + eBx)^2 - \frac{1}{2m}p_z^2 + \Delta E_c(z). \qquad (6)$$

The H_z part remains the same as in Eq. (1), which led to the electric subbands. The solution to $H_\parallel$ is simple; it is a harmonic oscillator with the center coordinate displaced by $X = \hbar k_y/eB$. Defining the cyclotron energy by $\hbar\omega_c = \hbar eB/m$ and the cyclotron radius of the lowest oscillator orbit by $\ell = (\hbar/eB)^{1/2}$, the displacement is given by $\ell^2 k_y = \hbar k_y/m\omega_c$.

The orbital motion on the $k_\parallel$ - plane is magnetically quantized into Landau levels N. The energy, Eq. (2), becomes

$$E_{n,N} = (N + 1/2)\hbar\omega_c + E_{nz.} \tag{7}$$

It can be shown readily that the electron states for a given energy at zero field, $A = \pi k_{\parallel}^2$ are accommodated after quantization by $(2\pi e/\hbar)B(N+1/2)$. The states for each Landau level is the degeneracy factor ζ,

$$\zeta = \frac{2}{(2\pi)^2} \cdot \frac{2\pi eB}{\hbar} = \frac{eB}{\pi\hbar}, \tag{8}$$

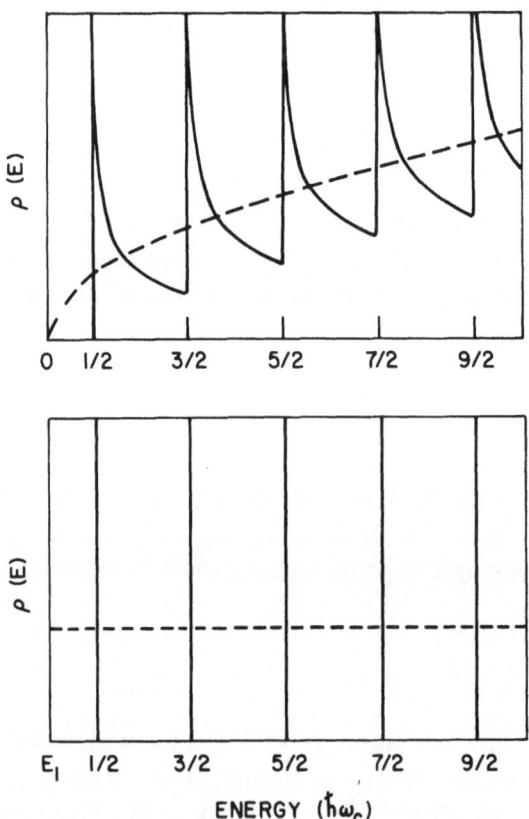

Figure 3 - Schematic density of states under a magnetic field for three-dimensional (upper, reference to zero) and two-dimensional (lower, reference to E_1) electron systems. The energy is in units of cyclotron energy. Dotted curves represent the cases without magnetic field.

where we have included the effect of spin. Recall that the two-dimensional density of states is $\rho = m/\pi\hbar^2$, the degeneracy is equal to the product of ρ and $\hbar\,\omega_c$. In other words, each Landau level created by the field collects the states of the contiuum throughout an energy range $\hbar\omega_c$. The situation is illustrated schematically in Fig. 3 where the corresponding case of three-dimensionality is also included for comparison. Although there are similarities, major differences exist: the two-dimensional density of states is of equal strength for each Landau level and it vanishes between levels. In reality, level broadening Γ_N is always present. The actual shape of the density has been shown to be elliptic, $1-[(E-E_N)/\Gamma_N]^{2\,1/2}$ where Γ_N has to be treated in a self-consistent manner.[10] The broadening depends on the range of scattering potentials. For short-range scattering, $\Gamma^2_N = 2\hbar^2\omega_c/\pi\tau$ where τ is the relaxation time. Here, the broadening is essentially the lifetime broadening, independent of the Landau level.

The inseparability of coordinates of Eq. (5) indicates that the electric and magnetic quantizations are in general mixed. For the other extreme case of B along the layer plane, say B(x) or A(0, Bz, 0), the system can be dealt with by making a transformation of $z' = z + \hbar k_y/eB$, and solved numerically. If B is not large so that ℓ is greater than the characteristic well width or, equivalenty, $\hbar\omega_c$ is smaller than the subband spacing, the influence of the magnetic field can be treated as a perturbation. The main effect is a diamagnetic shift of the ground-state energy.[5] In intermediate cases for which the field deviates only slightly at an angle θ from the surface normal, the B_x component is usually negligible. The quantized, constant energy surface perpendicular to B can be expressed as $A_\theta = \pi k_{\parallel}^2/\cos\theta = (2\pi eB/\hbar)(N + 1/2)$. This is equivalent to having an effective $B_z = B\cos\theta$ acting on the density of states. This cosine dependence is unique to the two-dimensional electron system, and has been commonly used to demonstrate its observation.[12]

To consider the effect of spin explicitly, an additional term, $\pm 1/2\,g\mu B$, should be included in Eq. (7), where g is the Zeeman spin factor, μ is the Bohr magneton given by $e\hbar/2m$, and the signs take into account up and down spins. The magnetic levels now include both Landau and spin levels, and the degeneracy factor for each level, Eq. (8), becomes eB/h. Unlike the Landau splittings which result from quantization of the orbital angular momentum, the spin splittings depend only on the magnitude of B regardless of its orientation. In addition, since the effective g-factor depends on the exchange energy of both types of spins, its value is varied when the Fermi level is located at different positions relative to the magnetic levels.[13]

3. MAGNETO-RESISTANCE

Measurements of transverse magento-resistance have been used extensively in the investigation of subbands in heterostructures, for such measurements probe conveniently the oscillatory density of states through electron scattering at the Fermi energy. For a typical Hall arrangement with current flow in the x-direction and a strong magnetic field along the z-direction, the conductivity coefficient σ_{xy} is usually much greater than σ_{xx}. The magneto-resistance, defined along x, is given by $\rho_{xx} \simeq \sigma_{xx}\rho_{xy}^{-2}$. The Hall resistance, ρ_{xy}, measured from the Hall voltage along the y-direction, can be directly related to $\sigma_{xy}^{-1} \simeq$ B/en, where n is the electron concentration. Since the electron scattering reaches a maximum when E_f coincides with a Landau level, the magneto-resistance oscillates with B, known as the Shubnikov-de Haas oscillations. From Eq. (8), the period of oscillations in inverse field ΔB^{-1} vs. the Landau index N is given by $e/\pi\hbar n$ and, consequently, by $2\pi e/\hbar A_f$ or $e\hbar/mE_f$. This period thus provides a direct measurement of the carrier density or the Fermi surface, or E_f if the effective mass is known. The magnitude of the oscillations depends in general on the mass, the temperature and the level broadening, which can only be obtained from a detailed treatment in transport theory. The behavior of the two-dimensional electron system[5] is somewhat different from that of the three-dimensional case,[14] reflecting the different density of states of the two systems under magnetic fields as illustrated in Fig. 3.

Results from the first magneto-resistance measurement, demonstrating the two-dimensional Shubnikov-de Haas oscillations in a GaAs-GaAlAs superlattice,[15] are shown in Fig. 4. The inset plots the linear variation of the period of oscillations with respect to N at perpendicular magnetic field ($\theta = 0°$), from which the electron density can be deduced. A large number of superlattices were investigated,[16] as shown in Table 1 which lists the experimental configurations and the theoretically calculated subband energies and widths from Eq. (3). The last columns compare E_f defined with respect to E_1: The experimental values were derived directly from the oscillations of the ground states, the only oscillations that were clearly observable; while the theoretical

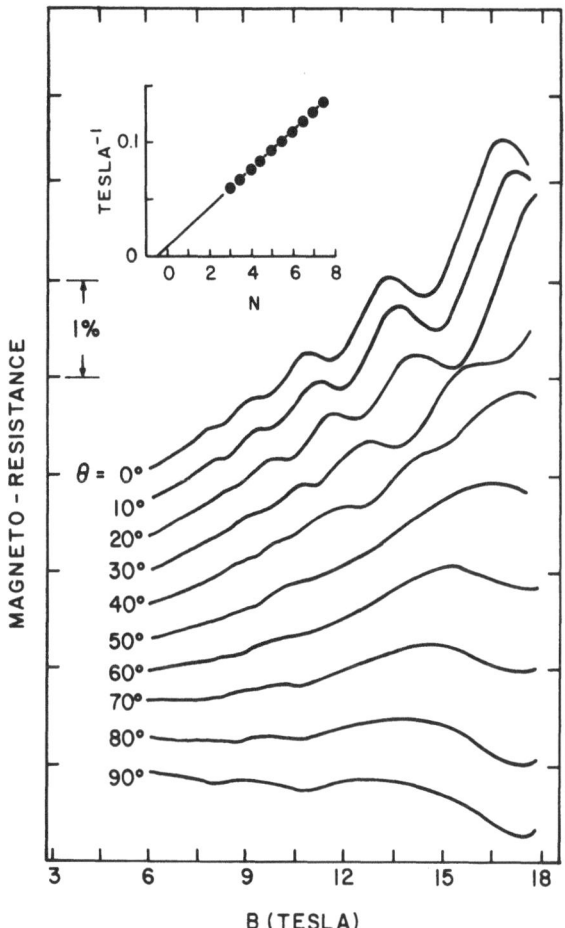

Figure 4 - Transverse magneto-resistance at 4.2K to illustrate Shubnikov-de Haas oscillations in a GaAs(90Å)-Ga$_{0.89}$Al$_{0.11}$As (90Å) superlattice with an electron concentration of 1.2 x 10^{18} cm^{-3}. The orientation of the field is varied from perpendicular ($\theta = 0°$) to parallel ($\theta = 90°$) to the sample surface. Extremal positions in inverse fields for the perpendicular case are plotted against Landau index in the inset.

values were calculated from the subband energies, knowing the total electron concentrations from the Hall measurements and taking into account the occupation of multiple subbands when applicable. The generally good agreement between the two sets of values was used to demonstrate in a systematic way the establishment of the superlattice subband structure. The measurements were made with two terminals on heavy doped superlattices, for which the mobility was only a few thousands. With the technique of modulation doping to separate spa-

Table 1. GaAs - Ga$_{1-x}$Al$_x$As superlattices. The first five columns list the experimental parameters. The next three columns list the calculated subband energies and widths. The last two columns compare the Fermi levels determined theoretically and experimentally. A bulk GaAs sample is also included for comparison.

d_1 (Å)	d_2 (Å)	x	#Pd.	n (10^{18}cm^{-3})	$E_1(\Delta E_1)$ (meV)	$E_2(\Delta E_2)$ (meV)	$E_3(\Delta E_3)$ (meV)	E_f(cal) (meV)	E_f(exp) (meV)
GaAs		0	-	0.17	0(-)	-	-	15	17
40	30	0.15	285	1.48	51.6(73.2)	209.5(-)	-	63	68
40	30	0.15	285	4.34	51.6(73.2)	209.5(-)	-	133	118
90	75	0.17	127	1.20	31.0(2.0)	109.0(-)	-	69	69
90	90	0.11	111	1.90	26.6(2.3)	87.0(20.2)	136.8(-)	96	97
90	50	0.18	143	3.09	29.8(5.8)	107.4(31.9)	204.4(-)	123	118
230	220	0.12	50	1.14	6.8(0)	27.0(0)	58.9(0)	84	97

tially the electrons and their donor impurities,[17] described elsewhere in this volume, enhanced mobilities and, consequently, improved oscillatory characteristics were achieved.[18]

As can be seen in Fig. 4, the extrema of oscillations shift progressively to higher field with θ as the field is tilted away from the surface normal. The angular dependence of the period follows the $\cos\theta$ relation, characterizing the two-dimensional nature of the electron system. Figure 5 illustrates explicitly the angular dependence with different superlattice configurations to illustrate the controllability of dimensionality in superlattices. In one extreme, represented by bulk

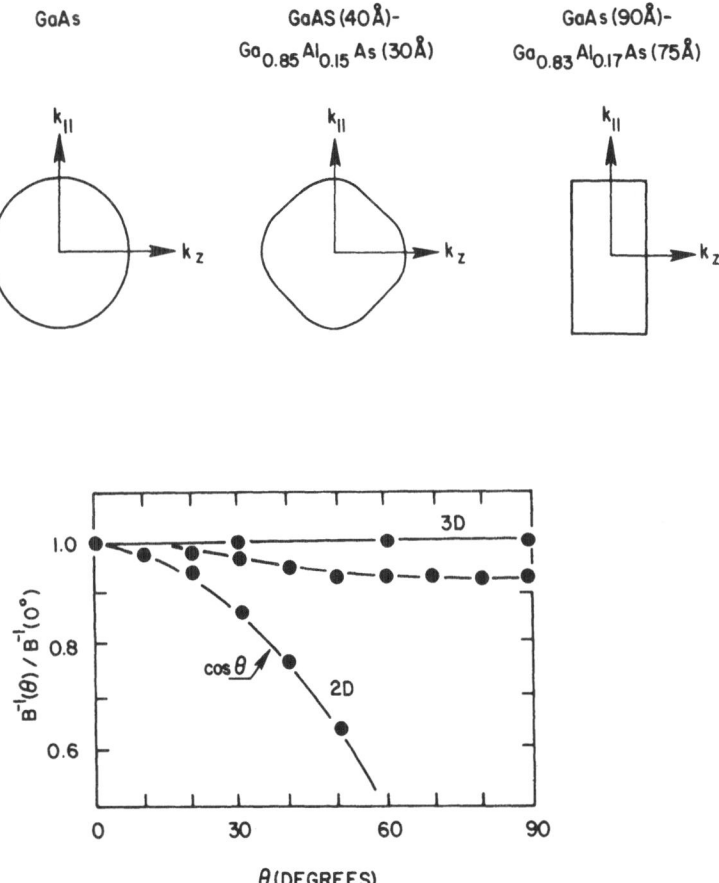

Figure 5 - Angular dependence of the period of oscillations for three sample configurations specified on the upper part. From left to right are three-dimensional bulk GaAs, intermediate and two-dimensional superlattices, with their corresponding Fermi surfaces.

GaAs, the electron system is three-dimensional with a spherical Fermi surface, the period is independent of the field orientation. The other extreme has the cylindrical surface of the two-dimensional system, hence the $\cos\theta$ dependence. The variation falls between the two extremes for the intermediate case. For parallel fields in the two-dimensional system, the magneto-resistance is usually featureless. The weak oscillations in Fig. 4 were interpreted as due to magnetic breakdown for electrons occupying the second subband.[15] Effects of inter-subband mixing have also been considered in this connection.[19]

Magneto-resistance has been employed systematically to probe the carrier variations with layer thickness in the InAs-GaSb superlattice, a system having the unique property of semiconductor-semimetal transitions.[6] As shown in Fig. 3, the superlattice energy gap decreases and, eventually vanishes when $d_1=d_2=85\text{Å}$. Beyond this point, the subbands of heavy holes in GaSb lie in energy above those of electrons in InAs, and electron transfers occur. The superlattice becomes semimetallic, with an equal increase of electrons in InAs and holes in GaSb.[20] With a series of configurations of different layers, the subbands were calculated and E_f compared with those obtained from the Shubnikov-de Haas oscillations,[21] as shown in Fig. 6. The behavior can be readily understood. The value of E_f first increases as electron transfer starts. The space charge thus created leads to band bendings which tend to push toward each other the subbands of electrons and holes. This effect, together with the occupation of multiple subbands as their energy spacings are narrowed with increasing thickness, reduced E_f. Eventually, the space-charge potentials localize electrons near the interface regions, and the superlattice behaves in essence as a series of isolated heterojunctions. The results shown in Fig. 6 cover the entire semimetallic regime, from the semiconductor-semimetal transition on the one end to the heterojunctions limit on the other.

The transition just described originated from the fact that the subbands of holes moved above those of electrons in different samples with increasing layer thickness. The situation can be reversed in the same sample under magnetic fields. In the semimetallic state, the fields cause the sweeping of Landau levels of both electrons and holes across E_f, resulting in a reverse transfer. In the limit when ground Landau levels are crossed, the semimetallic carriers are depleted and the superlattice is back in the semiconductor regime. This in principle is expected to be a general property in semimetals.[22] The criterion for its

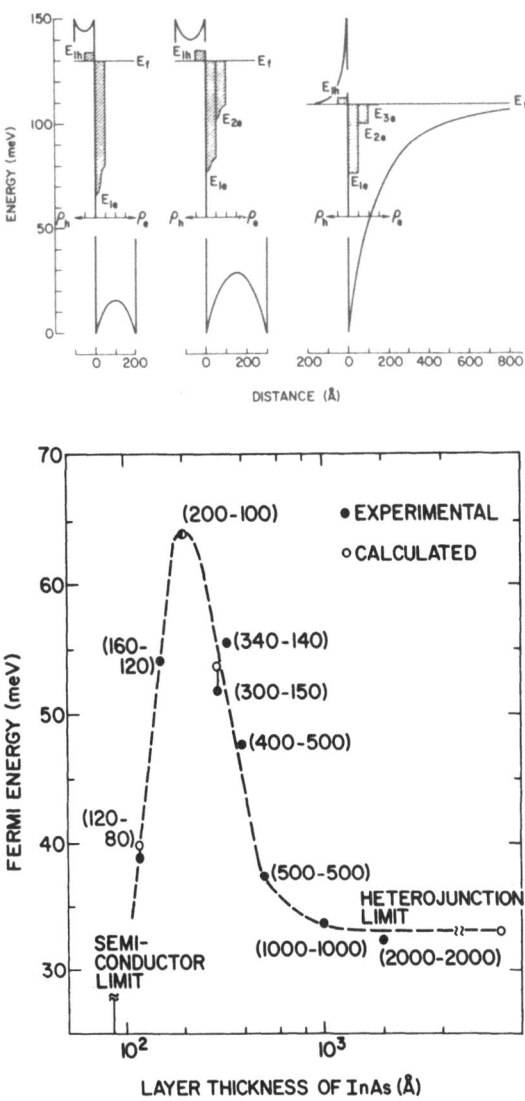

Figure 6 - InAs-GaSb superlattices with different layer thickness to demonstrate the semiconductor-semimetal transition. Upper diagrams shows the calculated electron subbands and density of states from charge transfer. The low curve shows the Fermi energy vs. InAs layer thickness; the GaSb layer thickness is indicated as the second value in the parenthesis.

observation requires that $\hbar\omega_c/2 = E_f$ can be reached with available fields. For superlattices with controllable subbands, this condition can be readily met with structures in the semimetallic region not far from the point of semiconductor-semimetal transition. A drastic increase in magneto-resistance was observed experimentally at the critical field corresponding to that expected theoretically.[23]

As the Shubnikov-de Haas effect, in general, reflects the extremal cross sections in the Fermi surface, additional oscillations may arise at the superlattice zone boundaries, $k_z = \pi/d$, where such extrema exist with proper dispersion. The observed magneto-resistance showed two distinct sets of oscillations and their periods themselves varied with the tilting angle in an oscillatory fashion.[24] The results are shown in Fig. 7 where the two sets are identified with the extrema at the zone center and edge in the inset. The oscillations with θ arise as the cyclotron

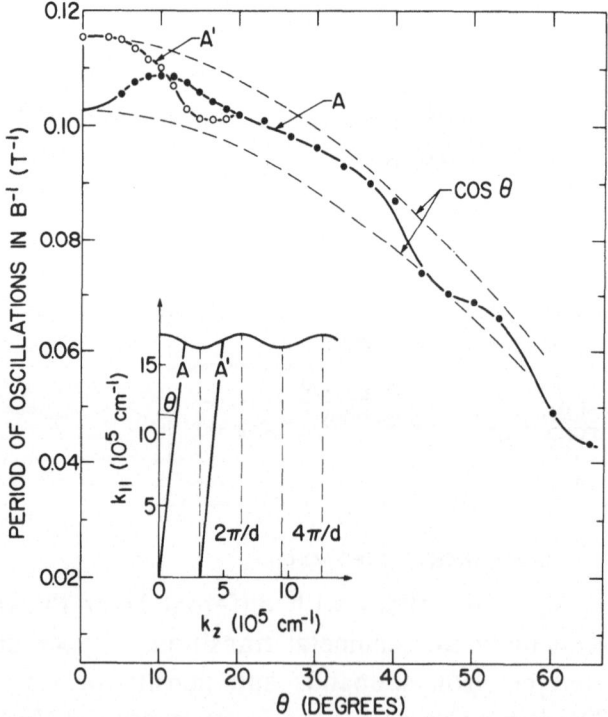

Figure 7 - Angular dependence of the period of Shubnikov-de Haas oscillations of an InAs(500Å)-GaSb (500Å) superlattice. The oscillatory periods correspond to the two extrema of the Fermi surface at the zone center (A) and edge (A′) as shown in the insert.

orbits traverse into the extended zone of the superlattice, switching their roles as being a maximum or minimum in the Fermi surface periodically. This observation provides another manifestation of the intermediate dimensionality of the superlattice.

In the characteristics of Shubnikov-de Haas oscillations, extra extrema were observed at high fields, which arose from Eq. (7) with the addition of the spin term $\pm 1/2g\mu B$. That the addition splittings are due to spin can be verified by tilting the field since they do not depend on the orientation as mentioned earlier. The g-values obtained, however, varied with the field and were much enhanced from the bulk values which are small in most semiconductors. These observations were explained from a systematic investigation of the temperature dependence of the characteristics in GaAs-GaAlAs[25] in accordance with the enhancement by the self-energy exchange interaction.[13] Since such interaction is maximum (or minimum) when E_f lies between spin levels (or Landau levels), the effective g-value oscillates as a function of B, approaching its fully enhanced value with decreasing level broadening. An enhancement of the order of 10 was obtained in GaAs-GaAlAs.[25,26] In other heterostructures such as InAs-GaSb[24] and InP-InGaAs,[27] effective g-values larger than those in the bulk have also been reported.

Of great recent interest is the quantum Hall effect, first observed in Si-inversion layers[28] and subsequently in GaAs-GaAlAs heterojunctions.[29] Results from the latter case are shown in Fig. 8. It is seen that, over certain ranges of the field when the magnetoresistance becomes vanishingly small, the Hall resistance exhibits pronounced plateau. This corresponds to the situation of E_f located between magnetic levels with $\rho_{xx} \sim \sigma_{xx} = 0$ and $\rho_{xy} = B/en = h/e^2\nu$ by substituting n with the product of ζ, the level degeneracy, and ν, the level occupation. Thus ρ_{xy} depends only on the fundamental constants which are related to the fine structure constant, $\alpha = (e^2/h)\mu_o c/2$ where μ_o is the vacuum premeability and c is the speed of light. Careful measurements[30] have demonstrated that the fine constant can be determined to a precision of a fractional part in 10^6. That ρ_{xy} exhibits plateaus can be understood on the basis of localized states, due to impurities or other disorders, which are present in the gap between the magnetic levels.[31,32] As long as E_f is located in the gap region, the remaining extended states will automatically adjust with the magnetic field to carry the entire Hall current.

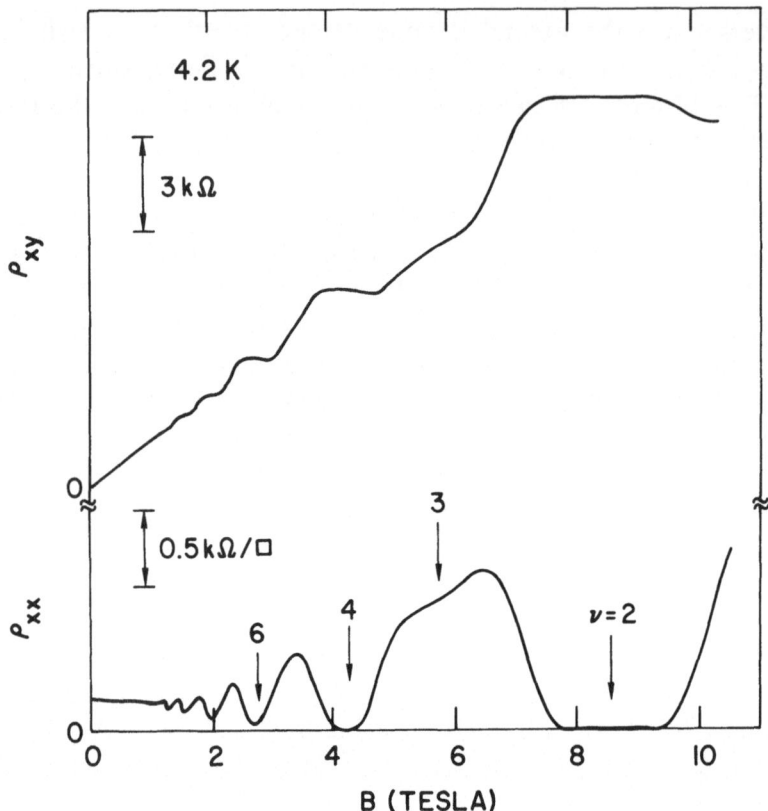

Figure 8 - Magneto (ρ_{xx}) and Hall (ρ_{xy}) resistance of a GaAs-Ga$_{0.7}$Al$_{0.3}$As heterojunction, with an electron density of 4.2 x 10^{11} cm^{-2} and a mobility of 7.9 x 10^4 cm^2/V•sec, to illustrate the quantum Hall effect. Arrows indicate the levels of occupation. (After Ref. 29).

Further experiments to low temperatures and high fields with samples having extremely high mobilities led to the striking observation that ν can assume fractional numbers,[33] notably $\nu = 1/3$ and $2/3$ as shown in Fig. 9. Other fractions such as 2/5, 3/5, and 4/5 were also reported subsequently.[34,35] In these cases, the electron system is in the extreme quantum limit where only the lowest magnetic level is occupied, and there exist no obvious gaps in the energy spectrum. A large number of theoretical considerations have been advanced to provide an explanation, of which the formulation of new, collective ground state wavefunctions describing an incompressible quantum fluid appears to be the most promising.[36] The theory predicts a series of ground states whose elementary excitations are fractionally charged. The series starts

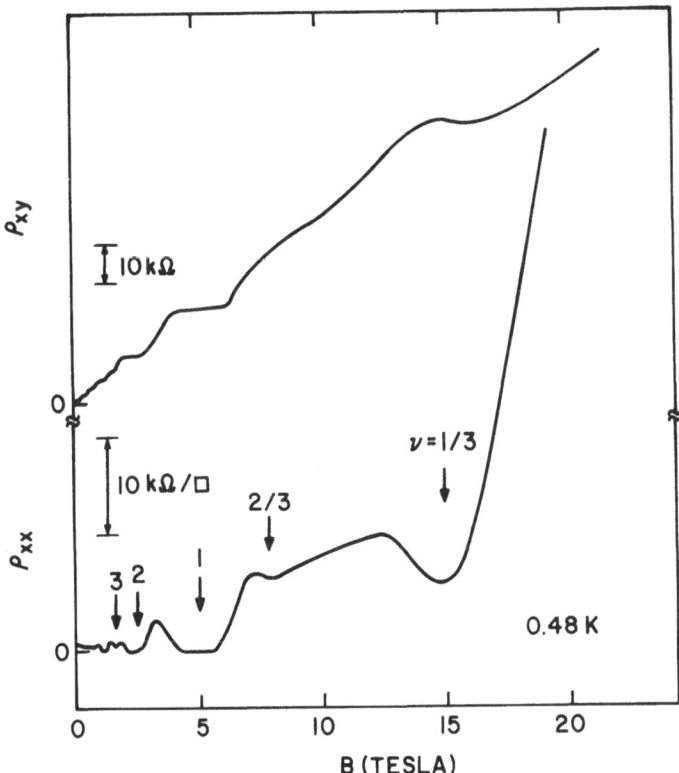

Figure 9 - Magneto (ρ_{xx}) and Hall (ρ_{xy}) resistance of a GaAs-Ga$_{0.7}$Al$_{0.3}$As heterojunction, with an electron density of 1.23 x 10^{11}cm^{-2} and a mobility of 9x10^{4}cm^2/V•sec, to illustrate the fractional quantum Hall effect. Arrows indicate the levels of occupation. (After Ref. 33).

with 1/3, progresses to other fractions with odd denominations, and end eventually in a crystalline state. Recent measurements with dilute electron densities have provided evidence to this trend.[37] The ρ_{xx} showed a rather weak structure corresponding to ν = 1/5, but otherwise featureless characteristics down to 1/11. It should also be mentioned that the fractional quantum Hall effect is not restricted to the extreme limit alone ($\nu<1$). Similar fractions were observed in the $\nu>1$ and $\nu>2$ regimes as well.[35]

Magneto-transport measurements have been performed in a number of other heterostructure systems,[38-40] where two-dimensional electron states were observed. The PbTe-PbSnTe superlattice was of special interest, for the oscillatory characteristics reflect the multi-valley energy ellipsoids.[40]   In terms of the quantum Hall effect, only integral ν values were observed,[41-43] because of the high electron densities which prevented access to the extreme quantum limit with available fields.   The InAs-GaSb quantum-well structure showed additional features in both ρ_{xx} and ρ_{xy} with unusual temperature dependence which could not be explained by electrons alone.[42]   This was attributed to interactions between electrons and holes which were also present in the system.

4. MAGNETO-ABSORPTION

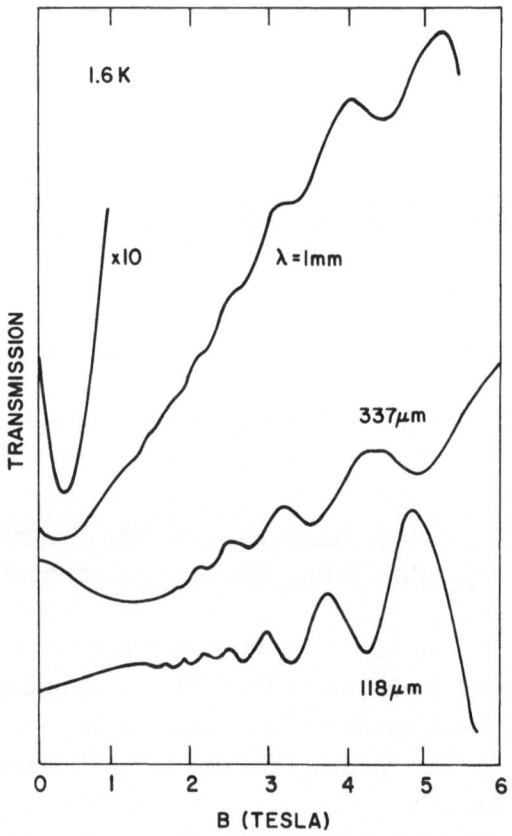

Figure 10 - Transmission spectra with different radiation wavelengths in an InAs(120Å)-GaSb(80Å) superlattice under magnetic fields. Minima correspond to inter-subband and cyclotron absorption.

Measurements of magneto-absorption provide a means of probing the oscillatory density of states of the Landau levels through transitions by photo excitation. Typically, different radiation sources are used and, for a given wavelength, the field is scanned to obtain the absorption or transmission spectrum. Two types of transitions are generally of interest. One is inter-subband absorption arising from Landau levels between the valence and conduction bands; the energy of absorption peaks extrapolated to zero field provides information about the energy difference, usually for the ground states E_{1e}-E_{1h}. The other is cyclotron resonance, or sometimes referred to as intra-subband absorption, between Landau levels across the Fermi energy within the same band of electrons or holes. The variation of absorption energies with field, in this case, gives a direction determination of the effective mass through $\hbar\omega_c$. In considering the various energies involved, the effect of band nonparabolicty should in general be taken into account.

Early measurements of cyclotron resonance were made in both GaAs-GaAlAs[44] and InAs-GaSb,[45] and, for the latter, inter-subband transitions were also reported in the semiconducting regime.[46] But most extensive experiments were performed for semimetallic InAs-GaSb, which provided a number of interesting observations.[47] We show in Fig. 10 the transmission curves in this case, where both types of magneto-absorption are present; the initial dips correspond to the cyclotron resonance. By use of different infrared sources with increasing energies, the oscillations move to higher fields, resulting from larger Landau level spacings. The situation is illustrated in Fig. 11 where the energies at absorption peaks or transmission dips are plotted against B in the standard way. The CR represents the cyclotron resonance associated with E_{1e}. The inter-subband transition between E_{1h} and E_{1e} consists of a series of transitions, each branch corresponding to a specific Landau index. These transitions are made possible by the magnetic field only when the Landau levels of electrons are lifted in energy across E_f and above those of holes, as illustrated in the inset. All the branches converge to a single, negative energy, $E_{gs} = E_{1e}$-E_{1h}. That the energy gap of the superlattice is negative is a direct demonstration of the semimetallic nature.

Subsequently, more systematic measurements provided other information. By used of more radiation sources and higher magnetic fields, additional data points were obtained which not only corroborated the results shown in Fig. 11 but revealed another set of

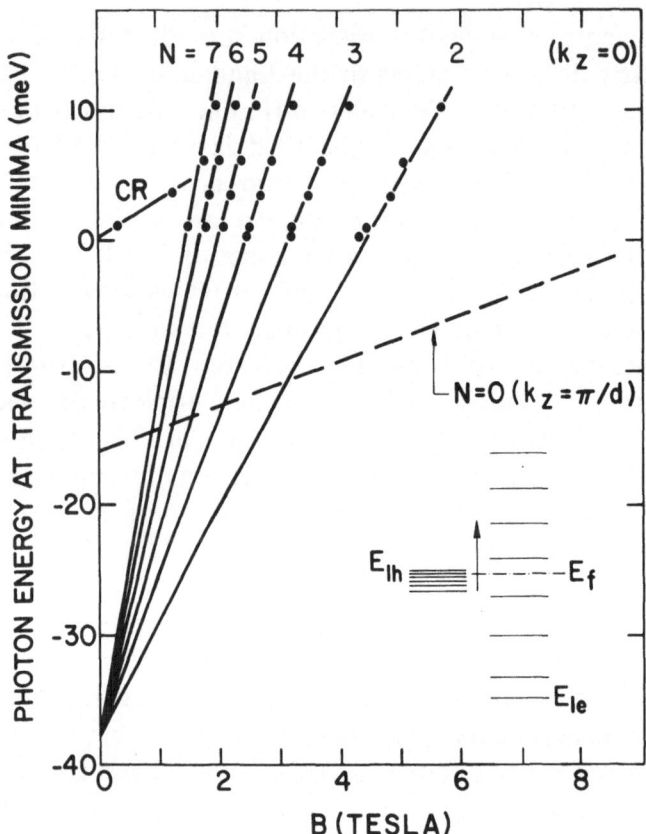

Figure 11 - Energies of transmission minima vs. fields for the superlattice in Fig. 10. CR denotes cyclotron resonance of electrons. Solid lines indicate inter-subband transitions at the zone center involving the ground states of electrons and holes as illustrated in the inset. The dotted line represents inter-subband transition at the zone edge, involving the same states. Landau level indices are identified, as shown.

transitions.[48] This is shown in dotted lines in the figure, and corresponds to transitions between the same bands but at the zone edge, $k_z = \pi/d$ where the density of states also peaked, similar to the situation at $k_z = 0$ with which the original set of transitions was identified. The energy difference at B=0 between the two sets is a direct measure of the subband width of ΔE_{1e}, the deviation of the superlattice from strict two-dimensionality.

The effect of nonparabolicity is particularly significant for electrons in the conduction band of InAs. For the energy range under

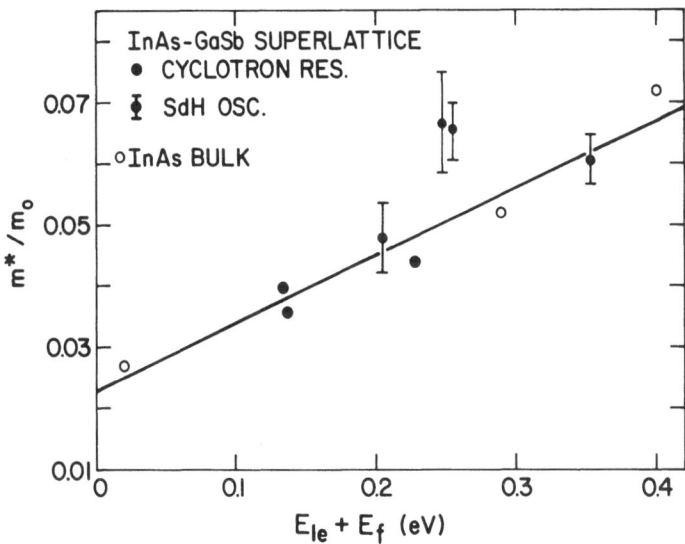

Figure 12 - Effective mass of electrons obtained from cyclotron resonance and Shubnikov-de Haas oscillations in InAs-GaSb superlattices. The conduction bandedge of InAs is the reference energy. The solid line represents the simple two-band parabolicity model. Some bulk values are included for comparison.

consideration, the two-band approximation has been found to be quite adequate, which involves the replacement of the electron energy E by $E(1+E/E_g)$ where E_g is the gap of InAs. The solid lines in Fig. 11 are calculated results from this approximation to fit the experimental data. They deviate from straight lines because of the nonparabolic electron mass. The increase of mass with energy from this approximation can be expressed explicitly by $1+2(E_{1e}+E_f)/E_g$, where E_{1e} is defined with respect to the conduction bandedge, and E_f to E_{1e}. Figure 12 shows the cyclotron masses from different structures, which are seen to be consistent with the theoretical expression. The measured, enhanced mass, corresponding to a high electron energy in the conduction band, was in fact used early in demonstrating the formation of the subband. Also included in Fig. 12 are masses obtained from the temperature dependence of the Shubnikov-de Haas oscillations. The large scattered values presumably reflect the difficulties known for such measurements.[49]

Another case of interest is the superlattice in the heterojunction limit shown above in Fig. 6 where more than one electron subband

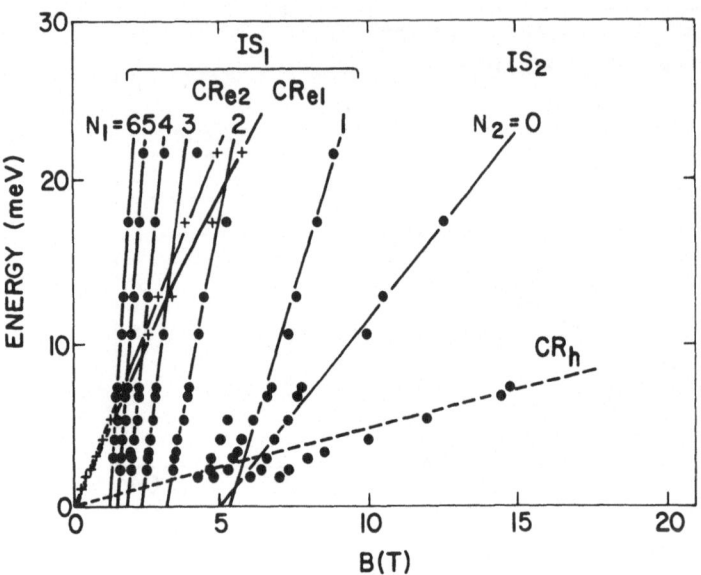

Figure 13 - Energies of transmission minima in an InAs(1000Å)-GaSb(1000Å) superlattice in the heterojunction limit. The transitions involve two electron subbands and one hole subband: two inter-subband transitions (IS$_1$ and IS$_2$), and three cyclotron transitions (CR$_{e1}$ and CR$_{e2}$ for electrons, and CR$_h$ for holes). Landau level indices are indicated.

exist. The magneto-absorption spectra showed a large number of peaks,[50] as plotted in Fig. 13. The interband series of the ground states, E$_{1e}$-E$_{1h}$, designated as IS$_1$, is similar to that in Fig. 11. But additional transitions are present. The IS$_2$ branch corresponds to the ground Landau level transition from the same hole subband, E$_{1h}$, to the second electron subband, E$_{2e}$; the difference of IS$_2$ and IS$_1$ in energy at B=0, in this case, equals the spacing between the two electron subbands. Two cyclotron resonances of electrons are resolvable, reflecting the two different electron masses associated with the two subbands as a direct consequence of the nonparabolicity effect. The fact that the higher-lying subband has a lighter mass, $m_2 < m_1$, contrary to intuitive thinking, is because the nonparabolic enhancement is less for E$_{2e}$ than E$_{1e}$ at the Fermi energy where cyclotron absorption is measured. The CR$_h$ branch in the figure is attributed to the cyclotron resonance of holes, whose mass is given correctly by the slope. The results shown in Fig. 13 demonstrate the ability of magneto-absorption for elucidating the various transitions arising from the band structure in heterostructures.

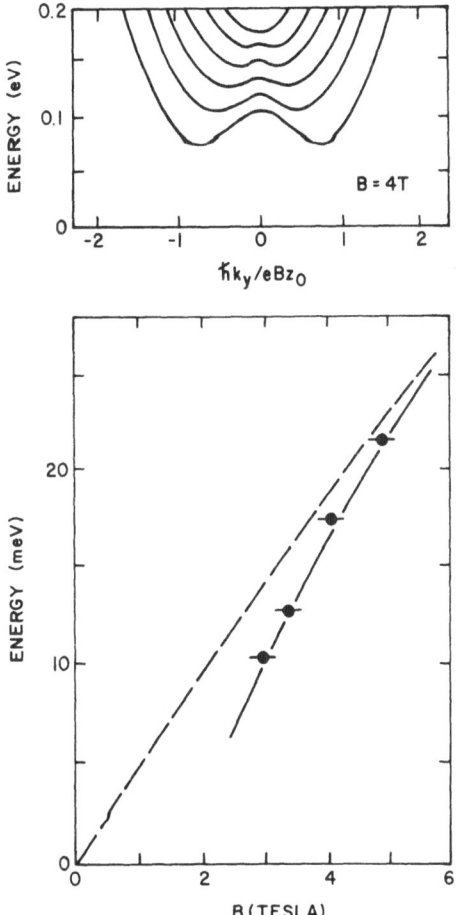

Figure 14 - InAs-GaSb superlattice of the same configuration as in Fig. 13 under parallel magnetic field. The upper diagram shows the calculated energy dispersion vs. the center coordinate of the cyclotron orbit normalized to the InAs layer thickness. The lower diagram shows the magneto-absorption data together with calculated hybrid subband transition (solid) and cyclotron resonance (dotted).

Tilting the magnetic field in a two-dimensional electron system has the same effect on absorption as on transport in that only the perpendicular field component is of importance.[51]   At large tilting angles and high fields, however, mixed electric and magnetic quantization occurs, as mentioned earlier. In the extreme case of a parallel field along x, the energies can be solved numerically from Eq. (5) by a proper transformation.[11] Results from both theoretical calculations and experimental absorption are shown in Fig. 14 for the InAs-GaSb heterojunction case.[52] The energy depends on the center coordinate position of the cyclotron orbit $\hbar k_y/eB$, as illustrated for a specific field. Here the magnetic field contribution becomes gradually dominant, leading to Landau-like states with energy spacings $\hbar\omega_c$ near the center $k_y=0$ and $2\hbar\omega_c$ at the interface $\hbar k_y = eBz_0$, the familiar magnetic surface states.[53] The InAs layer thickness, z_0, is the characteristic length parameter. Experimentally, well-defined transitions, which were not resolvable at low fields when magnetic and electric contributions are comparable, were observed as the field was increased. The transitions correspond to those near the center with an energy approaching $\hbar\omega_c$ as shown in Fig. 14.

As demonstrated in Si-inversion layers,[5] cyclotron resonance is a complex phenomenon, depending on the field, the frequency of excitation and the electron density. Crude analyses to the resonance lineshape in early work yielded scattering times, which were consistent with those estimated from mobilities.[44,51] Recently, interest has been focused on cyclotron interactions with impurities and phonons. The linewidth as a function of the field was studied extensively in GaAs-GaAlAs.[54,55] It was found to vary with $\sqrt{B}$ at low fields in accordance with the short-range scattering theory[56] but deviate below this dependence as the field was increased.[55] An oscillatory linewidth correlated with the occupation of the Landau levels was also reported.[57] The interaction between electrons and LO-phonons was considered theoretically,[58] predicting an enhanced polaron mass in a two-dimensional system. The effect of carrier screening, which has the opposite influence, however, plays a dominant role; the enhancement of the cyclotron mass has been shown to be adequately described by the nonparabolic effect alone.[59] In the InP-InGaAs system, the cyclotron linewidth was recently reported to depend on the field in a similar fashion, but increase dramatically at fields corresponding to the LO-phonons in InAs.[60] We should mention, in this connection, that interband- phonon collective excitations have been extensively investigated from inelastic light scattering experiments,[61] as treated thoroughly elsewhere in this volume. Magnetic fields, however, are usually not involved.

5. OTHER PROPERTIES IN MAGNETIC FIELDS

5.1 Magneto-Phonon Oscillations

Magneto-phonon resonance results from electron scattering with resonant absorption of phonons. With an increase in the magnetic field, the magneto-resistance, or σ_{xx}, goes through a minimum whenever the resonant condition, $\hbar\omega_{ph} = N\hbar\omega_c$, is satisfied. In this expression, $\hbar\omega_{ph}$ is the phonon energy, N is an integer, being the Landau index difference, and $\hbar\omega_c$ is the usual cyclotron energy, containing, strictly speaking, the polaron mass. Unlike Shubnikov-de Haas oscillations, the present effect is a direct consequence of electron interaction with the lattice, requiring $\hbar\omega_{ph} > \hbar\omega_c$ instead of $E_f > \hbar\omega_c$. In addition, since low temperature decreases the phonon population which is needed and high temperature broadens the Landau levels as usual, the magneto-phonon oscillation is observable only through an intermediate range of temperatures. Also, the effect is rather small, and can usually be resolved experimentally only through derivative techniques.

The magneto-phonon oscillation was first observed in GaAs-GaAlAs.[62] The results verified the relation between the period in $1/B$ and N through $e/m^*\omega_{ph}$, demonstrated the involvement of two-

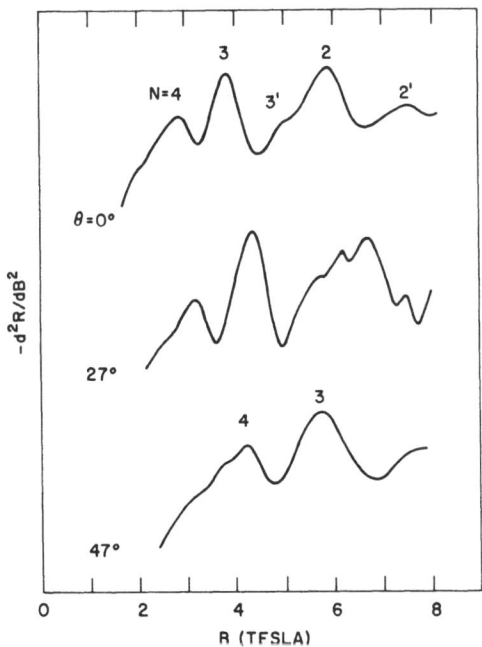

Figure 15 - Second derivative of magneto-resistance vs. field in an InGaAs(150Å)-InP(300Å) superlattice to illustrate magneto-phonon oscillations at 150K. Numbers and prime numbers indicate, respectively, LO phonon modes in InGaAs (GaAs-like) and InP. (After Ref. 63).

dimensional electrons with the cosθ dependence by tilting the field, and identified the phonon to be LO of GaAs. More recently, similar observations were made in the InGaAs-InP and InGaAs-InAlAs systems.[63,64] The oscillatory characteristics for the InGaAs-InP superlattice are shown in Fig. 15. The dominant phonon is LO(GaAs-like) in InGaAs layers where electrons are largely concentrated. But additional peaks associated with LO(InP) are also seen, as indicated by prime numbers in the figure. Mutual penetration, electrons into InP and phonons into InGaAs, was believed to cause their interaction. In contrast, the phonons involved in the heterojunctions of InGaAs-InP and InGaAs-InAlAs are GaAs-like and InAs-like LO phonons, respectively.[64] Again, the various peaks shift to higher fields with the tilting angle, as can be seen in Fig. 15.

5.2. Magneto-Plasmon Effect

There has been a growing interest in collective excitations involving an electron plasma. For the superlattice where the dimensionality can be varied, theoretical considerations have led to plasmon dispersion relations which depend on the coupling strength of the electron layers.[65,66] The frequency varies with the square-root of the wave-vector in the weakly coupled case but linearly when the coupling becomes strong, approaching the two and three-dimensional limits, respectively. Under magnetic fields, a number of collective modes were predicted, including magneto-plasmons, helicon and Alfren waves. Experimentally, the technique of infrared absorption has been commonly used to investigate the collective modes, as demonstrated in Si-inversion layers where both plasmons and magneto-plasmons have been observed.[67] The other technique is that of inelastic light scattering, mentioned earlier in connection with collective interband-phonon excitations. This latter technique was used to provide definitive evidence for the observation of plasma oscillations in GaAs-GaAlAs. With a strongly-coupled superlattice, the predicted linear dispersion relation was demonstrated by varying the scattering angle.[68]

For magneto-plasmons in heterostructures, the subject of our interest, only preliminary measurements have been reported. In GaAs-GaAlAs, light-scattering spectra exhibited additional peaks with energies up-shifted from those of the cyclotron resonance.[69] They were attributed to the magneto-plasmon mode with frequencies ω_{mp}, which appeared to follow the usual relation, $\omega^2_{mp} = \omega^2_p + \omega^2_c$. The deduced plasmon frequency, ω_p, however, was much lower than that expected theoretically. In InAs-GaSb, a series of maxima were observed in the infrared transmission spectra under magnetic fields, and they were explained on the basis of helicon wave propagation.[70] The magneto-plasma modifies the dielectric constant by $\omega_p^2/\omega\omega_c$ under proper conditions. With favorable polarization, resonant transmission of the helicon wave occurs at $\omega = \omega_h$ when the superlattice thickness becomes integral multiples of the half-wavelength of the radiation in the sample. The experiments verified the polarization conditions and the functional relations between ω_h and B. The electron mass obtained was, however, higher than that expected from the effect of nonparabolicity. In both cases, the plasmons are three-dimensional in nature, as the superlattices have strongly coupled layers for the wavelengths under consideration.

5.3 Cyclotron Emission

Under the application of an electric field, electrons are heated, and emission may occur through their radiative decay. Using this technique in GaAs-GaAlAs heterostructures with the field parallel to the layers, a variety of emission processes were observed, involving plasmons, interband and cyclotron transitions.[71] Of these, the former two do not inherently require a magnetic field, although it is commonly used to tune the detector to a particular energy window for the purpose of detecting the response signal. In cyclotron radiation, a GaAs detector is typically used with a fixed detecting energy. The emission signal is registered whenever $\hbar\omega_c$ coincides with this energy by varying the magnetic field.

Figure 16 shows the response vs. B for a GaAs-GaAl heterojunction as B is varied from the surface normal.[72] Both emissions from the bulk GaAs and the two-dimensional electron gas have been observed, and they can be distinguished as the energies of the latter shift to higher values with increasing θ, following the usual angular dependence of $\cos\theta$. The electron-mass obtained was slightly enhanced from the bulk mass by the small nonparabolicity in GaAs. The linewidth of the emission was narrowed with respect to that of the bulk, consistent with the effect of electron screening of the impurities. Similar experiments were also performed in superlattice structures. In all cases, the emission intensities follow approximately the square of the electric field, which is proportional to the input power. The efficiency of heating,

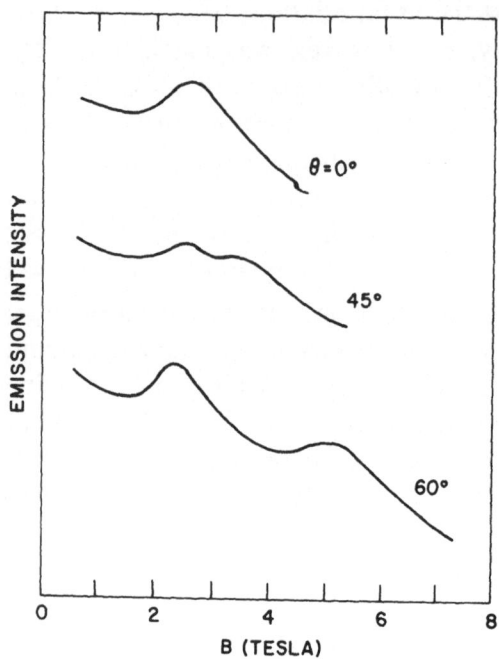

Figure 16 - Cyclotron emission in a GaAs-$Ga_{0.74}Al_{0.26}$As heterojunction obtained by using a GaAs detector at 4.4meV. The peak shifting to higher values with field is identified with the two-dimensional electron gas. (After Ref. 72).

however, decreases from the three-dimensional electrons in the bulk to the two-dimensional electrons in heterojunctions and, further, in super-lattices.

5.4 de Haas-van Alphen Effect

The de Haas-van Alphen effect has been used extensively in studying the Fermi surface of metals. Unlike magneto-transport which involves scattering, this effect deals with the measurement of magnetic moment or susceptibility, thus providing a direct means of probing the density of states. In a two-dimensional system, the energy of electrons fully filling up to the Nth Landau level is given by $\Sigma \zeta (N + 1/2)\hbar \omega_c$ or $(1/2)\zeta (N + 1)^2 \hbar \omega_c$. The electron energy for the next partially filled level is $(N + 3/2)[n - \zeta (N + 1)]\hbar \omega_c$ where n is the electron density. As can readily be seen that the free energy of the system, the sum of the two, is periodic in B^{-1}, similar to the Shubnikov-de Haas effect, and, in the ideal case of density of states of Fig. 3, shows a cusp each time E_f coincides with the Landau level. Under the same condition, the magnetic moment, defined as the first derivative of energy with respect to B, shows a discontinuity of magnitude $n(e\hbar/m^*)$, and the magnetic susceptibility, as the second derivative, exhibits a singularity. The sharpness of the oscillations is smeared by level broadening in reality. By measuring the magnetic moment, information with regard to the actual shape of the density of states can, in principle, be obtained.

While the de Haas-van Alphen effect is fundamental and elegant conceptually, experimental difficulty arises in two-dimensional systems as the number of electrons is limited, and it must be enhanced by increasing the area to get a detectable signal. This was achieved experimentally by stacking together in sequence a large number of superlattice layers of GaAs-GaAlAs.[73] Oscillatory magnetic moment with respect to B was observed with a SQUID magnetometer. The oscillations were consistent with those from Shubnikov-de Haas measurements, but the amplitude was much smaller than expected. This would indicate considerable level broadening, inconsistent with independent mobility and cyclotron resonance measurements. Inhomogeneities appeared to be the apparent cause for the discrepancy. We should mention, in this connection, that the deHaas-van Alphen effect was also observed recently in Si-inversion layers by a modulation technique of varying the electron density.[74] In both cases, the experiments were preliminary, but they demonstrated the feasibility of the de Haas-van Alphen measurement.

5.5. Weak Localization

There have been considerable theoretical and experimental activities in weak localization with voluminous reports in the literature, since the early work in this field.[75,76] In a two-dimensional electron system such as thin metallic films and Si-inversion layers,[77,78] the conductance at low temperatures decreases logarithmically with decreasing temperature; the proportional constant is given by $c(e^2/2\pi^2\hbar^2)$. Two mechanisms can contribute to this behavior: localization effect in which $c = \alpha p$ where α is the scattering parameter and p is the temperature exponent of the inelastic scattering rate;[75] and interaction effect in which $c = 1-F$ where F is the electron-electron screening factor.[79] The role the magnetic field plays here is to provide a means to distinguish between the two mechanisms. While the interaction effect is generally enhanced by the field as it reduces screening to produce a positive magneto-resistance, the localization effect is depressed by the field to give rise to a negative magneto-resistance as the cyclotron radius is shrunk and may become smaller than the inelastic scattering length, thus providing the limiting scale in the localization process.[80] In addition, it has been shown that the correction to the Hall coefficient with respect to that of the resistance is either 2 or zero in the interaction or localization regime, respectively. Furthermore, the negative magneto resistance from localization is an orbital effect, depending on the orientation of the field. In contrast, the interaction effect is primarily spin-related in nature. The ability to separate the various contributions in general depends on a number of parameters, including the electron densities and mobilities in the sample as well as the ranges of fields and temperatures in the measurement. Typically, localization is readily observable in relatively low-mobility samples in the low field range before this effect is quenched by the magnetic field.

The conductivity behavior in the weak localization reime has been investigated as functions of temperatures and fields in the GaAs-GaAlAs heterojunction.[81] Negative magneto-resistance as a result of the localization effect was observed, from which the inelastic scattering time was derived, giving $\alpha p \simeq 1$ as predicted.[75] This negative magneto-resistance was absent with parallel fields, and it turned positive as the field was increased; both were consistent with theoretical expectations. The existence of the interaction effect was evidenced both from the correction of the Hall coefficient[86,82] and, more recently, from the temperature dependence of the conductivity after the localization was completely quenched by the magnetic field.[83] Preliminary experiments have also been performed recently in InAs-GaSb quantum wells.[84] Negative magneto-resistance was observed, and was ascribed to the weak localization of holes.

ACKNOWLEDGMENTS

I would like to thank many of my collaborators who have contributed to the work covered in this Chapter, in particular, C. A. Chang, L. Esaki, and E. E. Mendez of this laboratory, G. Bastard, Y. Guldner, and M. Voos of l'ENS, Paris, and J. C. Maan of MPI, Grenoble. I am also grateful to many of my colleagues who have made available to me their work prior to publication. This research was partially sponsored by the U.S. Army Research Office.

REFERENCES

1. See, for example, A. G. Milnes and D. L. Feucht, Heterojunctions and Metal-Semiconductor Junctions, Academic Press, New York, 1972.
2. See, for example, H. Kroemer, Surf. Sci. 132 (1983).
3. See, for example, L. L. Chang, in Handbook of Semiconductors, ed. by T. S. Moss (North Holland, Amsterdam, 1980), Vol. 3, Chap. 9.
4. See, for a recent review, L. L. Chang, J. Vac. Sci. Technol. B1, 120 (1983).
5. See, for a recent review, T. Ando, A. B. Fowler and F. Stern, Rev. Modern Phys. 54, 437 (1982).
6. L. L. Chang and L. Esaki, Surf. Sci. 98, 70 (1980).
7. G. Bastard, Phys. Rev. B24, 5693 (1981).
8. J. N. Schulman and T. C. McGill, Phys. Rev. Lett. 39, 1680 (1977).
9. E. Caruthers and P. J. Lin-Chung, Phys. Rev. Lett. 38, 1543 (1977).
10. T. Ando and Y. Uemura, J. Phys. Soc. Japan 36, 959 (1974).
11. J. C. Maan, in Proc. Int. Conf. Application of High Magnetic Fields in Semicond. Phys., Grenoble, 1982, p. 164.
12. F. F. Fang and P. J. Stiles, Phys. Rev. 174, 823 (1968).
13. T. Ando and Y. Uemura, J. Phys. Soc. Japan 37, 1044 (1974).
14. See, for example, L. M. Roth and P. N. Argyres, in Semiconductors and Semimetals, Vol. 1, ed. by R. K. Willardson and A. C. Beer (Academic Press, New York, 1966) Chap. 6.
15. L. L. Chang, H. Sakaki, C. A. Chang and L. Esaki, Phys. Rev. Lett. 20, 1489 (1977).
16. L. L. Chang, presented at the 2nd Int. Conf. Electron Properties Two-Dimensional Systems, Berchtesgaden, Sept. 1977 (unpublished).
17. R. Dingle, H. L. Störmer, A. C. Gossard, and W. Wiegmann, Appl. Phys. Lett. 33, 665 (1978).
18. R. Dingle, H. L. Störmer, A. C. Gossard and W. Wiegmann, Surf. Sci. 98, 90 (1980).
19. T. Ando, J. Phys. Soc. Japan 50, 2978 (1981).
20. L. L. Chang, N. J. Kawai, G. A. Sai-Halasz, R. Ludeke, and L. Esaki, Appl. Phys. Lett. 35, 939 (1979).
21. L. L. Chang, N. J. Kawai, E. E. Mendez, C. A. Chang, and L. Esaki, Appl. Phys. Lett. 38, 30 (1981).

22. N. B. Brant and E. A. Svistova, J. Low Temp. Phys. 2, 1 (1970).

23. N. J. Kawai, L. L. Chang, G. A. Sai-Halasz, C. A. Chang, and L. Esaki, Appl. Phys. Lett. 36, 369 (1980).

24. L. L. Chang, E. E. Mendez, N. J. Kawai, and L. Esaki, Surf. Sci. 113, 306 (1982).

25. Th. Englest, D. C. Tsui, A. C. Gossard, and Ch. Uihlein, Surf. Sci. 113, 295 (1982).

26. S. Narita, S. Takeyama, W. B. Luo, S. Hiyamizu, K. Nanbu and H. Hashimoto, Surf. Sci. 113, 301 (1982).

27. R. J. Nicholas, M. A. Brummell, J. C. Portal, M. Razeghi, and M. A. Poisson, Solid State Commun. 43, 825 (1982).

28. K. von Klitzing, G. Dorda, and M. Pepper, Phys. Rev. Lett. 45, 494 (1980).

29. D. C. Tsui and A. C. Gossard, Appl. Phys. Lett. 38, 550 (1981).

30. D. C. Tsui, A. C. Gossard, B. F. Field, M. E. Cag, and R. F. Dzinba, Phys. Rev. Lett. 48, 3 (1982).

31. R. E. Prange, Phys. Rev. B 23, 4802 (1981).

32. R. B. Laughlin, Phys. Rev. B23, 5632 (1981).

33. D. C. Tsui, H. L. Störmer, and A. C. Gossard, Phys. Rev. Lett. 48, 1559 (1982).

34. H. L. Störmer, A. Chang, D. C. Tsui, C. M. Huang, A. C. Gossard, and W. Wiegmann, Phys. Rev. Lett. 50, 1953 (1983).

35. E. E. Mendez, L. L. Chang, M. Heiblum, L. Esaki, M. Naughton, K. Martin, and J. Brooks, to appear in Phys. Rev.

36. R. B. Laughlin, Phys. Rev. Lett. 50, 1395 (1983).

37. E. E. Mendez, M. Heiblum, L. L. Chang, and L. Esaki, Phys. Rev. B28, 4886 (1983).

38. J. E. Schirber, I. J. Fritz, L. R. Dawson, and G. C. Osburn, Phys. Rev. B28, 2229 (1983).

39. A. Kastalsky, R. Dingle, K. Y. Cheng, and A. Y. Cho, Appl. Phys. Lett. 41, 274 (1982).

40. H. Kinoshita, S. Takaoka, K. Murase, and H. Fujiyasu in Collected Papers 2nd Int. Symp. MBE and Related Clean Surface Techniques, Tokyo, 1982 (Japan Soc. Appl. Physics) p.61.

41. Y. Guldner, J. P. Hirtz, J. P. Vieren, P. Voisin, M. Voos, and M. Razeghi, J. Phys. 42, L613 (1982).

42. E. E. Mendez, L. L. Chang, C. A. Chang, L. F. Alexander, and L. Esaki, presented at the 5th Int. Conf. Electron Properties Two-Dimensional Systems, Oxford, 1983 (to appear in Surf. Sci.).

43. C. A. Chang, E. E. Mendez, L. L. Chang, and L. Esaki, presented at the 5th Int. Conf. Electron Properties Two-Diemsnional Systems, Oxford, 1983 (to appear in Surf. Sci.).

44. H. L. Störmer, R. Dingle, A. C. Gossard, W. Wiegmann, and M. D. Sturge, Solid State Commun. 29, 705 (1979).

45. H. Bluyssen, J. C. Maan, P. Wyder, L. L. Chang, and L. Esaki, Solid State Commun. 31, 35 (1979).

46. L. L. Chang, G. A. Sai-Halasz, L. Esaki, and R. L. Aggarwal, J. Vac. Sci. Technol. 19, 589 (1981).

47. Y. Guldner, J. P. Vieren, P. Voisin, M. Voos. L. L. Chang, and L. Esaki, Phys. Rev. Lett. 45, 1719 (1980).

48. J. C. Maan, Y. Guldner, J. P. Vieren, P. Voisin, M. Voos, L. L. Chang, and L. Esaki, Solid State Commun. 39, 683 (1981).

49. F. F. Fang, A. B. Fowler, and A. Hartstein, Surf. Sci. 73, 269 (1978).

50. Y. Guldner, J. P. Vieren, P. Voisin, M. Voos, J. C. Maan, L. L. Chang, and L. Esaki, Solid State Commun. 41, 755 (1982).

51. H. Bluyssen, J. C. Maan, P. Wyder, L. L. Chang, and L. Esaki, Phys. Rev. B25, 5364 (1982).

52. J. C. Maan, Ch. Uihlein, L. L. Chang, and L. Esaki, Solid State Commun. 44, 653 (1982).

53. M. Wanner, R. E. Dozema, and U. Strom, Phys. Rev. B12, 2883 (1975).

54. K. Muro, S. Narita, S. Hiyamizu, K. Nanbu, and H. Hashimoto, Surf. Sci. 113, 321 (1982).

55. P. Voisin, Y. Guldner, J. P. Vieren, M. Voos, J. C. Maan, P. Delescluse, and N. T. Linh, Physica 117-118BC, 634 (1983).

56. T. Ando, J. Phys. Soc. Japan 38, 989 (1975).

57. Th. Englert, J. C. Maan, Ch. Uihlein, D. C. Tsui, and A. C. Gossard, Solid State Commun. 46, 545 (1983).

58. S. Das Sarma, presented at the 5th Int. Conf. Electron Properties Two-Dimensional Systems, Oxford, 1983 (to appear in Surf. Sci.).

59. W. Seidenbush, G. Lindemann, R. Lassnig, J. Ellinger, and E. Gornik, presented at the 5th Int. Conf. Electron Properties Two-Dimensional Systems, Oxford, 1983 (to appear in Surf. Sci.).

60. M. A. Brummell, R. J. Nicholas, L. C. Brunel, S. Huyant, M. Baj, J. C. Portal, M. Razeghi, M. A. Di Forte-Poisson, K. Y. Cheng and A. Y. Cho, presented at the 5th Int. Conf. Electron Properties Two-Dimensional Systems, Oxford, 1983 (to appear in Surf. Sci.).

61. A. Pinczuk and J. M. Worlock, Physica 117-118BC, 637 (1983).

62. D. C. Tsui, Th. Englert, A. Y. Cho, and A. C. Gossard, Phys. Rev. Lett. 44, 341 (1980).

63. R. J. Nicholas, M. A. Brummell, J. C. Portal, M. Razeghi, and M. A. Poisson, in Proc. Int. Conf. Application of High Magnetic Fields in Semicond. Phys., Grenoble, 1982, p. 130.

64. J. C. Portal, G. Gregoris, M. A. Brummell, R. J. Nicholas, M. Razeghi, M. A. DiFort-Poisson, K. Y. Cheng, and A. Y. Cho, presented at the 5th Int. Conf. Electron Properties Two-Dimensional Systems, Oxford, 1983 (to appear in Surf. Sci.).

65. S. Das Sarma and J. J. Quinn, Phys. Rev. B12, 7603 (1982).

66. W. L. Bloss and E. M. Brody, Solid State Commun. 43, 423 (1982).

67. T. N. Theis, Surf. Sci. 98, 515 (1980).

68. D. Olego, A. Pinczuk, A. C. Gossard, and W. Wiegman, Phys. Rev. B25, 7867 (1982).

69. J. M. Worlock, A. C. Maciel, C. H. Perry, Z. J. Tien, R. Aggarwal, A. C. Gossard, and W. Wiegmann, in Proc. Int. Conf. Application of High Magnetic Fields in Semicond. Phys., Grenoble, 1982, p. 154.

70. J. C. Maan, M. Altarelli, H. Sigg, P. Wyder, L. L. Chang, and L. Esaki, Surf. Sci. 113, 347 (1982).

71. R. Hopfel, G. Lindemann, E. Gornik, G. Stangl, A. C. Gossard, and W. Wiegmann, Surf. Sci. 113, 118 (1982).

72. E. Gornik, R. Schawarz, D. C. Tsui, A. C. Gossard, and W. Wiegmann, Solid State Commun. 38, 541 (1981).

73. H. L. Stormer, T. Haavasoja, V. Narayanamurti, A. C. Gossard, and W. Wiegmann, J. Vac. Sci. Tech. B1, 423 (1983).

74. F. F. Fang and P. J. Stiles, presented at the 5th Int. Conf. Electron Properties of Two-Dimensional Systems, Oxford, 1983 (to appear in Surf. Sci.)

75. E. Abrahams, P. W. Anderson, D. C. Licciardello, and T. V. Ramakrishnan, Phys. Rev. Lett. 42, 673 (1979).

76. G. J. Dolan and D. D. Osheroff, Phys. Rev. Lett. 43, 721 (1979).

77. D. J. Bishop, D. C. Tsui and R. C. Dynes, Phys. Rev. Lett. 46, 360 (1981).

78. R. C. Dynes, Surf. Sci. 113, 510 (1982).

79. B. L. Altshuler, A. C. Aronov and P. A. Lee, Phys. Rev. Lett. 44, 1288 (1980).

80. H. Fukuyama, Surf. Sci. 113, 489 (1982).

81. D. A. Poole, M. Pepper and R. W. Glew, J. Phys. C 14, L995 (1981).

82. R. A. Davies, C. C. Dean and M. Pepper, presented at the 5th Int. Conf. Electron Properties of Two-Dimensional Systems, Oxford, 1983 (to appear in Surf. Sci.).

83. M. A. Paalanen, D. C. Tsui, B. J. Line and A. C. Gossard, presented at the 5th Int. Conf. Electron Properties of Two-Dimensional Systems, Oxford, 1983 (to appear in Surf. Sci.).

84. S. Washburn, R. A. Webb, E. E. Mendez, L. L. Chang, and L. Esaki, to appear in Phys. Rev. B.

MODULATION DOPING OF SEMICONDUCTOR HETEROSTRUCTURES

A.C. Gossard

Bell Laboratories, Murray Hill, N.J. 07974

ABSTRACT: Modulation-doped heterostructures are materials containing nonuniform distributions of dopant atoms and embodying heterostructures to guide carriers and separate them from impurities. The modulation-doping technique can produce extremely high carrier mobility, which is interesting both for device application and for observation of new two-dimensional effects. Growth and characteristics of modulation doped structures are described and dependences of carrier densities and mobilities on growth and structure parameters are developed. Two-dimensional phenomena, including the quantized Hall effect, are presented. Techniques for controlling carrier densities are discussed.

1. INTRODUCTION

A major goal of solid state physics and solid state technology is the perfection of materials and structures in which charge carriers have long lifetimes, low scattering, high mobilities, and controlled densities. Such materials have allowed the elucidation of the electronic structure of solids and the development of semiconductor electronics and photonics. In materials such as semiconductors where carriers are introduced by incorporation of impurities, however, the improvement of lifetimes and scattering rates is generally limited by scattering by impurities necessarily incorporated in order for carriers to be present or confined. The technique of modulation doping of semiconductor heterostructures provides a means for avoiding these limits on the performance of semiconductor materials and allows access to new regimes of physical effects and device performance.

The modulation doping (MD) technique traces its origins to studies of semiconductor heterostructures and superlattices, which demonstrated charge carrier confinement by heterojunctions[1] and showed that in layers which are sufficiently thin the carriers occupy quantized energy states.[2,3] The quantization occurs in the dimension perpendicular to the heterostructure, whereas the motion parallel to the interface is free. With only two degrees of free motion available, the charge carriers in each quantum state form a two-dimensional system. Deviations from two-dimensionality are produced only by scattering between quantum states or by tunnelling between different layers.

Carriers can be introduced into the two-dimensional quantum states by optical pumping of electrons between valence and conduction bands or by use of chemical doping. The first experimental observation of the two-dimensionality was the detection of Shubnikov-de Haas magnetoresistance oscillations of the conduction electrons in uniformly Sn-doped molecular beam epitaxy (MBE)-grown GaAs-$Al_xGa_{1-x}As$ superlattices.[4] With thick (>100Å) barrier layers of $Al_xGa_{1-x}As$, motion of the electrons was demonstrated to be only in the planes of the layers while with thinner barriers, evidence of barrier penetration was found. The observations required intense magnetic fields in order to make the cyclotron-orbit times of the carriers less than their scattering times. Although Esaki and Chang had speculated that carrier mobilities might be enhanced by use of nonuniform doping[5] they did not use modulation doping in preparation of the structures.[4]

   The first modulation doped heterostructures were MBE-grown GaAs-Al$_x$Ga$_{1-x}$As superlattices with modulated silicon doping.[6] The structures were designed to separate donors from mobile electrons in superlattices in order to increase the time between scattering events and thus observe more clearly any two-dimensional resistance effects. Silicon was selected as the dopant because it could be incorporated with abrupt concentration profiles in MBE growth. The initial experiments showed that efficient introduction of carriers into the superlattice was produced by either uniform doping or by doping of only barrier layers, but that substantially increased mobilities were produced when only barrier layers were doped. The source of transfer of charge from the barriers to the GaAs wells is the lower energy of electron states in the wells relative to electrons residing in the barriers or bound to donors in the barriers. The principal mechanism for the mobility enhancement is the reduction in scattering of the carriers from impurities by means of the spatial separation of the inpurities and the mobile electrons.

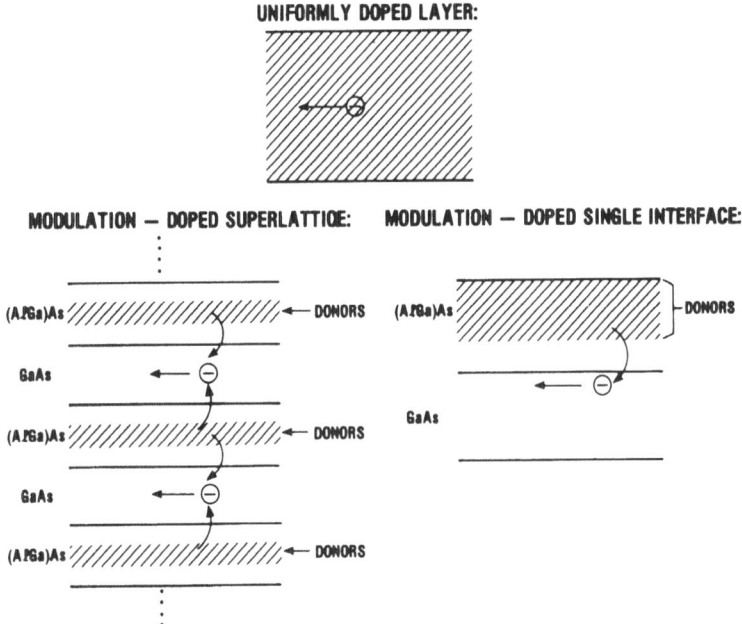

Figure 1 - Spatial arrangement of layers in GaAs modulation- doped structures. Shaded areas represent doped layers. Electrons transfer to lower energy mobile states in undoped GaAs layers.

   In a modulation doped heterostructure superlattice, electrons confined to the quantum layers are bounded on each side of each well by the heterojunction potential step between the well materials and the barrier material. It is also possible with modulation doping to confine electrons at a single interface between an undoped channel material and a doped barrier material. This was achieved both with MBE[7] and with LPE (liquid phase epitaxy)[8] grown structures, although the MBE structures are most widely used and developed.

In the single interface structures, the carriers in the channel are confined at the heterojunction potential step by the electric field of the ionized donors from which the carriers came. The resultant potential well is essentially triangular in shape and confines carriers in quantum states much as in the superlattice case. Motion along the heterointer-face occurs with enhanced mobility, also much as in the modulation-doped superlattice. These systems are illustrated schematically in Figure 1. They have provided an excellent means for the realization of the two-dimensional electron gas and have shown many remarkable new physical phenomena. They also have provided an opportunity for exploring novel device phenomena and for developing high performance device structures.

In Section 2 the principles of modulation doping are developed, particularly with regard to the various available parameters for design and construction of modulation-doped structures. Section 3 reviews the growth and measurement techniques. Section 4 presents the experimentally observed conductivity and mobility behavior of modulation-doped structures. Section 5 treats the various phenomena of the two-dimensional electron gas in these structures, including the quantized Hall effect.

2. PRINCIPLES OF MODULATION DOPING

2.1 Charge Transfer

The charge transfer and carrier confinement which form the basis for most modulation doping effects in heterostructures are produced by the step in band edge energies occuring at the heterojunction.[1] This step can be probed in several ways. Optical determination of confined particle energy levels permits a determination of confining potential barrier heights. The energy levels can be deduced from interband optical absorption spectra,[3] photoluminescence excitation spectra,[9] and from intersubband Raman scattering spectra.[10] In the GaAs/Al$_x$Ga$_{1-x}$As system, both electrons and holes can be confined in the GaAs layers, whereas in the InAs/GaSb system, they might be confined in separate layers, with electrons contained in the InAs layers and holes in the GaSb layers.[11] In the GaAs/Al$_x$Ga$_{1-x}$As system about 85% of the difference in bandgap between the two species occurs in the conduction band, corresponding essentially to the electron affinity difference between the materials, and the remaining 15% occurs in the valence band.[12] These discontinuities are illustrated in Figure 2. The bandgap difference between GaAs and Al$_x$Ga$_{1-x}$As is 1.247x electron volts, so that for x = 0.30, electrons experience a 0.32 eV conduction band-edge discontinuity and holes are subject to a 0.06 eV valence band-edge discontinuity.

These band-edge discontinuities form an upper limit for the difference in energy between electrons bound to a donor in the barrier

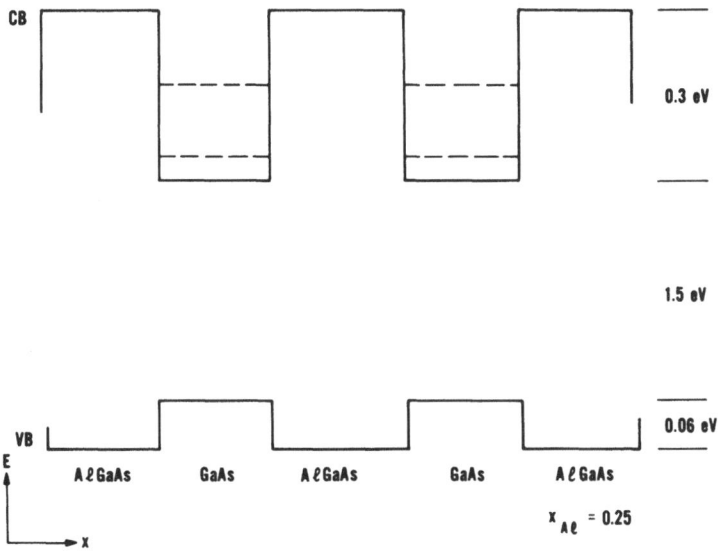

Figure 2 - Conduction-band and valence-band edges in undoped GaAs/(Al,Ga)As heterostructure. Electrons introduced into GaAs layers occupy quantum energies shown by dashed lines.

layer near the interface and electrons in the (GaAs) channel layer. The actual energy difference will be reduced by several effects.

2.1.1. Donor binding energy. The electrons bound to donors in barrier material will lie below the bottom of the barrier conduction band. This binding energy is greater in $Al_xGa_{1-x}As$ than in GaAs, reaching values on 50 to 100 meV in the range of x = 0.20 to 0.40.[13] It reduces the energy gained on dropping into the channel layer.

2.1.2. Conduction electron quantum energy. The lowest conduction electron energy in the quantum confined state in the channel is raised by the size quantization effect. For thin wells, the quantum state energies E_n are approximated by the energy of a particle in a square box of the same width, L with infinite barriers. $E_n = n^2h^2/8m^*L^2$, which has a value of $E_1 = 55meV$ for electrons in the lowest state in GaAs ($m^*/m_o = 0.0665$) at a well width L = 100Å. More exact values of the quantum energies may be calculated which take into account the actual potential profiles.[12] This energy also reduces the energy gained by an electron transferring into the channel.

2.1.3. Conduction electron Fermi energy. For a nonzero concentration of electrons in the channel layer, the electron energies will occupy a Fermi distribution. Electron states up to a Fermi energy of $II\hbar^2n/m^*$ above the quantum energy E_1 will be occupied for a two-dimensional gas of concentration n carriers/cm^2 with only one transverse quantum state occupied, and spin degeneracy of two. For $m^* = 00665m_o$ and $n = 1 \times 10^{12}cm^{-2}$, this is 35 meV and also acts to reduce the energy gained by transferring additional electrons into the channel layer.

2.1.4. Spatial transfer Coulomb energy. Transfer of charge into the channel creates a negatively charged channel region and leaves behind a positive space charge of the ionized donors. The electric field and potential drop associated with this space charge are also appreciable. The potential drop for separation d of charge density σ is $\sigma d/\varepsilon\varepsilon_o$, which is 138 meV for separation of 10^{12} electrons/cm^2 by a distance of 100Å in GaAs of dielectric constant 13.1. Of course, for continuous or extended distributions of carriers and ions, appropriate integral expressions are to be employed. The Coulomb potentials alter the shape of the potential wells and barriers and thus alter the quantum energies, so that an exact treatment requires a self-consistent calculation. The essential result of all of these considerations, however, is that it is energetically possible to transfer electrons up to densities of $\approx 10^{12}cm^{-2}$ over distances of ≈ 100Å into wells as thin as 100Å in modulation doped GaAs/Al$_x$Ga$_{1-x}$As layers with x$\gtrsim$0.2. These various effects are illustrated in the band diagram of Figure 3.[14]

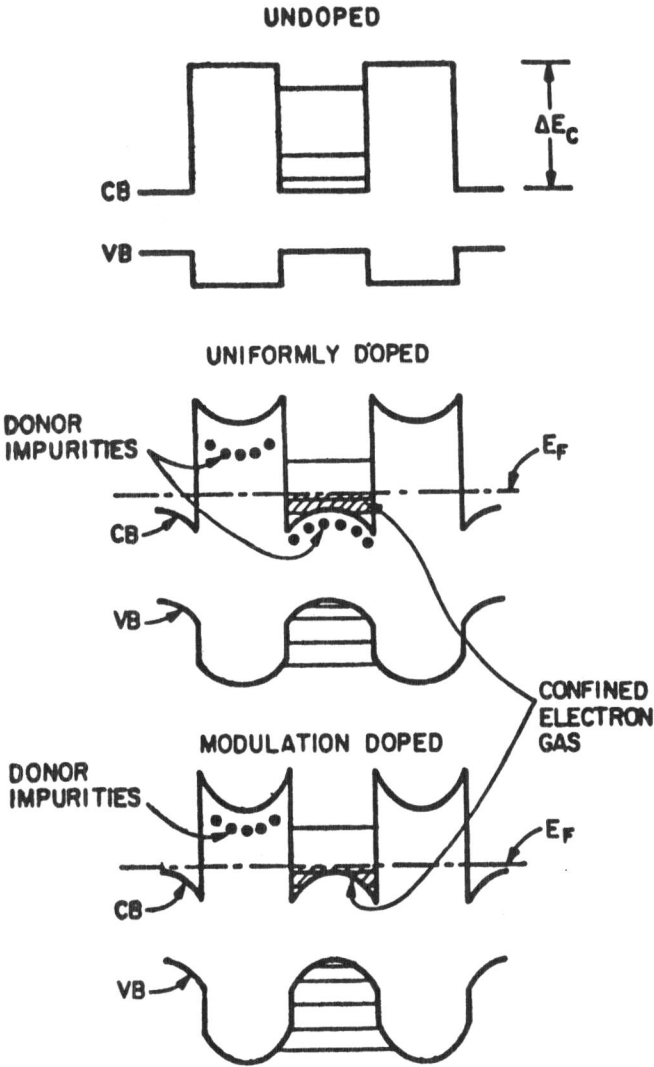

Figure 3 - Band edges, donor states, and electron states in undoped, uniformly doped and modulation doped multi-layered semiconductor system. (After Ref. 6).

2.2 Mobility Determining Process

The dominant sources of carrier scattering in uniformly doped semiconductors are phonons and impurities, with the phonon processes

SCATTERING MECHANISM

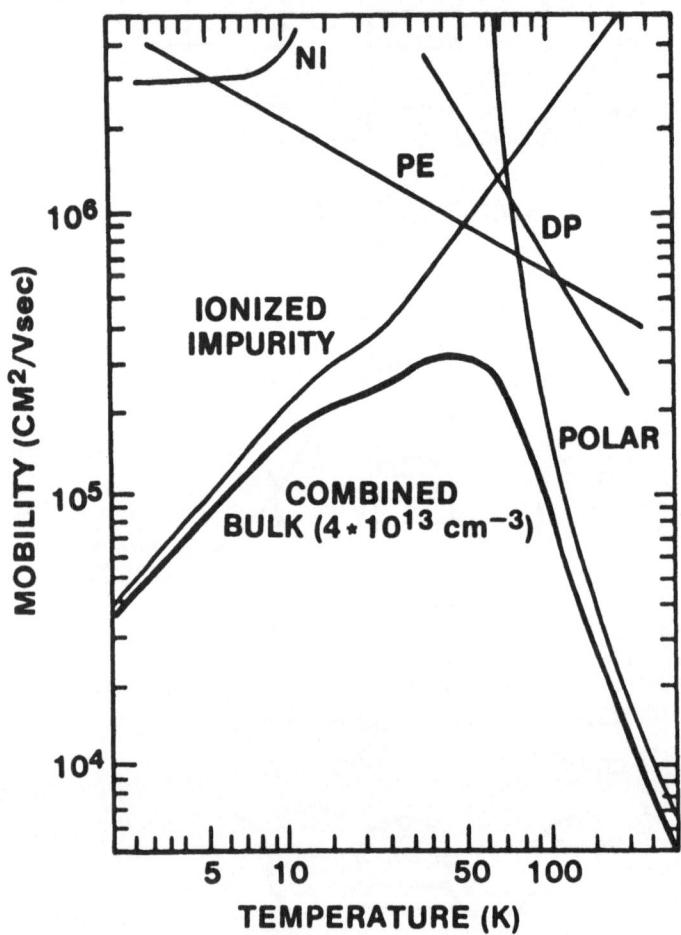

Figure 4a - Theoretical mobility limits for main scattering processes in uniformly-doped GaAs (Ref. 15). NI refers to neutral-impurity scattering, while PE indicates scattering by piezoelectrically active acoustic phonons and DP indicates deformation-potential scattering by acoustic phonons. In modulation-doped material with no impurity scattering. upper limit on mobility would be $\approx 5 \times 10^6 \text{cm}^2/\text{Vsec}$.

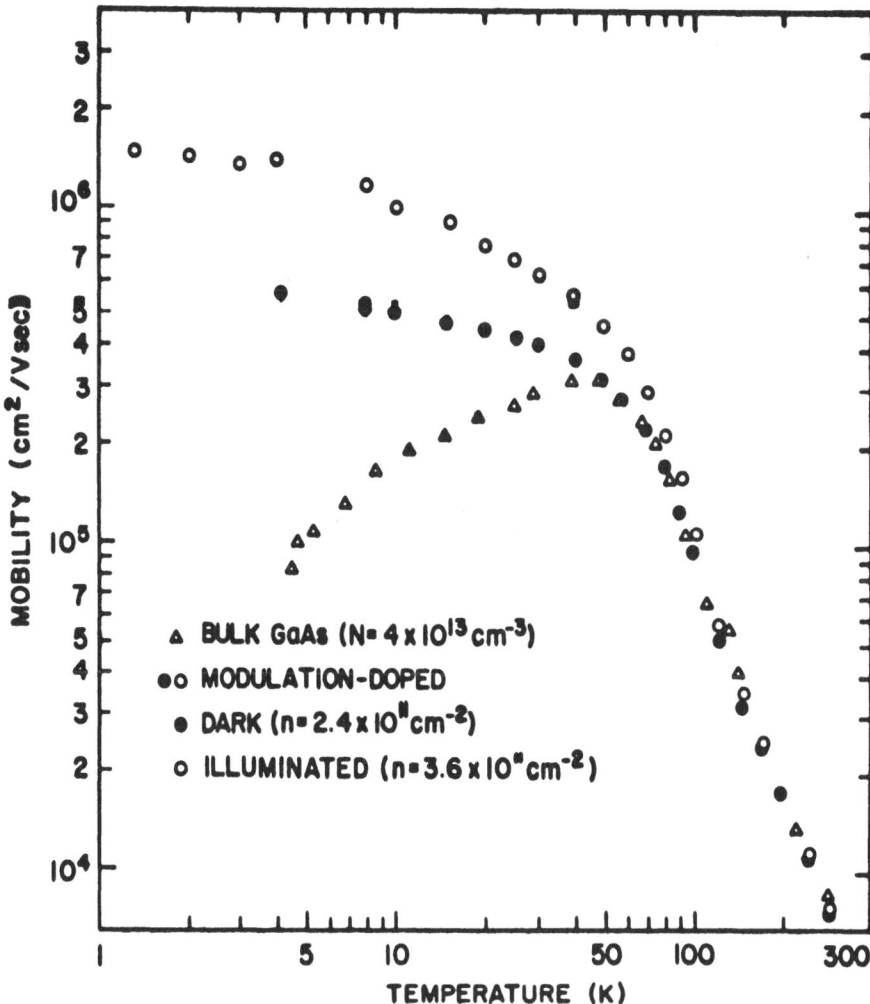

Figure 4b - Experimentally observed mobilities of high- mobility selectively-doped GaAs-(Al,Ga)As single interface structure and high-mobility lightly uniformly doped bulk GaAs.

dominating scattering at high temperatures and the impurity processes predominant at low temperature. A plot of the limits to the mobility from these processes for very weakly doped GaAs ($\approx 10^{13}\text{cm}^{-3}$) is shown in Figure 4.[15] In modulation-doped layers we may expect the phonon processes and rates to be roughly similar to the phonon scattering in uniformly doped material. The impurity scattering is drastically altered, however. When the impurity scatterers are separated from the carriers, the impurity scattering decreases both because of the fall-off in Coulomb fields with distance and because the short-wavelength (large wavevector) potential fluctuations, which are the most important source of scattering by impurities, fall off with distance from the impurity distribution. These effects have been calculated in detail by Price[16] and Ando.[17] A simple expression for the resultant mobility is obtained[16] for a degenerate two-dimensional sheet of n carriers/cm^2 separated by d from a two-dimensional sheet of $n_I$ions/cm.^{2}

$$\mu = 16\sqrt{\text{II}}\,ed^3n^{3/2}/\hbar n_I$$

The increase in mobility with carrier concentration n is especially noteworthy, because it more than compensates the customary decrease of μ with n_I in uniformly-doped cases where $n = n_I$. The increasing mobility with high n is a result of the increase in Fermi wavevector with n. With large Fermi wavevectors, small-k elastic ion scattering is less effective in relaxing the electron momentum distribution. The calculated mobility reaches 60,000cm^2/Vsec at $n = n_I = 10^{12}$cm^{-2} and $d = 100$Å. The mobility increase continues until a second quantum state becomes occupied, which occurs when the Fermi energy becomes equal to the difference between the first and second quantum states. The second quantum state forms the basis for a second subband of states and permits scattering of electrons between the subbands, which then produces a decrease in mobility, as first calculated by Ando.[17]

In the case of the incomplete electron transfer to the quantum well channels, electron states in both channel and barrier layers will be occupied. The conductivity will then consist of parallel conduction in the channels and barriers. For $Al_xGa_{1-x}As$ barriers, however, donor states in the barriers are sufficiently deep that carriers are typically frozen out at cryogenic temperatures in the barriers and conduction is dominated entirely by the quantum well channel carriers.

3. GROWTH AND MEASUREMENT TECHNIQUES

3.1 Growth Techniques

Modulation-doped structures with enhanced mobilities have now been grown with liquid phase epitaxy,[8] metal-organic chemical vapor deposition,[18] and molecular beam epitaxy.[6] In liquid phase epitaxy, the structures are produced by successively sliding a seed crystal under gradually cooled melts containing doped and undoped barrier and well material, whereas in metal-organic chemical vapor deposition, the structures are obtained by successively introducing and purging the requisite reacting gases from the reactor containing the crystal substrate. In molecular beam epitaxy, the modulation-doped structure is produced by shuttering on and off the constituent molecular beams needed for the respective layers. The smoothest and sharpest layers[19] and the highest mobilities[20,21] have been demonstrated with the MBE technique, but the mobility enhancements observed with the other two techniques[8,18] are also very appreciable. Consequently, all three techniques are potentially useful, and in less-demanding applications of modulation doping, considerations of economics or convenience may determine the choice of crystal growth technique.

The essential factors which must be present to achieve a high mobility structure are 1) absence of scattering centers in the conducting channel regions, 2) smooth and pure interfaces bounding these regions, 3) minimization of traps and compensating centers in the barriers, and 4) separation of the dopant atoms from the channel regions. The absence of unintentional scattering centers in the channel and barriers requires a low dislocation density and a low impurity density, whereas the smooth and pure interfaces require growth under conditions giving a smooth crystal surface with low interdiffusion between layers. The separation of dopant atoms and channel regions requires dopants which do not diffuse and which incorporate directly in the lattice without requiring accumulation of an enriched surface layer before incorporation in the lattice.

In MBE of modulation-doped GaAs-(Al,Ga)As structures, the pure conducting channel regions are best obtained by 1) use of substrate temperatures during epitaxial growth which are sufficiently high (600 C to 700 C) to reduce incorporation of impurities, but not roughen or decompose the sample surface; 2) minimization of unintentional contamination from sources and hot surfaces; 3) substrate introduction by vacuum interlock to prevent UHV vacuum contamination; and 4) use of arsenic rich growth conditions. As emphasized in the following section, the smoothness and purity of interfaces generally depend on the order in which the elements of the interface are grown, with inter-faces grown on GaAs surfaces being smoother and purer than interfaces grown on (Al,Ga)As. Absence of traps and compensating centers in (Al,Ga)As barrier layers is also favored by high substrate temperatures and good vacuum conditions, with optimal substrate temperatures being higher than for GaAs. The density of carriers measured in the channel for a selectively-doped GaAs/(Al,Ga)As single interface is generally lower than that calculated for complete ionization of donor levels above the Fermi level and suggests that impurities or defect centers in (Al,Ga)As barriers trap charges in deep states, preventing their entry into the GaAs channels. This presumably accounts for much of the variability in charge transfer and mobility which is observed from sample to sample in MD structures grown in different systems or at different times. It also presents a limitation on the lowest amount of charge which can be reliably transferred to a conducting channel, (currently in the high $10^{10}cm^{-2}$ range).

Silicon and beryllium are presently the most favorable n- and p-type dopants for creation of abrupt doping profiles in GaAs structures. Tin, although it can be incorporated at high concentrations, does not make a suitable dopant because of surface concentration effects which make it difficult to abruptly start and stop incorporation of tin in a GaAs structure. Diffusion of dopants occurs more rapidly between surface and near-surface layers than between layers further beneath the surface. In order to avoid diffusion of dopants from the doped barrier toward the undoped channel during growth of a selectively doped single interface the undoped channel layer may be grown before the doped barrier. In this way, dopant diffusion toward the growing surface will not move dopant atoms toward the channel.

3.2 Measurement Techniques

Basic characterization of modulation-dope structures has general-ly been based on 1) Hall effect and resistivity measurements which together can determine the carrier concentration and mobility of a single charge layer or an assembly of identical charge layers, and 2) oscillatory magnetoresistance data, which provides an alternate way for characterizing carrier densities and relaxation times. Each of the techni-ques requires Ohmic contacts to the electron gas, which for the GaAs case can often be achieved with alloyed indium contacts or, for smaller contacts, gold-germanium-nickel alloyed contacts. Capacitance measurements are also frequently useful and sometimes can be made with liquid electrodes which do not alloy with the sample. Optical measurements especially luminescence and Raman scattering observations, have also helped elucidate the behavior of modulation doped structures and will be discussed in the section on excitations of two-dimensional electron gases in modulation doped structures.

4. CARRIER MOBILITIES

The maximum carrier mobilities and lifetimes which have been obtained with the MBE technique have increased steadily as growth techniques have improved and the relevant structural parameters have become understood.[6,20,21] The highest carrier mobility which has been reported thus far is a value of $2.1 \times 10^6 \mathrm{cm}^2/\mathrm{Vs}$ for electrons at a modulation-doped $GaAs/Al_xGa_{1-x}As$ interface.[20] This is approximately three orders of magnitude greater than in bulk structures of comparable local density. Comparison between the mobilities observed in an MD single interface with extremely high mobility and a very lightly uniformly doped bulk sample are shown in Figure 4b. The similarity in the values at temperatures over 100 K illustrates the dominance of phonon processes in both systems at the higher temperatures. The superior mobility of the MD structure at low temperatures illustrates the reduced impurity scattering in MD structures, even when compared to bulk material of only $4 \times 10^{13} \mathrm{cm}^{-3}$ carrier density. We now describe the experimental results which demonstrate the factors influencing mobilities.

4.1 Influence of Undoped Spacer Layer

The separation between mobile carriers in the conducting channels and ionized impurities in the barriers of modulation doped structures can be increased by inserting undoped spacer layers in the barriers between the carriers and the ionized impurities. Monotonically increased mobilities have been demonstrated in $GaAs/Al_xGa_{1-x}As$ with progressively increasing setback thicknesses from 0Å to 150Å[14,22] In order to demonstrate the effect, series of samples were grown in succession in which spacer layer thicknesses were increased. The temperature dependences of mobilities observed in a series of such samples are shown in Figure 5. Throughout the range of temperature and spacings up to room temperature, the Hall mobilities increased with increasing width of undoped $Al_{0.12}Ga_{0.88}As$ spacer layers. The average Hall density decreased from $1.5 \times 10^{17}cm^{-3}$ to $0.7 \times 10^{17}cm^{-3}$ with the increase in spacer layer width. These results agree with the theoretical expectation that an undoped spacer layer progressively decreases ionized impurity scattering. However, the observed dependence is less than

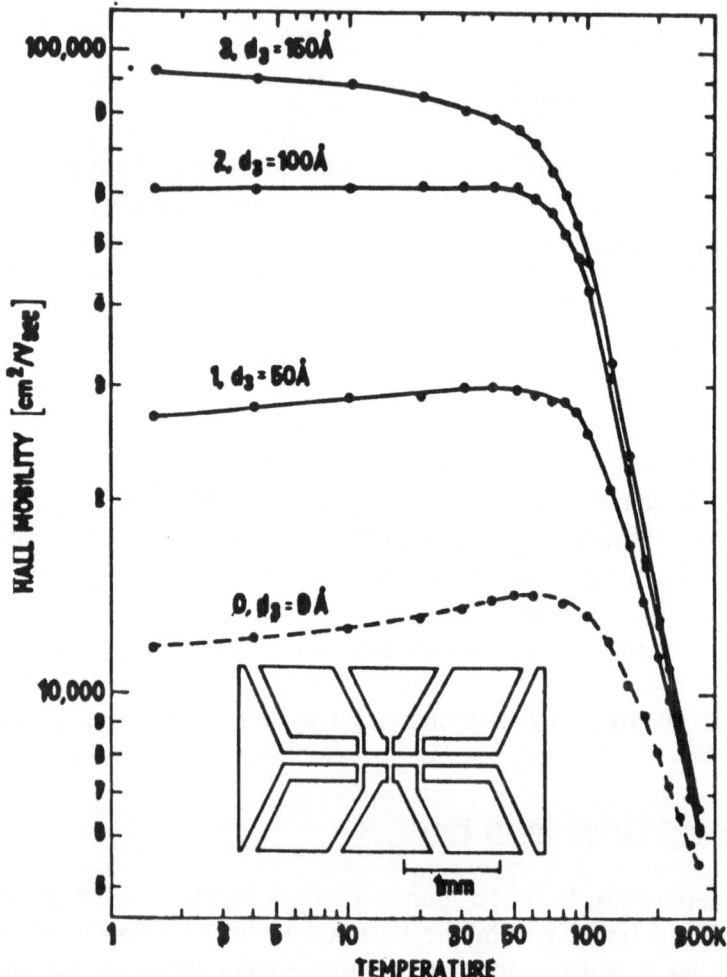

Figure 5 - Electron mobilities vs. temperature for a series of modulation-doped multilayer samples with different undoped (Al,Ga)As spacer layer between GaAs channel and doped (Al,Ga)As regions. (After Ref. 14).

the expected d^3 dependence and samples studied with $d > 150\text{Å}$ did not show further mobility increase. This occurs because other scattering mechanisms, perhaps associated with unintentional impurities or residual scattering from the heterojunction interface, become dominant at larger setback spacings of the dopant layer and because the carrier density in the layers was less when greater setbacks were present.

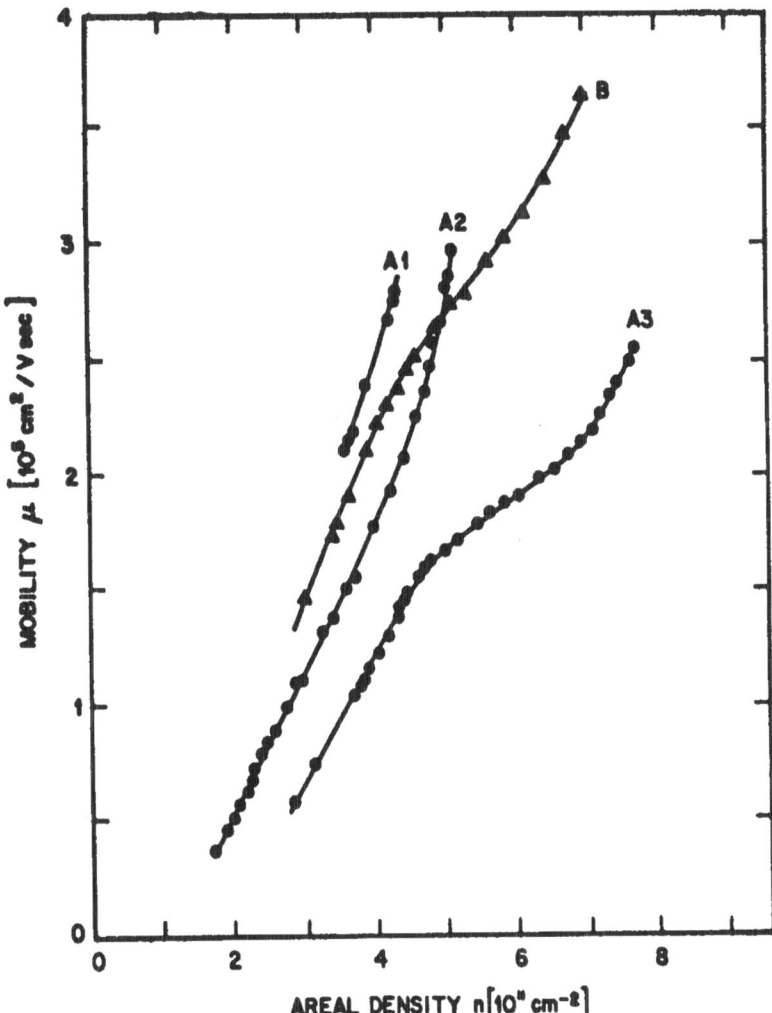

Figure 6 - Electron mobilities vs. carrier density in several selectively-doped single-interface GaAs/(Al,Ga)As samples. Carrier density is varied by persistent photoconductivity after illumination. (After Ref. 23).

4.2 Effects of Carrier Density

Several methods can be used to vary the number of carriers per unit area in a given modulation-doped structure. These techniques, which permit variation of the carrier density without the need to produce series of different samples, are particularly useful in clarifying the role of the carrier density in the conduction process as well as in modifying conductivity as desired in devices. The techniques to be discussed involve photoexcitation of carriers, gate control of carriers by a metal gate electrode on the sample surface, and gate control by a backside gate beneath an insulating substrate on the opposite side of the substrate from the epitaxial conducting layers.

In the photoexcitation technique, a low-temperature persistent photoconductive effect can be used to vary the electron concentration continuously over a large range. The origin of the persistent photoconduction is not understood in detail, but may be connected with deep traps in $Al_xGa_{1-x}As$ which can be optically activated.[23] Optically excited electrons transfer from the traps by tunnelling to the GaAs channel, where they increase the carrier concentration.

Using this effect, a number of single interface modulation-doped specimens have been studied. Light pulses emitted from light emitting diodes onto $GaAs/Al_xGa_{1-x}As$ single interface Hall bridges increased electron densities by amounts which persisted in the dark after the pulses ended. In this manner, charge densities in the 10^{11} to $10^{12} cm^{-2}$ range could be varied by up to a factor three. This produced increases in mobilities up to nearly one order of magnitude as measured by Hall effect, as shown in Figure 6.[23] The increases are monotonic in each case and actually somewhat greater than expected from the expression $\mu = 16\Pi^{1/2}ed^3n^{3/2}/\hbar n_I$. It is possible that the optical technique modifies the distribution of ions near the channel as well as modifying the more distant ions and the carrier density and that this is also affecting the observed mobilities in a positive way with increasing total illumination.

The density of carriers can also be varied by the field effect induced by voltages on gate electrodes in proximity to the channel. An unusual but useful configuration involves a metal electrode fixed to the back of an insulating substrate on which is grown a modulation doped single interface.[24] The gate voltage may be applied between ohmic contacts to the channel electrons on the front and the back electrode, which may be a metal plate. Using the technique of Shubnikov-de Haas oscillations to determine carrier density and four terminal resistance measurement to determine the resistivity, the carrier concentration was found to vary linearly with gate voltage by nearly a factor five, while the conductivity varied by a factor 250 and the mobility by a factor 50, as shown in Figure 7.[24] The mobility increases with increasing positive gate voltage following an approximate $\mu \approx n^{5/2}$ law. The increased electron density improves screening of ionized impurity fields while the positive backside bias increases the distance between the electrons in the channel and the impurities in the barrier by pulling electrons away from the interface. Both of these mechanisms increase mobility at positive biases and decrease mobilities under the negative biases.

A more widely employed configuration for gate control of carrier concentration involves metal Schottky gates on the top surface of a modulation-doped single interface structure. In this case, the gate voltage may be applied between the Schottky gate on the surface and the electrons in a channel approximately a depletion depth below the surface, which may be contacted by alloyed Ohmic contacts. This configuration forms the basis for selectively-doped heterostructure transistors as discussed in the accompanying paper by H. Morkoc.[25]

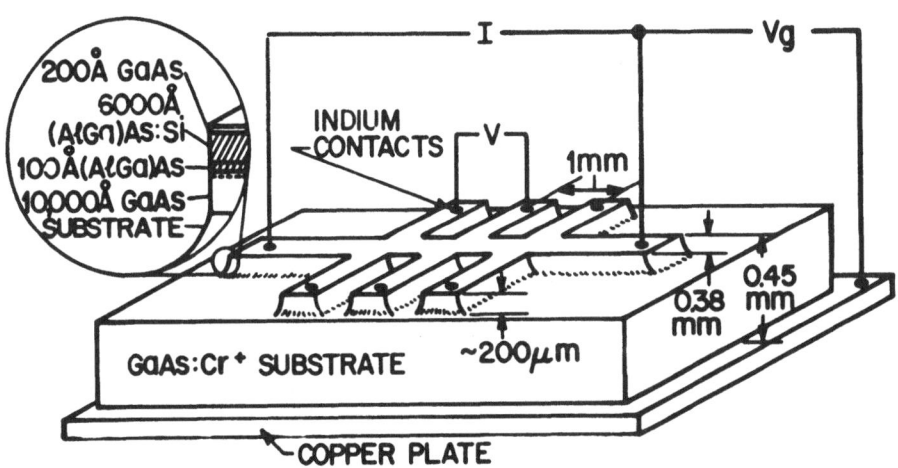

Figure 7a - Sample configuration for application of backside gate voltage, V_G, and measurement of resistance and Hall effect in selectively-doped single interface structure.

Strong variations in electron mobility with the gate-controlled carrier concentration are also seen in this configuration, as shown in Figure 8, which illustrates low temperature mobilities in gated structures with and without an undoped spacer layer.[26,27] On devices cover-

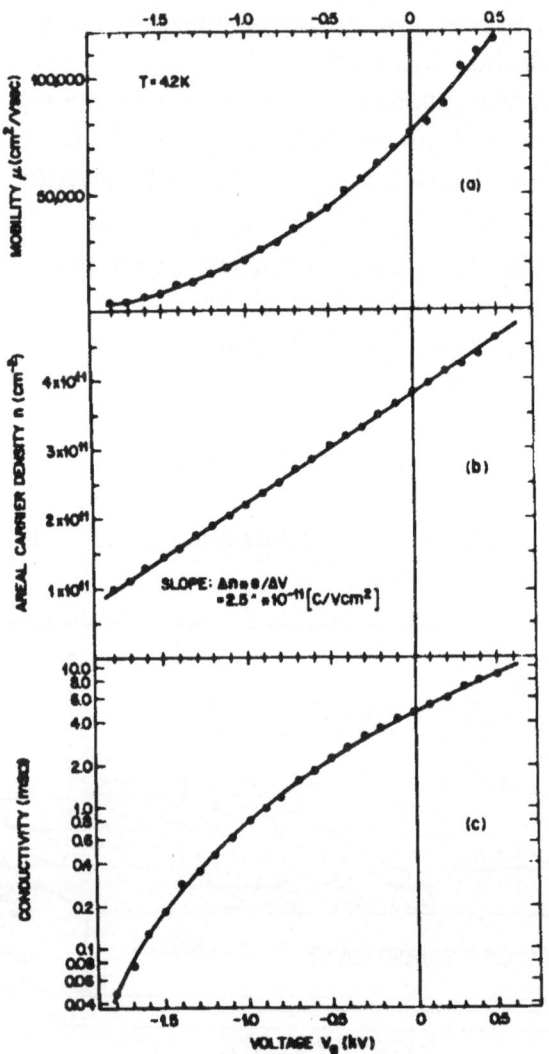

Figure 7b - Variation in mobility, carrier density and conductivity at a backdside-gated selectively-doped GaAs-(Al,Ga)As single interface. (After Ref. 24).

ing a density range of $7 \times 10^{10} \text{cm}^{-2}$ to $6 \times 10^{11} \text{cm}^{-2}$, mobilities increased with carrier concentrations according to power laws ranging from $\mu \approx n^{0.45}$ to $\mu \approx n^{1.4}$. The carrier density in these devices, with approximately 700Å of $Al_x Ga_{1-x}As$ and 200Å of GaAs over the channel, varied linearly with gate voltage, in contrast to the density in

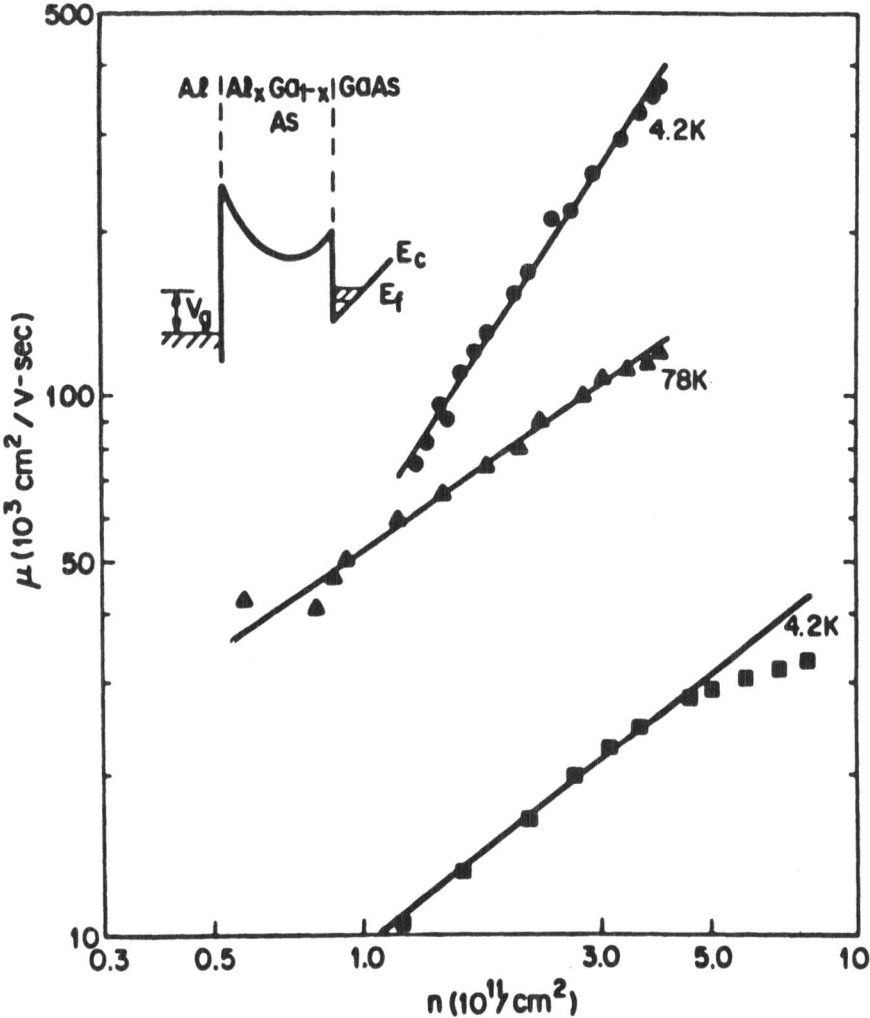

Figure 8 - Dependence of electron mobility on carrier density at a selectively-doped GaAs-(Al,Ga)As single interface with carrier density controlled by Al metal gate on top of (Al,Ga)As layer. (Ref. 26). The lowest curve refers to a device with no undoped (Al,Ga)As spacer layer. The upper curves are for a device with a 165Å-thick undoped spacer layer.

uniformly doped layers, which varies as the square root of gate voltage. The linear change in carrier concentration is produced by the insulating action of the (Al,Ga)As barrier layers. The weaker dependence of mobility on carrier concentration, relative to the backside gate configuration, reflects the motion of carriers toward the interface on applying a positive voltage above the interface as opposed to motion of carriers away from the interface on application of positive voltage below the substrate.

The separation of carriers from the surface of the modulation doped structure also provides the possibility of incorporating a layer of an insulator on the surface without having carriers in direct contact with the defects and impurities of the insulator and its surfaces. This provides a basis for a hybrid metal/insulator/semiconductor and semiconductor/semiconductor heterojunction technology.[28,29]

With increase in carrier density beyond $6 - 8 \times 10^{11} \text{cm}^{-2}$ in GaAs modulation-doped single interfaces, it becomes possible to populate a second band of two-dimensional electrons. The second band consists of electrons in the second confined quantum state of the potential well formed by the heterojunction band-edge potential step and the attractive potential of the ionized impurity. The electrons are, of course, free to move parallel to the interface, but they introduce and are subject to new scattering mechanisms which reduce mobility. Coincident with populating the second subband, decreases in Hall density and Hall mobility have been seen, the clearest results having been established for the backside gate samples mentioned above.[30,24] At positive gate voltages, the occupation of the second subband is clearly observed in Shubnikov-de Haas measurements by appearance of a new magnetoresistance oscillation period. The sum of the occupation of the first and second subbands remains linearly dependent on gate voltage. Combination of Hall and Shubnikov-de Haas data show that on occupying the new subband, mobility of electrons in the first subband drops by about 25% and the new (second) subband has carrier mobility $\approx$50% less than the first subband (Figure 9). The drop in mobility of

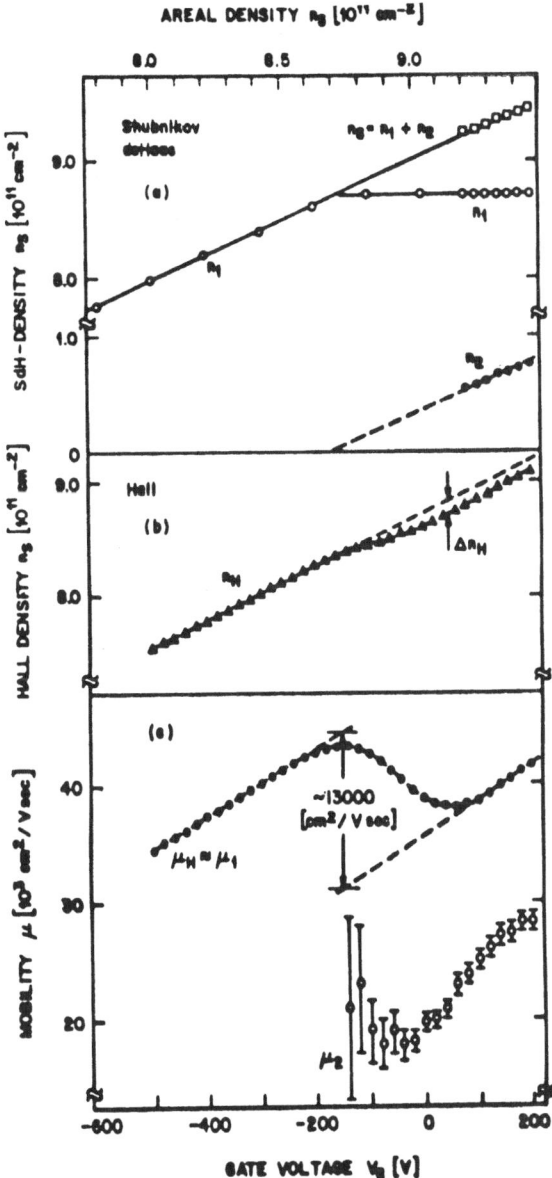

Figure 9 - Effects of occupation of upper quantum subband at selectively-doped single interface.Data are shown for electron mobilities, Hall effect electron density, and electron subband densities from Shubnikov-de Haas magnetoresistance oscillations as a function of gate voltage at a backside- gated single interface, where 1 and 2 refer to first and second quantum subbands. (After Ref. 30).

the first subband is produced by intersubband scattering, while the lower mobility of the second subband is a result of its lower Fermi wavevector magnitude. Closely related effects occur in the modulation-doped superlattice, as calculated by Ando and Mori.[17]

4.3 Dependences on Crystal System, Growth Conditions and Sequences

The electron mobilities which can be obtained with the modulation doping technique are sensitively dependent on the crystal system and compositions examined as well as on the conditions of epitaxial growth and the sequence in which the layers are grown. The dependence on extrinsic parameters is especially noticeable in the absence of the dominant impurity scattering in uniformly doped matter. In addition to the widely studied $GaAs/Al_xGa_{1-x}As$ (100)-crystal-face superlattices and single-interfaces, enhanced mobilities have been seen in $In_xGa_{1-x}As/In_yAl_{1-y}As$ on InP heterostructures,[31] where higher MD electron mobilities might be expected because of a lower conduction band effective mass, but where alloy scattering in the $In_xGa_{1-x}As$ channel might limit mobilities. Enhanced mobilities up to $10^5 cm^2/Vs$ have been produced in this system. In the $GaAs/Al_xGa_{1-x}As$ system, enhanced hole mobilities and two-dimensional hole-gas conduction have also been observed,[32,33] but no reports of modulation-doping of other crystal face orientations have yet appeared. Morkoc et al. have found and determined the effects of substrate temperature, AlAs mole fraction and background impurity concentration on mobility enhancement in single-period and multiple-period structures.[34]

Another factor which, somewhat surprisingly, produced a strong influence on the properties of modulation doped heterojunction is the order in which layers are grown. Almost universally, an MBE-grown heterojunction in which an undoped GaAs channel layer is grown before a doped $Al_xGa_{1-x}As$ barrier layer will show better behavior, i.e. more carriers and higher mobility, than the same layers grown in the reverse order.[35] This was surprising at first sight because of the initial expectation that layer properties and purity would be equivalent for layers grown with the same parameters but at different places in layered structures. Furthermore, carrier scattering in modulation-doped multilayers is more than twice as strong as at modulation doped single interfaces on GaAs, as evidenced by the fact that the highest single interface mobilities are significantly more than two times the highest multilayer mobilities. Subsequent work on the optical and structural properties of single and multiple quantum well structures demonstrated that the properties of an $Al_xGa_{1-x}As$ interface with GaAs grown on top of it are sensitively dependent on the thickness of the $Al_xGa_{1-x}As$ and the composition of underlying layers.[36,37] The purity and smoothness of interfaces on $Al_xGa_{1-x}As$ under common conditions of MBE growth are poorer than for interfaces on GaAs, and furthermore, the quality of the $Al_xGa_{1-x}As$ interface degrades with increasing thickness of the $Al_xGa_{1-x}As$ layer. A mechanism by which this could occur is an increased incorporation of impurities produced by an enrichment of surface impurity concentrations and associated increases in surface roughness during $Al_xGa_{1-x}As$ growth. In the presence of such mechanisms, carrier compensation and carrier scattering would be increased at interfaces of $Al_xGa_{1-x}As$ with overlying GaAs layers, accounting for the observed asymmetry in the electrical characteristics of modulation-doped single interface structures.

4.4 Mobilities of Hot Carriers

Electron mobilities in modulation doped heterojunction structures have been found to depend strongly on the strength of the electric field applied to the carriers. The effect of the electric field is to heat the carriers and thus to allow new momentum relaxation processes, including launching of phonons and electron scattering into higher, normally unoccupied, low-mobility valleys. Decreases in mobility as large as a factor five were seen in relatively modest fields of 200 V/cm in MD heterojunction structures at a temperature of 10K.[38] The modification of the distribution function of electrons in an applied field has been observed directly by measurement of shifts in photoluminescence spectra under application of the electric fields to a modulation-doped superlattice.[39] Simultaneous measurement of the drift mobility allows the correlation between electron temperature and electron mobility to be established for carriers in the wells. The electron temperature changed measurably in fields as low as 0.3 V/cm and for 10 V/cm the electron temperature exceeded 50K for a sample in a 2K helium bath. The mobility decreased rapidly for carrier temperature above 70K. The decrease in mobility with electric field and electron heating means that the full mobility enhancement observed at low fields in modulation doped structures is not available in devices or test structures employing high electric fields. Moderate device performance improvement has been obtained with modulation doping, however, and further study is certainly required.

Another potentially important property related to device application of modulation doped structures is resistance to radiation damage. Changes in transport properties under gamma-ray and electron-beam radiation have recently been studied for modulation-doped GaAs single interfaces.[40] The MD GaAs structures were found to be highly resistant to radiation damage, particularly in comparison to silicon MOS structures. The apparent reason for this is that the comparatively perfect GaAs/(Al,Ga)As interface lacks the defect centers and dangling bonds which act as trapping centers under electron-hole pair creation in the relatively amorphous Si/SiO_2 interface. Applications of this feature would include resistance to energetic beam lithography processes and operation in radiation environments.

5. TWO-DIMENSIONAL PHENOMENA IN MODULATION-DOPED STRUCTURES

The confinement of electrons to two-dimensional motion with long scattering lifetimes makes modulation doped structures fruitful media for study of the physics of two-dimensional electrons.

5.1 Excitations Between Quantum Subbands

The source of the two dimensional electron confinement in MD layers and interfaces, as presented above, is the quantization of motion perpendicular to the layer planes. The spacing between the quantized levels has been probed directly by Raman scattering excitation of electrons between the associated quantum[41] subbands. In this excitation, an electron is promoted from an occupied subband to a higher subband, and the energy of the scattered photon is reduced relative to the incident beam by the intersubband energy. Both single particle and collective electron-longitudinal optic phonon energies may be measured with the technique. The line shape of the Raman scattered spectral peaks also contains information on electron state lifetimes and a correlation between line shape and mobility is observed in MD superlattices.

5.2 Plasma Excitations

In a layered electron gas, as occurs in modulation doped structures, electrons are constrained to move along the planes of the channels, and this alters the spectrum of plasma charge oscillations from that of an unconstrained system of electrons. In the unconstrained, three-dimensional plasma of density n_V, effective mass m^* and dielectric constant ϵ, a uniform mode with frequency $n_V e^2 / \epsilon m^*$ exists, along with modes of higher wavevector with an additional energy proportional to wavevector, k, squared. In a single charge layer, as studied for electrons on liquid helium and electrons in a silicon MOS channel, the zero-wave-vector mode frequency goes to zero and the dispersion to a square-root dependence on k. In a multiple layer electron gas, a zero uniform mode frequency and a linear dependence on mode frequency on wavevector parallel to the planes is predicted.[42] This behavior has been quantitavely confirmed in modulation-doped GaAs-Al$_x$Ga$_{1-x}$As multilayer structures, as measured by inelastic light scattering.[43]

On exciting the plasmons, the possibility exists for emission of radiation by radiative deexcitation. This was observed originally in silicon MOSFETs[44] and more recently from single interface modulation doped GaAs heterostructures excited by passing a current along the channel.[45] A grating structure in close proximity to the two-dimensional electrons is necessary to couple out the far infrared radiation. The radiation intensity is apparently dependent on electron mobility, and high mobility samples with undoped barrier spacer layers show weaker plasmon emission than lower mobility samples. Intensities up to $\approx 3 \times 10^{-8}$ watts at 3 to 5 μm wavelengths were measured. The dependence on mobility suggests that impurity scattering is instrumental in exciting the plasmons.

5.3 Magnetic Field Effects

Application of a magnetic field perpendicular to a plane of electrons quantizes the two remaining degrees of translation freedom into cyclotron (Landau) orbits. The spacing between Landau levels $\hbar \omega_c = eB / m^* c$ has been measured directly in modulation-doped structures by cyclotron resonance absorption in MD multilayers,[7] and by excitation between Landau levels with Raman scattering.[46] The Landau level spectrum has also been probed by magnetophoton scattering, in which resistance singularities are observed in a single interface structure when the separation between Landau levels is equal to or is a submultiple of an LO phonon energy.[47] The Landau level separation is also responsible for Shubnikov-de Haas magnetoresistance oscillations which occur when Landau levels pass through the Fermi energy as magnetic field is increased. The period of the oscillations accurately measures the carrier density.[24,26] The cyclotron resonance measures an effective electron mass at the Fermi energy, which is slightly larger than the effective mass at the bottom of the conduction band because of non-parabolicity in the conduction band dispersion. The magnetophonon resonance measures a polaron mass, which is not measurably altered from the cyclotron mass. The Raman scattering from excitation between Landau levels had been unexpected because no theory predicts scattering from in-plane motion of carriers. The cyclotron mass deduced from the Raman excitation is nearly equal to the GaAs conduction band mass at fields up to 14 T. At higher magnetic fields, where the extreme quantum limit of all electrons in the lowest Landau level is reached, anomalies were observed in the frequency, intensity, and selection rules of the transitions.[46] Cyclotron resonance observations by far infrared transmission through modulation doped $GaAs/Al_xGa_{1-x}As$ single interfaces have shown oscillations of the cyclotron resonance linewidth which correlate with Landau level filling.[48] Both single interface and multilayer MD samples have shown far infrared emission at the cyclotron resonance frequency when excited by passing current through the layers.[45]

5.4 Quantum Hall Effect

At fields above ~4T and temperatures below ~4.2K, the Shubnikov-de Haas oscillations referred above in the electrical resistance vs. perpendicular megnetic field of high mobility modulation-doped $GaAs/Al_xGa_{1-x}As$ heterojunctions exhibit broad resistance minima approaching zero resistance.[49] Some samples have shown resistances below 10^{-6} ohms per square, corresponding to a resistivity in the current-carrying volume of the heterostructure of less than 10^{-12}ohm-cm, which is less than the resistivity of any known normal-state material. At the same fields as the minima of parallel resistance, the Hall resistance, defined as Hall voltage divided by the current along the layer, assumes wide, flat plateaus[50](Figure 10). The value of Hall resistance at these plateaus is accurately h/ne^2 , where h is Planck's constant and n is an integral number of filled Landau levels, including spin, at the operating field. This effect is the "Quantized Hall Effect," which had been observed initially in silicon MOSFET structures.[51] The Hall resistance plateaus and resistance minima occur at fields where an integral number of Landau levels are filled, although the explanation of the extreme width and flatness of the plateaus is still incomplete. At temperatures below 100 mK, the width of the plateaus has been measured to be more than 90% of the center-to-center field difference between plateaus.[52]

Relative to silicon structures, the GaAs heterostructure has a lower effective mass, requiring a lower magnetic field to produce a given separation between Landau levels. Furthermore, because the density of interface states is less at the $GaAs/Al_xGa_{1-x}As$ junction than at a silicon-oxide interface, the quantized Hall effect can be observed at lower carrier density in GaAs heterojunctions, allowing use of lower magnetic fields to reach a given plateau and allowing access to

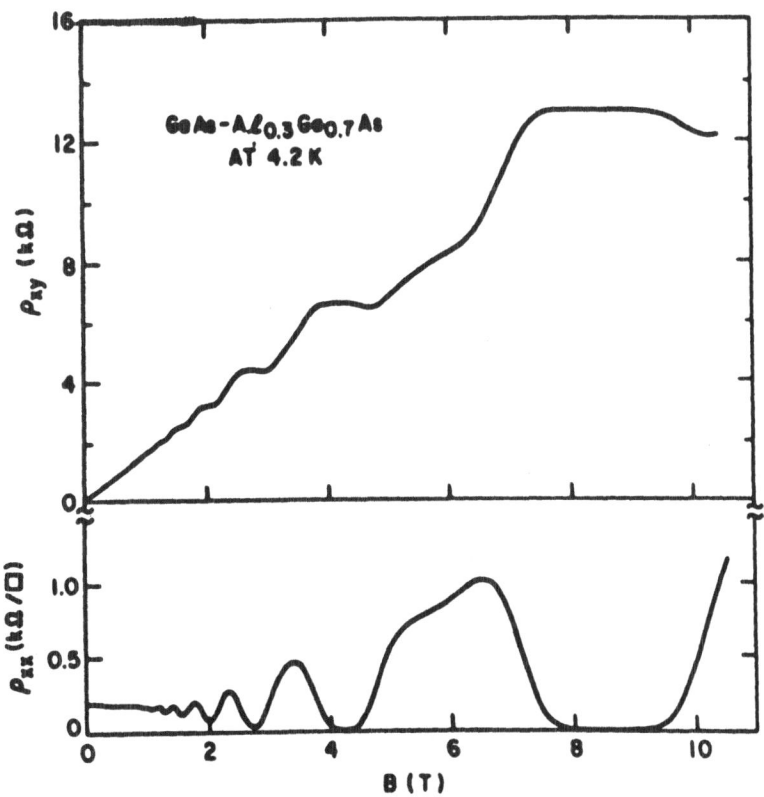

Figure 10 - Quantum Hall effect observed in Hall resistance, ρ_{xy}, and parallel resistance, ρ_{xx}, of selectively-doped GaAs-(Al,Ga)As single interface. Plateau in ρ_{xy} between 8 and 9 Tesla has value $h/2e^2$, corresponding to occupation of single, spin-degenerate Landau level. (After Ref. 50).

the extreme quantum limit in which all electrons occupy the lowest Landau level.

The Hall resistance plateaus have been found to be flat to at least one part in 10^6 over field ranges greater than 0.1 T in samples with the flattest plateaus. This enables use of the GaAs QHE for a fundamental standard of resistance and as a means for determination of e^2/h and thus the fine structure constant.[53] The flatness and width of the Hall resistance plateaus is not equal among samples of comparable mobility, however, and the mechanism for occurrence of the flat plateaus is apparently dependent on details of the structure of the junction. A possible source of the plateau width is impurity-produced potential fluctuations which produce electron paths along lines of intersecting Landau and Fermi levels.[54] These lines would form a connected cluster only for magnetic fields in ranges between the plateaus; the current would flow in globally extended nondissipating states parallel to the bridge and perpendicular to the electric field, whence the vanishing resistivity described above.

On increasing the magnetic field acting on a modulation-doped GaAs/$Al_xGa_{1-x}As$ interface into the region of the extreme quantum limit, where all electrons are in the lowest Landau level, additional new features have unexpectedly been seen[55](Figure 11). In the vicinity of magnetic fields corresponding to one-third and two-thirds occupation of the lowest Landau level, additional minima are observed in the resistance, and additional plateaus are seen in the Hall resistance, approaching values of $h/(e^2/3)$ and $h/(2e^2/3)$. The features are observable at T = 1.6 K and are tending toward zero resistance and a flat Hall plateau below T = 0.5 K. Whereas peaks and gaps in the energy spectrum of Landau levels can be invoked to explain the locations of QHE features at the lower fields, no such structures had been anticipated for the high-field case of the partially filled Landau level. No theoretical explanation of the new effects in the extreme quantum limit has yet been found, but the existence of plateaus and minima suggests the formation of new energy gaps, analogous to the gaps between Landau levels and possibly associated with ordering of the electrons, such as charge density wave or an electron crystal formation produced by electron-electron interactions.

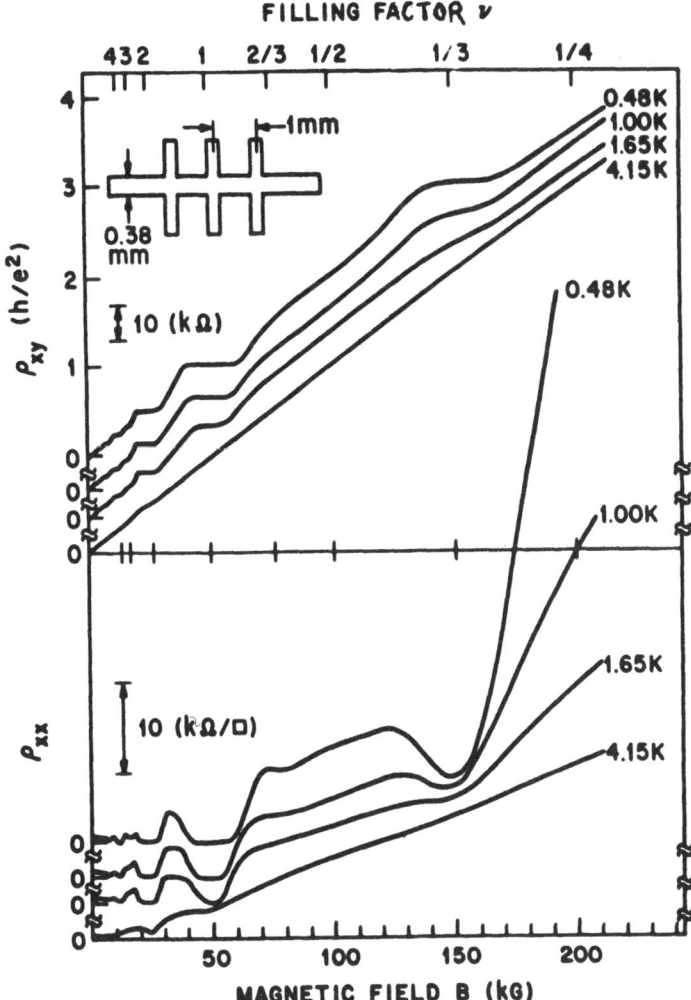

Figure 11 - Quantum Hall effect in extreme quantum limit. Plateau in ρ_{xy} at 15 T at 0.48 K has value $3h/e^2$, corresponding to one-third occupation of lowest Landau level. (After Ref. 55).

In principle, the quantized Hall effect could also be observed in multiple-layer structures as well as in the previously studied single-interface MOSFET and GaAs structures. This has been attempted successfully recently[56] by growth of a multiple layer GaAs-Al$_x$Ga$_{1-x}$As structure. Structures with multiple layers also allow study of extensive properties of interface and two-dimensional systems which might otherwise be too weak to measure. For example, in recent work, the first measurement of the magnetic susceptibility of the two dimensional electron gas was accomplished by means of a stack of over 4000 epitaxial multilayers assembled from the modulation doped structure described above. Using a SQUID magnetometer, oscillations of the field-dependent susceptibility (de Haas-van Alphen oscillations) were observed with the same period as the electrical resistance oscillations, giving information on the electronic density of states.[56]

6. CONCLUSIONS

The ability to selectively dope semiconductor heterojunctions has greatly expanded the range of performance and phenomena accessible in doped semiconductors. The modulation doping technique permits wide flexibility and innovation in structure design but simultaneously puts strict demands on the crystal growth parameters associated with control, smoothness, stability and purity. Contemporary molecular beam epitaxy technique is showing success in meeting these challenges and it is not unreasonable to expect continued progress in development and production of modulation-doped structures.

ACKNOWLEDGEMENTS

I wish to acknowledge the collaboration of W. Wiegmann, H. L. Stormer, D. C. Tsui, R. Dingle, M. A. Paalanen, M. E. Cage, R. F. Dziuba, B. F. Field, S. Luryi, R. F. Kazarinov, V. Narayanamurti, T. Haavasoja, R. C. Miller, P. M. Petroff, W. T. Tsang, M. B. Panish and J. Brooks in the work described here.

REFERENCES

1. I. Hayashi, M. B. Panish, and P. W. Foy, IEEE J. Quantum Electron. 5, 211 (1969).
2. L. L. Chang, L. Esaki, and R. Tsu, Appl. Phys. Lett. 24, 593, (1974).
3. R. Dingle, W. Wiegmann, and C. H. Henry, Phys. Rev. Lett. 33, 827 (1974).

4. L. L. Chang, H. Sakaki, C. A. Chang, and L. Esaki, Phys. Rev. Lett. 38, 1489 (1977).

5. L. Esaki and L. L. Chang (unpublished).

6. R. Dingle, H. Störmer, A. C. Gossard, and W. Wiegmann, Appl. Phys. Lett. 33, 665 (1978).

7. H. L. Störmer, R. Dingle, A. C. Gossard, W. Wiegmann, and M. D. Sturge, Solid State Comm. 29, 705 (1979).

8. D. C. Tsui and R. A. Logan, Appl. Phys. Lett. 35, 99 (1979).

9. C. Weisbuch, R. C. Miller, R. Dingle, A. C. Gossard, and W. Wiegmann, Solid State Comm. 37, 219 (1981).

10. G. Abstreiter and K. Ploog, Phys. Rev. Lett. 42, 1308 (1979); A. Pinczuk, H. L. Störmer, R. Dingle, J. M. Worlock, W. Wiegmann, and A. C. Gossard, Solid State Comm. 32, 1001 (1979).

11. G. A. Sai-Halasz, R. Tsu, and L. Esaki, Appl. Phys. Lett. 30, 651 (1977).

12. R. Dingle, in Advances in Solid State Physics, H. J. Queisser, ed., (Pergamon, Oxford, 1975), Vol. XV, p. 21.

13. D. M. Collins, J. Appl. Phys. (to appear).

14. H. L. Störmer, A. Pinczuk, A. C. Gossard, and W. Wiegmann, Appl. Phys. Lett. 38, 691 (1981).

15. C. M. Wolfe, G. E. Stillman, and W. T, Lindley, J. Appl. Phys. 41, 3088 (1970).

16. P. J. Price, Surf. Sci. 113, 199 (1982).

17. S. Mori and T. Ando, Phys. Rev. B 19, 6433 (1979).

18. S. D. Hersee, J. P. Hirtz, M. Baldy, and J. P. Duchemin, Electronics Letters 18, 1076 (1982).

19. A. C. Gossard, Thin Solid Films 57, 3 (1979); P. M. Petroff, A. C. Gossard, W. Wiegmann, and A. Savage, J. Cryst. Growth 44, 5 (1978).

20. S. Hiyamizu, in Collected Papers of 2nd Int. Sym. on MBE and Related Clean Surface Techniques, (Japan Soc. Appl. Phys., Tokyo, 1982), p. 113.

21. J. C. M. Hwang, A. Kastalsky, H. L. Störmer, and V. G. Keramidas, Postdeadline paper, 2nd Int. Sym. on MBE and Related Clean Surface Techniques, Tokyo, August 1982.

22. T. J. Drummond, H. Morkoc, and A. Y. Cho, J. Appl. Phys. Lett. 52, 1380 (1981).

23. H. L. Störmer, A. C. Gossard, W. Wiegmann, and K. Baldwin, Appl. Phys. Lett. 39, 912 (1981).

24. H. L. Störmer, A. C. Gossard, and W. Wiegmann, Appl. Phys. Lett. 39, 493 (1981).

25. H. Morkoc, this volume.

26. D. C. Tsui, A. C. Gossard, G. Kaminsky, and W, Wiegmann, Appl. Phys, Lett. 39, 712 (1981).

27. Takashi Mimura, Surf. Sci. 113, 454 (1982).

28. C. Y. Chen, A. Y. Cho, A. C. Gossard, and P. A. Garbinski, Appl. Phys. Lett. 41, 360 (1982).

29. Takashi Hotta, Hiroyuki Sakaki, and Hideo Ohno, Jap. Jour of Appl. Phys. 21, L122 (1982).

30. H. L. Störmer, A. C. Gossard, and W. Wiegmann, Solid State Comm. 41, 707 (1982).

31. A. Kastalsky, R. Dingle, K. Y. Cheng, and A. Y. Cho, Appl. Phys. Lett. 41, 274 (1982).

32. H. L. Störmer and W. T. Tsang, Appl. Phys. Lett. 36, 685 (1980).

33. H. L. Störmer, A. C. Gossard, and W. Wiegmann, (submitted for publication).

34. T. J. Drummond, R. Fischer, H. Morkoc, and P. Miller, Appl. Phys. Lett. 40, 430 (1982).

35. T. J. Drummond, H. Morkoc, S. L. Su, R. Fischer, and A. Y. Cho, Electronics Lett. 17, 870 (1981).

36. R. C. Miller, W. T. Tsang, and O. Munteanu, Appl. Phys. Lett. 41, 374 (1982).

37. A. C. Gossard, W. Wiegmann, R. C. Miller, P. M. Petroff, and W. T. Tsang, in Collected Papers of 2nd Int. Sym. on MBE and Related Clean Surface Technique, (Japan Soc. Appl. Phys., Tokyo, 1982), p. 39

38. T. J. Drummond, W. Kopp, H. Morkoc, and M. Keever, Appl. Phys. Lett. 41, 277 (1982).

39. Jagdeep Shah, A. Pinczuk, H. L. Störmer, A. C. Gossard, and W. Wiegmann, Appl. Phys. Lett. 42, 55 (1983).

40. D. C. Tsui, A. C. Gossard, and G. J. Dolan, Appl. Phys. Lett. (to appear).

41. A. Pinczuk and J. M. Worlock, Surf. Sci. 113, 69 (1982).

42. P. B. Visscher and L. M. Falicov, Phys. Rev. B 3 2541 (1971).

43. Diego Olego, A. Pinczuk, A. C. Gossard, and W. Wiegmann, Phys. Rev. B 26 7867 (1982).

44. D. C. Tsui, E. Gornik, and R. A. Logan, Solid State Comm., 35, 875 (1980).

45. R. Hopfel, G. Lindemann, E. Gornik, G. Stangl, A. C. Gossard, and W. Wiegmann, Surf. Sci. 113, 118 (1982).

46. · Z. J. Tien, J. M. Worlock, C. H. Perry, A. Pinczuk, R. L. Aggarwal, H. L. Störmer, A. C. Gossard, and W. Wiegmann, Surf. Sci. 113, (1982).

47. D. C. Tsui, Th. Englert, A. Y. Cho, and A. C. Gossard, Phys. Rev. Lett. 44, 341 (1980).

48. Th. Englert, J. C. Maan, Ch. Uihlein, D. C. Tsui, and A. C. Gossard, in Proc. of Int. Conf. on Physics of Semiconductors, Montpellier, Sept. 1982 (to appear).

49. H. L. Störmer, D. C. Tsui, and A. C. Gossard, Appl. Phys. Lett. 113, 32 (1982).

50. D. C. Tsui and A. C. Gossard, Appl. Phys. Lett. 39, 550 (1981).

51. K. von Klitzing, G. Dorda, and M. Pepper, Phys. Rev. Lett. 45, 494 (1980).

52. M. A. Paalanen, D. C. Tsui, and A. C. Gossard, Phys. Rev. B, Rapid Communication 25, 5566 (1982).

53. D. C. Tsui, A. C. Gossard, B. F. Field, M. E. Cage, and R. F. Dziuba, Phys. Rev. Lett. 48, 3 (1982).

54. R. F. Kazarinov and S. Luryi, Phys. Rev. B25, 7626 (1982).

55. D. C. Tsui, H. L. Störmer, and A. C. Gossard, Phys. Rev. Lett. 48, 1562 (1982).

56. H. L. Störmer, T. Haavasoja, V. Narayanamurti, A. C. Gossard, and W. Wiegmann, in Proc. of Int. Conf. on Metastable and Modulated Semiconductor Structures, Pasadena, Dec. 1982 (to appear).

DOPING SUPERLATTICES

Klaus Ploog

Max-Planck-Institut für Festkörperforschung
D-7000 Stuttgart 80, Federal Republic of Germany

ABSTRACT: Doping ("NIPI") superlattices represent a new type of artificial periodic semiconductor structures with tunable carrier concentration and tunable energy gap. The prototype structure, recently prepared by molecular beam epitaxy (MBE), consists of a periodic sequence of thin (5<d<300 nm) n(Si)- and p(Be)-doped GaAs layers. The space charge field of the ionized impurities varying in the direction of layer sequence produces a parallel periodic modulation of the energy bands which determines the unusual electrical and optical properties of the material. The 1-dimensional periodic potential induces a splitting of the CB and VB into subbands similar to those in the compositional superlattices. In contrast to the AlAs/GaAs superlattice, GaAs doping superlattices exhibit an indirect energy gap in real space, with the electrons and holes separated by half a superlattice period. Deviations from thermal equilibrium are thus quasi-stable, and excess-carrier lifetimes become very large (in the order of 10^3 sec). We demonstrate the tunability of bipolar conductivity, absorption coefficient, electro-

and photoluminescence, and subband spacing by external voltage and by optical excitation in this new class of semiconductor materials.

1. INTRODUCTION

In the two types of compositional superlattices AlAs-GaAs and GaSb-InAs, the fascinating idea of "tailoring" or "designing" semi-conductor materials with a new type of band structure (mini-or subbands),[1] which yield novel optical and transport properties, has been proven feasible.[2,3] In this paper we present a new type of artificial semiconductor superlattice originally proposed by Döhler[4,5] which does not contain any of the typical material interfaces and which exhibits unusual properties due to its <u>indirect energy gap in real space</u>. The structure which was successfully prepared by molecular beam epitaxy (MBE)[6] consists of a periodic sequence of thin ($5 \leq d \leq 300$ nm) n- and p-doped GaAs layers, possibly with interspersed intrinsic (i-) layers. This "doping" superlattice is also called "NIPI" crystal or structure[4] and is schematically shown in Fig. 1. The relatively small amounts of impurities required for doping (typically 10^{17} to 10^{19} cm^{-3}) induce only

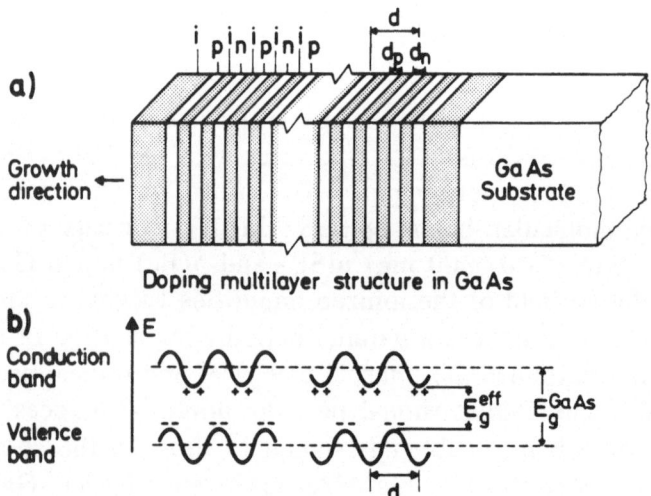

Figure 1 - Schematic illustration of GaAs doping ("NIPI") superlattice of periodicity d having layer thicknesses d_n, d_p, and d_i and doping concentrations n_D, n_A, and n_i, respectively (a), and modulation of conduction and valence band edges (b).

a minor distortion of the lattice of the homogeneous semiconductor material.

The superlattice potential in GaAs NIPI structures originates from the positive and negative space charge of the ionized impurities periodically varying in the direction of layer sequence. This produces a parallel periodic modulation of the energy bands (Fig. 1b) which determines the unusual electrical and optical properties of the material. The 1-dimensional periodic potential induces a splitting of the CB and VB into subbands similar to that in the well-known compositional superlattices. The effective bandgap in GaAs doping superlattices, defined as the difference between the bottom of the CB in the n-layers and the top of the VB in the p-layers, may have any value between the gap of unmodulated GaAs and zero, depending on the choice of constituent layer thickness and doping concentration ("design parameters"). The i-region in this superlattice may have a thickness close to zero without substantially modifying its peculiar electronic properties.

In contrast to the AlAs/GaAs superlattice, GaAs NIPI structures exhibit an indirect bandgap in real space, since the electron and hole states are spatially separated by half a period through the space charge potential. There is only a small overlap between CB and VB wavefunctions localized in the n- and p-layers. Deviations from thermal equilibrium of free-carrier concentration are thus quasi-stable, and excess-carrier lifetimes become very large. As a consequence, the effective bandgap and the carrier concentration are no longer fixed but become variable quantities for a given GaAs doping superlattice which may be tuned from the outside. In this paper we describe the tunability of conductivity, absorption coefficient, luminescence and subband spacing by external voltage or by optical excitation for this new class of semiconductor material.

2. PREPARATION OF GaAs DOPING SUPERLATTICES

The most useful growth technique for doping superlattices is molecular beam epitaxy (MBE).[7,8,9] For the present study we used superlattice configurations with i-regions close to zero thickness. We investigated two methods to accomplish abrupt changes in the doping type (i.e., p-n junctions) during MBE growth of GaAs:

1) In addition to the gallium and arsenic sources, only a single source containing an amphoteric group-IV-doping element (e.g., Ge) was used. The dopant incorporation on either Ga-sites resulting in n-type material or on As-sites resulting in p-type material was regulated by an intentional change of the growth surface composition that can be monitored via the abrupt change of surface reconstruction in the RHEED pattern.[10,11,12,13]

2) Two sources containing different doping elements (e.g., Be as acceptor and Si as donor impurities, respectively) were used together with the gallium and arsenic effusion cells. For a change of doping type, only the shutters in front of the dopant cells were operated in a precisely controlled manner at otherwise constant growth conditions.[6]

The dopant elements Si and Be as well as Ge have unity sticking coefficients on (100) GaAs over a wide range of growth conditions, i.e., substrate temperature up to 630°C, impurity flux 10^6 - 10^{11} atoms/cm^2 sec, As$_4$/Ga flux ratio varied widely. Therefore, the observed doping level is simply proportional to the dopant arrival rate.[7,8,9,14] During growth of all of the Si- and Be-doped material and of n-type Ge-doped GaAs the substrate temperature was typically maintained at 550°C, yielding excellent photoluminescence response and Hall mobilities of the layers as described in Refs. 6, 15 and 16.

The growth of p-type Ge-doped GaAs was achieved by carefully changing the As surface population during growth towards a more Ga-stabilized (4×2) surface reconstruction.[12,13] We were able to accomplish this variation either by reducing the As$_4$/Ga ratio to about one by an additional aperture in front of the As$_4$ cell at a constant growth temperature of 530°C, or - somewhat easier - by increasing the growth temperature from 530°C to >610°C at a constant As$_4$/Ga flux ratio of about two.[12,15] In both cases the site occupancy of Ge onto As- sites under Ga-stabilized (4×2) reconstruction was monitored in the RHEED pattern.

Both (Be,Si)- and (Ge,Ge)- doped abrupt p-n junctions in GaAs were grown, and detailed I-V and C-V measurements on isolated mesa diodes showed improved diode characteristics.[6,12] These encouraging results on the diodes formed the basis to realize the new GaAs doping superlattice by sequentially combining many of such p-n junctions in the same growth run. Figure 2 displays scanning electron micrographs

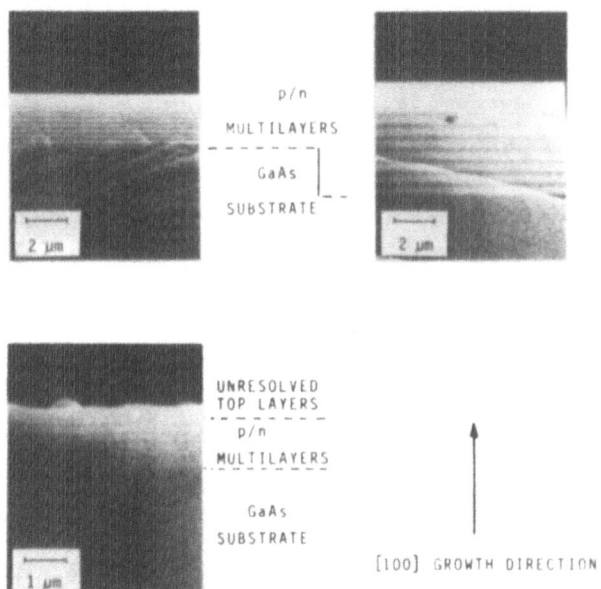

Figure 2 - Scanning electron micrographs of (110) cleavage plane of different periodic n-p doping multilayer structures in GaAs; (a) (Si,Be)-doped with $d_n = d_p = 0.12$ μm; (b) (Si,Be)-doped with $d_n = d_p = 0.27$ μm; (c) (Ge,Ge)-doped with $d_n = 0.16$ μm and $d_p = 0.04$ μm. No intrinsic (i-) layer was interspersed between the constituent intentionally doped layers.

of the (110) cleavage plane of three different periodic n-p multilayer structures in GaAs recently grown by MBE. In Fig. 2a and 2b, a total of 20 alternating n(Si)- and p(Be)-doped GaAs layers corresponding to 10 periods are clearly resolved. The thickness of the i-layers in these samples was chosen to be close to zero by opening the Si shutter immediately after the Be shutter has been closed and vice versa.

Inspection of Fig. 2c reveals, however, that the use of Ge for n- as well as p-type doping of MBE GaAs in these doping multilayer structures has been less successful. On top of the 12 clearly observable alternate n-p GaAs layers 4 additional n-p layers were deposited in the same growth run, which are no longer resolved due to the increasingly unstable Ga (4x2) growth conditions during growth of the p-type layers.[12,14] Growth had to be stopped after 16 layers because of the onset of 3-dimensional growth of Ga droplets. In this particular structure, the n-type layers were obtained at an As_4/Ga flux ratio >2 leading to the As-stabilized (2×4) surface structure, and the p-type material was grown at an As_4/Ga flux ratio of 1 resulting in a Ga-stabilized (4×2) surface structure. The width of the Ge incorporation on donor or acceptor sites, respectively, was accurately determined by the time interval of placing the aperture into the As_4 beam path, and it was monitored in the RHEED pattern. The occurrence of a wavelike bending in n-p GaAs <u>multilayer</u> structures, doped with Ge only, could also not be suppressed completely, when we performed the required variation of As surface population via a change of the substrate temperature from 530°C for n-type growth to 625°C for p-type growth at a <u>constant</u> As_4/Ga flux ratio of 2. Therefore, at present, only the application of Si as donor and Be as acceptor impurities yields a uniform doping within each individual layer and a controlled thickness with constant periodicity also for a large number of layers in sequence. Only in this case the transition regions ("interfaces") between the individual GaAs layers remain extremely smooth down to atomic steps also over large substrate areas.[6]

In our experiments described in Section 4, we found that 10 periods of the doping multilayer structures, as shown in Figs. 2a and 2b, are sufficient to produce the expected superlattice effects in this material. In order to study these effects, we have now grown a large number of (Si,Be)-doped GaAs doping superlattices, where we systematically varied the layer thicknesses d_n, d_p, and thereby the periodicity d as well as the individual doping levels over wide ranges according to Fig. 6 and our theoretical considerations in Section 3.

Finally, the question of the ultimate abruptness of the constituent (Be,Si)-doped p-n junctions in these GaAs doping superlattices arises. Since the depth resolution combined with the required high elemental sensitivity of the available profiling techniques (e.g., SIMS) is not yet sufficient, we have combined the results obtained from C-V profiling measurements of periodic doping profiles in n-GaAs:Si and in p-GaAs:Be, in order to derive the expected free-carrier profile of an abrupt p-n junction in (Be,Si)-doped GaAs by point-by-point subtraction of actually measured profiles.[6] The profile of a typical junction extracted in this way, where electron-hole recombination across the junction was totally neglected, is shown in Fig. 3. The dotted line indicates the ideal impurity profile calculated from the Be-and Si-effusion-cell temperature and from the shutter setting. With a final

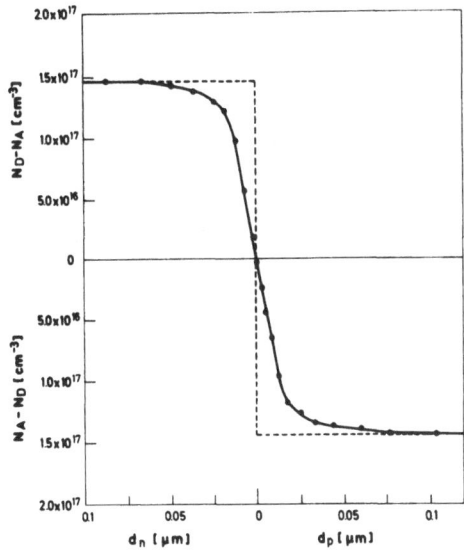

Figure 3 - Expected free-carrier profile (solid line) of an abrupt p-n junction in (Be,Si)-doped GaAs grown by MBE as derived from a point-to-point subtraction of actually measured periodic profiles of one carrier type (no correction for majority carrier diffusion). The dotted line indicates the ideal profile extracted from the Be and Si dopant beam intensity.[6]

free-carrier concentration reaching 1.45×10^{17} cm^{-3} on both sides of the junction (not stable because of recombination!), the width necessary to change the free-carrier concentration from n-type 1×10^{17} cm^{-3} to the same level p-type is found to be less that 200Å. In this simulation procedure, the problems arising from C-V profiling measurements have not been taken into account. Since the measured free-carrier profile may differ by several Debye lengths ($\lambda_{Debye} \sim 450$Å for $n = 1 \times 10^{16}$ cm^{-3} [17]) from the actual dopant profile due to majority carrier diffusion,[18] this value of 200Å represents an upper limit, and the width required to change the actual dopant impurities Si and Be is much smaller. Furthermore, our experiments have shown that any noticeable interdiffusion of the dopants Si and Be does not occur at growth temperatures below 600°C.

3.   ELECTRONIC PROPERTIES OF GaAs DOPING SUPERLATTICES

The unusual electrical and optical properties of doping (NIPI) superlattices are a direct consequence of the 1-dimensional periodic potential in this structure originating from the varying space charges of relatively small concentrations of ionized impurity atoms and charge carriers in an otherwise undistorted homogeneous bulk material. In this section we briefly discuss the schematic band structure in real space in the direction of layer sequence. This band structure can be tailored by an appropriate choice of the design parameters of the superlattice. In addition, the space charge potential and thereby the band structure may be tuned by electrical or by optical excitation of the crystal.

3.1 Band Structure of GaAs Doping Superlattices

We consider a periodic GaAs structure as illustrated in Fig. 1a, which consists of homogeneously doped n- and p-layers of thicknesses d_n and d_p and donor concentrations n_D and acceptor concentrations n_A, respectively, interspersed by intrinsic layers of thickness d_i. For the calculation of the periodic space charge potential in the crystal we make two approximations. (i) If the constituent layers are sufficiently thick (>100 nm) any subband effects resulting from the quantization of the motion in the direction of layer sequence may be neglected. (ii) For clarity we do not calculate the charge carrier distribution self-consistently. Instead, we assume that the impurity space charge is exactly neutralized in the central regions of width d_n^0 and d_p^0 in the respective layers by free carriers, whereas the impurity space charge is uncompensated in the remaining fractions of the doping layers of width (see Fig. 4 for illustration)

$$2d_n^+ = d_n - d_n^0 \text{ and } 2d_p^- = d_p - d_p^0 \tag{1}$$

The 2-dimensional density of electrons per n-layer and of holes per p-layer, respectively, is

$$n^{(2)} = n_D d_n^0 \text{ and } p^{(2)} = n_A d_p^0 \tag{2}$$

Since the condition of macroscopic neutrality requires that

$$n_D d^+ = n_A d_p^- \tag{3}$$

we obtain a relation between the 2-dimensional carrier densities

$$n^{(2)} = p^{(2)} + n_D d_n - n_A d_p. \tag{4}$$

The solution of Poisson's equation directly yields the internal electrical

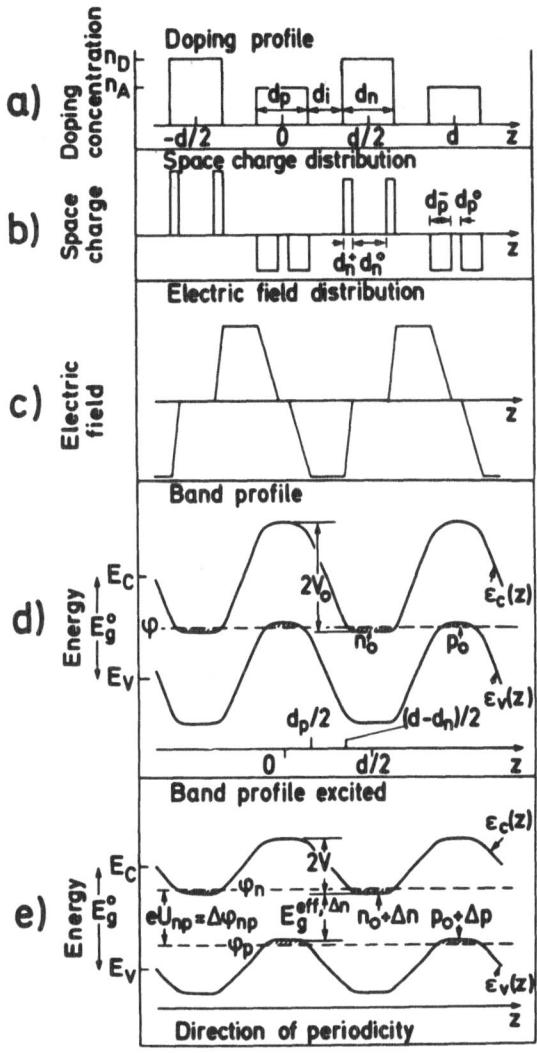

Figure 4 - Schematic illustration of the origin of periodic band edge modulation in GaAs doping superlattices; (a) doping profile; (b) space-charge distribution; (c) electric field distribution; (d) and (e) spatial modulation of the valence and conduction band edges. The effective bandgap of the superlattice is reduced compared to that of bulk GaAs. (d) Ground state with n_0 and p_0 free carriers in the constituent layers, vanishing bandgap ("semimetal"), and a common Fermi level ϕ; (e) excited state with $n = n_0 + \Delta n$, $p = p_0 + \Delta p$, and a splitting of the quasi-Fermi levels $\phi_n - \phi_p = \Delta\phi_{np} = eU_{np}$.

field F(z), as depicted in Fig. 4c,

$$\frac{dF(z)}{dz} = \frac{4\P}{\varepsilon}\rho \tag{5}$$

where ρ denotes the space charge, and the space charge potential V(z) (see Fig. 4d)

$$\frac{dV(z)}{dz} = -eF(z). \tag{6}$$

The periodic space charge potential V(z) for $-d/2 < z < d/2$ is then

1) constant in the neutral part of the doping layers, i.e.,

$$V(z) = \begin{cases} 0, \text{ for } z < d_n^0/2 & (7) \\ 2\,V_0, \text{ for } (d-d_p^0)/2 < z < d/2, & (8) \end{cases}$$

2) parabolic in the ionized impurity regions, i.e.,

$$V(z) = \begin{cases} (2\P e^2 n_D/\varepsilon)(|z|-d_n^0/2)^2 \text{ for } d_n^0/2 < |z| < d_n/2 \\ 2\,V_0-(2\P e^2 n_A/\varepsilon)(|z|-(d-d_p^0)/2)^2 \text{ for} \end{cases} \tag{9}$$

$$(d-d_p)/2 \;<\; |z| \;<\; (d-d_p^0)/2 \tag{10}$$

3) linearly increasing in the intrinsic layers (if present), i.e.,

$$V(z) = (4\P e^2 n_D d_n^+)(d_n^+/2 + |z| - d_n/2)/\varepsilon, \text{ for}$$

$$d_n/2 < |z| < (d-d_p)/2 \tag{11}$$

where ε is the static dielectric constant of the semiconductor. The amplitude V_0 of the space charge potential is given by[21]

$$2\,V_0 = V(d/2) - V(0) = (4\P e^2/\varepsilon)$$

$$[\bar{n}_D(d_n^+)^2/2 + n_A(d_p^-)^2/2 + n_D d_n^+ d_i] \tag{12}$$

The potential V(z) is modulating the CB and VB (see Fig. 4d). For the band edges we thus obtain

$$\varepsilon_c(z) = E_c + V(z) \tag{13}$$

and

$$\varepsilon_v(z) = E_v + V(z) \tag{14}$$

with

$$E_c = E_v + E_g^0. \tag{15}$$

As a consequence, the effective energy gap E_g^{NIPI} of the doping superlattice, i.e., the energy difference between the lowest electron states in the CB and the uppermost hole states in the VB, is reduced by $2V_0$ compared to the unmodulated bulk value E_g^0

$$E_g^{NIPI} = \varepsilon_c(z = 0) - \varepsilon_v(z = \frac{d}{2}) = E_g^0 - 2 V_0. \tag{16}$$

The periodic space charge potential in the direction of layer sequence results in a quantization of the energy spectrum. The free carriers thus represent a dynamically quasi-2-dimensional system. Motion of the free carriers parallel to the layers is free with a kinetic energy of

$$E_{kin} = \hbar^2 k_{||}^2 / 2m_c \tag{17}$$

at sufficiently large values of free-electron density $n^{(2)}$ near the Fermi level. In Eq. (17), m_c is the effective mass (including 2-dimensional many body corrections) and $k_{||}$ is the momentum parallel to the layers.[19] Any scattering by ionized impurity potentials as well as weak potential fluctuations arising from the random distribution of the impurities are neglected here.

The motion of free carriers in the direction of layer sequence, however, is quantized with discrete energy levels E_ν which are characterized by their subband index $\nu = 0,1,2...$. The energy of electrons in the νth subband is given by

$$E_\nu(k_{||}) = E_\nu + \hbar k_{||}^2 / 2m_c. \tag{18}$$

Detailed self-consistent calculations of the subband energies in the Hartree approximation have been described in Refs. 19 and 20.

The density of states is obtained as a superposition of the contribution of all subbands. In the effective mass approximation each subband contributes a constant amount of $m_c/\pi\hbar^2$. . This situation is schematically displayed in Fig. 5 for equal CB and VB effective masses.

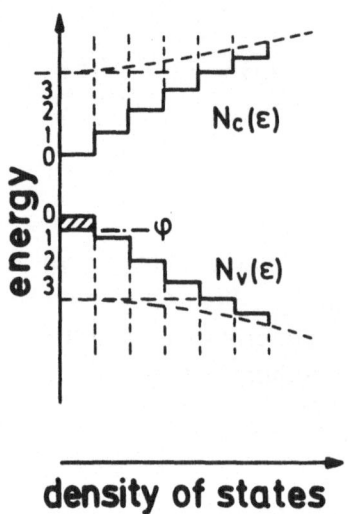

density of states

Figure 5 - Density of subband states in a GaAs doping superlattice (full lines) after Ref. [19] assuming m_{VB} equal to m_{CB}. The dotted line indicates the density of states in an unmodulated bulk GaAs crystal having the same number of impurities distributed randomly.

In the limit of long superlattice periods a quasi-continuum is achieved. In this case, the energetic separation between the distinct subbands becomes very small compared to the potential fluctuations arising from the random distribution of the impurities. The density of states smears out and no stepwise changes can be expected. However, considerable differences compared to the density of states of homogeneous bulk material persist.

3.2 Tailoring of the Band Structure of Doping Superlattices by the Choice of the Design Parameters

The periodic space charge potential in doping superlattices is determined by the doping concentration and by the thickness of the constituent layers according to Eqs.(7)-(12). Therefore, different types of superlattices may be produced intentionally by preselecting their design parameters. This enables us to vary the periodicity, amplitude and shape of the potential as well as the free-carrier population of the subbands. Three types of GaAs doping superlattices have so far been prepared by MBE.

In the case of low doping concentrations and a small superlattice period d, the space charge potential amplitude $2V_0$ is smaller than the band gap of the homogeneous bulk material. With equal doping densities in the constituent n- and p-layers, i.e., $n_D d_n = n_A d_p$, all excess electrons of the donors recombine with the holes of the acceptors. This structure represents a compensated intrinsic superlattice having no free carriers and with the Fermi level ϕ in the middle of the effective gap.

Next we assume the same value of $n_A d_p$ as before, however, we increase $n_D d_n$. Then the n-layers contain excess electrons of density

$$n_0^{(2)} = n_D d_n - n_A d_p \tag{19}$$

whereas the p-layers are totally depleted. This structure is n-type having the Fermi level in the conduction band. Correspondingly, the crystal is p-type, if $n_A d_p > n_D d_n$.

If the values of $n_D d_n$ and $n_A d_p$ are large, the space charge potential amplitude $2V_0$ has reached its maximum value of slightly more than the band gap of the homogeneous bulk, and the effective band gap has totally disappeared. Then, in the ground state, there are free electrons in the n-layers and free holes in the p-layers, and the doping superlattice resembles a semimetal. Also this structure may be n-type, as the example shown in Fig. 4d, or p-type, depending on the choice of the doping density in the respective layer type.

For GaAs multilayer structures with $n_D = n_A$ and $d_n = d_p$, the dependence of the material type on the design parameters is depicted in Fig. 6a using realistic values for the doping concentration and for the layer thickness. Curve (a) corresponds to a compensated intrinsic superlattice structure, where the space charge potential amplitude $2V_0$ is equal to half the band gap of bulk GaAs. Curve (b) shows the transition from a compensated structure to a semimetal structure, where $2V_0 \simeq E_g^0$. In the third case (curve (c)), there are 1×10^{12} cm^{-2} free carriers per layer in the semimetal structure. For this semimetal structure the width of one space charge layer and the total space charge per n- or p-type layer have been calculated, and the results are shown in Fig. 6b. These examples demonstrate that semiconductor structures

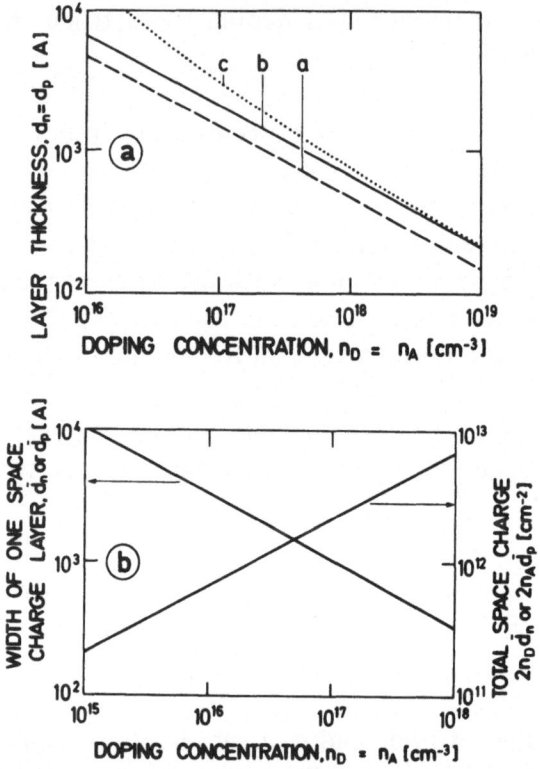

Figure 6 - Schematic illustration of the electronic properties of GaAs doping superlattices as a function of the design parameters; (a) dependence of conductivity and thereby "type" of doping superlattice (for details see text); (b) dependence of one space charge width and of space charge density per n- or p-layer, respectively.

with exactly tailored electronic properties may be produced by a suitable growth technique like MBE.

3.3 Tunability of the Band Structure of GaAs Doping Superlattices

The periodic 1-dimensional space charge potential in doping superlattices yields a band gap that is indirect in real space, since the CB minimum and the VB maximum are shifted by half the superlattice period d. Because of the small overlap of electron and hole states, electron-hole recombination lifetimes may exceed the corresponding bulk values by many orders of magnitude. As a consequence, even large deviations of the free-carrier concentration from thermal equilibrium become quasi-stable. The band profile of this "excited" state is depicted in Fig. 4e. Excess electrons and holes, where Δn equals Δp because of macroscopic neutrality, lead to a reduction of the original space charge density. Hence, the reduced space charge potential amplitude results in a widening of the effective energy gap as indicated in Fig. 4e. A rough estimate yields

$$\Delta E_g^{eff} \simeq (e^2 d/\varepsilon)\Delta n^{(2)}. \tag{20}$$

Simultaneously, a splitting of electron and hole quasi-Fermi levels is induced. The position of the Fermi level for electrons relative to the CB edge in the flat portions of the periodic potential is given by the corresponding bulk value for doping concentration n_D and the temperature T within our approximations

$$\varepsilon_c(z = 0) - \phi_n = (E_c - \phi)_{n_D};T. \tag{21}$$

A similar expression applies for the hole Fermi level ϕ_p

$$\phi_p - \varepsilon_v(z = d/2) = (\phi - E_v)_{n_A};T. \tag{22}$$

The expressions (1) to (4) and (12) to (16) relate the carrier concentrations $n^{(2)}$ and $p^{(2)}$ with the effective gap E_g^{NIPI} and, by (21) and (22), with the quasi Fermi level difference

$$\Delta\phi_{np} = \phi_n - \phi_p. \tag{23}$$

For the values of $n^{(2)}$ and $p^{(2)}$ as a function of $\Delta\phi_{np}$ we obtain the relations

$$n^{(2)} = n_D d_n - n_0^{(2)}[\{1 + (V_{bi} - \Delta\phi_{np})/(2V_0)_0\}^{1/2} - 1] \tag{24}$$

and

$$p^{(2)} = n_A d_p - n_0^{(2)}[\{1 + (V_{bi} - \Delta\phi_{np})/(2V_0)_0\}^{1/2} - 1] \tag{25}$$

with

$$n_0^{(2)} = [n_A n_D/(u_A + n_D)]d_i, \tag{26}$$

$$(2V_0)_0 = (2\P e^2/\varepsilon)n_0^{(2)}d_i, \tag{27}$$

and

$$V_{bi} = E_g - (E_c - \phi)_n - (\phi - E_v)_p. \qquad (28)$$

As the carrier concentrations may never become negative, a minimum value of the quasi Fermi level difference $\Delta\phi_{np}^{th}$ can be calculated from the threshold condition that the "minority layers" are totally depleted. The expressions (24) and (25) yield[21]

$$n_0^{(2)}[\{1 + (V_{bi} - \Delta\phi_{np}^{th})/(2V_0)_0\}^{1/2} - 1] = \text{Min}[n_D d_n, n_A d_p], \qquad (29)$$

whence

$$\Delta\phi_{np}^{th} = V_{bi} - (2V_0)_0 \begin{cases} [\{1 + (1 + n_A/n_D)d_p/d_i\}^2 - 1] \, ; \\ \quad \text{if } n_D d_n > n_A d_P \qquad\qquad (30) \\ \\ [\{1 + (1 + n_D/n_A)d_n/d_i\}^2 - 1] \, ; \\ \quad \text{if } n_D d_n < n_A d_p \qquad\qquad (31) \end{cases}$$

The condition of macroscopic charge neutrality requires that also the carrier concentration in the "majority layers" can never be reduced below their value at $\Delta\phi_{np}^{th}$, i.e.,

$$n_{min}^{(2)} = n_D d_n - n_A d_p; \text{ if } n_D d_n > n_A d_p \qquad (32)$$

or

$$p_{min}^{(2)} = n_A d_p - n_D d_n; \text{ if } n_A d_p > n_D d_n. \qquad (33)$$

The value of $\Delta\phi_{np}^{th}$ may be positive or negative, depending on the design parameters n_D, n_A, d_n, d_p, and d_i. Because of $\Delta\phi_{np} \approx E_g^{NIPI}$, negative values of $\Delta\phi_{np}^{th}$ imply the possibility of having a <u>negative</u> effective gap (!). If $\Delta\phi_{np}^{th} > 0$, the minority layers will be depleted in the ground state of the system. If, however, $\Delta\phi_{np}^{th} < 0$, there will be a finite carrier concentration in both types of layers at thermal equilibrium which is obtained from (24) and (25) with $\Delta\phi_{np} = 0$.

There are several possibilities to achieve deviations from the ground state in GaAs doping superlattices, two of which are discussed here and demonstrated experimentally in Sec. 4. First, the injection or extraction of free carriers may be accomplished via selective lateral electrodes to the constituent layers, as displayed in Fig. 8. A potential difference eU_{np} applied between the n- and p-layers induces charges of

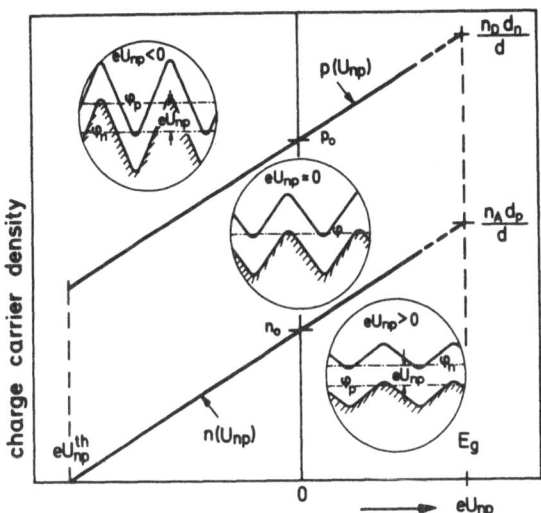

Figure 7 - Concentration of free electrons and free holes in a semimetallic p-type doping superlattice as a function of external potential
difference eU_{np}. The insets indicate the modulation of the band edges
and the position of the electron and hole quasi-Fermi levels for the case
$eU_{np} > 0$ and $eU_{np} < 0$, respectively.

free-carrier concentration such that the splitting of the quasi-Fermi
levels corresponds directly to the externally applied potential difference.

$$eU_{np} = \phi_n - \phi_p = \Delta\phi_{np} \qquad (34)$$

In Fig. 7 the relation between $n^{(2)}$ and $p^{(2)}$ as a function of eU_{np}
is displayed graphically for a p-type semimetal structure. The space
charge density in the constituent layers may be increased or decreased
depending on the sign of the applied potential difference. As a result,
the free-carrier concentration and thereby the conductivity can directly
be tuned in the constituent n- and p-layers. For $eU_{np} < eU_{np}^{th}$, the
n-layers are totally depleted and the free-hole concentration in the
p-regions remains constant according to Eq. (33). The upper value for
injection of free carriers is given by the bulk band gap energy. Increasing electron-hole recombination currents limit the splitting of the quasi
Fermi levels.

The second possibility to inject free carriers into GaAs doping
superlattices is given by the electron-hole pair generation during ab-

sorption of electromagnetic radiation. Two important aspects should be noted; first, the internal space charge fields result in a spatial separation of electrons which are collected in the n-layers and holes drifting into the p-layers, and secondly, the reduced effective band gap E_g^{eff} allows for absorption even far below the gap of homogeneous bulk material. In principle, optical absorption in a NIPI crystal is possible once the photon energy $\hbar\omega$ exceeds the effective energy gap E_g^{NIPI} (or, more accurately, for the case of finite $n^{(2)}$ and $p^{(2)}$, if $\hbar\omega > \overset{\circ}{\Delta}\phi_{np}$). In practice, however, the interband matrix elements for transitions between the uppermost valence and the lowest conduction subbands are extremely small for large superlattice periods and moderate doping concentrations used in the experiments described below. The matrix elements, however, increase exponentially as $\hbar\omega$ approaches the gap of the unmodulated bulk E_g^o. Consequently, an exponential increase of the absorption coefficient with $\hbar\omega$ is expected and values comparable with the unmodulated bulk are expected in the photon energy range above E_g^o. This absorption below E_g^o is comparable to the Franz-Keldysh effect[22,23] which results from an externally applied uniform electric field.

Also the absorption coefficient of a doping superlattice as a function of photon energy $\alpha^{NIPI}(\omega)$ depends on the effective energy gap E_g^{NIPI} and, therefore, on the quasi-Fermi level splitting $\Delta\phi_{np}$, since the optical matrix elements are strongly influenced by variations of the periodic space charge potential. In NIPI structures with large superlattice periods we can not expect to observe quantum oscillations. Therefore, a continuum approach represents a sufficiently good approximation, since the spatial variation of the internal fields is rather slow and the envelope

$$eF_i(z) = (dV/dz)_z \qquad (35)$$

wave functions of the true subbands do not differ significantly from the Airy functions obtained for homogeneous fields near the band edges.

The absorption coefficient of a NIPI crystal $\alpha^{NIPI}(\omega;\Delta\phi_{np})$ in this continuum approximation is given by the bulk value $\alpha^b(\omega; F_i)$ averaged over the internal field distribution. Thus one has

$$\alpha^{NIPI}(\omega;\Delta\phi_{np}) \approx d^{-1}\int_0^d dz\ \alpha^b(\omega;F_i(z)). \qquad (36)$$

The shape of $F_i(z)$ follows directly from the expressions (7) to (11) for $V(z)$ by differentiation. For the case of constant doping in the n- and p- layers, $F_i(z)$ increases and decreases linearly from zero to its maximum value in the ionized impurity regions, whence one obtains[21]

$$\alpha^{NIPI}(\omega;\Delta\phi_{np}) \approx d^{-1}\left[d_n^o\alpha_n^b(\omega) + d_p^o\alpha_p^b(\omega) + \right.$$

$$2d_i\alpha^b(\omega;F_{i,\ max}) + 2(d_n^+ + d_p^-)$$

$$\left. F_{i,\ max}^{-1} \int_0^{F_{i,\ max}} dF\alpha^b(\omega;F)\right]\Delta\phi_{np}$$

$$(37)$$

The subscript on the r.h.s. of (37) is a reminder that for the quantities $d_n^o, d_p^o, d_n^+, d_p^-$ and $F_{i_b max}$ their respective values at $\Delta\phi_{np}$ should be taken. $\alpha_n^b(\omega)$ and $\alpha_p^b(\omega)$ are the absorption coefficients of an unmodulated n- and p-type bulk material of the corresponding doping strength at zero electric field. In Ref. 21 we have calculated these quantities in detail using the analytical expression obtained for the effective mass approximation under the assumption of isotropic bands.

The tunability of the absorption coefficient is a direct consequence of the dependence of the space charge density on the population of the CB and VB states. The injection of free carriers leads to a reduction of the space charge density, a widening of the effective band gap, a reduction of the internal electrical field $F(z)$, and hence to a reduction of optical absorption below E_g^o. In addition, the tunability of the band structure has a strong influence on the luminescence properties of GaAs doping superlattices. Any variation of the (photoexcited) steady-state carrier concentration results in a strong shift of the position and intensity of the emission lines. (For details see Sec. 4.2.2.)

4. TUNABILITY OF FREE-CARRIER CONCENTRATION AND EFFECTIVE BANDGAP IN GaAs DOPING SUPERLATTICES

In this section we describe selected experimental results to demonstrate that in fact deviations from thermal equilibrium of free carriers are quasi-stable in GaAs doping superlattices and excess-carrier lifetimes become thus very large. Our studies clearly reveal that both the effective bandgap and the concentration of free carriers are variable quantities for a given doping superlattice, which can easily be tuned from the outside by external voltage or by optical excitation.

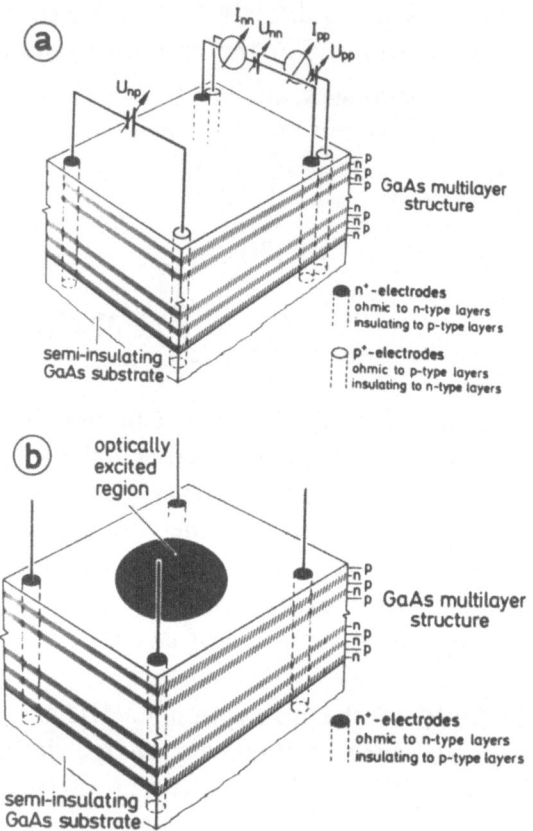

Figure 8 - GaAs doping multilayer device provided with selective electrodes to the constituent n- and p-doped layers; (a) multilayer device for simultaneous measurements of electron and hole conductance as function of external bias U_{np}; (b) multilayer device for Hall effect measurements in the n-doped regions during optical excitation.

4.1 Modulation of Free-Carrier Concentration under Negligibly Small Recombination Losses

A simultaneous modulation of electron and hole concentration in GaAs doping superlattices and hence of the bipolar conductivity parallel to the layers can be achieved by an external bias U_{np} applied between the constituent n- and p-doped layers or by the generation of electron-hole pairs via the absorption of light. We have demonstrated this modulation on multilayer structures provided with selective electrodes (Fig. 8a) to measure the variation of conductivity simultaneously in the constituent n- and p-layer.[24]

In a representative sample (#2282) the n- and p-doped GaAs layers have the same thickness of $0.19\mu m$. The thickness of the i-layers is zero. The doping concentration in the n-layers is $n_D = 3 \times 10^{17}cm^{-3}$, whereas the concentration in the p-layers is lower ($n_A = 1.9 \times 10^{17}cm^{-3}$). The thickness and the doping concentration of the constituent layers are chosen such that even in the ground state there are already free electrons and free holes in the respective layer regions. In this case, the effective band gap $E_g^{eff,o}$ is zero because twice the maximum space charge potential amplitude exceeds the band gap E_g^o (n-type "semimetal"). The electron mobility μ_n and the 2-dimensional electron concentration $n^{(2)}$ of the gound state of this structure are directly obtained from Hall effect measurements in the dark. We find $n^{(2)} = 2.99 \times 10^{12}cm^{-2}$ per layer, $\mu_n = 1798$ $cm^2V^{-1}sec^{-1}$, and $\sigma_{nn}^{(2)} = 8.63 \times 10^{-4}Ohm^{-1}$, where the notation $\sigma_{nn}^{(2)}$ denotes the conductivity in each n-layer parallel to the layers. The Hall coefficient does not yield the volume electron density n directly but only the product of the density $n^{(2)}$ and the width of the conducting layers.

If the electron concentration $n^{(2)}$ and the hole concentration $p^{(2)}$ are varied by the external voltage U_{np}, a corresponding modulation of the quasi-2-dimenstional conductivities $\sigma_{nn}^{(2)}$ and $\sigma_{pp}^{(2)}$ in the n- and p-layers, respectively, can be observed.[24] In the steady state, the potential difference eU_{np} applied between the n+- and the p+- electrode results in the splitting of the Fermi level into quasi-Fermi levels ϕ_n for electrons and ϕ_p for holes. At positive bias, $+U_{np}$, the injected carriers partly compensate the space charges resulting in a reduction of the space charge potential V_o and thus a widening of the effective band gap E_g^{eff} (see Fig. 4e). At negative bias, $-U_{np}$, free carriers are extracted as long as one layer type (here the p-layers) is totally depleted leading to an increase of the space charge potential.

In order to determine the conductivity in the respective doped regions, the current I_{nn} in the n-layers and I_{pp} in the p-layers generated by a small voltage U_{nn} and U_{pp} of less than 50 mV was measured as a function of the controlling bias U_{np} (Fig. 8a) yielding directly the conductance of the sample which still includes a geometrical factor. The observed conductances $G_{nn}(U_{np}) = I_{nn}(U_{np})/U_{nn}$ and

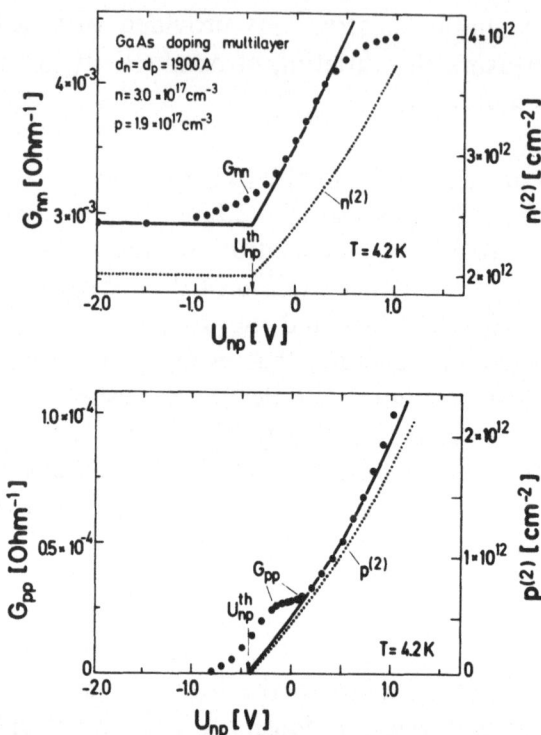

Figure 9 - Simultaneous variation of the conductances G_{nn} and G_{pp} in a GaAs doping superlattice as a function of applied bias U_{np} (full points). The full lines represent the expected conductance behaviour as calculated from the design parameters of the device. The dotted lines show the variation of free-carrier concentration with bias derived from the conductance values.

$G_{pp}(U_{np}) = I_{pp}(U_{np})/U_{pp}$ for our sample #2282 at a temperature of 4.2K are displayed in Fig. 9.[24] The design parameters of the sample (n_D, n_A, d_n, d_p) inserted in Eqs. (30-34) yield a threshold voltage U_{np}^{th} of -0.42V. Below this voltage, there are no free carriers in the p-regions, whereas the concentration of electrons in the more heavily doped n-layers remains constant. The experimental data of Fig. 9 indicate that a simultaneous enhancement of electron and hole concentration by more than 10^{12} carriers per cm^2 per layer has been obtained by the external potential difference eU_{np}. With increasing bias, the measurement is finally limited by the increasing leakage current I_{np} across the individual p-n junctions.

The intriguing properties of this versatile bulk multi junction FET structure result from a suppression of the direct electron-hole recombination by a space charge potential periodically modulated in the direction of layer growth. The experimental data are in good agreement with the calculated characteristics using the preselected design parameters of the structure. An alternative method to modulate the conductivity in these periodic GaAs doping multilayer structures is provided by the generation of electron-hole pairs via the absorption of light.[25]

The photoexcitation of the GaAs doping superlattice was achieved by continuous illumination with monochromatic light at 4.2K using the geometry of Fig. 8b. Typical light intensities on the sample were in the order of 10^{12} - 10^{13} photons/cm^2.sec. Electron-hole pairs generated by the absorption of photons are effectively separated by the space charge potential and collected in the respective doping layers. During excitation the carrier concentration in the constituent layers increases with duration of illumination because of the very low recombination rate. Simultaneously, the impurity space charges are becoming more and more compensated by the generated free carriers of concentration $\Delta n^{(2)}$ per layer resulting in a decrease of the width of the individual space charge layers. As a consequence of this reduction of the effective potential barrier width, the recombination of free electrons from the n-layers with free holes from the p-layers increases.[21] Finally, a steady state of the whole system is reached when the recombination losses balance the free-carrier generation rate during optical excitation. In principle, optical excitation should be possible if the photon energy $\hbar\omega$ exceeds the effective band gap E_g^{eff}. In practice, however, the absorption coefficient may be very small at $\hbar\omega \approx E_g^{eff}$. Below the gap of the homogeneous semiconductor material, E_g^o, the absorption coefficient decreases exponentially because of the reduced overlap between valence and conduction band wave functions involved in this process.[21] The photon energy dependence of the absorption, which will be described in Section 4.2.1, can be interpreted quantitatively as internal Franz-Keldysh effect.[22,23]

The results of Hall effect measurements to characterize the photoexcited steady state of sample # 2282 are displayed in Fig. 10. These four-point measurements do not only allow determination of the conductivity. They also yield the two factors which govern the conductivity, i.e: both free-carrier concentration and mobility can be deduced

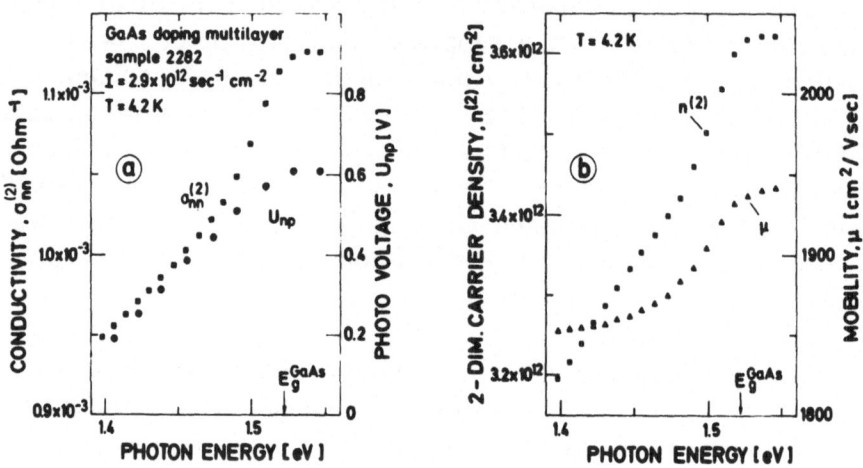

Figure 10 - Photon-energy dependence of the steady-state electronic properties of a strongly populated GaAs doping superlattice with design parameters as given in Figure 9; (a) variation of n-type conductivity parallel to the layers and photovoltage; (b) variation of free-electron concentration and mobility.

independently.[25] In Fig. 10a we only plotted the photon-energy dependence of the measured parameters in the constituent <u>n-layers</u>. As required from the macroscopic neutrality, an increase of the hole concentration in the p-layers by the same amount was observed. Using the same photon flux of $I = 2.9 \times 10^{12} \text{sec}^{-1} \text{cm}^{-2}$ at 1.52eV ($= E_g^o$), the dark value of $\sigma_{pp}^{(2)} = 9.3 \times 10^{-6} \text{Ohm}^{-1}$ was increased to $\sigma_{pp}^{(2)} = 7.8 \times 10^{-5} \text{Ohm}^{-1}$ by photoexcitation. This optically enhanced conductivity as a function of $\hbar\omega$ (see Fig. 10a) directly reflects the previously mentioned dependence of the absorption coefficient on the photon energy (see Sec. 4.2.1).

A second method to detect optically induced changes of the carrier concentration in doping multilayer systems is the measurement of the photovoltaic response. The optically induced splitting of the quasi-Fermi levels of electrons ϕ_n and holes ϕ_p (see Fig. 4c), can be measured directly as a voltage U_{np} if selective electrodes to the n- <u>and</u> the p-layers are applied to the sample. We have inserted the results of such experiments into Fig. 10a. A maximum value of $U_{np} > 0.6V$ was achieved for $\hbar\omega$ close to E_g^o. Besides its photon energy dependence, the photovoltage strongly relies on the photon flux. At high excitation levels, a value of eU_{np} in the order of E_g^o can be achieved.

Inspection of the results depicted in Fig. 10b reveals that the observed increase of the conductivity in the studied sample is partly due to an increase of mobility, but mainly due to an increase of the free-electron concentration $n^{(2)}$. The maximum increase of $n^{(2)}$ obtained at $\hbar\omega = E_g^o$ is $\Delta n^{(2)} = 21\%$. This value also depends strongly on the applied photon flux. The carrier mobility μ_n, on the other hand, only increases with optical excitation by about 8% at $\hbar\omega = E_g^o$. Therefore, as a result, the increase in carrier concentration and mobility both contribute to the measured conductivity enhancement.[25]

4.2 Modulation of Effective Bandgap

The validity of the concept of a tunable bandgap in GaAs doping superlattices has next been examined by measurements of the optical absorption and of the luminescence at photon energies $\hbar\omega$ below the forbidden gap of the unmodulated semiconductor material. Both quantities were found to depend strongly on the optical excitation of the samples.

4.2.1 Optical absorption at photon energies below the gap of bulk GaAs. In Sec. 4.1.2 we briefly mentioned that the optical absorption $\alpha(\omega)$ at photon energies $\hbar\omega$ close to the effective bandgap E_g^{eff} is very low and that $\alpha(\omega)$ increases exponentially with enhanced photon energy. The absorption coefficient for photon energies $\hbar\omega < E_g^o$ is thus no longer a constant quantity in GaAs doping superlattices, and it varies over a wide range if the effective bandgap is modulated by changing the carrier concentration. The strong photoconductive response observed in this material (see Sec. 4.1) makes feasible measurements of extremely low absorption coefficients in the tail far below the gap of bulk GaAs. The concentration of electrons and holes generated by photon absorption increases with the absorbed doses due to the extremely long recombination lifetime. Only when the rate of electron-hole generation becomes comparable with the (very low) recombination rate, the photo-response saturates. As a result, $\alpha(\omega)$ can easily be determined from the time dependence of the photoconductivity via two-point conductance measurements on selective electrodes to the respective layer types (see Fig. 8a).

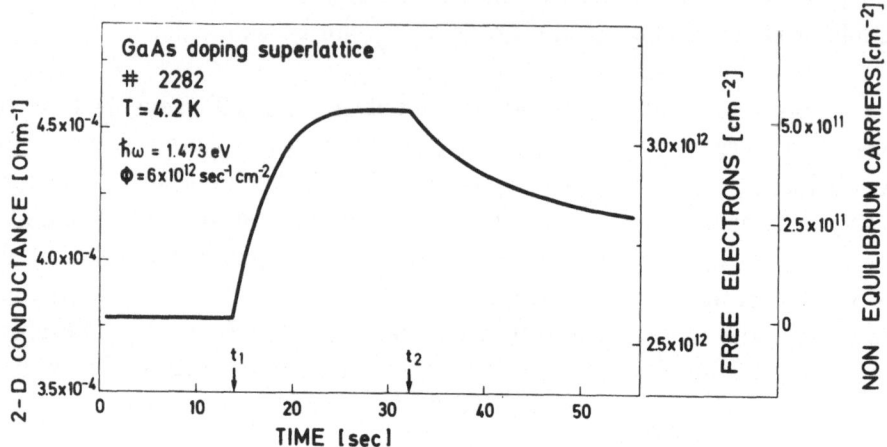

Figure 11 - Observed photoconductive response as a function of time in a GaAs doping superlattice used to determine the absorption coefficient. The photon flux of $\phi = 6 \times 10^{12}$ photons sec⁻¹ cm⁻² is switched on at t = t_1 and turned off at t = t_2.

A typical electron-conductance-versus-time diagram obtained for the constituent n-layers of sample #2282 is shown in Fig. 11. We can divide the diagram into three parts.[21] At t ≤ t_1, the conductance corresponding to the ground-state value of the sample without optical excitation is constant. The relatively high conductance ($G_{nn} = 3.8 \times 10^{-3} \Omega^{-1}$) in the n-layers arises from the higher doping concentration and carrier mobility as compared to the p-layers. The corresponding p-layer conductance is only $G_{pp} = 2.9 \times 10^{-5} \Omega^{-1}$. At t = t_1, the illumination of the sample by monochromatic light of defined photon energy is switched on. A strong enhancement of the conductance is observed which reflects the increasing free-carrier concentration due to the generation of electron-hole pairs by the absorption of photons. The slope of G_{nn} (t) decreases monotonously until G_{nn} (t) becomes practically flat. This continuous flattening is caused by both the decreasing absorption coefficient $\alpha^{NIPI}(\omega, \Delta\phi_{np})$ and the increasing recombination rate R ($\Delta\phi_{np}$). The absorption coefficient decreases with excitation since the overlap between valence and conduction band wave-functions differing in energy by $\hbar\omega < E_g^o$ decreases when the periodic space charge potential becomes flatter due to enhanced $\Delta\phi_{np}$. This decrease of $\alpha(\omega ; \Delta\phi_{np})$ is in marked contrast to the behaviour of the recombination rate which increases with $\Delta\phi_{np}$ because of enhanced overlap between relaxed electron and hole states.   The constant (steady-state) value of increased conductance is reached when the rate of both countercurrent processes balance each other.

After switching off the light at t = t_2, the optically induced enhancement of the conductance decreases due to recombination of the excess carriers.   The decrease of the photoconductance, however, does not show an exponential behaviour; it slows down very soon since the recombination rate becomes smaller with decreasing $\Delta\phi_{np}$ as shown in Fig. 12.

In order to determine the absorption coefficient as a function of

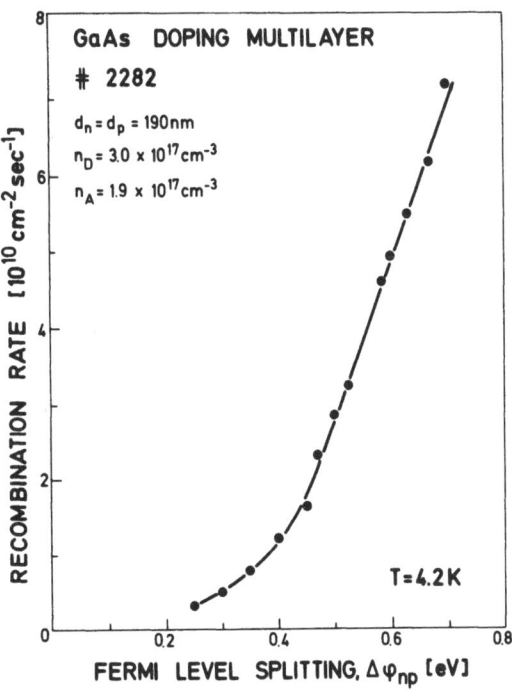

Figure 12 - Observed recombination rate in a photoexcited GaAs doping superlattice as a function of excitation level expressed as splitting of the quasi-Fermi levels $\Delta\phi_{np}$.

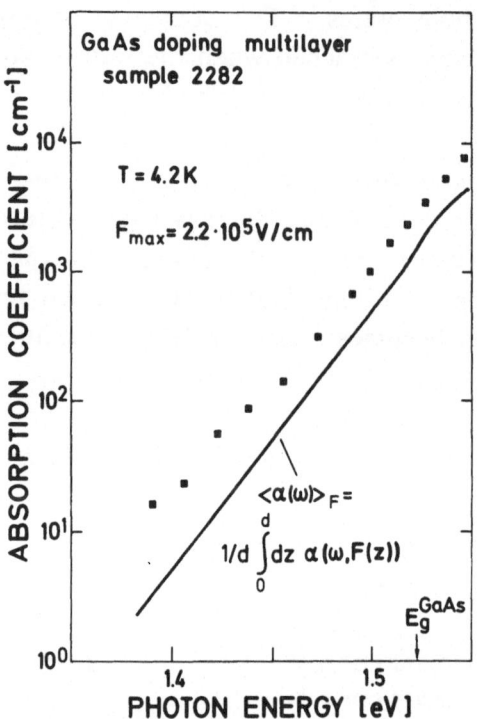

Figure 13 - Photon-energy dependence of the absorption coefficient of GaAs doping superlattice in the ground state ($\Delta\phi_{np} = 0$). The experimental square points are obtained from the measured increase of photoconductive response at time $t = t_1$. The theoretical curve is calculated with the design parameters $d_n = d_p = 190$ nm, $n_D = 3.0 \times 10^{17}$ cm^{-3} and $n_A = 1.9 \times 10^{17}$ cm^{-3} using Eq. (37).

frequency ω and potential difference $\Delta\phi_{np}$, we used the equation

$$\dot{n}^{(2)} = \dot{p}^{(2)} = (I_\omega/\hbar\omega)\alpha(\omega,\Delta\phi_{np})d - R(\Delta\phi_{np}) \qquad (38)$$

where the variation of carrier concentration with time, $\dot{n}^{(2)} = \dot{p}^{(2)}$, is given as the rate of electron-hole pair generation by the absorption of $(I_\omega/\hbar\omega)\alpha(\omega,\Delta\phi_{np})$ d photons per layer at a light intensity I_ω and the rate of electron-hole recombination. In practice, conductance versus time curves are measured systematically during and after illumination at different photon energies, and the deduced quantities $\dot{n}^{(2)}(\Delta\phi_{np})$ and $R(\Delta\phi_{np})$ are inserted into Eq. 38.[21]

Figure 13 displays the photon-energy dependence of the absorption coefficient in the ground state at $\Delta\phi_{np} = 0$. The experimental data (full squares) were obtained from the observed initial conductance increase at the time t_1 (Fig. 11) when the optical excitation starts. At this time, recombination can be neglected as it is strongly suppressed at very small $\Delta\phi_{np}$ values by the large spatial separation of free electrons and holes, through the potential barriers. The theoretical curve in Fig. 13 (full line) is calculated with the design parameters of sample #2282 using an extended version of the equation given in the figure. Inspection of Fig. 13 clearly demonstrates a considerable exponential tail of the absorption coefficient $\alpha(\omega)$ at photon energies <u>below</u> the gap of the unmodulated semiconductor. The finite absorption for $\hbar\omega < E_g^o$ observed in GaAs doping superlattices has been analysed quantitatively in terms of the Franz-Keldysh effect caused by the internal space-charge fields.[22,23] The nearly energy-independent shift of the experimental

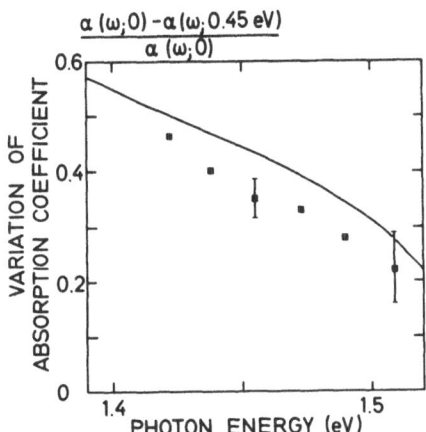

Figure 14 - Observed relative changes of the absorption coefficient vs. photon energy of an excited GaAs doping superlattice (design parameters as in Figure 13). The applied optical excitation yields a splitting of the quasi-Fermi levels of $\Delta\phi_{np} = 0.45$ eV. The full line represents the calculated results using the design parameters of the sample.

results to higher $\alpha(\omega)$ values arises from systematic errors in the determination of the incident light intensity.

We next evaluate the modulation of the absorption coefficient by optical excitation for a splitting of the quasi-Fermi levels of $\Delta\phi_{np} = +0.45eV$.[21] $\alpha(\omega, 0.45\ ev)$ was determined from the difference in the values of G_{nn} (0.45eV) measured during optical excitation $(t_1 < t < t_2)$ and after excitation $(t > t_2)$ via Eq. (38). The observed relative changes of the absorption coefficient between $\Delta\phi_{np} = 0.45eV$ and $\Delta\phi_{np} = 0$ eV are displayed in Fig. 14 and compared with calculations based on the design parameters. The values of $\alpha\ (\omega, 0.45eV)$ are considerably smaller at $\hbar\omega < E_g^o$ as compared to the ground state $(\Delta\phi_{np} = 0$ eV) values. With increasing $\Delta\phi_{np}$, i.e., optical excitation, the periodic space charge potential decreases and thus the internal electric fields are reduced (see Fig.4c and e).

Table 1 - Numerical values of the design parameters of GaAs doping superlattices used for luminescence measurements. V_o is the amplitude of the periodic space charge potential as calculated for zero electron and hole concentration in the constituent layers (ground state). $2V_o$ gives the value of the band edge modulation in the ground state. In the sample #2434 the value of $2V_o$ exceeds the bandgap E_g^o of the unmodulated semiconductor.

Sample	$d_n = d_p$ (nm)	$n_D = n_A$ (cm^{-3})	$2V_o$ (eV)
2227	20	1×10^{18}	0.145
2432	40	1×10^{18}	0.579
2229	60	1×10^{18}	1.305
2431	40	2×10^{18}	1.158
2434	40	4×10^{18}	2.316

A considerable increase of the absorption coefficient can be achieved with negative values of $\Delta\phi_{np}$ values by applying a negative external potential difference $eU_{np} = \phi_{np} < 0$ to the doping superlattice via selective electrodes (see Fig. 8a). The use of selective electrodes on the sample makes also feasible an electrical modulation of the absorption coefficient by varying $\Delta\phi_{np}$ through the external bias U_{np}. Our experiments on this subject have not yet been finished and will be reported elsewhere.

4.2.2 Tunable luminescence at photon energies below the gap of bulk GaAs. The spontaneous radiative recombination of excess electrons and holes across the indirect gap in GaAs doping superlattices has been analyzed by means of photo- and electroluminescence measurements (Refs. 26 - 29) on samples cooled to T ≤ 4K. From the reduced effective gap we expect that luminescence occurs also at photon energies far below the gap of bulk GaAs. The excess electrons and holes were generated by photoexcitation or by carrier injection over long distances via selective electrodes. For these experiments we selected superlattice configurations with high doping concentrations and with small superlattice period (Table 1), in order that the radiative recombination lifetimes τ_{rec} become relatively small and differ "only" by a few orders of magnitude from typical bulk GaAs values. Under these conditions, the radiative electron-hole recombination becomes easily detectable with sufficient intensities in the luminescence spectra.

Luminescence in GaAs doping superlattices arises from radiative recombination of electrons populating CB subbands below the quasi-Fermi level for the electrons, ϕ_n, with holes in the acceptor impurity band at energies above the quasi-Fermi level for the holes, ϕ_p,[26] as indicated in Fig. 15. In this figure, the 2-dimensional subbands are drawn neither for the electrons in the CB nor for holes in the VB, for clarity. Instead, the photoexcited or injected electrons, which are actually populating these CB subbands up to ϕ_n, are symbolized by the dotted area. They are partly compensating the positive donor space charge in the central part of the constituent n-layers, as found in self-consistent calculations.[20] The excess holes, however, are not populating the uppermost VB subband, but instead the acceptor impurity band in the constituent p-layers. They are partly neutralizing the negative charge of the acceptors in the central part of these layers.

In Fig. 16 four photoluminescence spectra covering the range of $1.20 < \hbar\omega < 1.50$ eV obtained from a representative GaAs doping superlattice (#2432) are displayed. The position of the asymmetric luminescence line is shifting strongly as a function of excitation intensity I_{exc}. The observation confirms our theoretical prediction that the

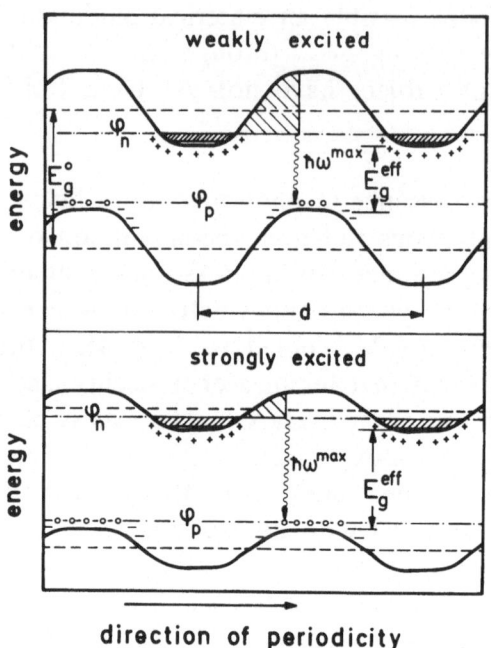

Figure 15 - Schematic real-space energy band diagram of a GaAs doping superlattice at two different excitation levels (not to scale), indicating the radiative electron-hole recombination across the indirect gap. The positively charged donors in the n-layers and the negatively charged acceptors in the p-layers are indicated by + and - signs, respectively.

dominant luminescence process in this superlattice is the radiative recombination of thermalized electrons in the conduction subbands with thermalized holes in the acceptor impurity band across the indirect gap in real space, as indicated in Fig. 15. With increasing I_{exc}, the enhanced steady-state concentration of (spatially separated) excess electrons and holes is screening more and more effectively the bare space charge potential of the ionized impurity atoms. The modulation of the band edges is thus reduced and the effective band gap finally approaches the gap of bulk GaAs (Fig. 15). Further examination of Fig. 16 shows that in addition to the shifting photoluminescence peak a narrow symmetric peak at $\hbar\omega = 1.455$ eV is detected, which does not shift with excitation. In all samples investigated we found, however, that the most intense luminescence line is the excitation-dependent broad asymmetric line. The observation confirms that most of the photoexcited carriers are separated in real space before they have a

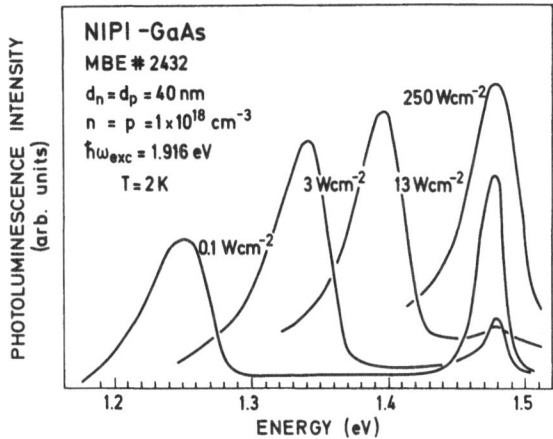

Figure 16 - Low-temperature photoluminescence spectra of a GaAs doping superlattice obtained with different laser excitation intensities, I_{exc}, expressed in W cm^{-2}.

chance to recombine via a "vertical transition in real space" (with $\hbar\omega$ $=E_g^0$).[29]

In Fig. 17 the energies of the PL peaks as a function of excitation intensity are shown for different GaAs doping superlattices with the design parameters of Table 1. For the three samples differing in their superlattice period d by a factor of 2 and 3, respectively, the excitation intensity required for a given position of the PL peak becomes larger for samples with longer superlattice period. It should be noted that at very low excitation intensities, the PL peak of sample #2227 approaches the theoretical minimum value of 1.435eV.

For samples with identical superlattice period but different doping concentrations, the calculated self-consistent potential becomes steeper and the ground-state effective gap becomes smaller with increased doping concentration. From the data of Fig. 17b it is evident that with increasing doping concentration, more and more excitation intensity is required to observe the PL line at a given peak energy. In addition, for a given peak energy, the width of the "tunneling barrier" decreases with increased doping level leading to a strongly enhanced recombination probability. Accordingly, the PL intensity is still high at very large red-shifts.[27]

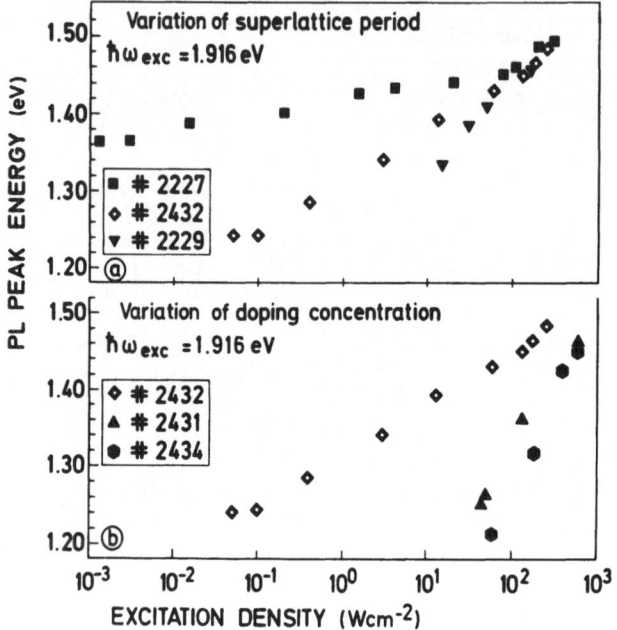

Figure 17 - Energies of photoluminescence peak, $\hbar\omega_{peak}$, as a function of excitation intensity for various GaAs doping superlattices with different design parameters as summarized in Table 1 (T = 2K).

The results of the photoluminescence measurements demonstrate that the observed photon energies of radiative recombination between thermalized electrons and holes directly reflect the effective gap E_g^{eff} of GaAs doping superlattices which depends strongly on the concentration of photoexcited excess carriers. Note that even at low excitation intensity the steady-state carrier concentration in this structure is already unusually high (of the order of the doping concentration). This is due to the reduced e-h recombination probabilities in these superlattices.[21]

For electroluminescence experiments[28] the as-grown wafers were cleaved into 0.4 cm[2] rectangular pieces. Selective ohmic contacts to the constituent GaAs layers were achieved by carefully alloying small Sn and Sn/Zn balls as n+- and p+-electrodes, respectively (see insert in Fig. 18). Both electrodes are selective since they are ohmic to one type of layer, but form blocking n-p junctions with respect to the layers of opposite doping. The injected current with variable pulse height was supplied by a pulse generator. The onset of electroluminescence from

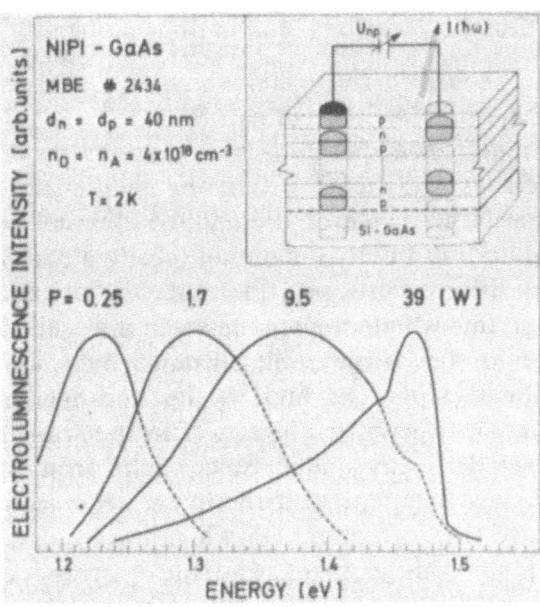

Figure 18 - Observed electroluminescence lines as a function of energy for various injection powers (relative luminescence intensities not drawn to scale). The inset shows a schematic cross-section of the multilayer device provided with two selective electrodes to the constituent n- and p-doped layers for current injection.

the three samples studied was observed at injection powers below 0.02W in the direction normal to the device surface with an IR viewer. In order to obtain sufficient radiation for the detection system of our luminescence experiments, however, the injection current had to be increased by a factor of five. In Fig. 18 four electroluminescence spectra covering the range $1.18 < \hbar\omega < 1.50$ eV obtained from a representative sample (#2434) are shown. The energetic position of the rather broad luminescence line is shifting strongly as a function of applied injection current. When we reduced the injection current from 40mA to 3 mA per constituent p-n junction, corresponding to a reduction of the power from 39 W to 0.1 W, the luminescence peak energy could be tuned from 1.48 to 1.18 eV, where the low-energy value is given by the spectral limitation of our detection system.

As in the case of photoluminescence, the observed electroluminescence of GaAs doping superlattice also arises from radiative recombination of electrons populating subbands below ϕ_n with holes in acceptor impurity band states above ϕ_p. This again implies a transition across the tunable indirect gap (Fig. 15). In addition, however, the electroluminescence spectrum is determined by the dynamics of the injected electrons and holes within the multilayer device.[28] We found that a much larger value of the applied external potential difference eU_{np} is required in order to maintain locally a certain value of the quasi-Fermi level difference $(\phi_n-\phi_p)$, because of the continuous drop of ϕ_n and ϕ_p within the whole region between the selective electrodes. This drop is due to the considerable distance between the two electrodes of our device (Fig. 18) and to the non-negligible resistance within the constituent superlattice layers. The spectra shown in Fig. 18 display the detected electroluminescence light from a rather large sample area, and they thus represent the integrated spectral distribution of the luminescence from a region over which the local value of $(\phi_n-\phi_p)$ varies considerably. This explains why the observed electroluminescence spectra are much broader than the corresponding photoluminescence spectra of GaAs doping superlattices and do not exhibit the typical asymmetric shape.

In addition to the broad tunable electroluminescence peak, we detected a narrow peak at $\hbar\omega = 1.475$ eV which increases in intensity particularly at high injection currents. This peak, which does not shift with current (see upper trace in Fig. 19), is due to "vertical" (in real space) radiative bulk GaAs transitions dominating in highly excited doping superlattices.[29] In all samples investigated, however, the broad shifting line is by far the most intense luminescence transition at moderate injection currents. This observation confirms our theoretical expectation that the dominant electroluminescence process in GaAs doping superlattices is the recombination of electrons injected into the n-layers and occupying conduction subbands with holes simultaneously injected into the p-layers and occupying a narrow acceptor impurity band.[20] As a consequence, the measured photon energies of electroluminescence also reflect the effective energy gap of the sample which depends strongly on the concentration of injected carriers (see Fig. 15).

In Fig. 19 the energies $\hbar\omega_{peak}$ of the electroluminescence peaks are shown as a function of injection power for three different GaAs doping superlattices with identical superlattice period but different

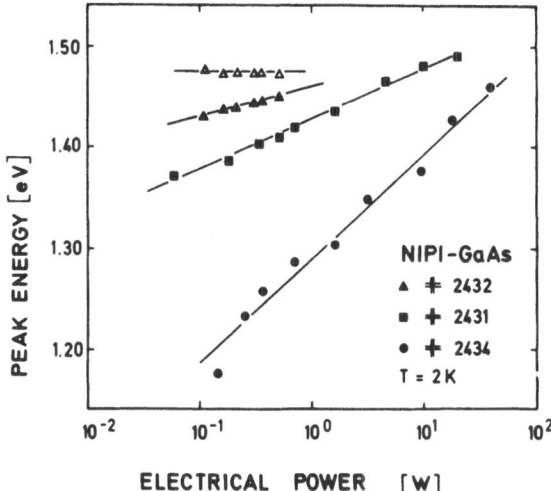

Figure 19 - Energies of electroluminescence peak, $\hbar\omega_{peak}$, as a function of injection power for varios GaAs doping superlattices with different design parameters as given in Table 1. The upper trace shows the non-shifting luminescence line which is due to the radiative bulk GaAs transition dominating in highly excited doping superlattices.

doping concentrations (see Table 1). With increased doping concentration, the calculated self-consistent potential becomes steeper and the ground-state effective gap becomes smaller. As a result, the width of the "tunneling barrier" for radiative recombination decreases with increased doping level for a given peak energy, thus leading to the strongly enhanced recombination probability.[20] Inspection of Fig. 19 reveals that exactly this behavior is observed in the three samples. With increasing doping concentration, more and more injection current is required to observe the luminescence line at the given peak energy. Thus, a high electroluminescence intensity even at very large red-shifts can readily be achieved by increasing the doping level of the constituent superlattice layers.

A step-like structure in the luminescence spectra, as expected from the formation of subbands in these space-charge-induced potential wells, has not been observed in the (photo) luminescence experiments, although the spectra show a strong asymmetric shape with a low energy tail (Fig. 16). This behaviour is attributed to inhomogeneous excitation of the layered structure, even during photoexcitation, due to the limited penetration depth of the incident laser light. The excitation intensity decreases by a factor of two within the superlattice structure. Therefore, the steady-state carrier concentration and the screening of the space charge potential and, hence, the luminescence frequencies decrease slightly from layer to layer. Since the observed spectrum is a superposition of all these contributions, the step-like structure cannot be detected.[26]

4.3 Quantization of Excess Carriers and Formation of Subbands

Direct information on quasi-2-dimensional excess carriers in the space charge induced potential wells of GaAs doping superlattices has been obtained from additional Raman and Shubnikov-de Haas measurements on samples with design parameters comparable to those of Table 1. In these superlattice configurations, the spacing between the subbands vary between ~20 meV and ~40 meV, depending strongly on the degree of excitation, and are then considerably larger than the level broadening which is of the order of ~10 meV.[20] A detailed discussion of our recent studies is beyond the scope of this paper and we only present a brief summary of the results.

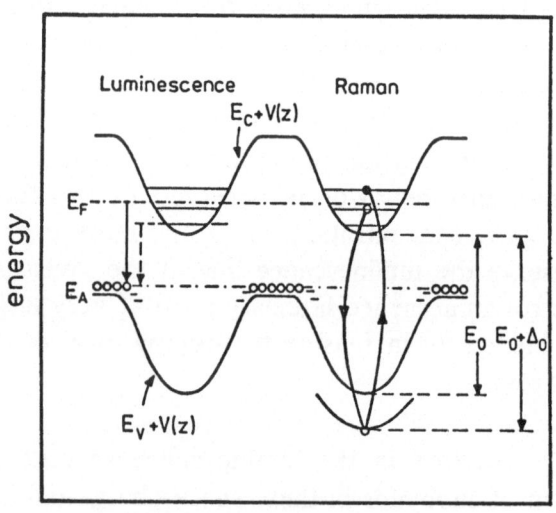

Figure 20 - Real-space energy band diagram of an excited GaAs doping superlattice indicating schematically the Raman and luminescence processes. (After Ref. 31)

The luminescence and Raman process in GaAs doping superlattices are schematically shown in Fig. 20. Light scattering signals from the quasi-2-dimensional photoexcited carriers can be obtained under resonance conditions where the scattering cross-section is stronly enhanced.[30] At the $(E_0 + \Delta_0)$ gap of GaAs, the resonance condition is fulfilled for electrons in the CB subbands, but not for holes in the VB. From the experiments,[26] the spin-flip single particle excitation provided direct information on the subband splittings, which are in good agreement with theoretical calculations.[20,31] Our studies further showed that with increasing excitation intensities, the subbands merge and a continuous change from a 2-dimensional to a 3-dimensional system occurs.[31] The striking result of these Raman measurements is that the statistical impurity potential fluctuations in GaAs doping superlattices are so effectively screened by the mobile electrons that they do not prevent the formation of subbands.

Finally, we briefly discuss the observed quantum transport in GaAs doping superlattices.[32] In Fig. 21, the measured resistivity of the

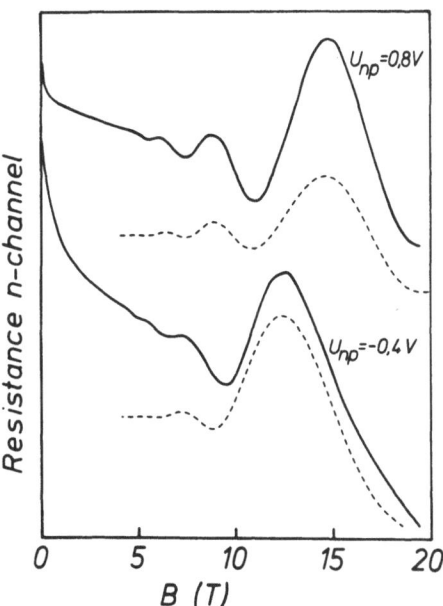

Figure 21 - Measured resistance in the n-channels of a GaAs doping superlattice ($d_n = d_p = 90$ nm, $n_D = n_A = 7 \times 10^{17}$ cm^{-3}) as a function of the magnetic field for two different p-n bias U_{np} (full line). The calculated resistivity is indicated by the dotted lines. (After Ref. 32)

n-channels at 4.2K as a function of the magnetic field perpendicular to the superlattice layers is shown for two different voltages U_{np}. Clear quantum oscillations are observed which shift to higher magnetic fields with increasing U_{np}. In addition, the magneto-quantum oscillations exhibit a distinct anisotropy for the orientation of the magnetic field with respect to the superlattice layers. From these experiments, the energetic spacing between the subbands could be deduced in the region of low excess-carrier densities and compared with theoretical evaluations.[32]

5. CONCLUDING REMARKS

In this paper we reviewed the molecular beam epitaxial growth of a new artificial doping ("NIPI") superlattice in GaAs, and we briefly described its fascinating electronic properties which result from the efficient spatial separation of electron and hole states (underline: indirect gap in real space) by the periodic space charge potential. Furthermore, we presented a few selected experiments which confirm the most crucial predictions of the theory of NIPI superlattices.[4,5,19,20] In particular, we could demonstrate that even large non-equilibrium carrier concentrations are metastable under weak excitation conditions and that the concept of tunable carrier concentrations, a tunable effective energy gap, and the formation and tunability of subbands is valid in this new class of semiconductor materials.

The original concept of doping superlattices has now been extended, and a new artificial semiconductor superlattice with tunable electronic properties and simultaneously with significant mobility enhancement of both 2-dimensional electrons and 2-dimensional holes has recently been prepared by MBE.[33] The structure consists of a periodic sequence of $n-Al_xGa_{1-x}As/i-GaAs/n-Al_xGa_{1-x}As/p-Al_xGa_{1-x}As/i-GaAs/p-Al_xGa_{1-x}As$ stacks with undoped $Al_xGa_{1-x}As$ spacers between the intentionally doped $Al_xGa_{1-x}As$ and the nominally undoped i-GaAs layers. In this new heterojunction doping superlattice we have for the first time achieved a spatial separation of electrons and holes by half a superlattice period as well as simultaneously a spatial separation of both types of free carriers from their parent ionized impurities. These unique properties were demonstrated by the strongly increased tunability of bipolar conductivity with bias. In addition, the observed temperature dependence of Hall mobilities provided direct evidence for a strong mobility enhancement of both electrons and holes in the spatially separated 2-dimensional accumulation channels formed in the lower band gap material.

ACKNOWLEDGMENT

The author wishes to thank many of his colleagues at the Max-Planck-Institut für Festkörperforschung who have contributed to this work, particularly G. H. Döhler, A. Fischer, H. Jung, J. Knecht, H. Künzel, P. Ruden, and Miss H. Willerscheid. Thanks are also due to H. J. Queisser and A. Rabenau for helpful discussions and encouragement. This work was sponsored by the Bundesministerium für Forschung und Technologie of the Fed. Rep. of Germany.

REFERENCES

1. L. Esaki and R. Tsu, IBM J. Res. Dev. 14, 61 (1970).
2. L. Esaki, J. Cryst. Growth 52, 227 (1981).
3. A. C. Gossard, in Thin Films: Preparation and Properties, ed. K. N. Tu and R. Rosenberg, Academic Press, New York, 1982,
4. G. H. Döhler, Phys. Status Solidi B52, 79 (1972).
5. G. H. Döhler, Phys. Status Solidi B52, 533 (1972).
6. K. Ploog, A. Fischer and H. Künzel, Epitaxy, J. Electrochem. Soc. 128, 400 (1981).
7. K. Ploog, in Crystals: Growth, Properties, and Applications,, ed. H. C. Freyhardt, Springer-Verlag, Berlin/Heidelberg, Vol. 3, 1980, p. 73.
8. C. T. Foxon and B. A. Joyce, in Current Topics in Materials Science, ed. E. Kaldis, North-Holland Publishing Company, Amsterdam/New York, Vol. 7, 1981, p. 1.
9. K. Ploog, Ann. Rev. Mater. Sci., 11, 171 (1981).
10. A. Y. Cho, J. Appl. Phys. 47, 2841 (1976).
11. K. Ploog, J. Vac. Sci. Technol. 16, 838 (1979).
12. K. Ploog, A. Fischer, and H. Künzel, Appl. Phys. Lett. 18, 353 (1979).
13. C. E. C. Wood, J. Woodcock, and J. J. Harris, Inst. Phys. Conf. Ser., 45, 1979, p.28.
14. M. Ilegems, J. Appl. Phys. 48, 1278 (1977).

15. H. Künzel, A. Fischer, and K. Ploog, Appl. Phys. 22, 23 (1980).
16. H. Künzel and K. Ploog, Inst. Phys. Ser., 56, 1981, p. 519.
17. D. P. Kennedy, P. C. Murley, and W. Kleinfelder, IBM J. Res. Dev. 12, 399 (1968).
18. W. C. Johnson and P. T. Panousis, IEEE Trans. Electron Dev. ED-18, 965 (1971).
19. G. H. Döhler, Surf. Sci. 73, 97 (1978).
20. P. Ruden and G. H. Döhler, Phys. Rev. B27, 3538 (1983).
21. G. H. Döhler, H. Künzel, and K. Ploog, Phys. Rev. B25, 2616, (1982).
22. W. Franz, Z. Naturforsch. A13, 484 (1958).
23. L. V. Keldysh, Sov. Phys.-JETP 7, 788 (1958).
24. K. Ploog, H. Künzel, J. Knecht, A. Fischer, and G. H. Döhler, Appl. Phys. Lett. 38, 870 (1981).
25. H. Künzel, G. H. Döhler, and K. Ploog, Appl. Phys. A27, 1 (1982).
26. G. H. Döhler, H. Künzel, D. Olego, K. Ploog, P. Ruden, H. J. Stolz, and G. Abstreiter, Phys. Rev. Lett. 47, 864 (1981).
27. H. Jung, G. H. Döhler, H. Künzel, K. Ploog, P. Ruden, and H. J. Stolz, Solid State Commun. 43, 291 (1982).
28. H. Künzel, G. H. Döhler, P. Ruden, and K. Ploog, Appl. Phys. Lett. 41, 852 (1982).
29. H. Jung, H. Künzel, G. H. Döhler, and K. Ploog, J. Appl. Phys. 54 (1983) to be published.
30. G. Abstreiter, K. Ploog, Phys. Rev. Lett. 42, 1308 (1979).
31. Ch. Zeller, B. Vinter, G. Abstreiter, and K. Ploog, Phys. Rev. B26, 2124 (1982).
32. J. C. Maan, Th. Englert, Ch. Uihlein, H. Künzel, K. Ploog, and A. Fischer, J. Vac. Sci. Technol. B1, 289 (1983).
33. H. Künzel, A. Fischer, J. Knecht, and K. Ploog, Appl. Phys. A30, 73 (1983).

SEMICONDUCTOR LASERS AND PHOTODETECTORS BY MOLECULAR BEAM EPITAXY

W. T. Tsang

Bell Laboratories Murray Hill, NJ   07974

ABSTRACT: The present review contains results on the performance of $Al_xGa_{1-x}As/Al_yGa_{1-y}As$ DH lasers for optical communication systems. In this respect, MBE-grown (AlGa)As DH lasers with threshold current densities, J_{th}, as low as, if not lower than, those prepared by liquid-phase epitaxy (LPE) over the entire wavelength range from infra-red to visible (0.88 μm - 0.7 μm) were obtained. Highly reliable DH lasers with $Al_{0.08}Ga_{0.92}As$ active layers were demonstrated. Median CW laser lifetimes $> 10^6$h at room temperature were projected for 70°C CW constant power accelerated aging. Optical transmitters containing MBE-grown lasers were installed in 45 Mbit/sec lightwave transmission systems and have been under field-test since 1980. More recently, high quality InP similar to the high purity InP layers grown by vapor phase epitaxy (VPE) and LPE were also prepared by MBE. 1.3μm wavelength InP/GaInAsP DH lasers having averaged J_{th} of 3.5 kA/cm^2 (1.8 kA/cm^2 lowest) and 1.5μm wavelength InP/GaAlInAs DH lasers having J_{th} of 3 kA/cm^2 were successfully prepared in a specially designed MBE system. $Al_{0.2}Ga_{0.8}Sb/GaSb$ DH lasers lasing at 1.78 μm were successfully prepared by MBE for the first time.

Meanwhile, the unique ability of MBE to grow atomically smooth ultra-thin ($\lesssim$ 200Å) (AlGa)As layers free of alloy clusters and layers with any desired compositional and doping profiles resulted in a new generation of electronic and photonic devices. These new devices yielded significant improvements in performance not generally achievable in conventional counterparts. In the photonic area, some examples are: multiquantum well (MQW) heterostructure lasers, double-barrier double-heterostructure lasers, graded-index waveguide separate-confinement heterostructure lasers (GRIN-SCH), multi-wavelength transverse-junction-strip lasers, superlattice avalanche photodetectors, graded-bandgap avalanche photodetectors and phototransistors, majority-carrier photodetectors, and superlattice etalon for optical bistability. With the modified MQW heterostructure and GRIN-SCH lasers extremely low J_{th} of 250 A/cm^2 were obtained. Buried heterostructure GRIN-SCH lasers have threshold of 2.5 mA under CW operation. A lowest value ever reported for semiconductor laser. With the superlattice APD and graded bandgap APD an impact ionization rates ratio enhancement of ~10 was measured.

1. INTRODUCTION

Molecular beam epitaxy (MBE) is an epitaxial crystal film deposition technique that involves the reaction of one or more thermal-energy beams of the constituent elements with a crystalline substrate surface held at a suitable temperature under ultra-high vacuum (UHV) conditions.[1] Gunther[2] in 1958, using multiple beams, described the growth of III-V materials. Stoichiometric films of both binary and ternary III-V compounds were grown by using his "three-temperature" technique. The group V element source oven was kept at a temperature T_1 so that it maintained a steady pressure in a static vacuum chamber, while the group III element source oven at a higher temperature T_3 was providing atoms incident on the substrate surface which was maintained at a temperature T_2. T_2 was higher than T_1 but lower than T_3. However, Gunther's films were grown on glass substrates and hence polycrystalline. In 1968, epitaxial growth of monocrystalline GaAs films were obtained by employing clean single crystalline GaAs substrates under highly improved vacuum conditions.[3,4] It is Arthur[4] who conducted fundamental studies of the kinetic behavior of Ga and As_2 species on GaAs surfaces, provided the first general understanding of the growth mechanism.   Subsequently, it was primarily Cho[1,5] whose continuing research in this epitaxial crystal growth technique led to the preparation of device-quality epitaxial GaAs/AlGaAs layers. By the mid-70's, the research of MBE can be divided into four major areas: (1) semiconductor microwave devices and current injection lasers primarily led by Cho and his coworkers;[5,6] (2) semiconductor superlattices primarily led by Dingle and Gossard and their coworkers,[7] Esaki and Chang and their coworkers,[8,9] and Ploog and his coworkers;[10] (3) growth kinetic studies primarily led by Joyce and Foxon and their coworkers;[11,12] and (4) preparation of II-VI and IV-VI compounds primarily carried out by Holloway[13] and Walpole[14] and their coworkers. At the same time, because of the growing importance and the rapid development of the lightwave communication and integrated opto-electronic technologies, a great deal of interest was generated and focused particularly in the preparation of (AlGa)As double-heterostructure current injection lasers by MBE. Traditionally, the quality of the injection laser prepared by a particular epitaxial growth technique is largely used as a yard-stick for calibrating the development of the epitaxial growth technique itself. This is true not only for MBE but also for liquid-phase epitaxy (LPE), vapor phase epitaxy (VPE) and metalorganic chemical vapor deposition (MO-CVD).

Excellent reviews on MBE growth technique may be found in references (1,10,15,16) and superlattices in references (7). The present review will cover new devices and device physics by MBE especially in the area of optoelectronics. In the case of MBE, the first current injection (AlGa)As DH laser was prepared by Cho, et al.[17] in 1974. In 1976, they also obtained the first CW MBE-grown stripe-geometry laser.[18] However, with these initial lasers, the threshold current densities of the best wafers were always twice that of similar-geometry DH laser wafers prepared by LPE, and the room-temperature CW operation lasted only on the order of hours.[17,18] It was not until 1979 that Tsang[19-21] first obtained MBE-grown (AlGa)As DH lasers with threshold currents as low as, if not lower than, those prepared by LPE over the entire wavelength range from infrared to visible (0.88 μm - 0.7 μm). In 1980, he and his coworkers also obtained highly reliable DH lasers with $Al_{0.08}Ga_{0.92}As$ active layers. Median CW laser lifetimes >

10^6h at room temperature were projected for 70°C CW constant power accelerated aging.[22-23] Optical transmitters containing MBE-grown lasers were installed in 45 Mbit/sec lightwave transmission systems and have been under field-test since 1980.[24] More recently, the unique ability of MBE to grow atomically smooth ultra-thin ($\leq$ 200Å) (AlGa)As layers free of alloy clusters[25] and layers with any desired compositional and doping profiles resulted in a new generation of electronic and photonic devices. These new devices yielded significant improvements in performances not generally achievable in conventional counterparts.

2. MBE LASER FOR OPTICAL COMMUNICATION SYSTEMS

For applications in lightwave communication systems highly reliable lasers are required. Consequently, the proton-bombarded stripe-geometry lasers fabricated from several MBE grown DH wafers containing $Al_{0.08}Ga_{0.92}As$ active layers were studied both for power reliability[22] and for functional reliability as 45 Mbit/s transmitters.[24] The laser diodes were formed in 5μm wide shallow proton-irradiated stripe geometry and 380μm long and operated without mirror coatings in dry nitrogen 70°C ambient at constant power outputs of 2.5 mW and 3.0 mW/mirror. Aged lasers from three early wafers yielded a median 70°C lasing lifetime of 8,800 h, a standard deviation of 1.5, and an extrapolated mean room temperature lifetime in excess of 10^6h. The long-term degradation rate of the operating current was as low as 0.7 mA/kh at 70°C. The results for the wafers studied show that the rate of change of the operating current with aging for the MBE diodes was more uniform than for typical LPE diodes fabricated with the same technology. However, this increased uniformity does not seem to result in a significantly smaller standard deviation σ in the failure contribution as determined from these three initial wafers. This indicates that a similar, "run-away" mechanism determined the end of life for both MBE and LPE lasers. Figure 1 shows a comparison of MBE lasers with three (best and sequential grown) of the ten LPE wafers used in the Hartman, Schumaker and Dixon[26] study, it is seen that the MBE lasers demonstrate a 70°C reliability comparable to those obtained for the LPW 12μm wide deep proton-bombarded stripe-geometry lasers.

We have also characterized the performance and studied the functional reliability of 45 Mb/s lightwave transmitters containing

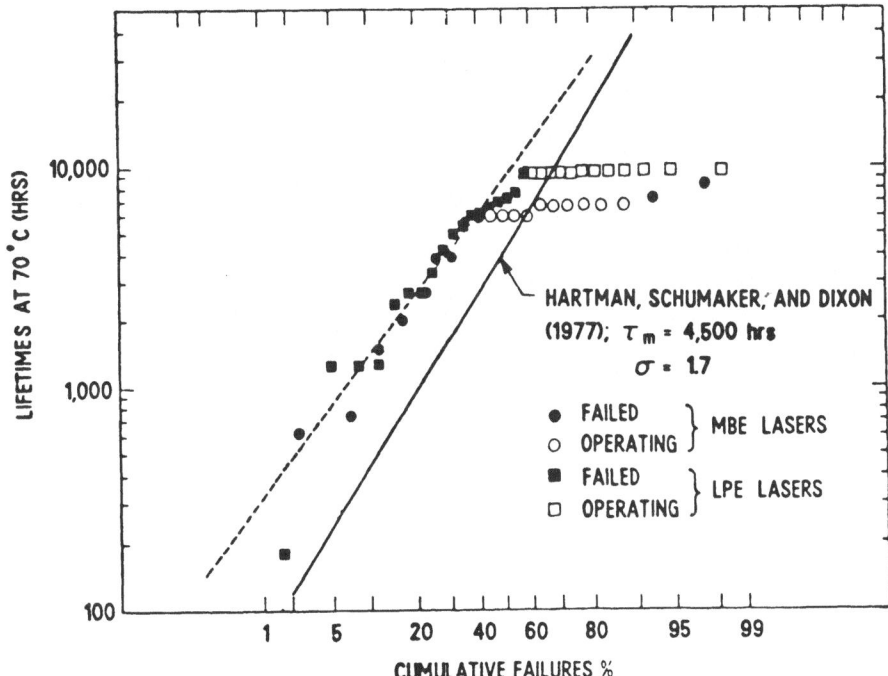

Figure 1 - Log-normal plot of the cw operating lifetime at 70°C of the three early best MBE wafers.

MBE-grown proton-bombarded stripe lasers.[24] In this study, the reliability of the lasers in the transmitters will depend not only on the degradation of the device characteristics e.g, linearity, pulsations, and symmetry of outputs, but also on other factors such as the laser packaging and its mechanical stability and the design of the transmitter circuit and its operating conditions. We observe that the transmitters containing MBE lasers have superior performance, i.e., low modulation current, small temperature dependence of lasing threshold, low noise below 20 MHz and high self-pulsation frequencies, compared to transmitters containing lasers of similar geometry grown in production facilities by liquid phase epitaxy (LPE), and subsequently processed using the same technology. The reliability data of over 10,000h indicate that the functional life-time of the MBE lasers will be limited ultimately by extinction ratio degradation in our circuit strategy (see Fig. 2). The origin of the degradation is the increase of the spontaneous light at the bias current.

It has generally been assumed that molecular beam epitaxy (MBE) is limited to slow growth rates ($\lesssim 2\mu m/h$) in the case of III-V

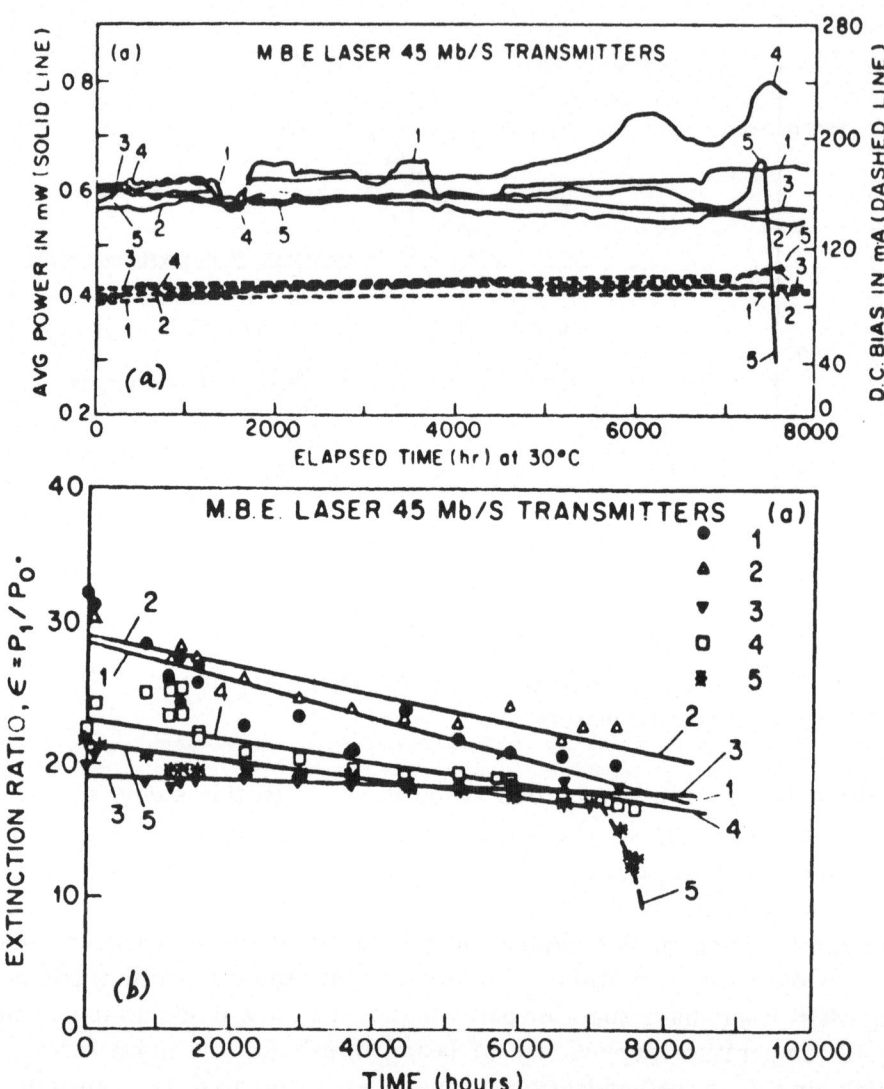

Figure 2 - (a) The aging behavior of the average power output and d.c. bias current of five randomly picked MBE laser transmitters at 45 Mb/s and 30°C; (b) the degradation of the extinction ratio as a function of time of the same five MBE laser transmitters.

compounds. As a result, it is also generally assumed that MBE will not be an economical way for high throughput mass-production of epitaxial layers for optoelectronic and microwave devices. Recently, we have also shown that high throughput, high yield, and highly reproducible (AlGa)As DH laser wafers can be grown by MBE with properly designed multi-chamber MBE systems.[27] This is demonstrated by growing two series of (AlGa)As DH laser wafers and measuring their broad-area threshold current density, J_{th}, distributions across the wafers (3.5 cm diameter). The first series consists of four different wafers grown consecutively at growth rates 2.9, 4.2, 7.4 and 9.5μm/h. The results show the J_{th}'s are essentially unaffected by accelerated growth rates. In the second series, four DH laser wafers having the same layer structures were grown under the same conditions without interruption at 11.5 μm/h. The results show that even at such high growth rates the qualities of the DH laser wafers are still highly reproducible, as shown in Fig. 3.

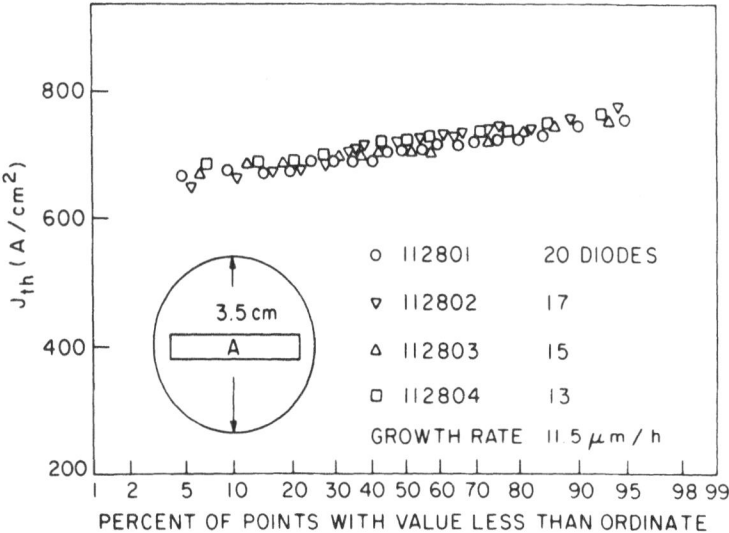

Figure 3 - Distributions of the threshold current densities of broad area lasers fabricated from four DH wafers of the same layer structures grown at an accelerated growth rate of 11.5 m/h by MBE under the same growth conditions.

The low averaged J_{th}'s ($\sim$ 700 A/cm^2) of the present DH wafers in both series also show that the material and heterojunction qualities of these wafers are as good as those previously grown at lower growth rates ($\sim$ 1.5μm/h). The half-peak full-width of the J_{th} distributions across an area of 3 cm width of the wafers (3.5 cm diameter) are about

50-60 A/cm². Such narrow range of distributions ensures high device
yield.

3. NEW DEVICES REALIZED WITH MBE

3.1 Important Capabilities and Unique Features of MBE Process

Because MBE is an ultra-high vacuum evaporation process in
which one or more species of atoms or molecules are impinging and
adsorbing on the substrate growth surface (see Fig. 4) and migrating to
suitable lattice sites as a result of the thermal energy provided by the
heated substrate, the resulting epitaxial layers are uniform over a large
area. Such large area (2 inch diameter) uniformity in layer thickness,
chemical composition and doping concentration and profiles is further
guaranteed by the use of the continuously rotating substrate scheme
employed during growth.[28,29] An example of such large are uniformity
is illustrated in Figure 5 by the doping profiles obtained at different
locations on an MBE-grown GaAs epilayer. This excellent uniformity

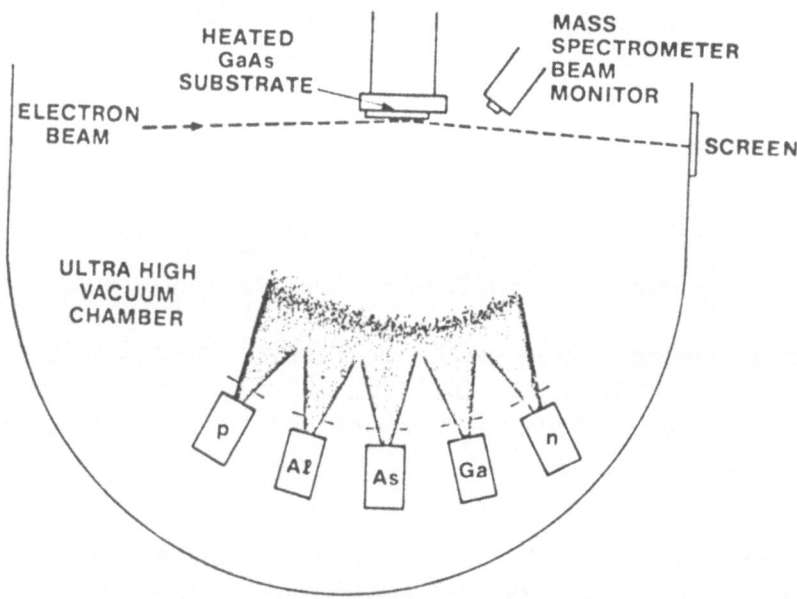

Figure 4 - Schematic diagram illustrating the MBE growth process.

results in narrow distributions of device properties as already shown in Fig. 3 for the threshold current densities of the lasers.

In growing the III-V compound semiconductors, the growth rate is solely determined by the arrival rate(s) of the atoms of the group III element(s). Thus, by controlling the arrival rate(s), the growth rate can be conveniently varied from extremely slow ($\lesssim$ 1Å/sec)[30] to as high as (35 Å/sec).[27] The slow growth rate of extremely precise control of layer thickness and the growth of extremely thin layers. An example is given in Fig. 6 in which a superlattice of alternating GaAs and AlAs layers were prepared by MBE. In fact 10^4 monolayers of alternating GaAs and AlAs have been prepared.[30] This opens the possibility of synthesizing a new class of alloys.

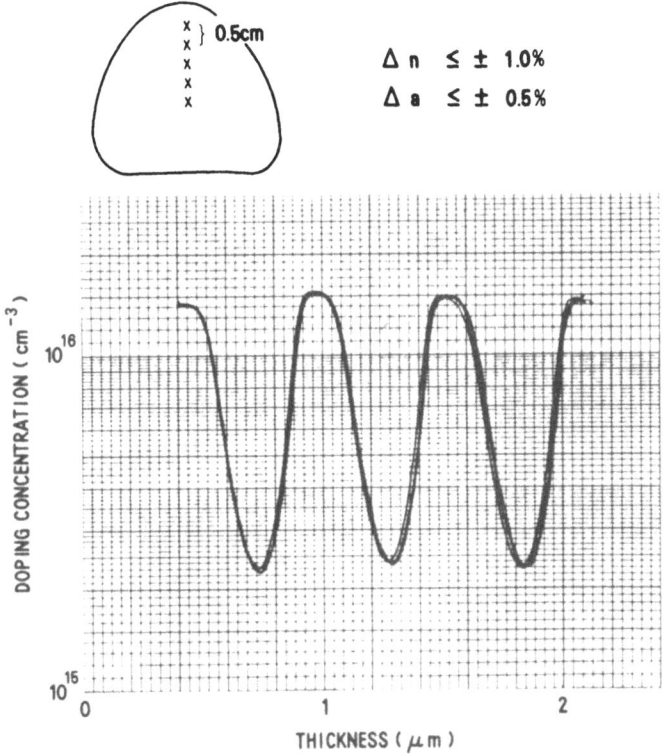

Figure 5 - Doping concentration as a function of depth from the surface of a GaAs layer grown with the substrate rotated at 4 rpm and doped periodically with Sn. Five traces were measured on different spots of the wafer as indicated (After Ref. 29).

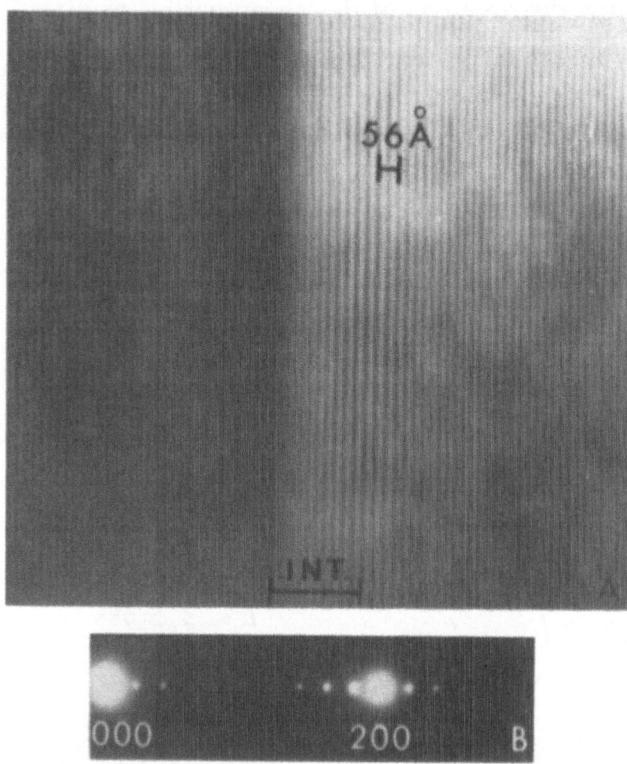

Figure 6 - Cross-sectional TEM image of $(GaAs)_n(AlAs)_m$ superlattices, where each period consists of n monolayers of GaAs and m monolayers of AlAs on a (100) GaAs substrate. For layers at left, n = 8.0
and m = 1.3. For layers at right, n = 6.1 and m = 3.4. Transmission
electron diffraction pattern of right side is seen at bottom. (000) and
(200) spots are fundamental diffraction lines of GaAs and remaining
spots are superlattice diffraction spots (After Ref. 30).

Since mechanical shutters are employed to switch on and off the
various individual beams (except the high vapor pressure group V
beams) and the shutter operating time for opening and closing by
pneumatic or electromechanical means can readily be made much less
than the deposition time for a monolayer, heterointerfaces and impurity
doping profiles can be obtained with monolayer abruptness. These are
demonstrated by the composition superlattice shown in Fig. 6 and the
preparation of doping superlattices.[31] Sharp doping profiles require the
use of dopant atoms which not only can be interupted by the shutter

but which incorporate completely in the growing layer without surface segregation or concentration.

In addition to being able to switch the various beams on and off independently and abruptly, the various beam flux intensities can also be varied over any arbitrary period of time following any arbitrary intensity profiles. This can be achieved by two means. In the first, the beam flux intensity can be varied by varying the effusion cell temperature. In the second, the flux intensity can also be varied by varying the opening of the aperture of the effusion cell at desired rate while keeping the oven temperature constant. The former technique is usually preferred because it gives more precise and accurate control. For ovens with large thermal mass, the growth rate of the epilayer has to be lowered in order to accommodate the thermal delay of the oven. This ability makes possible the growth of epilayers with any arbitrary desired composition and doping profiles.

Since MBE essentially is a controlled UHV evaporation process, mechanical shadow masking can be employed to achieve selective area growth[32] or lateral epitaxial writing of different materials and layer structures.[33] Such ability can be useful for integrated opto-electronics. Comparing with other growth techniques such as LPE and CVD, MBE possesses one very important and unique feature. Because molecular beam epitaxy takes place in an ultra high vacuum, the films are available for surface analysis, even during the growth process itself. Electron diffraction patterns obtained by the glancing incidence of kilovolt (5-10 keV) electron beam from the growing surface is especially useful. Surface reconstructions and smoothness can be monitored insitu.[34] The chemical composition of the surface can be monitored by briefly interrupting the growth to examine the Auger spectra from the surface. The residual gasses in the UHV growth chamber can be monitored and the chemical composition of the epilayers can be calibrated by using mass spectrometer, collimated ion gauge detectors, or optical fluorescence beam monitors.

One further important and unique feature is that the growth chamber is always maintained at UHV even during source recharge of the group V elements. The group V elements need much more frequent recharge that the group III elements. As a result, the UHV growth environment is maintained over long periods of times only interrupted by the necessity to recharge group III elements or accidental repairing of parts in the growth chamber. Such growth environments are particularly important in achieving high degree of controllability and reproducibility.

In the following, we give some illustrative examples of new devices and the device physics involved and which are based upon the above unique capabilities of MBE.

3.2 New Photonic Devices by MBE

3.2.1. Multiquantum well heterostructure lasers. The ability of MBE to prepare ultra-thin ($\lesssim$ 200Å) GaAs and $Al_xGa_{1-x}As$ layers with the latter free of alloy clusters[25] leads to the preparation of high quality multiquantum well (MQW) lasers, [35,36] the barriers and the cladding layers have the same AlAs composition X $\gtrsim$ 0.3 (Fig. 7a).

With the MQW lasers an extensive study has been made on the device characteristics.[37] Wafers with different numbers of wells, and different well and barrier thicknesses were investigated. The results show that threshold current densities J_{th} as low as the lowest J_{th} (800 A/cm^2) obtained for standard double-heterostructure (DH) lasers with approximately the same AlAs composition in the cladding layers were obtained in spite of the reduced optical confinement factor Γ and the increased number of interfaces (Fig. 7b).

Significant beam width reduction in the direction perpendicular to the junction plane was obtained. Half-power full-width as narrow as 15° was measured for some MQW wafers. Theoretically, because of the modification of density of states from the parabolic distribution in bulk material as in conventional DH lasers to the stair-case distribution in the MQW heterostructure as shown in Fig. 8a, the injected carrier distribution and hence the gain spectra will be different in both cases as depicted in Fig. 8b and 8c respectively.

For the laser to lase, the overall round-trip gain should overcome the overall round-trip loss. If such overall cavity losses are the same in both the DH and MQW lasers, the modification of the density of states in the MQW lasers should require fewer carriers to be injected for the laser to reach threshold. This means the threshold current for the MQW laser should be lower than the conventional DH laser. A detailed theoretical study of this can be found in Ref. 38. However, the

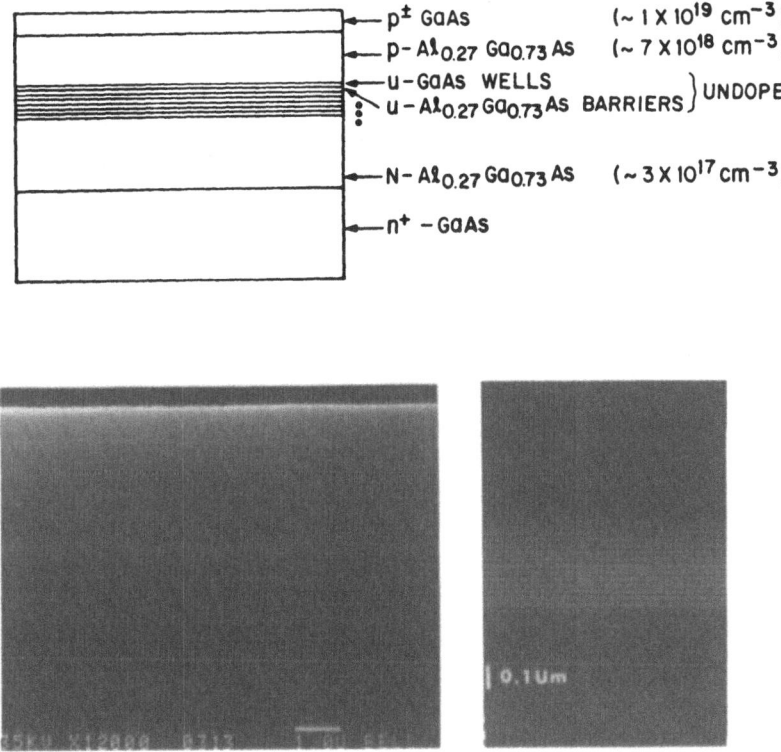

Figure 7a - Schematic diagram showing the layer structure and doping levels of the MQW lasers. The multilayers were unintentionally doped. The SEM photographs of the cleaved cross-sectional view of the actual MQW laser structure at high magnifications, respectively. There are 14 GaAs quantum wells ~136Å thick and 13 $Al_{0.27}Ga_{0.73}As$ barriers each ~130Å thick.

experimental results shown in Fig. 7 do not reflect such improvement. Recently, this was found to be related to the injection efficiency of the carriers over the various barrier in the MQW laser.[39] Subsequently, it was shown that by modifying the layer structures of the conventional MQW lasers (Fig. 9) extremely low J_{th} of 250 A/cm^2 (averaged value) for broad-area Fabry-Perot diodes of 200 μ x 380 μ was obtained.

This was achieved as a result of utilizing the beneficial effects of the two-dimensional nature of carrier confinement, the improved injection efficiency of the carriers into the GaAs wells, and an increased

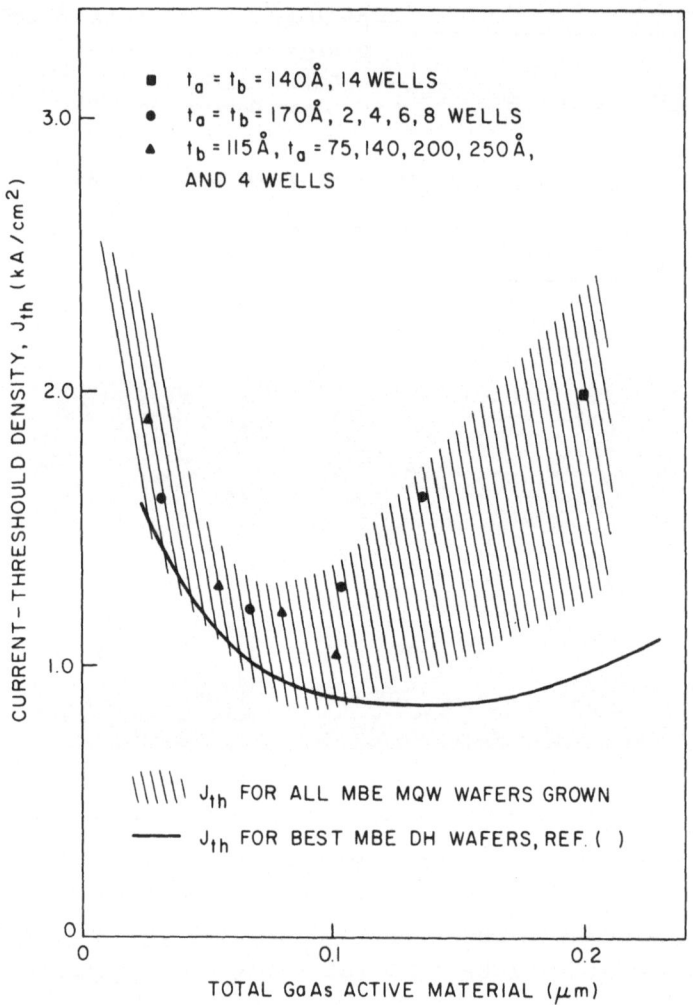

Figure 7b - Summarizes the distribution, as represented by the shaded region, of the J_{th}'s of all the MQW wafers grown by MBE during a period of about one and half year. The solid circles and triangles represent J_{th}'s of two systematic consecutive series of MQW wafers. The solid curve represents the best averaged J_{th} of standard DH lasers having $Al_{0.3}Ga_{0.7}As$ cladding layers grown also by MBE.

optical confinement factor in these modified MQW lasers. It was also determined that for low threshold operation in the optimal AlAs composition in the $Al_xGa_{1-x}As$ barrier layers is about x = 0.19 when GaAs wells are used. Such extremely low J_{th} of 250 A/cm^2 is to be com-

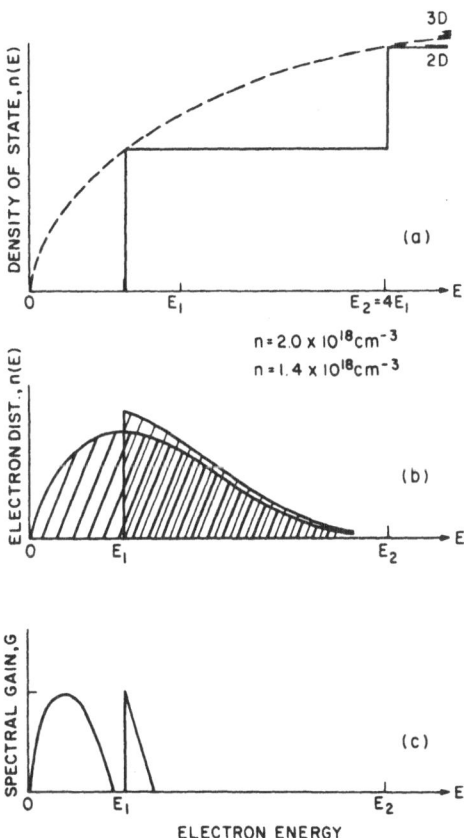

$$n = 2.0 \times 10^{18} cm^{-3}$$
$$n = 1.4 \times 10^{18} cm^{-3}$$

ELECTRON ENERGY

Figure 8 - Schematic diagrams of density of states for bulk material and QW heterostructures. The distribution of injected carriers in bulk QW structures needed to achieve the same peak gain spectra as shown in (c).

pared with ~ 800 A/cm² for the previous conventional MQW lasers[37] and for otherwise similar geometry double-heterostructure (DH) lasers.[2] Gain-guided proton bombarded stripe-geometry lasers fabricated from these MQW wafers have a cw threshold current of ~ 30 mA instead of ~ 80 mA which is typical of conventional MQW lasers and DH laser prepared by MBE.[23] CW accelerated aging in dry nitrogen 70°C ambient at constant power output of a 3 mW/mirror of 5μm shallow proton-bombarded uncoated stripe geometry lasers fabricated from conventional MQW wafers with GaAs wells has also been studied.

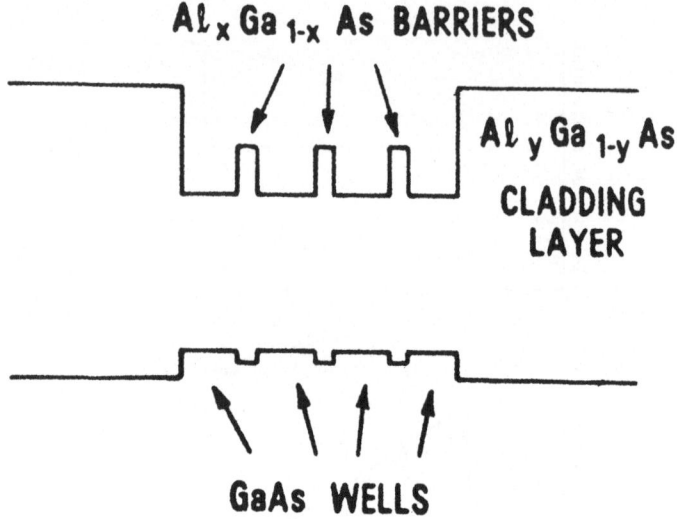

Figure 9 - The schematic energy band diagram from the modified multiquantum well laser.

Figure 10 shows the log-normal plot of the aging results of MBE grown DH lasers containing $Al_{0.08}Ga_{0.92}As$ active layers, MBE grown conventional MQW lasers with GaAs active layers without Sb or In. All lasers have the same proton-bombarded geometry and were processed under the same technology. Even though the MQW lasers have pure GaAs wells and more interfaces, a median laser lifetime of ~ 5,000h at 70°C was obtained. Such lifetime represents the longest-lived MQW lasers ever prepared.

3.2.2. Double-barrier double-heterostructure (DBDH) lasers. As sources in optical communication systems, lasers should be reliable and their device characteristices, e.g. threshold current I_{th} and external differential quantum efficiency η_D, should be relatively insensitive to temperature. It is also desirable that the beam divergence be narrow, particularly in the direction perpendicular to the junction plane, in order to achieve efficient coupling of energy into the optical fibers. With the commonly used (DH) lasers, good temperature stability of I_{th} and η_D can be achieved by using a large change in the atomic fraction of AlAs, Δx between the active layer and the cladding layers ($\Delta x >$ 0.03). Such a large value of Δx increases the barrier height and reduces

POWER LIFETIME OF ~0.87μm LASERS

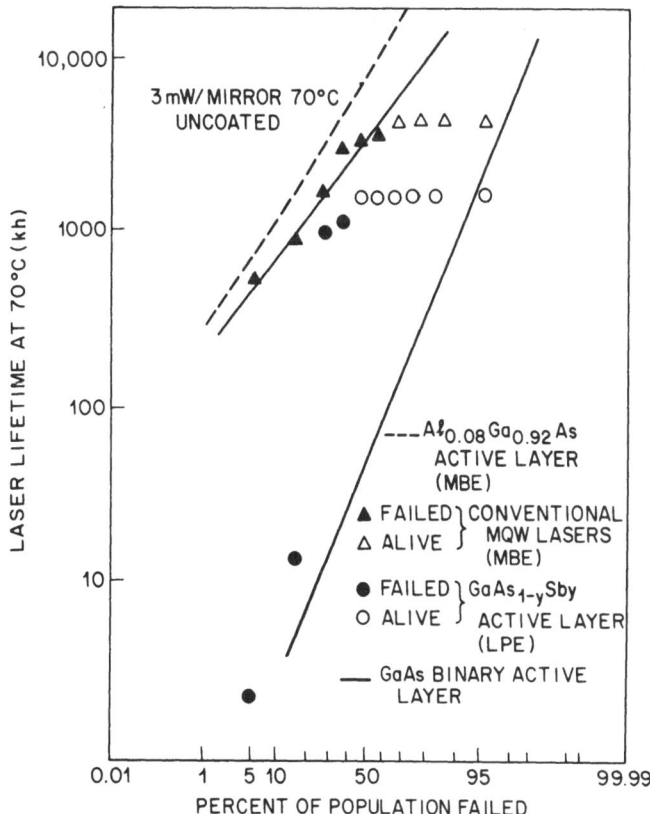

Figure 10 - The log-normal plot of 70°C CW aging results of MBE-grown conventional MQW lasers.

carrier leakage into the cladding layers especially at high temperatures. Furthermore, it has also been shown by Goodwin, et al.[41] that lasers with good temperature threshold stability also exhibit better aging reliability than those that have poor temperature stability. On the other hand, narrow beam divergence can be achieved by having a small Δx (a small refractive index step) and/or by having very thin active layers. The latter, however, tends to result in an undesirable increase in threshold. Since in DH lasers, both carrier and optical energy confinement are provided by the same heterojunctions, a trade-off is thus always made between threshold and beam divergence. For the DH lasers commonly used in lightwave transmission systems (active layer thickness 0.15 - 0.2μm, $\Delta x \sim 0.3$) the angle of divergence at half-power full-width $\theta_\perp$ is $\sim 50°$.

The ability to profile the AlAs composition of the epilayers make possible the preparation of a new semiconductor current injection heterostructure laser: the double-barrier double-heterostructure (DBDH) laser.[42] In this new heterostructure laser a pair of very thin (250 - 450Å) uni-directionally graded barriers of very wide bandgap (> 2.00 eV) between the active layer and the uniform cladding layers of the conventional double-heterostructure (DH) laser is incorporated (Fig. 11). As a result, the beam divergence of this new heterostructure laser can be independently varied by varying the AlAs composition of the uniform cladding layers without affecting the temperature stability of the threshold I_{th}, the external differential quantum efficiency η_D and possibly the reliability of the laser. With these lasers, excellent temperature stability of I_{th}, η_D, spontaneous emission level, and abrupt transition from LED mode to laser mode were achieved and maintained up to temperatures as high as 276°C. A comparison of the pulsed light-current characteristics at various temperatures for a typical DH laser

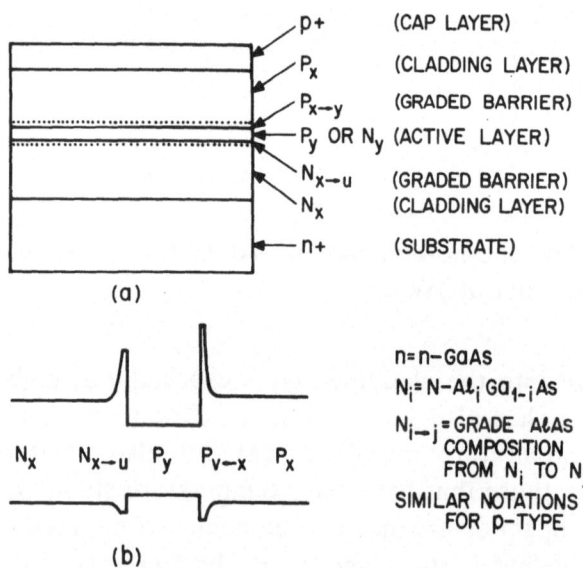

Figure 11 - (a) A schematic diagram of the cross-section of the DBDH laser. (b) The energy band diagram under lasing conditions.

and a typical DBDH laser is shown in Fig. 12. Narrow beam divergence ~26° and low I_{th} ~ 1 kA/cm^2 were also obtained simultaneously from the same lasers.

3.2.3. Graded-index waveguide separate-confinement heterostructure (GRIN-SCH) lasers. The ability to profile AlAs composition of the epilayers also made possible the preparation of a heterostructure semiconductor laser with graded-index waveguide and separate carrier and optical confinement (GRIN-SCH)[43] (Fig. 13).

Figure 14 shows the result of a theoretical and experimental comparison of the relative importance of the contributions to the threshold current density J_{th} by the intrinsic, the internal loss and the mirror loss terms for both regular (AlGa)As double-heterostructure (DH) and graded-index waveguide separate confinement heterostruc-

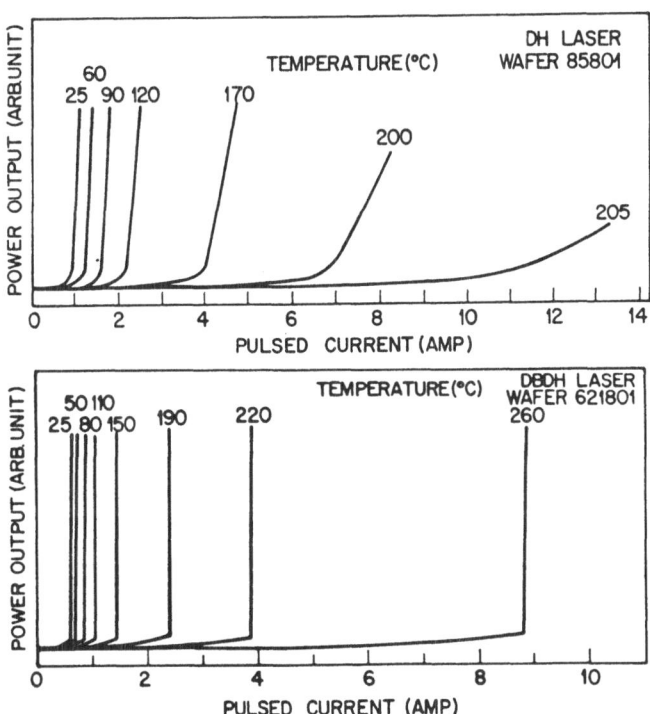

Figure 12 - (a) and (b) show the pulsed light-current characteristics at various temperatures for a typical DH laser and a typical DBDH laser, respectively.

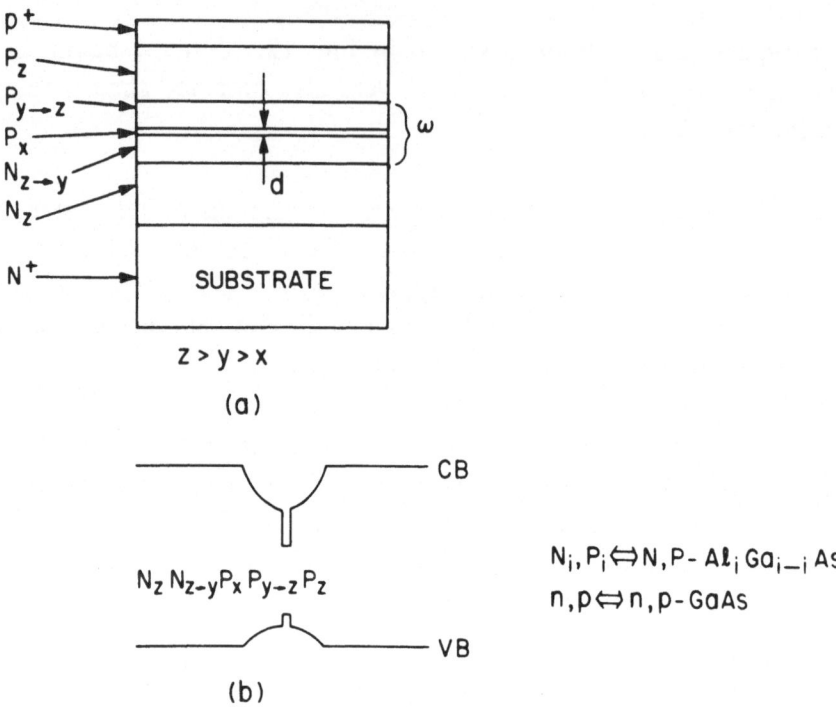

Figure 13 - (a) Shows schematically the layer structure of a graded-index waveguide SCH laser and one of the corresponding energy band diagram (b).

ture (GRIN-SCH) lasers in the very thin active layer regime,[43] $d \lesssim$ 700Å. In this regime and for both laser structures, both the mirror and internal loss terms become dominant. However, in contrast to the rapid increase of these two terms with decreasing d in DH lasers, they remain almost constant in the GRIN-SCH lasers, thus, enabling the J_{th} to continue decreasing with decreasing d even for $d \lesssim$ 700Å. Experimentally, extremely low threshold GRIN-SCH lasers with single and double active layers were prepared by molecular beam epitaxy as a result of an increased optical confinement, a significant reduction in the internal loss α_i, and an increased gain constant β. Averaged J_{th} of 250 A/cm^2 and 160 A/cm^2 for broad-area Fabry-Perot diodes of cavity lengths 380μm and 1125μm respectively, and averaged external differential quantum efficiency η_D of 65 - 80% were obtained. As determined, the internal quantum efficiency η_i is 0.95, the internal loss α_i is $< \sim 3$ cm^{-1}, and the gain constant β is 0.08 - 0.12. The dotted curves in Fig. 14 were calculated with these experimentally determined parameter values.

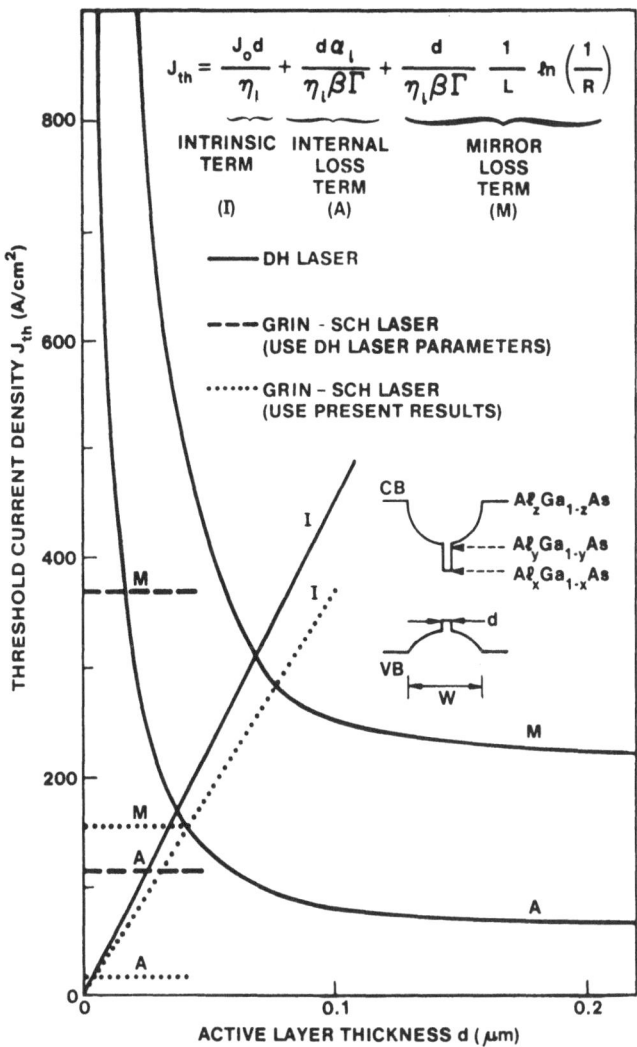

Figure 14 - The relative importance of the contributions to J_{th} by the intrinsic, the internal loss and the mirror loss terms. The solid curves were calculated for regular DH lasers, while the dashed curves were calculated for GRIN-SCH lasers. Both use previously determined parameter values as described in the text. The dotted curves were calculated for GRIN-SCH lasers using the parameter values determined in this experiment.

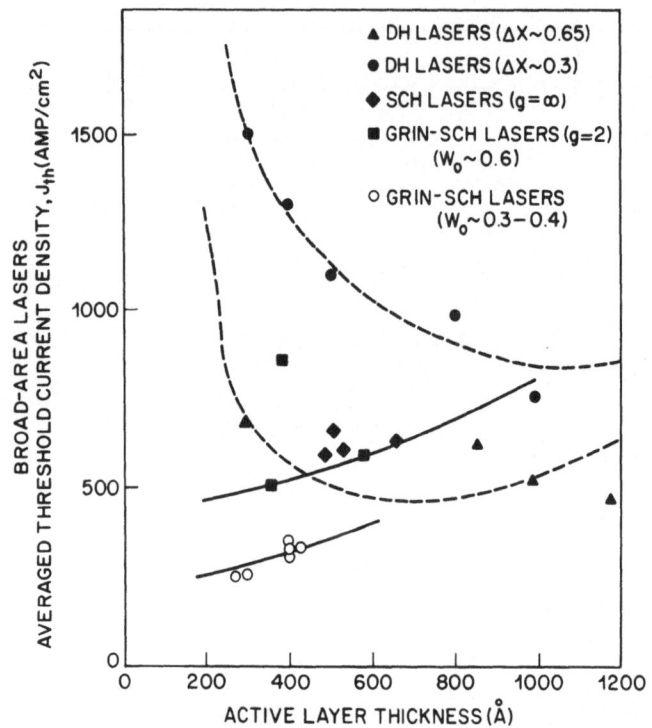

Figure 15 - Comparison of J_{th} as a function of active layer thickness for regular DH and GRIN-SCH lasers.

Figure 15 shows a comparison of J_{th} as a function of active layer thickness for regular DH lasers and GRIN-SCH lasers. Buried heterostructure lasers operating in the fundamental transverse mode fabricated from these wafers have cw thresholds as low as 2.5mA, the lowest ever reported in any lasers. CW heat-sinking is simply achieved by bonding the diode p-side up to Cu-block with silver-epoxy.[44] Pulsed output power as high as 20 mW/mirror and external quantum efficiency as high as 80% were obtained. (see Fig. 16)

Gain-guided proton-bombardment stripe-geometry lasers also show a significant improvement in device performances over lasers fabricated from conventional DH laser wafers also grown by MBE especially in the symmetry of outputs and threshold.

3.2.4. Multiwavelength transvers e-junction-stripe (MW-TJS) lasers. It is well established that transverse-junction-stripe (TJS) lasers[45] operate

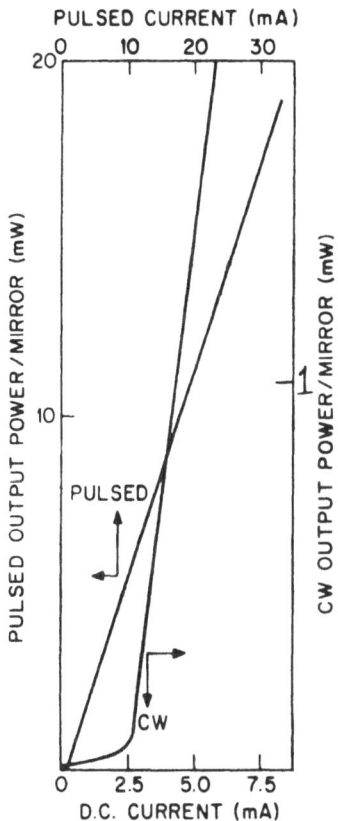

Figure 16 - The light-current characteristics of a GRIN-SCH BH laser under CW and pulsed operation.

3.2.4. Multiwayelength transvers e-junction-stripe (MW-TJS) lasers. It is well established that transverse-junction-stripe (TJS) lasers[45] operate in single-fundamental-transverse modes and tend to operate predomi- nantly in a single-longitudinal mode even at injection current levels higher than twice current thresholds I_{th}. With the TJS lasers, low current thresholds have also been obtained. The new TJS laser capable of emitting multiwavelength emissions was simultaneously demonstrated.[46] At each emission eavelength, the laser power is almost completely concentrated into one single-longitudinal mode. The struc- ture of this multiwavelength transverse-junction-stripe (MW-TJS) laser is shown schematically in Fig. 17a. In this example, there are four active layer (represented by the shaded layers), $Al_{x1}Ga_{1-x1}As$, $Al_{x2}Ga_{1-x2}As$, $Al_{x3}Ga_{1-x3}As$ and $Al_{x4}Ga_{1-x4}As$. These active layers are separated from each other by the intervening $Al_yGa_{1-y}As$ layers (y > x1, x2, x3, x4) and sandwiched on top and bottom by the thick

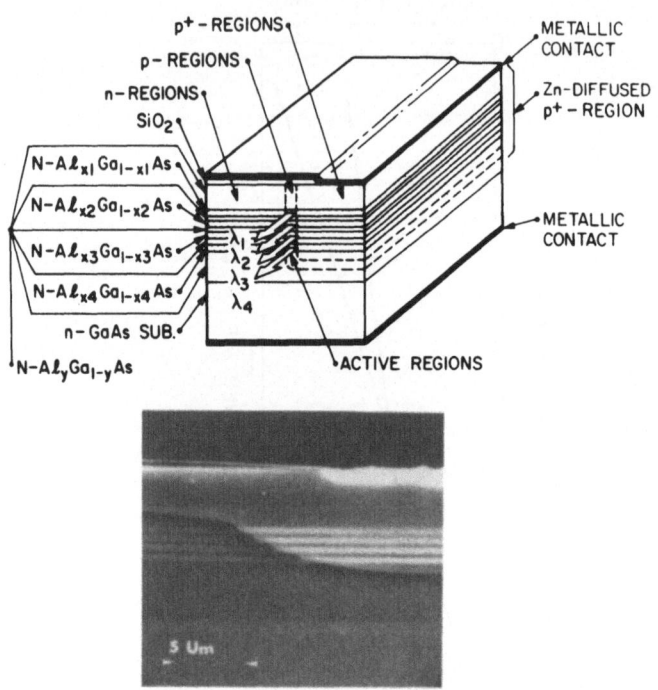

Figure 17(a) - Schematic diagram showing the laser structure of a multi-wavelength transverse-junction-stripe (MW-TJS) laser having four different AlGaAs active layers. The results in four different lasing wavelengths, λ_1, λ_2, λ_3, and λ_4. (b) Shows the SEM photograph of the cross-sectional view of an actual MW-TJS laser. Lateral Zn diffusion beyond the Si_3N_4 stripe edge also occurred.

$Al_yGa_{1-y}As$ confinement layers. The two-step Zn-diffusion process produces three different vertical regions the p^+-p-n regions. When current is injected through the p contact, it flow laterally across the four p-n junctions formed in the $Al_{x1}Ga_{1-x1}As$, $Al_{x2}Ga_{1-x2}As$, $Al_{x3}Ga_{1-x3}As$, and $Al_{x4}Ga_{1-x4}As$ active layers to the n side. Since the $Al_yGa_{1-y}As$ confinement and intervening layers have much wider bandgap than the active layers (y > x1, x2, x3 and x4), carriers are injected predominantly across the four p-n junctions formed in the four active layers. Because both the n- and p^+- regions are heavily doped ($\sim 3 \times 10^{18}$ cm^{-3} and $\sim 2 \times 10^{19}$ cm^{-3}, respectively) and because the intervening $Al_yGa_{1-y}As$ layers are made thin, the current is expected to be approximately equally divided among the four active p-n junctions.

Since the Al mole fractions, x1, x2, x3 and x4 in the four active layers are made different, the resulting lasing emissions from each p-n junction are at different wavelengths, λ_1, λ_2, λ_3 and λ_4. Thus by controlling the Al mode fractions in the various active layers, preselected multiwavelength lasing emissions with desired separations can be obtained simultaneously from such MW-TJS lasers. Furthermore, the number of lasing lines is simply determined by the number of active layers present in such lasers. More importantly, since it is characteristic for regular TJS lasers to lase predominantly in a single-longitudinal mode, we expect also that at each emission wavelength the lasing power will almost completely concentrate into one single-longitudinal mode at that particular wavelength. Figure 17b shows the SEM photograph of a MW-TJS laser prepared by MBE.

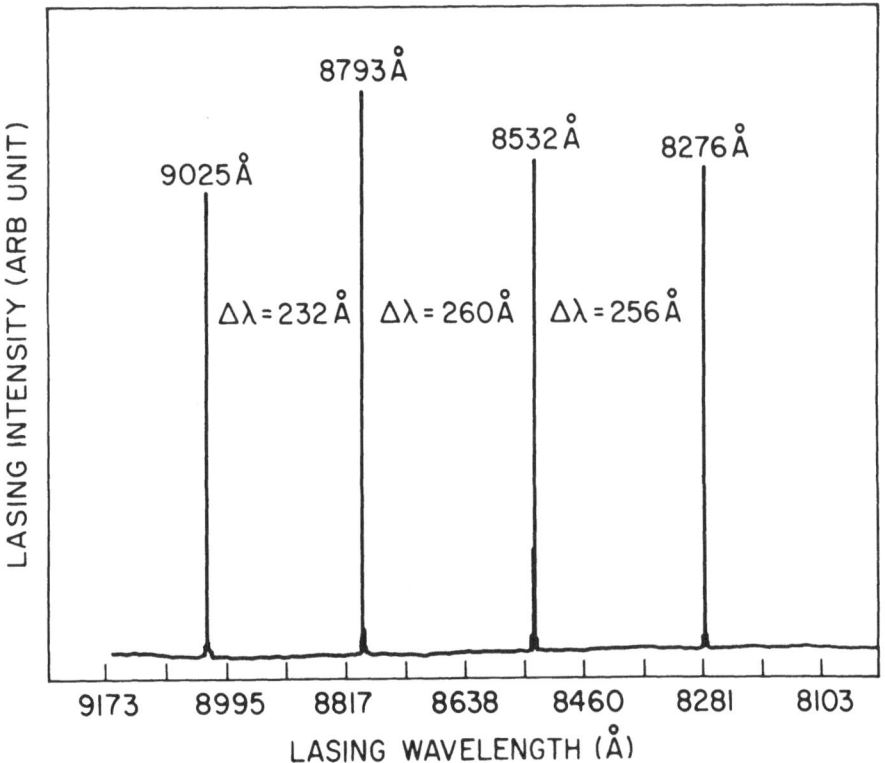

Figure 18 - Shows the lasing spectrum from a 4-wavelength TJS laser at ~ $1.2I_{th}$. Such spectral qualities were maintained up to ~ $1.7I_{th}$ tested. The pulsed I_{th} of this diode is 245 mA.

As an example, four different lasing lines at 9025, 8793, 8532 and 8276 Å were obtained simultaneously from a single 4-wavelength TJS laser as shown in Fig. 18. The pulsed and cw current thresholds I_{th} for such a 4-wavelength TJS laser with cavity length of 375 μm and each active layer thickness of 0.5μm are 245 and 252 mA, respectively. In the case of regular TJS lasers grown by MBE, the pulsed and cw I_{th}'s are 37-50 and 40-54 mA, respectively.

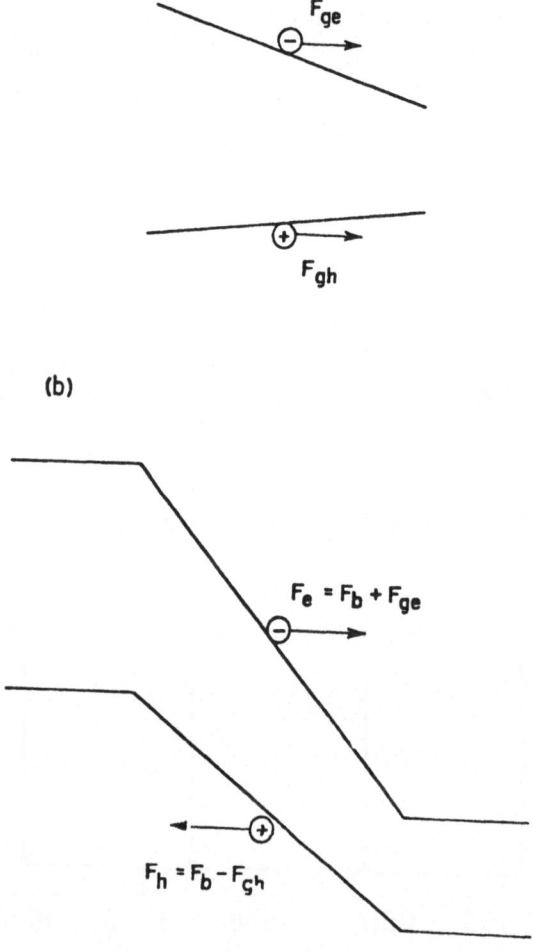

Figure 19 - (a)   Effects of quasi-electric fields (F_g) in a graded gap device.   (b)   combined effects of real (F_b) and quasi-electric fields (F_g).

3.2.5. Graded bandgap avalanche diode. The unique capabilities of MBE also spurred the conception and demonstration of new photodetectors. As demonstrated by McIntyre,[47] a large difference in the ionization rates for electrons and holes is essential for a low noise avalanche photodiode (APD). Silicon has an ionization rate ratio k = $\alpha/\beta \simeq 20$ and therefore is an ideal APD at wavelengths $\leq 1.06\ \mu$m. Unfortunately, most III-V semiconductors, including alloys of interest for long wavelengths detectors (InGaAsP, etc.) have α nearly equal to β. It is therefore of great interest to explore the possibility of "artificially" increasing (or decreasing) α/β in these materials by using new device structures. A novel device, the graded bandgap avalanche

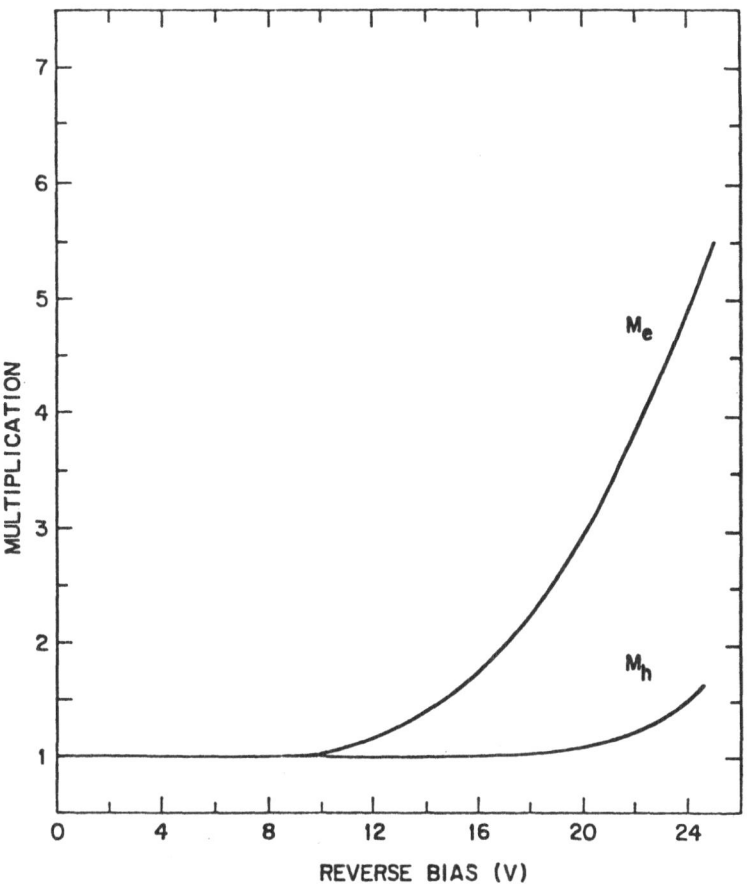

Figure 20 - Electron (M_e) and hole (M_h) initiated multiplications for p-i-n diodes with a 0.4μm wide graded region.

diode, was investigated.[48] By grading the composition of the avalanche region over distances $\leq$ 1μm, the ionization rates ratio α/β can be made significantly greater than one. This occurs because electrons have a lower ionization energy and experience a higher quasi-electric field than holes (see Fig. 19). The graded gap diode can also have a "softer" breakdown and thus a greater gain stability then a non-graded diode. The reason is that breakdown starts first in the low gap region and then gradually proceeds towards higher gap regions as the voltage is increased. An effective ionization rate ratio as high as 10 has been measured in MBE grown $Al_xGa_{1-x}As$ graded diodes. Figure 20 shows the electron (Me) and hole (Mh) indicated multiplications measured versus reverse voltage in the p-i-n diode having the 0.4μm wide graded region. We note a large difference in the electron and hole multiplications. This is conclusive evidence of a large effective ionization rates ratio. The turn-on voltage of the electron initiated multiplication corresponds to a field $\simeq$ 2 x 10^5 V/cm. The highest microplasma-free gain was 3900.

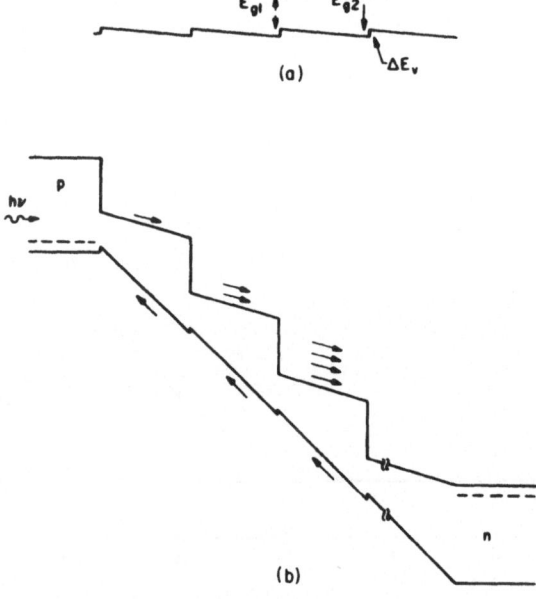

Figure 21 - Band diagrams of (a) the unbiased graded multilayer region and (b) the complete detector under bias.

Multi-stage graded bandgap APD[49] with the energy band discontinuity in the conduction band edge comparable to or greater than the ionization energy in the low bandgap material will result in impact ionization for electrons only when they cross the heterojunction with the device under low bias (Fig. 21). Such structure will mimic a photomultiplier and can possibly be constructed with the AlGaAsSb or AlGaInAs material system.

3.2.6. Superlattice avalanche photodiode. The first superlattice avalanche photodiode[50] has been reported. The high field region of the p-i-n structure consists of 50 alternating $Al_{0.45}Ga_{0.55}As$ (550 Å) and GaAs (450 Å) layers. (Fig. 22). A large ionization rate ratio has been measured in the field range (2.102.7) x 10^5 V/cm, with $\alpha/\beta \simeq 10$ at a gain of 10 giving a McIntyre noise factor $F_n \simeq 3$. The ionization rate ratio enhancement with respect to bulk GaAs and AlGaAs is attributed to the large difference in the band edge discontinuities for electrons and holes at the heterojunction interfaces.

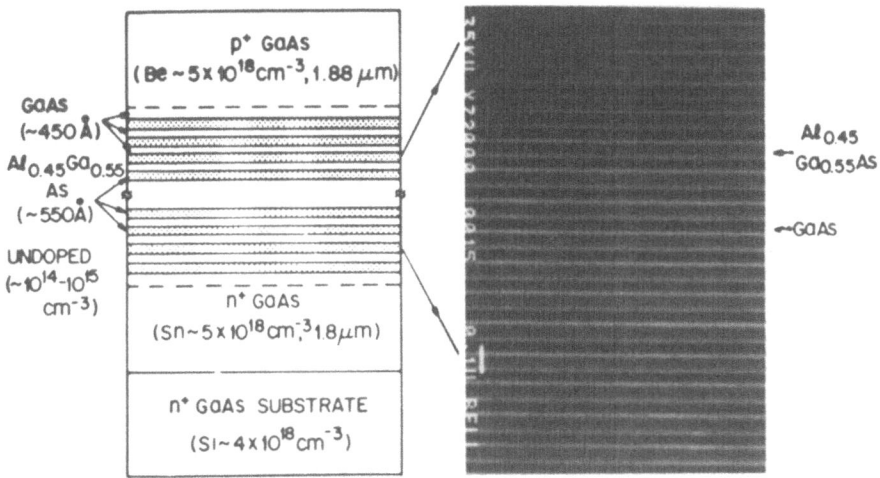

Figure 22 - Schematic diagram showing the layer structure and doping concentrations of the superlattice APD grown by MBE. The scanning electron micrograph shows the stained (H_2O_2 + NH_3OH, pH = 7.05) cross section of the superlattice. The bright images of the interfaces result from the chemical staining.

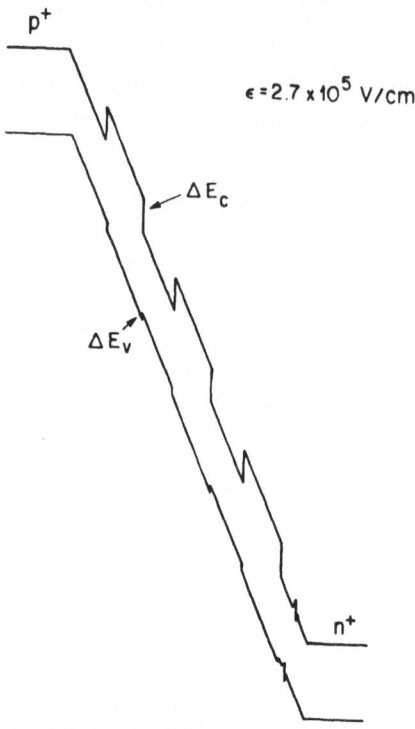

Figure 23 - Energy-band diagram of the superlattice APD. The band edge discontinuities are ΔE_c = 0.48eV, ΔE_v = 0.08 eV.

To understand the superlattice APD consider the energy-band diagram in Figure 23. Because of the very low doped material, the field is constant across the 2.5μm long depletion layer. This consists of 50 alternating GaAs (450 A) and AlGaAs (550 A) layers. For illustrative purposes, assume a field ϵ = 2.7 x 10^5 V/cm; at this value a sizeable multiplication was observed. For ϵ > 10^5 V/cm, electrons gain between collisions and energy greater than the average energy lost per photon scattering even ($\simeq$ 21 meV). Thus carriers are strongly heated by the field and can gain the ionization energy.

Consider now a hot-electron accelerating in an AlGaAs barrier layer. When it enters the GaAs well it abruptly gains an energy equal to the condition band edge discontinuity ΔE_c = 0.48 eV. The effect is that the electron "sees" an ionization energy (E_{th} $\simeq$ 1.5 eV) reduced by ΔE_c with respect to the threshold energy in bulk GaAs (E_{th} = 2.0

eV). The impact ionization rate α with respect to bulk GaAs is expected to increase, when the electron enters the next AlGaAs barrier region, the threshold energy in this material is increased by ΔE_c thus
decreasing α in the AlGaAs layer. However, since $\alpha_{GaAs} > \alpha_{AlGaAs}$,
the exponential dependence on the threshold energy ensures that the
average α (see later in text)

$$\bar{\alpha} = (\alpha_{GaAs}L_{GaAs} + \alpha_{AlGaAs}L_{AlGaAs})/(L_{GaAs} + L_{AlGaAs})$$

is largely increased (L denotes layer thicknesses).

Electrons that have impact ionized in the GaAs easily get out of
the well; the voltage drop across the well is > 1 V. In addition at field
$\geq 10^5$ V/cm in GaAs the average electron energy is ≥ 0.6 eV, so that
electron trapping effects in the wells are negligible.

In contrast, the hole ionization rate β is not substantially increased because the reduction in the hole ionization energy is only the
valence-band discontinuity of 0.08 eV. The net result is a large enhancement in the α/β ratio. Figure 24 shows the measured effective
ionization coefficients for electrons ($\bar{\alpha}$) and holes ($\bar{\beta}$) in the superlattice APD. The solid lines represent least squares fits to the data.
These data were reproduced in many diodes on two different wafers.

3.2.7.   Graded bandgap picosecond photodetector.   Previously,
Kroemer[51] proposed the use of a graded bandgap transistor to reduce
the base transit time which is normally diffusion limited. Recently
Levine,[52] et al. directly measured the electron drift velocity in compositionally graded p$^+$ Al$_x$Ga$_{1-x}$As and determined to be $\simeq 2 \times 10^6$ cm/s
in a quasi-electric field of 1.2×10^3 V/cm.[52] Capasso, et al.[53] also
reported on the first graded gap phototransistor structure.

The device was grown by molecular beam epitaxy on a Si-doped
($\sim 4 \times 10^{18}$ cm^{-3}) n$^+$ GaAs substrate. A buffer layer of n$^+$ GaAs ($\simeq$
μm) was subsequently grown followed by an Sn doped n type ($\simeq 10^{15}$
cm^{-3}) 1.5μm thick collector layer. The 0.4μm thick base layer was
compositionally graded from GaAs on the collector layer side to
Al$_{0.20}$Ga$_{0.80}$As (E$_g$ = 0.8 eV) and heavily doped with Be (p$^+$ $\simeq$ 5 x
10^{18} cm^{-3}). The wide gap emitter consists of an Al$_{0.45}$Ga$_{0.55}$As 1.5μm
(E$_g$ = 2.0 eV) thick window layer n doped with Sn to $\simeq 2 \times 10^{15}$ cm^{-3}.
A final 1000A thick GaAs layer heavily doped with Sn (n$^+$ $\simeq$ 5 x 10^{18}

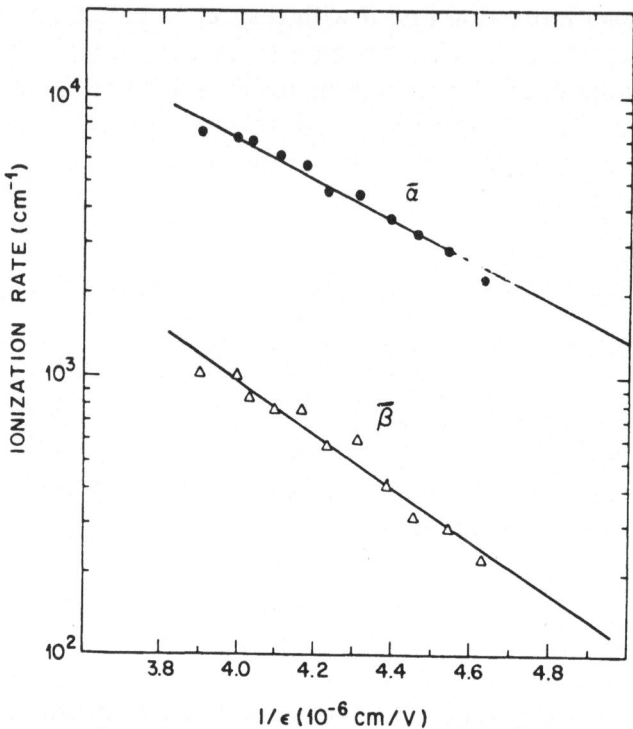

Figure 24 - Measured effective ionization coefficients for electrons ($\bar{\alpha}$) and holes ($\bar{\beta}$) in the superlattice APD. See text for the definition of $\bar{\alpha}$ and $\bar{\beta}$. The solid lines represent least-square fits to the data. These data were reproduced in many diodes on two different wafers.

cm^{-3}) was grown to facilitate ohmic contact. The emitter and collector doping were kept intentionally low to minimize the base emitter and base collector capacitances. The total series capacitance at zero bias measured at 1 MHz is $\simeq$ 0.30 pF; the device area is 10^{-4}/cm^2. Figure 25 shows the typical pulse response of the detector at zero bias.

Here the peak power of the laser was $\lesssim$ 500 mW. The sampling scope was operating with marginal sensitivity, so that the pulse from the S4 sampling head was signal averaged (150 sweeps) with a Nicolet 1170 multichannel analyzer. This improved the signal-to-noise and eliminated trigger jitter effects. Note again the ultra-fast ($t_r \simeq$ 30 ps, FWHM = 50 ps) and symmetric pulse response. The complete absence of ringing indicates a near perfect impedance matching. The devices were also operated with the base-collector-junction reverse biased as a phototransistor. The pulse response for a reverse bias of 2 volts and a

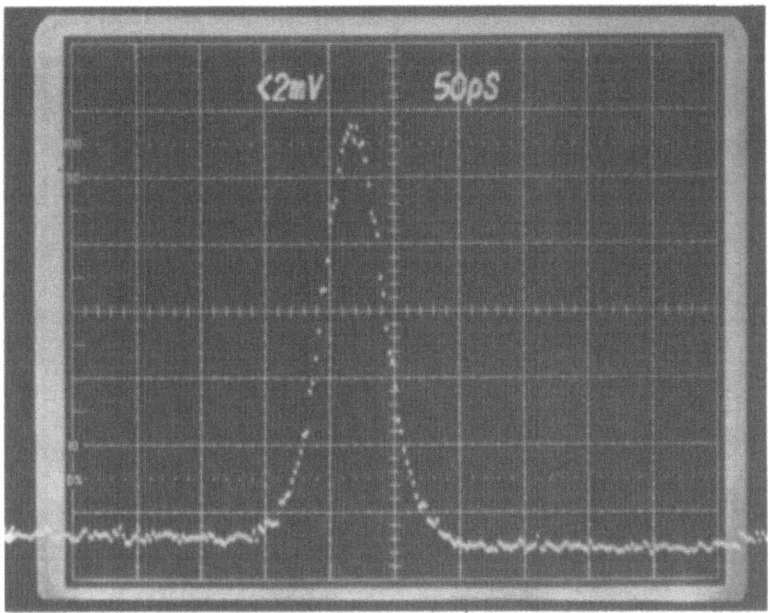

Figure 25 - Pulse response of a typical graded gap detector operated at zero bias, to a 5 ps laser (α = 62000Å). The pulse from the S4 sampling head was signal averaged.

peak power of $\simeq$ 0.1 watt has a rise time and a FWHM comparable to the observed response time at zero bias, while the sensitivity is much higher. However, a long tail with a time constant of a few hundreds of ps is clearly discernible after the fast decay of the pulse. The pulse response of the phototransistor was examined at laser peak powers in the range from 10mW to 10 watt; the rise and FWHM did not vary with power but it was clearly observed that the magnitude of the tail became smaller at lower incident power.

Figure 26 shows the phototransistor dc characteristics at different incident power levels. The light source was a 4 mW HeNe laser (λ = 6328Å) which was successively attenuated with 0.1, 0.3 and 0.5 neutral density filters. The flatness of the characteristics confirms the lack of any substantial base depletion (Early effect). The sensitivity is $\simeq$ 0.2A/W. The operation of the detector at zero bias is illustrated by the energy band diagrams of Fig. 27. Most of the incident radiation is absorbed in the graded region, since the absorption length $1/\alpha$ at λ =

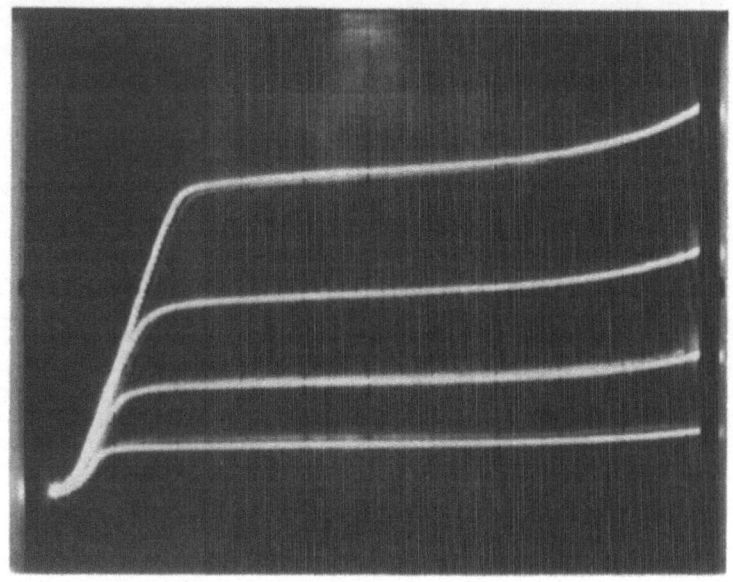

Figure 26 - Typical dc characteristics of the phototransistor under different light intensities. The upper curve correspond to illumination with a 4 mW HeNe laser, while the other curves are obtained by successively attenuating the beam with 0.1, 0.3 and 0.5 ND filters.

6200Å is $0.3\mu m$ and smaller than the base width. The absorption coefficient α was obtained from recent experimental data of Aspnes and Studna for AlAsGa ($0 \leq x \leq 0.2$). Under the action of the quasielectric field of $\simeq 7 \times 10^3$ V/cm the photogenerated electrons and holes drift towards the collector with an ambipolar group drift velocity estimated to be $\simeq 5 \times 10^6$ cm/s and reach the edge of the p^+n base collector junction in a time comparable to the laser pulse duration (Fig. 27a). Electrons are injected in the collector while the holes are stopped by the base collector built-in potential and accumulate. Thus a time-dependent photovoltage develops at the unbiased base collector p-n junction and is divided between the base emitter junction and the load resistor. As a result photoelectrons will flow from the collector to the emitter via the external load resistor until the electron quasi-fermi levels on both sides of the p^+ base are lined up and both the emitter-base and collector base capacitances are charged. This charging process occurs in a time of the order of the RC time constant ($\simeq 15$ ps).

The subsequent return to the equilibrium configuration occurs mostly via thermonic emission processes of holes over the reduced base-emitter and base-collector barriers, followed by recombination with electrons in the emitter and collector as at the contact regions.

These processes effectively discharge the two capacitances and are very fast because of the large base-emitter and base-collector barrier lowering (≥ 1 eV) caused by the relatively high laser power. When instead the base collector junction is reverse biased, the device behaves as a phototransistor with the base-collector diode acting as a current source amplified by transistor action via lowering of the base

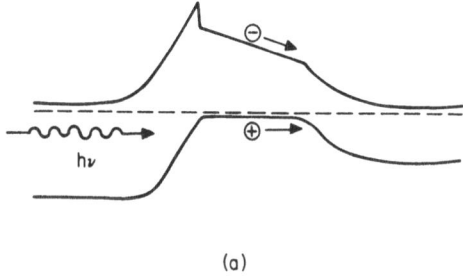

(a)

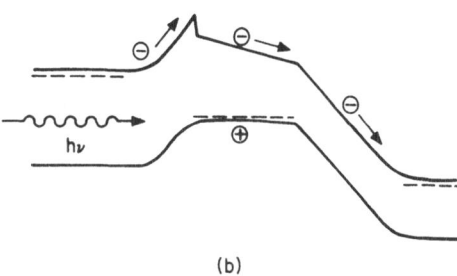

(b)

Figure 27 - Energy band diagram illustrating the different phases of the detection process at zero bias. The emitter and collector are connected by the 50 ohm load. (a) Absorption and electron hole separation mechanism. (b) Band configuration after the charging of the base emitter and base collector capacitors. Return to equilibrium via thermionic emission of the accumulated holes.

emitter potential (Fig. 27b). Thus a large increase in sensitivity is observed, as expected.

3.2.8. Optical bistability in superlattice etalon. In the bulk GaAs material, bound electron-hole pairs give rise to a sharp absorption feature, the excitonic resonance, which enhances the change in refractive index of the material. Previously, pure GaAs[54] (~ 4μm thick) was used as the core of the etalon with thin dielectric reflecting coatings applied on both surfaces. When light in incident on the etalon at the excitonic resonance, it changes the refractive index of the GaAs. The coating above and below the core produces internal reflection, as in Fabry-Perot resonator. The combination of the nonlinear refractive index and the optical feedback (resonance) gives rise to the nonlinear transmission characteristic of the device, or the optical ability of the device. However, the binding energy of the excitons in pure GaAs is small (4.2 meV) and thus easily broken by thermal energy at room temperature. Consequently, optical bistability with pure GaAs is difficult to obtain at room temperature.[54,55] On the other hand, the binding energy of the excitons in GaAs/AlGaAs superlattice is greater depending on the thickness of quantum wells.[56] This makes its nonlinear refractive index larger at room temperature than that of pure GaAs. Etalon with the pure GaAs replaced by a superlattice consisting of 61 periods; each contains a 336Å GaAs layer and a 401Å $Al_{0.27}Ga_{0.73}As$ layer is shown in Fig. 28. Room temperature optical bistability was obtained.[57] The intensity required are about 1 mW/(μm)^2 and the switching times are 20-40 ns, similar to those of low temperature pure GaAs values.

4. MATERIALS SYSTEMS

Figure 29 shows the composition diagram for III-V compounds, while Table 1 gives the possible quarternary-to-binary III-V lattice-matched systems. Since MBE is an ultra-high vacuum evaporation process with beams derived from elemental sources, no chemical reactions take place either in the intermediate gas phase or on the substrate surface unlike those ocurring in VPE or CVD. Consequently, in principle, all of the alloy III-V compound semiconductors can be grown epitaxially by MBE. However, the main difficulty encountered with growing alloy III-V compound semiconductors is the requirement for precise lattice-matching. This is particularly difficult for systems in

TABLE 1. Binary to quaternary III-V lattice matched systems for heterostructure lasers (From Heterostructure
Lasers by H.C. Casey, Jr., and M.B. Panish)

Quaternary	Lattice Matching Binary	Comments
$Al_xGa_{x-x}P_yAs_{1-y}$	GaAs	No binary lattice match except at low y (0.01). Used to adjust Al_xGa_{1-x} lattice constant
$Al_xGa_{1-x}P_ySb_{1-y}$	GaAs, InP, and InAs	Mostly indirect energy gap and probably miscibility gaps.
$Al_xGa_{1-x}As_ySb_{1-y}$	InP, InAs	Regions of probably miscibility gaps at compositions where lattice matched to InP. Step-graded DH lasers grow on GaAs for 1 micron emission. Lattice match to InAs for emission between 1.2 and 1.6 micron.
$Al_xIn_{1-x}P_yAs_{1-y}$	InP	InP active region not at an interesting emission wavelength
$Al_xIn_{1-x}P_ySb_{1-y}$	GaAs, InAs, AlSb, GaSb	Probably miscibility gaps.
$Al_xIn_{1-x}As_ySb_{1-y}$	InP, GaSb, AlSb	Very similar in lattice constant variation with composition to $Ga_xIn_{1-x}As_ySb_{1-y}$. Less interest than that system because of greater growth problems and indirect energy gap regions.
$Ga_xIn_{1-x}P_yAs_{1-y}$	InP, GaAs	For laser emission in the 2 to 1.5 micron region with lattice match to InP. Low threshold cw lasers have been prepared.
$Ga_xIn_{1-x}P_ySb_{1-y}$	GaAs, InP, InAs, AlSb	Probable extensive miscibility gaps.
$Ga_xIn_{1-x}As_ySb_{1-y}$	InP, GaSb, AlSb	For low temperature DH lasers at wavelengths greater than 2 micron. Miscibility gap over part of the InP lattice match region.
$(Al_xGa_{1-x})_yIn_{1-y}P$	GaAs, $Al_xGa_{1-x}As$	Heterostructure lasers with visible emission to 2.1 eV.
$(Al_xGa_{1-x})_yIn_{1-y}As$	InP	Heterostructure lasers with emission between 0.8 and 1.5 micron
$(Al_xGa_{1-x})_ySb_{1-y}$	InP	Heterostructure lasers with emission between 1.1 and 2.1 micron. Problems with AlSb surface oxidation
$Al(P_xAs_{1-x})_ySb_{1-y}$	InP	All Indirect gap, probable miscibility gap.
$Ga(P_xAs_{1-x})_ySb_{1-v}$	InP	Probable miscibility gap in desired composition range.
$In(P_xAs_{1-x})_ySb_{1-y}$	AlSb, GaSb, InAs	Heterostructure lasers with emission between 1.51 and 31 micron. Expected difficulties with miscibility gaps over part of the desired composition ranges.

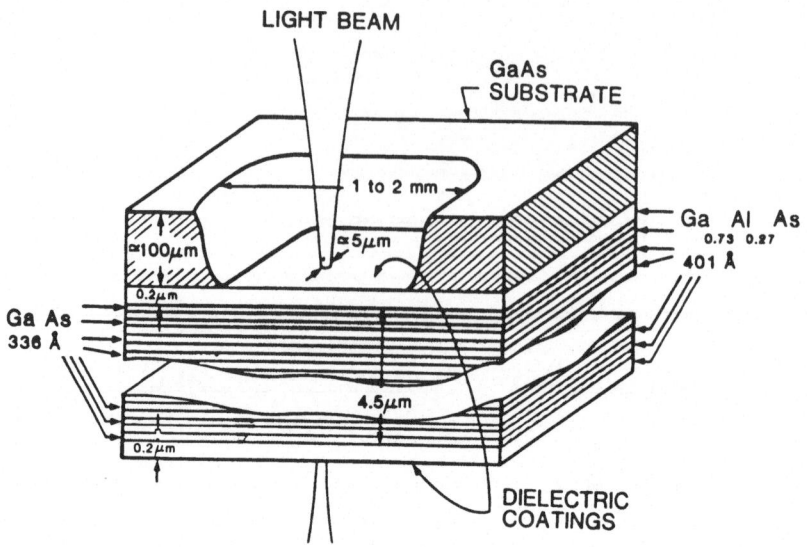

Figure 28 - The "superlattice" of the core consists of 61 layers of alternating GaAs (0.0336μm thick) and GaAlAs (mostly 0.0401μm thick but 0.2&m7.m on the top and bottom). After the superlattice has been formed, the substrate is etched away to expose the multiple layers, and a dielectric reflective coating is them deposited on top and bottom, forming the Fabry-Perot resonator (After Ref. 57).

volving two group V elements, which usually have small sticking coefficients on the substrate surface and their sticking coefficients depend on the substrate growth temperature. However, with specially designed oven systems[35] or using gaseous source,[58-60] their relative ratio can be also easily controlled for lattice-matching. For growing lattice-matched alloy systems, e.g. III-III-V or III-III-III-V, the control is simpler. In this case, because the group III elements have unity sticking coefficients at the usual substrate growth temperatures, their ratios in the epitaxial layers can be controlled by their relaitve beam flux intensities, while the group V element is overpressured.

At present, research in MBE growth of III-V compound semiconductors other than the AlGaAs, GaInAsP, InP, GaSb, AlSb, InAs, AlGaSb and AlGaAsSb in areas of heterostructure lasers and superlattices. An excellent review has recently been given by Wood.[61]

Lattice-matched $Al_{0.48}Ga_{0.52}As/Ga_{0.47}In_{0.52}As/$ $Al_{0.48}In_{0.52}As$ double heterostructure lasers lasing at $1.65\mu m$ grown on InP substrates has been prepared by molecular-beam epitaxy.[62] The result represents the first current injection semiconductor laser ever fabricated from the $Al_{0.48}In_{0.52}As/Ga_{0.47}In_{0.53}As/$ $Al_{0.48}In_{0.52}As$ double heterostructure. Broad-area Fabry-Perot laser (0.15 μm thick $Ga_{0.47}In_{0.53}As$ active layer) with pulsed current threshold density as low as 3.3 kA/cm² has been obtained at room temperature and operated at heat-sink temperatures as high as 115°C studied. The threshold-current temperature dependence has a characteristic temperature T_0 of 70°C in the temperature range of 25° - 115°C.

Recently, CW operation at up to 60°C at $1.65\mu m$ has been achieved in MBE-grown InGaAs/InP buried-heterostructure lasers with LPE InP burying layers.[63] Threshold current was as low as 35 mA at room temperature.

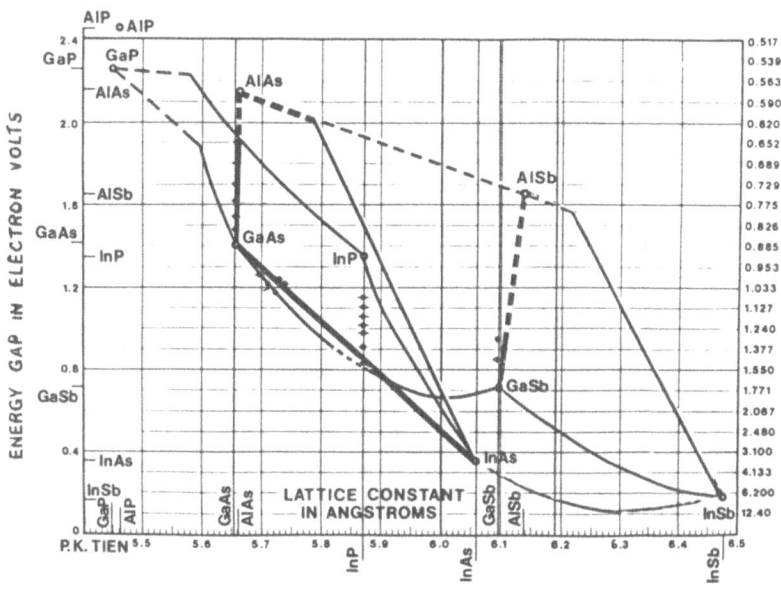

Figure 29 - The composition diagram for III-V compound semiconductors (Courtesy P.K. Tien).

The first high quality InP has been grown by molecular beam epitaxy (MBE).[64] The undoped InP layers are n-type with residual impurity concentrations $\sim 5 \times 10^{14}$ - $\sim 5 \times 10^{15}$ cm^{-3}. Fine structure attributed to polariton, neutral donor-exciton (D$^\circ$-X), neutral donor-hole (D$^\circ$-h), neutral acceptor-exciton (A$^\circ$-X) transitions at the exciton edge and neutral donor-neutral acceptor (D$^\circ$-A$^\circ$) transitions are clearly resolved in the low temperature (5K) photoluminescence spectra with a line width of < 1 meV for D$^\circ$-X as has been observed with high purity InP layers grown by other methods. Elemental In and P (red phosphorus) were used as the primary molecular beam sources. The P_4 is dissociated into P_2 before use. The growth temperature has a very significant effect on the quality of the InP layers. The present MBE system employed for growing III-V compound semiconductors containing P from elemental red phosphorus is of an advanced design.

Furthermore, the As and P ovens were equipped with recharge interlock systems. As a result, the growth chamber was always maintained in ultra-high vacuum condition even during As and P recharge. A detailed description of the system is given in Ref. 64. With this system, as shown schematically by the vertical cross-sectional view in Fig. 30, the cryopanel 2 condenses most of the unused arsenic and phosphorus (P_2 + P_4) during the growth can be transferred into the phosphorus removal chamber. Subsequent bake-out by the four in-situ heaters removes the volatile P_4 by pumping with a turbomolecular

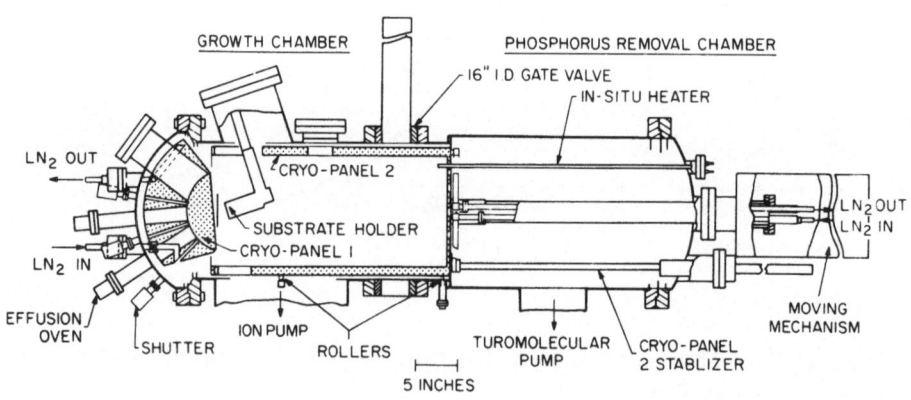

Figure 30 - The vertical cross-sectional view of an advanced MBE system employed for growing materials containing P from red phosphors.

pump. Before growth starts the cryopanel 2 is cooled with LN and then transfers into the growth chamber. The first preparation of current injection GaInAsP/InP double-heterostructure laser lasing at 1.3μm by molecular beam epitaxy has been achieved. The layer structures are shown in Fig. 31.

The averaged threshold current density J_{th} is 3.5 kA/cm^2 while the lowest J_{th} is 1.8 kA/cm^2 for broad-area Fabry-Perot diodes of 380 μm x 200 μm and an active layer thickness of 0.2μm. The threshold current-temperature dependence is described very closely by exp (T/T_0) with T_0 as high as 87K in the temperature range of 10° - 55°C as shown in Fig. 32.

Recently, Tsang also prepared and characterized for the first time a new current injection double-heterostructure (DH) laser with a $Ga_xAl_yIn_{1-x-y}As$ as the active layer and InP as the cladding layers operating at 1.5μm wavelength.[66] Figure 33 shows the layer structures.

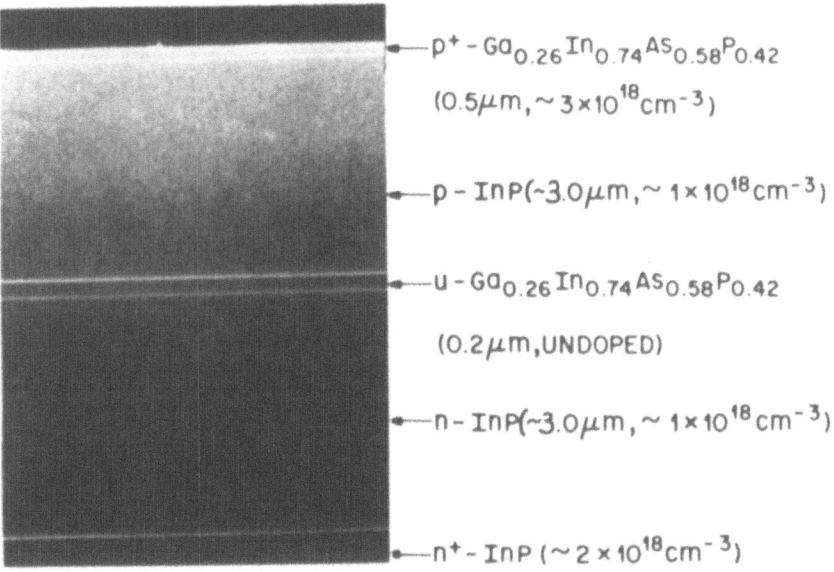

$$p^+ - Ga_{0.26}In_{0.74}As_{0.58}P_{0.42}$$
$$(0.5\mu m, \sim 3\times10^{18}cm^{-3})$$

$$p - InP(\sim 3.0\mu m, \sim 1\times10^{18}cm^{-3})$$

$$u - Ga_{0.26}In_{0.74}As_{0.58}P_{0.42}$$
$$(0.2\mu m, UNDOPED)$$

$$n - InP(\sim 3.0\mu m, \sim 1\times10^{18}cm^{-3})$$

$$n^+ - InP (\sim 2\times10^{18}cm^{-3})$$

Figure 31 - The multilayer structure of 1.3μm wavelength GaInAsP/InP DH laser grown by MBE.

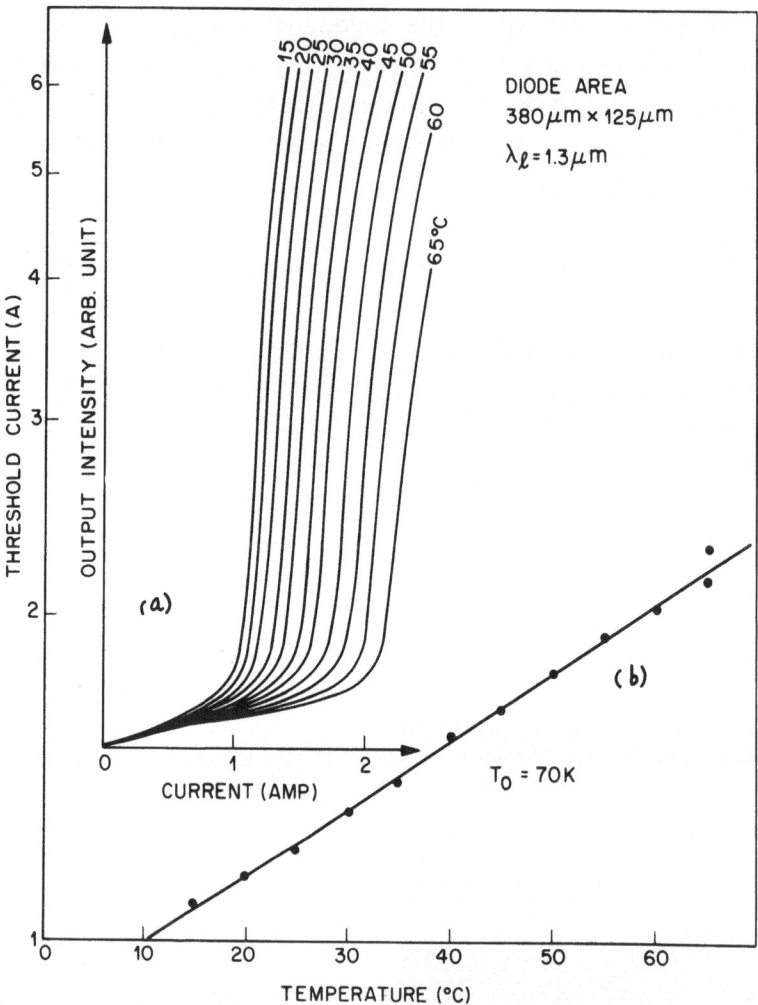

Figure 32 - (a) The light-current characteristics for an MBE-grown 1.3μm wavelength GaInAsP/InP DH laser at various heat-sink temperatures. (b) The threshold current-temperature dependence of the diode is closely described by exp (T/T$_0$) with T$_0$ = 70K.

In this new heterostructure, there is only one group V element involved in every layer. This eases the precise control of lattice matching during MBE growth. Excellent lattice-matching was obtained as shown by the x-ray diffraction results given in Fig. 34. At the same time, it eliminates the use of Al$_{0.48}$In$_{0.52}$As, which at present is of lower quality than InP when grown by MBE, as the cladding layers. Broad-area Fabry-Perot diodes of 380μm x 200μm have a threshold current density of

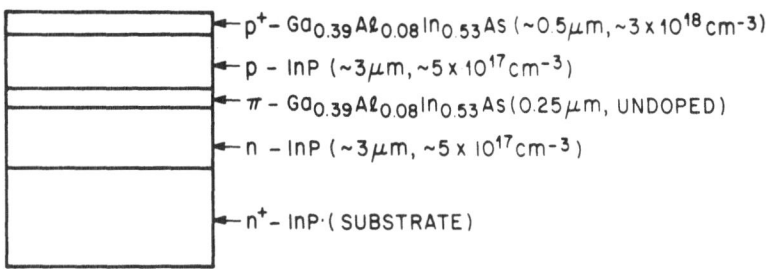

Figure 33 - A schematic diagram of the GaAl InAs/InP DH laser structure lattice-matched to InP substrate together with typical layer thicknesses and approximate doping concentrations.

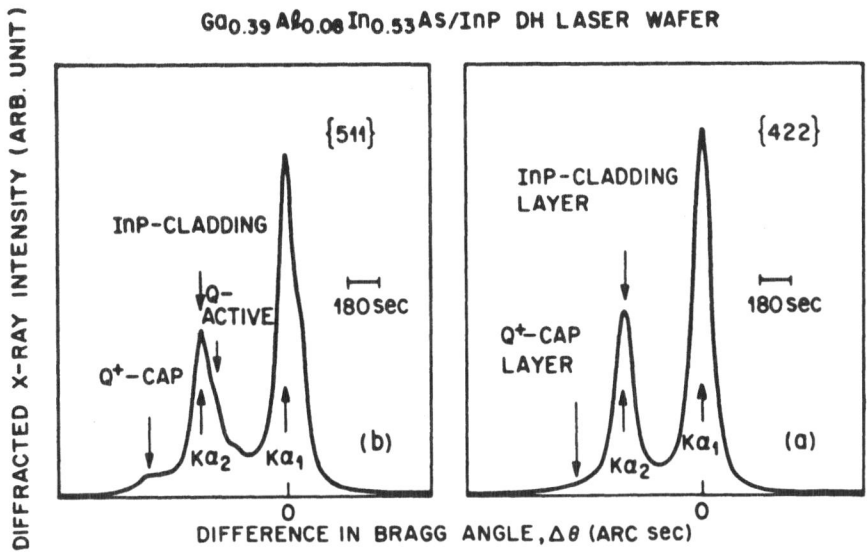

Figure 34 - (a) and (b) are single x-ray rocking curves from 442 and 511 planes diffracted from the surface of GaAl InAs/InP DH laser wafer, respectively.

~ 3.2 kA/cm^2 for active layer thickness of 0.25μm. In the temperature range of $\sim 15\text{-}50°$C, the threshold temperature dependence coefficient T_o is typically ~ 40K. Above $\sim 50°$C, T_o decreases to $\sim 25\text{-}35$K. These results are given in Fig. 35. The present laser also represents the first current injection DH with GaAlInAs active layer emitting at 1.5μm ever prepared by MBE. In the present experiment, As$_2$ instead of As$_4$ was also used for the first time in growing the Ga$_x$As$_y$In$_{1-x-y}$As layers by MBE in order to improve their quality. This becomes particularly important as the substrate growth temperatures employed are relatively low, ~ 625°C.

   As the losses due to Rayleigh scattering decreases strongly with increasing wavelength λ, the future generation of optical fibers, light sources, and detectors may well be operating at still longer wavelength beyond 1.55μm. Tsang reported the first preparation of $Al_{0.2}Ga_{0.8}Sb$/GaSb double-heterostructure (DH) lasers by molecular beam epitaxy (MBE) operating at 1.78 μm.[67] For $Al_xGa_{1-x}Sb$ with x ≲ 0.1, room temperature photoluminescent intensity and linewidth similar to those of bulk GaSb substrate of similar carrier concentration were obtained. The $Al_{0.2}Ga_{0.8}Sb$/GaSb DH laser wafers grown by MBE have smooth, featureless, mirror-reflecting surfaces. Reflection high energy electron diffraction study shows that the abrupt

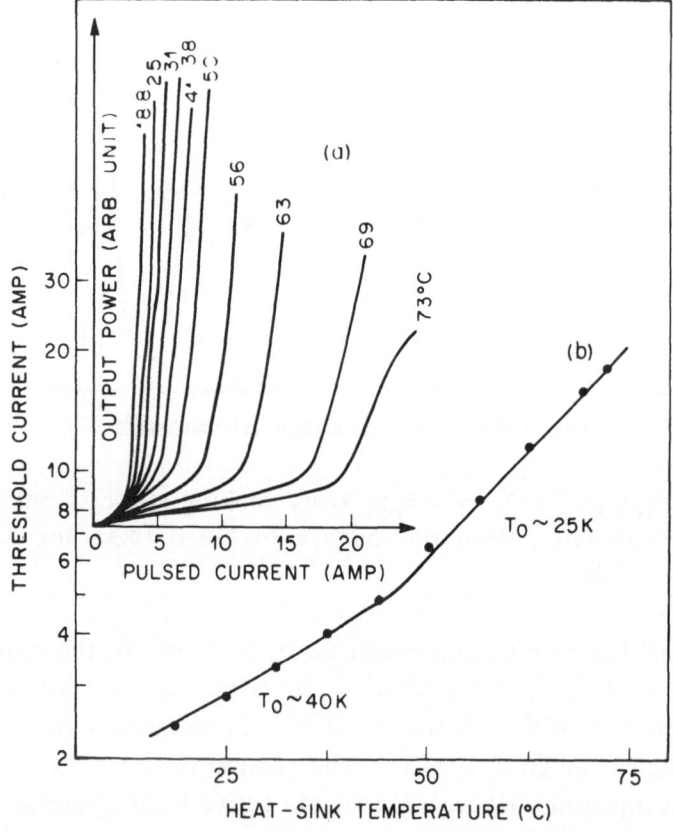

Figure 35 - (a) the L-I characteristics at various heat-sink temperatures of a typical diode. (b) The threshold-temperature dependence of the same diode.

$Al_{0.2}Ga_{0.8}Sb/GaSb$ interfaces were atomically smooth. Initial threshold current measurements gave a pulsed threshold current density of 3.4 kA/cm^2 for a diode of 380 μm x 200 μm and an active GaSb layer of 0.33 μm.

5. CONCLUDING REMARKS ON MBE, LPE, AND CVD

From the above discussions and the examples given, it is very clear that MBE is a very versatile and powerful thin-film crystal growth technique and the advantages of MBE over others are obvious. Though the present discussion is limited to the preparation of III-V compound semiconductors, MBE can be and has been extended to the growth of II-VI and IV-VI compound semiconductors as well as elemental semiconductors such as silicon and Germanium, magnetic material, superconducting superlattices and many heteromaterial systems, e.g. Ge-GaAs, GaP-Si, and InP or GaInAsP on GGG substrate. MBE is an UHV process and hence is the only growth technique that is compatible with most of the semiconductor processing technologies used today, e.g. ion-beam milling, ion implantation, sputtering deposition and etching, electron beam and x-ray lithographies, etc. Such compatibility may have important future implications in integrating various processing steps in-situ. One obvious and already carried out integration is the incorporation of ion-implantation in an MBE system. Because MBE is an UHV crystal growth technique, it is thus accessible to in-situ on-time study of crystal growth process by various diagnostic techniques, e.g. HEED, LEED, Auger, SEM, SIMS, etc.

Liquid-phase epitaxy is more limited in the number of materials that can be grown because of the presence of, for example, miscibility gaps, or the distribution coefficient is too large, and others. However, the LPE process seems to be more tolerant in that fair material quality can be obtained quite easily in general. As a result, it is quite commonly employed. However, because of the limitation in achieving large-area material or thickness uniformities, LPE is generally most suited for discrete devices and where high-yield large volume production are not crucial. For ICs, the LPE process is not likely to be utilized.

The various chemical vapor deposition process are attractive because they can be scaled up to large volume production as in silicon CVD. However, thus, far, no such demonstration has been carrier out with the III-V compound semiconductors. Unlike in Si-CVD, in the case of III-V CVD, there is the added complication of uniformity of gas mixture of the various gaseous specied across even a single 2-inch diameter GaAs wafer. To date, in fact, the best uniformity has been obtained with MBE. Multi-substrate loading cassette interlock system, high growth rate, continuously rotating substrate design, source recharge and interlock systems are all demonstrated in MBE. In the ease of CVD's there are also present a large number of intermediate species in the gas mixture as illustrated by the example of VPE growth of InGaAsP given in Table 2. A comparison is also given for MBE. It is with admiration that such a complicated system is brough under control. However, simplicity is also the merit of a successful technology.

TABLE 2 - Comparison of VPE and MBE processes for growing InGaAsP

InGaAsP VPE

$$H_2 + 2MC\ell + 1/2P_4 \rightarrow 2MP + 2HC\ell$$

Inputs	Intermediate			Output
AsH_3	AsH_3	As_2	As_4	$Ga_xIn_{1-x}As_yP_{1-y}$
PH_3	AsP	AsP_3	As_2P_2	$In_{1-x}Ga_xAs$
$Hcl(In)$	AsP	As_3P	PH_3	$P_2In_{1-x}Ga_xP$
$Hcl'(Ga)$	P_4	$InCl$	In_2Cl_2	$InAs_xP_{1-x}$
H_2	$InCl_3$	$GaCl$	Ga_2Cl_2	$GaAs_xP_{1-x}$
	Ga_2Cl_4	In_2Cl_4	$GaInCl_2$	GaP
	$GaInCl_4$	$GaClCl_3$	HCl	$GaAs$
				InP
				$InAs$

InGaAsP MBE

Inputs	Intermediate	Output
In	In	$Ga_xIn_{1-x}As_yP_{1-y}$
Ga	Ga	
As_4	As_4, As_2	
P (red)	P_4, P_2	

REFERENCES

1. A.Y. Cho, J.R. Arthur, Progr. Solid State Chem. 10, 157 (1975).
2. K.G. Gunther, U.S. Patent No. 2,938,816 (1956); Naturwissen-Shaften, 45, 415 (1958).
3. J.E. Davey, T. Pankey, J. Appl. Phys. 39, 1941 (1968).
4. J.R. Arthur, J. Appl. Phys. 39, 4032 (1968).
5. A.Y. Cho, Jpn. J. Appl. Phys. 16 (suppl. 16) 435 (1977).
6. A.Y. Cho, J. Vac. Sci. Technol. 16, 275 (1979) and references therin.
7. A.C. Gossard, in Thin Films: Preparation and Properties, edited by K.N. Tu and R. Rosenberg, Academic Press (to be pressed).
8. L.L. Chang, L. Esaki, W.E. Howard, R. Ludeke, J. Vac. Sci. Technol. 10, 11 (1973).
9. L.L. Chang, R. Ludeke, in Epitaxial Growth edited by J.W. Mathews, (Academic Press, New York 1975) p. 57.
10. K. Ploog, in Crystals: Growth, Properties and Applications, edited by Freyhardt, 1980, Vol. 3, p. 73.
11. B.A. Joyce, C.T. Foxon, J. Crystal Growth 31, 122 (1975).
12. C.T. Foxon, B.A. Joyce, Surf. Sci. 50, 434 (1975).
13. H. Holloway, J.N. Walpole, Prog. Cryst. 2, 49 (1979).
14. J.N. Walpole, A.R. Calawa, T.C. Hamman, S.H. Groves, Appl. Phys. Lett. 28, 552 (1976).
15. C.E.C. Wood, in Physics of Thin Films Vol 11, edited by C. Hoff and M. Francombe, Academic Press (1980).
16. A.Y. Cho, H.C. Casey, Jr., Appl. Phys. Lett. 25, 288, (1974).
17. A.Y. Cho, R.W. Dixon, H.C. Casey, Jr., R.L. Hartman, Appl. Phys. Lett. 28, 501 (1976).
18. W.T. Tsang, Appl. Phys. Lett. 34, 473 (1979).
19. W.T. Tsang, Appl. Phys. Lett. 36, 11 (1980).
20. W.T. Tsang, J. Appl. Phys. 51, 917, (1980).
21. W.T. Tsang, R.L. Hartman, B. Schwartz, P.E. Fraley, W.R. Holbrook, Appl. Phys. Lett 39, 683 (1981).
22. W.T. Tsang, J. Cryst. Growth 56, 464 (1982).
23. W.T. Tsang, M. Dixon, B.A. Dean, (to be published in IEEE J. Quant. Electron.)
24. R.C. Miller, W.T. Tsang, Appl. Phys. Lett. 39, 334 (1981).
25. R.L. Hartman, N.E. Schumaker, R.W. Dixon, Appl. Phys. Lett 31, 756 (1977).
26. W.T. Tsang, Appl. Phys. Lett. 38, 587 (1981).

27. A.Y. Cho, K.Y .Cheng, Appl. Phys. Lett. 38, 360 (1981).

28. J.C.M. Hwang, T.M. Brennan, H. Temkin, A.Y. Cho, to be published.

29. A.C. Gossard, P.M. Petroff, W. Wiegmann, R. Dingle, A. Savage, Appl. Phys. Lett 29, 323 (1976); P.M. Petroff, A.C. Gossard, W. Wiegmann, A. Savage, J. Cryst. Growth 44, 5 (1978).

30. G.H. Dohler, K. Ploog, Prog. Crystal. Growth Charact. 2, 145 (1979).

31. W.T. Tsang, M. Illegens, Appl. Phys. Lett 31, 301 (1977).

32. W.T. Tsang, A.Y. Cho, Appl. Phys. Lett. 32, 491 (1978).

33. A.Y. Cho, J. Appl. Phys. 41, 2780 (1970).

34. R. Dingle, C.H. Henry, U.S. Patent 3982207, Sept. 21, 1976.

35. W.T. Tsang, C. Weisbuch, R.C. Miller, R. Dingle, Appl. Phys. Lett. 35, 673 (1979).

36. W.T. Tsang, Appl. Phys. Lett 38, 204 (1981).

37. N.K. Dutta, to be published.

38. W.T. Tsang, Appl. Phys. Lett. 39, 786 (1981).

39. W.T. Tsang, R.L. Hartman, Appl. Phys. Lett. 38, 502 (1981).

40. A.R. Goodwin, J.R. Peters, M. Pion, G.H.B. Thompson, J.G.A. Whiteaway, J. Appl. Phys. 46, 3126 (1975).

41. W.T. Tsang, Appl. Phys. Lett. 38, 835 (1981).

42. W.T. Tsang, Appl. Phys. Lett. 39, 134 (1981).

43. W.T. Tsang, R.A. Logan, J.A. Ditzenberger, to be published.

44. H. Namizaki, IEEE J. Quant. Electron. QE-11, 427 (1975).

45. W.T. Tsang, Appl. Phys. Lett. 36, 441 (1980).

46. R.J. McIntyre, IEEE Trans. Electron Devices ED-13, 164 (1966).

47. F. Capasso, W.T. Tsang, A.L. Hutchinson, P.W. Foy, 1981, Inst. Phys. Conf. Ser. No. 63, 463 (1982).

48. G.F. Williams, F. Capasso, W.T. Tsang, IEEE Electron Device Lett. ED-3, 71 (1982).

49. F. Capasso, W.T. Tsang, A.L. Hutchinson, G.F. Williams, Appl. Phys. Lett. 40, 38 (1982).

50. H. Kroemer, RCA Rev. 18 33 (1957).

51. B.F. Levine, W.T. Tsang, C.G. Bethea, F. Capasso, to be published.

52. F. Capasso, W.T. Tsang, C.G. Bethea, A.L. Hutchinson, B.F. Levine, to be published.

53. H.M. Gibbs, S.L. McCall, T.N.C. Venkatesan, A.C. Gossard, A. Passner, W. Wiegmann, Appl. Phys. Lett 34, 511 (1979).

54. S.M. Jensen, SPIE 34, paper 10 (1982).

55. D.A.B. Miller, P.W. Smith, D.S. Chemla, D.J. Eilenberger, A.C. Gossard, W.T. Tsang, , to be published.
56. H.M. Gibbs, S.S. Taing, J.L. Jewell, D.A. Weinberger, K. Tai, A.C. Gossard, S.L. McCall, A. Passner, W. Wiegmann, to be published in Appl. Phys. Lett.
57. A.R. Calawa (private communication).
58. A.R. Calawa, Appl. Phys. Lett. 38, 701 (1981).
59. M.B. Panish, J. Electrochem, Soc. 127, 2729 (1980).
60. C.E.C. Wood, edited by T. Pearsall, John Wiley and Sons (New York 1982), p. 87.
61. W.T. Tsang, J. Appl. Phys. 52, 3861 (1981).
62. Y. Kamamura, Y. Noguchi, H. Asahi, H. Nagai, Electron Lett. 18, 91 (1982).
63. W.T. Tsang, R.C. Miller, F. Capasso, W.A. Bonner, Appl. Phys. Lett. 41, (1982).
64. W.T. Tsang, F.K. Reinhart, J.A. Ditzberger, to be published.
65. W.T. Tsang, N.A. Olsson, Appl. Phys. Lett. in press, (1983).
66. W.T. Tsang, N.A. Olsson, Appl. Phys. Lett. in press, (1983).

MODULATION DOPED $Al_xGa_{1-x}As$/GaAs FIELD EFFECT TRANSISTORS (MODFETS): ANALYSIS, FABRICATION AND PERFORMANCE

Hadis Morkoç

University of Illinois Urbana, IL   61801

ABSTRACT: Lattice matched heterojunction structures have received a great deal of attention owing to their applications to high speed and opto-electronic devices.   In designing devices with ultra small dimensions, it is necessary to obtain high carrier concentrations in a controllable manner in a very narrow region of the epitaxial structure. Modulation doped $Al_xGa_{1-x}As$/GaAs structures meet this requirement without degrading the quality of the semiconductor. This is done by doping the larger bandgap $Al_xGa_{1-x}As$ layer quite heavily with donors. Electrons associated with donors in $Al_xGa_{1-x}As$ layer transfer to GaAs where they are confined in a quantum well about 100Å wide at the heterointerface. Since the ionized donors are located in the $Al_xGa_{1-x}As$ layer, the GaAs layer having an interface sheet carrier concentration of about 10^{12} cm^{-2} is virtually free of ionized carriers. The high field transport parallel to the interface then becomes similar to pure GaAs crystal with peak electron velocities of about 2.1 x 10^7 cm/s and 3.3 x 10^7 cm/s at

300 and 77K respectively. Compared to GaAs doped at comparable levels, these figures represent substantial improvements.

Field effect transistors with a 1μm gate length and a 3μm channel length were fabricated and characterized under dc operating conditions at both 300 and 77K. A model was also developed to analyze the device operation and performance. Using this model, the heterojunction structures were optimized ($Al_xGa_{1-x}As$ doping level and the donor-electron separation) for high performance field effect transistors. Transconductances of as high as 275 mS/mm and over 400 mS/mm were obtained at 300 and 77K respectively. Normally-off devices exhibited power gains of over 10 dB at 10 GHz. A simple prediction shows that these devices should exhibit switching speeds of about 10 ps and 5 ps at 300 and 77 K respectively in a digital circuit with fan out of 1, e.g. ring oscillators. The experimental switching times obtained in ring oscillators are about 17 and 12 ps at 300 and 77 K respectively which are about a factor 2 away from the expected performance.

1. INTRODUCTION

The subject of high speed devices is a rather wide one and encompasses any device capable of operation above about 1 GHz. Depending on their use, these devices can be grouped into two categories, analog and digital devices. In general the analog devices are much faster than the digital ones and, needless to say, both types have different applications. This manuscript will be concerned with only the special high speed field effect transistors that can be best prepared by molecular beam epitaxy (MBE).

Devices intended for high speed applications need to have very short transit times between the input and the output. This can be obtained by choosing a semiconductor or a special structure which exhibits high carrier velocity and by reducing the distance that the carriers have to travel. In vertical devices, e.g. bipolar junction transistors, the important distance is the base thickness. This means that the control of the base thickness is extremely important. Thicknesses below 1000Å and extremely sharp doping profiles are needed and can best be obtained by molecular beam epitaxy.

Lateral three terminal devices, such as FETs, must have correspondingly small lateral dimensions. Small lateral dimensions also require small epitaxial layer thicknesses and larger doping. The drawback of very high carrier concentrations is that the electronic properties of the semiconductor degrade with increased doping. The point should be made very clear that for current conduction one needs electrons (preferred) or holes. In conventional structures, electrons and donors, and holes and acceptors are present in the same space. Using the heterojunction concept that can be engineered by MBE it is possible to separate the electrons needed for current conduction from donors, minimizing their adverse effects.

Electrons associated with donors placed in the $Al_xGa_{1-x}As$ layers of $Al_xGa_{1-x}As/GaAs$ heterojunction structures transfer to the GaAs layers. The resulting space charge sets up a strong electric field (over $10^5 V/cm$) at each interface and leads to the formation of a triangular potential well. Aided by the energy band gap discontinuity, the electrons are confined to the heterointerface and are spatially separated from the donors. This is a very convenient way of obtaining electron concentrations of over 10^{18} cm^{-3} essentially in a plane. Since the donors are separated from the electrons, the electron transport properties of undoped GaAs are preserved. This phenomenom is called modulation doping, the FET applications of which will be the main topic of this text.

In this chapter both the theoretical and experimental aspects of modulation doped field effect transistors (MODFETs) will be described. The interface sheet carrier concentration will be calculated in terms of both the triangular potential barrier, related energy levels and the device terminal voltages. Next, the drain saturation current will be calculated. This will be followed by the development of expressions for the transconductance and device capacitances. Following the device fabrication procedure, experimental and theoretical results as well as the effects of the dominant structural parameters will be discussed. Finally the high speed small signal and large signal results will be presented.

2. MODULATION DOPED FET

2.1 Materials Considerations

Modulation doped structures consist of single or multiple periods of doped $Al_xGa_{1-x}As$ layer(s) and undoped GaAs layer(s). The electrons transferring from the $Al_xGa_{1-x}As$ layer(s) into the undoped GaAs layer(s) are confined at the heterointerface as proposed by Esaki and Tsu in 1969[1] and experimentally observed by Dingle *et al.* in 1978.[2] Being spatially separated from the donors, the electrons, even at extremely high concentrations, are not subjected to ionized impurity scattering and thus can exhibit very high mobilities. This is especially pronounced at cryogenic temperatures where the ionized impurity scattering would have been dominant. The absence of ionized impurity scattering changes the dependence of mobility on lattice temperature in that it increases steadily as the temperature is decreased. In a perfect heterointerface the remaining scattering processes are the Coulombic interaction between the donors and electrons and the scattering due to background (in GaAs layers) ionized impurities. The results on MBE grown GaAs indicate that the total background ionized carrier concentration is in the high 10^{14} cm^{-3} range which should lower the low temperature mobility of modulation doped structures below about 50K. The lack of such a drop in experimental mobilities is a result of high mobility electrons screening background ionized impurities very effectively. The details of such a screening mechanism have been treated by Price[3].

The coulombic interaction between the donors and electrons across the heterointerface has also received a great deal of attention. This interaction can be reduced by the incorporation of a thin undoped $Al_xGa_{1-x}As$ spacer layer between the doped $Al_xGa_{1-x}As$ layer and the undoped GaAs layer, as first demonstrated by the University of Illinois group[4-6] and adopted by every laboratory involved in this area of research. Using a spacer thickness of about 200Å, electron mobilities of over 10^6 cm^2/Vs at 4.2K have been obtained. Later on we will see that devices fabricated from structures exhibiting extremely high electron mobilities do not necessarily perform as well as those with thinner undoped $Al_xGa_{1-x}As$ layers and, naturally, lower mobilities. As we will show in the following sections, the channel conductance of the modulation doped FETs is the parameter that should be optimized.

2.2 Structural Parameters

We have so far briefly discussed the multiple period and single period structures but did not address the question of which one is better for FETs. Even though the multiple period structures are a means of producing many parallel paths and thus a large conductance these structures are not used for devices. This is a result of an asymmetry between the properties of GaAs grown prior to and after $Al_xGa_{1-x}As$ growth. The heterointerface when the GaAs is grown on $Al_xGa_{1-x}As$ is not of high quality which makes the "inverted", and "multiple period" modulation doped structures unattractive[7,8]. In this text then we will solely consider the normal single period structures where the doped $Al_xGa_{1-x}As$ layer is grown on an unintentionally doped GaAs buffer layer. It should, however, be remembered that a thin undoped $Al_xGa_{1-x}As$ layer still exists at the heterointerface.

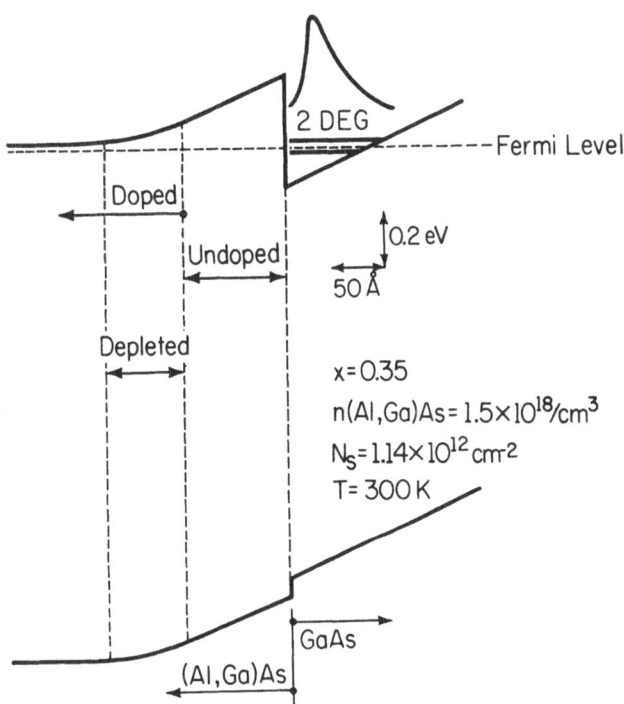

Figure 1 - Band diagram (drawn to scale) of a single period $Al_xGa_{1-x}As$/GaAs modulation doped heterostructure having an un-doped $Al_xGa_{1-x}As$ spacer layer thickness of 100Å. The electron concentration in the $Al_xGa_{1-x}As$ is 1.5 x 10^{18} cm^{-3}.

The band diagram of a normal modulation doped structure is shown in Figure 1 where the electrons confined at the interfacial trian-

gular potential will have the characteristics of a two dimensional electron gas.[9,10] In addition, higher order subbands formed in such a narrow (~50Å) quantum well may be populated if the electron concentration exceeds about high 10^{11} cm^{-2}. In such cases, the electron mobility
decreased, but improvements in the current carrying capability of the
devices more than compensates for this reduction in mobility.

2.3 Mobility and Velocity Considerations

In many instances we have talked about electron mobility in the
context of modulation doped FETs. In this section we will quantitatively relate the device performance to the mobility of carriers while
keeping in mind that the electric fields present in short channel devices
of this kind can be many times those where the mobility is no longer a
constant with electric field. A schematic representation of such a
surface oriented device (FET) having a gate length of 1μm is shown in
Fig. 2. For such short channel devices, the electric field is high, therefore the carrier velocity is the determining factor instead of the mobility. This text will treat the short channel FET with the understanding
that the saturation velocity determines the FET performance. To do
such a treatment, we need to know the sheet carrier concentration at
the heterointerface as well as its dependence on the gate voltage that is
used to control it.

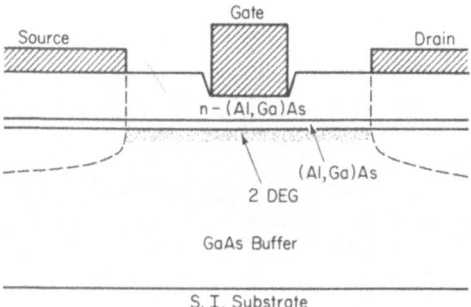

Figure 2 - A schematic representation of a cross-sectional view of a
MODFET with the $Al_xGa_{1-x}As$ on top of the GaAs. Source and drain
contacts are diffused in to reach the 2DEG plane. The gate metal (Al)
is deposited after the channel has been recessed by chemical etching.

3. MODFET MODELLING

3.1 Interface Sheet Carrier Concentration

To relate the sheet carrier concentration to energy levels, we must assume that the potential well at the heterointerface is triangular and that the electric field is quasi-constant. The solution of the longitudinal quantized energy is given by[11]

$$E_i = \left(\frac{\hbar^2}{2m^*}\right)^{1/3} \left(\frac{3}{2}\pi q E_o\right)^{2/3} \left(i + \frac{3}{4}\right)^{2/3} \tag{1}$$

where E_i is the energy, E_o the electric field at the interface and i an integer from 0 and up. The other parameters have their usual meanings. The energies of the two lowest subbands are thus

$$E_o = 1.83 \times 10^{-6} E_o^{2/3}$$

$$E_1 = 3.23 \times 10^{-6} E_o^{2/3} \tag{2}$$

$$E_o \text{ in } V/m$$

Experimental results should be used to modify the constants to minimize the adverse effects of the assumption that the electric field is constant.

The interface carrier concentration can be related to the subband energies if it is expressed in terms of the electric field. To do that, Poisson's equation must be solved. In the depletion approximation and assuming no impurities in the GaAs layer[12]

$$\frac{dE_1}{dx} = -\frac{q}{\epsilon_1} \bar{n}(x) \tag{3}$$

where $\bar{n}(x)$ and ϵ_1, are the free electron concentration and the dielectric constant in the small bandgap material respectively. Integration within the depletion region results in

$$\epsilon_1 E_1 = q n_{so} \tag{4}$$

where E_1 and n_{so} are the interface electric field and carrier concentration respectively. Using equations (2) and (4)

$$E_0 = \lambda_0 (n_{so})^{2/3} \text{ and } E_1 = \lambda_1 (n_{so})^{2/3} \tag{5}$$

where λ_0 and λ_1, are adjustable parameters used to yield a good agreement with experiments.

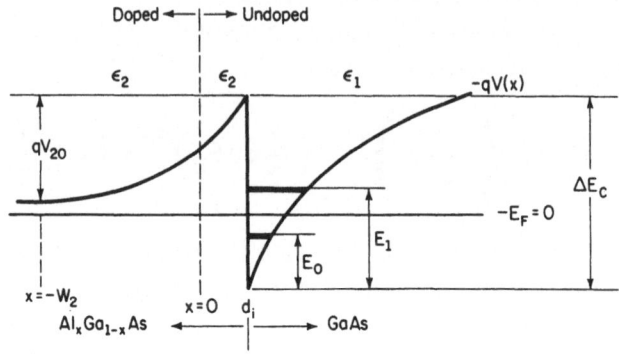

Figure 3 - Conduction band edge diagram of a single period modulation doped heterostructure.

Next we must establish a relationship between n_{so} and $qV(d_i^+)$. E_F is taken as the reference energy so that the bottom of the conduction band $qV(d_i^+)$ represents the Fermi level (Fig. 3). The density of states associated with a single quantized energy level is given by (in m^{-2} eV^{-1})

$$D = \frac{m^*}{\pi \hbar^2} \tag{6}$$

and the interface carrier concentration can be calculated using the Fermi Dirac distribution (with a spin degeneracy of 2)

$$n_{so} = D \int_{E_0}^{E_1} \frac{dE}{1 + \exp\left(\frac{E - qV(d_i^+)}{kT}\right)} \tag{7}$$

$$+ 2D \int_{E_1}^{\infty} \frac{dE}{1 + \exp\left(\frac{E - qV(d_i^+)}{kT}\right)}$$

Solving the integral (using $\int \dfrac{dx}{1+e^x} = -\ell n\,(1 + e^{-x})$) yields

$$n_{so} = DkT\ell n\left[\left(1 + \exp\left(\frac{qV(d_i^+) - E_o}{kT}\right)\right)\right.$$
$$\left.\left(1 + \exp\left(\frac{qV(d_i^+) - E_1}{kT}\right)\right)\right] \tag{8}$$

which for very low temperatures reduces to

$$n_{so} = D(qV(d_i^+) - E_o) \tag{9}$$

for an empty second subband or reduces to

$$n_{so} = D\left[qV(d_i^+) - E_o\right] + D\left[qV(d_i^+) - E_1\right] \tag{10}$$

for an occupied second subband. Using the experimental Shubnikov de Haas results[9] or cyclotron resonance measurements, the factors λ_o and λ_1 can be estimated as

$$\begin{aligned} \lambda_o &= 2.5 \times 10^{-12}\;(Jm^{4/3}) \\ \lambda_1 &= 3.2 \times 10^{-12}\;(Jm^{4/3}) \end{aligned} \tag{11}$$

and from the measured cyclotron mass

$$D = 3.24 \times 10^{17}\;(m^{-2}\,eV^{-1}) \tag{12}$$

The interface sheet carrier concentration we just calculated must be provided by the larger bandgap semiconductor. Under equilibrium, the charge depleted from the larger bandgap material must equal the inter-face charge density. A solution is then found such that the Fermi level is constant across the heterointerface. Since the doping level in the $Al_xGa_{1-x}As$ layer is quite large (approaching the density of states), full depletion approximations can not be used[13]. One must then use a charge which is a function of distance even with a constant doping. Therefore, the modified Poisson's equation is

$$\frac{d^2V}{dx^2} = \frac{-\rho(x)}{\epsilon_2} \tag{13}$$

Here, V is the electrostatic voltage and x is the perpendicular distance away from the heterointerface. The space charge density $\rho(x)$ is given by

$$\rho(x) = q[N_d^+(x) - n(x)] \tag{14}$$

where $n(x)$ is the free electron concentration and

$$N_d^+(x) = \frac{N_d}{1 + g \ \exp \ [(E_d + qV)/kT]} \tag{15}$$

is the ionized donor concentration. Here N_d is the total donor density, g is the degeneracy factor of the donor level and E_d is the donor activation energy. On the other hand[14]

$$n(x) = N_c \frac{e^{qV/kT}}{1 + \exp \ (qV/kT)/4} \tag{16}$$

where N_c is the density of states of the conduction band in the $Al_xGa_{1-x}As$ and the Fermi level E_F is chosen as the origin of the energy scale ($E_F = 0$).

Combining equations (13) through (16), we obtain

$$\frac{d^2V}{dx^2} = -\frac{qN_c}{\epsilon_2}\left(\frac{N'_d}{1 + g' \exp \dfrac{qV}{kT}} - \frac{\exp \ (qV/kT)}{1 + \exp \ (qV/kT)/4}\right) \tag{17}$$

where $N'_d = N_d/N_c$ and $g' = g \ \exp \ (E_d/kT)$. The integration of equation (17) from V $(-W_2)$ to V (0) with respect to V using the boundary condition $E_2 \ (-Wd_2) = 0$ where E_2 is the electric field in the $Al_xGa_{1-x}As$ layer as shown in Fig. 3 and W_2 is the edge of the depletion region, yields

$$E_2^2(0) = \frac{2kTN_c}{\epsilon_2}\left[N'_d \ell n \ \frac{g' + \exp \ \{-qV(0)/kT\}}{g' + \exp \ \{-qV(-W_2)/kT\}}\right.$$

$$\left. + \ 4\ell n \ \frac{4 + \exp \ \{qV(0)/kT\}}{4 + \exp \ \{qV(-W_2)/kT\}}\right] \tag{18}$$

The constant V $(-W_2)$ can be found from the space charge neutrality that exists at $x = -W_2$, i.e.

$$\rho(-W_2) = -\frac{qN_c}{\epsilon_2}$$

$$\frac{N'_d}{1+g' \exp \dfrac{qV(-W_2)}{kT}} - \frac{\exp \left(\dfrac{qV(-W_2)}{kT} \right)}{1+ \exp \left(\dfrac{qV(-W_2)}{kT} \right)/4} = 0 \qquad (19)$$

The solution of equation (19) is given by

$$y = \exp \frac{qV(-W_2)}{kT} = \frac{-(1-\dfrac{N'_d}{4}) + \sqrt{(1-\dfrac{N'_d}{4})^2+4g'N'_d}}{2g'} \qquad (20)$$

As can be seen from Fig. 3, V $(-W_2)$ is simply equal to the difference between the Fermi level and the bottom of the conduction band edge away from the heterojunction. This values is shown in Fig. 4 as a function of the doping density at 300 and 77K for the classical Boltzmann statistics or depletion approximation (solid line) and for the approximate Fermi-Dirac statistics (Equation (19) dotted line). This comparison demonstrates that the deviation from Boltzmann statistics is quite noticeable even at a relatively low doping level, $N_d = 10^{17}$ cm^{-3} and becomes quite substantial at doping densities commonly used (1 - 2 x 10^{18} cm^{-3}). The equilibrium interface density is determined by the interface field as

$$q_{so} = \frac{\epsilon_2}{q}E_2(0) = \frac{\epsilon_2}{q}E_2(d_i^-) \qquad (21)$$

Equation (21) follows from Gauss' law if the doping density in the GaAs layer is small enough so that the total bulk charge in the depletion layer of GaAs is much small than qn_{so}. The expression for $E_2(0)$ is given by equation (18) which may be simplified when the inequality

$$\exp \left(-\frac{qV(0)}{kT} \right) >> 1 \qquad (22)$$

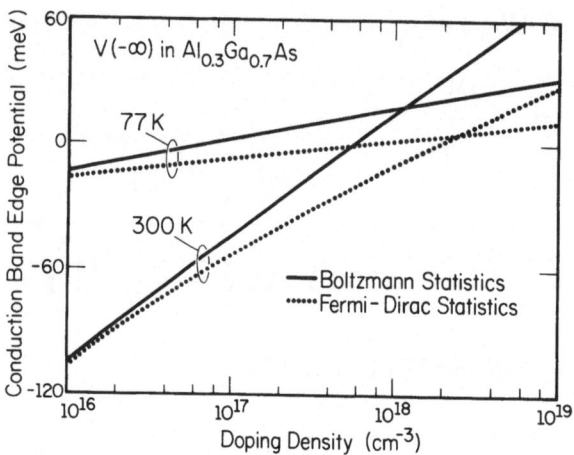

Figure 4 - Conduction band edge energy in the $Al_xGa_{1-x}As$ away from the heterointerface with respect to the doping density for 300 and 77 K lattice temperatures.

is taken into account, e.g.

$$E_2^2(0) = \frac{2qN_d}{\epsilon_2}$$
$$-V(0) + V(-W_2) - \frac{kT}{q}\left[\ell n(1 + g'y) + \frac{4}{N'_d}\ell n(1 + \frac{y}{4})\right] \quad (23)$$

Substituting equation (23) into equation (21) yields

$$n_{so} = \sqrt{\frac{2\epsilon_2 N_d}{q}[-V(d_i^-) + V(-W_2) + \delta] + N_d^2 d_i^2 - N_d d_i} \quad (24)$$

where

$$\delta = -\frac{kT}{q}\left\{\ell n(1 + g'y) + \frac{4}{N'_d}\ell n(1 - \frac{y}{4})\right\} \quad (25)$$

The relationship

$$V(d_i^-) = V(0) - E_2(0)d_i \quad (26)$$

has been incorporated to derive equation (24). The coordinate $x = d_i^-$ corresponds to the $Al_xGa_{1-x}As$ side of th heterointerface.

Equation (24) differs from the depletion approximation case by δ

in the right hand side which is substracted from the total band bending

$$- V(d_i) + V(-W_2)$$

This contribution to the band bending is shown in Fig. 5 as a function of the doping density, N_d, in $Al_xGa_{1-x}As$ for 77 and 300K. As can be seen from this figure, the correction is quite important because it is comparable to ΔE_c.

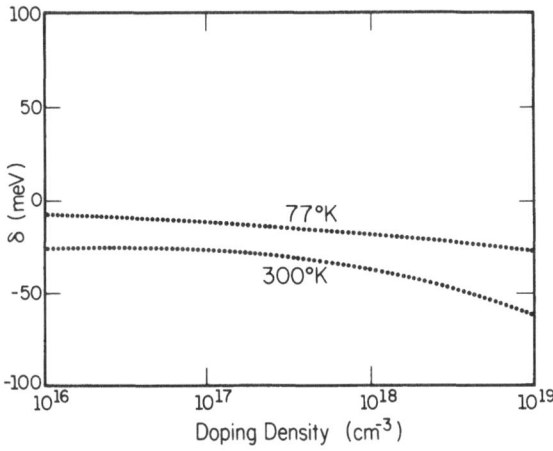

Figure 5 - The energy correction term arising from the degeneracy in the $Al_xGa_{1-x}As$ (without depletion approximation) as a function of the donor concentration in the $Al_xGa_{1-x}As$ layer.

In order to determine n_{so}, we should solve equation (24) with equation (8) which expresses n_{so} in terms of the density of states D and the two lowest energy levels in the potential well in GaAs as described earlier in equation (5).

Solving equations (24) and (8) simultaneously using (5), (25), (26) and

$$V(d_i^+) = \frac{1}{q} \Delta E_c + V(d_i^-) \tag{27}$$

with numerical techniques yields n_{so}, (dotted lines in Fig. 6).

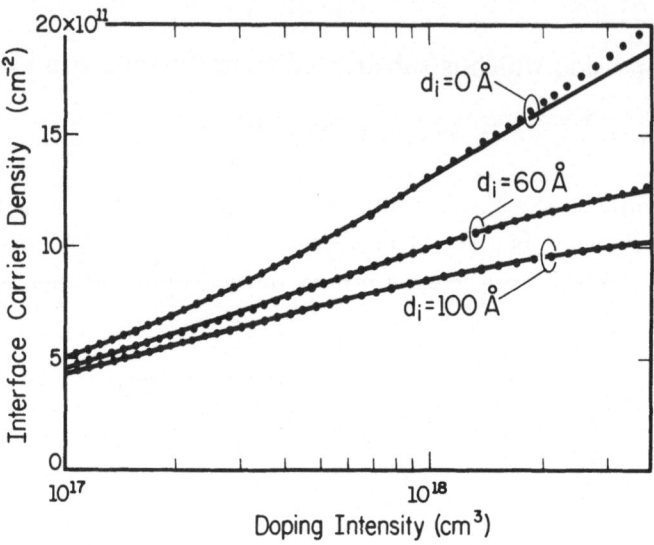

Figure 6 - Interface sheet electron 2DEG concentration vs. the doping level in the $Al_xGa_{1-x}As$ layer.

However, a very accuracte analytical approximation can be obtained if equation (5) and (8) are linearized with respect to $V(d_i+)$ as reported earlier.[15,16] It should also be pointed out that $V(d_i+)$ depicts the difference between the Fermi level (taken as reference) and the bottom of the conduction band in the GaAs at the heterointerface. It would be more appropriate to term $V(d_i+)$ as E_{Fi}/q. Repeating equation (8) with this replacement, we obtain

$$(28)$$

$$n_{so} = DkT \, \ell n \left[\left(1 + \exp \left(\frac{E_{Fi} - E_o}{kT} \right) \right) \left(1 + \exp \left(\frac{E_{Fi} - E_1}{kT} \right) \right) \right]$$

With a little modification of equation (28), the linearization we discussed above can be realized. For large values of n_{so} we obtain

$$-\frac{1}{q} E_{Fi} = \frac{1}{q} \Delta E_{Fo}(T) + a \, n_{so} \qquad (29)$$

where

$$a = 0.125 \times 10^{-16} (V \, m^2)$$

and $\Delta E_{Fo} = 0$ at 300K and 25 meV at 77K and below. Noting $qV(d_i-) = \Delta E_o - E_{Fi}$ equation (24) leads to the following simple formula for

n_{so}:

$$n_{so} = \sqrt{\frac{2\epsilon_2 N_d}{q}\left[\frac{\Delta E_c - \Delta E_{Fo}(T)}{q} + \delta + V(-W_2)\right] + N_d^2(d_i + \Delta d)^2}$$
$$- N_d(d_i + \Delta d) \tag{30}$$

where

$$\Delta d = \frac{\epsilon_2 a}{q} \approx 80\mathring{A} \quad \text{(Ref. 15 and 16)}$$

If one were to use the depletion approximation in the $Al_xGa_{1-x}As$ layer, the expression for the interface electron concentration can be arrived at by setting

$$\delta = 0 \text{ in equation (30)}$$

$$n_{so} = \sqrt{\frac{2\epsilon_2 N_d}{q}[V_{20}] + N_d^2(d_i + \Delta d)^2} - N_d(d_i + \Delta d). \tag{31}$$

where $V_{20} = V(-W_2) - V(d_i^-)$. At 77K and 300K, the depletion approximation underestimates the voltage term by 25 and 50 mV respectively (see Fig. 7).

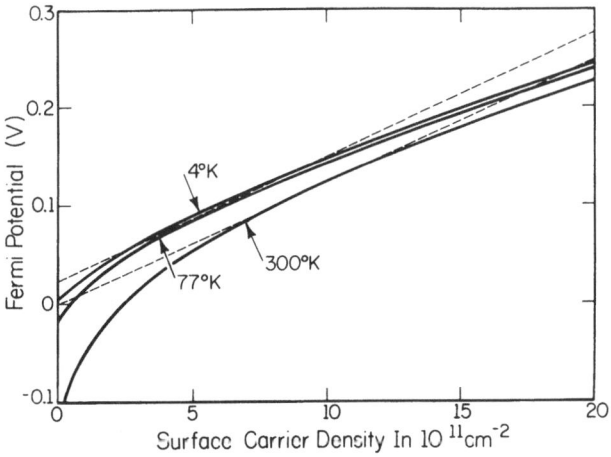

Figure 7 - Interface Fermi potential (E_{Fi}) vs. the sheet carrier concentration.

In Fig. 8 the interface carrier concentration is plotted with respect to the thickness of the undoped interfacial layer. For comparison, experimental values obtained in our laboratory are also included.[17] The large change in the measured n_{so} with respect to temperature not predicted by the theory is a result of the relatively thick doped $Al_xGa_{1-x}As$ layer.

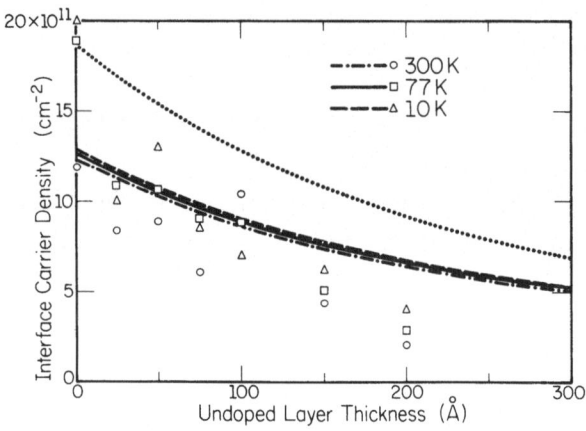

Figure 8 - Interface sheet carrier concentration vs. the undoped spacer layer thickness.

3.2 Charge Voltage Relation

We have so far derived the expressions relating n_{so} to the parameters of the heterojunction structure. Since our intent is to fabricate and understand the operation of modulation doped FETs, we must derive the necessary expressions showing how n_{so} is modulated by the applied gate bias. A band diagram of such a heterojunction with a Schottky barrier (biased negatively) is shown in Fig. 9. As can be seen there are two depletion regions, one caused by the heterointerface and another by the Schottky barrier which will be referred to as the "gate". In the schematic diagram the doped $Al_xGa_{1-x}As$ layer is shown to be depleted entirely. This can be achieved by a combination of a sufficiently large gate voltage and/or a sufficiently thin doped $Al_xGa_{1-x}As$ layer.

Using equations (21) and (24), and assuming that the depletion approximation is accurate enough, we find

$$qn_{so} = \epsilon_2 E_2(0) =$$

$$\overline{\sqrt{2q\epsilon_2 N_d[-V(d_i^-)+V(-W_2)]+q^2 N_d^2 d_i^2 - qN_d d_i}} \tag{32}$$

which can then be used to calculate the electric field strength at the heterointerface. The same result can be obtained by the use of equation (23) in the light of depletion approximation.

Since the transport through $Al_x Ga_{1-x} As$ is not as good as it is through GaAs, the structure and operational parameters are chosen such that the $Al_x Ga_{1-x} As$ layer is depleted entirely. The electrostatic potential in such a case can be calculated simply by integrating the area under the E-field curve. One must, however, remember that the electric field is constant in the undoped space layer. If we assume that the thickness of the doped $Al_x Ga_{1-x} As$ is d_d, we find

$$V_2 = \frac{qN_d}{2\epsilon_2} d_d^2 - E_2(0)d \tag{33}$$

where $d = (d_d + d_i)$ and

$$\epsilon_2 E_2(d_i^-) = \epsilon_2 E_2(0) = qn_{so} = \frac{\epsilon_2}{d}(V_{p2} - V_2) \tag{34}$$

where

$$V_{p2} = \frac{qN_d}{2\epsilon_2} d_d^2 \tag{35}$$

is the voltage that is necessary to pinch off the doped $Al_x Ga_{1-x} As$ layer. Examination of Fig. 9 shows that

$$V_2 = \phi_b - V_G + \frac{1}{q}(E_{Fi} - \Delta E_c) \tag{36}$$

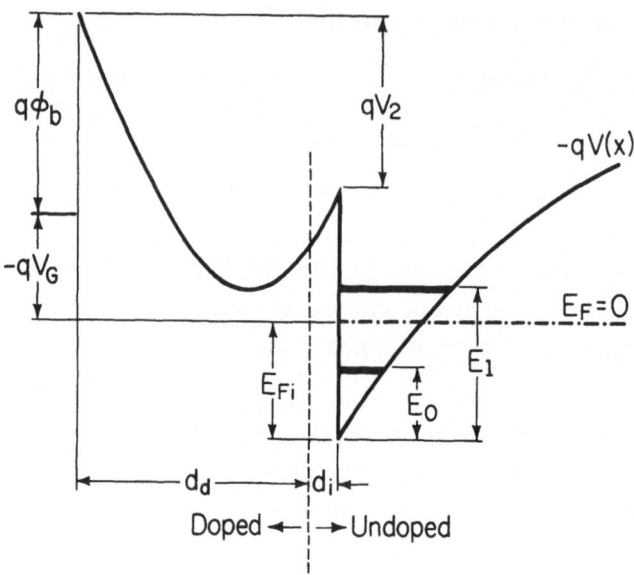

Figure 9 - Conduction band edge diagram of a single period modulation doped structure with a Schottky barrier deposited on the $Al_xGa_{1-x}As$ layer. Note that the $Al_xGa_{1-x}As$ layer is shown to be depleted entirely by a combination of the Schottky gate and the space charge balancing the 2DEG.

Combining equations (34) and (36) leads to

$$n_{so} = \frac{\epsilon_2}{qd}\left[V_G - (\phi_b - V_{p2} + \frac{1}{q}E_{Fi} - \frac{1}{q}\Delta E_c)\right] \tag{37}$$

We must, however, remember that E_{Fi} is a function of n_{so} as well as the gate voltage as depicted by equation (29). Substituting equation (29) into equation (37)

$$n_{so} = \frac{\epsilon_2}{qd}\left[V_G - (\phi_b - V_{p2} - \frac{1}{q}\Delta E_c + \frac{1}{q}\Delta E_{Fo}(T) + a\, n_{so})\right] \tag{38}$$

or

$$n_{so} = \frac{\epsilon_2}{q(d+\Delta d)}(V_G - V_{off}) \tag{39}$$

where

$$V_{off} \equiv \phi_b - \frac{1}{q}\Delta E_d - V_{p2} + \frac{1}{q}\Delta E_{Fo} \tag{40}$$

and

$$\Delta d = \frac{\epsilon_2 a}{q} \approx 80 \overset{\circ}{A} \tag{41}$$

The charge is then given by

$$Q_s = qn_{so} = \frac{\epsilon_2}{(d+\Delta d)}(V_G - V_{off}) \tag{42}$$

If we were to assume that E_{Fi} is not a function of the gate voltage, we would not have had the correction term Δd ($\approx 80 \overset{\circ}{A}$) which cannot be neglected next to d_d which is about $300 \overset{\circ}{A}$. Thus Δd gives an important correction factor and dropping it can lead to overestimation of n_{so} for a particular V_G and thus to overestimation of current or underestimation of electron velocity.

3.3 Current Voltage Relation

In an FET configuration such as the one shown in Fig. 10, application of a drain voltage in addition to the gate voltage gives rise to a potential distribution varying from zero at the source end to V_D' at the drain (see Fig. 10). This potential then leads to an effective charge control voltage which is different from the applied gate voltage and is a function of distance under the gate. The potential distribution shown in Fig. 10 is obtained when the ohmic drop across the source and drain resistances are neglected. Later on we will incorporate these resistances, but for now the expressions and concepts become much simpler to understand without them.

At a distance x from the source, the effective voltage controlling the charge is

$$V_{eff}(x) = V_G - V_c(x) \tag{43}$$

where V_G and $V_c(x)$ are the external applied gate voltage, and the channel voltage respectively. Equation (42) must read

$$Q_s(x) = \frac{\epsilon_2}{(d + \Delta d)}(V_g - V_c(x) - V_{off})$$
(44)

The channel current expression at x is

$$I = Q_s(x)\, Zv(x)$$
(45)

where Z denotes the width of the gate and v(x) is the electron velocity at x. Since the gate dimensions we are concerned with are on the order of a 1μm or less one must be concerned with high field effects such as velocity saturation.

We know that the electron mobility in the 2DEG is comparable to what is expected of pure GaAs.[18] The electron saturation velocity also was reported to be what one can expect from a pure GaAs crystal without degrading effects of ionized impurities. In this treatment we will use a two piece linear approximation (and briefly mention the three piece linear approximation) for the velocity field characteristic as shown in Fig. 11. Mathematically

and
$$\begin{aligned} v &= \mu E \quad \text{for } E < E_c \\ v &= v_s \quad \text{for } E \geq E_c \end{aligned}$$
(46)

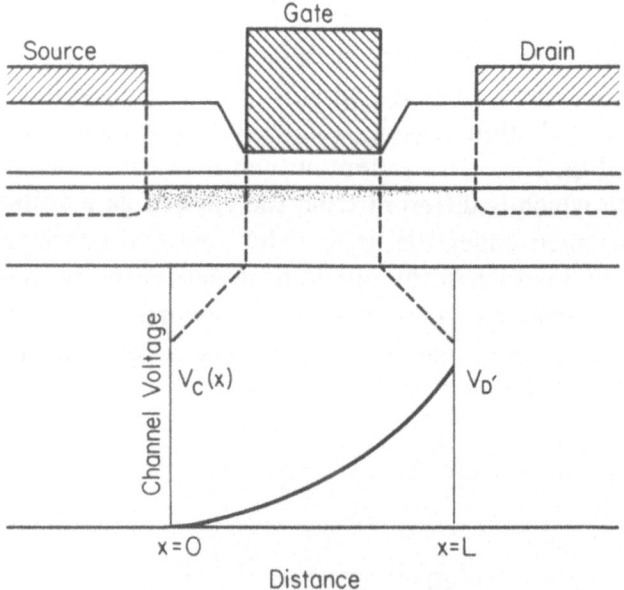

Figure 10 - Channel potential along the gate of a field effect transistor.

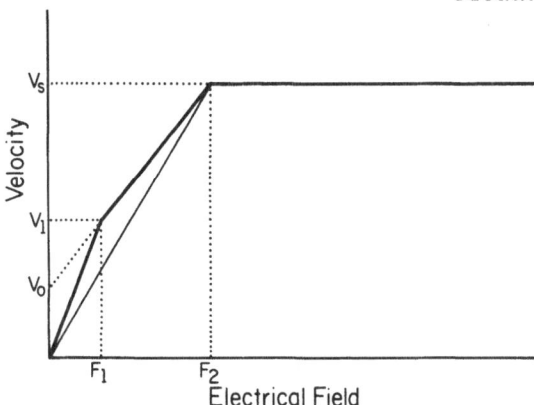

Figure 11 - Assumed velocity vs. electric field characteristics, two piece and three piece models. The three piece model is for closely matching the cryogenic temperature velocity-field characteristic. Here F_2 corresponds to E_c in the text.

At fields less than E_c, from Eqs. (44) and (45), we obtain

$$I = \mu Z \frac{\epsilon_2}{d + \Delta d}[V_G - V_c(x) - V_{off}]\frac{dV_c}{dx} \qquad (48)$$

where $dV_c/dx \equiv E$ is the electric field. As can be seen from Fig. 10, dV_c/dx is maximum near the drain. For small drain voltages below saturation, one can assume that the linear region of the $v(E)$ curve is applicable.

For the case where $E < E_c$, integration of equation (48) with the use of

$$V_c(x = 0) = R_s I \qquad (49)$$

$$V_D' = V_c(x = L) = V_D - (R_D + R_s)I \qquad (50)$$

where R_s and R_D denote the source and drain parasitic resistances and V_D depicts the external drain voltage, leads to

$$V_c(x) = V_G' - \sqrt{(V_G' - R_s I)^2 - 2\frac{(d + \Delta d)Ix}{\epsilon_2 \mu Z}} \qquad (51)$$

where

$$V_G' = V_G - V_{off} \qquad (52)$$

Differentiation of equation (51) with respect to x results in the electric field

$$(53)$$

$$E(x) = \frac{dV_c(x)}{dx} = \frac{(d + \Delta d)I}{\mu Z \epsilon_2} \left[(V_G' - R_s I)^2 - 2\frac{(d + \Delta d)Ix}{\epsilon_2 \mu Z} \right]^{-1/2}$$

As we increase the drain voltage, the electric field near the drain end at $x = L$ will reach the critical value first, called E_c or E_s;

$$E(x)|_{x=L} = E_c = \left(\frac{I_s}{\beta L}\right) \left[(V_G' - R_s I_s)^2 - \frac{2I_s}{\beta} \right]^{-1/2} \quad (54)$$

where

$$\beta = \frac{\mu \epsilon_2 Z}{(d + \Delta d)L}$$

and I_s is the saturation current. Letting

$$V_{sl} = E_c L = \left(\frac{I_s}{\beta}\right) \left[(V_G' - R_s I_s)^2 - \frac{2I_s}{\beta} \right]^{-1/2} \quad (55)$$

then

$$\left(\frac{I_s}{\beta}\right)^2 \left[(V_G' - R_s I_s)^2 - \frac{2I_s}{\beta} \right]^{-1} - V_{sl}^2 = 0. \quad (56)$$

or

$$\left(\frac{I_s}{\beta}\right)^2 - V_{sl}^2 \left[(V_G' - R_s I_s)^2 - \frac{2I_s}{\beta} \right] = 0 \quad (57)$$

Solving for I_s leads to

$$I_s = \frac{\beta V_{sl}^2}{1 + (\beta R_s V_G')^2}$$

$$(58)$$

$$\left[\sqrt{1 + 2\beta R_s V_G' + (V_G'/V_{sl})^2} - (1 + \beta R_s V_G') \right]$$

If R_s were to be set to 0, then

$$I_s = \beta V_{sl}^2 \left[\sqrt{1 + (V_G'/V_{sl})^2} - 1 \right] \quad (59)$$

or

$$I_s = \frac{\mu\epsilon_2 Z}{(d+\Delta d)L}V_{sl}\left[\sqrt{(V_G{}')^2 + V_{sl}^2}-V_{sl}\right].$$ (60)

Using $v_{sl} = E_c$ and $\mu E_c = v_s$, one finds

$$I_s = \frac{Z\epsilon_2 v_s}{(d+\Delta d)}\left[\sqrt{(V_G{}')^2 + V_{sl}^2}-V_{sl}\right]$$ (61)

From Equation (51) with $R_s = 0$ and $x = L$

$$V_c(L) = V_G{}' - \sqrt{(V_G{}')^2 - 2\frac{I_sL(d+\Delta d)}{\mu\epsilon_2 Z}}.$$ (62)

On the other hand

$$I_s = \frac{Z\epsilon_2 v_s}{(d+\Delta d)}(V_G{}' - V_{DS}{}')$$ (63)

where $V_{DS}{}'$ denotes the intrinsic drain to source voltage. Using equations (61) and (63), we find

$$V_{DS}{}' = V_G{}' + V_{sl} - \sqrt{(V_G{}')^2 + V_{sl}^2}$$ (64)

Extrinsic drain to source voltage is then given by

$$V_{DS} = V_G{}' + V_{sl} - \sqrt{(V_G{}')^2 + V_{sl}^2} + I_s(R_s + R_D)$$ (65)

3.4 Transconductance

For short channel devices, V_{sl} can be small compared to $V_G{}'$ and thus the expression for I_s can be reduced to

$$I_s = \frac{\epsilon_2 Z v_s}{d+\Delta d}V_G{}'$$ (66)

The intrinsic transconductance in the saturation regime is given by

$$g_m|_{max} = \frac{\partial I_s}{\partial V_G}|_{V_D'} = constant$$

$$= \frac{dI_s}{dV_G'} \cdot \frac{dV_G'}{dV_G} = \frac{\epsilon_2 Z v_s}{(d+\Delta d)} \cdot \frac{d}{dV_G}(V_G - V_{off}) \quad (67)$$

$$g_m|_{max} = \frac{\epsilon_2 Z v_s}{(d+\Delta d)}. \quad (68)$$

For long channel devices, using equation (59)

$$g_m = \frac{\partial I_s}{\partial V_G} = \beta \frac{V_G'}{\sqrt{1 + (V_G'/V_{sl})^2}} \quad (69)$$

$g_m|_{max}$ is obtained at $V_{G|max}'$ which is the pinch-off voltage of the 2DEG

$$V_{G}'|_{max} = V_{po}|2d = \frac{qn_{so}(d+\Delta d)}{\epsilon_2}. \quad (70)$$

Substituting equation (70) into equation (69) we obtain

$$g_m|_{max} = \frac{q\mu Z n_{so}}{L}\left[1 + \left(\frac{qun_{so}(d+\Delta d)}{\epsilon_2 v_s L}\right)^2\right]^{-1/2} \quad (71)$$

which reduces to equation (68) for small values of L.

4. SMALL SIGNAL GATE CAPACITANCE

In order to find the various device capacitances the total charge under the gate needs to be calculated. In the Shockley approximation the total charge is given by

$$Q_T = Z\int_0^L qn_{so}dx = Z\int_{V_S'}^{V_D'} qn_{so}\frac{dx}{dV_c}dV_c \quad (72)$$

Using equation (48), we find

$$I = Zq\mu n_{so}\frac{dV_c}{dx} = Z\mu \frac{\epsilon_2}{d+\Delta d}(V_G' - V_c)\frac{dV_c}{dx} \qquad (73)$$

Integrating both sides of equation (73)

$$\int_0^L I dx = Z\mu\frac{\epsilon_2}{d+\Delta d}\int_{V_S'}^{V_D'}(V_G' - V_c)dV_c \qquad (74)$$

which leads to

$$I = \frac{Z\mu}{L}\frac{\epsilon_2}{d+\Delta d}\cdot\frac{1}{2}\left[(V_G' - V_S')^2 - (V_G' - V_D')^2\right] \qquad (75)$$

From equation (73)

$$\frac{dx}{dV_c} = \frac{Zq\mu n_{so}}{I}. \qquad (76)$$

Substituting equation (76) into equation (72) and using equation (75)

$$Q_T = Z\int_{V_S'}^{V_D'}\frac{Z\mu\left[\dfrac{\epsilon_2}{d+\Delta d}(V_G' - V_c)\right]^2}{\dfrac{Z\mu}{L}\dfrac{\epsilon_2}{d+\Delta d}\dfrac{1}{2}\left[(V_{GS}')^2 - (V_{GD}')^2\right]}dV_c \qquad (77)$$

where $V_{GS}' = V_G' - V_S'$ and $V_{GD}' = V_G' - V_D'$ Then

$$Q_T = ZL\frac{\epsilon_2}{d+\Delta d}\frac{2}{(V_{GS}')^2 - (V_{GD}')^2}\int_{V_S'}^{V_D'}(V_G' - V_c)^2 dV_c \qquad (78)$$

$$= \frac{2}{3}C_o\frac{(V_{GS}')^3 - (V_{GD}')^3}{(V_{GS}')^2 - (V_{GD}')^2}$$

where

$$C_o = \frac{\epsilon_2 ZL}{d+\Delta d} \qquad (79)$$

$$C_{GS} = \frac{\partial Q_T}{\partial V_{GS}'} = \frac{2}{3}C_o\frac{V_{GS}'(V_{GS}' + 2V_{GD}')}{(V_{GS}' + V_{GD}')^2} \qquad (80)$$

and similarly the gate-drain capacitance is

$$C_{GD} = \frac{\partial Q}{\partial V_{GD}'} = \frac{2}{3}C_o\frac{V_{GD}'(V_{GS}' + 2V_{GD}')}{(V_{GS}' + V_{GD}')^2} \qquad (81)$$

In Fig. 12 and Fig. 13, the normalized capacitances C_{GS}/C_o and C_{GD}/C_o are plotted against the normalized drain-to-source voltage V_{DS}'/V_{sl} using the normalized gate voltage V_{GS}'/V_{sl} as the parameter. These capacitance values are calculated for the region below current saturation. Above saturation both capacitances should stay constant at about C_o. In addition, for sufficiently large voltages, e.g. 0.5 V, both C_{GS} and C_{DG} can be assumed constant having values of about $C_o/2$.

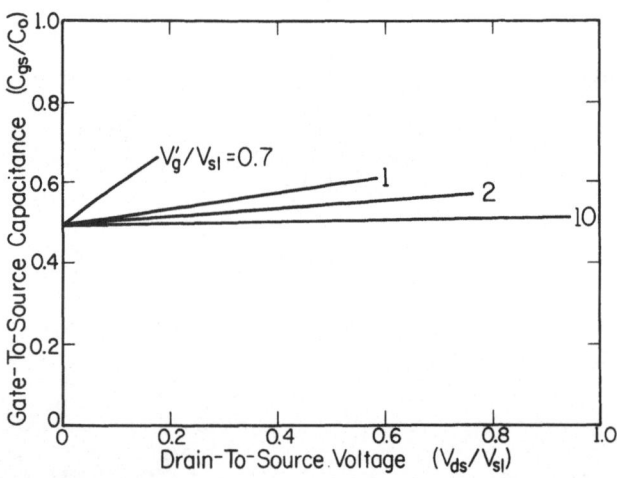

Figure 12 - The normalized gate-to-source capacitance vs. the normalized drain voltage.

5. DEVICE FABRICATION

Shown in Fig. 14 is the schematic cross-sectional view of a single interface modulation doped field effect transistor (MODFET). Multiple interface MODFETs are not going to be covered here because of the asymmetry in the interface quality of GaAs on $Al_xGa_{1-x}As$ and $Al_xGa_{1-x}As$ on GaAs.[19,20] Inverted modulation doped structures with the binary on top of the ternary have also been fabricated[7] but their performance is still inferior because of a lower quality $GaAs/Al_xGa_{1-x}As$ interface.[19]

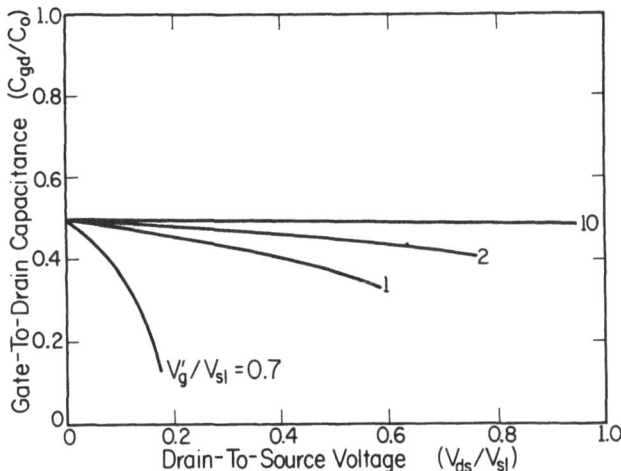

Figure 13 - The normalized gate-to-drain capacitance vs. the normalized drain voltage.

Using the normal MODFETs with $Al_xGa_{1-x}As$ on top of GaAs, the devices reported here were fabricated by first etching mesas which provide isolation of the devices and allows the gate contact pads to be placed on the Cr-doped semi-insulating substrate. The source and drain regions are then defined in positive photoresist and AuGe/Ni/Au metallization is evaporated. Following the lift-off process, this metallization is alloyed at 500°C for 1 minute in a H_2 ambient. The gate pattern is then defined with positive photoresist and the top $As_xGa_{1-x}As$ layer is thinned by etching, the amount of etching being dependent on whether a normally-off or a normally-on FET is desired, and the Al gate metallization is evaporated and lifted-off.

Figure 15 shows a top view of a MODFET having a gate length of 1 μm and a source-drain spacing of 3 μm. The width of the device is 2 x 145 μm. The source, drain and gate pads are built up to a total thickness of about 4000Å to facilitate bonding. The wafer is then thinned down to about 200 μm by polishing the back side. Following an electroless Pd plating process on the back side, the wafer is scribed and diced into individual devices. For dc testing the devices are first

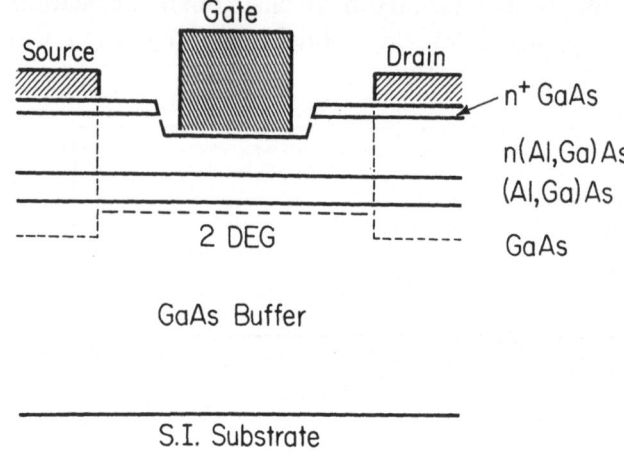

Figure 14 - Cross-sectional view of a MODFET with a 1μm gate length and a 3μm source-drain (channel) length.

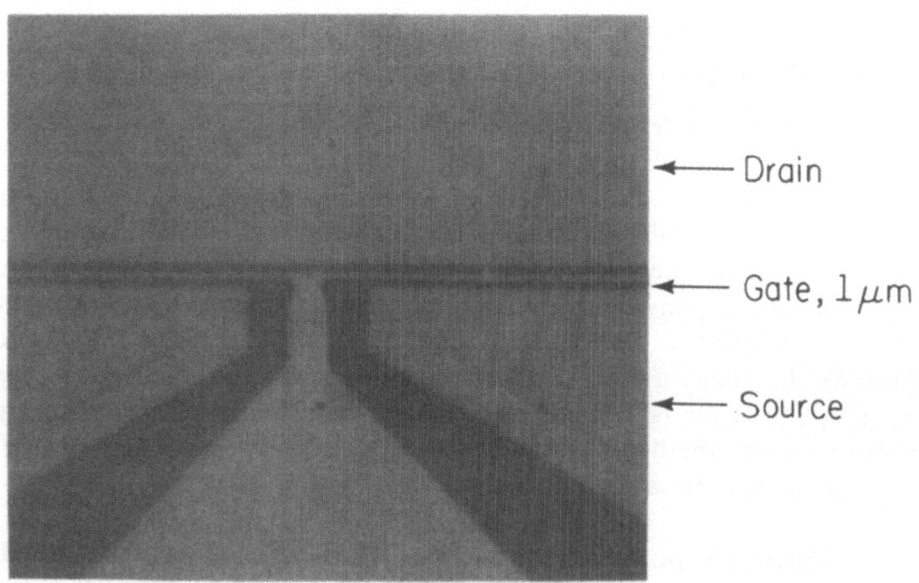

Figure 15 - Top view of a MODFET having dimensions indicated in the Fig. 14 caption.

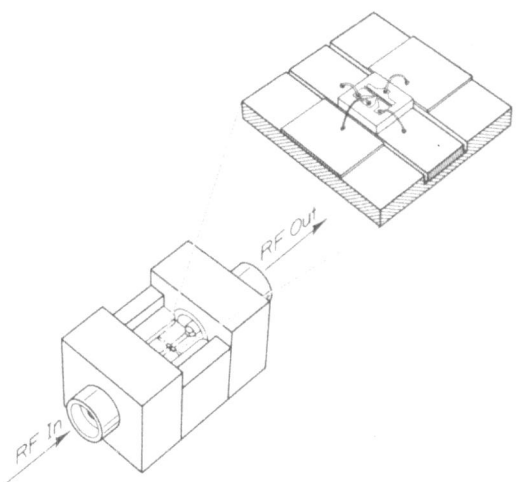

Figure 16 - Schematic diagram of a microwave carrier with 50Ω input
and output microstrip lines used for rf measurements.

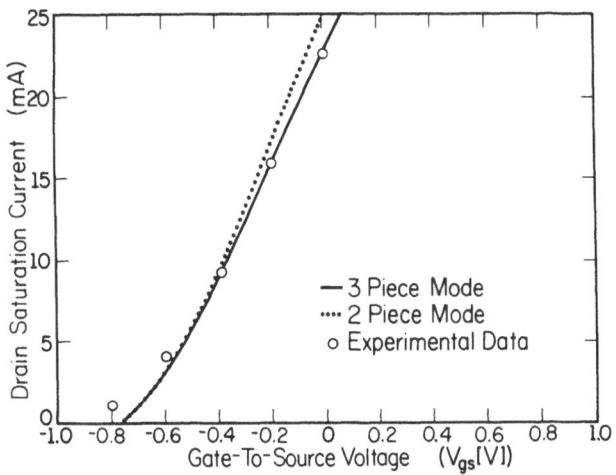

Figure 17 - Drain saturation current (for a device width Z = 145μm)
vs. gate voltage of a N-ON MODFET.

bonded onto TO-18 headers and then wire bonded. For rf testing, the device is bonded to a microwave carrier and wire bonded to 50Ω microstrip lines as shown in Fig. 16.

6. COMPARISON OF THE THEORY WITH EXPERIMENTAL RESULTS

To test the accuracy of the model, a normally-off and a normally-on modulation doped FET fabricated in our lab were selected and characterized. The low field mobility was obtained from Van der Pauw-Hall measurements of the particular wafer from which the FETs were fabricated. The gate width is 145 μm and the length is 1 μm. The structure consisted of a 1 μm undoped GaAs buffer layer, a 60Å undoped $Al_xGa_{1-x}As$ layer and a 600Å n-type $Al_xGa_{1-x}As$ layer doped with Si to a level of 1 x 10^{18} cm^{-3}. The 600Å dimension was used to allow fabrication of either normally-on or normally-off devices by recessing the gate. For x = 0.3, N_d = 1 x 10^{18} cm^{-3} and d_i = 60Å, the doped (Al,Ga)As remaining beneath the gate should be about 250Å and 350Å thick to obtain normally-off and normally-on devices respectively.

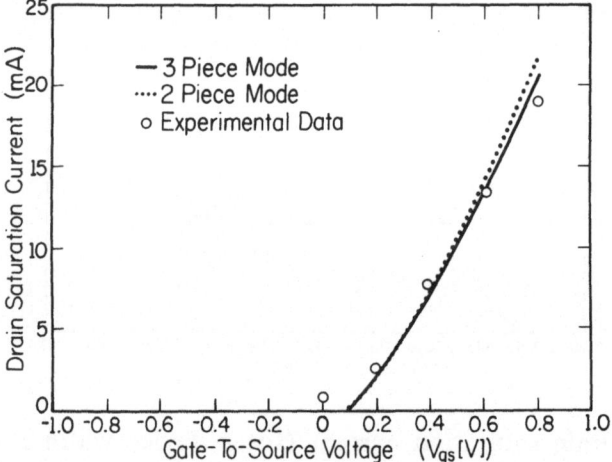

Figure 18 - Drain saturation current (for a device width Z = 145μm) vs. gate voltage of a N-OFF MODFET.

The calculated and experimental drain saturation currents at room temperature as a function of the gate voltage are shown in Figs. 17 and 18. N_d = 1 x 10^{18} cm^{-3}, μ = 6800 cm^2/V$_s$ and R_s = 7Ω for the normally-on and 10Ω for the normally-off FETs were measured.

v_{sat} = 2 x 10^7 cm/sec[18], F_1 = 1 kV/cm and F_2 = E_c = 3.5 kV/cm are used for our calculation in Figures 17 and 18 (solid line: three piece velocity model; dotted line: two piece velocity model). The experimental values are marked by the dots. The measured transconductance of 275 mS/mm for the N-ON device at 300K is the highest value ever reported. Even though 10Ω is measured as R_s for the N-OFF FET, 12Ω is used for our model to have the best fit to the experimental data. A possible explanation for this is given below. As can be seen from Fig. 17, the agreement between our model and the experimental data is quite good except near the threshold. (This region is not described well by the model.) The two piece model overestimated the current predicted by the three piece model by approximately 10-20%. The I_s-V_D characteristics for the N-on and N-off FETs are shown in Fig. 19 a) and b). As can be seen from the figure, the agreement between the measurement and our calculation is very good.

For the N-off FET, the agreement is also good but the value of R_s has to be adjusted to give a best fit. The slightly smaller current measured at V_{GS} = 0.8V may be due to the fact that our model becomes invalid at gate voltage higher that V_{off} + $(V_{po})_{2d}$. The justification for using R_s = 12Ω instead of 10Ω may be as follows: The fabrication FET has 1μm spacing between the source and gate. This region can be thought of as ungated FET indicating that R_s should increase as the current increases. This effect should be more pronounced for N-off FET's.

7. EFFECT OF DONOR-ELECTRON SEPARATION

From equation (68) it is clear that any reduction in the thickness of the $Al_xGa_{1-x}As$ layer under the gate leads to increased transconductances. This can either be done by increasing the doping in the $Al_xGa_{1-x}As$ (thereby reducing the thickness required under the gate) and/or reducing the undoped $Al_xGa_{1-x}As$ spacer layer.[21] Reduction of the spacer layer thickness also has the added advantage of leading to increased two dimensional electron concentrations. The resulting increased current carrying capability leads to a faster charging time of device and parasitic capacitances and thus to higher speeds. High speeds can be obtained at small voltages swings and thus low power consumption. This argument basically leads to the conclusion that the

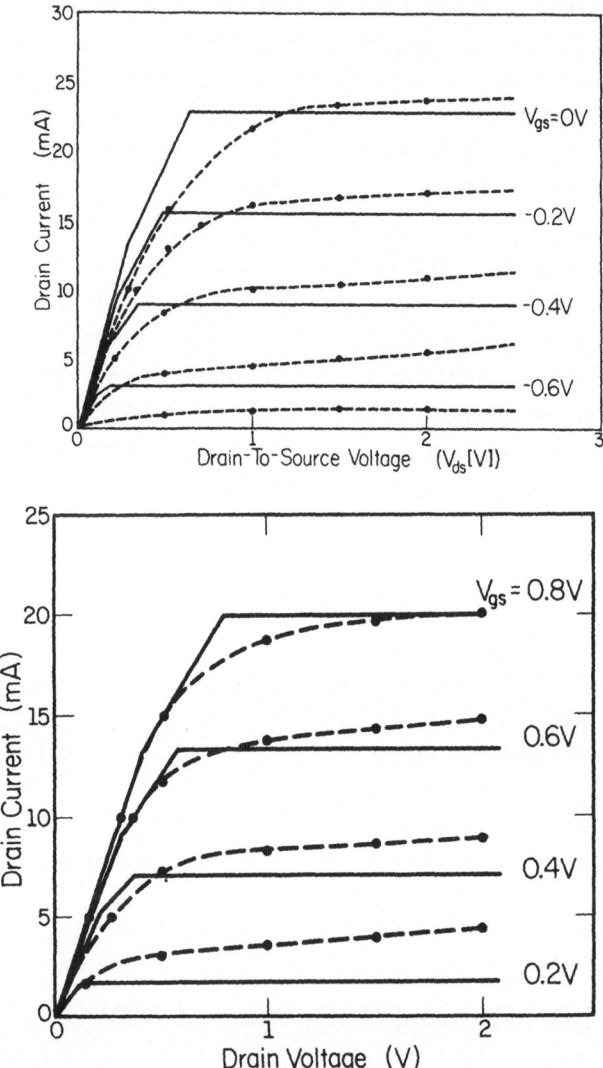

Figure 19 - Three terminal output characteristics of the N-on and N-off MODFETs. Solid lines indicate the results of the three piece model, the dots and the broken lines represent the experimental drawn I-V.

doping level in the $Al_xGa_{1-x}As$ layer should be increased as much as possible and that the spacer layer should be made as small as possible.

The maximum net electron concentration that can be obtained in $Al_xGa_{1-x}As$ doped with Si is about 2×10^{18} cm^{-3} which is smaller than

Sn and Be doping limits possible by MBE. The limit in Si doping is thought to be a result of the high Si effusion cell temperature required. Any impurities present in the effusion cell assembly can easily be released at these high temperatures and incorporated in the growing $Al_xGa_{1-x}As$ film. The Si doping limit in GaAs is about 5×10^{18} cm^{-3} which is basically comparable to the aforementioned figure for the $Al_xGa_{1-x}As$ layer. The use of a Si ribbon heated by passing current through it in place of an effusion cell has tentatively been shown to result in carrier concentrations of over 10^{19} cm^{-3} in GaAs.[22]

Even if higher doping concentrations were possible, non-tunneling Schottky barriers are difficult if not impossible to obtain on $Al_xGa_{1-x}As$ with a doping level greater than about 2×10^{18} cm^{-3}, which sets the practical limit.

The second parameter, the space layer, can be made as thick or as thin as one desires, meaning that there is no technological limit by the MBE process, except that a very thin layer may have to be incorporated to avoid Si diffusion into GaAs. Depending on the growth temperature, the required thickness to prevent Si atom diffusion for a conventional modulation doped FET structure is about 5-10Å. When the spacer layer is made this thin, the Coulomb interaction between the Si donors and the 2DEG is stronger which then may lead to a degradation of electron velocity. The electron velocity in undoped GaAs and doped (10^{17}cm^{-3}) GaAs layers is 2.1×10^7 cm/s and 1.8×10^7 cm/s respectively. This means that advantages gained in reducing spacer layer thickness must be weighed against the degradation in electron velocity. The experimental results obtained in our laboratory indicate that the optimum thickness is about 30Å.

To investigate the role of the undoped $Al_xGa_{1-x}As$ spacer layer, a series of MODFET structures were grown and devices were fabricated[21] as described earlier. The doping level in the $Al_xGa_{1-x}As$ was chosen to be about 10^{18} cm^{-3} so that adequate Schottky gates could be obtained.

As the thickness of the spacer layer was decreased from 100Å to 20Å a number of trends were observed. The decrease in the undoped $Al_xGa_{1-x}As$ spacer layer thickness was closely matched by an increase in the doped $Al_xGa_{1-x}As$ thickness needed to accommodate the increased charge transfer. As a result, the gate capacitance remained almost constant while the maximum saturation current almost doubled. On the other hand, the maximum transconductance scaled with the current. The drain I-V characteristics of the best normally on and normally off devices are shown in Fig. 20 indicating transconductances of 230 and 250 mS/mm respectively.

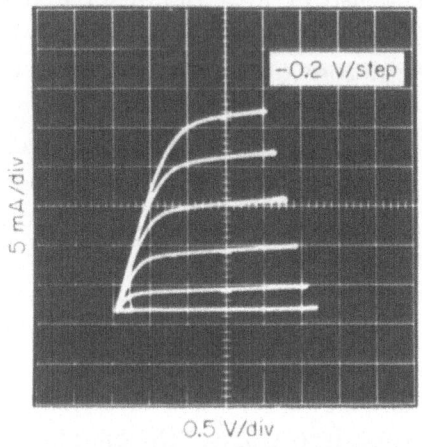

0.5 V/div

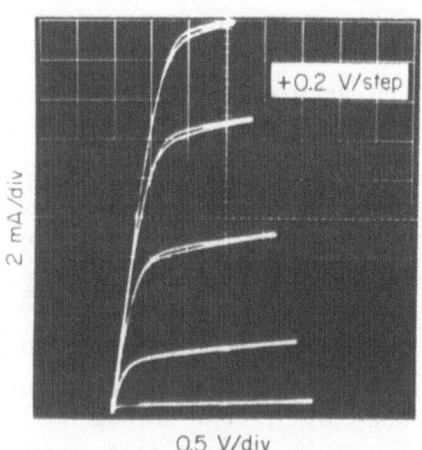

0.5 V/div

Figure 20 - Output drain characteristics of a N-ON MODFET with d_i = 40Å and a N-OFF MODFET with d_i = 20Å. Both characteristics are for Z = 145μm.

The best previously reported values were 193 mS/mm for a normally-off device and 117 mS/mm[23] for a normally-on device. The maximum currents obtained are 12-20% higher than the best previously reported values. Recently optimized devices have shown improved current carrying capabilities up to twice those of the ones reported here.

The transistor characteristics were modeled using the equations derived earlier with the exact dependence of the Fermi level on the gate voltage accounted for. The saturation velocity was chosen to be 2 x 10^7 cm/s, independent of the mobility, on the basis of earlier work on modulation doped heterostructures.[18] The two parameters used to fit the data were the doped $Al_xGa_{1-x}As$ layer thickness beneath the gate, d_d, and the source resistance, R_s.

At 300K, the mobility of the 2DEG is relatively independent of d_i. Therefore we assume that $\mu = 7000$ cm^2/Vs and is independent of d_i. The transconductance g_m vs. d_i for a 1μm gate device is shown in Fig. 21 for the doping levels most commonly used. $N_d = 2 \times 10^{19}$ cm^{-3} is also included (theoretical) to show the degree of dependence. It was pointed out earlier that non-tunneling Schottky barriers cannot be obtained at these doping levels. The dotted lines are obtained when

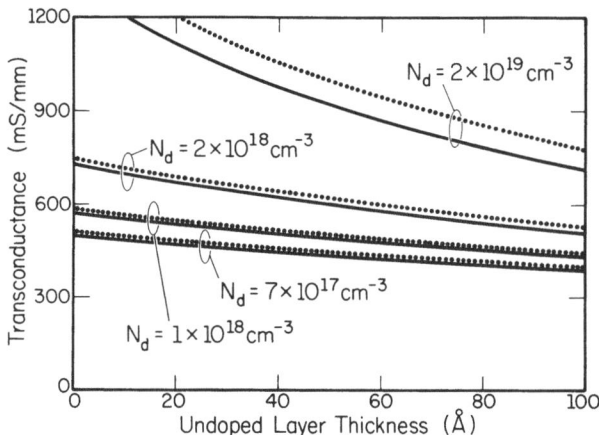

Figure 21 - Intrinsic transconductance per mm of gate width given by $g_m/(1-g_mR_s)$ as a function of the undoped $Al_xGa_{1-x}As$ layer thickness. Dotted lines assume non-degeneracy and solid lines assume degeneracy of doping in the $Al_xGa_{1-x}As$. $V_{off} = 0.2V$, $\mu = 7,000$ cm^2/Vs and L $= 1\mu$m.

Boltzmann statistics are used in the $Al_xGa_{1-x}As$ and the solid lines take degeneracy into account.

Once the I_s vs. V_s characteristic was determined, the transconductance and gate capacitance were calculated numerically as $g_m = dI_s/dV_G$ and $C_g = dQ/dV_G$. While the values obtained for the transconductance are an accurate reflection of the experimental data, the capacitance values are good only as a first approximation.

Table 1 lists the source resistances used to model each device as well as the maximum saturation currents and transconductances. The data represent parameters for the largest value of I_s for which the data are accurately fit by the theory. These would be +0.6V for the normally-off device and -0.2V for the normally-on device in Fig. 22 which shows the I_s vs. V_G characteristics of the best normally-on and normally-off devices fabricated. The maximum theoretical values of the saturation current and transconductance are given in parenthesis.

The thicknesses of the undoped $Al_xGa_{1-x}As$ spacer, d_i, the doped $Al_xGa_{1-x}As$ beneath the gate, d_d, the equilibrium junction depletion depth, d_j, and the maximum gate capacitance are given in Table 2. The large discrepancy in the parameter values of the 20Å N-on FET relative to the other N-on devices results from it being only quasi-normally-on.

TABLE 1 - Measured and theoretical maximum values of transconductance and saturation current. Theoretical values are in parentheses. The values of source resistances are those employed in fitting the theoretical curve to the data.

d_i (Å)	g_m (mS/mm)		I_{sat} (max) (mA)		R_s (Ω)	
---	N-on	N-off	N-on	N-off	N-on	N-off
20	210 (290)	250 (275)	7.5 (27)	19 (19)	5	5.5
40	230 (245)	155 (175)	24 (2.75)	13 (14.5)	5	16.5
60	235 (240)	205 (210)	23 (26.5)	18.8 (16)	5.5	12.5
80	210 (225)	160 (170)	14 (19)	8.4 (11)	4	13
100	145 (170)	125 (135)	7.5 (11)	8.5 (8.5)	13	22

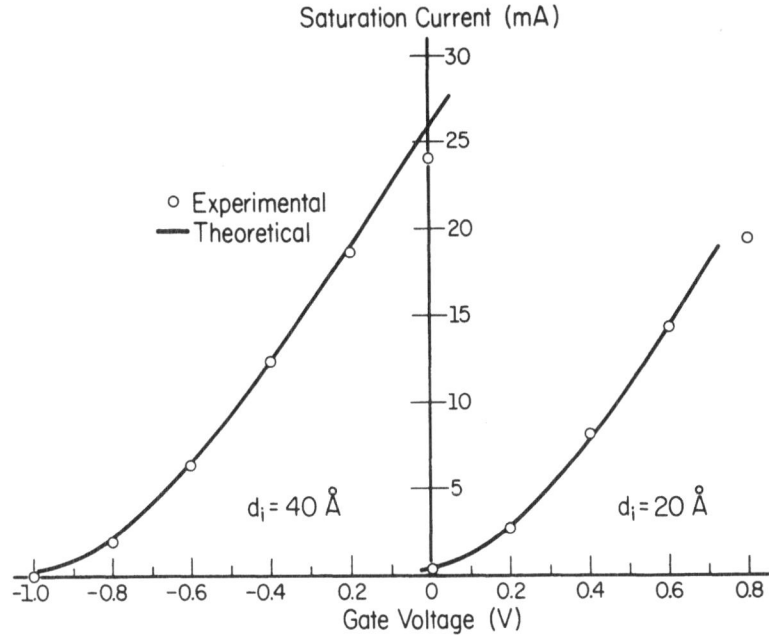

Figure 22 - Saturation current vs. gate voltage characteristics for N-on and N-off MODFETs with $d_i = 40\text{Å}$ and $d_i = 20\text{Å}$. The measured data are presented as open circles ($Z = 145\mu m$).

The gate voltage swing ran from -0.4V to +0.6. The current and transconductance values for this device are listed in Table 2 for $V_g = 0V$. The normally-off device with $d_i = 20\text{Å}$ is anomalous in that the doped layer is too thin and the source resistance too small to be consistent with the other normally-off devices.

In Table 2 the gate capacitance is nearly independent of spacer layer thickness. This is a result of the particular parameters used rather than being a characteristic of MODFETs. The maximum capacitances tabulated represent an upper limit for devices operating a current saturation regime. This implies that maximum operating frequencies in excess of 10 GHz can be obtained.

8.   CRYOGENIC TEMPERATURE PERFORMANCE

Conventional bulk GaAs layers doped to a level of 10^{17} cm^{-3} or higher do not show any appreciable mobility or velocity enhancement when cooled to 77K. Undoped GaAs, however, shows a considerable

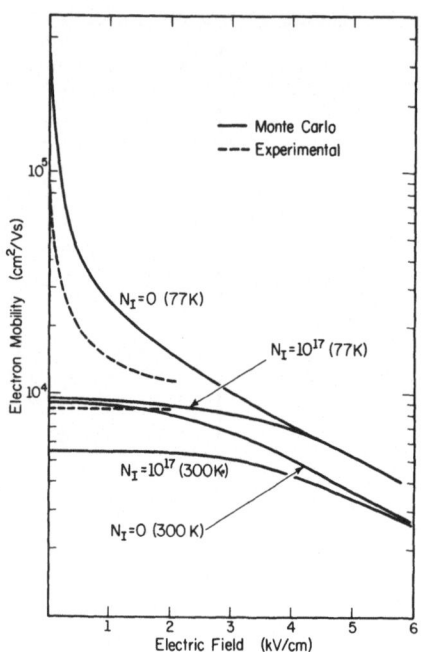

Figure 23 - Static electron mobility vs. electric field in a modulation doped GaAs/Al$_x$Ga$_{1-x}$As heterostructure deduced from pulse measurements up to an electric field of 2 kV/cm. Monte Carlo calculation up to 6 kV/cm are also shown for comparison for pure GaAs and GaAs doped to 10^{17} cm^{-3}.

TABLE 2 - The function depletion depth, d_i, for each spacer layer thickness and the doped layer thickness, d_d, used in modelling each device is given as well as the total Al$_x$Ga$_{1-x}$As layer thickness (d_i + d_d). The calculated gate capacitance represents an upper limit.

d_i (Å)	d_j (Å)	d_d (Å)		$d_i + d_d$ (Å)		C (pF)	
---	---	N-on	N-off	N-on	N-off	N-on	N-off
20	138	358	283	378	293	0.73	0.88
40	127	460	307	500	347	0.55	0.78
60	117	441	281	501	341	0.57	0.80
80	109	420	268	500	348	0.56	0.78
100	101	382	245	482	345	0.58	0.78

enhancement when cooled to 77K. This is a result of the absence of ionized impurities. Ironically, one must have ionized donors to obtain free electrons and thus current carrying capability. Modulation doped structures as described earlier provide electrons in sufficient numbers in the GaAs while maintaining its freedom from ionized impurities and, therefore, one would expect mobility-field and velocity -field characteristics similar to those of pure GaAs. Pulse measurements between 200 V/cm and 2kV/cm and dc mesurements below 200 V/cm show that the transport in modulation doped structures is similar to that in pure GaAs.[18] Static mobility-field and velocity-field characteristics of modulation doped structures along with undoped GaAs are shown in Figs. 23 and 24 respectively.

Based on the above argument the transconductance of a short channel MODFET, being proportional to the electron velocity, should register an enhancement scaling with the velocity when cooled to 77K. For this experiment, the devices with an undoped $Al_xGa_{1-x}As$ layer thickness of 20Å were mounted on TO-18 headers and tested at 300K and, by immersing into LN_2, at 77K. The drain I/V characteristics of a MODFET with a gate width of 290μm are shown in Fig. 25. For a gate voltage of +0.6V the transconductance is 225 mS/mm at 300K and 400 mS/mm at 77K. At V_G = +0.8V, the 2DEG is entirely depleted and an additional conduction path through the low quality

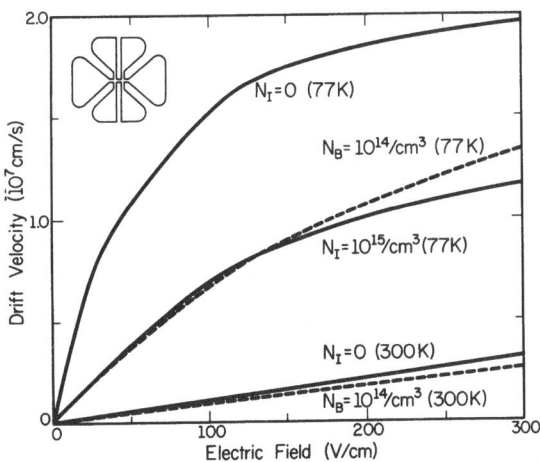

Figure 24 - Drift velocity vs. electric field in a modulation doped heterostructure up to an electric field of 300 V/cm. GaAs with an ion concentration of 10^{15} and 10^{17} cm^{-3} is also shown (Monte Carlo calculations).

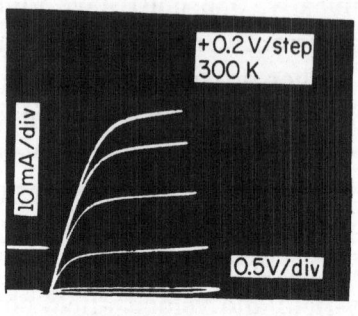

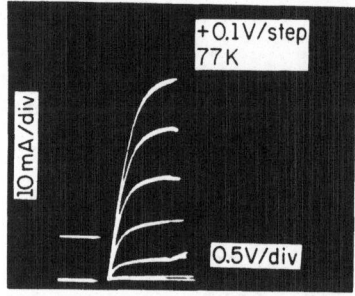

Figure 25 - Drain I/V characteristics of a modulation doped FET having gate dimensions of 1 μm x 290μm at 300K (a) and 77K (b).

$Al_xGa_{1-x}As$ is formed resulting in the observed decrease in transconductance.

Using the model described earlier, the drain saturation current vs. gate voltage was calculated at 300 and 77K and plotted in Fig. 26. At 300K a source resistance of 4Ω and a saturation velocity of 2 x 10^7 cm/s were used. Such an excellent match in the entire range of currents can only be obtained for a specific combination of source resistance and saturation velocity making the values reported here very reliable. At 77K, R_s decreased to 2.5Ω and v_s increased to 3 x 10^7 cm/s. These results are in extremely good agreement with the predictions based on pulse measurements.

Also characterized in Fig. 26 is the best previously reported MODFET operating at 77K (Fujitsu Laboratories, Ltd., Ref. 25). The data, represented as triangles, were taken from the published Fujitsu

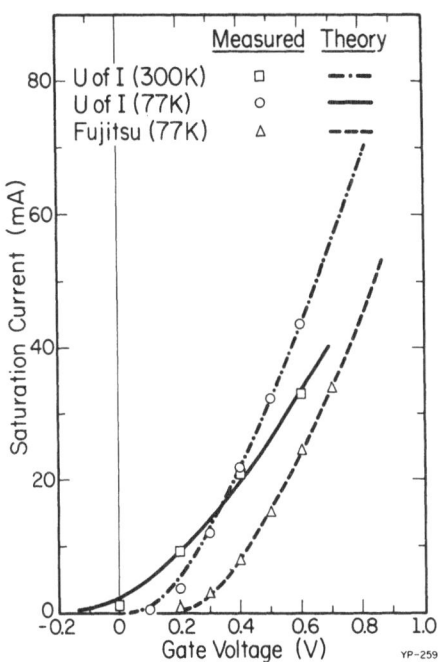

Figure 26 - Drain saturation as a function of gate voltage of the device shown in Fig. 25. For comparison, the saturation current of a high performance Fujitsu device is also shown.

drain I-V characteristic. The dotted line is then calculated to fit the data. The Fujitsu device had a gate length of 2μm, s doping level of N_d = 2 x 10^{18} cm^{-3} in the $Al_{0.3}Ga_{0.7}As$ and no undoped spacer layer. To model the device at 77K the reported mobility of 20,000 cm^2/Vs was used with R_s = 3.8Ω and v_s = 3 x 10^7 cm/s.

In all devices it is necessary to assume that 25% of the electrons in the $Al_{0.33}Ga_{0.67}As$ are frozen out at 77K to account for the shift in the threshold voltage. This figure was initially determined from measurements made on bulk $Al_{0.33}Ga_{0.67}As$ doped to 1 x 10^{18} cm^{-3} with Si. When modelling MODFETs, the thickness of the doped $Al_{0.33}Ga_{0.67}As$ beneath the gate is adjusted to result in a match between experimental and theoretical drain currents at a given gate voltage. At 300K a thickness of 305Å was required to fit the data; at 77K, assuming 25% freeze-out, the required thickness was 308Å. A consequence of the freeze-out is that the best parameters for a device intended for operation at cryogenic temperatures may not be the same as those optimizing the 300K performance.

The transconductance of each device is calculated numerically from the curves in Fig. 26 ($g_m = dI_s/dV_G$) and plotted in Fig. 27. As can be seen, the potential maximum transconductance of our device at 77K for a gate voltage of +0.8V is 450 mS/mm at 77K, as compared to a value of 400 mS/mm obtained for the Fujitsu device at 77K. Calculating the maximum intrinsic transconductance (zero source resistance) for our device we find $g'_m = 352$ mS/mm at 300K and 668 mS/mm at 77K.

9. HIGH FREQUENCY PERFORMANCE

For digital applications, inverters with an odd number of stages can be fabricated in a way to be connected in series. By bringing a feedback signal from the last stage to the first one, an oscillation, the frequency of which is a function of number of stages as well as the delay time associated with each device is obtained. In addition, parasitics such as line capacitances can add to the delay time. An output buffer stage, active or passive, is used to measure the oscillation frequency with a fast oscilloscope or a spectrum analyzer. Spectrum

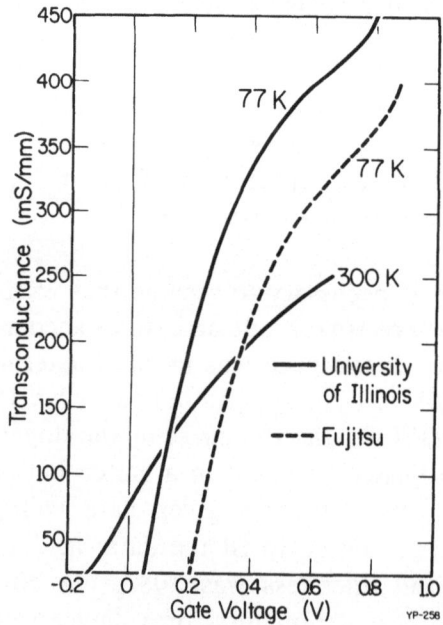

Figure 27 - Calculated transconductance as a function of gate voltage using drain current vs. gate voltage characteristics shown in Fig. 26.

analyzers have the advantage that they can detect very weak signal levels.

The ring oscillators mentioned above have been fabricated and their performance has been reported.[26,27] The best propagation delay time result obtained so far is 17 ps at 300K with a power dissipation of slightly under 1 mW per gate with a gate length of $0.7\mu m$. At 77K, the best performance is 12 ps with a power dissipation of about 10 mW per gate which must be trimmed down to about 1 mW to make refrigeration to 77K practical (see Fig. 28).

High frequency small signal measurements made in our laboratory indicated a maximum available gain of about 12 dB at 10 GHz which compares well with published results by other laboratories.[23] The noise measurements on individual devices[23] as well as cascaded three stage amplifiers have also been performed.[28]

10. SUMMARY

Modulation doped $Al_xGa_{1-x}As/GaAs$ field effect transistors (MOSFETs) whose operation is similar but performance is superior to Si/SiO_2 metal oxide semiconductor field effect transistors (MOSFETs)

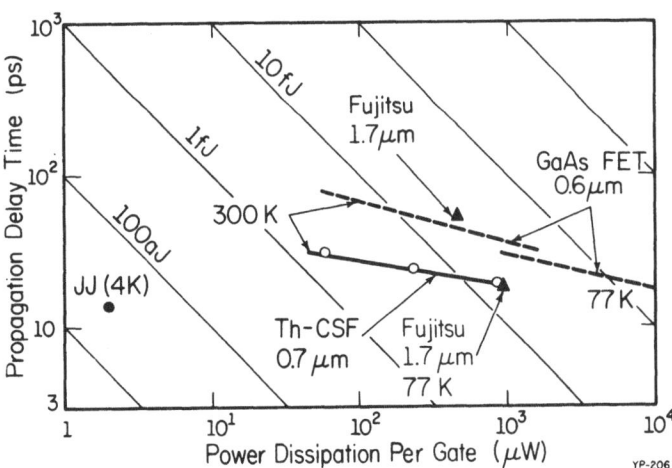

Figure 28 - Propagation delay vs. power consumption chart obtained from ring oscillators made from devices competing for high speed including the MODFETs.

were described. A complete numerical model and very accurate analytical model with closed form expressions were developed and used to predict and improve the performance. The key aspect of the model is the use of the triangular potential well approximation to relate the two dimensional electron concentration (2DEG) to the subband energies. Using Gauss' law and Poisson's equation, the 2DEG is also related to the gate voltage as well as other heterojunction parameters. The drain current of the MODFET was calculated using a two piece and also a three piece velocity-field model. From the charge under the gate the device capacitances were also calculated. Expressions for the transconductance were derived which indicate that, in terms of charge balance, the 2DEG can effectively be assumed to be 80Å away from heterointerface and if this correction is neglected it can lead to overestimation of current or underestimation of the electron velocity and also to overestimation of gate capacitance.

Devices with 1μm gate lengths showed extremely high transconductance and high current capabilities. The transconductances were observed to scale up with decreasing undoped spacer layer thickness. Maximum transconductances of about 275 mS/mm at 300K and over 400 mS/mm at 77K were obtained. By applying the model developed, the electron saturation velocity has been estimated to be 2 x 10^7 cm/s and 3 x 10^7 cm/s at 300 and 77K respectively.

Microwave measurement on N-off MODFETs intended for logic circuit applications showed power gains of about 12 dB at 10 GHz. If cooled to 77K, better gains and very promising noise performance should be obtained. Ring oscillator data indicate switching speeds of about 17 and 12 ps at 300 and 77K, respectively.

Finally, improved fabrication techniques should lead to improved performance as well as applications to functional high speed integrated circuits. In a short time, switching times of 10 ps at 300K should be possible. While the device itself should be capable of delivering an internal switching time of about 5 ps at 77K, it may take a while to optimize the circuitry and device fabrication to deliver this expected performance.

ACKNOWLEDGEMENTS

This work was made possible by funds from the United States Air Force Office of Scientific Research.

Analysis of MODFETs presented here would not have been possible without Professor M.S. Shur and his student K. Lee who helped develop most of the analytical model. The author would also like to acknowledge the constant support and encouragement of Dr. A.Y. Cho and discussions with Dr. N.T. Linh. Finally, he is indebted to his graduate students, T.J. Drummond, W. Kopp, R. Fischer, R.E. Thorne, S.L. Su, W.G. Lyons, D. Arnold, P. Tejayadi and J. Klem for their dedication, hard work and the accomplishment of results

APPENDIX

In this section we will derive the expressions for the interface electric field and the sheet carrier concentration in the case where the depletion approximation is assumed applicable. Poisson's equation states

$$\frac{d^2 V_2}{dx^2} = -\frac{q}{\epsilon_2} N_2(x) \tag{A1}$$

where the parameters have their usual meanings and $N_2(x)$ is the donor concentration in the depletion region of the larger band gap material. Equation (A1) can be expressed in terms of the electric field. Referring to Fig. 3, we find

$$\frac{dE_2(x)}{dx} = \frac{q}{\epsilon_2} N_2(x) \tag{A2}$$

or

$$\frac{dE_2(x)}{dx} = \frac{q}{\epsilon_2} N_d \text{ for } -W_2 < x < 0 \tag{A3}$$

where W_2 is the edge of depletion region. For $0 \leq x \leq d_i$

$$\frac{dE^2(x)}{dx} = 0 \tag{A4}$$

Solving equations (A3) and (A4) and using the boundary conditions that the electric field is continuous and is zero at $x = -W_2$, we obtain

$$E_2(x) = \frac{qN_d}{\epsilon_2}(x + W_2) \quad \text{for} \quad -W_2 < x < 0 \tag{A5}$$

and

$$\epsilon_2 E_2(x) = qN_d W_2 \quad \text{for} \quad 0 \leq x < d_i \tag{A6}$$

These plots are shown in Fig. A1. Integration of the electric field leads to the electrostatic voltage drop in the space charge region of the $Al_xGa_{1-x}As$ layer. Starting with

$$V = -\int E \, dx \tag{A7}$$

we obtain

$$V_{20} = -\int_{d_i}^{0} \frac{qN_d}{\epsilon_2} W_2 \, dx - \int_{0}^{-W_2} \frac{qN_d}{\epsilon_2}(x + W_2) \, dx \tag{A8}$$

where $V_{20} = V(-W_2) - V(d_i^-)$. (see Fig. A2)

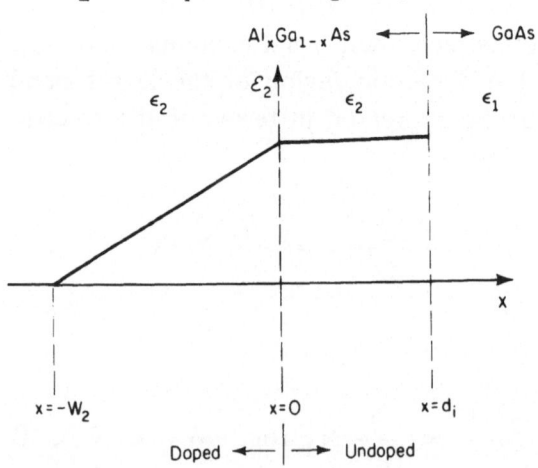

Figure A1 - Electric field vs. distance, x, in the space charge region.

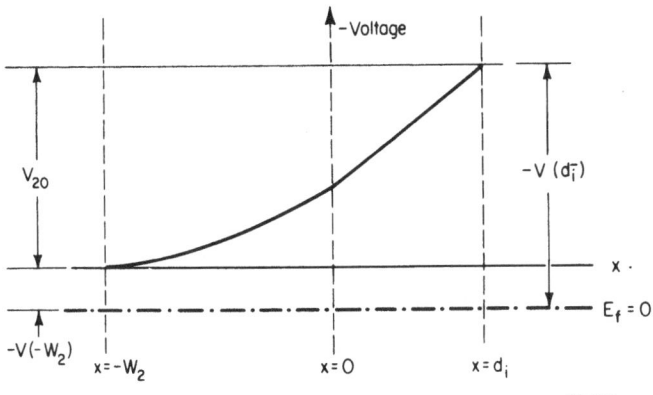

Figure A2 - Potential vs. distance in the space charge region.

From equation (A6),

$$E_2 \mid d_i = E_2(d_i^-) = \frac{qN_dW_2}{\epsilon_2} \qquad (A9)$$

and

$$\epsilon_2 E_2(d_i^-) = qN_d = qn_{so} \qquad (A10)$$

From equation (A8), the depletion depth can be found as

$$W_2^2 + 2W_2d_i - \frac{2\epsilon_2 V_{20}}{qN_d} = 0 \qquad (A11)$$

and thus

$$W_2 = -d_i + \sqrt{d_i^2 + \frac{2\epsilon_2 V_{20}}{qN_d}} \qquad (A12)$$

Using equation (A10) and (A12), and noting $E_2(0) = E_2(d_i^-)$ one finds

$$\epsilon_2 E_2(d_i^-) = -qN_dd_i + \sqrt{q^2N_d^2d_i^2 + 2\epsilon_2 V_{20}qN_d} = qn_{so} \qquad (A13)$$

This expression is identical to equation (32).

LIST OF SYMBOLS

E_i:	Energy in the ith subband
$\hbar$:	Reduced Planck's constant
q:	Electronic charge
m^*:	Effective mass in the conduction band
E_0:	Electric field perpendicular to the heterointerface in the triangular potential well
i:	Integer
E_0:	1st subband energy w.r.t. the bottom of the conduction band
E_1:	2nd subband energy
ϵ_1:	Dielectric constant in GaAs
n(x):	Charge concentration in the triangular potential well
E_1:	Electric field (perpendicular to the heterointerface) at the interface in GaAs
n_{so}:	Equilibrium areal density of two dimensional electron gas
γ_0:	Adjustable parameter relating E_0 to n_{so}
γ_1:	Adjustable parameter relatin E_1 to n_{so}
D:	Two dimensional density of states in the triangular potential well
$V(d_i{}^+)$:	Potential at the heterointerface when approached from the GaAs side
k:	Boltzmann constant
T:	Absolute temperature
V:	Potential depicting the conduction band edge w.r.t. the Fermi level
$\rho(x)$:	Charge density in the quasi space charge region of $Al_xGa_{1-x}As$
ϵ_2:	Dielectric constant of $Al_xGa_{1-x}As$
$N_d{}^+(x)$:	Ionized donor concentration in $Al_xGa_{1-x}As$
n(x):	Electron concentration in the quasi space charge region
N_d:	Total donor concentration in $Al_xGa_{1-x}As$
g:	Degeneracy factor of the donor level in $Al_xGa_{1-x}As$
E_d:	Donor binding energy in $Al_xGa_{1-x}As$
N_c:	Density of states in the conduction band of $Al_xGa_{1-x}As$
N'_d:	Quantity equal to N_d/N_c
g':	Quantity equal to $g \exp(E_d/kT)$
V(0):	Potential at $x = 0$, boundary between the doped and undoped $Al_xGa_{1-x}As$

$E_2(-W_2)$:   Electric field in $Al_xGa_{1-x}As$ at the edge of the depletion region

W_2:   Distance from the doped-undoped boundary to the edge of the depletion region in $Al_xGa_{1-x}As$

y: Quantity equaling exp $qV(-W_2)/kT$

$E_2(0)$:   Electric field at the boundary between the doped and undoped $Al_xGa_{1-x}As$

$E_2(d_i^-)$:   Electric field at the heterointerface in the $Al_xGa_{1-x}As$

d_i: Thickness of the undoped $Al_xGa_{1-x}As$ spacer layer

δ: Correction term arising from the degeneracy of the $Al_xGa_{1-x}As$ (see Equation 25)

ΔE_c: Conduction band edge energy discontinuity at the heterointerface

$V(d_i^-)$:   Potential at the heterointerface on the $Al_xGa_{1-x}As$ side

E_{Fi}:   Fermi level at the heterointerface

ΔE_{Fo}: Asymptotic value of the Fermi level for very small 2DEG concentrations

a: Slope of the Fermi level vs. 2DEG concentration for large concentrations

Δd: Effective distance of the 2DEG from the heterointerface

V_2:   Electrostatic potential across the $Al_xGa_{1-x}As$ if all of the doped $Al_xGa_{1-x}As$ were depleted

d: Total thickness of the $Al_xGa_{1-x}As$ under the gate

d_d: Thickness of the doped $Al_xGa_{1-x}As$ under the gate

V_{p2}:   The voltage necessary to deplete the doped $Al_xGa_{1-x}As$ layer

ϕ_b: Barrier height of the Schottky gate

V_G: Applied gate voltage

V_{off}:   Voltage necessary to deplete the 2DEG

Q: Charge of the 2DEG

$V_{eff}(x)$:   Effective voltage along the channel modulating the 2DEG

$V_c(x)$:   Channel voltage under the gate

I: Drain current

Z: Width of device

$v(x)$:   Electron velocity

μ: Electron mobility

E: Electric field parallel to the heterointerface in the potential well

E_c: Critical field above which velocity saturation occurs

R_s: Source resistance

V_D:	Applied drain voltage
V'_D:	Intrinsic drain voltage
R_D:	Drain resistance
V'_G:	Defined in equation (52)
x:	Distance from the source end of the gate towards the drain
L:	Gate length
I_s:	Drain saturation current
β:	Defined in equation (54)
V_{sl}:	Voltage in the channel under the gate at the onset of velocity saturation
v_s:	Saturation velocity
g_m:	Transconductance
$g_m|max$:	Maximum transconductance
Q_T:	Total charge under the gate
V_{GS}:	Gate-to-source voltage
V_{GD}:	Drain-to-gate voltage
C_o:	Characteristic maximum capacitance under the gate
C_{GS}:	Gate-to-source capacitance
C_{GD}:	Gate-to-drain capacitance
$(V_{po})_{2d}$:	Electrostatic potential equilibrating the 2DEG $(d + \Delta d)$ away
V_{20}:	Electrostatic potential depleting the doped $Al_xGa_{1-x}As$ to a depth equal to W_2

REFERENCES

1. L. Esaki and R. Tsu, Internal Report RC 2418, IBM Research, March 26, 1969.
2. R. Dingle, H.L. Stormer, A.C. Gossard and W. Wiegmann, Appl. Phys. Lett. **37**, 805 (1978).
3. P.J. Price, Annals of Physics **133**, 217 (1981).
4. L.C. Witkowski, T.J. Drummond, C.M. Stanchak and H. Morkoç, Appl. Phys. Lett. **37**, 1033 (1980).
5. L.C. Witkowski, T.J. Drummond, S.A. Barnett, H. Morkoç, A.Y. Cho and J.E. Green, Electron Lett. **17**, 126 (1981).
6. T.J. Drummond, H. Morkoç, and A.Y. Cho, J. Appl. Phys. **52**, 1380 (1981).

7. R.E. Thorne, R. Fischer, S.L. Su, W. Kopp, T.J. Drummond and H. Morkoç, Jap. J. Appl. Phys. Lett. 21, L223 (1982).

8. T.J. Drummond, M. Keever, and H. Morkoç, Jap. J. Appl. Phys. Lett. 21, L65 (1982).

9. H.L. Störmer, R. Dingle, A.C. Gossard, and W. Wiegmann, Solid State Commun. 29, 705 (1979).

10. D.C. Tsui and R.A. Logan, Appl. Phys. Lett. 35, 99 (1979).

11. L.D. Landau and E.M. Lifshitz, Quantum Mechanics (Non-Relativistic Theory) Oxford: Pergamon Press, 74 (1977).

12. D. Delagebeaudeuf and N.T. Linh, IEEE Trans. Electron Dev. ED-28, 790 (1981). The same authors treated the normal MOD-FET case as well, IEEE Trans. Electron Dev. ED-29, 955 (1982).

13. K. Lee, M.S. Shur, T.J. Drummond and H. Morkoç, J. Appl. Phys. 54, 2093, (1983).

14. R.A. Smith, "Semiconductors", Cambridge University Press, Second Edition, p. 83 (1978).

15. T.J. Drummond, H. Morkoç, K. Lee and M.S. Shur, IEEE Electron Dev. Lett. EDL-3, 338 (1982).

16. K. Lee, M.S. Shur, T.J. Drummond and H. Morkoç, IEEE Trans. Electron Dev. 30, March (1983).

17. T.J. Drummond, W. Kopp, M. Keever, H. Morkoç and A.Y. Cho, J. Appl. Phys. 53, 1023 (1982).

18. T.J. Drummond, W. Kopp, H. Morkoç and M. Keever, Appl. Phys. Lett. 41, 277 (1982).

19. H. Morkoç, T.J. Drummond and R. Fischer, J. Appl. Phys. 53, 1030 (1982).

20. T.J. Drummond, R. Fischer, P. Miller, H. Morkoç, and A.Y. Cho, J. Vac. Sci. Technol. 21, 684 (1982).

21. T.J. Drummond, R. Fischer, S.L. Su, W.G. Lyons, H. Morkoç, K. Lee and M.S. Shur, Appl. Phys. Lett. 42, 262 (1983).

22. J.S. Harris and D.L. Miller, private communication.

23. M. Laviron, D. Delagebeaudeuf, P. Delescluse, P. Etienne, J. Chaplart and N.T. Linh, Appl. Phys. Lett. 40, 530 (1982).

24. T.J. Drummond, S.L. Su, W.G. Lyons, R. Fischer, W. Kopp, H. Morkoç, K. Lee and M.S. Shur, Electron Lett. 18, 1057 (1982).

25. T. Mimura, S. Hiyamizu, K. Joshin and K. Hikosaka, Jap. J. Appl. Phys. Lett. 20, L317 (1981).

26. S. Hiyamizu, T. Mimura and T. Ishikawa, Jap. J. Appl. Phys. 21, Supplement 21-1, 161 (1982).

27. P.N. Tung, D. Delagebeaudeuf, M. Laviron, P. Delescluse, J. Chaplart and N.T. Linh, Electron Lett. 18, 109 (1982).
28. M. Niori, T. Saito, S. Joshin, and T. Mimura, presented at ISSCC February 23-25, 1983, New York, NY.

METAL ORGANIC CHEMICAL VAPOR DEPOSITION

J.P. Duchemin, S. Hersee, M. Razeghi, M.A. Poisson

Thomson - CSF Orsay, France

ABSTRACT: In this paper it is shown that the low pressure organometallic chemical vapor deposition (LP-MOCVD) technique of growing semiconductors is generally applicable to most of the III-V compounds that are currently of interest. The principles of the technique and five selected specific applications are detailed: 1) GaAlAs/GaAs quantum well lasers; 2) GaAlAs/GaAs modulation doped structures; 3) Two dimensional electron gas layer at GaInAs/InP interfaces; 4) Low threshold-current density lasers operating at $1.3\mu m$; 5) Multiquantum well GaInAs/InP heterostructures.

1. INTRODUCTION

There is an ever inceasing demand for submicron monocrystalline silicon, gallium arsenide, and other related III-V compounds for the applications of both microwave and optoelectronics devices. Thin GaAs layers are grown for making microwave Schottky diodes or high frequency FET's; thin GaAlAs layers are used to make low current threshold laser diodes, operating in the 0.85μm wavelength range or TEGFET; and thin GaInAsP layers are used for the active layers of the 1.3 - 1.5μm wavelength range laser diodes and in the future a new generation of FET's.

In all these cases, it is essential for the different layers to be clearly defined, i.e., for the substrate-layer or layer-layer interfaces to be sharp, and for the grading in impurity concentration or composition to be confined to a region of the order of 100Å. For alloys, where the lattice parameter depends on composition, the interfaces may need to be even sharper to avoid high interfacial mismatch strains. This is the case, for example, in the GaInAsP/InP system.

Impurity grading at a heterojunction is generally a result of either solid state diffusion or of "autodoping". The latter effect can be particularly significant when the dopant is volatile and the dopant concentration which is sustained by the stagnant layer is high. Compositional grading can also occur by solid state diffusion, although it is generally more likely to be a result of the finite time that is taken to change between two steady state gas compositions.

Use of the lower pressure (LP) MOCVD technique to grow III-V compounds allows one to act on both impurity and compositional grading. At low pressure, the concentration gradient across the stagnant layer is greater for species diffusing out from the substrate, thus less volatile dopant is contained in the stagnant layer, and impurity grading due to autodoping is reduced. Compositional grading is reduced in the LP MOCVD either by stopping the growth completely while establishing the new gas flow required for the subsequent layer or by using the higher gas flow rate, a feature of LP MOCVD, to more quickly establish the new gas composition.

The method of LP MOCVD was originally perfected for the growth of silicon[1] and GaAs[2], and has since been extended in InP[3]

GaAlAs[4] and GaInAsP[5-7] for microwave or optoelectronic applications. For silicon growth, the starting materials is silane (SiH_4), while for the compounds listed above, group III metal alkyls with either "triethyl" or "trimethyl" molecular structure are used as indicated in Table 1.

In common with silicon growth, it was found that for GaAs, InP, and GaAlAs, parasitic reactions in the gas phase (smoke formation) could be eliminated by growing at low pressure. Also, as already described, the effects of autodoping were reduced and much sharper impurity profiles were obtained at layer-layer or substrate-layer interfaces.

A further feature of the LP MOCVD process is that it generally operates at much higher gas velocity than conventional CVD or MOCVD. To achieve the high gas velocity that is necessary for a good uniformity of growth it is better to resort to a low pressure in the reactor than to simply increase the flow of gas at atmospheric pressure. The latter results in an increased cooling effect of the increased mass flow rate of cold gas, whereas by use of a low pressure it is possible to maintain the same mass flow rate while achieving a higher gas velocity. This will tend to improve the thickness uniformity of layer growth, because, without exception the growth of the III-V host lattice was found to be controlled by diffusion of the group III element through the stagnant layer.

TABLE 1. Materials used in the growth of various III-V compounds by LP MOCVD

Materials Grown	Starting Materials Group III	Group V	Substrate
GaAs	$Ga(CH_3)_3$	AsH_3	GaAs
GaAlAs	$Ga(CH_3)_3, Al(CH_3)_3$	AsH_3	GaAs
GaInAs	$Ga(C_2H_5)_3, In(C_2H_5)_3$	AsH_3	InP
GaInAsP	$Ga(C_2H_5)_3, In(C_2H_5)_3$	AsH_3-PH_3	InP
InP	$In(C_2H_5)_3$ PH_3	InP	

The paper discussed the LP MOCVD processes for growing GaAs, GaAlAs, InP, GaInAs, GaInAsP materials or heterostructures and gives some examples of application. We will start the paper in trying to explain how a MOCVD reactor works. Because the MOCVD technique is similar to the silicon epitaxy we will report on the characterization of a silicon reactor.

2. GENERAL MECHANISMS OF A LP MOCVD TECHNIQUE

2.1 Analogy with the Silicon Epitaxy

Certain epitaxial growth methods, such as MBE which take place under ultrahigh vacuum permit the characterization of the structure (RHEED and LEED) and the chemical composition (AUGER electrons) of the surface during growth. Relationship can then be established between the properties of the growing surface and the electric and crystallographic properties of epilayers.

The surface of epilayers grown by chemical vapor deposition (CVD) cannot be characterized by the same method. However, in many cases the electric and crystallographic properties of epilayers are related to the motionless gas layer called boundary layer located next to the growth surface.

In the section, it will be shown how samples of the boundary layer can be probed as a function of the distance above the growth surface and analyzed by gas phase chromatopgraphy in the case of silicon epitaxy in order to measure concentration and concentration gradient of pertinent chemical species close to the growth surface. It will be shown that this analysis allows us to explain the behavior of the growth rate as a function of the growth parameters. At last, general results will be used to explain the different regimes of impurity incorporation in the case of the CVD epitaxy of Si, GaAs and InP.

2.1.1. Mechanism of the Deposition - The gas phase of a chemical vapor deposition reactor has been characterized in several ways: V. Ban[8-12] used gas phase chromatography. He was able to analyse quantitatively most of the stable and unstable species of the gas phase by using a specially designed experimental device for taking small samples of gas inside or outside the boundary layer. T. Sedgwick[13,15] used the Raman scattering method to analyse the composition of the gas phase in the growth area of an Si growth reactor. He was able to get to know the variation in composition of the gas phase as a function of the distance above the growth surface or along this latter by using an optical focalisation device placed outside the reactor.

We have made the same investigations in a vertical reactor usually called pancake reactor.[16] In such a system, the boundary layer is notably thicker than in a horizontal reactor (20 mm instead of 4 mm) which makes it easier to observe the phenomena which take place in the gas phase in the neighborhood of the growth surface. By using a special device shown in Figure 1, it has been possible to take samples of gas at various distances above the growth surface. These samples were then submitted to chemical analysis: gas phase chromatography, mass spectroscopy, infrared spectroscopy, to define the stable species present in the gas phase.

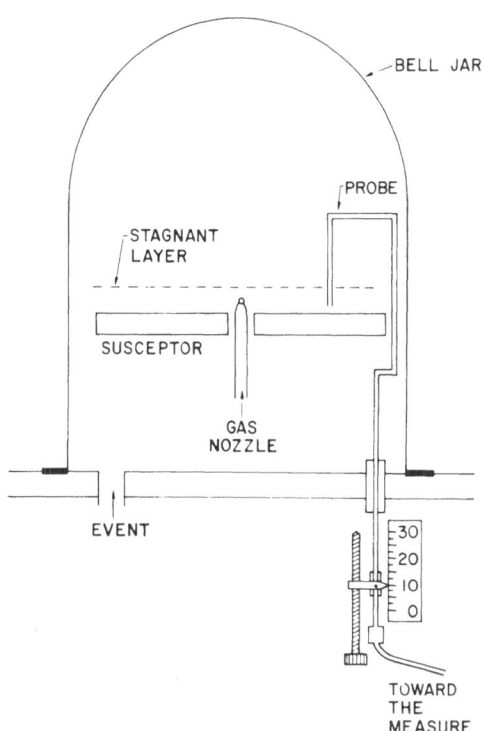

Figure 1 - Probe used to take samples of the gas phase in a CVD reactor.

The observations made by these various methods have shown that the composition of the gas phase was uniform in the convective gas area which is maintained at a lower temperature than the growth temperature and which therefore remains, for this reason, relatively stable. In the neighborhood of the growth surface, there is therefore a gas layer whose characteristics change gradually from those of the convective gas

phase to those of the interface growth surface - gas phase. This transition layer is called stagnant layer or boundary layer. The temperature
of this layer and its composition are function of the distance above the
growth surface.

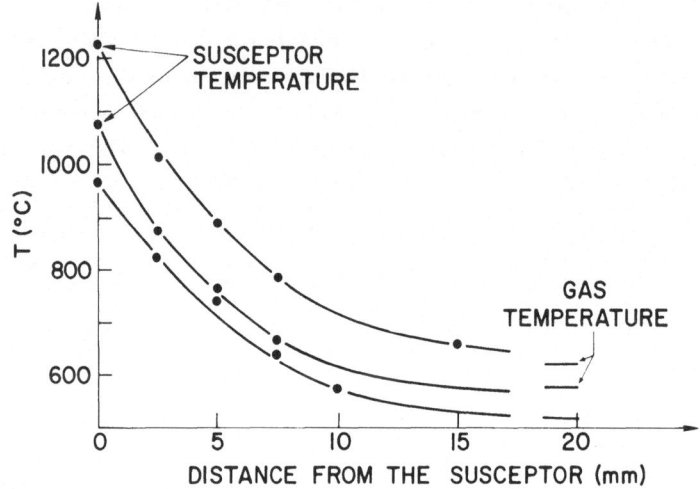

Figure 2 - Temperature profile of the gas phase in a CVD reactor
versus distance above the susceptor.

An example of the temperature profile in the neighborhood of
the growth surface is given in Fig. 2. The temperature of the growth
surface was obtained by means of an optical pyrometer and then the
temperature gradient between two points situated at various distances
above the growth surface was linked to the variation in concentration
of a small quantity of argon inserted in the hydrogen of the reactor.
This variation in concentration is a result of the thermal separation of
species of different molecular weights under the effect of a temperature
gradient.

For various temperature on the growth surface, Figure 3a shows
the concentration gradient of the $SiCl_4$ source which diffuses towards
the growth surface. Figure 3b shows the concentration gradient of the
HCl compound formed on the growth surface which diffuses in the
opposite direction to the convective gas phase. At a relatively low
temperature, T = 960°C, the mole fraction of $SiCl_4$ on the growth
surface represents a notable fraction of that contained in the gas phase
which shows that the rate of decomposition reaction on the surface is a

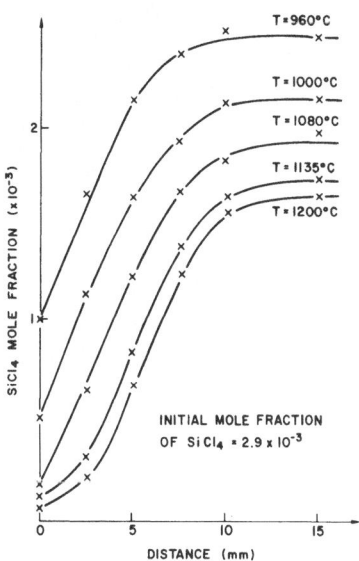

Figure 3a - Mole fraction of SiCl$_4$ close to the growth surface during the silicon growth occurring in a pancake reactor for different growth temperatures.

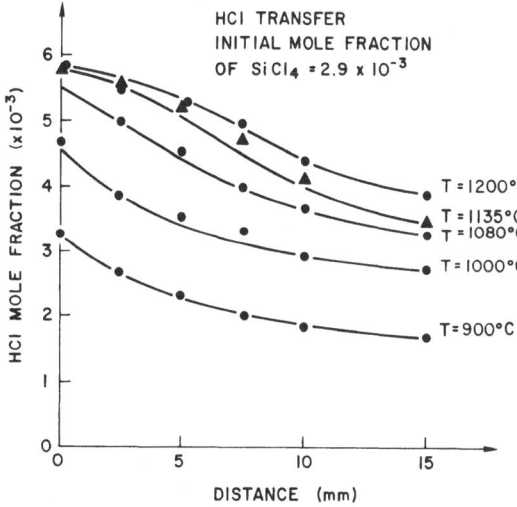

Figure 3b - Mole fraction of HCl close to the growth surface during the silicon growth occuring in a pancake reactor for different growth temperatures. SiCl$_4$ diffuses from the gas phase to the growth surface. HCl is formed on the surface and diffuses toward the gas phase. At high temperature the shape of the curves changes which shows that the starting compounds has been decomposed before reaching the growth surface.

slow process as is also that of the diffusion through the boundary layer. When the temperature rises, T = 1000°C and T = 1080°C, the rate of reaction on the surface becomes more and more rapid and progressively the concentration of the SiCl₄ source lessens on the surface. The rate of reaction increases and the concentration of SiCl₄ in the convective gas phase also lessens.

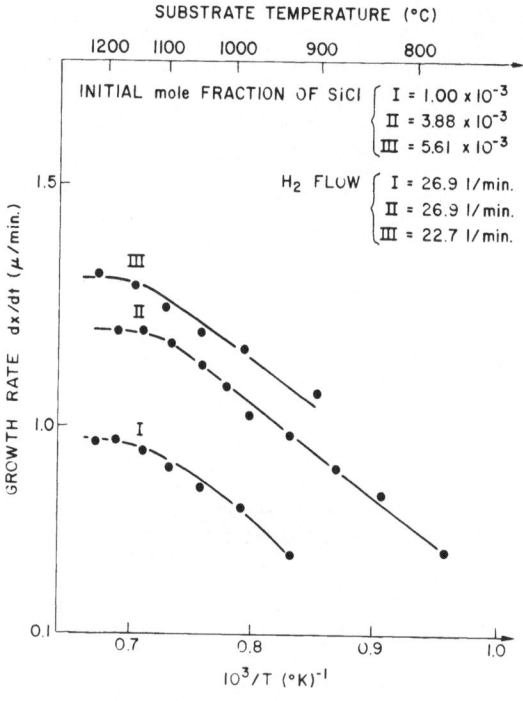

Figure 4 - Growth rate as a function of the temperature for a given SiCl₄ mole fraction in the case of the silicon growth occurring in a pancake reactor.

It can be seen in Figure 4 that the growth rate effectively increases when the temperature changes from 960° to 1000°C then to 1080°C. Beyond 1100°C, the growth rate no longer increases and the concentration of SiCl₄ on the surface is very weak. The kinetics of growth are then controlled by the diffusion rate of SiCl₄ through the boundary layer. At temperature close to 1100°C, the aspect of the profile of SiCl₄ concentration undergoes a transformation showing that a large part of the initial compound undergoes chemical transformation before reaching the surface. The SiCl₄ compound is transformed into SiCl₄ following the reaction

$$SiCl_4 + H_2 \rightarrow 2HCl + SiCl_2$$

2.1.2. Mechanisms of the Incorporation of impurities. The preceding observations have therfore shown that the growth rate was controlled by the properties of the boundary layer. Let us now see how the boundary layer also controls the kinetics of incorporation of impurities during the growth.

The compounds which are used as source of impurities are present in the gas phase in a concentration too weak (10^{-6} to 10^{-10}) to be observed by chemical analysis. The study of the incorporation mechanisms of the impurities must therefore consist, in most cases, of a phenomenological analysis of the results.

An experiment made with a compound being used as a source for the growth does however let us imagine a general mechanisms which could explain all the doping behavior observable.[16,17]

In this experiment, we sought to effect a silicon growth using a $SiCl_4$ source at the concentration 10^{-3} on a growth surface non initially covered with silicon in presence of a fairly high concentration of HCl (10^{-2}). No silicon deposit was observed on the growth surface but chemical analysis of the gas phase showed that the $SiCl_4$ compound diffused towards the growth surface on which the silicon should have been deposited. This silicon was immediately re-attacked by the HCl compound and move away from the growth surface in the form of the $SiCl_4$ compound (Figure 5).

By analogy, the following doping mechanism can be imagined:

- the compound being used for the doping is inserted into the reactor where its mole fraction is Xg in the convective gas phase situated outside the boundary layer;

- the "g" compound diffuses in the boundary layer towards the growth surface without undergoing decomposition in the gas phase (like $SiCl_4$);

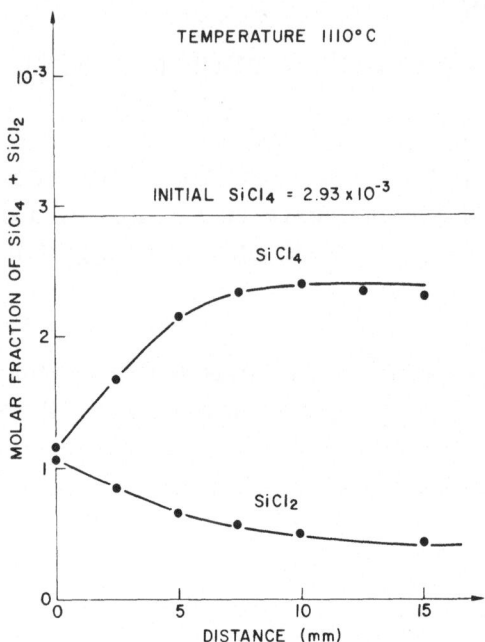

Figure 5 - $SiCl_4$ and $SiCl_2$ mole fraction profiles close to the growth surface. In this experiment $SiCl_4$ (10^{-3}) and HCl (10^{-2}) were initially introduced in the gas phase. No deposition was observed. Nevertheless, $SiCl_4$ diffuses toward the surface where it was decomposed. The free silicon was immediately etched by HCl and diffused back toward the gas phase.

- the "g" compounds, on the growth surface or in its immediate neighborhood, undergoes a chemical transformation, changing it into "g'" (as $SiCl_4$ is changed into $SiCl_2$);

- the "g'" compound contained in the gas in contact with the surface is in equilibrium with the doping element present in the solids according to Henry's law;

- the excess "g'" compound diffuses towards the convective part of the gas phase where it is then drawn along by the general circulation of gases outside the reactor (Figure 6a)

Some border-line cases can then be observed: in Figure 6b, the "g" compound is totally decomposed on the surface. Its decomposition rate is limited by its diffusion rate through the boundary layer. A very

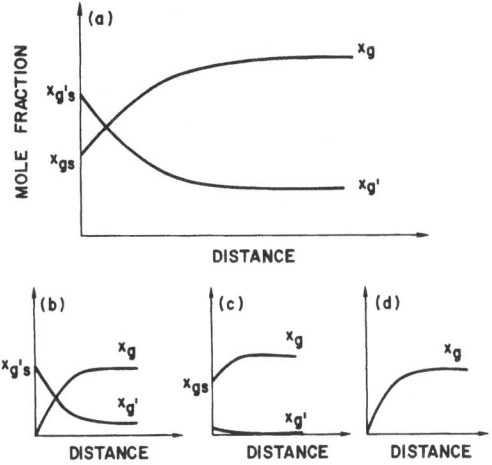

Figure 6 - (a) General mechanisms explaining the kinetics of the incorporation of impurities during material deposition in a cold wall CVD (found from results of Fig. 5). The compound g which contains the doping element diffuses toward the surface on which it is decomposed to form the compound g'. The compound g' is in equilibrium on the surface with the growing film. The compound g' which is not incorporated inside the growing film diffuses toward the gas phase. Three limit cases can be observed. (b) The decomposition rate of g on the surface is fast, (it is limited by the diffusion rate of g) and a small quantity of g' is incorporated inside the growing film the larger quantity diffuses toward the gas phase. In that case, the impurity level does not depend on the growth rate but it decreases when the temperature increases. That is the case with the phosphorus incorporation in the silicon growth or the zinc incorporation in the GaAs growth using the Ga $(CH_3)_3AsH_3$ system or the sulfur or zinc incorporation in the InP growth using $In(C_2H_5)-PH_3$ system. (c) The decomposition rate of g on the surface is slow and the larger quantity of g' formed on the surface is incorporated inside the growing film. The impurity level is then inversely proportional to the growth rate and increases when the temperature rises. That is the case with the boron incorporation in the silicon growth or the case with the silicon or germanium incorporation in the GaAs growth using the $Ga(CH_3)_3-AsH_3$ system. (d) The decomposition rate of g on the surface is fast (it is limited by the diffusion rate) and the larger quantity of g' formed on the surface is incorporated inside the growing film. The impurity level is then inversely proportional to the growth rate and does not depend on temperature. That is the case with the germanium incorporation in the silicon growth.

small quantitity of "g'" compound is incorporated in the solid which is growing, the greater part of the "g'" compound formed, diffuses towards the gas phase. In this case, the incorporation of impurities decreases exponentially with the temperature because of the displacement of the equilibrium on the surface between the "g'" compound and the growth surface. It is the case, for example, with the doping of silicon with phosphorus by using PH_3 (Figure 7).[17]

In Figure 6c, the decomposition rate of the "g" compound is a slow step. The greater part of the compound formed g' is incorporated in the crystal which is growing. The concentration of impurities in the solid deposited is then inversely proportional to the growth rate and increases exponentially with the temperature according to the same law

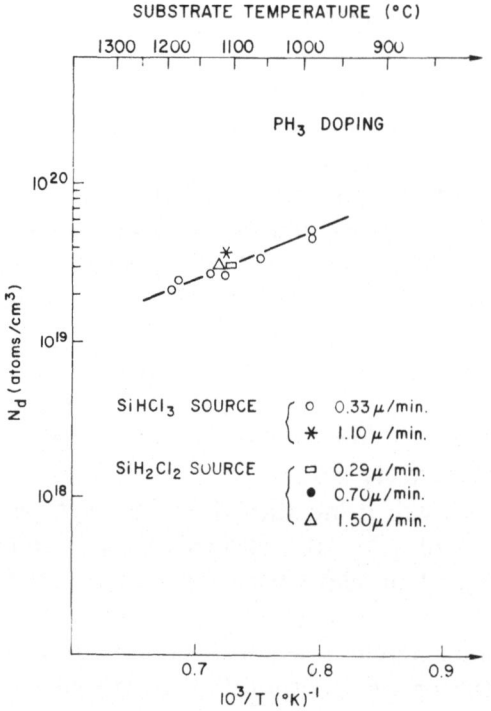

Figure 7 - Phosphorus incorporation during the silicon growth. The free carrier concentration was plotted as a function of $1/T$. The free carrier concentration does not depend on the growth rate but is decreased when the temperature rises because phosphorus becomes more volatile at high temperature.

as the kinetics of decomposition of the "g" compound. It is the case with the doping of silicon with boron by using B_2H_6 (Figures 8a and 8b).[17]

In Figure 6d, the rate of diffusion through the boundary layer of the "g" compound is the slowest step compared to its rate of decomposition which is rapid. The greater part of the "g'" compound is incorporated in the crystal which is growing. The concentration of impurities in the solid does not depend on the temperature. It is inversely proportional to the growth rate. This case represents the incorporation of germanium in silicon using GeH_4 (Figures 9a, b).[17]

2.2 MOCVD Growth of III-V Alloys

The MOCVD technique is similar to the silicon epitaxy using $SiCl_4$, $SiHCl_3$ or SiH_2Cl_2. The growth of III-V compounds is then limited by the diffusion through the boundary layer of the organometallic compounds of gallium $[Ga(CH_3)_3$ or $Ga(C_2H_5)_3]$, aluminum $Al(C_2H_5)_3$ or $Al(CH_3)_3$ or indium $In(C_2H_5)_3$. At the growth surface the starting source is pyrolised the group III elements react with the group V elements to grow III-V compounds. The alkyles (CH_4 or C_2H_6) formed at the surface diffuse away from the surface toward the gas phase in the same way as HCl in the silicon epitaxy according to the following chemical reactions:

$$Ga(CH_3)_3 + AsH_3 \rightarrow GaAs + 3CH_4$$
$$In(C_2H_5)_3 + PH_3 \rightarrow InP + 3C_2H_5$$
$$Al(CH_3)_3 + AsH_3 \rightarrow AlAs + 3CH_4$$

In the same way ternary or quaternary alloys can be grown by mixing different starting materials:[18-26]

$$1-xGa(CH_3)_3 \quad +xAl(CH_3)_3 \rightarrow Ga_{1-x}Al_xAs$$
$$(1-x)Ga(C_2H_5)_3xIn(C_2H_5)_3 + yPH_3 + (1-y)AsH_3$$
$$\rightarrow Ga_{1-x}In_xAs_{1-y}P_y$$

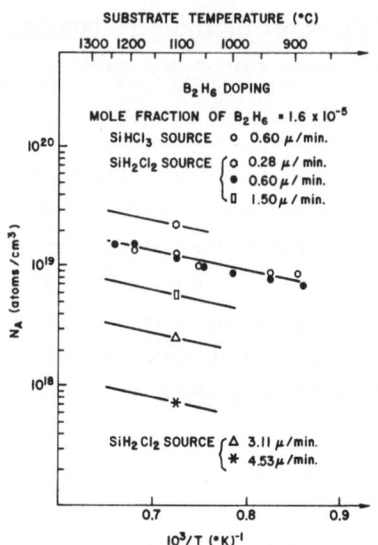

Figure 8a - Boron incorporation during the silicon growth. The free carrier concentration was plotted as a function of the growth rate. The free carrier is inversely proportional to the growth rate.

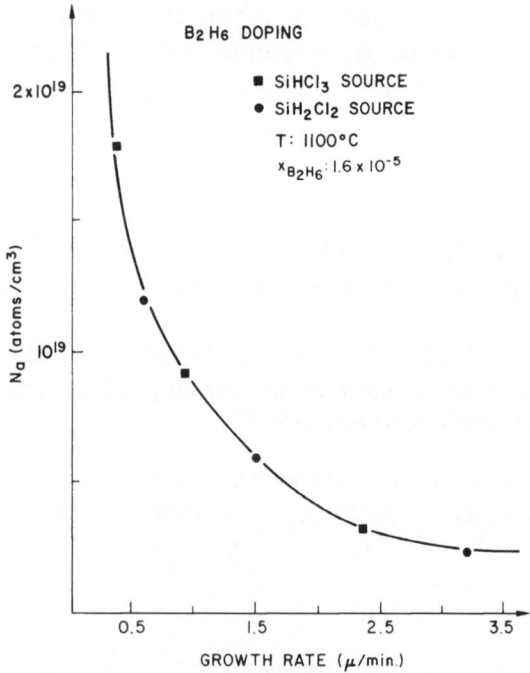

Figure 8b - Boron incorporation during the silicon growth. The free carrier concentration was plotted as a function of $1/T$. The boron incorporation is controlled by B_2H_6 decomposition rate.

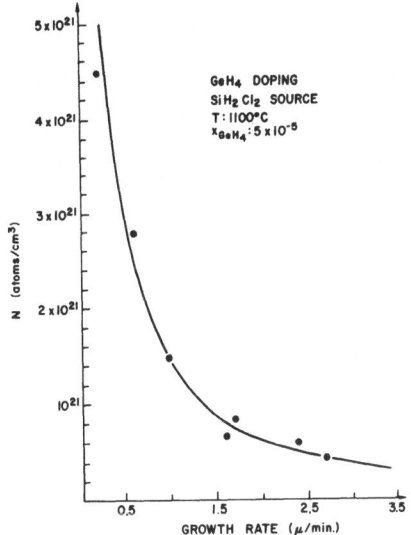

Figure 9a - Germanium incorporation during the silicon growth. The impurity incorporation was plotted as a function of the growth rate. The impurity incorporation is inversely proportional to the growth rate.

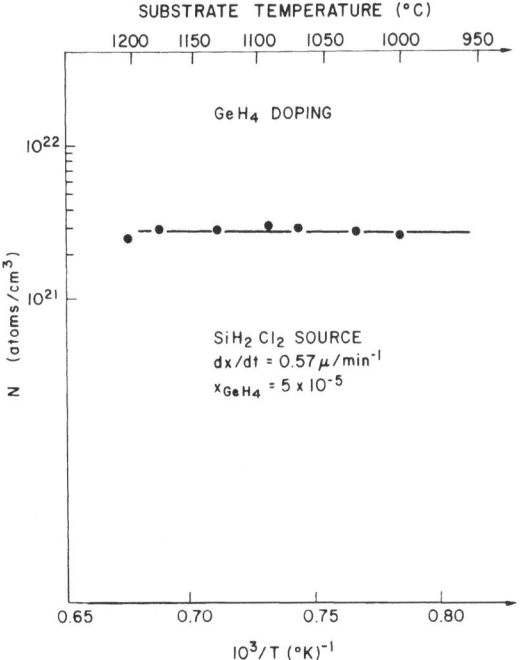

Figure 9b - Germanium incorporation during the silicon growth. The impurity concentration was plotted as a function of $1/T$. The germanium incorporation is controlled by GeH_4 diffusion rate.

In all cases the growth rate is controlled by the sum of the diffusion rates of group III elements through the boundary layer and the composition is proportional to the ratio of the same group III elements in the gas phase for alloys such as GaAlAs, GaInAs or AlInAs. For the group V elements there is a balance between the gas phase and the solid.

In every case, the growth rate is insensible to the temperature (Figure 10a), it does not depend on the partial pressure of arsine (Figure 10b), this latter must however be at least ten times greater than that of gallium to obtain smooth layers. The growth rate increass on the other hand proportionally to the mole fraction of the organic compound of gallium. All these points show that the growth rate is only controlled by the diffusion rate of the organic compound of gallium through the boundary layer.[23,24]

2.2.1. Growth of GaAs by the Organometallic Method. The GaAs layers obtained by this method can be intentionally doped to N type with Si or Ge by using SiH_4 or GeH_4. The kinetics of incorporation of these elements then follows the mechanisms corresponding to the border-line case of Figure 6c. The level of free carriers is inversely proportional to the growth rate and increases exponentially with the temperature. It is insensible to the partial pressure of AsH_3.

Bass[25] has shown that the GaAs layers obtained by this method could be doped P type with zinc. In this case, the level of free carriers decreases exponentially when the temperature rises, it increases when the partial pressure of AsH_3 increases and it does not depend on the growth rate as first approximation. The behavior corresponds to the border-line case described in Figure 6b.

2.2.2 Growth of InP by the Organometallic Method. In the same way, epitaxial layers of InP can be grown at atmospheric pressure or under reduced pressure by using an organic compound of indium $In(C_2H_5)_3$ and PH_3.[26] Unlike the $Ga(CH_3)_3$ - AsH_3 system, in the $In(C_2H_5)_3$ - PH_3 system it is necessary to pyiolyse the PH_3 compound before inserting into the reactor because PH_3 is too stable.

In this system, the growth rate is still insensible to the temperature, it does not depend on the partial pressure of phosphine as long as this latter is sufficient to ensure a good quality growth. The partial

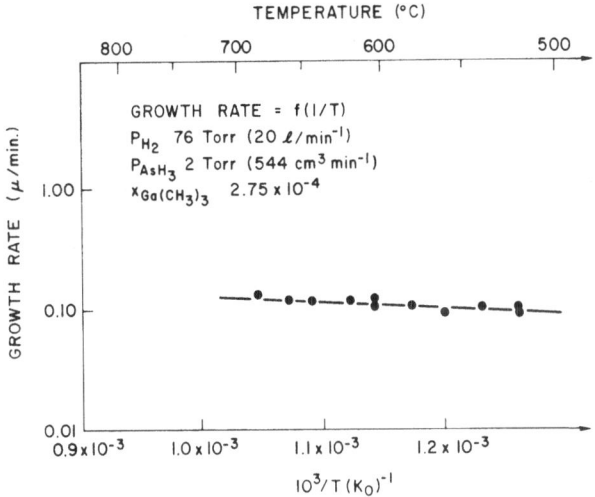

Figure 10a - GaAs growth rate as a function of the growth temperature.

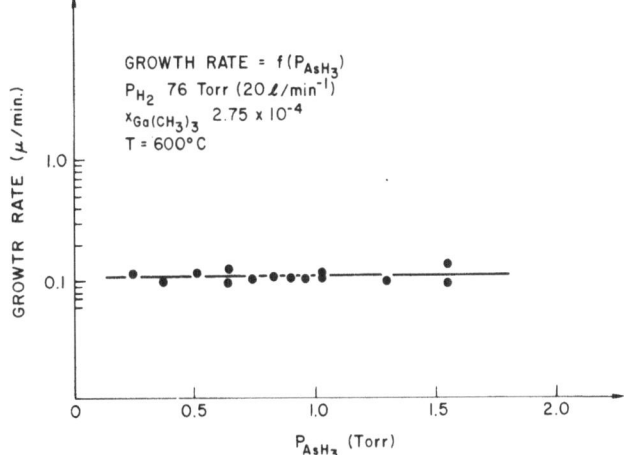

Figure 10b - GaAs growth rate as a function of the AsH_3 partial pressure. The growth rate of GaAs grown with the $Ga(CH_3)_3$-AsH_3 system is only controlled by the $Ga(CH_3)_3$ diffusion rate. It does not depend neither on temperature nor on AsH_3 partial pressure.

pressure of PH_3 must be equal to at least ten times that of $In(C_2H_5)_3$ The growth rate is linearly proportional to the mole fraction of triethylindium. All these points show that the growth rate is only controlled by the diffusion rate of the organic compound of indium through the boundary layer.

The InP layers can be intentionally doped N type by using H_2S. They can be intentionally doped P type by using diethyl-zinc. In both these cases, the level of free carriers in insensible to the growth rate, in first approximation, on the other hand it decreases exponentially when the temperature rises. In the case of doping with zinc, the level of free carriers increases when the quantity of PH_3 increases whereas the opposite effect is observed with sulphur. This behavior corresponds to the border-line case described in Figure 6b.

2.2.3. Growth of GaInAsP by the organometallic method. In the same way, epitaxial layers of GaInAsP have been grown by using under reduced pressure an experimental equipment previously used for growing InP. The gallium and indium sources were respectively the triethyl-gallium and triethyl-indium compounds whereas the sources of the elements of group V were AsH_3 and PH_3 but it seemed more interested to pyrolyse PH_3 before its insertion in the reactor as for the growth of InP.

This system behaves like the preceding system. The growth rate is insensible to the temperature and does not depend on partial pressures of arsine or phosphine which however must be greater than the partial pressures of the gallium and indium compounds. The growth rate is only controlled by the sum of the diffusion rates of the triethyl-gallium and of the triethyl-indium through the boundary layer. The composition in compound III is proportional to the ratio of the quantity of triethyl-gallium on the quantity of triethyl-indium which diffuses in the boundary layer. The composition in compound V is on the contrary proportional to the chemical reactivities of the AsH_3 and PH_3 compounds. AsH_3 is then about five times more reactive than PH_3. A schematic diagrams showing the principle of a MOCVD reaction is shown in Figure 11.

3. SOME APPLICATIONS OF THE LP MOCVD TECHNIQUES

In order to demonstrate that the MOCVD is capable of producing sophisticated heterostructures exhibiting sharp interfaces and leading to high characteristics devices. We have selected five examples:

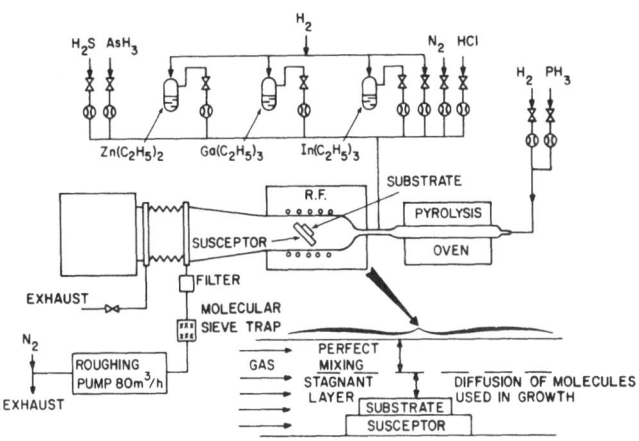

Figure 11 - Schematic diagram showing the principle of a MOCVD reactor.

- GaAlAs/GaAs quantum well laser diodes,

- High mobility selectivity doped GaAs/GaAlAs structures,

- Very low threshold current density laser diodes operating at 1.3 μm

- Two dimensional electron gas in a InGaAs-InP heterojunction,

-GaInAs/InP multiquantum well structures.

3.1 Very Low Threshold GaAs/AlAs GRIN-SCH
(Graded Refractive Index Separate Confinement Heterostructure)

3.1.1 Experimental details. The GRIN-SCH structure (see Figure 12) was grown in a horizontal low-pressure OM-VPE reactor as described in a previous publication[27] using the work of Tsang,[28,29] broad area lasers were fabricated by evaporating Au/Zn and Au/Ge contacts, annealing, then cleaving and sawing. The lasers were then tested, unmounted, under pulsed conditions at a pulse repetition rate of 10^4 Hz with a pulse length of 100 ns (duty cycle 0.1%).

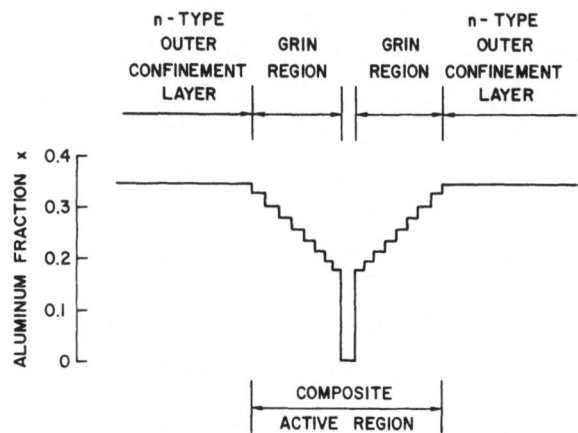

Figure 12 - Graded refractive index separate confinement heterostructure.

3.1.2.   Broad area laser characteristics.   The excellent uniformity of device characteristics demonstrated that the layer thickness remained virtually constant over the slice area in spite of the thinness of the active layer (L_z = 60Å) For example, on chips cleaved from the same bar the standard deviation in the lasing threshold current density was only 4%. The layer uniformity was further demonstrated by the experimental result that the threshold current density continued to fall with lasing cavity length even for chips as long as 1788μm (see Figure 13). Extrapolation of the data in Figure 12 yields a horizontal intercept $\bar{\alpha}$ of 16 cm^{-1} after expr. 2 shown inset in Figure 13.

The positive enhancement of carrier confinement offered by the GRIN region can be observed by comparing the value T_0 in simple SCH and GRIN-SCH structures. The value of T_0 is obtained from the expression

$$T_0 = \frac{T_1 - T_2}{\log{}_e(I_1/I_2)}$$

where T_1 and T_2 are temperatures and I_1 and I_2 are the threshold currents at temperatures T_1 and T_2.

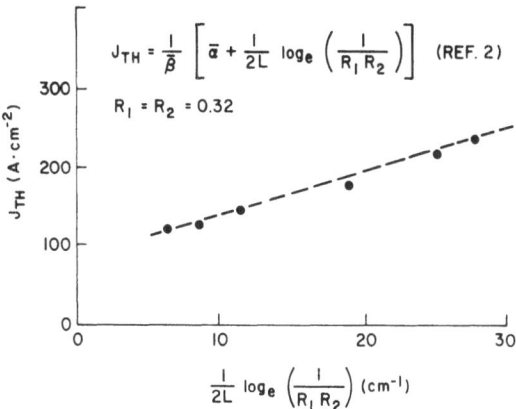

Figure 13 - Variation of threshold current density with the reciprocal of lasing cavity length.

SCH structure shown in Figure 14 is approximately 60K. For the GRIN-SCH structure shown in Figure 12, the value of T_0 is 159K.

The effect of the GRIN region is also seen in the variation of threshold current density (J_{TH}) with quantum well thickness. For the simple SCH structure of Figure 14, the optimum value of L_z is 140Å, and at 60Å the value of J_{TH} increases again. For the GRIN-SCH structure the value of J_{TH} decreases with L_z and still decreases at L_z = 60Å.

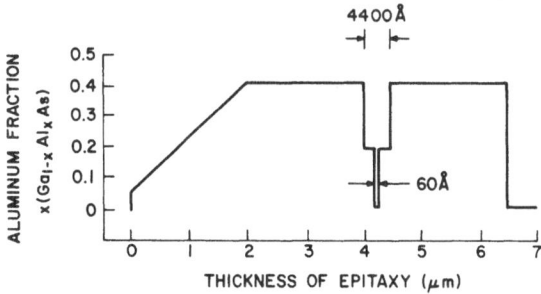

Figure 14 - Simple separate confinement heterostructure.

A high aluminum fraction (x = 0.6) was used in the outer confinement layers of this GRIN-SCH structure. Nevertheless, the forward voltage at threshold (V_s) was only slightly higher than in normal devices, V_s = 1.9V, compared to $1.2V < V_s < 1.5V$, and no series resistance problems were observed. We attribute these results to the low electrical compensation ratio which we now obtain for n-type GaAlAs as a result of eliminating all sources of water and oxygen from the reactor.

The theoretically predicted lasing wavelength[33] for recombination between an "n=1" conduction band electron and a heavy hole, in a 60Å quantum well with a barrier height of 250 meV, is 844 nm. There is a discrepancy between this value and the measured wavlength, which varied between 856 nm and 855 nm, depending on the injection level.

A possible explanation of this discrepancy is that the quantum well thickness is 80Å rather than 60Å, due either to a timing error in the 12 growth period of the well or to a change of growth rate for very thin layers. The discrepancy may also be due to recombination at an acceptor impurity level, which would involve an acceptor level at approximately 60 meV above the valence band.

We intend to resolve this point in the near future by studying the spectral behavior of single quantum wells under optical excitation.

As with the previously reported GRIN-SCH structure[27-29] this device exhibits a high internal quantum efficiency (e_{int}) of between 80% and 100% as predicted by extrapolation of the data in Figure 15.

Conclusions: Further optimization of the GRIN-SCH single quantum well laser structure has resulted in broad-area lasers having the lowest ever threshold current densities of 232 A/cm^{-2} at a cavity length of 413μm and 121 A/cm^{-2} at 1788μm. The device has good thermal and electrical properties and a high internal quantum efficiency; thus we are optimistic that very low threshold buried-heterostructure stripe lasers can be fabricated from this type of structure.

3.2 <u>High Mobility Selectively Doped GaAs/GaAlAs</u> Structures

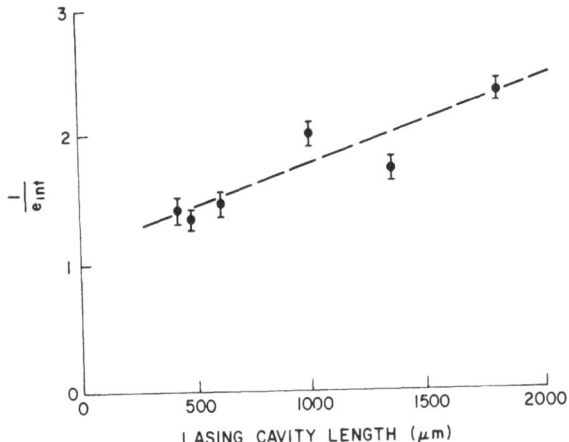

Figure 15 - Variation of reciprocal of differential quantum efficiency $(1/e_{int})$ with lasing cavity length.

3.2.1. Introduction. Since the demonstration of mobility enhancement in a modulation-doped multiheterojunction structure,[34] the transport properties of carriers confined at a heterointerface has become an area of much interest. Recently a mobility of 150 000 cm^{-2}V^{-1}s^{-1} (at 4K) was reported[35] for a selectively doped GaAs/GaAlAs heterojunction grown by MBE. The high-mobility interface has been incorporated into FET-type structures, and it has been demonstrated that in the case of the TEGFET[36] a higher speed and lower power consumption can be obtained even at room temperature.

Although most workers have used MBE to produce the heterojunctions, it is evident from studies of quantum well lasers[37-39] that the OM-VPE growth technique can also grow the very abrupt interfaces that are required. Two-dimensional electron gases (2DEG) have been observed at OM-VPE grown GaAs/GaAlAs heterojunctions[39,40] and GaInAs/InP heterojunctions[41,42] although in the case of GaAs/GaAlAs system the mobilities obtained were lower than those reported for MBE-grown structures.

In this letter we report on a preliminary study of the growth of selectively doped GaAs/GaAlAs structures by OM-VPE at reduced pressure. The structures exhibited a confined 2DEG and the number of confined carriers which transferred from the GaAlAs was varied between 4×10^{11} and 10^{12} cm^{-2}. In the best sample a mobility of 162 000 cm^{2}V^{-1}s^{-1} was obtained at 2K for 5×10^{11} cm^{-2} confined carriers.

3.2.2   Experimental details.   Following the previous MBE work we grew the selectively doped GaAs/GaAlAs structure shown in Figure 16 using the horizontal geometry reduced-pressure OM-VPE system that has been described previously.[38] The selectively doped heterojunction was grown after a 0.5μm to 3μm-thick undoped GaAs buffer layer, and the spacer thickness t_s was varied between 100Å and 200Å. The structures were grown continuously at a rate of 270Å min^{-1} (for GaAs) at 0.3 Atm and at a temperature of 620°C on chromium-doped semi-insulating and "undoped" semi-insulating GaAs substrates. The residual doping level of the undoped layers was adjusted by varying the V/III ratio during growth. The structures containing an n-type GaAs buffer layer were grown continuously at a constant V/III ratio which gave a residual p-type doping level in the GaAlAs spacer layer ($N_A - N_D \simeq 10^{16}$ cm^{-3}) and an n-type GaAs buffer layer ($N_D - N_A \simeq 3 \ 10^{15}$ cm^{-3}). In the samples containing a p-type GaAs buffer layer ($N_A - N_D < 5 \times 10^{15}$ cm^{-3}), it was necessary to interrupt the growth at the heterojunction while the V/III ratio was changed, in order to maintain a low residual acceptor concentration in the GaAsAl spacer layer.

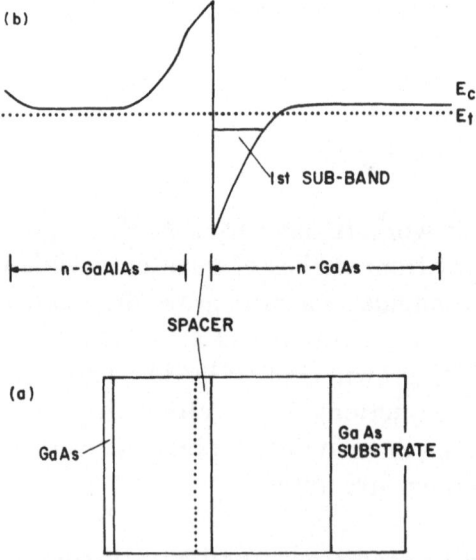

Figure 16 - (a) Selectively doped heterostructures. (b) Conduction band edge diagram of heterojunction.

A final layer of GaAs, 50Å thick, was grown on the structures to facilitate the production of ohmic contacts, which were made by annealing an evaporated film Au-Ge at 430°C for 1 min. Hall effect and Shubnikov de Haas (SdH) measurements were performed on Van der Pauw "clover leaf" samples at 2K. The magnetic field was provided by a superconducting coil and could be varied between 0 and 8 T.

3.2.3. Results. The characteristics of the selectively doped heterojunctions that have been studied are given in Table 2. The "spacer" layer of undoped GaAlAs, which was interposed between the 2DEG and the ionized impurities, was varied between 100Å and 200Å while the donor concentration in the doped GaAlAs layer was within the range 5×10^{17} cm^{-3} to 10^{18} cm^{-3}. We investigated modulation-doped structures with both p-type and n-type buffer layers, and demonstrated that the higher mobility values were related to a two-dimensional electron gas (2DEG) by studying the angular dependence of the Shubnikov de Haas (SdH) oscillations (see Figure 17). In the SdH measurement, each oscillation in resistance indicates a Landau level in the 2DEG. For a fixed angle θ, where θ is the angle between the magnetic field (B) and the perpendicular to the plane of the 2DEG, the separation between these levels was proportional to B^{-1}, giving the straight line dependence shown in Figure 17. As θ was increased from zero the separation between the levels increased following the expected $(\cos\theta)^{-1}$ dependence until $\theta = 90°$ no SdH oscillations were observed.

Sample 235 was typical of our structures having a p-type (p < 5×10^{15} cm^{-3}) buffer layer. The poorer mobilities that were measured for this type of structure were believed to be related to the interrupted growth, and were probably related to a contamination or higher defect density at the heterojunction.

The charge transfer in structures having an n-type buffer was found to be related to the doping level in the n-GaAlAs layer, and fell into one of the following three categories: (a) no transfer under dark conditions (sample 229); (b) one electron subband populated (samples 236, 237); (c) two electron subbands populated (sample 228). In structure 229, which had a doping level of 5×10^{17} cm^{-3}, no 2DEG was observed when the sample was tested under dark conditions. After exposure to LED light, however, we observed a 2DEG with a poor mobility (20 000 $cm^2V^{-1}s^{-1}$) at 2K and a charge transfer 4×10^{11} cm^{-2}

TABLE 2   Parameters of the selectively doped GaAs/GaAlAs structures.

Sample number	GaAs buffer layer	AlGaAs			2 K	
		Al	Doped concentration	Undoped "spacer" interface layer	Mobility	Electron transfer
	m	%	cm^{-3}	$\mathring{A}$	cm^2/Vs	cm^{-2}
228	3(n-type)	30	1.1×10^{18}	100	40 000	1.5×10^{12} 2 sub-band
229	3(n-type)	30	5×10^{17}	100	20 000 after light exposure	4×10^{11} 1 sub-band
235	0.5(p-type)	30	5×10^{17}	100	11 000	4×10^{11} 1 sub-band
236	3(n-type)	30	8×10^{17}	150	115 000	7×10^{11} 1 sub-band
237	3(n-type)	30	8×10^{17}	200	160 000	5×10^{11} 1 sub-band

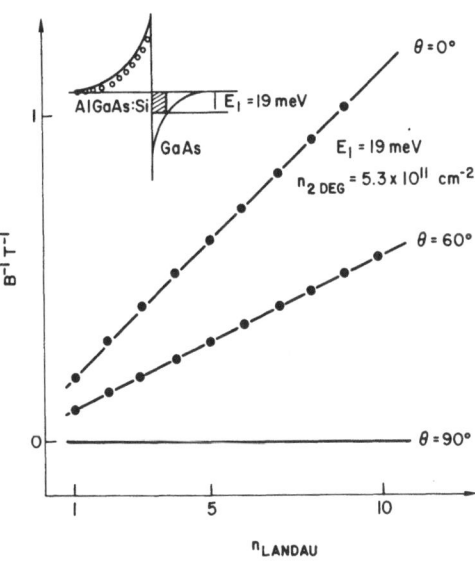

Figure 17 - Reciprocal field at maxima of shubnikow de Haas oscillation against Landau quantum number of different values of θ (sample 237).

which persisted when the light source was removed. Other workers[43] working with MBE-grown materials, have observed this persistent photoconductivity effect and have attributed it to the ionization of electrons from deep traps in the GaAlAs layer.

The typical Hall mobilities at 300K and 77K for samples with an n-type buffer layer were 6000 $cm^2V^{-1}s^{-1}$ and between 60 000 and 75 000 $cm^2V^{-1}s^{-1}$, respectively. These values contained contributions from conduction in the GaS and GaAlAs layers adjacent to the 2DEG, as carrier "freeze-out" was incomplete at these temperatures. Ignoring conduction in the GaAlAs layer and using previously measured values for the doping level ($5x10^{15}$ cm^{-3}) and bulk mobility at 77K (60 000 $cm^2V^{-1}s^{-1}$) for the GaAs buffer material, we estimate the mobility of the 2DEG in sample 237 to be approximately 100 000 $cm^2V^{-1}s^{-1}$ at 77K in this sample (237) the measured mobility was 162 000 cm^2V^{-1} s^{-1} at 2K.

The characteristics of the charge transfer and the 2DEG were independent of the type of semi-insulating substrated used, showing that under these growth conditions a buffer layer of between 0.5μm and 3μm was sufficient to avoid substrate effects.

3.2.4 Conclusion. In this preliminary study of the OM-VPE growth of selectively doped heterostructures we have observed 2DEGs with mobilities as high as 162 000 $cm^2V^{-1}s^{-1}$ at 2K. It was possible to control the charge transfer by varying the "spacer" layer thickness and the doping concentration in the n-type GaAlAs layer.

These results confirm that very abrupt heterojunctions can be obtained by continuous growth in an OM-VPE systems, and we expect to further improve the mobility of the 2DEG by optimization of the characteristics of the spacer layer.

3.3 Very Low Threshold GaInAsP/InP Double Heterostructure Lasers

3.3.1. Introduction. We have recently reported room-temperature operation of InP/ $Ga_xIn_{1-x}As_yP_{1-y}$/InP double-heterostructure (DH) lasers emitting in the 1.2 - 1.6μm spectral region grown by low-pressure metalorganic chemical vapor deposition (LP MOCVD).[44-48] In this section, we report the LP MOCVD growth of very low threshold InP/GaInAsP/InP DH lasers having room-temperature (300K) pulsed threshold current densities lower than the best values reported for comparable devices grown by liquid phase epitaxxy (LPE).[50]

Experimental Details: The epitaxial structures were grown by the LP MOCVD process on (100) InP (Sn) substrate misoriented 2° toward (110) at 650°C, at growth rates ≃ 200Å/min. The growth process has been described elsewhere.[47] The layers grown for these devices were as follows: (i) 1μm S-doped InP (n ≃ 10^{18} cm^{-3}); (ii) 0.22μm GaInAsP (no intentional doping); (iii) 2μm Zn-doped InP (P ≃ $3x10^{17}$ cm^{-3}); (iv) 0.2μm Zn-doped GaInAsP (P ≃ $2x10^{18}$ cm^{-3}).

Quaternary GaInAsP layers were lattice-matched to

$$\left(\frac{\Delta a}{a}\right) = \frac{a_{GaInAsP} - a_{InP}}{a_{InP}} \leq 5x10^{-4}$$

at room temperature. The grown wafers were then lapped to a thickness of 100μm and Au-12% Ge and Au-8% Zn contact metallizations were deposited on the n and P sides, respectively. The contacts were then annealed at 400°C for 5 min in an argon ambience. The devices were cleaved and sawn, producing chips of 150μm wide with cavity lengths in the range 300μm to 1000μm. The laser chips were tested, unmounted, under pulsed conditions at a pulse repetition rate of 10^4 Hz with a pulse length of 100 ns.

3.3.2. Results and discussion. For chips cleaved from the same bar the standard deviation in the lasing threshold current density was only ±5%. On larger areas of slice (6 cm^2) the standard deviation in lasing threshold was typically less than 20%.[48] A good layer uniformity was also demonstrated by the experimental result that the threshold current density continued to fall with lasing cavity length for cavity lengths up to 950μm. The variation of lasing threshold current density J_{th} with cavity length "L" is shown in Figure 18. The value of the absorption coefficient $\bar{\alpha}$ was 16 cm^{-1} and was obtained by fitting the standard expression[49] to the data in Figure 18.

$$J_{th} = \frac{1}{\beta}\left[\bar{\alpha} + \frac{1}{2L}\left(\log_e\left(\frac{1}{R_1 R_2}\right)\right)\right]$$

where R_1 and R_2 are the facet reflectivities ($R_1 = R_2 = 0.32$) and β is a constant equal to 0.058 cm.A^{-1}. In a series of eight laser slices the average emission wavelength was 2.284μm with a standard deviation of only ± 1%. For a broad area laser of cavity length 400μm (width 150μm) the average threshold current density is 800 A.cm^{-2} and this

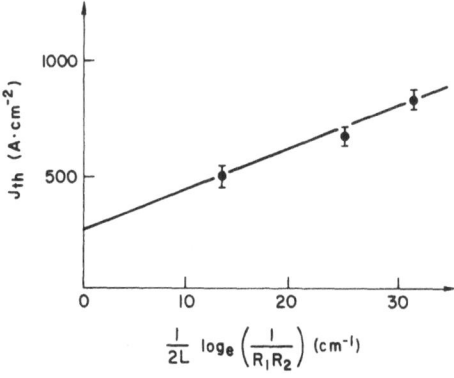

Figure 18 - Variation of threshold current density J_{th} with reciprocal of cavity length.

decreases to 500 A.cm^{-2} for a cavity length of 950μm. We believe that these are the lowest lasing threshold current densities that have yet been reported.

The threshold current density of DH lasers generally depends on temperature as

$$J_{th}(T) = J_{th}(T_0) \exp T/T_0$$

where T_0 described the characteristic temperature. The average T_0 value is 52°C.

3.3.3 Conclusion. We have prepared GaInAsP/InP DH lasers by LP MOCVD having power J_{th} than the best values reported for comparable LPE devices. The structure has an average J_{th} of 800 A.cm^{-2} for a chip size (400 μm x 150μm) which decreases to 430 A.cm^{-2} for a cavity length of 950μm.

3.4. <u>Two Dimensional Electron Gas in a In$_{0.53}$Ga$_{0.47}$ As-InP Heterojunction</u>

The two dimensional electron gas (TEDG) occuring at the interface in modulation-doped[51] GaAsAl$_x$Ga$_{1-x}$As heterojunctions has recently received great attention. Indeed, due to the spatial separation of electrons in the GaAs layer from their parent donor impurities in the Al$_x$Ga$_{1-x}$As layer, these structures exhibit a high electron mobility[52], so that fruitful applications[53] in the field of fast devices can be expected.

We wish to report here Shubnikov-de Haas and cyclotron resonance measurements done in a selectively doped In$_{0.53}$Ga$_{0.47}$As-InP heterojunction grown by metalorganic-chemical vapor deposition (MOCVD). These measurements give the first evidence for the formation of a TEDG at the interface between In$_{0.53}$Ga$_{0.47}$As and InP. From these investigations we obtain several parameters such as the electron cyclotron mass and mobility, which is found to be good. We derive also from our results the value of the conduction band discontinuity at the interface. We believe that this study shows the MOCVD can be used to grow In$_{0.53}$Ga$_{0.47}$As-InP heterojunctions with good interfacial properties. This is important because In$_{0.53}$Ga$_{0.47}$As in an interesting material due to its potential applications in integrated optics for optical communications.

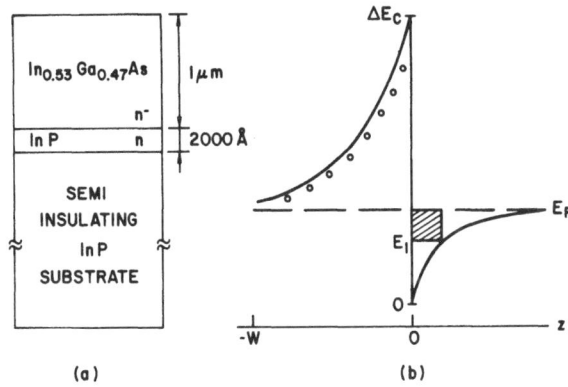

(a)   (b)

Figure 19 - (a) Schematic configuration of the hèterojunction used here. (b) Energy band diagram for a selectivity doped heterojunction.

The heterojunctions used here was grown by low-pressure metalorganic chemical vapor deposition[51] (see Table 3) on a (100) semi-insulating Fe-doped InP substrate, and is schematized in Figure 19a. The InP epilayer, 2000Å thick, was n type with $N_D - N_A \simeq 3 \times 10^{16}$ cm^{-3}. The In$_{0.53}$Ga$_{0.47}$As epilayer, which was nonintentionally doped, was also n type with $N_D - N_A \lesssim 2 \times 10^{15}$ cm^{-3}, its thickness being equal to 1μm. Note that, at low temperature, the band gaps[55] of InP and In$_{0.53}$Ga$_{0.47}$As are 1.42 and 0.81 eV, respectively.

TABLE 3. Growth parameters. The growth temperature and background pressure are 550°C and 100 mbar, respectively.

Sources	Partial Pressure (mbar)	Temperature (°C)
Triethylindium	0.0082	30
Triethylgallium	0.0106	0
Phosphine	1.584	25
Arsine	0.792	25

Standard techniques were used to do Shubnikov-de Haas measurements and, for cyclotron resonance experiments, the far infrared sources used here were carcinotrons, and the detector was a carbon resistance at 2K. The magnetic field was provided by a superconducting coil, and could be continuously varied from 0 to 8.5T. It is worth adding that Hall measurements done on this heterojunction gave electron mobilities equal to 8800, 42 000, and 51 000 cm^2 V^{-1}s^{-1} at 300, 77 and 4.2K respectively.

Figure 20a shows pronounced Shubnikov-de Haas oscillations observed at 4.2K with the magnetic field B perpendicular to the heterojunction interface, and some plateaus[56] characteristic of 2D systems can be seen. Figure 20b gives the reciprocal magnetic field corresponding to the magneto-oscillation maxima as a function of the Landau index for different values of θ, which is the angle between B and the perpendicular to the interface. The oscillations are periodic in $1/B$, and they follow the expected $(\cos\theta)^{-1}$ dependence[57] of a TDEG. This manifests the two-dimensionality of the electron gas under consideration. From these data, we deduce, using standard procedures, an electron density n_s equal to 4.3×10^{11} cm^{-2}, in agreement with the Hall data. In the structure, as in modulation-doped GaAs-Al$_x$Ga$_{1-x}$As

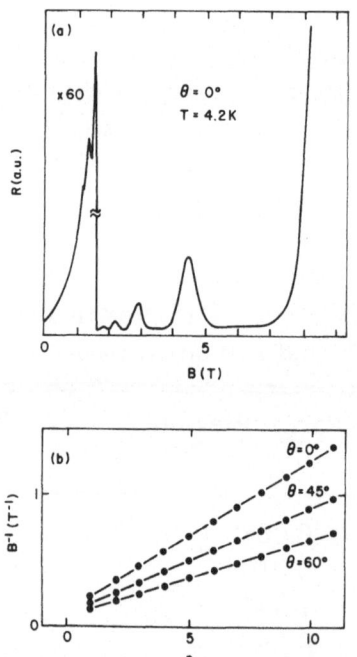

Figure 20 - (a) Magnetoresistance oscillations as a function of the magnetic field B. (b) Reciprocal field at maxima of these magneto oscillation versus Landau quantum number for different values of θ.

heterojunctions,[51] electrons from the donors in the wide band gap material (InP) are transferred across the interface in the narrow gap material ($In_{0.53}Ga_{0.47}As$) to maintain a constant Fermi level in the system. This leads to the formation of the observed TEDG which is confined in the potential well resulting from the strong band bending occurring in the vicinity of the interface and due to the spatial separation of positive (ionized donors) and negative (electrons) charges, as schematized in Figure 19b.

Figure 21 presents typical cyclotron resonance data obtained at 2K for $\theta = 0$ and an infrared photon wavelength equal to $630\mu m$. The cyclotron frequency is found to vary as $(\cos \theta)^{-1}$, as it should for a TDEG.[58] The corresponding electron cyclotron mass is $m_c^* = (0.047 \pm 0.001)m_0$. It is larger than the band edge electron effective mass in bulk[55] $In_{0.53}Ga_{0.47}As$, $m_e^* = 0.041 m_0$, and this effect is certainly due to the nonparabolicity of the conduction band of $In_{0.53}Ga_{0.47}As$. Using this value for m_c^* and the Shubnikov-de Haas data, we get a Fermi energy $E_F = 22.4$ meV measured from the lowest energy level E_1 in the potential well (Figure 19b). Besides, from the model developed by Stern E_1[59] is given by

$$E_1 = \frac{5}{8}\left(33e^2hn_s/8\epsilon_0\epsilon_r m_c^{*1/2}\right)^{2/3},$$

where $\epsilon_r = 14$ is the statis dielectric constant of $In_{0.53}Ga_{0.47}As$.[55] This yields $E_1 = 57$ meV, measured from the bottom of the $In_{0.53}Ga_{0.47}As$ conduction band at $z = 0$, so that $E_1 + E_F \simeq 80$ meV. Now, using the model of Antcliffe, et al.[10] to take into account the nonparabolicity in a

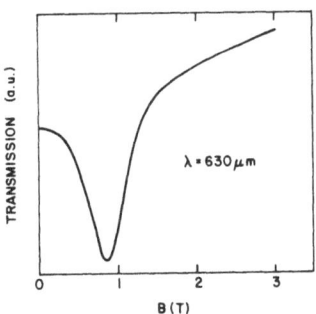

Figure 21 - Cyclotron resonance data at 2K for an infrared wavelength equal to $630\mu m$.

triangular well, one gets 0.0465 m_0 for the electron effective mass at an energy equal to $E_1 + E_F$, in good agreement with the measured electron cyclotron mass.

From the width of the cyclotron line shown in Figure 21, we can obtain the electron mobility μ_e. In a TDEG[61] $\Delta B_{1/2}$ $\mu_e^{-1/2} B_r^{1/2}$, where $\Delta B_{1/2}$ is the half-width at half-amplitude of the cyclotron resonance line and B_r is the resonance magnetic field. It follows that μ_e is found to be equal to $(60\ 000 \pm 5000) cm^2 V^{-1} s^{-1}$ at 2K, which compares favorable with the Hall mobility.

Considering that the electrons in the TDEG come from InP, the width ω of the depletion charge layer in InP (see Figure 19a) is $\omega = n_s/N_D \simeq 1400\text{Å}$, which allows us to determine the conduction-band discontinuity ΔE_c at the interface. Indeed, ΔE_c is given by

$$\Delta E_c = E_1 + E_F + e^2 N_D \omega^2 / 2 \epsilon_r \epsilon_o,$$

where e is the electron charge and $\epsilon_r = 12.4$ for InP.[55] In the heterojunction studied here this yields $\Delta E_c \simeq 530$ meV, namely 87% of the difference between the gaps of InP and $In_{0.53}Ga_{0.47}As$, comparable to the result obtained in $GaAs/Al_xGa_{1-x}As$ structures[62] with x ~ 0.2.

In summary, we have reported here the first observation of a two-dimensional high mobility electron gas in a selectivity doped $In_{0.53}Ga_{0.47}As$-InP heterojunction grown by MOCVD. As in $GaAs$-$Al_xGa_{1-x}As$ structures, the mobility can certainly be improved by having an undoped InP spacer layer between the n-type InP and the $In_{0.53}Ga_{0.47}As$ layer, increasing the potential technological applications of these heterostructures.

3.5 Growth of GaInAs/InP Multi-Quantum Well Structures

The potentially important application of $Ga_{0.57}In_{0.53}As$ in optical communications make it a semiconductor worthy of detailed study. One possible line of application involve the fabrication of quantum well injection lasers incorporating this material. As recently demonstrated in systems such as $GaAs$-$Al_xGa_{1-x}As$, high quality semiconductor injection lasers can be fabricated from quantum well structures.[63]

We describe here the growth of single- and multi-quantum well $Ga_{0.47}In_{0.53}As$-InP structures by low-pressure metalorganic chemical vapor deposition (LP MOCVD). Results obtained from photoluminescence measurements performed with these structures are presented, with specific reference to the experimentally determined and theoretically predicted luminescence line positions.

The processes involved in the growth of $Ga_{0.47}In_{0.53}As$-InP and the growth apparatus have been described in detail in previous publications.[64,66] Triethlyindium (TEI) and triethylgallium (TEG) are used as sources of In and Ga, while arsine (AsH_3) and phosphine (PH_3) provide As and P, respectively. Hydrogen (H_2) and nitrogen (N_2) are used as carrier gases, the presence of N_2 slowing down the parasitic reaction between Te In and PH_3 or AsH_3 and that of H_2 being necessary to avoid the deposition of carbon.

Growth was carried out at 76 Torr and at a temperature of 540°C. The InP/$Ga_{0.47}In_{0.53}As$ interfaces were realized by turning off the PH_3 flow and turning on both the TEG and AsH_3 flows; in the same way the reverse procedure was used to obtain $Ga_{0.47}In_{0.53}As$/InP interfaces. The growth rate was small ($\sim 3\text{Å sec}^{-1}$) and stabilized to its new steady-state value in less than one second after switching. The optimum conditions for growing at reduced pressure InP and $Ga_{0.47}In_{0.53}As$, as determined during these investigations, are presented in Table 4.

The quantum well samples grown for this study were not intentionally doped. Residual impurity concentrations are assumed to be of the order of those determined for bulk layers grown under identical conditions. Capacitance-voltage measurements in 3-5μm thick layers grown under such conditions yield carrier values (N_D - N_A) of 3×10^{14} cm^{-3} for $Ga_{0.47}In_{0.53}As$ and 6×10^{14} cm^{-3} for InP.

The exact composition of the quantum well samples is difficult to measure directly. X-ray diffraction measurements carried out on thick ($\sim 1\mu$m) layers of GaInAs grown under conditions identical to those used for the quantum well growth indicate (0.47 $\pm$ 0.01) Ga and (0.53 $\pm$ 0.01) In.

In order to facilitate the study of more than one quantum well simultaneously, a multilayer structure consisting of 25, 50, 100 and 200

TABLE 4. Optimum growth parameters for reduced pressure (76 Torr) growth as determined during this study

		InP	$Ga_{0.47}In_{0.53}As$
Substrate temperature during growth	^{o}C	540	540
Total gas flow	ℓmin^{-1}	6	6
N_2 flow through TEI bubbler	$cm^3 min^{-1}$	225	225
H_2 flow through TEG bubbler	$cm^3 min^{-1}$	–	90
PH_3 flow	$cm^3 min^{-1}$	300	–
AsH_3 flow	$cm^3 min^{-1}$	–	90
Growth rate	$\overset{o}{A} min^{-1}$	100+10	200+10

Å thick $Ga_{0.47}In_{0.53}As$ layers (wells) separated by 500Å thick InP layers (barriers) was grown on an InP[52] substrate (Figure 23 inset). The quantum well layer thickness was deduced from the steady-state growth rate, which in turn was calibrated by step measurements on thick layers of InP and $Ga_{0.47}In_{0.53}As$ in which steps had been selectively etched.

As a preliminary test of these multilayer structures, Auger scans were carried out. Samples were prepared by chemically etching a linear bevel through all layers and interfaces. Figure 22 indicates the nature of the bevel for the two outermost $Ga_{0.47}In_{0.53}As$ layers (25 and 50Å) and also shows the Auger line scan of the bevel. The four $Ga_{0.47}In_{0.53}As$ layers are clearly visible and the interfaces are abrupt.

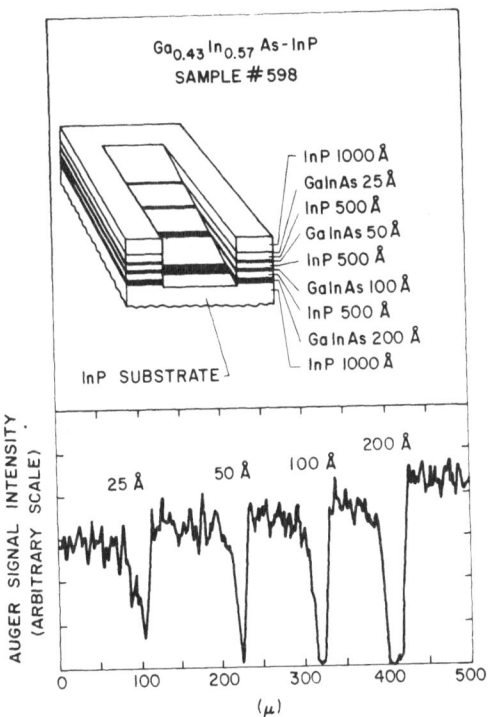

Figure 22 - Down: Auger spectrum of a chemically etched bevel which cuts all $Ga_{0.47}In_{0.33}As$ layers of the four well samples. The abscissa indicates the distance along the bevel in microns. The zero value is determined by the start of the bevel in the outer-most InP layer. Layer assignments are indicated above the trace. Up: The bevel is schematically depicted in the inset - the actual bevel angle with respect to the outmost InP surface is between 0.02 and 0.06 of a degree.

The trace of the 25Å layer is less distinct than the others because the 25Å layer, being the outermost layer comprises the most non-linear section of the bevel. The bevel angle is therefore less well defined in this region and Auger results correspondingly more noisy. Although the Auger results indicate the four distinct $Ga_{0.47}In_{0.53}As$ layers with abrupt interfaces exist, they do not yield accurate values for layer thickness or interface abruptness.

Photoluminescence measurements were carried out at 2K, using a focussed Nd YAG laser beam (E_{exc} = 1170 meV, 20 mW) to excite only the $Ga_{0.47}In_{0.53}As$ (Eg (bulk, low T) = 810 meV) layers and not the InP (Eg (bulk, low T) = 1420 meV) layers[67] The photolumines-

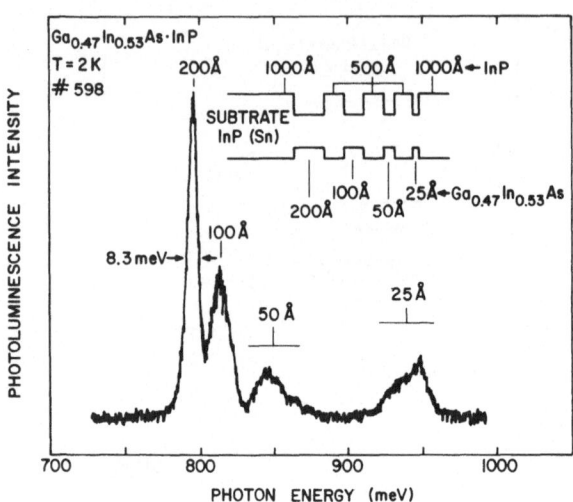

Figure 23 - Photoluminescence spectrum of $Ga_{0.47}In_{0.33}As$-InP sample measured at 2K with excitation at 1170 MeV (Nd yag laser 20 mW focussed beam). The inset schematically illustrates the sample structure.

cence was analyzed with a Jobin-Yvon HRD-2 monochromator in conjunction with a PbS detector, using conventional clock-in techniques.

The quantum well samples luminesce well - the luminescence intensity is at least comparable to that of bulk $Ga_{0.47}In_{0.53}As$. Figure 23 shows the photon luminescence spectrum of the four well structure and indicates the peak assignments. Peak positions shift by up to 15 meV from those indicated in Figure 23, depending on the point of excitation of the sample. These shifts are probably caused by compositional inhomogeneities within any given samples.

Figure 24 is a plot of luminescence line energies as a function of the associated layer thickness (well widths) for the four well sample and for three single well samples grown under similar conditions. Also shown on the same plot are theoretical curves calculated within the framework of the envelope function approach, taking into account band non-parabolicity and assuming transitions between the first conduction and valence states in $Ga_{0.47}In_{0.53}As$ quantum wells[68-69]. The magnitude band discontinuity has been determined to be $\Delta E_c \simeq 530$ meV.

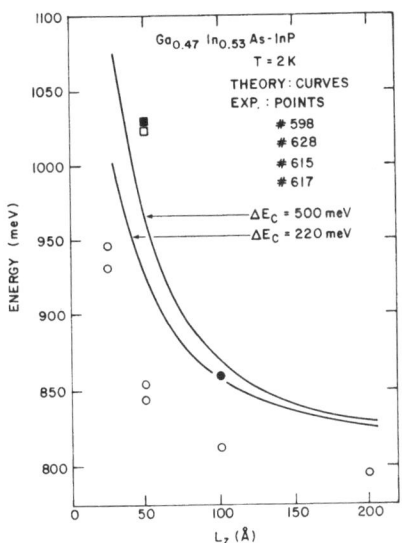

Figure 24 - Plot of measured photoluminescence line energies against well widths (L_z) for the four well sample 598 and for single well samples 628 (well width 100Å), 615 and 617 (well width 60Å). Theoretical curves are shown for two values of the conduction band offset, ΔE_c.

Note that all measurements of the four well sample fall consistently below the theoretical curves. The 50Å single well points fall above the theoretical curves and the 100Å single well point is in agreement with theory. It is probably that a number of factors contribute jointly to create the observed discrepancies between experiment and theory. Compositional variation within the stated limits (GA: 0.47 ± 0.01; In: 0.53 ± 0.01) and the resultant increase in lattice mismatch (which gives rise to strained quantum wells) probably contribute to the observed discrepancies. Layer thickness is not known exactly (accuracy is ± 3Å), and deviations from the quoted thickness may also contribute to the observed discrepancies. Participation of impurities in the recombination process may be responsible for the consistently low line energies measured in the four well sample. This hypothesis is consistent with the situation observed in n-type $Ga_{0.47}In_{0.53}As$ bulk, where all photoluminescence has been identified as impurity assisted luminescence (donor to valence band and donor to acceptor transitions).[70]

In summary, we have described the growth by LP MOCVD of single and multi-quantum well $Ga_{0.47}In_{0.53}As$-InP structures. As verified by Auger measurement, the multi-well structure consists of four wells ($Ga_{0.47}In_{0.53}As$ layers) of different thicknesses separated by InP barriers of uniform thickness. All interfaces are observed to be abrupt. Both single and multi-well samples luminesce at least as well as bulk samples, however, discrepancies exist between the measured and the theoretically predicted photoluminescence energies in most of the samples studies. Possible explanations, including variation of well composition, variation of well thickness and participation of impurities, in the recombination process have been suggested.

REFERENCES

1. P. Duchemin, M. Bonnet, F. Koelsch, J. Electrochem Soc. 125, 637 (1978).

2. P. Duchemin, M. Bonnet, F. Koelsch, D. Huyghe, J. Electrochem. Soc. 126, 1134 (1979).

3. P. Duchemin, M. Bonnet, G. Beuchet, F. Koelsch, GaAs and Related Compounds, edited by C.M. Wolfe (Institute of Physics, London, 1978), p. 10.

4. P. Hirtz, B.D. Voung, J.P. Duchemin, P. Hirtz, B. De Cremoux, R. Bisaro, P. Merenda, M. Bonnett, E. Duda, G. Messsssquida, J.C. Carballes, Appl. Phys. Lett. 36, 15 (1980).

5. P. Hirtz, P. Larivain, J.P. Duchemin, T.P. Pearsall, Electron. Lett. 16, 415 (1980).

6. P. Hirtz, Electron Lett. 16, 275 (1980).

7. P. Hirtz, M. Razeghi, S. Hersee, J.P. Larivain, J.P. Duchemin, Electron. Lett. (to be published).

8. V.S. Ban, J. Electrochem Soc. 118, 1473 (1971).

9. V.S. Ban, J. Electrochem Soc. 122, 1389 (1975).

10. V.S. Ban, S.L. Gilbert, J. Crystal Growth 31, 284 (1975).

11. V.S. Ban, S.L. Gilbert, J. Electrochem Soc. 122, 1382 (1975).

12. V.S. Ban, J. Electrochem Soc. 125, 317 (1978).

13. T.O. Sedgwick, J.E. Smith, R. Ghez, M.E. Cowher, J. Crystal Growth 31, 264 (1975).

14. T.O. Sedgwick, Proc. of the Sixth International Conf. on CVD, ed. L.F. Donaghery et al., The Electrochem Soc., Princeton, NJ 59 (1977).

15. T.O. Sedgwick, G.V. Arbock, R. Ghez, Proc. of the Sixth International Conf. on CVD, ed. L.F. Donaghery et al., The Electrochem Soc., Princeton, NJ 66 (1977).

16. J.P. Duchemin, Revue Thomson-CSF Vol. 9 n° 1 mars 1977.

17. J.P. Duchemin, Revue Thomson-CSF Vol. 9, n° 2 juin 1977.

18. H.M. Manasevit, W.I. Simpson, J. Electrochem Soc. 116, 1725 (1969).

19. H.M. Manasevit, F.M. Erdmann, W.I. Simpson, J. Electrochem Soc. 118, 1864 (1971).

20. H.M. Manasevit, J. Electrochem. Soc. 118, 1865 (1971).

21. H.M. Manasevit, A.C. Thorsen, J. Electrochem. Soc. 119, 99 (1972).

22. H.M. Manasevit, J. Crystal Growth 22, 125 (1974).

23. J.P. Duchemin, M. Bonnet, F. Koelsch, D. Huyghe, Proceeding of the Fourth International Conf. on Vapor Growth and Epitaxy Nagoya (Japan). J. Crystal Growth 45, 1 (1978).

24. J.P. Duchemin, M. Bonnet, D. Huyghe, Revue Thomson-CSF Vol. 9, December 1977.

25. S.J. Bass, Third International Conference on Vapor Growth and Epitaxy. August 18-25, 1975, Amsterdam.

26. J.P. Duchemin, M. Bonnet, G. Beuchet, F. Koelsch, Int. Symposium on Gallium Arsenide and Related Compounds. Saint Louis, USA, September 24, 1978.

27. S.D. Hersee, M. Baldy, P. Assenat, B. De Cremoux, J.P. Duchemin, Electron. Lett. 18, 618 (1982).

28. W.T. Tsang, Appl. Phys. Lett. 39, 134 (1981).

29. W.T. Tsang, Appl. Phys. Lett. 40, 217 (1982).

30. H. Kressel, J.K. Butter, Semiconductor Lasers and Heterojunctions LED's, (Academic Press, New York, 1977), p. 270.

31. S.D. Hersee, M.A. Di Forte-Poisson, M. Baldy, J. Duchemin. J. Crystal Growth 55, 53 (1981).

32. S.D. Hersee, M. Baldy, J.P. Duchemin, presented at Electronic Materials Conference, Colorado, USA, June 1982. (to be published in J. Electron. Mat.)

33. R. Dingle, Festkorperprobleme XV/Advances in Solid State Physics (Pergamon-Vieweg, Stuttgart, 1975), p. 21.

34. R. Dingle, H.L. Stormer, A.C. Gossard and W. Wiegmann, Appl. Phys. Lett. 33, 665 (1978).

35. S. Hiyamizu, Collected paper of MBE-CST2, Tokyo, 113 (1982).

36. P.N. Tung, D. Delagebeaudef, M. Laviron, P. Delescluse, J. Chaplart, N.T. Linh, Electron. Lett. 18, 109 and 517 (1982).

37. N. Holonyak, R. Kolbas, R.D. Dupuis, P.D. Dapkus, IEEE J. Quantum Electron QE-16, 170 (1980).

38. S.D. Hersee, M. Baldy, P. Assenat, B. De Cremoux and J.P. Duchemin, Electron Lett. 18 896 (1982).

39. P. Frijlink, J. Malvenda, Les Editions de Physique (1982), to be published.

40. J.J. Coleman, P.D. Dapkus, J.J.J. Yang, Electron. Lett 17, 606 (1981).

41. Y. Guldner, J.P. Vieren, P. Voisin, M. Voos, M. Razeghi, M.A. Poisson, Appl. Phys. Lett. 40, 877 (1982).

42. M. Razeghi, M.A. Poisson, J.P. Larivain, B. DeCremoux, J.P. Duchemin, M. Voos, Electron Lett. 18, 339 (1982).

43. T.J. Drummond, W. Kopp, R. Fischer, H. Morkoc, R.E. Thorne, A.Y. Cho, J. Appl. Phys. 53, 1238 (1982).

44. M. Razeghi, J.P. Hirtz, P. Hirtz, J.P. Larivain, R. Blondeau, B. DeCremoux, J.P. Duchemin, Electron Lett 17, 597 (1981).

45. M. Razeghi, P. Hirtz, J.P. Larivain, R. Blondeau, B. DeCremoux, J.P. Duchemin, Electron. Lett. 17, 643 (1981).

46. M. Razaghi, P. Hirtz, R. Blondeau, J.P. Larivain, L. Noel, B. DeCremoux, J.P. Duchemin, Electron Lett. 18, 132 (1982).

47. M. Razeghi, Revue Technique Thomson-CSF 15, 1 (1983).

48. M. Razeghi, M.A. Posiion, J.P. Larivain, J.P. Duchemin, J. Electronic Material, 1983 (to be published).

49. H. Kressel, J.K. Butler, Academic Press 1977.

50. R.J. Nelson, Appl. Phys. Lett. 35, 654 (1979).

51. See, for example, H.L. Stormer, R. Dingle, A.C. Gossard, W. Wiegmann, M.D. Sturge, Solid State Commun 29, 705 (1979).

52. L.C. Witkowski, T.J. Drummond, S.A. Barnett, H. Morkoc, A.Y. Cho, J.E. Green, Electron Lett. 17 126 (1981); H.L. Stormer, A.C. Gossard, W. Wiegmann, K. Baldwin, Appl. Phys. Lett 39, 914 (1981); S. Hiyamizu, T. Mimura, T. Fujii, K. Nanbu, H. Hashimoto, Jpn. J. Appl. Phys. 20, L455 (1981); J.J. Coleman, P.D. Dapkus, J.J.J. Yang, Electron Lett 17, 606 (1981).

53. T. Mimura, K. Joshin, S. Hiyamizu, K. Hirosaka, M. Abe, Jpn. J. Appl. Phys. 20 L598 (1981); Nguyen T. Linh (private communication).

54. P. Hirtz, J.P. Larivain, J.P. Duchemin, T.P. Pearsall, Electron Lett 16, 415 (1980); M. Razeghi, P. Hirtz, J.P. Larivain, R.

Blondeau, B. DeCremoux, J.P. Duchemin, Electron Lett. 17, 643 (1981).

55. See, for example, Y. Takeda, M.A. Littlejohn, J.R. Hauser, Appl. Phys. Lett 39, 620 (1981).

56. D.C. Tsui, A.C. Gossard, Appl. Phys. Lett. 38, 550 (1981).

57. F. Stern, W.E. Howard, Phys. Rev. 163, 816 (1967).

58. R.J. Nicholas, S.J. Sessions, J.C. Portal, Appl. Phys. Lett. 37, 178 (1980).

59. F. Stern, Phys. Rev. B12, 4891 (1972).

60. G.A. Antcliffe, R.T. Bate, R.A. Reynolds, In Proceedings of the Conference on the Physics of Semimetals and Narrow-gap Semiconductors, Dallas, 1970, edited by D.L. Carter and R.T. Bate (Pergamon, New York, 1971) p. 499.

61. P. Voisin, Y. Guldner, J.P. Vierner, M. Voos, P. Delescluse, Nguyen T. Linh, Appl. Phys. Lett. 39, 982 (1981).

62. R. Dingle, In Festkorperproblemr (Advances in Solid State Physics), edited by H.J. Queisser (Pergamon Viewag, Braunschweig, 1975), Vol. XV, p. 21.

63. See, for example, S.D. Hersee, M. Baldy, P. Assenat, B. DeCremoux, J.P. Duchemin, Electron. Lett. 18 896 (1982); W.T. Tsang, Appl. Phys. Lett. 40, 217 (1982).

64. M. Razeghi, M. Poisson, J.P. Larivain, J.P. Duchemin, J. Electron Mat. 12, 371 (1983).

65. M. Razeghi, J.P. Duchemin, J. Vac. Sci. Tech. (1983), to be published.

66. M. Razeghi, J.P. Duchemin, J. Cryst. Growth (1983), to be published.

67. See, for example, Y. Takeda, M.A. Littlejohn, J.R. Hauser, Appl. Phys. Lett. 39, 620 (1981).

68. G. Bastard, this volume.

69. Y. Guldner, J.P. Vieren, P. Voisin, M. Voos, M. Razeghi, M.A. Poisson, Appl. Phys. Lett. 40, 877 (1982).

70. Y.S. Chen, O.K. Kim, J. Appl. Phys. 52, 7392 (1981); J.Y. Marzin, J.L. Benchimol, B. Sermage, B. Etienne, M. Voos, Solid State Commun. 45, 79 (1983).